OXYGEN TRANSPORT
TO TISSUE XI

ADVANCES IN EXPERIMENTAL MEDICINE AND BIOLOGY

Recent Volumes in this Series

OXYGEN TRANSPORT TO TISSUE XI

Edited by

Karel Rakusan and George P. Biro

University of Ottawa
Ottawa, Ontario, Canada

Thomas K. Goldstick

Northwestern University
Evanston, Illinois

and

Zdenek Turek

University of Nijmegen
Nijmegen, The Netherlands

PLENUM PRESS • NEW YORK AND LONDON

Library of Congress Cataloging in Publication Data

International Society on Oxygen Transport to Tissue. Meeting (16th: 1988: Ottawa, Ont.)
    Oxygen transport to tissue XI / edited by Karel Rakusan . . . [et al.].
      p.   cm. — (Advances in experimental medicine and biology; v. 248)
    "Proceedings of the 16th Annual Meeting of the International Society on Oxygen Transport to Tissue, held August 7-11, 1988, in Ottawa, Ontario, Canada" — T.p. verso.
    Includes bibliographies and indexes.
    ISBN-13: 978-1-4684-5645-5   e-ISBN-13: 978-1-4684-5643-1
    DOI: 10.1007/978-1-4684-5643-1

    1. Oxygen — Physiological transport — Congresses. 2. Tissue respiration — Congresses. 3. Blood substitutes — Congresses. I. Rakusan, Karel. II. Title. III. Title: Oxygen transport to tissue 11. IV. Title: Oxygen transport to tissue eleven. V. Series.
QP99.3.O9I5 1988   89-3851
599′.012 — dc19   CIP

Proceedings of the 16th Annual Meeting of the International Society on Oxygen Transport to Tissue, held August 7-11, 1988, in Ottawa, Ontario, Canada

© 1989 Plenum Press, New York
Softcover reprint of the hardcover 1st edition 1989
A Division of Plenum Publishing Corporation
233 Spring Street, New York, N.Y. 10013

INTERNATIONAL SOCIETY ON OXYGEN TRANSPORT
TO TISSUE 1987–88

Officers:

President: K. Rakusan, Canada
President-Elect: J. Piiper, West Germany
Past Presidents: C. R. Honig, USA
                  M. Mochizuki, Japan
Secretary: N. S. Faithful, UK
Treasurer: J. Grote, West Germany

Executive Committee:

A. Eke, Hungary
W. Erdmann, The Netherlands
T. K. Goldstick, USA
T. Koyama, Japan
J. C. LaManna, USA
A. Mayevsky, Israel
M. McCabe, Australia
W. Mueller-Klieser, West Germany
E. M. Nemoto, USA

OTTAWA MEETING, AUGUST 7–11, 1988

Organizing Committee:

G. P. Biro
G. V. Forrester (Secretary)
W. J. Keon
B. Korecky
R. Masters
K. Rakusan (Chairman)
J. W. Reeves
Z. Turek

We gratefully acknowledge generous support from the following:

National Science Foundation (Washington, D.C.)
Canadian Heart Foundation
Heart and Stroke Foundation of Ontario
University of Ottawa Heart Institute
University of Ottawa
Ortho Pharmaceuticals Canada Ltd.
Parke-Davis Pharmaceuticals Ltd.
Pfizer-Shiley Canada Ltd.

PREFACE

The Ottawa '88 meeting of the International Society for Oxygen Transport to Tissue attracted a record number of participants and presentations. We were able to avoid simultaneous sessions and still keep the scientific program to four days by using poster sessions followed by plenary debate on each poster. To paraphrase the British physicist David Bohm, we tried to avoid an ordinary discussion, in which people usually stick to a relatively fixed position and try to convince others to change. This situation does not give rise to anything creative. So, we attempted instead to establish a true dialogue in which a person may prefer and support a certain point of view, but does not hold it nonnegotiably. He or she is ready to listen to others with sufficient sympathy, and is also ready to change his or her own view if there is a good reason to do so.

Our Society is in its "teen" years, and there are even some arguments about its exact age. Many newer members have raised questions concerning the history of the Society. For this reason, I have asked one of the "founding fathers", D. Bruley, to prepare a brief account of the birth and early history of the Society which appears on the following page.

A final note concerning the presented material in this volume: we have continued the policy of having all papers reviewed by both editors and anonymous referees. We had to compromise between the aim for rapid publication on the one hand, and the need for thorough review on the other. As we would be using camera-ready manuscripts provided by the authors themselves, we were rather lenient concerning small format errors and typos. On the other hand, we tried not to compromise on scientific content; several manuscripts had to be rejected or returned for revision. Thus, the present volume represents a mid-course that we have navigated between the two opposing forces.

Karel Rakusan

ACKNOWLEDGEMENTS

We gratefully acknowledge generous support from the following:

National Science Foundation (Washington, D.C.)
Canadian Heart Foundation
Heart and Stroke Foundation of Ontario
University of Ottawa Heart Institute
University of Ottawa
Ortho Pharmaceuticals Canada Ltd.
Parke-Davis Pharmaceuticals Ltd.
Pfizer-Shirley Canada Ltd.

THE BIRTH OF ISOTT

Due to membership interest a request was made to recall the events
that took place in the structuring of our society. To the best
recollection of major contributors the sequence of events took place as
follows.

Several meetings addressing oxygen-tissue interaction were held in
workshop style over the years (Brompton Hospital, London, 1960; Queen
Elizabeth College, London, 1963; Bedford College, London, 1963). In 1971
a workshop on oxygen supply organized by Dr. Dietrich W. Lubbers and
Dr. Manfred Kessler at the Max-Planck-Institute in Dortmund, West Germany[1]
inspired Drs. Duane F. Bruley and Haim I. Bicher to initiate plans for a
subsequent workshop. The intended purpose of the meeting was to highlight
interdisciplinary research involving theoretical and experimental
investigations for oxygen transport in tissue. It was to bring life
scientists and engineers together to examine the many complex phenomena of
normal tissue growth and maintenance, and tissue survival and repair under
pathological conditions.

Immediately following the Dortmund workshop, Dr. Bruley informed the
Clemson University administration of the intended meeting. Shortly
thereafter approval was granted. The university agreed to sponsor a
workshop in honor of Dr. Melvin H. Knisely for his many scientific
contributions regarding the microcirculation.

Upon Dr. Haim I. Bicher's return to the Medical School of South
Carolina from an extended trip to Israel, similar support was obtained
from the Medical University of South Carolina at Charleston, S.C.,
Dr. Knisely's institution, and thus it was decided to hold the workshop
jointly in Charleston, S.C. and Clemson, S.C. Once this decision was made
both schools committed sizeable funds to support the meeting along with
resources generated from other companies and agencies.

After an intensive period of planning and preparation an initial
meeting announcement was sent out to sample community interest. The
results demonstrated enthusiasm far beyond projections and triggered
Drs. Bicher and Bruley to consider the meeting as a launching pad for a
very focused international society regarding oxygen transport to tissue.
After several discussions with other investigators it was decided that a
formal society would be in the best interest of groups around the world to
achieve research goals related to oxygen transport in tissue and that the
Charleston/Clemson meeting would be an appropriate forum to formalize and
begin an international society. The name "International Society of Oxygen
Transport to Tissue" and a society "logo" were then agreed upon. As the
meeting date of April 22-28, 1973 approached, charter members of the
International Committee were selected. The membership consisted of the
following scientists and engineers:

Dr. Melvin H. Knisely, Charleston, U.S.A.
Dr. Duane F. Bruley, Clemson, U.S.A.

Dr. Haim I. Bicher, Little Rock, U.S.A.
Dr. Gerhard Thews, Mainz, West Germany
Dr. Ian A. Silver, Bristol, England
Dr. Herbert J. Berman, Boston, U.S.A.
Dr. Britton Chance, Philadelphia, U.S.A.
Dr. Leland C. Clark, Jr., Cincinnati, U.S.A.
Dr. Lars-Erik Gelin, Goteborg, Sweden
Dr. Jurgen Grote, Mainz, West Germany
Dr. Manfred Kessler, Dortmund, West Germany
Dr. Dietrich W. Lübbers, Dortmund, Germany
Dr. José Strauss, Miami, U.S.A.
Dr. William J. Whalen, Cleveland, U.S.A.
Dr. Daniel D. Reneau, Ruston, U.S.A.

At the society organizational meeting the following slate of officers
was elected:

President - Dr. Melvin H. Knisely, Charleston, U.S.A.
President-Elect - Dr. Gerhard Thews, Mainz, West Germany
Secretary - Dr. Haim I. Bicher, Little Rock, U.S.A.
Treasurer - Dr. Ian A. Silver, Bristol, England

The first meeting of ISOTT surpassed all expectations and established
a society that has continued to meet annually at various locations around
the world. The Society is now incorporated in Holland. Society meetings
have been held at the following locations under the leadership of the
listed presidents:

1973	Charleston/Clemson, U.S.A.	Melvin H. Knisely, Honorary
1974	Atlantic City, U.S.A.	Melvin H. Knisely
1975	Mainz, West Germany	Gerhard Thews
1976	Anaheim, U.S.A.	Britton Chance
1977	Cambridge, U.K.	Ian A. Silver
1978	Atlantic City, U.S.A.	Jose Strauss
1979	La Jolla, U.S.A.	Jose Strauss
1980	Budapest, Hungary	Arisztid G.B. Kovách
1981	Detroit, U.S.A.	Haim I. Bicher
1982	Dortmund, West Germany	Dietrich W. Lübbers
1983	Ruston, U.S.A.	Duane F. Bruley
1984	Nijmegen, Holland	Ferdinand Kreuzer
1985	Raleigh, U.S.A.	Ian S. Longmuir
1986	Cambridge, UK	Ian A. Silver
1987	Sapporo, Japan	Masaji Mochizuki/Carl Honig
1988	Ottawa, Canada	Karel Rakusan

Dr. Johannes Piiper is the 1989 President of ISOTT. He will host the
meeting in Gottingen, West Germany.

In 1983 at the Ruston, Louisiana meeting the first Melvin H. Knisely
Award to a promising young investigator was made. This award was
initiated to express the spirit and willingness of Dr. Knisely to work
with and contribute to the growth of beginning scientists and engineers
addressing the problems of oxygen transport to tissue. The recipients,
through the 1988 meeting in Ottawa, Canada are as follows:

1983	Antal G. Hudetz/Hungary
1984	Andras Eke/Hungary
1985	Nathan A. Busch/U.S.A.
1986	Karlfried Groebe/West Germany
1987	Isumi Shibuya/Japan
1988	Kyung Kang/Korea-U.S.A.

The first society proceedings were published by Plenum Press[2a,b]. The number of total proceedings published has been confused by the mixing of two different publishers and the publishers mistaken use of two different names. Some of the first meeting proceedings were published under the Library of Congress Cataloging title of "International Symposium on Oxygen Transport to Tissue" rather than the official title of "International Society on Oxygen Transport to Tissue". Since the two title are listed separately the uninformed might not be aware of both sets of proceedings and many libraries do not have all volumes available.

D.F. Bruley

[1] OXYGEN SUPPLY - Theoretical and practical aspects of oxygen supply and microcirculation of tissue, edited by Manfred Kessler, Duane F. Bruley, Leland C. Clark Jr., Dietrich W. Lübbers, Ian A. Silver, and Jose Strauss, Urban & Schwarzenberg, 1973.

[2a] OXYGEN TRANSPORT TO TISSUE - Instrumentation, methods, and physiology, Edited by Haim I. Bicher and Duane F. Bruley, Advances in Experimental Medicine and Biology, Volume 37A, Plenum Press, 1973.

[b] OXYGEN TRANSPORT TO TISSUE - Pharmacology, mathematical studies, and nematology, Edited by Duane F. Bruley and Haim I. Bicher, Advances in Experimental Medicine and Biology, Volume 37B, Plenum Press, 1973.

CONTENTS

OPTICAL SPECTROSCOPY OF LIVING TISSUE

OTHER METHODS AND INSTRUMENTATION

MATHEMATICAL MODELS

SKELETAL MUSCLE

OTHER ORGANS AND TISSUES

TUMORS

INFORMATION

OPTICAL SPECTROSCOPY OF LIVING TISSUE

MULTICOMPONENT ANALYSIS OF NEAR-INFRARED SPECTRA OF ANESTHETIZED RAT HEAD:

(I) ESTIMATION OF COMPONENT SPECTRA BY PRINCIPAL COMPONENT ANALYSIS

Ryuichiro Araki and Ichiro Nashimoto

Department of Hygiene, Saitama Medical School
38 Morohongo, Moroyama, Iruma-gun, Saitama 350-04, Japan

INTRODUCTION

Near-infrared (NIR) spectrophotometry has been proposed for continuous and noninvasive monitoring of Hb oxygenation (So_2) and redox state of cyt. aa_3 in living tissues (Jöbsis, 1977 and others). In the previous articles, however, it has been assumed that only Hb and cyt. aa_3 contribute to changes in NIR spectra of the tissues. Since various biological materials exist in the living organs, spectral changes of the other kinds of biological chromophores may be involved in the changes in NIR spectra of the tissues. We attempted to establish a method for quantitating near-infrared spectra of anesthetized rat head using multiple regression analysis of the spectral data which is so-called the 'curve-fitting technique' (Araki and Nashimoto, this volume). However, the number and standard spectral curves of ingredients of the mixtures must be already known to establish quantitative analysis by this technique (Leggett, 1977).

A method for estimating the number and spectra of pure components from spectra of unknown mixtures with various relative concentrations has been developed by Sasaki et al. (1983). This method is based on principal component analysis and a constrained nonlinear optimization technique, and is capable of estimating the number of components and component spectral curves from a set of spectra of unknown mixtures with no use of a standard spectral library of the components. Even the principal component analysis is capable of indicating the number of spectral components included in the spectra of unknown mixtures.

By using this technique, we attempted to estimate the number of the components and component spectral curves from a set of NIR spectra of anesthetized rat head measured under acute hypoxia. The results will serve as a basis of quantitating near-infrared spectra in the living organs by means of the multiple regression analysis.

MATERIALS AND METHODS

Animals: Male albino rats of the Wistar strain (230-290 g body weight) fed on a commercial diet were used. Animals were anesthetized with sodium

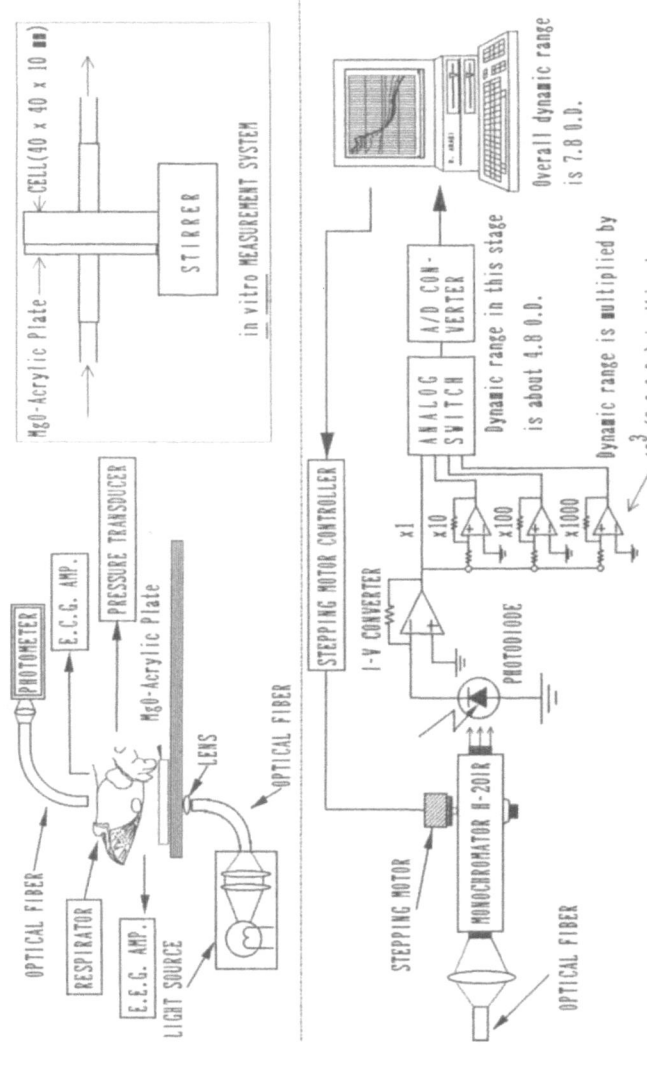

Figure 1. The experimental setup (upper) and the block diagram of the optical system (lower) used in this study. Optical density of the MgO-Acrylic plate was about 2 O.D.. Models of operational amplifiers used for the I–V converter and the gain amplifiers were AD549LH (Analog Devices Inc.) and OP27 (Precision Monolithic Inc.), respectively. Details are described in the text.

pentobarbital (50 mg/kg ip). After the trachea and femoral artery and vein were cannulated, the animals were immobilized with pancuronium bromide (0.5 mg/kg iv). Animals were then artificially respired (480 ml/kg/min) with a respirator (Harvard). The position of the head was fixed with a simple stereotaxis. ECG, EEG and arterial blood pressure were monitored continuously. When EMG got mixed in ECG, appropriate amounts of the anesthetics and the immobilizing agent were injected through the venous cannula to maintain anesthesia.

Near-infrared spectrophotometry: Near-infrared spectra of the light transmitted through the rat head was measured with a computer-controlled scanning spectrophotometer. Light longer than 680 nm from an air-cooled tungsten-iodine lamp (12 V, 100 W) was focused as a light spot of 2 mm diameter, and guided onto the top of the shaved head of the animal via a light guide of 5 mm diameter. An acrylic plate containing magnesium oxide powder (about 2 O.D.) served as an opal-glass was placed between the light guide and the head to diffuse the incident light completely. The mouth of the animal was opened widely with a forceps, and another light guide was inserted into the mouth to detect light transmitted through the brain. The transmitted light was then fed into a sensitive photodiode (Model S2386-8K, Hamamatsu Photonics, Japan) through a monochromator (Model H-20IR, Jobin-Yvon, France) equipped with a high-speed stepping motor. The photomultiplier used in the previous paper (Araki et al., 1988) was replaced with the photodiode to avoid deformation of the spectra in the wavelength range longer than 850 nm due to a steep decline of sensitivity of the photomultiplier and stray-light of the monochromator. The output signal of the photodiode via a current-voltage converter was then fed into an analog switch either directly or via 3 amplifiers of which gain were 10, 100 and 1000 to minimize errors due to A/D conversion of low signal level. The most appropriately amplified signals were used as the data, and changes in the amplifier gain were compensated in computation. An 8086-based computer (Model PC-9801M2, NEC) equipped with a stepping motor controller, an A/D converter, and a floating point math co-processor was used for wavelength scanning, data acquisition and calculations. Spectra in the wavelength range from 700 to 1000 nm was measured in 600 msec (500 nm/sec), and accumulation of 16 spectra were performed to obtain high S/N ratio. Real-time calculation of optical density was performed, and the spectra were displayed on a video display terminal at 10 sec intervals. All the data were also stored in floppy diskettes, and analyzed lately. A reference spectrum against an MgO-acrylic plate (about 2 O.D.) was measured before or after the experiment, and the data spectra were represented as absolute spectra. For in vitro measurements, an acrylic cell (40x40 mm, light path 10 mm) was used. An MgO-acrylic plate was placed between the light source and the cell to minimize effects of light-scattering. Fig. 1 summarizes the experimental setup and gives a block diagram of the optical system.

Resolution and accuracy of the optical system: Since resolution of the A/D converter was 12 bits, resolution of the optical signal was limited below 1/4095. However, 16 times of accumulation provides virtual resolution of 1/65520. In addition, the output signal of the photodiode was appropriately amplified by the amplifiers of the next stage. Therefore, overall dynamic range of 7.8×10^7 was achieved. The signal-to-noise ratio was less than 0.01% per full-scale, and the resolution of optical density was about 1×10^{-4} in the worst case. A wavelength resolution of 0.5 nm was obtained and half-band width was ±4 nm.

Computer analysis: Computer programs for the principal component analysis of the NIR spectra were written in C language according to the algorithm of Sasaki et al. (1983). The C source programs were then compiled with Microsoft C compiler (version 4.00, Microsoft Corp., U.S.A.) to obtain executable programs. Validity of the programs were checked with computer-simulated data and spectra of mixtures of dyes (bromocresol green, methyl orange, and indigo carmine) with various relative concentrations (data not shown).

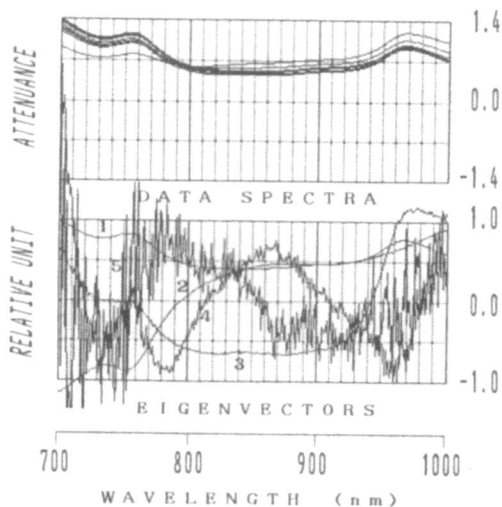

Figure 2. Data spectra measured during 100 sec of 5% oxygen inhalation (upper) and eigenvectors estimated from the data set by principal component analysis (lower).

RESULTS

Fig. 2 shows a set of NIR spectra of the rat head measured during 100 sec of 5% O_2 inhalation (upper) and eigenvectors estimated from the spectral data set by means of the principal component analysis (lower). The data set consisted of NIR spectra measured in 5 independent experiments. Since the eigenvectors were normalized by area to show the spectral characteristics of the eigenvectors obviously, care should be taken that the spectral intensities of the eigenvectors do not correspond to the concentrations of the components. The numbers attached to the eigenvectors represent the decreasing order of eigenvalues, which correspond to contribution factors. For instance, an estimated component of which eigenvalue is the largest one corresponds to the first principal component, and contributes to the spectral changes most significantly. It was shown that the 1st principal component in the NIR spectral changes caused by the short-term hypoxia was the mean of the data spectra, and the 2nd and 3rd principal components were attributable to Hb deoxygenation and reduction of cyt. aa_3. It should be noted that due to the limitations of the algorithm employed in the present study, our computer program is not capable of either giving the true spectra of the components or isolating component spectra of plural ingredients of which quantity changes in parallel. Thus the spectral curves and polarities of the 2nd and 3rd eigenvectors are not exactly true, and the spectra of the eigenvectors are represented as difference spectra of oxy- minus deoxy-Hb and oxidized-minus reduced- cyt. aa_3, respectively. The 4th and 5th eigenvectors were not reproducible and had negligible eigenvalues. The water absorption peak around 960-970 nm could not be isolated from the eigenvectors.

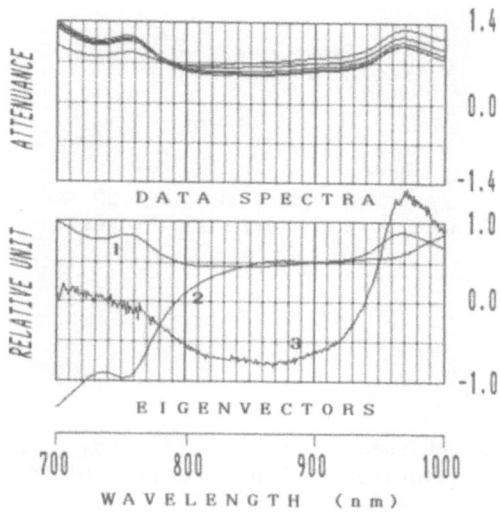

Figure 3. NIR spectra measured during the early 50 sec of hypoxia (upper) and eigenvectors estimated from the data set by principal component analysis (lower). The 4th and 5th eigenvectors were not determined because these eigenvectors were not reproducible and had negligible eigenvalues.

In Fig. 3, NIR spectra measured during the early 50 sec of 5% O_2 inhalation was used as data spectra. The 1st principal component corresponded to the mean of the data spectra, and the 2nd and 3rd principal components were Hb deoxygenation and reduction of cyt. aa_3 as seen in Fig. 2. The values of $\log_{10}$(eigenvalue) for the components 1, 2

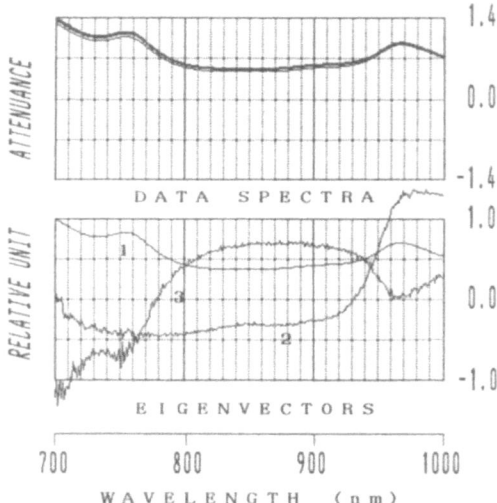

Figure 4. NIR spectra measured during the last 50 sec of hypoxia (upper) and eigenvectors estimated from the data set (lower).

Table 1. Eigenvalues obtained in the present study. The eigenvalues were represented as $\log_{10}$(eigenvalue). N.D., not determined.

$\log_{10}$(eigenvalue)

	Fig. 2	Fig. 3	Fig. 4
No. of eigenvector			
1	0.780	0.457	0.571
2	-1.359	-1.535(Hb)	-3.618(cyt)
3	-3.222	-3.865(cyt)	-4.066(Hb)
4	-5.092	N.D.	N.D.
5	-6.129	N.D.	N.D.

and 3 were 0.457, −1.535, and −3.865, respectively. Thus, the major part of the changes in NIR spectra during the early 50 sec of hypoxia was attributable to Hb deoxygenation. In Fig. 4, a spectral set measured during the last 50 sec of 5% O_2 inhalation was used as data spectra. In contrast to the results shown in Fig. 3, the eigenvector of the 2nd principal component corresponded to reduction of cyt. aa_3, followed by Hb deoxygenation in the latter half of the hypoxic period. The values of log_{10}(eigenvalue) for components 1, 2 and 3 were 0.571, −3.618 and −4.066, respectively. Table 1 summarizes the eigenvalues obtained in this study.

DISCUSSION

This is the first attempt to apply the principal component analysis to the NIR spectra of living organs for estimating the number and spectral characteristics of the components. It was found that meaningful eigenvectors estimated from the spectral data set measured during 100 sec of 5% O_2 inhalation corresponded to the mean of the data spectra, Hb deoxygenation and reduction of cyt. aa_3. The 4th and 5th eigenvectors were not reproducible, and their eigenvalues were 10^6 and 10^7-fold lower than that for the 1st principal component, respectively. The principle of the principal component analysis predicts that the number of eigenvectors which have positive eigenvalues is equal to the true number of components included in the data (Sasaki et al., 1983). The actual data, however, always contain instrumental noise components. The signal-to noise ratio of our instrument is 80 dB (10000:1 in the worst case). In addition, the resolution of optical density in the worst case is limited to be 10^{-4} due to the A/D conversion in spite of the effort to minimize such errors. Thus false components may be estimated from the data spectra. The 4th and 5th components shown in Fig. 2 are considered to be false components because these were not reproducible and had very small eigenvalues. Therefore, we conclude that principal components in the NIR spectral changes caused by acute hypoxia are Hb deoxygenation and reduction of cyt. aa_3, and contribution of the remaining biological materials to the changes in NIR spectra are negligible in quantitative multivariate analysis (curve-fitting method). Estimation of principal components in the changes in NIR spectra under the other experimental conditions (e.g., chronic hypoxia, hyperbaroxia, carbon monoxide inhalation and administration of cyanide) is under investigation, and will be published elsewhere.

In the early 50 sec of hypoxia, the 2nd principal component was Hb deoxygenation, and had 200-fold larger eigenvalue than that of the 3rd principal component corresponding to reduction of cyt. aa_3 (Fig. 3 and Table 1). In contrast, the eigenvalue for reduction of cyt. aa_3 was comparable as that of Hb deoxygenation in the last 50 sec of hypoxia (Fig. 4 and Table 1). Despite that the method used in this study is qualitative (Sasaki et al., 1983), the comparison of the contribution factors (eigenvalues) seems reasonable. These results suggest that Hb deoxygenation was almost completed within 50 sec of hypoxia, and reduction of cyt. aa_3 continued after completion of Hb deoxygenation. There seems two possible explanations for the experimental results: 1) Full-reduction of cyt. aa_3 occurs at extremely low oxygen tension in the brain in vivo. 2) There is difference in time course of Hb deoxygenation and that of reduction of cyt. aa_3 caused by acute hypoxia. With respect to this point, further studies by means of multivariate analysis of Hb and cyt. aa_3 are required.

SUMMARY

By measuring precise NIR spectra of anesthetized rat head, we determined principal components included in changes of the NIR spectra caused by acute hypoxia. There was apparent delay in reduction of cyt. aa_3 as compared with Hb deoxygenation in the hypoxic period. With regard to the eigenvalues determined by principal component analysis, principal components in the NIR spectral changes caused by acute hypoxia were Hb and cyt. aa_3, and contribution of the remaining biological materials to the changes in NIR spectra caused by acute hypoxia was considered to be negligible in quantitative multivariate analysis of Hb and cyt. aa_3 <u>in situ</u>.

REFERENCES

Araki R., Nashimoto I., and Takano T., 1988, The effect of hyperbaric oxygen on cerebral hemoglobin oxygenation and dissociation rate of carboxyhemoglobin in anesthetized rats: Spectroscopic approach, *Adv. Exp. Med. Biol.*, 222:375-381.

Jöbsis F. F., 1977, Noninvasive, infrared monitoring of cerebral and myocardial oxygen sufficiency and circulatory parameters, *Science*, 198:1264-1267.

Leggett D. J., Numerical Analysis of Multicomponent Spectra, 1977, *Anal. Chem.*, 49:276-281.

Sasaki K., Kawata S., and Minami S., 1983, Constrained Nonlinear Method for Estimating Component Spectra from Multicomponent Mixtures, *Appl. Opt.*, 22:3599-3603.

MULTICOMPONENT ANALYSIS OF NEAR-INFRARED SPECTRA OF ANESTHETIZED RAT HEAD: (II) QUANTITATIVE MULTIVARIATE ANALYSIS OF HEMOGLOBIN AND CYTOCHROME OXIDASE BY NON-NEGATIVE LEAST SQUARES METHOD

Ryuichiro Araki and Ichiro Nashimoto

Department of Hygiene, Saitama Medical School
38 Morohongo, Moroyama, Iruma-gun, Saitama 350-04, Japan

INTRODUCTION

There have been a number of attempts to estimate oxygen saturation in Hb (So_2) and redox state of cyt. aa_3 in the living tissues by means of near-infrared (NIR) spectrophotometry[3] (Jöbsis, 1977 and others). In the early stage of the investigation, the simple dual-wavelength technique has been used to estimate changes in So_2 and redox state of cyt. aa_3. However, one can not distinguish changes in O.D. due to changes in So_2 and in blood volume by means of the simple dual-wavelength method. In order to estimate changes in blood volume, several authors have used changes in O.D. at 805 nm which is the isosbestic point for oxygenation-deoxygenation of Hb. However, cyt. aa_3 has absorption band in this wavelength region, and thereby, quantitative analysis has been hampered (Hazeki and Tamura, 1988). In addition, there seems no appropriate wavelength-pair for estimating changes in redox state of cyt. aa_3 by means of this method without interferences resulted from changes in So_2 and/or in blood volume because both of the wavelengths must be isosbestic points for oxygenation-deoxygenation of Hb, and O.D. at these isosbestic points must be equal for such purpose.

Recently, the limitations of the simple dual-wavelength technique for quantitating these chromophores of biological interest have been widely recognized. Several authors (Hazeki et al., 1987; Cope et al., 1988; Hazeki and Tamura, 1988; Seiyama et al., 1988) have attempted to establish quantitative analysis of NIR data. However, most of these attempts were based on triple-wavelength equations derived from three simultaneous equations in three unknowns, and aimed to estimate relative changes in So_2, blood volume and redox state of cyt. aa_3. NIR spectra of living tissues consist of 4 component spectra, namely, spectra of oxy- and deoxy-Hb, and of oxidized and reduced cyt. aa_3 as confirmed by our investigation on the basis of principal component analysis (Araki and Nashimoto, this volume). In addition, there is a scattering background due to optical turbidity of the tissue. Our final goal is to establish absolute quantitative analysis of So_2 and redox state of cyt. aa_3. For this purpose, O.D. at five different wavelengths must be measured[3] because five unknowns are involved in changes of optical attenuance in NIR region.

Quantitative multivariate analysis of multicomponent spectrophotometric data on the basis of multiple regression analysis has been developed and improved by several authors (Sternberg et al., 1960; Blackburn, 1965; Leggett, 1977). The so-called curve-fitting technique has been widely used by analytical chemists as a powerful tool for quantitating ingredients in the multicomponent mixtures simultaneously. Hoffmann and Lübbers (1986) employed the technique to quantitate reflection spectra obtained from the surface of the living organs by applying two-flux theory of Kubelka and Munk (Kubelka, 1948). We have developed a computer program on the basis of non-negative least squares and simplex optimization algorithms (Leggett, 1977). This program is applicable to the quantitative analysis of NIR spectra of living tissues measured with our optical system. By using this technique, we examined fundamental problems which must be solved to quantitate absolute value of So_2 and that of redox state of cyt. aa_3.

MATERIALS AND METHODS

NIR spectrophotometry of the anesthetized rat head. Near-infrared spectra of the anesthetized rat head was measured with a computer-controlled scanning spectrophotometer. The experimental conditions were essentially the same as described previously (Araki et al., 1988; Araki and Nashimoto, this volume) except that wavelength range was limited to be 700-900 nm to avoid interference by absorption band of water around 965 nm. Hyperbaric oxygen at 3 ATA (atmospheric absolute) was used to induce well-oxygenated state of the brain tissue.

Multiple regression analysis of spectrophotometric data. The computer program for multiple regression analysis (curve-fitting method) of NIR spectrophotometric data was written in C language according to non-negative linear least squares and simplex optimization algorithms (Leggett, 1977). Validity of the program was checked with computer-simulated data and actual spectral data measured with mixtures of dyes (bromocresol green, methyl orange, and indigo carmine) with various relative concentrations.

Standard spectra of Hb and cyt. aa_3. Standard spectra of Hb solution, suspension of red blood cells (RBC) and solution of purified cyt. aa_3 were measured with the same apparatus used for the measurement of the animals in situ. Heparinized blood of the rat was centrifuged and washed with cold physiological saline three times to collect RBC. Finally the packed RBC were lysed with distilled water, and 10% (final concentration in v/v) of hemolysate was obtained. A few drops of 10% NH_4OH was added into the hemolysate to establish complete hemolyzation. The ghosts were removed by centrifugation and filtration through a filter (0.22 μm). The hemolysate was used as Hb solution. 10% RBC suspension was obtained by suspending RBC in cold physiological saline containing 50 mM of phosphate buffer (pH 7.4). Purified cyt. aa_3 was obtained from Sigma Chemical Co., and dissolved in 50 mM of phosphate buffer (pH 7.4). Aerobic and anaerobic conditions were obtained by bubbling the fluids with oxygen and adding an adequate amount of sodium dithionite into the fluids, respectively. Fig. 1 shows the standard spectra used in the present study.

Assumptions and terminology. We assumed the following items in this study:

1) Spectral intensity of each component is proportional to the quantity of each one.
2) A spectrum of a mixture can be represented as a sum of spectra of the components in the mixture.
3) Quantity of each component does not have negative value.
4) Intensity of light scattering is wavelength-independent and the optical

attenuance spectrum can be represented as a flat spectrum in NIR region.

The assumptions 1 and 2 are descriptions of linear additivity of the spectral intensities of the components. The assumption 3 is self-evident, and this restricted condition was used in the algorithm of the computer program to establish more precise analysis (Leggett, 1977; the terminology "non-negative least squares" was derived from this restricted condition). Assumption 4 will be discussed in "DISCUSSION".

In this paper, the term "optical attenuance" will be used instead of "absorbance" because light transmitted through the tissue is attenuated by light scattering due to turbidity of the tissue as well as light absorption by the chromophores. Definition of optical attenuance is similar to that of absorbance ($\log_{10}(I_0/I_t)$, where I_0 and I_t are intensity of the light transmitted through the reference and the sample, respectively).

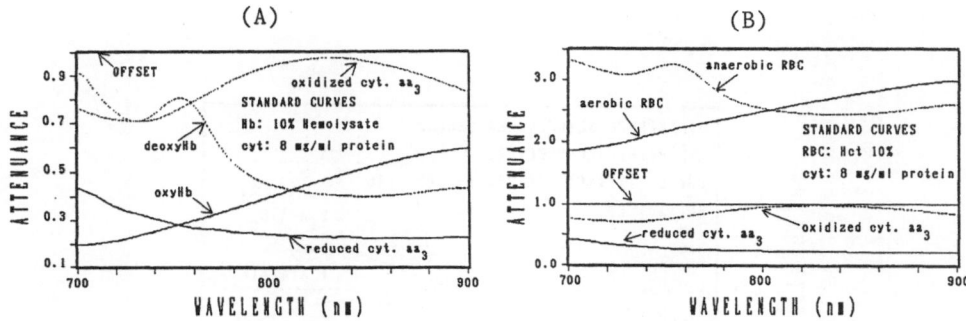

Figure 1. Standard spectra used for the multiple regression analysis. Part A: spectra of oxy- and deoxy-Hb were measured using hemolysate as Hb solution. Part B: spectra of oxygenated and deoxygenated RBC were measured using RBC suspension of 10% hematocrit.

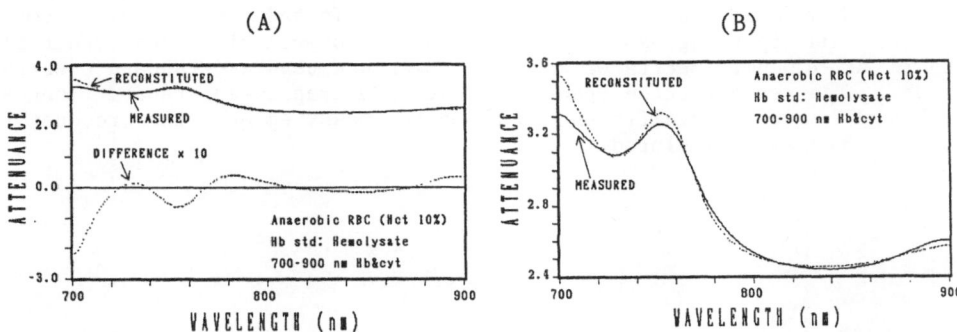

Figure 2. Part A: an NIR spectrum of RBC suspension (hematocrit 10%) measured under the anaerobic condition and a best-fit curve reconstituted from standard spectra including spectra of Hb solution. A difference spectrum between the measured and the reconstituted spectra was calculated and indicated in the figure. Part B: the ordinate of Fig. 2A was magnified.

RESULTS

Spectral properties of RBC suspension. Fig. 2A shows a spectrum of RBC suspension in anaerobic state and a best-fit curve reconstituted from spectra of hemolysate and cyt. aa₃ solution. Contents of both oxidized and reduced cyt. $_3$ were determined to be zero, and So$_2$ was estimated to be 0%. However, there was considerable difference between the measured and the reconstituted spectra. This phenomenon is resulted from the flattening of the absorption spectrum of suspensions as compared to that of solutions due to localization of light absorber in the restricted region (Duysens, 1956). In Fig. 2B, the ordinate of Fig. 2A was magnified to show the flattening effect obviously. The peak around 755 nm was apparently flattened, and the attenuance spectrum of the RBC suspension in 700-730 nm, in which Hb has greater absorption than in the other wavelength range, was apparently lowered. Thus, we conclude that spectra of Hb solution can not be used as standard spectra of oxy- and deoxy-Hb to quantitate So$_2$ and blood volume in the tissue by means of multicomponent analysis.

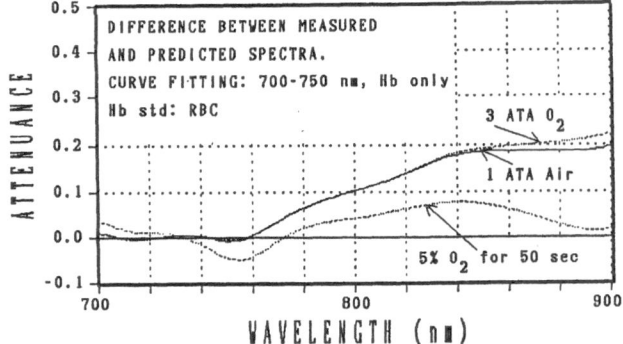

Figure 3. Extraction of the spectrum of cyt. aa₃ from the NIR spectra of the rat head measured under the control, hypoxic and hyperbaroxic conditions. Multiple regression analysis was performed using the spectral data in 700-750 nm. Spectra of RBC suspension was used as standard spectra of Hb. The reconstituted spectra were then extrapolated toward longer wavelength range, and difference between the measured and the reconstituted spectra was calculated.

Extraction of the spectrum of cyt. oxidase from NIR spectra __in situ__. Oxidation-reduction of cyt. aa₃ causes little changes in absorption in the wavelength range from 700 to 750 nm. Thus, most of the changes in the NIR spectra of the living tissues in this wavelength range are attributable to changes in So$_2$ and/or blood volume. We performed multiple regression analysis on NIR spectra of the rat head by using the data in this wavelength range, and a best-fit curve was calculated. In Fig. 3, the reconstituted spectra were then extrapolated toward longer wavelength range (750-900 nm), and difference between the measured and the reconstituted spectra were calculated. Optical attenuance of the difference spectrum calculated from the NIR spectrum measured under 1 ATA inhalation was about 0.2 in 840-900 nm. The spectral intensity of the difference spectrum was decreased when the animal inspired 5% O$_2$ for 50 sec, and decrease in O.D. at 840 nm was about 0.1. 3 ATA of hyperbaric oxygen caused only a little change in the difference spectrum as compared with that obtained under normal air-breathing.

Flattening effect of the spectrum of RBC in situ. Fig. 4 shows the NIR spectrum of the rat head measured under 1 ATA air breathing and the best-fit curves. When spectra of Hb solution was used as the standard spectra (Fig. 4A and B), apparent spectral flattening of the in situ spectrum in 700–760 nm was observed (Fig. 4B) as well as in vitro experiments (Fig. 2). When spectra of RBC suspension was used as standard spectra (Fig. 4C and D), the difference between the measured and the reconstituted spectra in the range from 700–760 nm was diminished. However, the measured spectrum was still slightly flattened as compared with the reconstituted one (Fig. 4D), and apparent difference around 840 nm, where absorption band of oxidized cyt. aa_3 exists, was observed. The flattening effect of the spectrum of RBC in situ appeared more apparently when the animal breathed 5% O_2 (Fig. 5A and B).

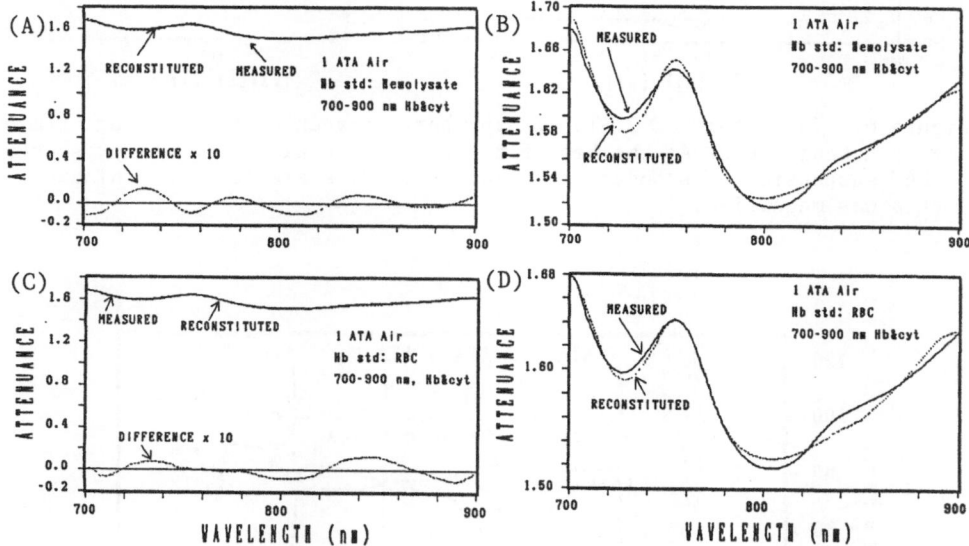

Figure 4. The NIR spectrum of the rat head measured under normal air-breathing and the best-fit curves. Part A: spectra of Hb solution were used as standard spectra. Part C: spectra of RBC suspension were used as standard spectra. Part B and D: scale for the ordinate of part A and C was magnified.

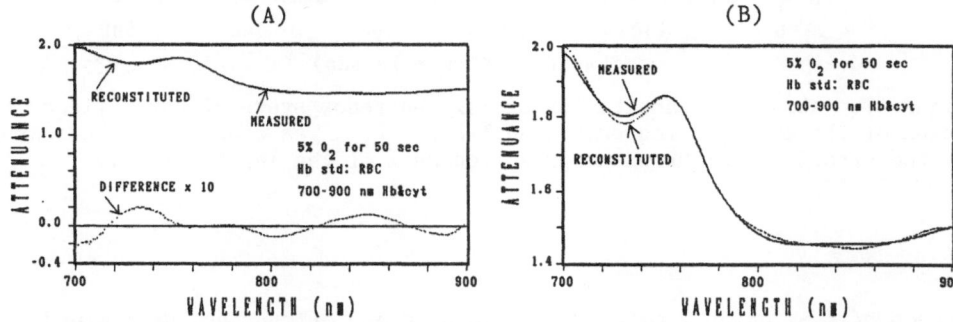

Figure 5. The effect of 5% O_2 inhalation for 50 sec on the NIR spectrum of the rat head. Part A: the best-fit curve was recalculated using spectra of RBC suspension as standard spectra of oxy- and deoxy-Hb, respectively. Part B: scale for the ordinate of part A was magnified.

Effect of hyperbaric oxygen on the NIR spectrum of the rat head. Fig. 6
shows the effect of 3 ATA of hyperbaric oxygen on the NIR spectrum of the
rat head. As compared with the spectra shown in Fig. 4 and 5, O.D. in
700-800 nm, where absorption of deoxy-Hb exists, was apparently decreased,
while O.D. in 800-900 nm was increased, indicating increase in So_2.
However, the spectral characteristics in 700-760 nm showed that Hb was not
fully oxygenated even under 3 ATA oxygen (cf. Fig. 1).

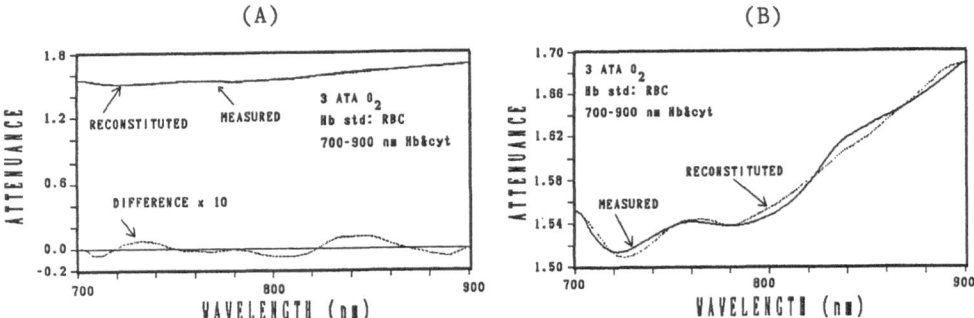

Figure 6. The effect of 3 ATA of hyperbaric oxygen on the NIR spectrum of
the rat head. Part A: the best-fit curve was recalculated using spectra
of RBC suspension as standard spectra. Part B: scale for the ordinate of
part A was magnified.

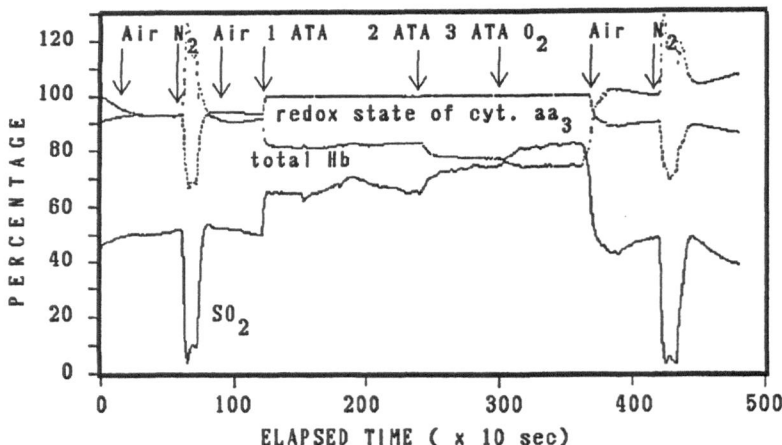

Figure 7. Absolute quantitation of So_2 and redox state of cyt. oxidase by
means of the multiple regression analysis. F_{IO2} was changed as indicated
by the arrows. Total Hb was represented as % of the initial level.

*Effect of F_{IO2} on So_2, total Hb and redox state of cyt. aa_3 determined by
the multiple regression analysis.* Fig. 7 shows the effect of F_{IO2} on So_2,
total Hb and redox state of cyt. aa_3 determined by means of the present
method. All the spectra measured at 10 sec intervals and stored in the
floppy diskette were subjected to the multiple regression analysis, and

the results were plotted against elapsed time. Total Hb was represented as % of the initial level. So$_2$ and redox state of cyt. aa$_3$ were estimated to be 50 and 90% under normal air-breathing, respectively. 5% O$_2$ inhalation caused almost complete deoxygenation of Hb and increase in total Hb (about 130% of control), whereas redox state of cyt. aa$_3$ did not fall below 65%. Inhalation of 1, 2 and 3 ATA O$_2$ induced stepwise increase in So$_2$ up to 80% and decrease in total Hb down to 70%, respectively. In contrast, cyt. aa$_3$ was fully oxidized when the animal breathed 1 ATA O$_2$, and further oxidation did not occur even under 3 ATA O$_2$.

DISCUSSION

By using the curve-fitting technique, we examined problems to be solved for establishment of absolute quantitative analysis of So$_2$ and redox state of cyt. aa$_3$ in the anesthetized rat head by means of NIR spectrophotometry. Advantages of measuring spectral data over multi-wavelength spectrophotometry are as follows: 1) Analysis can be performed more precisely by applying the least squares method. 2) Artifacts due to unexpected incidents (e.g., movement of the position of the measuring site, instrumental artifacts) can be easily recognized. 3) Estimation of absolute (not relative) value of So$_2$ and that of redox state of cyt. aa$_3$ can be performed. It should be noted that this method is not capable of calculating absolute quantity of components because of difficulties in estimating effects of changes in light path length and multiple scattering.

Flattening effect of the spectrum of Hb in vitro and in vivo. We observed apparent flattening effect of the spectrum of Hb in RBC suspension as compared to that in Hb solution in NIR region (Fig. 2). Shimada et al. (1984) reported that in NIR region the flattening effect due to localization of Hb within RBC is smaller than that in UV or visible region. Our results, however, demonstrated that there are considerable differences between the spectrum of Hb in solution and that in suspension which may affect the estimation of So$_2$ and changes in blood volume. Such errors may also interfere estimation of redox state of cyt. aa$_3$.

We also found that the spectrum of blood in the brain tissue was more flattened as compared with the spectrum of RBC suspension *in vitro*. As has been described by Duysens (1956), the flattening is enhanced when the degree of localization of the dye within the particles becomes higher. In addition, the flattening is stronger in suspension of bullet-like particles than in that of spherical particles. This phenomenon, thereby, seems resulted from localization of RBC within the vessels as well as the deformation of RBC in the microcirculatory system.

Wray et al. (1988) reported precise absorption coefficients of oxygenated and deoxygenated Hb solution in NIR region. By using the parameter obtained *in vitro* and solving three simultaneous equations, they proposed an algorithm for calculations of oxy- and deoxy-Hb, and redox state of cyt. aa$_3$. By using the blood-perfused rat head preparation, Hazeki et al. (1987) determined multiplying factors for the three simultaneous equations which consisted of changes in optical attenuance at 3 different wavelengths (700, 730 and 830 nm) for estimating changes in So$_2$, total Hb, and redox state of cyt. aa$_3$. The present results, however, suggest that the optical parameters *in vitro* can not be simply applied for estimating So$_2$, total Hb, and redox state of cyt. aa$_3$ and the multiplying factors for multi-wavelength equation can be varied due to changes in the degree of localization of RBC within the vessels and/or deformation of RBC in the microcirculatory system. Thus, even if the spectra of RBC suspension *in vitro* are used as standard spectra, the results of multiple regression analysis will contain considerable errors due to the flattening effect of the spectrum of Hb.

Wavelength-dependency of scattering intensity in NIR region. In the present study, intensity of light scattering was assumed to be wavelength-independent. The basis for this assumption is as follows: 1) Wendlandt and Hecht (1966) examined relationship between size of light-scattering particles and wavelength-dependency of scattering intensity in visible region, and found that scattering intensity is wavelength-independent when the particle size is larger than 2 µm. 2) We found that RBC suspension (10% hematocrit) containing MgO powder (3-6 µm, about 2 O.D./cm light path length) equilibrated with carbon monoxide gas gave an almost flat spectrum in 700-900 nm. Also, The NIR spectra of the rat head gave an almost flat spectrum in 700-900 nm when the animal inspired 80% O_2 + 20% CO under 3 ATA for 30 min and CO-Hb content in the venous blood became more than 97% (data not shown). Since CO-Hb has no significant absorption and the spectral intensity of reduced cyt. aa_3 is quite small in NIR region, the flat spectrum was considered to be the scattering background.

Redox state of cyt. aa_3 in situ. In the present study, cyt. aa_3 was fully oxidized when the animal inspired pure oxygen above 1 ATA (Fig. 7). By using dual-wavelength reflectance spectrophotometry in visible region, Hempel et al. (1977) reported that under resting air-breathing conditions, heme moiety in cyt. aa_3 is about 85% reduced in the cat cerebral cortex. Instead we found that copper in cyt. aa_3 is over 90% oxidized under the conditions (Fig. 7). This discrepancy might be explained as follows: 1) This is due to the limitation of dual-wavelength technique with regard to quantitating cyt. aa_3 signal without interferences due to changes in So_2 and/or in blood volume. 2) There might be differences in oxygen characteristics of heme moiety and copper in cyt. aa_3.

Problems in the present algorithm for computation. The algorithm used in this study still has the following theoretical problems.

1. Non-linear relationship between concentration and optical density of suspensions of light-scattering absorber predicted by Kubelka-Munk's theory. According to Kubelka-Munk's theory (1948), light-scattering absorber such as RBC does not obey Lambert-Beer's law. In cuvette experiments, this non-linear relationship has been shown by Steinke and Shepherd (1986). Instead Seiyama et al. (1988) found that a linear relationship existed between the absorbance and the concentration of blood in a perfused rat thigh muscle. Thus the non-linear component does not seem predominant in quantitating NIR spectra *in vivo*. Even if the linear relationship does not exist in the tissue, we can quantitate NIR spectra by using appropriate transformation equations determined experimentally.

2. Flattening effect of the spectrum of RBC in vivo. We found that the spectrum of RBC *in vivo* is more flattened than that *in vitro* (cf. Fig. 2 and 5). The extent of the flattening between RBC *in vitro* and *in vivo* was smaller than that between RBC suspension and Hb solution (Fig. 2), but not negligible (a sum of least squares was about 0.04 O.D. in the range from 700 to 750 nm, cf. Fig. 5). By using fluorocarbon-transfused rats, Hazeki et al. (1988) and Wray et al. (1988) have shown that the absorption change around 830-840 nm caused by the reduction of cyt. aa_3 alone was 0.05-0.1 O.D. in the brain, and the value was comparable with the error caused by the flattening effect of the spectrum of Hb *in vivo*. In fact, when we performed the analysis using the data of 700-750 nm measured under normal air breathing, So_2 was estimated to be 30%, which did not agree with the results shown in Fig. 7. Thus, parameters determined *in vitro* can not be simply applied to quantitate So_2 and redox state of cyt. aa_3 *in vivo* even if RBC is being used to determine the optical parameters of oxy-

and deoxy-Hb. In addition, it has been reported that the spectrum of RBC suspension, which is being stirred, is more flattened as compared with that in the stationary state (Tyuma, 1966). Thereby, changes in circulatory parameters may vary the extent of the flattening of the spectrum of RBC _in vivo_. In order to overcome this problem, we are planning to revise our computer program to consider the changes in the flattening factor for the spectra of oxy- and deoxy-Hb in order to establish more precise analysis.

3. *Difference of the spectrum of cyt. aa$_3$ in vitro and in vivo*. Keizer et al. (1985) have examined the NIR absorption band of cyt. aa$_3$ in purified enzyme, isolated mitochondria and in the intact brain of fluorocarbon-transfused animals, and found that there is an apparent difference between _in vitro_ and _in vivo_ spectra of cyt. aa$_3$. The differences between the NIR spectra around 840 nm measured under 1 ATA air (Fig. 4) or 3 ATA O$_2$ (Fig. 6) and recalculated spectra seem to be attributable to the difference of the spectrum of cyt. aa$_3$ _in vivo_ and _in vitro_. The details of spectral characteristics of cyt. aa$_3$ _in vivo_ and _in vitro_ are still controversial, and must be solved to establish quantitative analysis of redox state of cyt. aa$_3$ by means of the curve-fitting technique. The use of spectra of mitochondria or those of the fluorocarbon-transfused rat as standard spectra is another solution to this problem.

SUMMARY

We developed a method for quantitating absolute value of Hb oxygenation and that of redox state of cyt. aa$_3$ on the basis of multiple regression analysis of near-infrared spectrophotometric data. Flattening of the spectrum of Hb in both RBC suspension and the brain _in situ_ was observed. This phenomenon was explained by localization of Hb within RBC for _in vitro_ flattening, and by localization of RBC within the vessels as well as deformation of RBC in the microcirculation _in vivo_. Under resting air-breathing conditions, So$_2$ and redox state of cyt. aa$_3$ were estimated to be 50-70% and over 90%, respectively. Increase in F$_{IO2}$ up to 3 ATA O$_2$ caused stepwise increase in So$_2$, whereas cyt. aa$_3$ was fully oxidized when the animal inspired O$_2$ under 1 ATA. Problems which must be solved for more accurate estimation of absolute values of So$_2$ and redox state of cyt. aa$_3$ were described and discussed.

REFERENCES

Araki R., Nashimoto I., and Takano T., 1988, The effect of hyperbaric oxygen on cerebral hemoglobin oxygenation and dissociation rate of carboxyhemoglobin in anesthetized rats: spectroscopic approach, *Adv. Exp. Med. Biol.*, 222:375-381

Blackburn J.A., 1965, Computer program for multicomponent spectrum analysis using least squares method, *Anal. Chem.*, 37:1000-1003

Cope M., Delpy D.T., Reynolds E.O.R., Wray S., Wyatt J., and van der Zee P., 1988, Methods of quantitating cerebral near infrared spectroscopy data, *Adv. Exp. Med. Biol.*, 222:183-189

Duysens L.N.M., The flattening of the absorption spectrum of suspensions, as compared to that of solutions, *Biochim. Biophys. Acta*, 19:1-12

Hazeki O., Seiyama A., and Tamura M., 1987, Near-infrared spectrophotometric monitoring of haemoglobin and cytochrome a,a$_3$ _in situ_, *Adv. Exp. Med. Biol.*, 215:283-289

Hazeki O. and Tamura M., 1988, Quantitative analysis of hemoglobin oxygenation state of rat brain in situ by near-infrared spectrophotometry, *J. Appl. Physiol.*, 64:796-802

Hempel F.G., Jöbsis F.F., LaManna J.L., Rosenthal M.R. and Saltzman H.A., 1977, Oxidation of cerebral cytochrome aa_3 by oxygen plus carbon dioxide at hyperbaric pressures, *J. Appl. Physiol.*, 43:873-879

Hoffmann J. and Lübbers D.W., 1986, Estimation of concentration ratios and the redox states of the cytochromes from noisy reflection spectra using multicomponent analysis methods, *Adv. Exp. Med. Biol.*, 200:119-124

Hoffmann J. and Lübbers D.W., 1986, Improved quantitative analysis of reflection spectra obtained from the surface of the isolated perfused guinea pig heart, *Adv. Exp. Med. Biol.*, 200:125-130

Jöbsis F.F., 1977, Noninvasive, infrared monitoring of cerebral and myocardial oxygen sufficiency and circulatory parameters, *Science*, 198:1264-1267

Keizer H.H., Jöbsis F.F., Lucas S.S., Piantadosi C.A., and Sylvia A.L., 1985, The near infrared (NIR) absorption band of cytochrome aa_3 in purified enzyme, isolated mitochondria and in the intact brain in situ, *Adv. Exp. Med. Biol.*, 191:823-832

Kubelka K., 1948, New contributions to the optics of intensely light-scattering materials, part I, *J. Opt. Soc. Am.*, 38:448-457

Leggett D.J., 1977, Numerical analysis of multicomponent spectra, *Anal. Chem.*, 49:276-281

Seiyama A., Hazeki O., and Tamura M., 1988, Noninvasive quantitative analysis of blood oxygenation in rat skeletal muscle, *J. Biochem.*, 103:419-424

Shimada Y., Yoshiya I., Oka N., and Hamaguri K., 1984, Effects of multiple scattering and peripheral circulation on arterial oxygen saturation measured with a pulse-type oximeter, *Med. Biol. Eng. Comput.*, 22:475-478

Sternberg J.C., Stillo H.S., and Schwendeman R.H., 1960, Spectrophotometric analysis of multicomponent systems using the least squares method in matrix form, *Anal. Chem.*, 32:84-90

Steinke J.M. and Shepherd A.P., 1986, Role of light scattering in spectrophotometric measurements of arteriovenous oxygen difference, *IEEE Trans. Biomed. Eng.*, BME-33:729-734

Steinke J.M. and Shepherd A.P., 1986, Role of light scattering in whole blood oximetry, *IEEE Trans. Biomed. Eng.*, BME-33:294-301

Tyuma I., 1969, Spectrophotometrical studies on erythrocyte suspensions by opal glass method-An investigation on the nature of flattening of soret band, *Bunko Kenkyu*, 8:86-93 (*in Japanese*)

Wendlandt W.W. and Hecht H.G., 1966, *in*: "Reflectance Spectroscopy", Interscience, New York, pp65, Interscience, New York

Wray S., Cope M., Delpy D.T., Wyatt J.S., and Reynolds E.O.R., 1988, Characterization of the near infrared absorption spectra of cytochrome aa_3 and haemoglobin for the non-invasive monitoring of cerebral oxygenation, *Biochim. Biophys. Acta*, 933:184-192

TIME RESOLVED SPECTROSCOPIC (TRS) AND CONTINUOUS WAVE

SPECTROSCOPIC (CWS) STUDIES OF PHOTON MIGRATION IN HUMAN ARMS AND LIMBS

Britton Chance

Department of Biochemistry and Biophysics, University of
Pennsylvania, Philadelphia, PA, USA

INTRODUCTION

The optical study of tissues began with the spectroscopic studies of
Glenn Millikan in 1935 who proposed a "metabolic microscope" by which he
would follow metabolic demand as expressed by the deoxygenation of myoglobin and hemoglobin in tissue. This was beautifully demonstrated in his
studies of the cat soleus muscle during functional actvity (tetanic
contraction and ischemia) (1). While the optical changes could be attributed to both hemoglobin and myoglobin, the demonstration of the
effectiveness of the dual-wavelength technique using a differential
detector and color filters was established in his pioneer studies.
Applications to humans emerged in 1940 (2) with the "Millikan Oximeter"
which was applied to the lobe of the ear, and using the same princiles,
this presaged the popular "pulse oximeter" as applied to the human finger
tip (3). Neither of these approaches presumed to provide intracranial
homoglobinometry. Thus the results of Jobsis-Vander Vleit and later
Piantadosi on transcranial spectroscopy are noteworthy. They have evolved
a much more sophisticated instrument which attempts to deconvolute
cytochrome from hemoglobin and myoglobin changes in the exercising muscle
(4,5). Such studies have been vexed by an unknown optical path requiring
the need for either speculation or transfer of data from one model to
another in abortive attempts to convert what has been termed justifiably a
"trend indicator" to a quantitative spectroscopic technique.

The problem in essence is that photons, having traversed vastly
different optical pathlengths, are indiscriminately accepted by a CWS
photodetector and indeed such a photodetector is likely to emphasize the
higher intensity, short path, or even "short circuit" photons. Thus
attempts to assign concentrations, or concentration changes, to
observations in human tissues have been unsuccessful and available reports
of concentration changes should be accepted with caution (6).

Two approaches have generally been employed: the first is to employ
NADH fluorometry in which the pathlength is so short compared to the
tissue volume that no expectations of pathlength variation in transfer
from model to model is involved Thus, the NADH fluorometer may be used on

a relative basis or may be calibrated by using standard solutions (7). The pathlength in the standard solution and _in vivo_ must be matched.

The second approach is different: here we time resolve the photon migration in tissue to accept for measurement only a small segment of the total pathlength, made feasible by time selection. In this way, the optical pathlength of the system is clearly defined. As we shall show, concentration changes are readily calculated with a single beam method, and with a dual wavelength method, absolute concentrations appear to be determinable. The pathlength of photon migration in tissue is measured from the velocity of light. This velocity is not altered by scattering and depends upon the refractive index of the medium (8-11).

Two unexpected benefits of this approach have emerged, first is that the decay of intensity of emitted photons from tissue appears to be exponential over a wide range, i.e. 4-5 orders of magnitude of intensity, and thus photons having traveled short paths, 2 or 3 cm, appear to experience the same absorbance as those travelling 20-30 cm in normal hemogeneous brain tissue. However, spatial discontinuities of absorption might be expected to be detected by this method, for example in tumors. Secondly, simultaneous deployment of CSW and TRS at corresponding places in the brain and with similar input-output geometry, affords a calculation of the effective pathlength for the CSW, thereby allowing a quantitation of data from this technique hitherto not possible.

The application of these methods to brain tissue is presented elsewhere (9,12,13) and here we present a summary of current work on photon migration in muscle. This tissue is particularly appropriate because deoxygenation in ischemia and reoxygenation by reflow thereafter provides a calibration not possible with the human brain and requiring significant instrumentation in the animal model.

MODEL SYSTEMS

A model system for simulating a progressive deoxygenation of oxygen transport pigments due to tissue respiration is based upon a suspension of blood and yeast cells located at variable distances from the input output optics (Fig. 1). Scattering material intervenes between the two to simulate the skull, in the case of brain studies, or the adipose layer in the case of muscles of particular individuals, particularly females. This apparatus affords a basic demonstration in pulsed light studies of the effect of the absorber in converting a power law decay of exiting photons to an exponential function as required by photon migration theory (13,14) (Fig. 2).

INTERCELLULAR PO_2's

As expected under these conditions, the oxygen response of hemoglobin, myoglobin, and indeed cytochrome corresponds closely to their _in vitro_ values since the oxygen gradients are negligible in suspensions of cells that do not adhere or clump (Fig. 3). For example, in such a system the high oxygen affinity of yeast hemoglobin, 10^{-8} M _in vitro_, is reproduced precisely _in vivo_ verifying that within the single yeast cell, the oxygen gradients are negligible and that the apparently larger values observed in cells and tissues by various extracellular oxygen indicators can be attributed to diffusion gradients for oxygen.

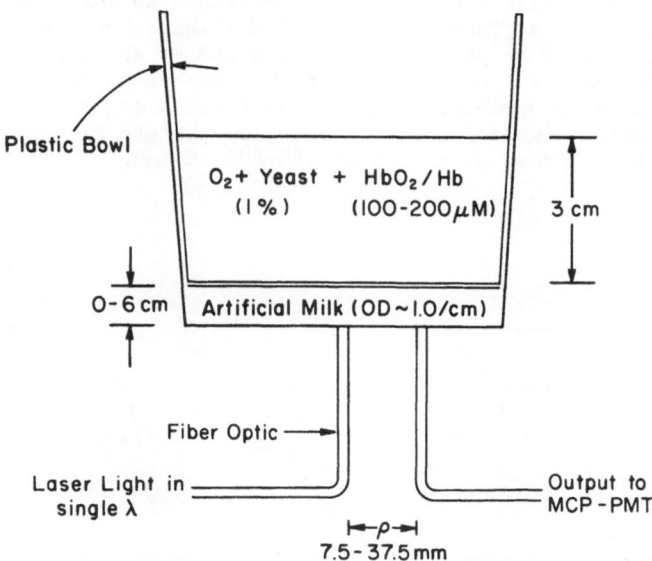

Test System for TCSPC

Figure 1. A model system appropriate to evaluating photon migration in the brain. The skull is simulated by a layer of milk substistute not containing hemoglobin while the brain tissue is simulated by hemoglobin added to a suspension of yeast. The respiration of the yeast causes oxy-deoxy transitions simulating cerebral ischemia.

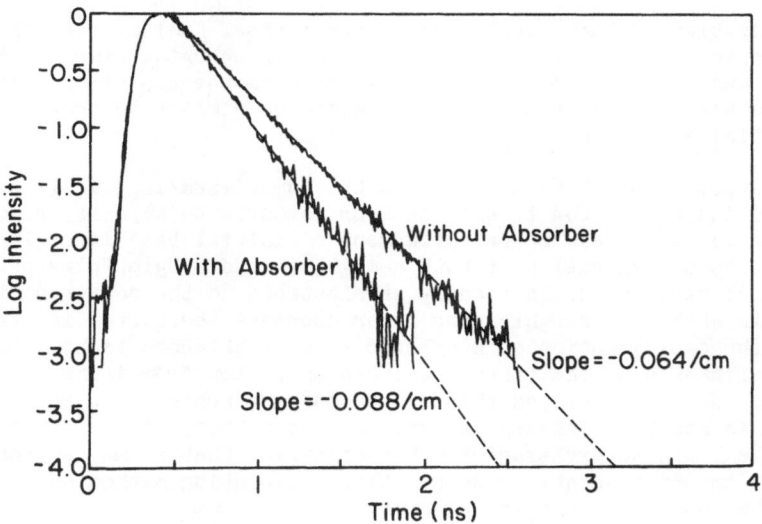

Figure 2. The effect of an absorber upon the kinetics of photon exit from the model of Figure 1.

The application of CWS to studies of the limb is illustrated by the photograph of the Millikan type oximeter attached to the adult leg, However, this device differs from that of Millikan; here the optical geometry is optimized by prior studies with TRS so as to secure adequate tissue penetration. The arrangement of the optical components as shown in Fig. 4 consists of a symmetrical location of flash light bulbs as light sources, and silicon diode detectors with appropriate filters for the wavelength pair desired, in this case 760-800 nm and.

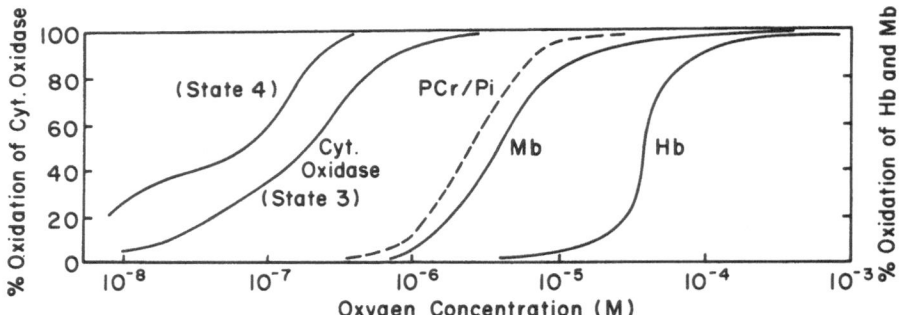

Figure 3. An indication of the oxygen affinity of cytochrome oxidase, intracellular yeast hemoglobin of high oxygen affinity, is deoxygenated at approximately the same oxygen tension as cytochrome in yeast cells, the oxygen gradient within the cell being negligble.

Equalization of the output signals for optimal dual wavelength operation is achieved by varying the capacitor charging times of the two signals, one charging and the other discharging the capacitor. This affords a high degree of common mode rejection similar to that of a differential amplifier.

The application of this device to the human (arm/leg) is illustrated in Fig. 5 indicating the transition from normoxia to hypoxia, recovery to hyperemia on reflow and re-establishment of initial baseline. The time for complete deoxygenation of both hemoglobin and myoglobin is prolonged. The initial rapid phase is readily attributable to the deoxygenation of hemoglobin while the roughly zero order decrease thereafter is attributed to show the deoxygenation of myoglobin. The continuous light instruments may be calibrated by the abrupt response on reflow from ischemia to hypermia. We have employed this criterion, the achievement of an endpoint in ischemia and the recovery in hyperemia as a "total signal change from deoxy hemoglobin to oxyhemoglobin," recognizing that it may uderestimate the total amount present. However, this calibration method does not afford absolute concentration changes.

CALIBRATION OF PATH LENGTH OF CWS BY TRS

It is preferable to employ the pathlength calibration obtainable from the pulse light method and employ this to measure the concentration change from normoxia to ischemia in recovery.

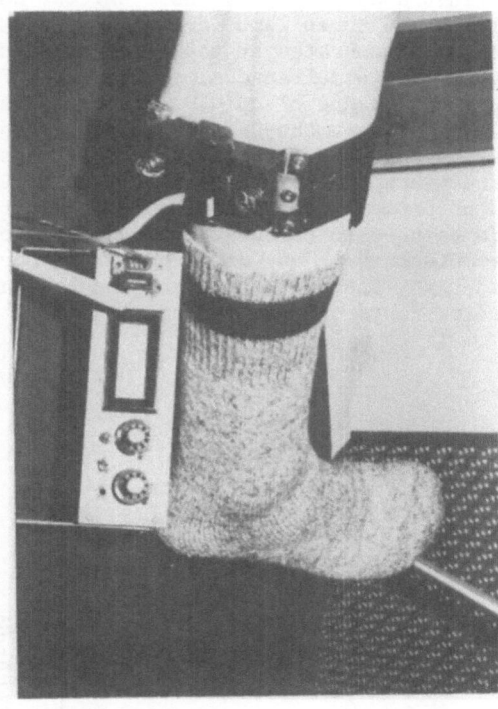

Figure 4. The use of a simple hemoglobinometer for measuring tissue hemoglobin-myoglobin deoxygenation.

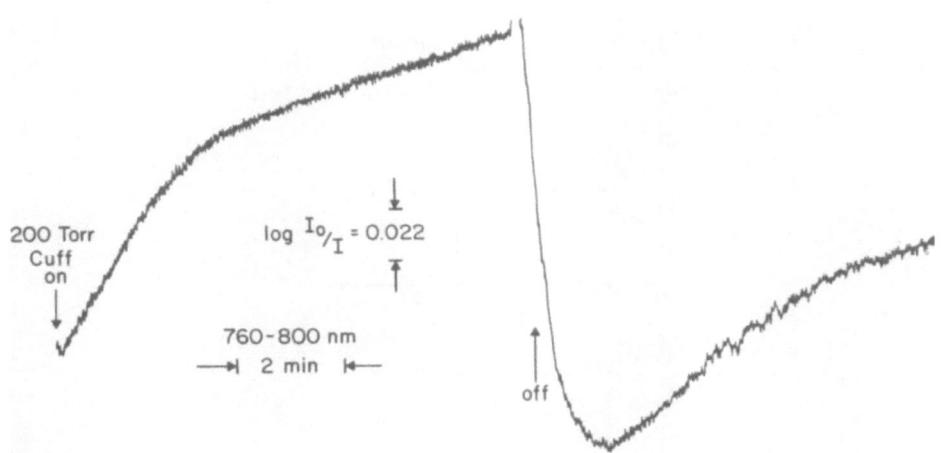

200 Torr
Cuff
on

log I_0/I = 0.022

760-800 nm

2 min

off

Figure 5. Application of simple hemoglobinometer to deoxygenation in the arm. In this case, the tourniquet ischemia is employed to cause the initial deoxygenation of hemoglobin and possibly subsequent deoxygenation of myolgobin.

The method is illustrated in Fig. 6 where the CWS and TRS devices are applied directly to the human limb (arm or leg) and the tourniquet ischemia is repeated as illustrated by the time course of absorbancy changes in the figure. The simultaneous measurements of the CWS absorbancy change and the change of u (O.D./cm) are clearly parallel in increasing hypoxia. If we plot the CWS and TRS data for increasing hypoxia one against the other, a straight line relationship is obtained (Fig. 7) in which the slope for the skeletal tissue is 2 cm. This is the mean pathlength of photon migration as observed in the continuous light device and can be used in calculating concentrations by CWS, or by calibrating the absorbance change by dividing the measured absorption change by the pulsed light optical pathlength and the known extinction coefficient to give its range of concentration.

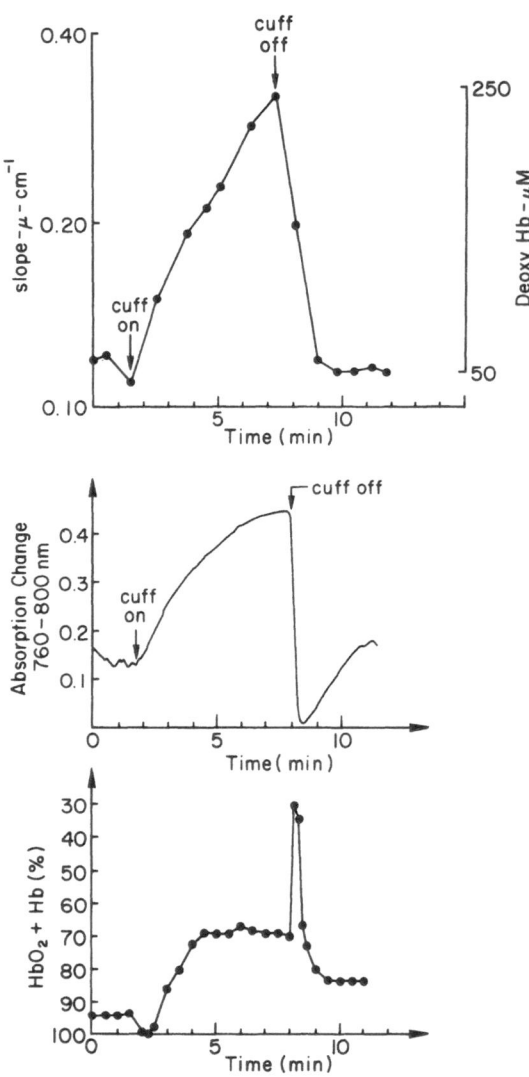

Figure 6. Simultaneous CWS and TRS measurements of hemoglobin deoxygenation in cuff ischemia of the human limb.

As might be expected, different individuals respond in vastly different ways to stress of exercise in the set point of hemoglobin oxygenation at the initiation of exercise and in the homeostasis of tissue oxygenation during exercise, ranging from elite performers, whose oxygen homeostasis is excellent, to patients with vascular disease who show significant deoxygenation at very light exercise (Fig. 8) (15,16). The percent deoxygenation in exercise performance has been calibrated as described above, from longer change from ischemia and hyperemia.

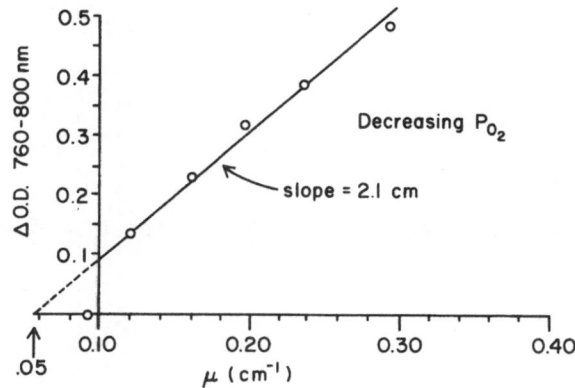

Figure 7. Comparison of CWS and TRS reponse to hypoxia in the human limb. The slope of the correlation function corresponds to a mean optical pathlength of 2 cm.

The kinetics of response to exercise are plotted in Fig. 9 for an individual who shows significant deoxygenation during work. Thus, the transition from the steady state of oxygen delivery at rest to that during work occurs in a few seconds and a steady state of oxygen delivery during steady state work is precisely maintained. On cessation of work, the recovery of hemoglobin oxygenation, of tissue oxygenation to the resting level, is significantly prolonged and occurs with a half time of 45 sec. The cessation of steady state exercise does not give a hyperemic response as noted in the release of an ischemic interval; i.e., the capillary immediately adjusts to the decreased oxygen utilization by the muscle but the recovery of tissue oxygenation to the steady state rest level occurs over an interval much longer than that required to reestablish the phosphocreatine level (approximately 1 min.). It is apparent that we are observing the response of the microvasculature as independent of the response to the needs of oxidative metabolism. In order, however, to justify that oxygenation of the working muscle was measured, an inflation pressure of 80 torr was employed to impede the venous return from the muscle with remarkable effects upon the degree of deoxygenation observed during exercise and corresponding percentage of MVC (Fig. 9).

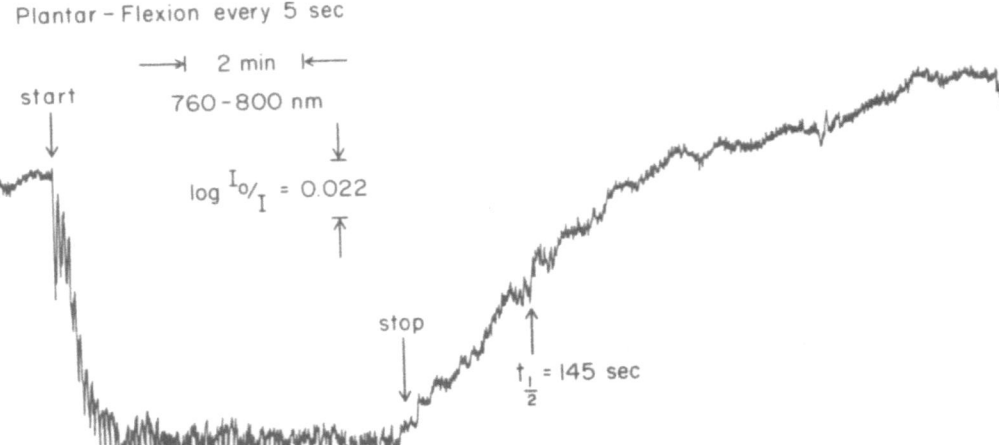

Figure 8. Hemoglobin deoxygenation in exercise in a vascular patient.

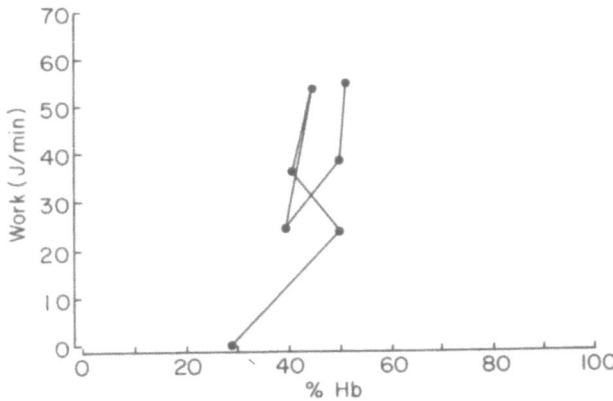

Figure 9. The work/hemoglobin deoxygenation profile for a vascular
patient.

CORRELATION WITH MRS

In order better to delineate the anaerobic threshold, magnetic resonnace spectroscopy (MRS) may be employed to identify the biochemical impact upon the work through measurements of the ADP value or its approximate equivalent, in NMR terms, of the inorganic phosphate and phosphocreatine ratio: P_i/PCr ($33 \times P_i/PCr$ = ADP to a first approximation and at pH 7). In this case, the optical and magnetic

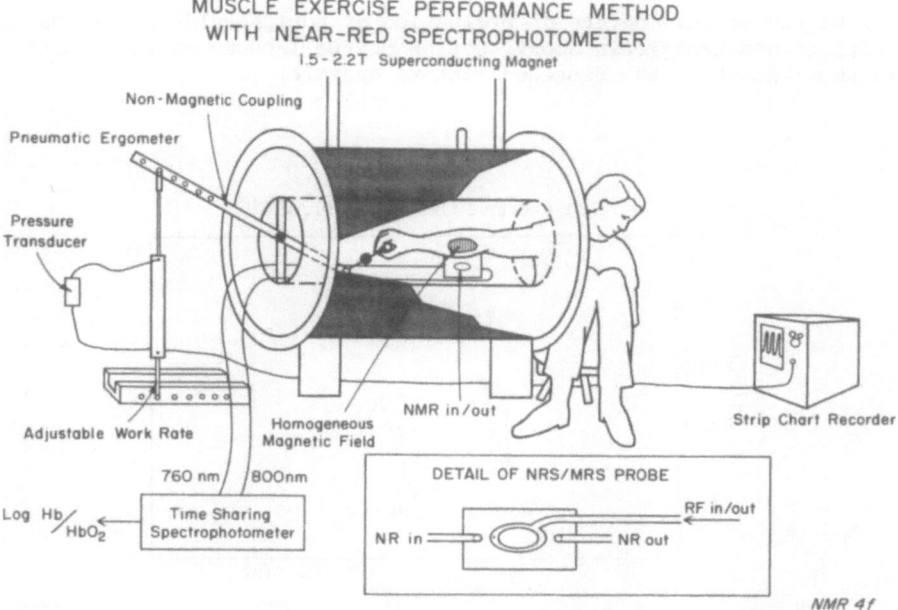

Figure 10. Technical arrangement for simultaneous near red spectroscopy (NRS) and magnetic resonance spectroscopy (MRS) of the human limb at rest and during exercise within the magnet bore. It should be noted that the fibers are placed strategically with respect to the surface coil of the NMR and have a geometry appropriate to a mean penetration into the tissue similar to that of NMR (2 cm).

resonance spectroscopies are coupled together as illustrated in Fig. 10 where fiber optics coupling into the bore of the magnet is necessary to avoid disturbing the field homogeneity. Furthermore, the function of the time-sharing spectrophotometer is identical to that described above. These fibers are located symmetrically with respect to the radio frequency coil of the NMR system and at its spacing as determined by the TRS system mentioned above.

SUMMARY

    The combination of continuous light spectrophotometry (CWS) and time resolved spectrophotometry (TRS) afford for the first time a quantitation of the optical path and the concentration changes detected by the CWS instrument.

    The application of these two techniques and magnetic resonnace spectroscopy (MRS) to muscle during exercise affords a correlation of the biochemical activation and the response of the peripheral circulation (NRS) to the exercise stress (MRS) (Fig. 11). In preliminary experiments, the well-trained endurance performance limb shows a near perfect homeostasis to exercise stress while ischemia will cause a significant deoxygenation and an impairment of the work output.

    The use of this device in evaluation of peripheral vascular disease is obvious and hemoglobin deoxygenation may well occur at work levels less than those at which the diseased limb is capable.

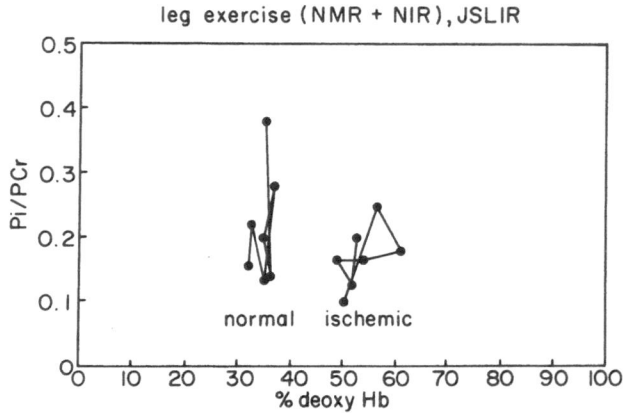

Figure 11. Correlation of biochemical stress (P_i/PCr) as measured by MRS and arterial or venelular capillary bed deoxygenation as indicated by NRS. In this diagram a normal subject with excellent microvascular control exercises with negligible change of hemoglobin deoxygenation in a protocol which demands 20, 40, 60%-20, 40, 60% MVC work output. If a venous occlusion, i.e. a pressure cuff at 80 cm of pressure, is applied then oxygen delivery is compromised and the work is accomplished at a greater level of deoxygenation.

REFERENCES

1. G. A. Millikan, Proc. Roy. Soc. London B 129:218 (1937)

2. G. A. Millikan, The oximeter, an instrument for measuring continuously the oxygen saturation of arterial blood in man. Rev. Sci. Instru. 13:434 (1942).

3. I. Yoshiya, Y. Shimada, and K. Tanaka, Spectrophotometric monitoring of arterial oxygen saturation in the finger tip. Med. Biol. Enz. Comp. 18:27 (1980).

4. F. F. Jobsis, J.H., Keizer, J.C. LaManna and M. Rosenthal, Reflectance spectrophotometry of cytochrome a,a_3 in vivo. J. Appl. Physiol. 43:858, 1977.

5. C. A. Piantadosi, and F.F. Jobsis-VanderVliet, Spectrophotometry of cerebral cytochrome a,a_3 in bloodless rats. Brain Res.305:89,1984

6. M. Cope, D. T. Delpy, E. O. R. Reynolds, S. Wray, J. Wyatt and P. van der Zee, Methods of quantitating cerebral near infrared spectroscoy data. Adv. Exp. Med. Biol. 222:183, 1988.

7. G. Renault, M. Sinet, M. Muffat-Joly, J. Cornillault, J. and J. J. Pocidalo, In situ monitoring of myocardial metabolism by laser fluorimetry: Relevance of a test of local ischemia. Lasers and Surgery & Medicine 5:111, 1985.

9. B. Chance, J.S. Leigh, Jr, H. Miyake, D. S. Smith, et al Comparison of Time Resolved and Unresolved Measurements of Deoxyhemoglobin in Brain. Proc. Natl. Acad. Sci. USA 85:4971, 1988.

9. B. Chance, J. S. Leigh, Jr. R. Greenfeld, H. Miyake, D. S. Smith and S. Nioka, Time Resolved Spectroscopy (TRS): A new approach to the spectroscopy of hemoglobin in Brain. New Eng. J.Med. Submitted 1988

10. B. Chance, S. Nioka, J. Kent, K. McCully, M. Fountain, R. Greenfeld, and G. Holtom, Time Resolved Spectroscopy of Hemoglonbin and Myoglobin in Resting and Ischemic Muscle. Anal. Biochem. In press.

11. B. Chance (ed.) Photon Migration in Tissues. A Workshop Proceedings. W. DeGruyter and Co. in press, 1988

12. T. Tamura, O. Hazeki, S. Nioka, and B. Chance, B. In vivo study of tissue oxygen metabolism using optical and nuclear magnetic resonance spectroscopies. Ann. Rev. Physiol. 51: in press, 1988.

13. R. F. Bonner, R. Nossal, S. Havlin and G.H. Weiss, Model for photon migration in turbid biological media. J. Opt. Soc. Am. A. 4:423 1987

14. B. C. Wilson and Adam, C. T. A monte carlo model for the dependent of cellular energy metabolism. Arch. Biochem. Biophys. 195:485, 1983.

15. J. H. Park, R. L. Brown, C .R. Park, M. Cohn, and B. Chance, A Genetic Endowment for Endurance Exercise: Energy Metabolism of the Untrained Muscle of Elite Runners as Observed by ^{31}P Magnetic Resonance Spectroscopy. Proc.Natl.Acad.Sci.USA 85:4971, 1988.

17. K. Kitagishi, L. Hao, and B. Chance, B. Heterogeneity Response of an Exercising Forearm as Studied by a Surface Coil Scan. 7th Annual Society of Magnetic Resonance in Medicine Mtg., San Francisco, CA.(Aug. 20-26), p. 339, 1988

A CCD SPECTROPHOTOMETER TO QUANTITATE THE CONCENTRATION OF CHROMOPHORES IN LIVING TISSUE UTILISING THE ABSORPTION PEAK OF WATER AT 975 nm.

Cope, M., Delpy, D.T., Wray S.[*], Wyatt J.S.[+], and Reynolds E.O.R.[+]

Departments of Medical Physics and Bioengineering, Physiology[*] and Paediatrics[+]
University College London,
11-20 Capper Street
London WC1E 6JA

Introduction

Optical spectroscopy has long been used to study the oxygenation of living tissue in-vivo. It is possible to detect concentration changes of naturally occurring chromophores whose absorption spectra are affected by the presence of oxygen, such as haemoglobin and the cytochrome enzymes.

Optical spectroscopy applied to tissue measurements is complicated however by the highly scattering nature of tissue. This has two major consequences. Highly sensitive spectrophotometers are required to overcome scattering losses and to measure the small changes in attenuation caused by the chromophores of interest. This problem is offset to some extent by a second effect. Scattering increases the optical pathlength of light in the tissue, producing a larger than expected attenuation change for a given chromophore concentration change. Thus determination of the optical pathlength is necessary before quantification of chromophore concentration can be obtained.

There is an inherent advantage in using a system which is as sensitive as possible. Increased sensitivity allows more accurate measurement of chromophore concentration (either by utilising a thicker section of tissue or by obtaining lower noise spectra). Alternatively the same accuracy in chromophore concentration can be obtained with an improved time resolution, enabling observation of rapid oxygenation changes.

Once a tissue spectrum has been collected the individual effects of all the chromophores present need to be identified. The technique of least squares multilinear regression from known absorption spectra allows the identification of each chromophore (Haaland, Easterling and Vopicka, 1985), including tissue water (from measurement of the height of the water absorption peak at 975 nm). If the concentration of water in tissue is known, the optical pathlength can then be determined (Wray et al., 1988) allowing the quantification of other chromophores. The purpose of this paper is to describe a new, highly sensitive, spectrometer system for the study of tissue oxygenation in-vivo and to present preliminary results from studies in-vivo in experimental animals.

System Description

A schematic diagram of the system is shown in Figure 1. The light source is a 100 watt quartz halogen lamp (current stabilised) with a fibre optic output (Oriel 77501, U.S.A.). A short pass filter with a sharp cutoff for wavelengths longer than 1000 nm (Oriel 58903, U.S.A.) removes unwanted heat from the light source. Various colour glass filters with long pass characteristics at visible wavelengths are employed to eliminate unwanted higher order spectra.

A 3 mm diameter glass fibre optic bundle of high numerical aperture (NA > 0.5) carries light to the sample. Transmitted light is collected by a second fibre optic bundle (Eurotec Optical Fibres, UK) consisting of 100 μm core silica glass fibres (NA 0.2) with a cross section changing from a circle of 1 mm diameter at the sample to a 5 mm high by 100 μm slit at the input to the spectrograph.

The spectrograph itself has a 30 cm focal length with an input aperture of f/4.2. Light enters through a variable width slit of micron resolution and is dispersed by a custom made ruled grating of 60 lines/mm blazed at 650 nm (Jobin Yvon HR320, France). Detection of the dispersed light is performed by an array detector at the focal point of the output mirror. The detector consists of a liquid nitrogen cooled silicon charge coupled device (CCD) camera (Wright Instruments, UK) using a 385 by 578 array of 22 μm square elements (EEV P8603, UK). The 385 horizontal elements are set across the wavelength dispersed axis while the 578 vertical elements record an image of the input slit. The CCD frame area is centred on the optic axis of the spectrograph and the storage area remains unilluminated.

The system provides a dispersion of 1.11 nm per CCD element and a total wavelength scan of 424 nm. Light incident on the CCD causes charge to accumulate in the elements. This charge is then read out and digitised by a 14 bit analogue to digital convertor (ADC, equivalent to 16 384 counts) with an additional programable gain

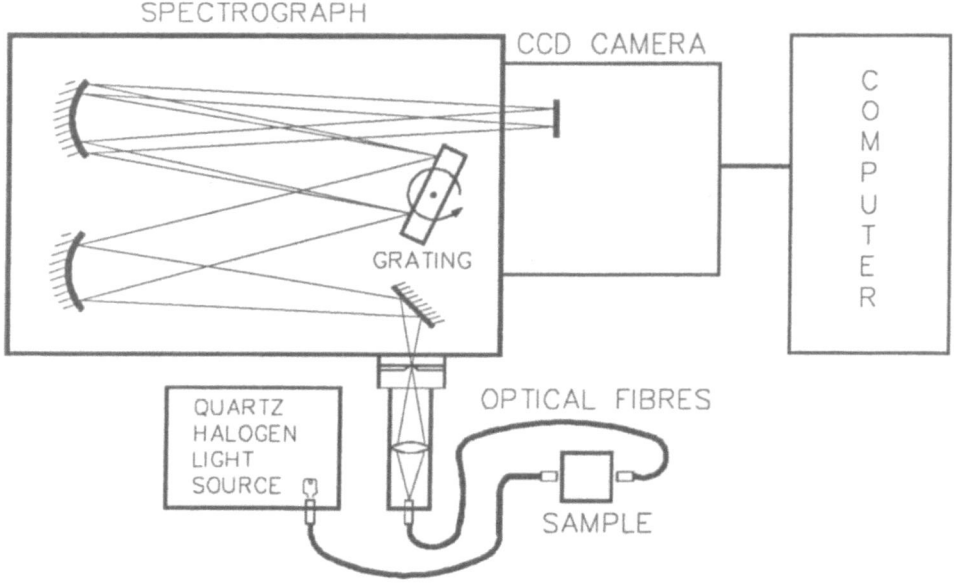

Figure 1. Schematic diagram of the CCD spectroscopy system used for tissue oxygenation measurements.

amplifier. Each count of the ADC can be set equivalent to 10 to 40 electrons.

Control, analysis and data acquisition are all performed by a IBM-AT compatible computer. An essential feature of the software is the ability for the time integrated charge to be binned (accumulated) either by row or column prior to read out.

Advantages of Liquid Nitrogen Cooled CCD Detectors in Spectroscopy

The general features of CCD devices have been described elsewhere (Janesick et al., 1987) but here their application in spectroscopy systems will be discussed.

Array based spectrometers allow observation of (n) multiple wavelengths simultaneously as opposed to scanning systems, which observe different wavelengths sequentially. This is of particular importance in the study of dynamic spectral changes. Here the array system will produce a spectrum correctly averaged over its integration time, whereas a scanning system will produce spectral distortion. Offset against this advantage is the size of each array element (A_a), which is typically much smaller than the detector area (A_s) of scanning systems of similar wavelength resolution. The ratio of the signal current from the two types of system for the same acquisition time is:

$$\frac{\text{Total signal current (array)}}{\text{Total signal current (scan)}} = \frac{S_a}{S_s} = \frac{n\,A_a}{A_s}$$

Similarly, the dark current ratio is:

$$\frac{D_a}{D_s} = \frac{n\,A_a}{A_s}$$

Hence:

$$\frac{S_a}{D_a} = \frac{S_s}{D_s}$$

Thus the signal current to dark current ratio is the same for both types of system. The improvement in performance of the array system comes from the ratio of the total signal currents $S_a/S_s = n\,A_a/A_s$. If a typical array system of 300 independent detectors with array element dimensions 5 mm by 22 µm is compared to a conventional scanning system with a 5 mm by 0.5 mm detector, then the ratio $S_a/S_s = 13.2$. The scanning system would thus take 13.2 times longer than the array system to produce a spectrum with the same signal current.

The quantity of dark charge produced by a liquid nitrogen cooled camera is less than 1 electron hr^{-1} pixel^{-1} and for integration times of less than 1 hour this may be neglected. The noise from a CCD array system is the sum of the readout noise (R) and the signal shot noise from the number of signal electrons (N). Hence the signal to noise ratio is given by:

$$\text{S/N ratio} = \frac{N}{R + \sqrt{N}} \quad \dots \dots \dots \dots \dots \dots \dots \dots \dots \dots \dots \dots (1)$$

Where $\sqrt{N}$ is the shot noise limited by the discrete nature of light.

Table 1 gives illustrative values for the number of photons at 600 nm required to achieve signal to noise ratios of 1, 10, 100 and 1000 for three different array detectors, a cooled CCD, a cooled photodiode array (PDA) and a cooled intensified photodiode array. The values have been calculated using equation (1).

Although the intensified PDA has the lowest readout noise, its lower quantum efficiency implies that more photons are required to achieve the same signal to noise ratio at 600 nm. Obviously quantum efficiency will affect system performance considerably, and for wavelengths longer than 500 nm the CCD system performs best,

Table 1. A comparison of the characteristics for three types of array detector used in spectroscopy systems.

	Cooled CCD	Photodiode array (PDA)	Intensified PDA
Wavelength range (nm)	400 - 1050	200 - 1050	200 - 850
Quantum effic. (600 nm)	0.50	0.65	0.10
Readout noise (e-)	10	1500	3
No. electrons S/N = 1	14	1 539	5
No. electrons S/N = 10	209	15 441	133
No. electrons S/N = 100	11 010	161 325	10 303
No. electrons S/N = 1000	1 010 010	2 501 498	1 003 003
No. photons S/N = 1	27	2 368	53
No. photons S/N = 10	418	23 755	1 329
No. photons S/N = 100	22 020	248 192	103 030
No. photons S/N = 1000	2 020 020	3 848 458	10 030 030

while the intensified PDA is better at wavelengths shorter than 500 nm. Photodiode arrays have some advantages despite higher noise. The number of elements can be greater, typically 1024, and the wavelength range is also larger. We compared experimentally the CCD system described here, with an unintensified peltier cooled photodiode array system using a different spectrograph.

Under identical low light conditions, the PDA system took 20 times longer than the CCD system to obtain a spectrum with a signal to noise ratio of 100. This compares reasonably with the factor of 11.3 predicted in Table 1. The discrepancy probably results from the relative efficiency of the spectrographs.

System Performance

The resolution and accuracy of wavelength against horizontal pixel number was checked using a low pressure neon lamp which produces many narrow spectral lines in the red and near infrared. The input slit was set at 22 µm, the same size as the detector elements. With careful focusing a full width half maximum of 1.5 pixels could be obtained at the maximum aperture of the spectrograph (f/4.2), corresponding to a wavelength resolution of 1.7 nm. A simple linear fit of wavelength against pixel number proved adequate using a dispersion of 1.11 nm/pixel. The precision of this fit was better than 1 nm over the entire 424 nm spanned by the array.

The relative sensitivity of the entire system against wavelength is shown in figure 2. The effects of the light source, fibre optics, spectrograph and detector are included. The effect of any additional filters are excluded. Any spectral band of range 424 nm can be placed on the array by manually rotating the grating to a new centre wavelength. The grating selected here is optimised for the 600 to 1000 nm region.

The maximum signal to noise ratio for a CCD is determined by the finite number of electrons which can be stored in a single pixel. For the P8603 device from EEV the maximum stored charge per frame or storage pixel is approximately 360 000 electrons giving a maximum signal to noise ratio of 590. A higher signal to noise ratio can be obtained if the charge in each wavelength column is binned (accumulated) into the output register. Approximately 1 080 000 electrons can be stored, providing a potential signal to noise ratio of 1 029. If even lower noise spectra are required then multiple spectra may be read out and summed arithmetically in the computer.

When rapid spectral changes are being observed (and sufficient light is available) the storage area of the CCD can be used for temporary storage of up to 288 spectra prior to readout. In this case, spectra can be transfered from the illuminated frame area by binning into the storage area in only 5.6 ms. The ability to store 288 spectra on chip makes this system ideal for the study of rapid spectral changes.

Experimental Results

Preliminary results from studies of the brain in experimental animals are now reported.

Male Wistar rats of approximately 300 g weight were anaesthetized with urethane (36% w/v solution, 0.5 ml/100 g body weight i.p.). A tracheostomy was performed and a femoral artery and vein cannulated. The head was secured in a stereotactic frame and superficial tissue together with the temporal muscles were removed. The optical fibres were fixed on either side of the skull against the parietal bones, and connected to the CCD spectrometer system. Integrated EEG was monitored from 3 electrodes placed in burr holes made in the skull, using a cerebral function monitor (Critikon, USA). Short severe periods of hypoxia were induced by changing the inspired gases from 100% oxygen to 100% nitrogen for periods of 5 to 20 seconds followed by a return to 100% oxygen. This manoeuvre consistently produced a significant reduction in EEG activity followed by recovery. This was repeated 3 to 6 times and the animals were then terminated in nitrogen.

A total of 4 animals were studied. In the first group (n = 2), animals were studied without exchange transfusion. In the second group (n = 2), a haemoglobin free preparation was obtained by exchange transfusion with 150 ml of a perfluorocarbon blood substitute (Green Cross Corp. FC-43, Japan). This permits absorption changes of

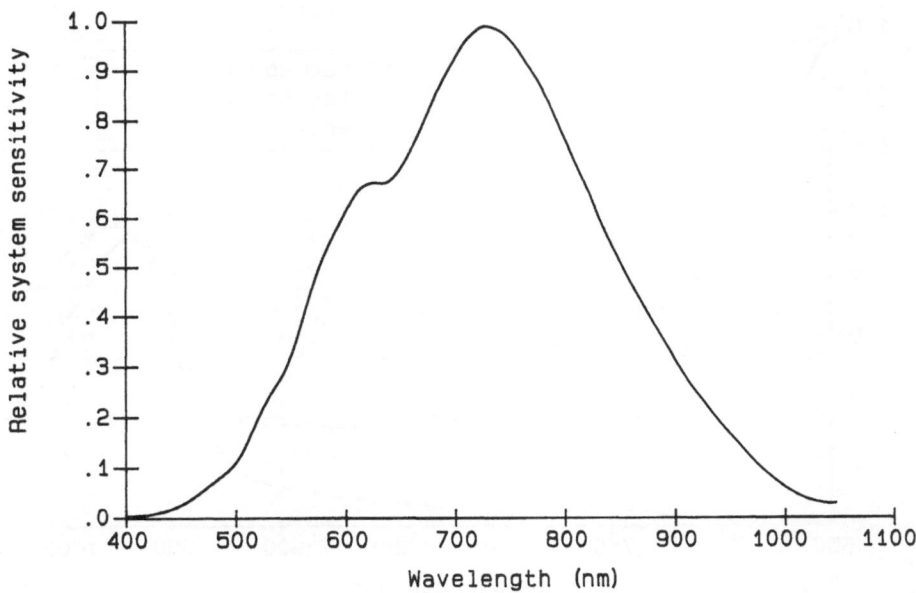

Figure 2. The relativity sensitivity of the CCD spectrophotometer system when no sample is placed between the optical fibres.

the cytochrome enzymes to be studied in the absence of haemoglobin. In all animals the height of the water peak at 975 nm was measured to allow estimation of the optical path length in the brain tissue.

The optical attenuation of the normal rat head meant that spectra with a peak signal to noise ratio of 280:1 were obtained with an acquisition time of 1 second. Following perfluorocarbon exchange, the attenuation across the head decreased leading to an improved signal to noise ratio of 480:1 in the same acquisition time.

Pathlength Determination

In order to determine the contribution of water to the composite tissue spectrum, the raw transmission spectra were first normalised to the system response measured in air (figure 2). Normalised spectra are plotted, by normal convention, on a minus log (base 10) scale. Due to the effects of scattering the zero is arbitrary.

A typical rat brain spectrum is shown in figure 3 together with a synthesised spectrum resulting from a least square multiple linear regression, performed between 650 and 1000 nm. The synthesised tissue spectrum was obtained by regression against known in-vitro spectra for oxygenated haemoglobin (HbO_2), deoxygenated haemoglobin (Hb), and water. A constant and a linear gradient are included in the fit to take account of tissue scattering. This simple regression provided a reasonable agreement with the original spectrum, but additional refinements are required to improve the accuracy. Figure 3 also shows a spectrum for 4.83 cm of pure water illustrating its spectral contribution determined by the regression.

Curve fits were performed on all 4 animals, using spectra which were obtained during respiration with 100% oxygen and following death. The height of the water peak was determined in each animal by a least squares regression and the optical path

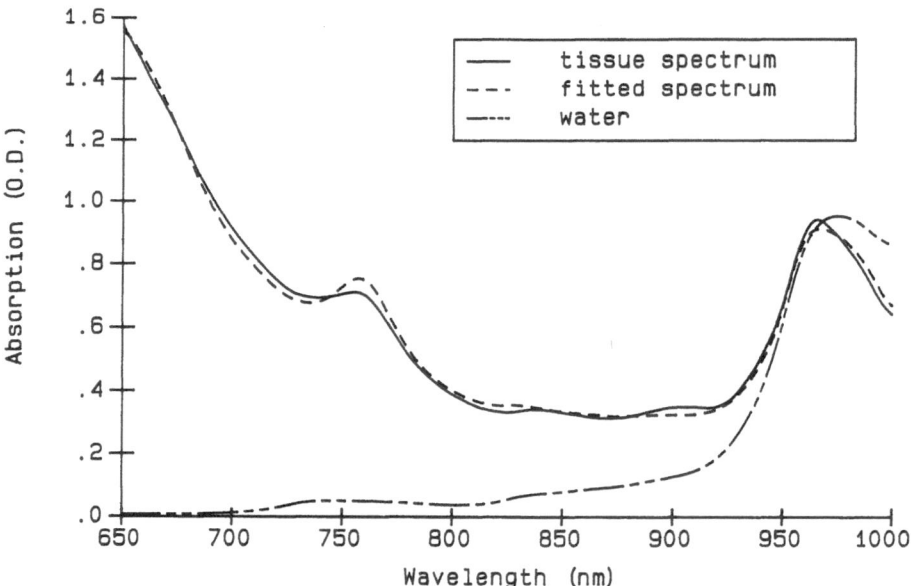

Figure 3. Least squares multilinear regression of tissue spectra to determine the optical pathlength from water absorption.

length was calculated, assuming a water content for the brain of 80% (Dobbing and Sands, 1970). The mean path length was 6.35 cm (±0.52 cm standard deviation). The mean interoptode spacing was 1.45 cm giving a pathlength factor of 4.4 times the interoptode spacing. this agrees reasonably well with previously published results (Wray et al., 1988) in which the size of the water peak was estimated by inspection.

Tissue oxygenation changes in FC-43 exchange transfused rats

In these experiments the grating was positioned to record spectra from 580 to 1004 nm. This allowed simultaneous observation of the visible absorption peak at 605 nm of reduced cytochrome aa_3 (Brunori, Antonini and Wilson, 1981), the near infrared peak at 830 nm of oxidised cytochrome aa_3 and the water peak at 975 nm. Spectra were obtained at 1 second intervals during respiration with 100% oxygen followed by periods of transient hypoxia. Figure 4 shows changes in the spectra during the onset of hypoxia. Peak signal to noise ratios recorded at 1 second acquisition times were 460:1 at 720 nm dropping to 25:1 at 580 nm and 1000 nm.

Two clear isobestic points at 594 nm and 617 nm can be detected during transient hypoxia. This suggests that chromophores other than cytochrome aa_3 make a negligible contribution to these spectra. It is also apparent that absorption changes at 830 nm and 605 nm occur simultaneously (within a time resolution of 1 second). Transient reduction of the cytochrome aa_3 enzyme correlated closely with reduction in the EEG activity. Following recovery of oxygenation, both the cytochrome aa_3 and the EEG returned rapidly to the baseline values.

During prolonged anoxia leading to death, a wavelength independent increase in attenuation was observed. This remains unexplained, and further studies are required.

In a final experiment, samples of circulating fluid and brain tissue were analysed following perfluorocarbon exchange transfusion to determine the residual haemoglobin

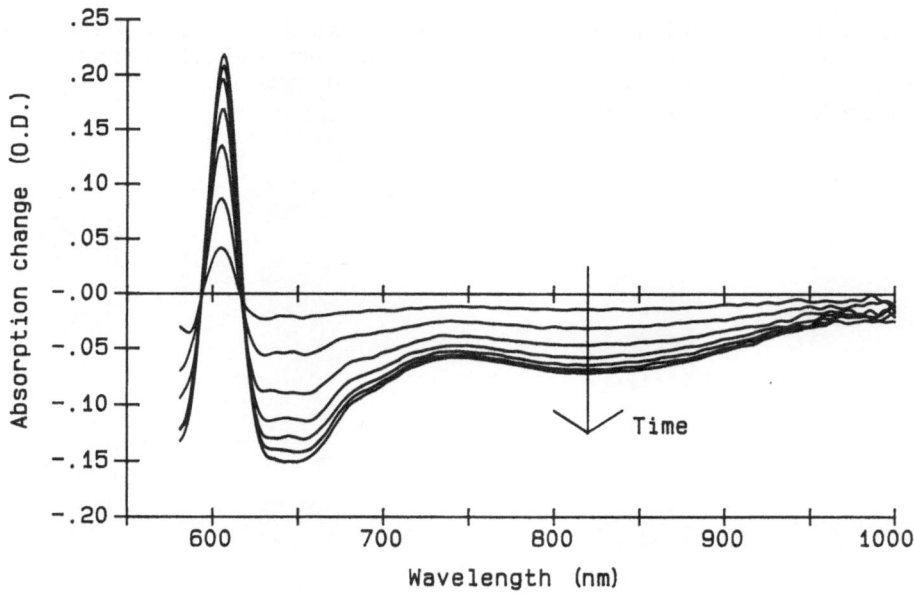

Figure 4: Difference spectra of the brain, relative to normoxia ($F_iO_2 = 100\%$), during the onset of hypoxia measured in a perfluorocarbon exchange transfused rat.

concentration. Both samples were disrupted mechanically and ultrasonically to lyse any red cells present, and microscopy was performed to confirm that complete lysis had occurred. Following centrifugation, spectrophotometric examination of the supernatant revealed a haemoglobin concentration of 2 micromolar in the sample of circulating fluid. The brain sample was estimated to have a haemoglobin concentration of less than 0.5 micromolar. These results confirmed that negligible concentrations of haemoglobin remained in the brain tissue and circulation following exchange transfusion, and thus no detectable interference in the absorption spectra could have occurred from residual haemoglobin.

Conclusion

The CCD spectrophotometer system provides the ability to monitor rapid spectral changes in the near infrared region up to 1000 nm. Observation of the tissue water peak at 975 nm permits determination of the optical pathlength and rapid changes in cytochrome aa$_3$ redox state can be identified and quantified.

Acknowledgments

We are grateful to The Wellcome Trust, Hamamatsu Photonics KK (Japan) and the Medical Research Council for supporting this work.

References

Brunori, M., Antonini, E., and Wilson, M.T., 1981, Cytochrome C Oxidase: An overview of recent work, in: "Metal ions in Biological Systems", H. Siegel, ed., Dekker, N.Y.

Dobbing, J., and Sands, J., 1970, Growth and development of the brain and spinal cord of the guinea pig, Brain Res., 17:115-123.

Haaland, D.H., Easterling, R.G., and Vopicka D.A., 1985, Multivariate least-squares methods applied to the quantitative spectral analysis of multicomponent samples, Appl. Spect., 39(1):73-84.

Janesick, J.R., Elliot, T., Collins, S., Blouke, M.M., and Freeman, J., 1987, Scientific Charge-Coupled Devices, Opt. Eng., 26(8):692-714.

Wray, S., Cope, M., Delpy, D.T., Wyatt, J.S., and Reynolds, E.O.R., 1988, Characterization of the near infrared absorption spectra of cytochrome aa$_3$ and haemoglobin for the non-invasive monitoring of cerebral oxygenation, Biochim. Biophys. Acta, 933:184-192.

QUANTITATION OF PATHLENGTH IN OPTICAL SPECTROSCOPY.

Delpy, D.T., Arridge, S.R., Cope, M.,[+]Edwards, D.
[+]Reynolds, E.O.R., Richardson, C.E.,[*]Wray, S.,[+]Wyatt
J.,and van der Zee, P

Department of Medical Physics,[*]Physiology,[+]Paediatrics
University College London
Shropshire House
Capper Street
London WC1E 6JA

Introduction

The relative transparency of tissues to near infrared light means that it is possible to transilluminate intact organs. In the infrared region, oxygen dependent absorptions due to haemoglobin and cytochrome aa_3 can be observed, and it is therefore possible to monitor changes in both the blood and tissue oxygenation of the organ.[1] This monitoring technique is particularly applicable to the study of the brain since there is no interfering absorption from myoglobin, and recent technical developments of the instrumentation have made it possible to transilluminate 8-9 cm of brain tissue.[2] However, once measurements of absorption change at several wavelengths are available, there are still considerable problems in converting this data into quantitative changes in the concentration of oxy and deoxy haemoglobin and of oxidised cytochrome aa_3.

These problems arise because of the scattering of light as it traverses the tissues. It has been shown previously that when transilluminating a thick section of highly scattering material, a simple "Beer Lambert" formula can to a first approximation be applied to calculate concentration changes, although the optical pathlength used must be increased to take into account the effects of multiple scattering.[3] This increased pathlength can be measured indirectly by using additional data from other physiological monitors, or directly by measuring the absorption due to water whose concentration in brain tissue is known.[4] It is however obvious that this pathlength is not a constant, but will vary slightly with changes in tissue absorption or scattering coefficient. In this paper this variation in pathlength is considered theoretically using a Monte Carlo model of light transport in tissue, and measured experimentally by timing the passage of ultrashort light pulses through tissue.

Background

Light travels through tissue at a speed of approximately
0.2 mm/picosecond. Using a synchronously pumped dye laser as a
source of ultrashort light pulses, and a synchroscan streak
camera as a detector it is now possible to measure the time of
flight of light pulses through tissue with a resolution of a
few picoseconds, and hence measure pathlength to a few
millimetres. However, the multiple scattering that occurs in
tissue means that the emerging photons are dispersed in time,
and have not travelled a single unique pathlength. This leads
to a problem when trying to choose a single transit time (and
hence pathlength) to use in the "Beer Lambert" calculation of
concentration. In a previous study[5] this problem has been
addressed using a Monte Carlo model of light transport in
tissue. The model predicted that the mean value of the time
dispersed pulse (the Temporal Point Spread Function) correlated
best with the pathlength in the "Beer Lambert" calculation.
This prediction was verified experimentally using a phantom of
known optical characteristics. Finally, transit time was
measured experimentally across the rat head, and a mean
pathlength of 5.3 + 0.3 times the head diameter calculated.
This work has now been extended, to study the way in which this
pathlength will change with absorption.

Model predictions and verification

A Monte Carlo model of light transport in tissue was used.
The scattering phase function in the model was the phase
function calculated from Mie theory for the phantom used for

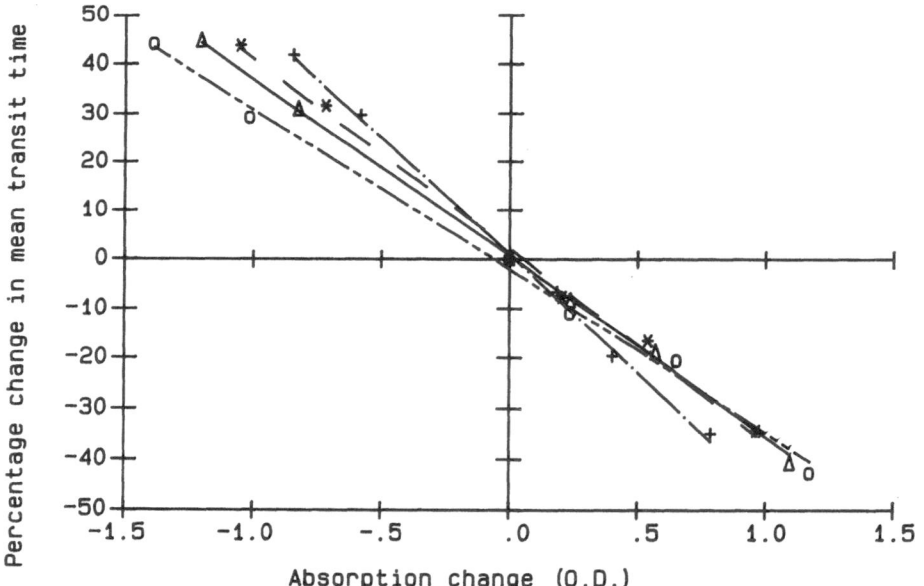

Figure 1. Monte Carlo predictions of the relationship
between percentage change in mean optical pathlength and
variation in optical density. Data is shown for four
different scattering coefficients (+) μs= 40 cm^{-1}; (*) μs=
60 cm^{-1}; (Δ) μs= 80cm^{-1}; (O) μs= 100 cm^{-1}.The slope for the
regression line for μs= 100 is 33%.

the experimental verification. Specular reflections at the slab
boundaries due to differences in refractive index were taken
into account. The time dispersion of a spatial and temporal
delta function transmitted through a 1 cm slab was modelled for
a range of absorption and scattering coefficients. Absorption
coefficients (base 10) of 0.456, 0.334. 0.263, 0.217, 0.0867
and 0.0434 cm^{-1} and scattering coefficients (base e) of 100, 80,
60, 40, and 20 cm^{-1} were simulated encompassing the range quoted
in the literature for brain tissue.
The mean value of each Temporal Point Spread Function (TPSF)
was then calculated. The percentage change in this value,
referenced to the value at an absorption coefficient of 0.217,
was then plotted against the calculated change in optical
density. Figure 1 shows this relationhip for scattering
coefficients of 40, 60, 80 and 100 cm^{-1}. It can be seen that a
linear relationship is obtained over a wide range in optical
density. The slope of the relationship is dependent upon the
scattering coefficients, an increase in scattering coefficient
causing a decrease in slope.

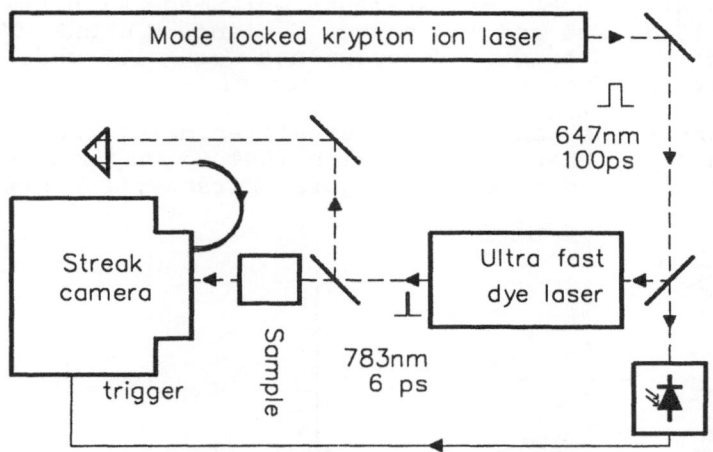

Figure 2. Experimental apparatus for the verification of
model predictions and in-vivo time of flight measurements.

The model predictions were verified using a phantom consisting
of polystyrene latex microspheres in water containing an
infrared absorbing dye.[5] Two sizes of polystyrene latex
microspheres were used in the phantom in an attempt to
reproduce the scattering phase function of brain tissue,
measured previously.[6] The solution contained 2.5% by volume of
1.0 μm diameter particles and 6.7% by volume of particles of
0.05 μm in diameter. Changes in scattering coefficient were
obtained by volumetric dilution of this mixture. Absorption
coefficient changes were obtained by adding known quantities of
infrared absorbing dye (ICI S109564). Measurements were made

using the experimental system shown in figure 2. with a sample cell 6 cm wide, 6 cm high and 1 cm thick. The absorption coefficient was varied over a range of 0.05 to 0.42 cm^{-1}. Scattering coefficients were 80, 40 and 20mm^{-1}. The percentage change in mean was again referenced to the value at an absorption of 0.21cm^{-1}. The results in figure 3 show a similar trend to those from the model predictions.

Pathlength measurement in brain

Experimental measurements have been made of the optical pathlength in the transilluminated rat brain. Nine adult wistar rats were studied. Following anaesthesia (urethane 36% w/V, intraperitoneal, 5ml/Kg), the temporoparietal muscles were reflected, the skull exposed and cleared of residual tissue. The animals were ventilated via a tracheal tube and the femoral artery was cannulated for blood sampling. The animal was placed in front of the streak camera in the sample position shown in figure 2, the head being immobilised in a stereotactic frame. Light pulses from the dye laser were incident upon the skull diametrically opposite the streak camera entry slit. Some of the light emerging from the far side of the head was sampled via a 1 mm diameter optical fibre coupled to a photomultiplier tube (Hamamatsu R928). Measurements were made with the animals breathing 100%, 21% and 12% oxygen (balance N_2) and 90% oxygen with 10% carbon dioxide. Measurements were also made immediately upon death by N_2 inspiration.

Figure 4 shows the data points for all animals, the reference pathlength being that obtained with 100% O_2 inspiration. Several points can be noted from this figure. Firstly, the overall

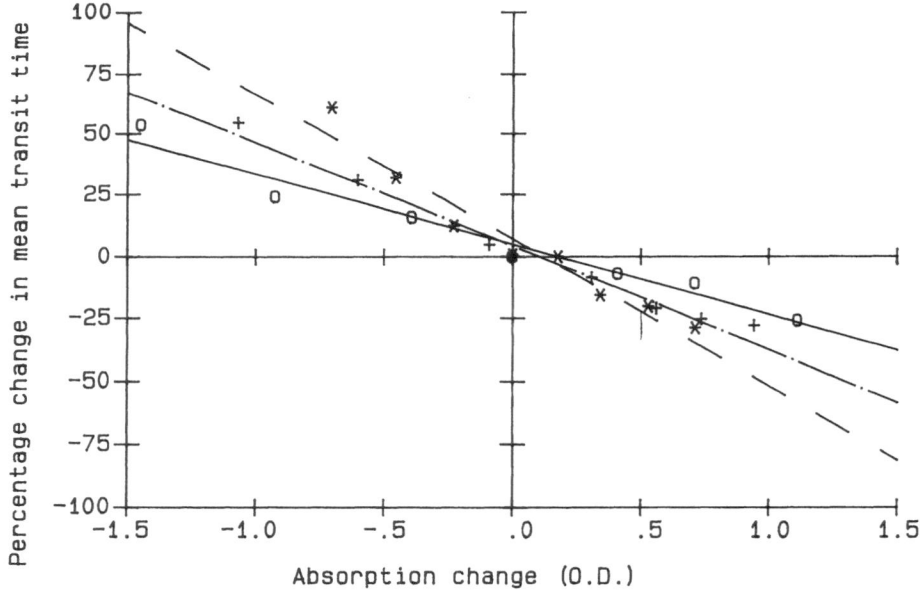

Figure 3. Measured percentage change in the mean optical path as a function of changing optical density for the phantom solution. Data is shown for three different values of scattering coefficients. (*) μs= 20 mm^{-1}; (+) μs= 40 mm^{-1}; (o) μs= 80 mm^{-1}.

relationship is very similar to that predicted by the Monte
Carlo model, and verified in the phantom studies. Secondly,
the relationship appears to be linear over a total change in
optical density of 0.5 OD induced both by a variation in blood
oxygenation (F_1O_2 = 100→12%) and blood volume ($FiCO_2$ = 10%).
The slope of the regression line gives a value for the change
in pathlength per OD of 30% for the data on live rats. Finally,
it can be seen that the data obtained during immediate death
does not follow the same relationship. The reasons for this are
as yet unclear, but may be due to changes in tissue scattering
coefficient post morten[7].

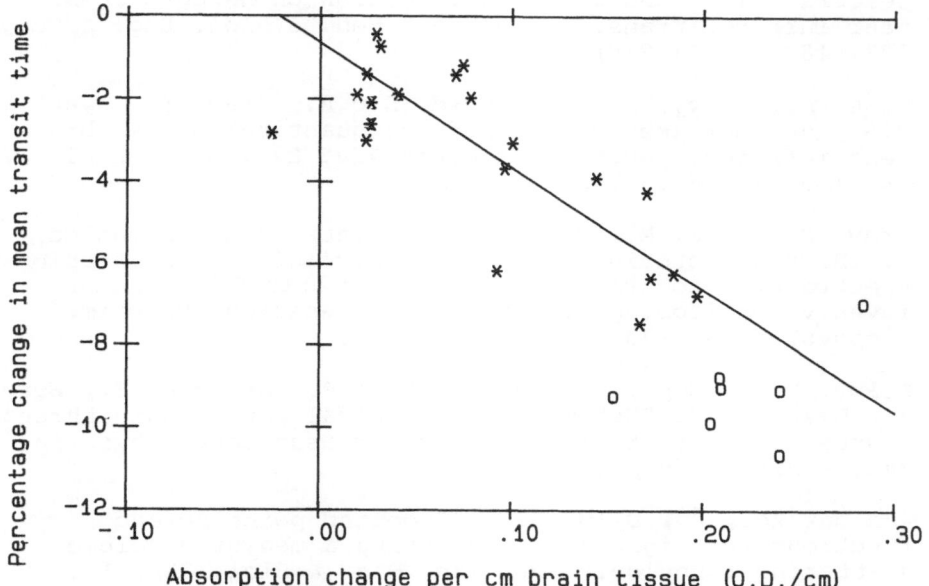

Figure 4. Percentage change in the mean optical path across
the rat head as a function of change in optical density. (*)
Data obtained on the live animal; (o) data obtained during
death. The slope of the regression line (excluding 'o' data
Points) is 30% /OD.

Conclusions

The optical pathlength through the transilluminated rat
head has been experimentally measured, and changes in the
pathlength have been shown to follow the predictions of a Monte
Carlo model of light transport in tissue. The mean pathlength
increased by approximately 3% for each 0.1 OD increase in
transmitted intensity. Incorporation of this pathlength change
into the Beer Lambert equation should significantly improve the
accuracy of estimation of chromophores at low concentration.

Acknowledgements

This work was carried out with funding provided by the Wellcome Trust, MRC, SERC and Hamamatsu Photonics KK. The authors also wish to thank Dr. B. Vincent and Dr. K. Ryan, Dept. of Physical Chemistry, University of Bristol for the generous supply of polystyrene-latex particles.

References

1) Jobsis F.F. "Non invasive infrared monitoring of cerebral and myocardial oxygen sufficiency and circulatory parameters" Science, 198, 1264-1267 (1977).

2) Cope, M., Delpy, D.T. "System for long term measurement of cerebral blood and tissue oxygenation on newborn infants by near infrared transillumination" Med. & Biol. Eng. & Comp. 222, 183-190, (1988).

3) Cope, M., Delpy, D.T., Reynolds E.O.R., Wray, S., Wyatt, J.S., van der Zee, P. "Methods of quantitating cerebral near infrared spectroscopy data" Adv. Exp. Med & Biol. 222, 183-190, (1988).

4) Wray, S., Cope, M., Delpy, D.T., Wyatt, J.S., Reynolds, E.O.R. "Characterisation of the near infrared absorption spectra of cytochrome aa_3 and haemoglobin for the non invasive monitoring of cerebral oxygenation "Biochim. Biophys. Acta. 933, 184-192, (1988).

5) Delpy, D.T., Cope, M., van der Zee, P., Arridge, S., Wray S., Wyatt, J.S. "Estimation of optical path length through tissue from direct time of flight measurement" Phys. in Med. & Biol. (in press).

6) van der Zee, P., Delpy D.T. "Computed point spread functions for light in tissue using a measured volume scattering function." Adv. Exp. Med. & Biol. 222, 191-198, (1988).

7) Svaasand, L.O., Ellinsen, R., "Optical properties of human brain" Photochem. & Photbiol. 38, 293-299, (1983).

DETERMINATION OF CEREBRAL VENOUS HEMOGLOBIN SATURATION BY DERIVATIVE NEAR INFRARED SPECTROSOCPY

Marco Ferrari*, David A. Wilson^, Daniel F. Hanley^, Jean F. Hartmann^, and Richard J. Traystman^

*Department of Biomedical Sciences and Technology, and Biometrics University of L'Aquila, 67100 L'Aquila, Italy and Physiopathology Laboratory, Istituto Superiore di Sanita, Viale Regina Elena 299 00162, Rome, Italy; ^Department of Anesthesiology/Critical Care Medicine, The Johns Hopkins University, Baltimore, MD 21205 USA

INTRODUCTION

A non invasive measure of brain tissue hemoglobin saturation could prove more useful than a systemic measure, since it reflects the balance between O_2 delivery and O_2 consumption and is organ specific. Near infrared spectroscopy (NIRS) was developed to monitor changes in brain hemodynamics and O_2 utilization (Jobsis, 1977; Wyatt et al., 1986; Ferrari et al., 1987). Quantification of the NIRS data has been difficult, therefore a new method was developed to quantitatively measure cerebral oxygenation using derivative NIRS (DNIRS), a secondary measurement method, which is calibrated to a primary conventional analytical method (Williams and Norris, 1987). DNIRS has been extensively applied in agriculture and its calibration involves establishing a mathematical relationship between the optical information and the primary species of interest. By using derivative spectra, scattering and non-specific absorption can be minimized while selection for the chromophores of interest can be optimized. This derivative method, well known in "in vitro" biochemistry, could be applied "in vivo" taking into account optical properties of the near IR transilluminated brain tissue and meninges (Wilson and Patterson, 1986; Eggert and Blazeck, 1987).

In this study, we determined if DNIRS could be utilized to measure non-invasively cerebral venous O_2 saturation (SvO_2). SvO_2 changes were obtained in dogs during graded levels of hypoxic hypoxia (HH) and the corresponding first derivative spectra were regressed against measured values of sagittal sinus (SS) SvO_2 taking into account that: 1) SS represents one of the largest components of the illuminated tissue volume and drains most of the venous blood of the detected cortical regions; 2) the SS drains most of the venous blood from cerebral cortical regions; and 3) venous blood volume is about 70% of total cerebral blood volume (Mchedlishvili, 1986). Therefore, in steady state conditions, SvO_2, calculated from the spectral data, should correspond to the SvO_2 of all the venous compartments of the transilluminated cerebral tissue. The resulting algorithm was tested in other dogs subjected to HH.

METHODS

In 8 barbiturate anesthetized, paralyzed, mechanically ventilated dogs, catheters were placed to collect arterial and SS blood. Spectral measurements

were made using a microprocessor-controlled scanning prototype spectrophotometer, interfaced to an IBM-AT computer, which could produce a full spectrum
from 730 to 960 nm at a rate of 3 Hz. The wavelength selection was obtained
by 3 interference filters mounted on a rotating filter wheel. A fiber
optic lightpipe (5 mm active diameter), carrying the IR photons, illuminated
the skull overlying the right frontal cortex one cm centrally to the part
of the SS previously cannulated. The same sized optic fiber, forming an
angle of 40-50 degrees with the first, was placed symmetrically on the
contralateral hemisphere (Figure 1). Near IR photons, passing through the
bone (a very low IR absorbing material) were partially absorbed and/or back
scattered by gray matter and the circulating hemoglobin that was present
in the SS as well as the illuminated vasculature. The photons were partially
collected by the receiver optic fiber. The intensity of the collected
light, defined as diffuse transmittance (T_d) (Williams and Norris, 1987),
was measured by silicon detector. The T_d measurement, expressed as log
($1/T_d$), was obtained by subtracting log ($1/T_{dr}$), where T_{dr} was the
detected radiation from the reference, from log ($1/T_{do}$), where T_{do} was
the detected diffuse transmittance from the brain. Plotting log ($1/T_d$) as a
function of wavelength gives a spectrum that depends on the oxy and deoxyhemoglobin contents, scattering effects of the optical field, and other
non oxygen specific absorbers. The stored spectra can be analyzed and
edited with software supplied by the manufacturer. The term derivative
spectroscopy refers to a spectral measurement technique in which the slope
of the spectrum, that is, the rate of change of absorbance with wavelength,
is measured. Mathematical transformation of normal spectra to first derivative spectra is standard in many modern IR spectrophotometers. The transformation has considerable potential in analytical spectroscopy because it is
uniquely insensitive to sources of error that can degrade the precision
and accuracy of spectrophotometric measurements. For example, because the
first derivative measures spectral slope, constants, such as baseline
differences between spectra, are not characterized. A linear, tilted baseline in the normal spectrum yields a flat, offset baseline in the first
derivative.

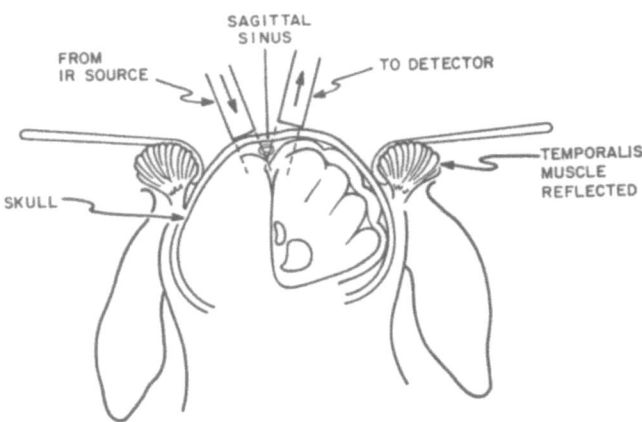

Fig. 1. Experimental layout.

In our experimental conditions, the first derivative could nullify baseline shifts inside each experiment due to cerebral blood volume changes, and eliminate the log $(1/T_d)$ difference from one experiment to another. Therefore, derivative spectra from different dogs could be pooled and correlated with the corresponding SvO_2 values by multiple linear regression analysis (Draper, 1981). In the regression program each wavelength across a specified spectral region was tested individually using a stepwise regression. The selection of the terms to enter a calibration equation were optimized by determining the wavelengths that provided the most information for improvement of the calibration equation. The calibration accuracy could be checked by the standard error (SE) of estimation and the correlation coefficient (R). Similar data treatments have been described in detail in a recent review (Williams and Norris, 1987).

Experimental Protocol

Inspired fractional O_2 concentration (FiO_2) was measured with an O_2 analyzer. A control spectrum (consisting of 100 consecutive spectra, averaged) was recorded in normoxic normocapnia. HH was induced by introducing N_2 into the inspired gas. After any FiO_2 change, spectral recording was repeated up to the time when consecutive spectra were unchanged. Since DNIRS technique is based on the correlation with an analytical method, the calibration samples, used for elaborating the method, are critically important; therefore, they include a large, equally distributed range of SvO_2 measurements. FiO_2 was decreased from normoxia to 6% at 1-3% intervals. Arterial and SS blood samples were collected just after the end of each collection and analyzed immediately with an Instrumentation Laboratory Model 282 CO-Oximeter (Lexington, MA) and a Radiometer ABL 3 gas analyzer (Copenhagen, Denmark). End-tidal CO_2 and arterial pH were maintained constant throughout the experimental procedure.

To test the ability of the DNIRS to predict SvO_2, the dogs were divided into two groups, the first for calibration and a second for prediction. Derivative spectra from the first five dogs were pooled and used to develop a calibration (prediction) equation. This equation was then used to predict SvO_2 from spectral changes of the other 3 dogs. Bias, determined as the difference between the average of the SvO_2 values and the average of predicted values, the SE of bias, the SD of difference, and the simple correlation (r) were then determined.

RESULTS

Eighty-three spectra were obtained from 8 dogs with SvO_2 ranging from 1.5% to 70%. Typical log $(1/T_d)$ spectra measured on one representative dog brain during normoxia and exposure to progressively more severe hypoxic breathing mixtures are shown in fig. 2. In the conventional scans, increased peak height around 755 nm reflects the relative increase in deoxyhemoglobin during hypoxia, whereas reductions in oxyhemoglobin content are expressed as decreases in the broad peak centered around 850 nm, consistent with published "in vitro" and "in vivo" spectra (Jobsis, 1977; Wray et al., 1988). The first derivative spectra reveal an increase in spectral slope below 758 nm that inversely correlates with SvO_2. By comparing to the "in vitro" hemoglobin spectrum, the increasing peak height seen in the normal spectrum is due to a progressive increase in deoxyhemoglobin. Conversely, oxyhemoglobin, prevalent at wavelengths higher than 830 nm, decreased during HH. The first derivative spectra enhanced the deoxyhemoglobin spectral features and damped, mostly at the highest wavelengths, the spectral shifts due to blood volume changes.

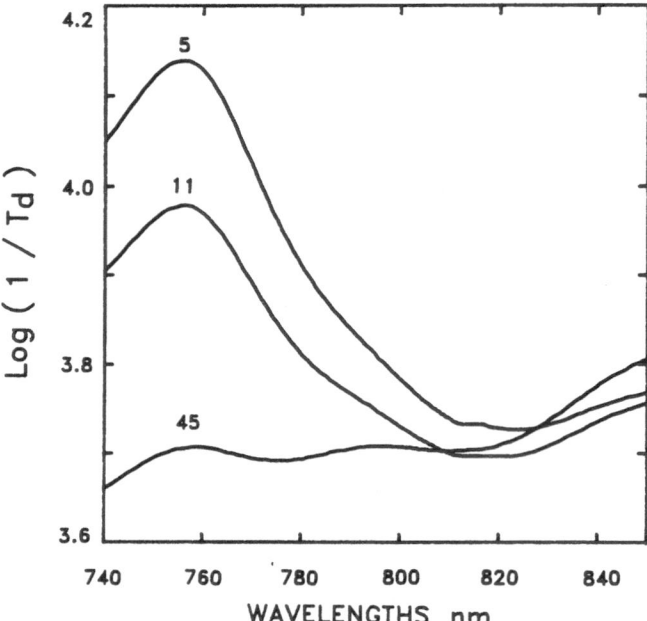

Fig. 2. Spectra obtained in one representative animal during normoxia
(SvO_2 45%) and two levels of hypoxic hypoxia corresponding to
SvO_2 = 11% and 5%. Each spectrum is the average of 100 scans
measured over a 30 sec period.

Fifty-two spectra, obtained from the first 5 dogs, were regressed
against the corresponding SvO_2 values. Log ($1/T_d$) values provided poor
correlations with SvO_2. Conversely, first derivative treatments provided
high correlations. The ratio of two first derivatives, obtained at two
different wavelengths, added to the first derivative at a third wavelength
were the terms of the regression equation:

$$SvO_2 = a + b(A_1/A_2) + c(A_3) \qquad \text{(eq. 1)}$$

which yielded the best correlation between the spectral data and SvO_2 (SE of
estimation = 3.8; multiple R = .98; a, b, c = regression coefficients;
A_1, A_2, A_3 = derivatives at different wavelengths (fig. 3). Equation 1
was used to predict SvO_2 in 33 spectra. The bias, the SE of bias and the
SD of differences were -.7, .5 and 3.2 respectively. The r value 0.97 and
highly significant (p < .001).

DISCUSSION

Tissue hypoxia is a major cause of morbidity and is ultimately the cause
of death in most humans. Non-invasive monitoring of arterial oxygenation

by pulse oximetry has become a standard in anesthetic and intensive care practice (Taylor and Whitman, 1987). Pulse oximetry is easy to use, and provides a rapid, continuous estimate of SaO_2. Unfortunately, its use does not provide information about the oxygen availability of the brain, one of the organs highly dependent on oxygen availability.

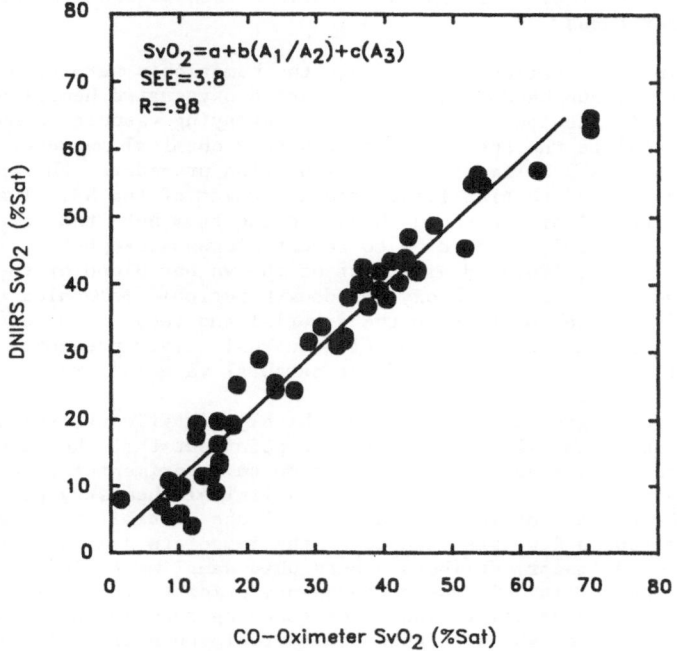

Fig. 3. Relationships in 5 dogs between SvO_2 calculated from a linear algorithm using the first derivative at 3 wavelengths and the SvO_2 value measured in the superior sagittal sinus. Each data point represents a different hypoxic hypoxia level. The line represents the first order regression that fits all data points. Inset: regression equation, correlation coefficient (R) and, standard error of estimation (SEE).

We tested the hypothesis that canine cerebral SvO_2 can be directly and non-invasively determined by DNIRS. This technique is based on the possibility of carrying out fast near IR spectra of brain tissues which are semi-transparent to near IR light. Since it is difficult experimentally to measure the amount of radiation transmitted through high density material such as the brain, the T_d, i.e. diffuse transmittance, was measured. Near IR log $(1/T_d)$ brain spectra are characterized by broad overlapping absorption bands (fig. 2). The shape of a log $(1/T_d)$ spectrum is a function of amount and type of chromophores present in brain, while the absolute value of signal is a function of optical scattering properties, the optic fibers geometry, and absorption. Special data treatments such as derivative transformations are required to obtain compositional information from such spectra. Although a number of studies dealing with the determination of constituents such as fat, protein, carbohydrates, and fibers on different biological materials and agricultural products have been published (Williams and Norris, 1987), only one paper has reported the application of this method to the living body (Conway, Norris and Bodwell, 1984).

We evaluated the merits of the use of zero derivative spectra, i.e. log $(1/T_d)$ spectra, as well as first and second derivative spectra in living animals. We conclude that first derivative spectra are better suited to this application because they diminish contributions from wavelength independent background absorbances.

No technique is available to measure the hemoglobin saturation in intact living cerebral tissue because oxygenated and deoxygenated hemoglobin are heterogenously distributed between several overlaying vascular compartments. In order to correlate the spectral features to a chemical measure, we measured the SS SvO_2. To improve the calibration procedure, the optic fibers were positioned to transilluminate a segment of the SS. The SS collects at least 40% of the venous blood coming from both the cerebral hemispheres and probably, according to recent microspectrophotometric data (Buchweitz and Weiss, 1987), drains most of the venous blood of the cortical regions. In addition, although oxy and deoxyhemoglobin molecules are present in different quantities in the arterial and venous side of the brain circulation, the venous circulation anatomically represents, in normoxic conditions, at least 70% of the cerebral vascular tree.

Although the regression equation was highly significant, some outliers (fig. 3) were obtained. In this regard, we point out that the mathematical treatment of these data gave no consideration to experimental error associated with measuring the reference SvO_2. The clinical chemistry procedure to measure SvO_2 had an intrinsic error even if the possibility of sampling error in the measured SvO_2 was reduced by the immediate availability of the reference CO-Oximeter, deliberate anaerobic sampling technique, and confirmed reproducibility of SvO_2 measurements prior to initiating the protocol. Despite these limitations, the scanning spectrophotometer was able to track measured SvO_2 with acceptable precision over a wide range of abnormal saturations in our animals. SvO_2 values were well estimated also at low FiO_2 values when the brain hemodynamics, as well as spectra, could abruptly change.

In conclusion, the applicability of DNIRS to quantitative measurement of SvO_2 during HH has been clearly established by this study. The generation of the algorithm suggests that a scanning instrument can non-invasively predict SvO_2 and may be useful in humans as an adjunct to conventional monitoring devices of anesthesia and intensive care.

SUMMARY

An "in vivo" method for non-invasive determination of cerebral venous hemoglobin O_2 saturation (SvO_2) was developed. A specially designed spectrophotometer recorded the Td near IR spectra of transilluminated brain tissue surrounding the SS. The accuracy of the method, based on the principle of DNIRS was tested on eight pentobarbital anesthetized dogs during hypoxic hypoxia (inspired O_2 6-21%). Spectral data were transformed into first derivative for correlation with SvO_2 data measured from the SS. Linear regression analyses were applied using data from 5 dogs, with SvO_2 ranging from 1.5%-70%, to build a 3 wavelength algorithm for predicting brain SvO_2. In three dogs, this regression equation was employed to predict SvO_2 in 31 separate spectra of varying HH intensity. The standard deviation of differences between SvO_2 and predicted values was 3.2%. The predicted values, when regressed against the sampled SvO_2, yielded an r value of 0.97. The results demonstrate that it is possible to noninvasively quantify SvO_2 utilizing IR spectroscopy.

ACKNOWLEDGEMENTS

Dr. M. Ferrari was supported by a PHS Fogarty International Fellowship #1 FO5 TWO 3384-01. This work was supported in part by USPHS NIH NS 20020. We would also like to thank Candace Berryman for her secretarial assistance.

REFERENCES

Buchweitz, E., and Weiss, H.R., 1987, Effect of MCA occlusion on brain O_2 supply and consumption determined microspectrophotometrically, Am. J. Physiol., 253:H454.

Conway, J.M., Norris, K.H., and Bodwell, C.E., 1984, A new approach for the estimation of body composition: infrared interactance, Am. J. Clin. Nutr., 40:1123.

Draper, N., and Smith, H., 1981, Applied Regression Analysis. 2nd Ed., John Wiley & Sons, New York.

Eggert, H.R., and Blazeck, V., 1987, Optical properties of human brain tissue, meninges, and brain tumors in the spectral range of 200 to 900 nm, Neurosurgery, 21:459.

Ferrari, M., Zanette, E., Sideri, G., Giannini, I., Fieschi, C., and Carpi, A., 1987, Effects of carotid compression, as assessed by near infrared spectroscopy, upon cerebral blood volume and hemoglobin oxygen saturation, J. Royal Soc. Med., 80:83.

Jobsis, F.F., 1977, Non-invasive, infrared monitoring of cerebral and myocardial oxygen sufficiency and circulatory parameters, Science, 98:1264.

Jobsis, F.F., 1985, Non-invasive, near infrared monitoring of cellular oxygen sufficiency in vivo, in: Advances in Experimental Medicine and Biology - Oxygen Transport to Tissue VII, Plenum Press, New York, 191:833.

Mchedlishvili, G., 1986, Arterial Behavior and Blood Circulation in the Brain, Plenum Press, New York.

Taylor, M.B., and Whitwam, J.G., 1987, The current status of pulse oximetry. Clinical value of continuous noninvasive oxygen saturation monitoring, Anaesthesia, 41:943.

Williams, P., and Norris, K., 1987, Near-infrared technology in the agricultural and food industries, American Association of Cereal Chemists, St. Paul.

Wilson, B.C., and Patterson, M.S., 1986, The physics of photodynamic therapy, Phys. Med. Biol., 31:327.

Wyatt, J.T., Cope, M., Delpy, D.T., Wray, S., and Reynolds, E.O.R., 1986, Quantification of cerebral oxygenation and haemodynamics in sick newborn infants by near infrared spectrophotometry, Lancet, II:1063.

Wray, S., Cope, M., Delpy, D.T., Wyatt, J.T., and Reynolds, E.O.R., 1988, Characterization of the near infrared absorption spectra of cytochrome aa_3 and haemoglobin for the non-invasive monitoring of cerebral oxygenation, Biochem. Biophys. Acta, 933:184.

NEAR INFRARED DETERMINED CEREBRAL TRANSIT TIME AND OXY- AND DEOXYHEMOGLOBIN

RELATIONSHIPS DURING HEMORRHAGIC HYPOTENSION IN THE DOG

Marco Ferrari*, David A. Wilson^, Daniel F. Hanley^, and Richard J.
Traystman^

*Department of Biomedical Sciences and Technology, and Biometrics,
University of L'Aquila, 67100 L'Aquila, Italy and Physiopathology
Laboratory, Istituto Superiore di Sanita, Viale Regina Elena 299,
00161, Rome, Italy; ^Department of Anesthesiology/Critical Care
Medicine, The Johns Hopkins University, Baltimore, MD 21205, USA.

INTRODUCTION

Many recent papers suggest that in patients with cerebrovascular
diseases, cerebral blood flow (CBF) measurement alone is a poor indicator of
hemodynamic reserve (Gibbs et al., 1984). A usual finding of positron emission tomography studies is a fall of the CBF/cerebral blood volume (CBV)
ratio, mathematically equivalent to an increase of the mean transit time
(MTT) (Zierler, 1965).

The goal of the present study was to investigate the relationships among
CBF, cerebral metabolic rate for oxygen ($CMRO_2$), MTT and tissue/superior
sagittal sinus hemoglobin saturations over a wide range of perfusion pressures (CPP) using a canine model of hypotension. MTT and brain oxy- and
deoxyhemoglobin (HbO_2, Hb) were measured by near infrared (IR) spectroscopy
(NIRS), a technique which offers a means of monitoring, in non-invasive
fashion, changes in brain hemoglobin content/oxygenation (Jobsis, 1977),
and quantifying MTT. The principles of NIRS have recently been reviewed
(Jobsis, 1985). Briefly, IR light of the 700-900 nm region is transmitted
sufficiently through the soft and hard tissues of the head to provide an
adequate signal for spectrophotometric purposes. The intensity of the
transmitted light depends on the wavelength, scattering effects, and upon
near IR absorption by the most abundant chromophore, i.e. the heme portion
of hemoglobin.

METHODS

Ten adult mongrel dogs were utilized in this study. Animals were anesthetized with sodium pentobarbital (30 mg/kg, i.v.). Following placement of
catheters, pancuronium bromide (2 mg/kg, i.v.) was administered for muscle
relaxation every 1 hr, and heparin (500 μg/kg, i.v.) was used as the anti-
coagulant. Femoral arteries and veins, omocervical arteries, right common
carotid artery and skull surface were exposed. A cardiac catheter was inserted into the left ventricle via a femoral artery for the injection of radiolabelled microspheres. The microsphere reference sample used for calculating
regional organ blood flow was obtained through a catheter placed into the
brachiocephalic artery via an omocervical artery. A femoral artery was

cannulated with a large bore catheter connected to a pressurized reservoir so that the animal's blood pressure could be set and maintained at any level via blood volume alterations. A catheter was inserted into the isolated cephalic thyroid artery for injection of indocyanine green dye (ICG) (Cardiogreen, Hynson, Wescott and Durring, Baltimore, MD) into the common carotid artery. A sagittal sinus catheter was placed in the superior sagittal sinus. Sagittal sinus and intracranial blood pressures, together with arterial pressure, were measured. Regional blood flow was measured with 15 ± 1.5 μm diameter, radiolabelled microspheres (^{153}Gd, ^{113}Sn, ^{103}Ru, ^{95}Nb, ^{46}Sc; Dupont-NEN Products, Boston, MA) using the reference sample method (Marcus et al., 1976). Approximately 4×10^6 microspheres were injected into the left ventricle over 20 sec, followed by a 15 sec flush of 20 ml saline. Simultaneous reference samples were withdrawn from the proximal aorta at a rate of 3.82 ml/min. Samples were counted by a scintillation spectrometer and the overlap of isotope activity among the five windows was subtracted by solving simultaneous equations using overlap coefficients determined from pure isotope spectra (Heymann et al., 1977). Blood flow was calculated as the product of these overlap-corrected counts in the tissue times the arterial reference withdrawal rate divided by the counts in the arterial reference sample.

Optical Methods

A three-wavelength spectrophotometer (OMNI-3 Monitor, Oxidative Metabolism Near Infrared; International Instrumentation Laboratories, Inc., Durham, NC), described in detail elsewhere (Piantadosi et al., 1986), was used to measure the near IR optical-absorption features of the brain. A fiber optic lightpipe (6 mm active diameter), carrying the infrared photons, illuminated the skull overlying the right parietal cortex and a same size optic fiber forming an angle of 40-50 degrees with the first, was placed 2.9 cm in front. Both fibers, roughly at 2.5 cm from the dog head midline, were firmly applied against the bone surface in order to avoid movements during the experiment as well as to optimize the optic coupling between the fiber endings and the skull surface.

The near IR photons, passing through the bone - a relatively low IR absorbing material - were partly absorbed and/or back scattered by the cortex, almost completely reflected by the white matter and, finally, partly collected by the receiver optic fiber. According to experimental data on optical properties of these different tissues (Wilson and Patterson, 1986; Eggert and Blezek, 1987) the biggest absorption contribution came from the first mm of cerebral tissue. Apparent extinction coefficients for the used wavelengths were experimentally obtained in animals. The multiplying factors for HbO_2 were -1, -1.6 and 1.64 at 775, 815 and 905 nm respectively; the corresponding factors for Hb were 1.6, -.61 and -.41. Because HbO_2 and Hb were generated by a 3 wavelength algorithm, their changes from control value can be expressed in relative terms as variation of density unit (v/d) (Jobsis, 1985).

The MTT was measured by a modification of a recently described technique (Tomita et al., 1983) calculating the passage into the brain optical field (BOF) of an intra-common carotid injected ICG bolus. ICG has a broad absorption band centered around 805 nm, OMNI 815 nm absorption signal only was used for MTT analysis. ICG doses from .05 to .2 ml (2.5 mg/ml) produced a delta OD bigger than .3, sufficient for ICG clearance analysis. IR monitoring allows for the observation of the appearance of the bolus in the BOF and determination of the rate of its washout from the BOF. ICG was collected at a sampling rate of 20 Hz. An acquisition time of 3 min was sufficient to carry out 1-3 injections. Mathematical analysis of the clearance curve conducted by several authors (Lassen and Perl, 1979) has suggested that the experimental data could be fitted by an exponential function. Therefore, the ICG clearance curves were represented within experimental error by the

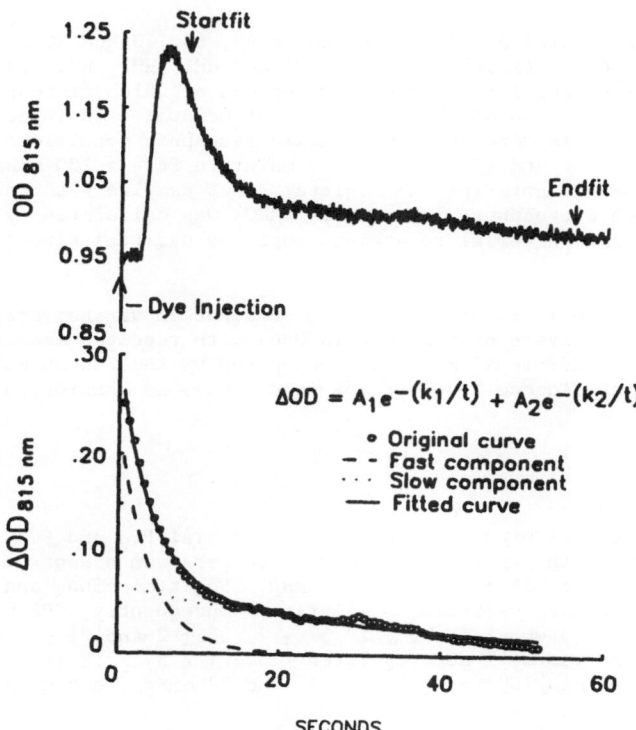

Fig. 1. Representative passage of intracarotid injected indocyanine green dye (ICG) bolus through a brain optical field (top panel). The curve is the sum of two exponential functions (bottom panel). The contribution of each component is determined by performing a least square fit of the sum of two exponential functions to the experimental data points.

sum of two exponential functions, a fast component and a slow component according to the equation:

$$\Delta \text{ OD} = A_1 e^{-(k_1/t)} + A_2 e^{-(k_2/t)}$$

where Δ OD was the 815 nm OD at any time and A_1 and A_2 were the initial delta OD due to the fast and slow components (fig. 1). The contribution of each one was determined by performing a least square fit of the sum of the two exponential functions to the experimental data points. Each exponential function yields a straight line when plotted versus time on a semilogarithmic scale. The slopes of these lines, K_1 and K_2, calculated by the biexponential fitting procedure, yielded the transit times, expressed in seconds, of the first and the second compartments, respectively, in the optical field.

Experimental Protocol

The dogs were bled until MAP was about 80, 60, 40 and 20 mmHg, respectively. Each step, defined as control (C) and HH1, HH2, HH3 and HH4 respectively, was maintained for 15-20 min at the end of which microspheres were injected. After the end of each reference withdrawal, ICG injections were carried out. Animals were kept in normoxic isocapnic conditions by changing the ventilation rate and tidal volume to maintain $PaO_2 > 100$ mmHg and $PaCO_2$ between 30-35 mmHg. Arterial and sagittal sinus samples were collected just before each microsphere injection. $CMRO_2$ was calculated by multiplying total CBF times arterial to venous (superior sagittal sinus) O_2 content difference.

Statistical analysis of the results obtained in various stages of HH was accomplished by analysis of variance (ANOVA) with repeated measures for effects of CPP. Individual means were compared by the Duncan multiple range test. A confidence level of 95% ($p < 0.05$) was considered statistically significant.

RESULTS

Results are presented as means $\pm$ SE. Arterial PO_2 and PCO_2 were unchanged throughout the experiment. MAP decreased from a control value of 136 $\pm$ 4 to 79 $\pm$ 1, 59 $\pm$1, 38 $\pm$ 1 and 16 $\pm$3 mmHg. Sagittal sinus and cerebrospinal fluid pressures decreased similarly. Consequently, CPP was decreased from control (133 $\pm$3 mmHg) to 76 $\pm$ 2, 57 $\pm$ 2, 39 $\pm$ 2 and 26 $\pm$ 4 mmHg. The optical field CBF was well autoregulated (25.2 $\pm$ 2.5, 27.2 $\pm$ 2.9, 33 $\pm$ 4.4, and 29.5 $\pm$ 3.6 ml/min/100g at 133, 76, 57 and 39 mmHg), but dropped to

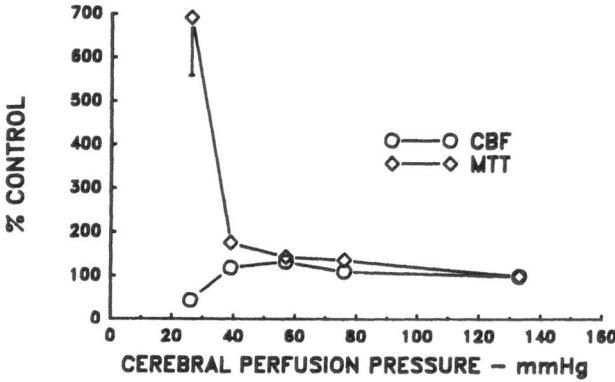

Fig. 2. Cerebral blood flow (CBF) and mean transit time (MTT) percentage changes vs cerebral perfusion pressure during normotension and four levels of hemorrhagic hypotension. Values are means $\pm$ SE.

10.4 $\pm$ 1.3 when CPP was 26 mmHg (fig. 2). $CMRO_2$ was maintained to the last tested CPP when it dropped under the control level (2.5 $\pm$ .3, 3.5 $\pm$ .5, 3.5 $\pm$ .5, 3.2 $\pm$ .3 and 1.6 $\pm$ .2 ml O_2/100g/min). MTT constantly increased from control (3.9 $\pm$ .4 sec) to 5.4 $\pm$ .6, 5.6 $\pm$ .7, 6.7 $\pm$ .7 and 24.9 $\pm$ 4.4 sec at 76, 57, 39 and 26 mmHg. HbO_2 decreased from control of 0 to -.17 $\pm$ .03, -.26 $\pm$ .05, -.31 $\pm$.06 and -43 $\pm$ .1 v/d at 76, 57, 39 and 26 mmHg. Hb increased from control of 0 to .18 $\pm$ .03, .27 $\pm$ .06, .34 $\pm$ .06 and .43 $\pm$ .1

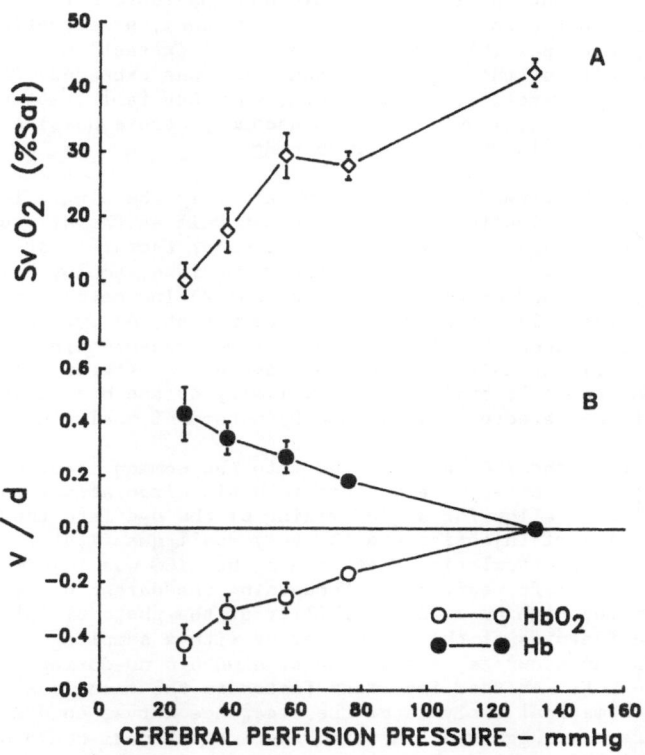

Fig. 3. Hb/HbO_2 (panel B) and superior sagittal sinus saturation (SvO_2) (panel A) vs cerebral perfusion pressure during normotension and four levels of hemorrhagic hypotension. Values are means $\pm$ SE.

v/d at 76, 57, 39 and 26 mmHg (fig. 3B). SvO_2 variations matched HbO_2 decreases (fig. 3A) confirming that $CMRO_2$ constancy could be maintained by means of oxygen extraction increases. At CPP of 26 mmHg, when presumably the vasodilatatory capacity was almost exhausted, CBF dropped and MTT drastically lengthened about 7 times control as final attempt to keep constant the O_2 availability to the cerebral tissue.

DISCUSSION

The cerebral circulation is capable of autoregulating during changing perfusion pressure (Heistad and Kontos, 1986). In this study, CPP reduction was produced, by graded steps of arterial hypotension, bleeding the dog into a pressurized reservoir connected to the femoral artery. According to the NIRS signal monitoring, a 20 min period was necessary to reach a steady state for HbO_2 and Hb. Autoregulation was maintained up to around 40 mmHg CPP (fig. 2). At the last stage of hemorrhagic hypotension when CPP was around 30 mmHg, CBF in all the measured cerebral areas, significantly decreased to about 50% of flow values measured inside the autoregulatory range.

The relative contribution of aerobic and anaerobic metabolism in hemorrhagic shock has not been elucidated. In our model, preservation of aerobic metabolism was made possible by the increased O_2 extraction. When the autoregulatory capacity of the cerebral vasculature was exceeded, CBF and $CMRO_2$ decreased in parallel. Furthermore, when CBF falls, metabolic factors maximally dilate cerebral vessels which become pressure passive as demonstrated by a further MTT increase at CPP 39 mmHg.

Zierler (1965) first described mathematically the single-injection method for the determination of blood flow as well as the MTT technique for calculating vascular volume. Different major factors that can affect markedly the accuracy of the ICG clearance technique, employed in this study are: 1) site and modality of injection; 2) indicator trapping in the vascular system; 3) extracerebral contamination; 4) dye recirculation; 5) relationship between the plasma transit time, measured in this study and red blood cells transit time; 6) correspondance between MTT measured in the BOF and the whole brain MTT; 7) validity of the dye passage model; 8) features of the detector system; and 9) method of analyzing the curves.

In this study the ICG was injected into the common carotid artery via the superior thyroid artery, very close to brain circulation, but also sufficiently far to allow for a good mixing of the dye into the blood stream. This site of injection and the very small quantity of injected dye prevented the recirculation that occurs when ICG was injected intravenously or in the left ventricle. Concerning the detection system employed in this study, the high near IR sensibility of the photomultiplier tube, as well as the features of the A/D converter with a sampling rate of 200 ms, allowed for an accurate measurement of Δ OD 815 nm during the dye transit in the BOF. Another important factor is the technique for curves analysis. Mathematical analysis of the clearance curves conducted by several authors has suggested that the experimental data could be fitted by an exponential function. We fitted successively a biexponential and a monoexponential function to the experimental data points. The monoexponential function was employed only for the curves in which the slow compartment was undetectable. The best fit was found in almost all curves with the biexponential function.

Recently, Tomita (1983) discussed in detail a similar photoelectric method to measure MTT on cats based on brain tissue carbon-black clearance curves. The reproducibility and the stability of our MTT method were quite satisfactory and comparable to Tomita's technique. It is not clear, however, if the relative MTT variability reflected the accuracy of the analysis system or actual physiopathological variation in hemodynamic conditions. In fact, in our study, the most consistent variability was observed at the lowest tested CPP when the hemodynamic conditions can quickly and dramatically change. On the contrary, MTT fluctuations were never related to the cardiac or respiratory cycles.

MTT increased through the hemorrhagic hypotension: considering the single experiments, only one animal had a MTT value lower than control at CPP 57 mmHg, and all MTT were longer than control at CPP 39 and 26 mmHg. Finally, the reversibility of MTT was witnessed with its recovery to control values in some animals in which the MAP was recovered by blood reinfusion.

Few cerebral MTT data are available from the literature. Moreover, using a different technique, studies carried out on cats, monkeys and goats reported MTT values in control conditions of $5.77 \pm .27$ SE, 4.33 ± 1.3 SD and $4.5 \pm .5$ SE sec respectively; very similar to our control value of $3.9 \pm .4$ sec. Recently, MTT for healthy humans of 3.2 ± 0.7 SE sec was reported (Carlsen and Hedegaard, 1987). A prolonged MTT was measured on patients older than 60 or with cerebral ischemia and stroke within the lesion (Szabo, 1986). In a cat middle cerebral artery occlusion, MTT increased progressively with longer period of occlusion (Little et al., 1981). Similar increases were calculated in cerebrovascular patients from CBF and CBV emission tomography data (Gibbs et al., 1984).

Due to the undefined tissue scattering properties, the NIRS data cannot be expressed in absolute terms and can be reported only in relative terms, i.e. v/d units. Moreover, since in each animal v/d units depend on the geometry of the optic fibers - distance between and the angle of placement - the NIRS results are not directly comparable. To circumvent this problem, the optic fibers geometry was almost the same in all the experiments. HbO_2 changes matched SvO_2 data, but drastically decreased at the lowest tested CPP. Currently, many methods are being explored to obtain a quantitative HbO_2 measurement (Hazeki and Tamura, 1988; Wray et al., 1988). In this framework, quantitative measurements of MTT and tissue HbO_2 by NIRS might give a more sensitive evaluation of the hemodynamic reserve than CBF measurements alone. This may be applicable for clinical situations in anesthesia and intensive care.

SUMMARY

The goal of the present study was to investigate the relationships among CBF, $CMRO_2$, MTT and tissue/superior saggital sinus hemoglobin saturations over a wide range of CPP using a canine model of hypotension. Microsphere-determined CBF was autoregulated in all tested cerebral regions over the 40-130 mmHg CPP range, but it decreased 50% around 30 mmHg. MTT, calculated by an indicator-dilution technique, progressively and proportionally increased over the 40-130 mmHg CPP range. The HbO_2 content, measured in the same area by an optical method, parallely decreased. Around 30 mmHg CPP, MTT disproportionately lengthened (700% of control). $CMRO_2$ maintenance was accompanied by a progressive decrease of tissue HbO_2 and sagittal sinus hemoglobin saturation. In this framework, MTT measurement and tissue HbO_2 monitoring by NIRS might give a sensitive evaluation of the hemodynamic reserve than CBF measurement in anesthesia and intensive care.

ACKNOWLEDGEMENTS

Dr. M. Ferrari was supported by a PHS Fogarty International Fellowship #1 F05 TWO 3384-01. We thank Candace Berryman for her secretarial help.

REFERENCES

Carlsen, O., and Hedegaard, O., 1987, Evaluation of regional cerebral circulation based on absolute mean transit times in radionuclide cerebral angiography, <u>Phys. Med. Biol.</u>, 32:1457.

Eggert, H.R., and Blezek, V., 1987, Optical properties of human brain tissue meninges, and brain tumors in the spectral range of 200 to 900 nm, Neurosurg., 21:459.

Gibbs, J.M., Wise, R.J.S., Leenders, K.L., and Jones, T., 1984, Evaluation of cerebral perfusion reserve in patients with carotid-artery occlusion, Lancet, 1:310.

Hazeki, O., and Tamura, M., 1988, Quantitative analysis of hemoglobin oxygenation state of rat brain in situ by near infrared spectroscopy, J. Appl.Physiol., 64:786.

Heistad, D.D., and Kontos, H.A., 1986, Cerebral circulation, in: Handbook of Physiology - The Cardiovascular System, sect. 2, vol. III, part I, p.137-182, J.T. Shepherd and F.M. Abboud, eds., American Physiological Society, Bethesda.

Heymann, M.A., Payne, B.D., Hoffman, J.I.E., and Rudolph, A.M., 1977, Blood flow measurements with radionuclide-labeled particles. Prog. Cardiovas. Dis., 20:55.

Jobsis, F.F., 1977, Noninvasive, infrared monitoring of cerebral and myocardial oxygen sufficiency and circulatory parameters, Science, 98:1264.

Jobsis, F.F., 1985, Non-invasive, near infrared monitoring of cellular oxygen sufficiency in vivo, in: Advances in Experimental Medicine and Biology, 191:833, Plenum Press, New York.

Lassen, N.A., and Perl, W., 1979, Tracer Kinetic Methods in Medical Physiology, Raven Press, New York.

Little, J.R., Cook, A., Cook, S.A., and MacIntyre, W.J., 1981, Microcirculatory obstruction in focal cerebral ischemia: albumin and erythrocyte transit, Stroke, 12:218.

Marcus, M.L., Heistad, D.D., Ehrhardt, J.C., and Abboud, F.M., 1976, Total and regional cerebral blood flow measurement with 7-, 10-, 15-, 25-, and 50-μm microsphers, J. Appl. Physiol., 40:501.

Piantadosi, C.A., Hemstreet, T.M., and Vandervliet-Jobsis, F.F. 1986 Near infrared spectrophotometric monitoring of oxygen distribution to intact brain and skeletal muscle tissue. Crit. Care Med., 14:698.

Szabo, Z., and Ritzl, F., 1983, Mean transit time image - a new method of analyzing brain perfusion studies, Eur. J. Nucl. Med., 8:201.

Tomita, M., Gotoh, F., Amano, T., Tanahashi, N., Kobari, M., Shinohana, T., and Mihara, B., 1983, Transfer function through regional cerebral cortex evaluated by a photoelectric method, Am. J. Physiol., 245:H385.

Wilson, B.C., and Patterson, M.S., 1986, The physics of photodynamic therapy, Phys. Med. Biol., 31:327.

Wray, S., Cope, M., Delpy, D.T., Wyatt, J.S., and Reynolds, E.O.R., 1988, Characterization of the near infrared absorption spectra of cytochrome aa$_3$ and haemoglobin for the non-invasive monitoring of cerebral oxygenation, Biochem. Biophys. Acta, 933:184.

Zierler, K.L., 1965, Equation for measuring blood flow by external monitoring of radioisotopes, Circ. Res., 16:309.

NEAR INFRARED QUADRUPLE WAVELENGTH SPECTROPHOTOMETRY OF THE RAT HEAD

Osamu Hazeki and Mamoru Tamura

Biophysics Division
Research Institute of Applied Electricity
Hokkaido University, Sapporo 060, Japan

INTRODUCTION

Cytochrome oxidase is the terminal enzyme in the respiratory chain, which ultimately transfers electrons to molecular oxygen. Its redox state can be used as a direct indicator of the sufficiency of oxygen supply to the site of cellular energy production, the mitochondria. The enzyme when oxidized has a broad absorption band in near infrared region and this absorption disappears when its copper component is reduced. Because the extinction coefficients for most biological materials are low in near infrared region, it is not difficult to detect near infrared light that penetrated intact tissues. Thus near infrared spectroscopy is expected to be a new tool for noninvasively monitoring tissue hypoxic injury. The technique was first introduced by Jobsis(1977), and has been applied in several clinical and physiological situations (e.g. Wiernsperger et al., 1981; Cairns et al., 1985; Ferrari et al., 1985; Wyatt et al., 1986). The main obstacle to quantitation of the cytochrome oxidase redox state in intact organ has been the overlap of the spectral changes due to variation in the amount and oxygenation state of hemoglobin. Algorithms previously reported include dual and triple wavelength analysis of difference spectrum. Wray et al. (1988) assumed the additive absorption of three major components, namely oxygenated and deoxygenated hemoglobin and the redox state of cytochrome oxidase, and calculated the change of each component by solving simultaneous equations at three separate wavelengths. The absorptivity constant of the cytochrome change was determined in fluorocarbon exchange transfused rat brain, while those of hemoglobin were measured in a clear solution. Hazeki et al. (1987) proposed a similar method where the standard spectra of hemoglobin were obtained directly in perfused rat head preparations.

A limitaion of the triple wavelength method is the lack of compensation for the light scattering change of tissue itself, and for instability of the photomultiplier or light source. For a one component system, classical dual wavelength analysis provides the adequate compensation for these factors. In the present paper, we attempt to apply a quadruple wavelength (three measuring and one reference, or three pairs of dual wavelength) method to the three component system of our interest.

METHOD

General

The experiments were performed in three series, in vivo, in situ and in vitro. Male Wistar rats were anaesthetized with urethane (0.6g/kg, intraperitoneally) and underwent tracheostomy for experiments in the live animal. For in situ experiments, external carotid arteries were occluded, common carotid arteries cannulated with polyethylene catheters, and perfusate (deoxygenated red blood cell suspension) infused into them. Circulated perfusate was recovered at the right atrium.

Spectroscopy

A halogen lamp illuminates through a lens system a rotating disc containing four interference filters. The successive monochromatic pulses of light thus obtained are guided to the rat head (in animal experiments) or to a quartz cuvette (1cm depth, in vitro experiments) via a flexible light pipe (ϕ= 5mm). Transmitted light is collected with an L shaped light pipe placed within the rat mouth or on the opposite side of the cuvette, and guided to a photomultiplier. The change in light intensity at each wavelength is measured by use of a sample and hold circuit and recorded after logarithmic transformation.

Analysis

The difference in optical density (OD) of a red blood cell suspension between two wavelengths can be approximated by a straight line over a limited range, as shown in Figs.1 and 2. By assuming that the total OD is a summation of that from the of oxygenated and deoxygenated species, the following expressions are obtained within the linear region.

$$OD_m - OD_r = a_m[HbO_2] + b_m [Hb] + x_m \qquad ---(1)$$

$$\Delta(OD_m - OD_r) = a_m \Delta[HbO_2] + b_m \Delta[Hb] \qquad ---(2)$$

where OD_m and OD_r are the optical densities at the measurement and reference wavelengths respectively, a_m and b_m are proportionality factors determined experimentally (Table 1). $[HbO_2]$ and $[Hb]$ are the concentrations of oxygenated and deoxygenated hemoglobin, respectively, multiplied by optical pathlength, and X_m is the scattering factor. For convenience , we define G_m to be the OD difference between the measuring and referece wavelengths.

$$\Delta G_m \equiv (OD_m - OD_r) \qquad ---(3)$$

Since the OD change in a rat head in the wavelength range 700-750nm can be attributed to hemoglobin (Hazeki and Tamura, 1988) , the following equations can be written for two measuring wavelengths (λ_1 and λ_2) and one reference wavelength (λ_r) within this range.

$$\Delta G_1 = a_1 \Delta[HbO_2] + b_1 \Delta[Hb] \qquad ---(4)$$

$$\Delta G_2 = a_2 \Delta[HbO_2] + b_2 \Delta[Hb] \qquad ---(5)$$

Because the in vivo spectra at longer wavelengths are affected considerably by the redox change of cytochrome oxidase, then for wavelength λ_3 in this range,

$$\Delta G_3 = a_3 \Delta[HbO_2] + b_3 \Delta[Hb] + c_3 \Delta[Cyt] \qquad ---(6)$$

Although cytochrome oxidase consists of two components (oxidized and reduced forms), the redox change ($\Delta[Cyt]$) can be treated as one component because the total tissue concentration of this enzyme can be regarded as constant during a short term experiment. From equations 4 and 5, we can obtain the following expressions for the changes of oxygenated and deoxygenated hemoglobin concentrations.

$$\Delta[HbO_2] = (-b_2 \Delta G_1 + b_1 \Delta G_2)/(b_1 a_2 - a_1 b_2) \qquad ---(7)$$

$$\Delta[Hb] = (a_2 \Delta G_1 - a_1 \Delta G_2)/(b_1 a_2 - a_1 b_2) \qquad ---(8)$$

By incorporating equations 7 and 8 into equation 6, the following expression for the redox change of cytochrome oxidase is obtained.

$$c_3\Delta[Cyt] = \Delta G_3 - \{(-b_2 a_3 + a_2 b_3) \Delta G_1 + (b_1 a_3 - a_1 b_3) \Delta G_2\}/(b_1 a_2 - a_1 b_2) \qquad ---(9)$$

RESULTS

Figure 1a shows the OD of deoxygenated red blood cell (RBC) suspension plotted against the hematocrit of the suspension. The relationship deviates systematically from linearity especially at lower hematocrit. The correlation coefficient of a linear regression line fitted in hematocrit range of 3–10% is 0.9923 at 700nm. Figure 1b shows the OD difference between two wavelength as a function of hematocrit. The relationship can be approximated by a straight line over the hematocrit range 3–10%, although there still is a slight but systematic deviation. The correlation coefficient for $(OD_{700} - OD_{805})$ is 0.9989. The total OD of whole blood can be expressed as a summation of absorbance and scattering terms.

$$ODtotal = ODabs. + ODscat. \qquad ---(10)$$

The first (absorbance) term is proportional to hematocrit and dependent on wavelength, reflecting the molar extinction coefficient of hemoglobin (Steinke and Shepherd, 1986). The scattering term is a highly nonlinear function of to hematocrit, and monotonically dependent on wavelength. Therefore the proportionality factors (slopes) obtained in Fig.1b should be regarded as the difference in absorbance terms between two wavelengths corrected empirically for the scattering terms. The relative values of

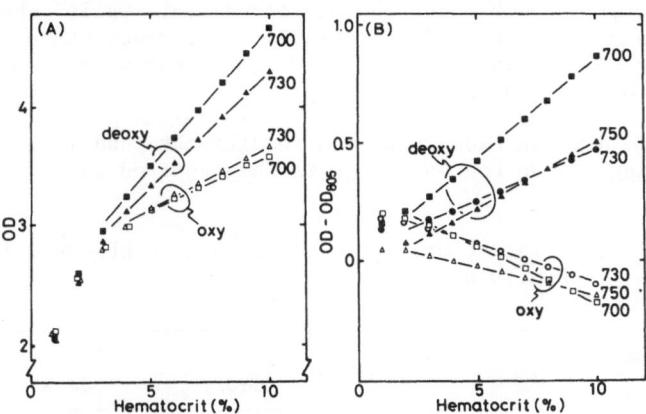

Fig.1 Optical properties of oxygenated and deoxygenated red blood cell suspensions in vitro at several wavelengths.(A) Optical density (OD) is plotted against hematocrit. (B) OD difference from 805 nm is plotted against hematocrit.

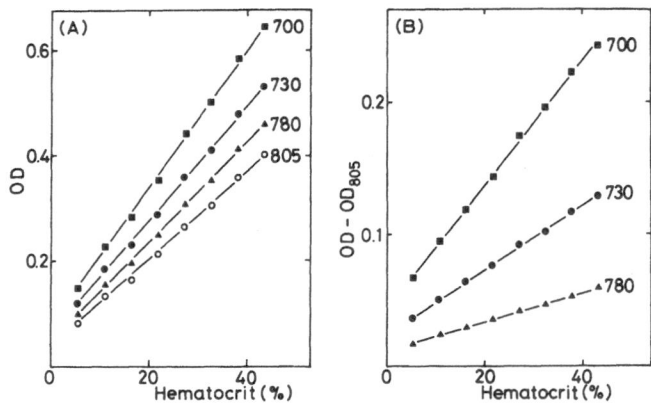

Fig.2   Optical properties of deoxygenated red blood cell
        suspension in situ. (A) Change in optical density (OD) is
        plotted against hematocrit. (B) OD difference from 805 nm
        is plotted against hematocrit.

the factors for oxygenated and deoxygenated suspensions are listed in Table 1.

Figure 2a shows the changes in OD of in situ rat head preparation which is perfused with deoxygenated RBC suspensions, plotted against the hematocrit of the suspension. ΔOD shows an approximately linear relationship function in the range 5-45%. This result validates one of the main assumptions in previously described triple wavelength methods (Hazeki et al., 1987; Wray et al., 1988), that a logarithmic relationship as described by Beer's law holds for living tissue. The OD difference between two wavelengths is shown as a function of hematocrit (Fig.2b). The slopes relative to that of (730-805)Hb are listed in Table 1. The values are similar to those obtained in vitro.

Equations 7,8 and 9 have been applied to the OD changes of the _in situ_ rat head preparation, employing the parameters determined _in vitro_. Figure 3a shows the calculated changes in oxygenated and deoxygenated hemoglobin in rat head, when fully reduced RBC suspension is used as the perfusate. Deoxygenated hemoglobin concentration linearly increases as the hematocrit of the suspension increases, while the oxygenated concentration does not change appreciably. The redox state of cytochrome oxidase (Fig.3b) fluctuates within 0.006 OD, but no systematic error is observed in the range 5-45%.

Figure 4 is an example of the results obtained in a live animal. With decreasing oxygen concentration in the inspired gas, oxygenated

Table 1   Relative proportionality factors observed in
          Fig.1b and Fig.2b

$\lambda m - \lambda r$	In vitro		In situ
	oxy	deoxy	deoxy
700-805	-1.15	2.00	1.96
730-805	-0.87	1	1
750-805	-0.62	1.30	1.41
780-805	-0.25	0.45	0.47

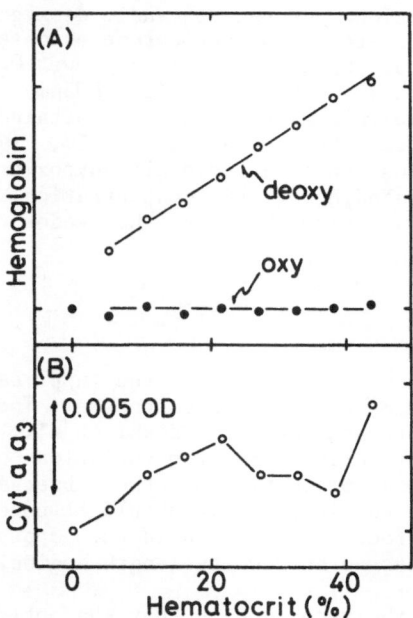

Fig.3 Calculated changes of oxygenated and deoxygenated hemoglobin concentration and the redox state of cytochrome oxidase in deoxygenated red blood cell suspension perfused rat head. 700, 730, 780 and 805 nm are employed as first, second, third and reference wavelengths respectibvely (see Analysis).

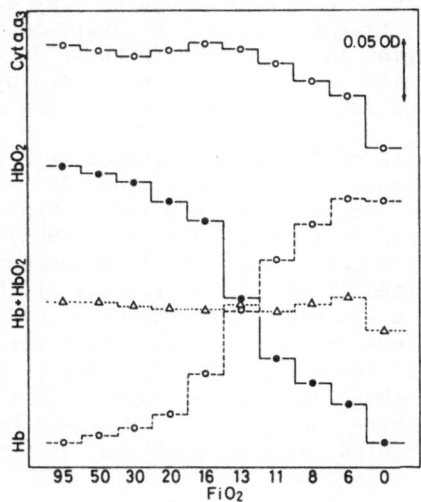

Fig.4 Calculated changes in concentration of oxygenated and deoxygenated hemoglobin and the redox state of cytochrome oxidase in a live animal. The oxygen concentration of the inspired gas is changed as shown in the abscissa. 700, 730, 750 and 805 nm are employed as the first, second, third and reference wavelengths respectively(see Analysis).

hemoglobin concentration decreases and deoxygenated concentration increases. The redox state of the cytochrome oxidase varies by only 0.01 OD unit above 13%O_2, but changes markedly at lower O_2 concentration. The fully reduced state is 0.08 OD unit (at 805nm) below the hyperoxic conditions. This value is comparable to that obtained in the fluorocarbon exchange transfused rat head (Hazeki et al., 1987). Bleeding from external jugular veins under a severely hypoxic condition decreases markedly the deoxygenated hemoglobin concentration, but does not affect appreciably oxygenated hemoglobin or the redox state of cytochrome oxidase.

DISCUSSION

In this report we have examined the applicability of quadruple wavelength analysis to near infrared spectra. The technique has two advantages over our previous method (Hazeki et al., 1987). Firstly, it can minimize the error derived from the variations of transmitted light due to the scattering properties of the observed tissue itself by employing a reference wavelength. It should be noted that the wavelength dependence of this factor is ignored in this approach, because it is expected to be small over a narrow wavelength region.

Another merit of the technique is that the optical characteristics of the RBC's within a tissue are the same as in a cuvette system. Thus, the determination of optical parameters, which is crucial for this type of analysis, can be performed in vitro. We can now re-estimate these parameters without having to repeat troublesome animal experiments, when a new apparatus is assembled and the center wavelength and half intensity bandwidth of the interference filters or aperture radii and mutual geometry of light guides are changed.

One important assumption in our algorithm is that the redox change of cytochrome oxidase gives no OD changes in the 700-780nm region. This is questioned by a recent spectrum obtained by Wray et al.(1988) in the fluorocarbon exchange transfused rat. However, because the wavelength dependence in this region is small even in their difference spectrum, the present method is valid as long as 750nm is employed as a reference wavelength and 700nm and 730nm for hemoglobin monitoring.

The basic principle of the present method is a simple extension of multiwavelength techniques commonly employed in analytical chemistry. Similar approaches have recently been applied by Wray et al.(1988) to the near infrared region and Kariman and Burkhart (1985) to the visible red region. The triple wavelength methods fail to compensate for non specific variations of transmitted light, the overcoming of which is the aim this paper. We feel such a approach is essential in order to evaluate the redox change of cytochrome oxidase in vivo, where variations in the amount or oxygenation state of hemoglobin severely affect the overall spectrum. The principle of the method will also be applicable to the measurement of muscle preparations where movement of the organs and the presence of myoglobin have presented severe problems in the monitoring of mitochondrial signals.

SUMMARY

A quadruple wavelength method to monitor the changes in concentration of oxygenated and deoxygenated hemoglobin and the redox state of cytochrome oxidase within a living tissue is presented. The expected advantages of this technique over the triple wavelength method are (i) that it can compensate for the light scattering change of tissue itself or for instabilities of light source and photomultiplier, (ii) that it can treat the optical properties of the red blood cell in a tissue in the same way as in an in vitro model system, and (iii) that it requires no estimation of the absorption coefficient of cytochrome oxidase in situ.

ACKNOWLEDGEMENT

This work is supported by a Grant-in-Aid from the Ministry of Education, Science and Culture of Japan, and in part by Kowa Life Science Foundation.

REFERENCES

Cairns, C.B., Fillipo, D. and Proctor, J. (1985) A noninvasive method for monitoring the effect of increased intracranial pressure with near infrared spectrophotometry. Surg. Gynecol. Obstet., 161: 145-148.

Ferrari, M., Giannini, I., Siderei, G. and Zanette, E. (1985) Continuous non invasive monitoring of the human brain by near infrared spectroscopy. Adv. Exp. Med. Biol., 191: 873-882

Hazeki, O., Seiyama, A. and Tamura, M. (1987) Near-infrared spectro-photometric monitoring of haemoglobin and cytochrome a,a_3 in situ. Adv. Exp. Med. Biol., 215: 283-289

Hazeki, O. and Tamura, M. (1988) Quantitative analysis of hemoglobin oxygenation state of rat brain in situ by near-infrared spectro-photometry. J. Appl. Physiol., 64: 796-802

Jobsis, F.F. (1977) Noninvasive, infrared monitoring of cerebral and myocardial oxygen sufficiency and circulatory parameters. Science, 198: 1264-1267

Kariman, K. and Burkhart, D.S. (1985) Non-invasive in vivo spectro-photometric monitoring of brain cytochrome a,a_3 revisited. Brain Res., 360: 203-213

Steinke, J.M. and Shepherd, A.P. (1986) Role of light scattering in whole blood oximetry. IEEE Trans. Biomed. Eng., BME-33: 294-301

Wiernsperger, N., Sylvia, A.L. and Jobsis, F.F. (1981) Incomplete transient ischemia: a non-destructive evaluation of in vivo cerebral metabolism and hemodynamics in rat brain. Stroke, 12: 864-868

Wray, S., Cope, M., Delpy, D.T., Wyatt, J.S. and Reynolds, E.O.R. (1988) Characterization of the near infrared absorption spectra of cytochrome aa_3 and haemoglobin for the non-invasive monitoring of cerebral oxygenation. Biochim. Biophys. Acta, 933: 184-192

Wyatt, J.S., Cope, M., Delpy, D.T., Wray, S. and Reynolds, E.O.R. (1986) Quantification of cerebral oxygenation and haemodynamics in sick newborn infants by near infrared spectrophotometry. Lancet, II(1986): 1063-1066

THE OXYGEN DEPENDENCY OF THE REDOX STATE OF HEME AND COPPER IN CYTOCHROME OXIDASE IN VITRO

Y. Hoshi, O. Hazeki and M. Tamura

Biophysics Division, Research Institute
of Applied Electricity
Hokkaido University, Sapporo, Japan

INTRODUCTION

Near infrared spectrophotometry has been proposed as a tool for non-invasive evaluation of the metabolic processes in living tissue. With this technique, it is possible to monitor the tissue blood content, hemoglobin saturation level, and the redox state of copper in cytochrome oxidase. The redox behaviour of heme $a+a_3$ in cytochrome oxidase and its use as an indicator of tissue oxygenation under low oxygen concentrations has already been studied by Oshino et al. (1974) However, there have been few studies upon the redox behaviour of copper under low oxygen concentrations corresponding to some mitochondrial energy states.

In the present study, we investigated the oxygen dependency of the redox state of copper under hypoxic conditions and compared it with that of heme $a+a_3$. It should be possible to apply the results of this in vitro study to in vivo data.

METHODS

Oxygen measurement

Leghemoglobin, present in the root nodules of leguminous plants, was used for an oxygen indicator in this study. The P50 of this compound, ie the oxygen concentration required for half deoxygenation is 0.373×10^{-7} M (Wittenberg et al, 1972). From Bergersen and Turner (1977), the following equation was used for calculation of the oxygen concentration.

$$[O_2 \text{ free }] = \frac{0.373 \times 10^{-7} Y}{1 - Y} \quad [M],$$

where Y is the fractional oxygenation of leghemoglobin as determined by the absorbance change at 430-460nm. Leghemoglobin was prepared according to the method of Bergersen and Turner. (1979)

Titration of the redox states of copper and heme $a+a_3$ with oxygen

A suspension of mitochondria(see later) and leghemoglobin were added to a reaction mixture consisting of 0.25M sucrose, 10mM Tris-HCl (pH7.2), 2mM KCl, 0.2mM EDTA and 10mM glutamate. The ratio of leghemoglobin to heme $a+a_3$ was chosen as 1:7 so that the absorbance change due to the oxygenation-deoxygenation of the leghemoglobin did not affect the absorbance changes due to the oxidation-reduction of either heme $a+a_3$ measured at 605-650nm or copper measured at 830-760nm. The contributions of leghemoglobin absorption were 26% and 37% of the total change at 430-460nm with an ADP and an ATP generating systems respectively. This is due to the large overlap of cytochrome absorbance change in this wavelength pair with that due to leghemoglobin. A Multiwavelength photometer was used for simultaneous monitoring of heme $a+a_3$, copper, and leghemoglobin.

Other procedures

Rabbit heart mitochondria were prepared according to a modification of the method using protease (Kagawa, 1976). Pyruvate Kinase (28 IU/ml), 3mM phosphoenolpyruvate, and 0.5mM ATP were used as an ATP generating system (State4), and hexokinase (15 IU/ml), 10mM glucose, 0.5mM ADP, and 0.5mM magnesium chloride were used for an ADP generating system (State3).

RESULTS AND DISCUSSION

Steady state responses

Fig.1 shows the absorbance changes caused by an aerobic-anaerobic transition of heme $a+a_3$. In State 4, the reduction of heme $a+a_3$ was biphasic. The rapid reduction phase (about half of the total absorption change) was followed by a relatively slow reduction phase. However, in both State 3 and the uncoupled state, the slow phase was reduced and only a rapid transition to full reduction was observed. As earlier workers have reported (e.g.Oshino et al, 1974, Lindsay and Wilson, 1972), these results indicate that the redox state of heme $a+a_3$ is dependent upon the mitochondrial energy state.

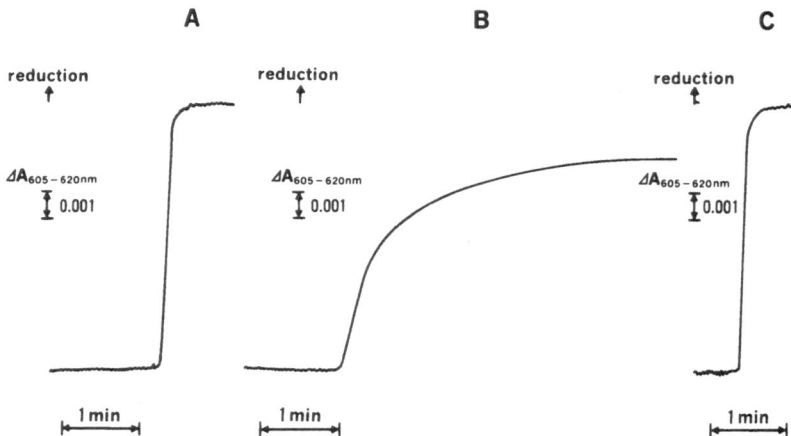

Fig. 1. The absorbance changes caused by an aerobic-anaerobic transition of heme $a+a_3$. Rabbit heart mithochondria, 0.8mg protein /ml, were suspended in the reaction mixture consisting of 0.25M sucrose, 10mM Tris-HCl(pH7.2), 2mM KCl, 0.2m MEDTA, 10mM KH_2PO_4, 10mM glutamate. (A), in State 3; (B), in State 4; (C), in the uncoupled state(2,4-dinitrophenol). The wavelength pair used was 605-620nm.

In contrast to heme a+a$_3$, the redox state of copper was independent of the mitochondrial energy state as can be seen in Fig.2A. The absorbance changes of the copper caused by an aerobic-anaerobic transition are compared to those of heme a+a$_3$ in Fig.2B. Here the maximum absorbance changes of heme a+a$_3$ at each mitochondrial energy states were taken as 100%. The corresponding values for the copper were 10% in State 3, 17% in State 4, and 5% in the uncoupled state.

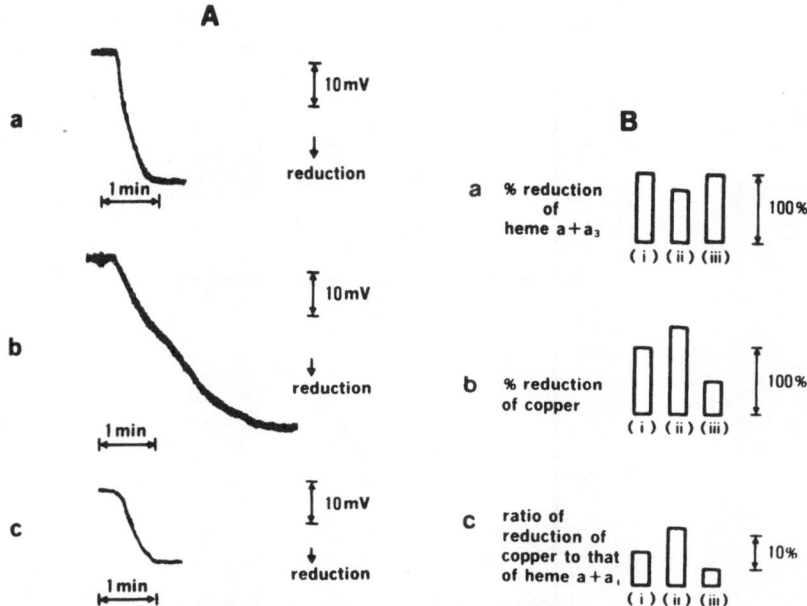

Fig. 2. (A): The absorbance changes of the copper caused by an aerobic-anaerobic transition. The experimental conditions are as described in Fig.1. except that the mitochondrial concentration was 1.8mg protein/ml. (a), in State 3; (b), in State 4; (c), in the uncoupled state. The wavelength pair used was 830-760nm. (B): % reduction of cytochrome oxidase. (a), heme a+a$_3$; (b), copper; (c), the ratio of the reduction of copper to that of heme a+a$_3$; (i), in State3; (ii), in State 4; (iii), in the uncoupled state. In both (a) and (b), the absorbance change in State 3 was taken as 100%, and those of State 4 and the uncoupled state were calculated. In(c), the maximum absorbance changes of heme a+a$_3$ at each mitochondrial energy states were taken as 100%, and the relative absorbance ratios of copper to heme a+a$_3$ were calculated.

The effect of oxygen concentration on the redox states of copper and heme a+a$_3$

Fig.3 shows the effect of oxygen concentration on the redox states of copper and heme a+a$_3$. The result for heme a+a$_3$ is in accordance with that observed by Oshino et al, who used photobacterium phosphoreum for the assay of oxygen concentration (Oshino et al, 1974). The oxygen concentration required for a half reduction of the heme $a+a_3$ (P50 $a+a_3$) varied with the energy state as well as the respiratory rate, and was 7.8×10^{-8}M and 1.6×10^{-7}M in State 3 and State 4, respectively. In contrast to heme $a+a_3$, the oxygen concentration required for the half-maximal reduction of copper (P50Cu) showed little dependence. (7.4×10^{-8}M and 8.0×10^{-8}M in State 3 and State 4, respectively.)

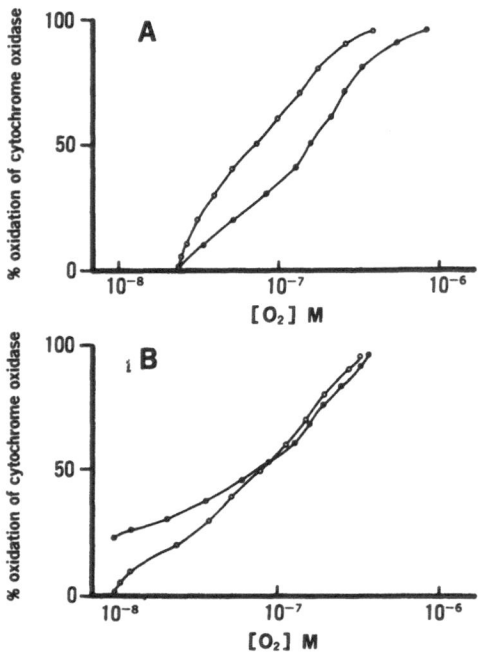

Fig. 3. The effect of oxygen concentration on the redox state of heme $a+a_3$ and copper. The experimental conditions are as described in Fig.2. (A), in State 3; (B), in State 4; ● ,heme $a+a_3$; ○ ,copper.

Relative oxidation-reduction states of copper and heme $a+a_3$ at low oxygen concentrations.

Fig.4 is a plot of the percentage reduction of copper with respect to heme $a+a_3$ under various conditions. The relationship deviates from the straight line which would be expected if both components had the same oxygen affinity. The deviation was larger in State 3 than in State 4.

These plots, obtained by the simultaneous monitoring of the redox states of copper and heme $a+a_3$, can be used to give us information about the oxygen concentration as well as the metabolic state of living tissue.

Fig.5 summarizes the oxygen dependencies of the near infrared active oxygen indicators which could be successfully used for non-invasive evaluation of metabolic processes in blood perfused living tissue (Jöbsis, 1977, Tamura, 1978).

SUMMARY

The oxidation-reduction state of cytochrome oxidase in isolated mitochondria at low oxygen concentrations was measured by the use of leghemoglobin as an oxygen indicator. P50 a+a$_3$ varied with energy state as well as the respiratory rate.

In contrast to heme a+a$_3$, copper was slower to reduce than heme a+a$_3$. The P50Cu of 8×10^{-8} M in State 4 and 7.4×10^{-8} M in State 3 was independent of both the energy state and the respiratory rate.

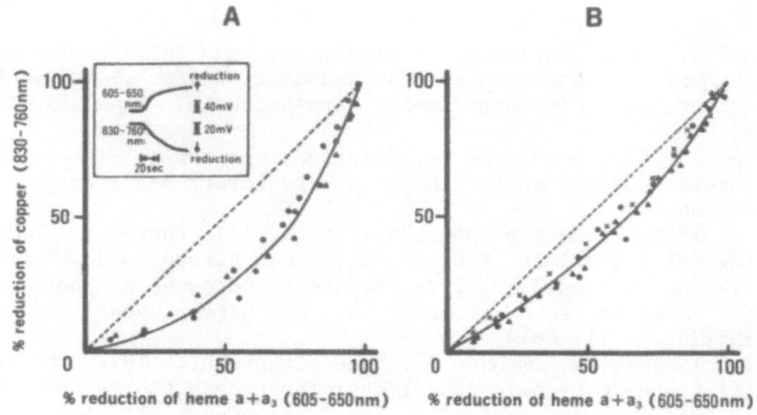

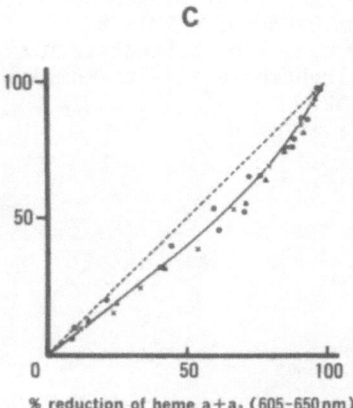

Fig. 4. Relative oxidation-reduction state of copper with respect to that of heme a+a$_3$. The experimental conditions are as described in Fig.2. (A), in State 3; (B), in State 4; (C), in the uncoupled state. The inset shows the absorbance changes caused by an aerobic-anaerobic transition of heme a+a$_3$ and copper in State 3.

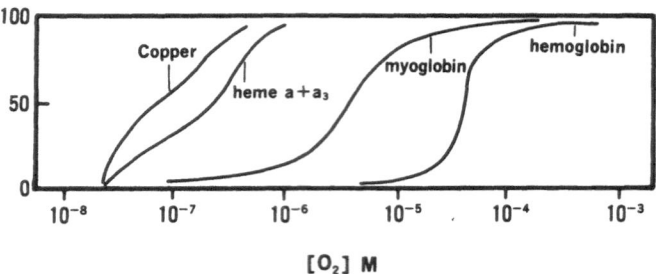

% O₂ saturation of hemoglobin and myoglobin
% oxidation of copper and heme a+a₃

Fig. 5. The titration with oxygen of copper, heme a+a$_3$, myoglobin, and hemoglobin. The redox states of copper and heme a+a$_3$ were measured in State 3.

REFERENCES

Bergersen,F.J. and Turner,G.L., 1979, Systems utilizing oxygenated leghemoglobin and myoglobin as a source of free dissolved O_2 at low concentrations for experiments with bacteria. Anal.Biochem., 96: 165.

Jöbsis,F.F., 1977, Non invasive, infrared, monitoring of cerebral and myocardial oxygen sufficiency and circulatory parameters. Science, 198: 1264.

Kagawa,Y., 1979, Oxidative phosphorylation, in: Energy metabolisms and biological oxidations, Naoki,A,ed., Tokyo Kagaku Dojin, Toko.

Lindsay,J.G. and Wilson,D.F., 1972, Apparent adenosine triphosphate induced ligand changes in cytochrome a$_3$ of pigeon heart mitochondria. Biochemistry, 11: 4613.

Oshino,N., Sugano,T., Oshino.R., and Chance,B., 1974, Mitochondrial function under hypoxic conditions: The steady states of cytochrome a+a$_3$ and their relation to mitochondrial energy state. Biochem.Biophys.Acta., 368:298.

Tamura,M., Oshino,N., Chance,B., and Silver,A.I., 1978, Optical measurements of intracellular oxygen concentration of rat heart in vitro. Arch.Biochem.Biophys., 191: 8.

Wittenberg,J.B., Appleby,C.A., and Wittenberg,B.A., 1972, The kinetics of the reaction of leghemoglobin with oxygen and carbon monoxide. J.Biol.Chem., 247: 527.

EXPONENTIAL ATTENUATION OF LIGHT ALONG NONLINEAR PATH

THROUGH THE BIOLOGICAL MODEL

Y. Nomura, O. Hazeki, and M. Tamura

Biophysics Division,
Research Institute of Applied Electricity
Hokkaido University, Sapporo 060, Japan

INTRODUCTION

Although exhibiting some transparency when illuminated with red or near infrared light(e.g. Jobsis, 1977; Hazeki and Tamura 1988), biological tissues are intensely light scattering within this range. As a result, the optical pathlength through the tissue is significantly different from the spacing between the points incidented and transmitted. Beer-Lambert law that stands for clear solution can't be simply applied to the tissue and various devices for the correction are done. Then two analytical approaches to the study of optical propagation in blood are investigated: a multiple scattering theory (Twersky, 1970) and a diffusion theory (Zdrojkowski and Pisharoty, 1970). We have studied the way that the spectroscopic data can be quantitatively treated by discriminating temporally and spatially the passage of an instantaneous light pulse through the tissue. In this report, assumption that light intensity along the non linear path taken by photons through the living tissue is exponentially attenuated has been tested in model system. Light path taken by photons is important for time resolved spectroscopy. The application to tissue has been performed by several investigators (e.g. Martin et al., 1980; Tamura et al., 1987; Chance et al., 1988).

MATERIALS AND METHOD

The scattering medium used is a suspension of 40g/1 defatted dry milk or no defatted milk (fat 14.4g/1 and protein 12.2g/1). Main scattererrs are casein micelle $0.5 \sim 2.8 \times 10^{-7}$ m diameter in defatted milk or fat globule $0.3 \sim 1 \times 10^{-5}$ m and casein micelle in no defatted milk. Hemoglobin(Hb) as an absorber coexisted in the medium. Laser beam (585nm wavelength, 30ps pulse width) is incidented to a sample cell with 1cm thickness and transmitted photons detected by streak camera.

RESULTS AND DISCUSSION

The experimental set up is illustrated by Fig.1. Io in Fig. 2A shows the probability distribution in highly scattering and no absorbing system as a function of time taken by photons to transmit through it. Addition of absorber changes the distribution. Assuming that the scattering property of each photon does not change by the addition of Hb, the

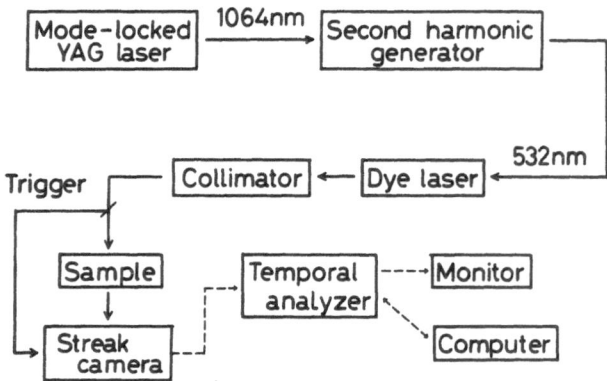

Fig.1. Block diagram of temporal analyzing system for the measurement of pulse profile.

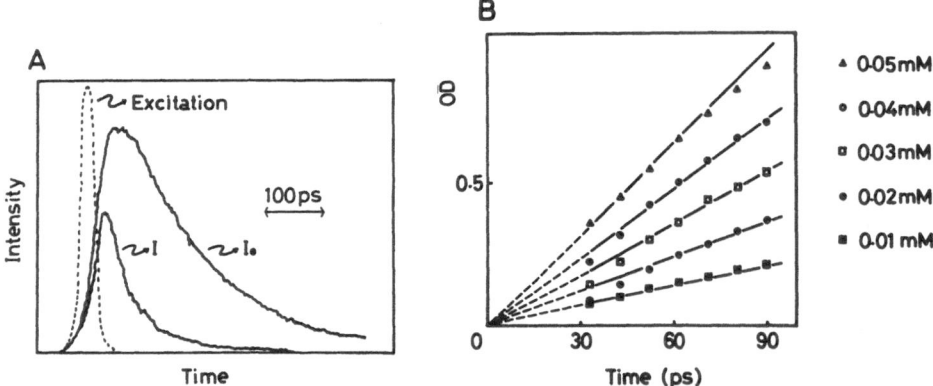

Fig.2. (A) Profile of the light intensity against time. Broken line, excitation pulse; Io, scattered material (milk) in the absence of Hb; I, presence of Hb. (B) Plot of OD against time of each sample (0.01 0.05mM oxy Hb). The origin and time range are determined by extrapolation and the sensivity to weak signals respectively.

following equation stands at a certain time.

$$OD = \log Io(t)/I(t) = ECD \qquad \text{---(1)}$$

E : molecular extinction coefficient
C : concentration
D : light pathlength

The appropriate light pathlength is simply related to, t, the "time of flight(TOF)" by:

$$D = Vot/n = Vt \qquad \text{---(2)}$$

Vo : velocity of light
n : the average refractive index ; for water n = 1.33
V : light velocity in water, V = 0.023cm/ps. Therefore

$$OD = \log Io(t)/I(t) = ECVt \qquad \text{---(3)} .$$

Fig.2B shows the relationship between the observed value of log Io(t)/I(t) and t at varying Hb concentration. This linear relationships, expected from the equation(3), are obtained within the time range 30∿90ps. The values at the time out of this range are omitted because I(t), light

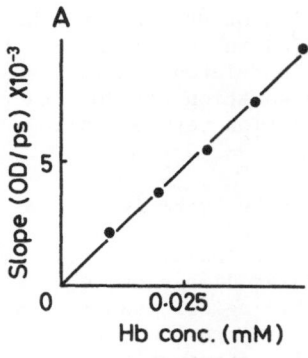 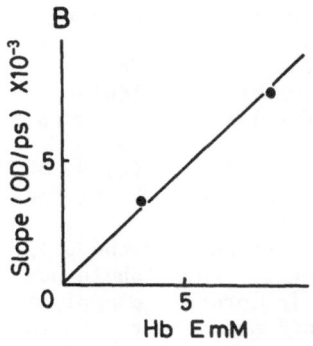

Fig.3. (A):The plot between the observed slope and concentration of oxyHb(0.01∿ 0.05mM). (B):The relationship between the slope and extinction coefficient at same concentration, oxy- and metHb, 0.04mM.

intensity in the presence of the absorber, is too weak to measure with reliable accuracy. The origin obtained by extrapolation nearly corresponds with the peak time of excitation pulse. To explain the relationship between OD and concentration, Fig.3A shows the slope in Fig.2B aginst Hb concentration. OD/t is proportional to Hb concentration. Similar experiment is performed at a constant Hb concentration (0.04mM) and in different extinction coefficient oftained by changing the oxygenation state of the pigment. The result is shown as a function of mM extinction coefficient in Fig.3B. Again we can find a linear relationship between OD/t and extinction coefficient, which can be predicted from the equation (3) at a constant concentration. Fig.4A shows the relationship between the observed slopes in Fig.2B and the calculated value (ECV). Broken line is a line of identity and experimental values fit well. This result evidences the validity of the equation(3) and the assumption prior to it.

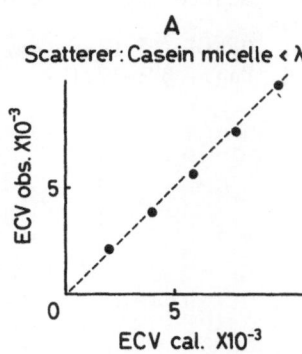 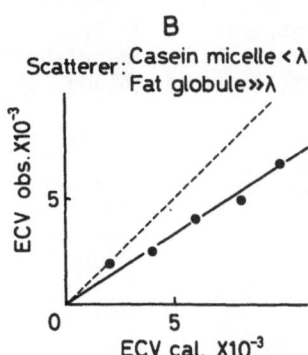

Fig.4 Relationship between experimental point OD/t and theoretical line ECV of different scatterers.(A) defatted milk media, (B) no defatted milk media.

In no defatted milk of Fig.4B, experimental points are shifted to small values. Because casein micelle($0.5∿2.8 \times 10^{-7}$m) in defatted milk is smaller than the wavelength(5.85×10^{-7}m), TOF in the particle is negligible and therefore all TOF is related to absorption of Hb. But in the case of no defatted milk, fat globule($0.3 \sim 1 \times 10^{-5}$ m) is much larger than the wavelength. And TOF in the particle, unrelated with absorption, overestimates the calculated value. As the equation stands with regard to only TOF related with absorption, the calibration term as follows is added to it.

$$OD = \log I_o(t)/I(t) = \alpha ECVt$$

α : Ratio of TOF related with absorption to total TOF

$0 > \alpha > 1$

α is determined by the property of particles (size, shape, refractive index etc). Fat globule has a common cellular dimension. e.g. red blood cell. In order to apply this method to biological sample, α must be determined separately with the tissue.

Present results evidence the exponential attenuation along the non linear light path. The combination of such temporal and spatial analysis can achieve the spectroscopic imaging of living tissue, other than that of computer tomography, where temporal analysis has been established as a first step. Now we can evaluate the absorption quantitatively along light path experienced by photons in highly scattering system like animal tissue.

REFERENCE

Chance, B., Greenfeld, R., Miyake, H., Smith, D., Nioka, S., Holtom, G., Kaufmann, K. and Leigh, J.S., 1988, Time resolved near infrared spectroscopy of the adult human brain, submitted
Hazeki, O. and Tamura, M., 1988, Quantitative analysis of hemoglobin oxygenation state of rat brain in situ by near-infrared spectrophotometry, J. Appl. Physiol., 64:796
Jobsis, F.F., 1977, Non invasive monitoring of cerebral and myocardial oxygen Sufficiency and circulatory parameters, Science, 198:1264
Martin, J.L., Lecarpentier, Y., Antonetti, A. and Grillon, G., 1980, Picosecond laser stereometry light scattering measurements on biological material, Med.&Biol.&Eng.&Comput., 18:250
Tamura, M., Nomura, Y. and Hazeki, O., 1987, Laser tissue spectroscopy - near infrared CT., Rev. Laser Eng., 15:657
Twersky, V., 1970, Absorption and multiple scattering by biological suspensions, J. Opt. Soc. Amer., 60:1084
Zdrojkowski, R.J. and Pisharoty, N.R., 1970, Optical transmission and reflection by blood, IEEE Trans. Biomed. Eng., BME-17:122

BEHAVIOR OF THE COPPER BAND OF CYTOCHROME C OXIDASE IN RAT BRAIN DURING FC-43-FOR-BLOOD SUBSTITUTION

Claude A. Piantadosi
Department of Medicine
Duke University Medical Center
Durham, North Carolina 27710

INTRODUCTION

Over the past few years, it has become possible to monitor tissue oxygenation in situ by measuring changes in oxygen-dependent absorption of near infrared (NIR) light by tissue chromophores. Soft tissues and bone are relatively translucent to near infrared light. Hemoglobin (Hb), oxyhemoglobin (HbO$_2$) and the oxidized copper band of cytochrome c oxidase (cytochrome a,a$_3$) have oxygen-dependent absorption spectra in the 700-900 nm wavelength region. These concepts were first reported when Jöbsis (1977) demonstrated transmission of NIR light through intact tissues and the feasibility of NIR monitoring of changes in both the oxygenation of hemoglobin and the oxidation level of cytochrome a,a$_3$ in living brain. NIR signals acquired from intact brain must be partitioned into absorption changes proportional to the concentrations of each of the three absorbers because of overlap of the NIR absorption spectra of Hb, HbO$_2$ and oxidized cytochrome a,a$_3$. The NIR contributions of Hb, HbO$_2$, and cytochrome a,a$_3$ to the spectrum of each of the other absorbers in brain tissue (t) can be accounted for by means of algorithms that calculate changes in the relative amounts of each component. The algorithms resolve changes in absorption at several wavelengths according to the relative contribution of each absorber at each wavelength. Since three overlapping absorption spectra must be deconvoluted, spectral data from at least three NIR wavelengths are necessary to correct for the contributions of the three molecular species. In some situations, four wavelength algorithms appear to provide more accurate approximations of the NIR absorption properties of living tissues.

Algorithm development for NIR monitoring has been enhanced greatly by the availability of oxygen-transporting blood substitutes that contain perfluorochemicals (PFCs). PFC preparations are fine emulsions of white particles that scatter light effectively. This optical property is useful in acquiring absorption spectra from highly scattering biological tissues that normally receive oxygen from red blood cells. PFC-for-blood substitution in rats also increases cerebral blood flow and cerebrovenous PO$_2$ without altering the cerebral metabolic rate for oxygen (Lee et al.,1988). These circulatory adjustments may reflect low O$_2$ content or low viscosity of PFC emulsions relative to whole blood. Such differences in O$_2$ delivery to tissue by blood and PFC may alter mitochondrial oxidation-reduction (redox) states in vivo. Redox effects of PFC circulation are difficult to ascertain with visible spectroscopy because the disappearance of hemoglobin unveils the visible spectra of the cytochromes and the appearance of the PFC may alter the absorption characteristics of tissue. In the NIR, these problems are somewhat less troublesome, particularly if the multiwavelength algorithms are free of crosstalk from the other absorbers in the tissue. This paper reports the results of continuous brain monitoring with NIR spectroscopy during the process of PFC-for-blood substitution in anesthetized rats. The NIR data indicate that the oxidation level of the copper band of cerebrocortical cytochrome $\underline{c}$ oxidase in the rat remains within a few percent of its baseline value during elimination of HbO_2 by exchange transfusion with PFC emulsion at high concentrations of inspired oxygen.

METHODS

Preparation of Bloodless Rats

Adult male Sprague Dawley rats, anesthetized with pentobarbital (50 mg/kg) were used in the experiments. The rats were prepared surgically with tracheostomy, arterial and venous femoral catheters and splenectomy. The animals were paralyzed and ventilated mechanically to prevent artifacts from respiratory motion. The arterial CO_2 tension ($PaCO_2$) was adjusted to approximately 30 Torr. The animals were monitored continuously for electroencephalogram (EEG) activity, arterial blood pressure and rectal temperature. Aliquots of arterial blood were sampled periodically for measurement of arterial blood gases, pHa and microhematocrit. The animals were prepared for continuous NIR monitoring (see below) and placed on 100% oxygen. After stable optical and physiological parameters were achieved, the animals underwent isovolemic, exchange transfusion with perfluorotributylamine emulsion (FC-43, Alpha Therapeutics, Los Angeles) as described previously (Piantadosi et al.,1985). The exchange volume was equivalent to five blood volumes based upon 60 ml/kg of circulating volume. The animals were killed with a lethal intravenous injection of KCl.

Continuous NIR Spectroscopy

The hair was removed from the scalp and two optical fiber bundles were pressed firmly against the cranium just behind the orbits using the transillumination mode. The transillumination arrangement provided a center-to-center distance of about 2.5 cm between the two optical fiber bundles. The optical path bridged both cerebral hemispheres across the sagittal suture. The transillumination mode and wide separation of the fiber bundles minimized detection of back-scattered incident light from superficial structures such as the cranium that would increase the ratio of scattered to transmitted light measured by the photodetector. Back-scattered light contains little absorption data and adversely affects the quality of NIR signals from deep cerebrocortical structures.

NIR spectroscopy was performed using a laser-based instrument designed and constructed in our laboratory. A description of the instrument has been published previously (Hampson and Piantadosi, 1988). Monochromatic light (1.5nm bandwidth) illuminated the tissue from a bank of GaAlAs laser diodes at 775, 810, 870 and 904 nm. Using time-domain multiplexing, the laser diodes were pulsed at a frequency of 1KHz and a pulse width of 200 ns. The intensity of the incident light was monitored with a reference (R) photodiode and the strength of the signals (S) from the tissue measured by a photomultiplier. The photocurrents from the photodetectors were integrated, demultiplexed and fed into a log ratio amplifier. The log S/R voltages were sent to a dedicated microprocessor where algorithms were applied. The $tHbO_2$, tHb and cytochrome $\underline{a,a_3}$ signals were displayed on a printer. At initial conditions, the signals were adjusted to S=R at each wavelength.

NIR Algorithms

The algorithms were generated from $tHbO_2$ and tHb spectra scaled to equimolar quantities using apparent in vivo extinction coefficients obtained independently during FC-43 exchange transfusion at hyperbaric pressures of oxygen. Although normal tissue contains more oxyhemoglobin than deoxyhemoglobin, the initial concentrations of $tHbO_2$ and tHb were equalized to insure that any change in the amount of one compound was offset exactly by an opposite change in the amount of the other. The ΔOD at the hemoglobin isosbestic point and the maximum ΔOD of the copper band of cytochrome a,a_3 measured after death in bloodless animals were used to compute a hemoglobin to cytochrome a,a_3 ratio for the algorithm derivation. A more detailed analysis of the absorption characteristics of the cytochrome a,a_3 copper band and procedures for stripping residual traces of hemoglobin from the spectrum of the copper complex have been published recently (Jöbsis-VanderVliet et al.,1988).

The four wavelength algorithms were based on the following relationships:

$$\Delta OD_{775nm} = a\Delta OD(HbO_2) + b\Delta OD(Hb) + c\Delta OD(cyt a,a_3) + d\Delta OD \text{ (fo)}$$
$$\Delta OD_{810nm} = e\Delta OD(HbO_2) + f\Delta OD(Hb) + g\Delta OD(cyt a,a_3) + h\Delta OD \text{ (fo)}$$
$$\Delta OD_{870nm} = i\Delta OD(HbO_2) + j\Delta OD(Hb) + k\Delta OD(cyt a,a_3) + l\Delta OD \text{ (fo)}$$
$$\Delta OD_{904nm} = m\Delta OD(HbO_2) + n\Delta OD(Hb) + o\Delta OD(cyt a,a_3) + p\Delta OD \text{ (fo)}$$

where a-p are fractional absorption values for each molecule at each wavelength. A correction to the 830 nm copper band of cytochrome a,a_3 was derived empirically using Lorentzian analysis of the line shapes as previously reported (Keizer et al.,1985). The curve fitting indicated that two functions described the copper band adequately; the dominant component had a wavelength maximum of approximately 820 nm and the minor component had a wavelength maximum at approximately 870 nm. Since the smaller band has not been identified, it was subtracted from the larger band and its algorithm was not used in these studies. The four expressions above were solved simultaneously by matrix inversion to obtain the algorithms. The sum of $tHbO_2$ and tHb was used to calculate tissue blood (hemoglobin) volume (tBV).

Multiwavelength algorithms relate changes in the relative quantities of $tHbO_2$, tHb and cytochrome a,a_3 to each other according to total absorption changes at each of several wavelengths. Minor differences in wavelength-dependent light scatter and pathlength and the absence of reliable measurements of optical pathlength under in vivo conditions prevents NIR multiwavelength data from being expressed either as changes in concentration or as true changes in optical density (OD). The term variation in density (vd) has been used instead of OD to indicate that in vivo weighting coefficients derived at several different wavelengths were employed to deconvolute the spectra. Variations in density, expressed in logarithmic form according to the Beer-Lambert relation, are proportional to concentration. One vd unit has been defined as a 10-fold change in a signal computed through the appropriate algorithm (Jöbsis-VanderVliet,1985). Changes in the relative amounts of $tHbO_2$, tHb or cytochrome a,a_3 were compared between experiments by expressing the NIR values from each experiment as fractions of the total labile signal (TLS) between 100% O_2 before PFC exchange and death after PFC exchange.

RESULTS

The physiological responses of the 12 rats to PFC-for-blood exchange are summarized in Table 1. There was a significant decrease in mean arterial pressure (MABP) during the first 2 blood volumes of exchange, however the MABP recovered later in the procedure. The hematocrit decreased from 39% at control to 1% after 5 blood volumes of exchange transfusion. There also was a significant increase PaO_2 during the PFC infusion, but no significant change in $PaCO_2$ or other measurements.

Table 1. Physiological Variables [1]

Variable	Blood Volumes of FC-43 Exchanged					
	0	1.0	2.0	3.0	4.0	5.0
MABP (mmHg)	108 ± 4	97 ±3	90[2] ±3	89[2] ±3	94[2] ±4	95 ±4
Hematocrit(%)	39.3 ±1.1	18.8[3] ±0.4	9.2[3] ±0.3	4.2[3] ±0.2	2.1[3] ±0.2	1.0[3] ±0.0
PaO_2 (Torr)	395[3] ±10	-	-	-	-	479[3] ±10
$PaCO_2$ (Torr)	32 ±2	-	-	-	-	36 ±1
pHa	7.48 ±0.02	-	-	-	-	7.42 ±0.02
T body (°C)	36.8 ±0.2	-	-	-	-	36.7 ±0.1

[1] Values are Mean ± S.E.M.
[2] Value significantly lower than control ($P < 0.05$) by one-way ANOVA and modified t statistics.
[3] Value significantly different than preceding value ($P < 0.05$) by paired t-test.

Cerebrocortical NIR responses to exchange transfusion are shown in Figure 1A for a representative animal. The NIR signals for $tHbO_2$, tBV and cytochrome a,a_3 expressed as changes from baseline in Figure 1A, have been converted to fractions of the TLS and displayed in Figure 1B. Note the stability of the NIR cytochrome a,a_3

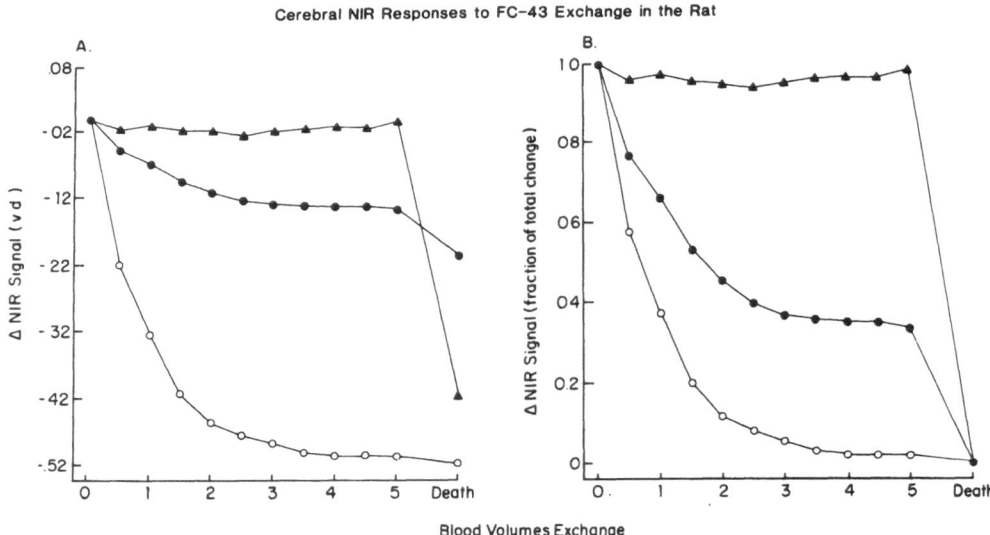

Cerebral NIR Responses to FC-43 Exchange in the Rat

Figure 1A and B

signal relative to the other two signals during the exchange process.

Figure 2 shows composite results of the hemoglobin washout experiments in 12 rats. In these animals, no significant change in cytochrome $\underline{a},\underline{a}_3$ oxidation level was detected by the algorithm for the copper band until the circulation was arrested with KCl. After arrest, the copper band disappeared with small changes in the $tHbO_2$ and tBV signals. These data indicate that the mean decrease in cytochrome $\underline{a},\underline{a}_3$ oxidation level after the 5 blood volumes of exchange was only 4% of the TLS, while 76% of the

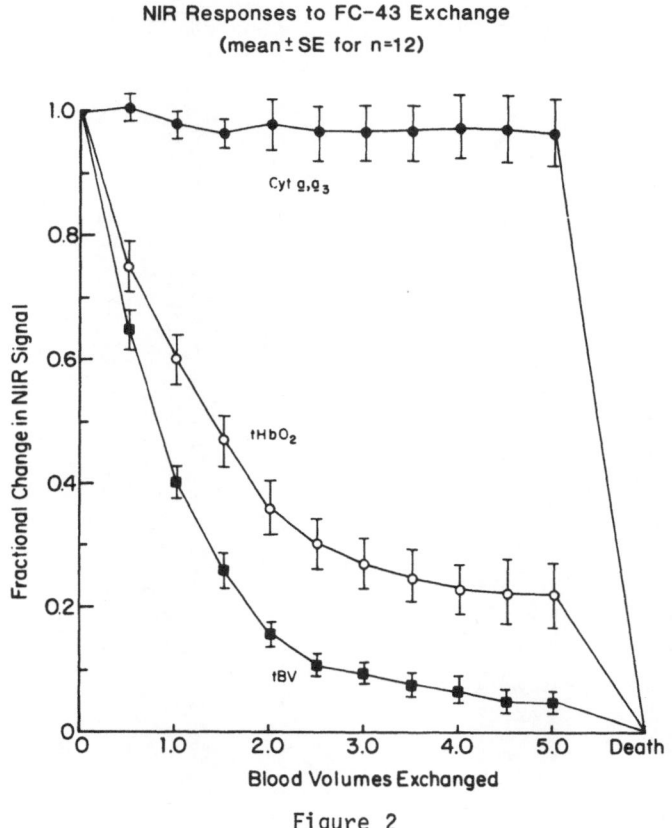

NIR Responses to FC-43 Exchange
(mean ± SE for n=12)

Figure 2

$tHbO_2$ and 95% of the tBV were washed out of the brain during the procedure. Thus, the maximum error in determination of cytochrome $\underline{a},\underline{a}_3$ was 5%, assuming any changes in cytochrome $\underline{a},\underline{a}_3$ calculated by the algorithm during the washout were due only to crosstalk from $tHbO_2$ and/or tHb. This assumption however, overestimates the amount of crosstalk because cytochrome $\underline{a},\underline{a}_3$ oxidation level usually changed only after elimination of most of the $tHbO_2$ and tHb.

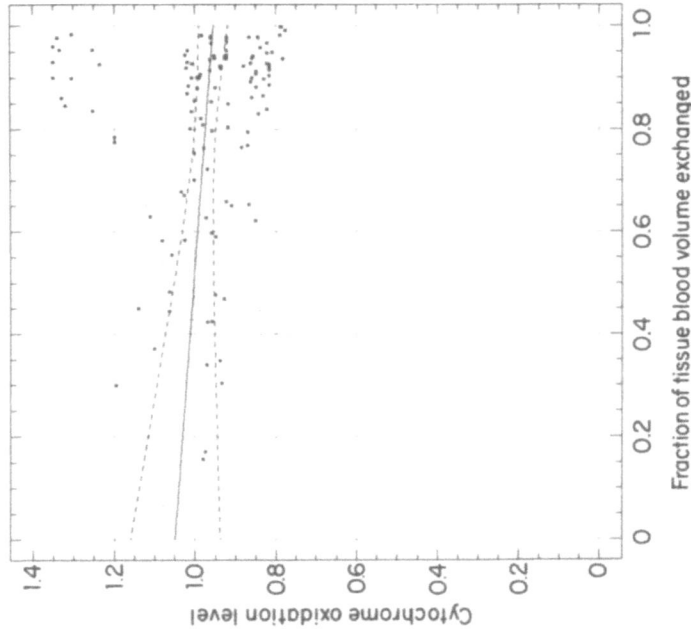

Figure 3B. Regression of Cytochrome Oxidation Level as a
Function of Tissue Blood Volume

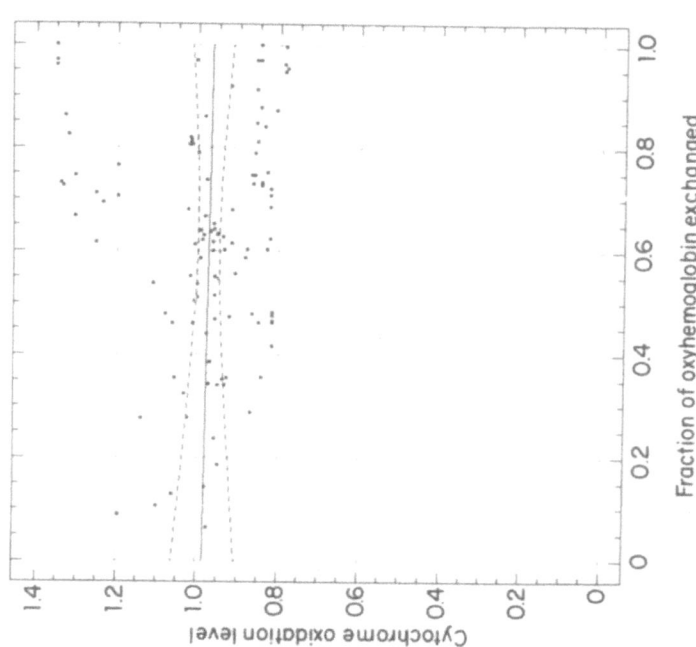

Figure 3A. Regression of Cytochrome Oxidation Level
vs Oxyhemoglobin Washout

86

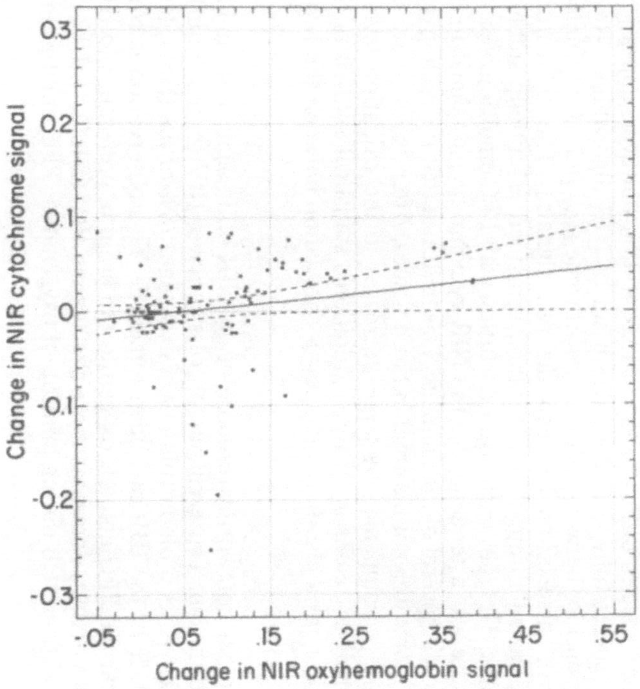

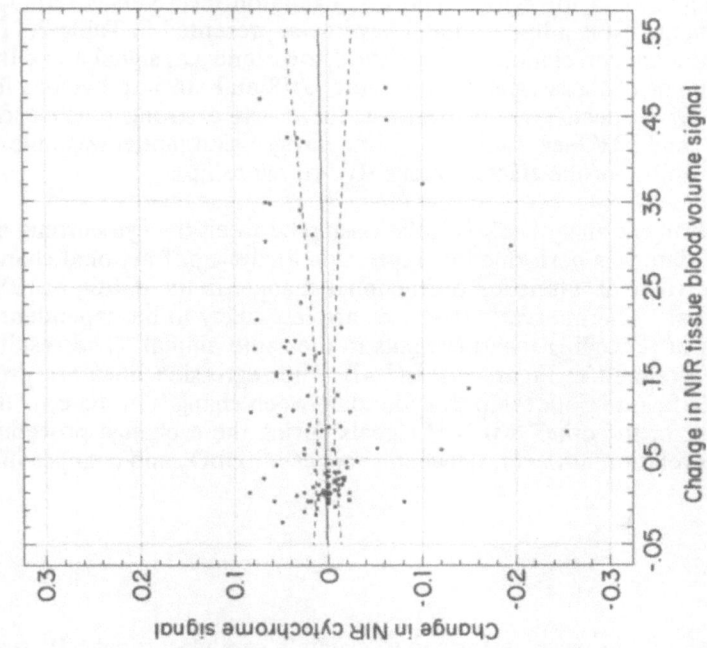

Figure 4A. Regression of Change In Cytochrome as a Function of Change in Oxyhemoglobin

Figure 4B. Regression of Change in Cytochrome as a Function of Change in Blood Volume

The relationships between the NIR signals for cytochrome a,a_3 oxidation level, $tHbO_2$ and tBV during the exchange are plotted in Figures 3A and 3B respectively. Regression analysis data for cytochrome a,a_3 oxidation level versus $tHbO_2$ and tBV during the exchange using a linear model have been presented in Table 2. These data indicate no significant correlation between the cytochrome a,a_3 signal and either $tHbO_2$ or tBV. The slopes of these relationships were -.018 and -.095 and values for r^2 were 0.08% and 1.65% respectively. In contrast, there was a strong positive correlation between $tHbO_2$ and tBV (see Table 2). A linear regression model was able to explain 60% of the variability of the $tHbO_2$ versus tBV relationship.

A more stringent analysis of the relationship between the cytochrome a,a_3 versus HbO_2 and tBV data was performed by regression analysis of fractional changes in the cytochrome a,a_3 value as a function of fractional changes in the $tHbO_2$ or tBV for each measured interval. The interval differences are less likely to be dependent upon the absolute values of preceding measurements in the same animal. The results of these regressions are graphed in Figure 4A and 4B. The regression lines are presented in Table 3. No significant relationship was found between changes in the cytochrome a,a_3 signal and either of the other two NIR signals during the exchange procedure. There was a strong correlation however, between changes in $tHbO_2$ and changes in tBV (See Table 3).

DISCUSSION

These experiments were performed to evaluate cerebrocortical NIR responses to PFC-for-blood exchange with particular attention to changes in the redox state of the oxidized copper moieties of cytochrome a,a_3. The optical responses during the exchange process also test how well the four wavelength algorithms assess independent changes in the amount of $tHb0_2$, tHb and cytochrome a,a_3. The NIR spectra of HbO_2 and the oxidized copper complex of cytochrome a,a_3 overlap considerably, thus making optical separation of the two species difficult. Optical separation however, appears to have been accomplished successfully with the present set of NIR algorithms. Regression analysis of the relationships between oxidized cytochrome a,a_3 and $tHbO_2$ indicated no statistically significant association between the two signals. The cytochrome a,a_3 oxidation level, measured through its algorithm, remained essentially stable during washout of most of the $tHbO_2$ by exchange transfusion with PFC emulsion. Near the end of the exchange, there was a tendency for the NIR cytochrome a,a_3 signal to move in the oxidized direction in several of the animals. This indicates that the resting oxidation level of the copper atoms was not maximal in those tissues. After exchange transfusion, circulatory arrest was accompanied by large decreases in the oxidation level of the oxidase, but only small decreases in $tHbO_2$. Additional statistical analysis of the NIR changes during brain anoxia also failed to show a correlation between the two signals (data not shown).

There is one other noteworthy response of the NIR brain signals to the PFC-for-blood exchange transfusion. The exchange process removes tBV (hemoglobin) out of proportion to $tHbO_2$. After 5 blood volumes of exchange, 76% of the $tHbO_2$ had been eliminated, while 95% of the tBV had been removed. This observation is consistent with concept the that the PFC unloads oxygen to the brain in preference to $tHbO_2$ at high PO_2. Therefore, $tHbO_2$ decreases more slowly than tBV ($tHbO_2$ +tHb) because $tHbO_2$ is replenished from tHb during infusion of oxygenated PFC emulsion into the circulation.

Table 2

REGRESSION ANALYSIS - LINEAR MODEL: y=a+bx

1. Dependent variable: NIR cytochrome Signal   Independent variable: NIR HbO$_2$ Signal

Parameter	Estimate	Standard Error	T Value	Prob. Level
Intercept	0.984	0.0392	25.067	.000
Slope	-0.018	0.0583	-0.3069	.759

Analysis of Variance

Source	Sum of Squares	Df	Mean Square	F-Ratio	Prob. Level
Model	.002	1	.002	.094	.759
Error	2.470	118	.021		
Total (Corr)	2.472	119			

Correlation Coefficient = -0.028   R squared = .08 percent
Stnd. Error of Est. = 0.145

2. Dependent variable: NIR cytochrome Independent variable: NIR tBV signal

Parameter	Estimate	Standard Error	T Value	Prob. Level
Intercept	1.049	0.056	18.73	.000
Slope	-0.095	0.067	-1.41	.162

Analysis of Variance

Source	Sum of Squares	Df	Mean Square	F-Ratio	Prob. Level
Model	.0408	1	.0408	1.982	.162
Error	2.431	118	.0206		
Total (Corr)	2.472	119			

Correlation Coefficient = 0.129   R squared = 1.65 percent
Stnd. Error of Est. = 0.144

3. Dependent variable: NIR HbO$_2$ Signal Independent variable: NIR tBV Signal

Parameter	Estimate	Standard Error	T Value	Prob. Level
Intercept	1.0967	0.562	19.517	.000
Slope	-0.9028	0.068	-13.371	.000

Analysis of Variance

Source	Sum of Squares	Df	Mean Square	F-Ratio	Prob. Level
Model	3.708	1	3.708	178.787	.000
Error	2.448	118	.021		
Total (Corr)	6.156	119			

Correlation Coefficient = -0.776   R squared = 60.24 percent
Stnd. Error of Est. = 0.144

Table 3

REGRESSION ANALYSIS - LINEAR MODEL : y=a+bx

1. Dependent variable: Δ Cytochrome Signal   Independent variable: Δ HbO$_2$ Signal

Parameter	Estimate	Standard Error	T Value	Prob. Level
Intercept	-.004	.006	-0.697	.487
Slope	0.093	0.049	1.894	.061

Analysis of Variance

Source	Sum of Squares	Df	Mean Square	F-Ratio	Prob.Level
Model	.0088	1	.0088	3.586	.061
Error	.277	113	.0025		
Total (Corr.)	.286	114			

Correlation Coefficient = 0.175   R squared=3.08 percent
Stnd. Error of Est. = 0.050

2. Dependent variable: Δ Cytochrome Signal   Independent variable: Δ tBV Signal

Parameter	Estimate	Standard Error	T Value	Prob. Level
Intercept	.002	.006	0.324	.746
Slope	0.014	0.038	0.366	.715

Analysis of Variance

Source	Sum of Squares	Df	Mean Square	F-Ratio	Prob.Level
Model	.0003	1	.0003	.134	715
Error	.2857	113	.0025		
Total (Corr.)	.286	114			

Correlation Coefficient = 0.034   R squared = .12 percent
Stnd. Error of Est. = -0.050

3. Dependent variable: Δ HbO$_2$ Signal   Independent variable: Δ tBV Signal

Parameter	Estimate	Standard Error	T Value	Prob. Level
Intercept	0.018	0.006	2.948	.0039
Slope	0.636	0.040	16.056	.0000

Analysis of Variance

Source	Sum of Squares	Df	Mean Square	F-Ratio	Prob.Level
Model	.7045	1	.7045	257.800	.000
Error	.3088	113	.0027		
Total (Corr.)	1.013	114			

Correlation Coefficient = 0.834   R squared =69.53 percent

Stnd. Error of Est. = 0.052

SUMMARY

FC-43-for-blood substitution experiments were conducted in anesthetized rats to evaluate NIR spectroscopic responses by living brain to exchange transfusion at very low hematocrits. The NIR responses to the exchange process also assess the ability of multiwavelength algorithms to measure independent changes in the relative amounts of Hb, tHbO$_2$ and oxidized cytochrome a,a$_3$ in the tissue. These data indicate that FC-43 circulation at high FIO$_2$ (1.00) delivers sufficient O$_2$ to the rat brain to maintain the oxidation state of the mitochondrial oxidase near pre-exchange values. The results also confirm the ability of four wavelength algorithms to distinguish changes in the oxidized copper band of cytochrome a,a$_3$ from changes in absorption by Hb and HbO$_2$.

REFERENCES

Hampson, N. B. and C. A. Piantadosi, 1988. Near infrared monitoring of human skeletal muscle oxygenation during forearm ischemia, J. Appl. Physiol.64:2449-2457.

Jöbsis-VanderVliet, F. F.,1985, Non-Invasive, near infrared monitoring of cellular oxygen sufficiency in vivo, In: Advances in Experimental Medicine and Biology, Vol. 191 (F. Kreuzer, S. M. Cain, Z. Turek and T. K. Goldstick, eds.). Plenum Press, New York. pp. 833-841.

Jöbsis-VanderVliet, F. F., C. A. Piantadosi, A. L. Sylvia, S. K. Lucas and H. H. Keizer,1988, Near infrared monitoring of cerebral oxygen sufficiency: I. Spectra of cytochrome c oxidase, Neurol.Res, 10: 7-17.

Jöbsis, F.F.,1977, Noninvasive, infrared monitoring of cerebral and myocardial and circulatory parameters. Science, 198:1624-1267.

Keizer, H. H., F. F. Jöbsis-VanderVliet, S. K. Lucas, C. A. Piantadosi and A. L. Sylvia, 1985, The near infrared (NIR) absorption band of cytochrome a,a$_3$ in purified enzyme, isolated mitochondria and in the intact brain in situ, In: Advances in Experimental Medicine and Biology, Vol 191 (F. Kreuzer, S. M. Cain, Z. Turek and T. K. Goldstick, eds.). Plenum Press, New York. pp. 823-832.

Lee, P.A., A.L. Sylvia and C.A. Piantadosi, 1988, Effects of fluorocarbon-for-blood exchange on regional cerebral blood flow in rats, Am J. Physiol, 254 (Heart Circ. Physiol 23): H 719-H729.

Piantadosi, C.A., A.L. Sylvia, H.A. Saltzman and F.F. Jöbsis-VanderVliet, 1985, Carbon monoxide cytochrome interactions in the brain of the fluorocarbon-perfused rat, J. Appl. Physiol.,58:665-672.

TOWARD ABSOLUTE REFLECTANCE OXIMETRY: I. THEORETICAL CONSIDERATION FOR NONINVASIVE TISSUE REFLECTANCE OXIMETRY

Setsuo Takatani

National Cardiovascular Center Research Institute
5-7-1 Fujishiro-dai, Suita, Osaka, Japan

INTRODUCTION

Methods of absorption and reflection photometry are sensitive enough to allow in vivo measurement of pigments such as hemoglobin, myoglobin and cytochromes which participate in oxygen transport to tissues. Since the influence of light on these pigments can be made really small, transmission and reflection photometry can be useful tools to noninvasively evaluate the oxygenation process that takes place in various tissues. Historically, the application of the spectrophotometric method for measurement of tissue oxygenation in situ has been made first in 1930-40's by Kramer(1934), Matthes(1934), Millikan(1942) and Brinkman(1949) in which two wavelengths were employed, one for the measurement of pigment concentration and the other for compensation of nonspecific light loss by tissue. In the 1950's, the two wavelength method was very much improved by Chance (1954).

In the last two decades, although the transmission method has made significant progress to yield clinical instrument such as transmission pulse oximeters(Yoshiya et al., 1980), reflection photometry did not follow in the same proportion, mainly due to difficulty in quantitatively relating the tissue optical reflectance to physiological parameters. In the reflection method, mostly two wavelength method or similar approach were employed utilizing the empirical analysis(Cohen and Wadsworth, 1972, Mendelson et al., 1983). Lubbers and Wodick (1969) introduced a multi-component method through analysis of reflection spectra obtained by a rapid scan photometer. Also, the one-dimensional Kubelka-Munk equation was used to construct tissue spectra by including both absorption and scattering by tissues(Lubbers and Hoffmann, 1980). The three-dimensional photon diffusion theory has also been applied for the analysis of reflectance from whole blood and tissues(Longini and Zdrojkowski, 1968, Johnson, 1970, Reynolds et al., 1976, Takatani and Graham, 1979). The appropriate theoretical approach should consider both scattering and absorption by pigments as well as by tissue entity. The adequate theory should also include the inhomogeneous effect of tissue. In addition, in order to test the applicability of the theory, development of tissue models that simulate the actual living tissues is required. In this study, theoretical consideration toward quantitative analysis of tissue reflectance spectra for noninvasive determination of hemoglobin oxygen saturation and design of optical sensor will be addressed based on the application of three-dimensional photon diffusion theory.

THEORETICAL BACKGROUND

Assumptions Underlying the Diffusion Theory

Based on the photon diffusion theory, the medium can be expressed by its scattering(S), absorption(K) and diffusion(D) constants. The underlying assumptions for valid application of the diffusion theory include 1) the medium must be diffusely scattering and 2) the medium must be semi-infinite in its extent. The diffusely scattering condition can be evaluated through the parameter called "Albedo" where it is defined as S/(S+K). In order for the medium to be diffusely scattering, the albedo should be close to 1.0 or at least greater than 0.5. The semi-infinite extent assumption can be evaluated by comparing the incident photon penetration depth defined as 1/(S+K) and the medium thickness. When the medium thickness is much larger than the incident photon penetration depth, then the incoming photons can diffuse in the medium, provided that the medium satisfies the first assumption. When we evaluate these two parameters for whole blood and tissue, as shown in Figs. 1 and 2, in the wavelength regions above 600 nm, both whole blood and tissue satisfy the diffusely scattering condition. The penetration depth in both whole blood and tissue is less than 1 mm, and thus this condition can be easily met if the medium thickness can be made larger than at least 3 mm. In the wavelength region below 600 nm, since the absorption by hemoglobin is large, the light can rarely propagate in the medium, but absorbed quickly after penetrating into the tissue. Thus, theories based on only absorption such as Beer-Lambert Law may find better application in these regions, though some error may result due to scattering by the tissue entity. In application of the diffusion theory, the third parameter that is important to understand propagation in tissue is the diffusion length, where it indicates how far the scattered photons can diffuse in the medium. Fig 3 shows the diffusion length in the tissue when tissue scattering constant is 1.0/mm with the tissue blood volume of 0.1. When the hemoglobin oxygen saturation is high, the maximum diffusion length would become about 5.0 mm at 665 nm. However, in the region of around 550 nm the diffusion length is very small, less than 1 mm. Hence, when the optical reflection from tissue is measured at this wavelength, the measured reflectance should come from very shallow layer of the tissue, nominally less than 1.0 mm.

Three-dimensional Diffuse Reflectance Equation

For the transducer model shown in Fig 4, the diffuse reflectance received by the receiver can be given as(Takatani et al., 1988);

$$R(r_a) = \frac{2}{d}\left[\frac{S}{S+K}\right]\sum_{n=1}^{\infty} A_n\left[1 - e^{-d/\delta_0}(-1)^n\right]$$

where

$$\times \left[1 - \frac{2r_a}{b} M_1(\gamma_n b) N_1(\gamma_n r_a)\right]$$

$R(r_a)$ = all the reflected light collected within the circle of radius r_a at the z=0 plane,
S, K = scattering and absorption constants of the medium,
d = thickness of the medium
b = radius of the light source,
δ_0 = penetration depth of the incident photons given by 1/(S+K),
γ_n = eigen value given by $[(1/\delta_d)^2+(n\pi/d)^2]^{1/2}$,
M, N = first-order Bessel's functions of the first and second kind respectively,
δ_d = diffusion length defined as $[K(2S+K)]^{-1/2}$.

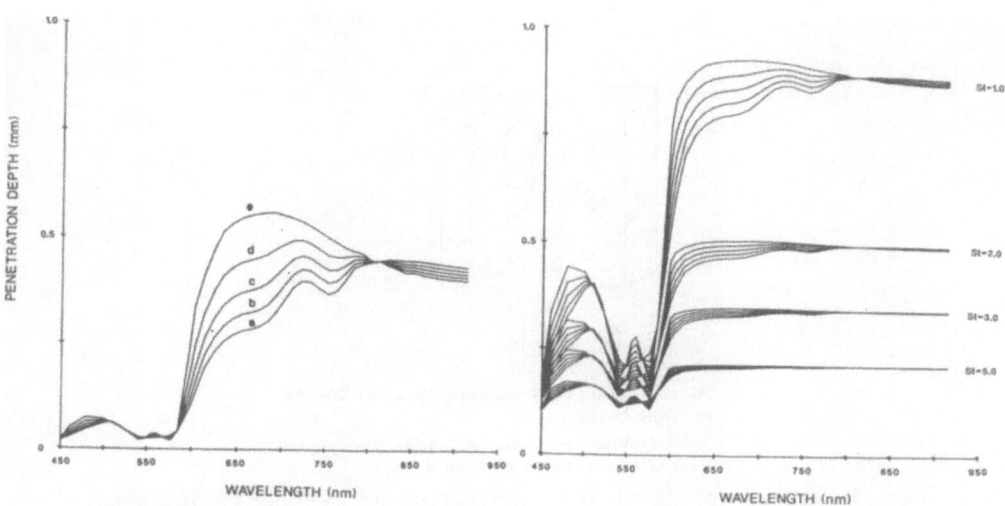

Fig. 1. Penetration depth in whole blood (left) and in tissue (right).

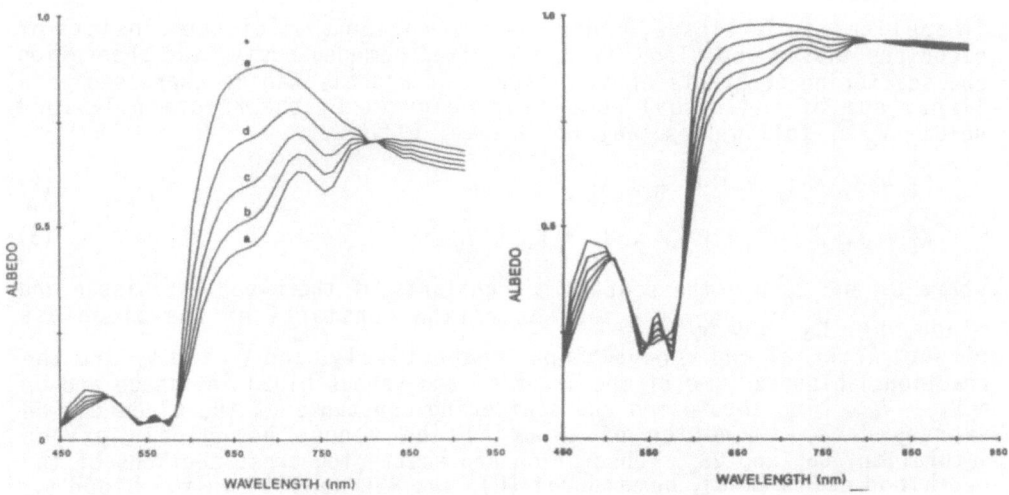

Fig. 2. Albedo of whole blood (left) and of tissue (right).

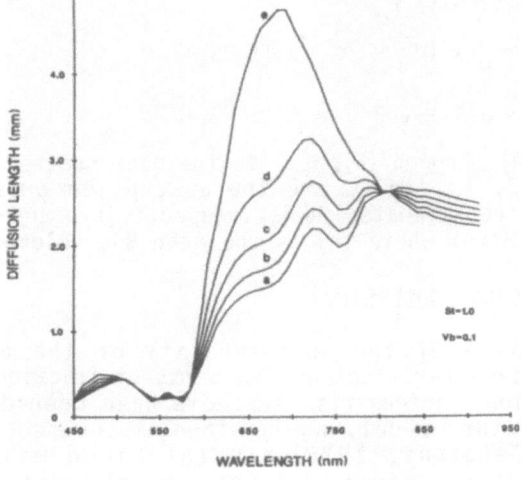

Fig. 3. Diffusion length in tissue for OS ranging from 20(a) to 100%(e).

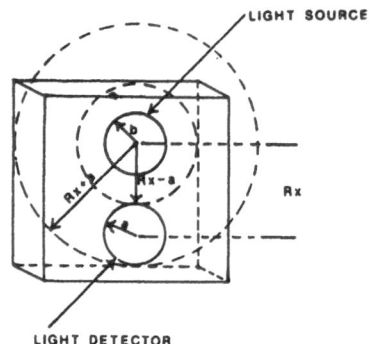

LIGHT SOURCE

R+b   Rx-a

Rx

LIGHT DETECTOR

**Rx: SEPARATION DISTANCE BETWEEN LIGHT SOURCE
AND DETECTOR**
a: RADIUS OF THE LIGHT DETECTOR
b: RADIUS OF THE LIGHT SOURCE

Fig. 4. Transducer model for computation of diffuse reflectance.

In application to tissue, under the assumption that tissue consists of bloodless tissue and blood which are mixed homogeneously, the absorption and scattering constants of the tissue, K_T and S_T, can be expressed as a linear sum of individual component weighted by the fractional blood volume V_b as follows(Takatani and Graham, 1979);

$$S_T = S_t \times (1-V_b) + S_B \times V_b \tag{2}$$

$$K_T = K_t \times (1-V_b) + K_a \times V_a + K_v \times V_v \tag{3}$$

where S_t and S_B are the scattering constants of the bloodless tissue and blood, K_t, K_a and K_v are the absorption constants of the bloodless tissue, arterial and venous blood, respectively, and V_a and V_v are the fractional blood volume of the arterial and venous blood in tissue and $V_b = V_a + V_v$. The absorption and scattering constants of the blood can be expressed as a function of arterial and venous hemoglobin oxygen saturation, OS_a and OS_v, absorption and scattering cross sections of the red blood cells (RBC), hematocrit (H), and RBC density in the blood (ρ) by the following equations(Takatani and Graham, 1979).

$$S_B = \sigma_s^- \times (1-H) \times \rho \tag{4}$$

$$K_a = \rho \times [\sigma_o \times OS_a + \sigma_r \times (1-OS_a)] \tag{5}$$

$$K_v = \rho \times [\sigma_o \times OS_v + \sigma_r \times (1-OS_v)] \tag{6}$$

In Equations (4) through (6), σ_s^- is the back scattering cross section of the single RBC, σ_o and σ_r are the absorption cross sections of the oxygenated and deoxygenated RBC's, respectively, and ρ is the density of RBC's given by H/Vol where Vol is the mean RBC volume.

HOMOGENEOUS TISSUE MODEL STUDY

In evaluation of the applicability of the above equation for prediction of tissue reflectance in terms of fractional blood volume and oxygen saturation, various tissue models have been developed including blood-milk mixture model, homogenized tissue model, and isolated gut tissue model(Takatani, 1978). In the blood-milk mixture model, homogenized milk was mixed with whole blood for different mixing ratio

94

and reflectance from the compound system was measured and compared with the theoretical reflectances as predicted by the photon diffusion theory. Fig. 5 shows the theoretical and experimental reflectances from the blood-milk mixture for different mixing ratios and oxygen saturation. Good agreement between the theory and experimental results can be seen. Thus, it can be concluded that the blood-milk mixture can be treated as a homogeneous system where optical constants can be expressed by a linear summation of each component using Equations (4) through (6). In the homogenized tissue model, the mucosal tissue of the canine gut was homogenized after the gut was made bloodless through perfusing it with physiological saline, and mixed with whole blood for different ratios. The result of this model was utilized in the third model where reflectance from the isolated gut loop was measured and compared against the theoretical reflectances. Since there are unknown distributions of arterial and venous blood in tissue, the tissue was cooled to minimize the A-V oxygen differences. Fig. 6 shows the experimental and theoretical reflectances of the isolated gut tissue with the tissue cooled and under normal condition. Again good agreement between the experimental and theoretical reflectances can be seen. Please note that under normal condition as determined from the spectral characteristics mean tissue hemoglobin saturation is closer to the OS_V, probably due to larger amount of venous blood volume in tissue.

Since the applicability of the 3-dimensional photon diffusion theory in combination with an linear analysis was shown in the model study, next a prototype sensor was constructed to sample tissue reflectance in the red and near-infrared regions at the wavelengths of 635, 665, 795, 910 and 940 nm and to predict tissue hemoglobin content and oxygen saturation based on the photon diffusion theory(Takatani et al., 1980). First of

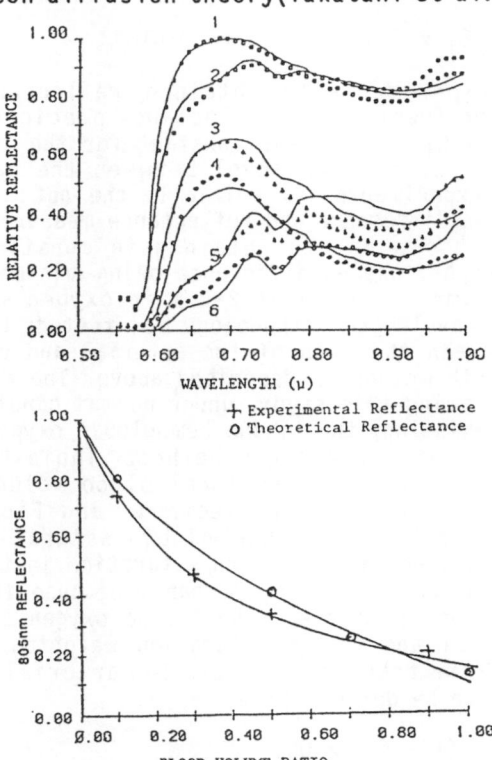

Fig. 5. Experimental and theoretical reflectance spectra (top) and 805 nm reflectance (bottom) of the homogeneous tissue model. #1,2:Vb=0.1, OS=98,40%, #3,5:Vb=0.5, OS=98,40%, #4,6:Vb=0.7, OS=98,40%.

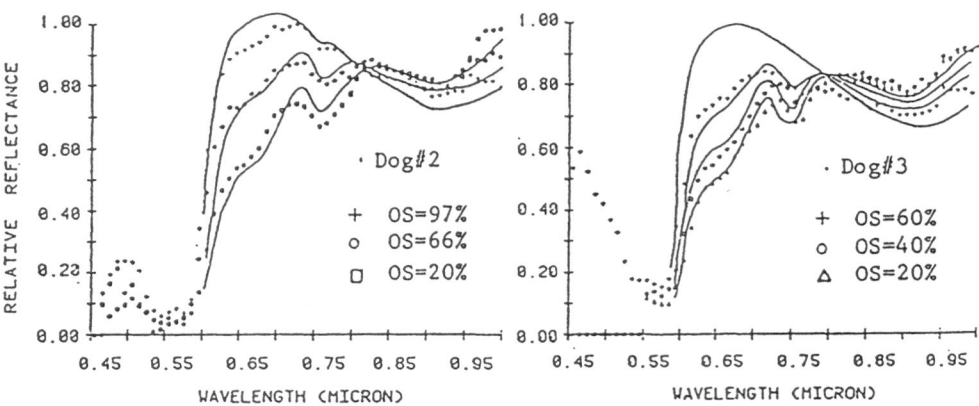

Fig. 6. Experimental and theoretical (solid lines) reflectance of the gut mucosa of dogs during tissue cooling (left) and under normal condition (right).

all, the tissue hemoglobin content Hb_T was derived from the isosbestic reflectance at 795 nm and used in the prediction of tissue hemoglobin OS, OS_T, through minimizing the sum of the squared error between the experimentally obtained tissue reflectance at the five wavelengths and theoretical reflectances from the photon diffusion theory.

$$\text{Error} = \sum_{i=1}^{5} [R_i - G_i \times \hat{R}_i]^2 \longrightarrow \text{MINIMUM} \qquad (7)$$

where R_i is the experimentally obtained reflectance at the ith wavelength, $\hat{R}_i$ is the theoretical reflectance predicted by the photon diffusion theory, and G_i is the gain constant for the ith wavelength to adjust the absolute reflectance level between the experimental and theoretical values. Experimentally, initially the gut tissue was cooled to minimize A-V oxygen difference and reflectance measurement was made at high OS value with known Hb_T to obtain gain constant for the five wavelengths. With the A-V oxygen difference being minimized (usually A-V oxygen difference came to be about 2-3 % in oxygen saturation), the accuracy of the five wavelength methodology was tested; Fig. 7 shows the correlation plot between the mean of the arterial and venous saturation vs. the five wavelength method as described above. The mean error ranged from 5 to 10%. Following this study, under normal condition with the A-V oxygen difference existing, the tissue hemoglobin oxygen saturation OS_T was estimated by the five wavelength method. Table I summarizes the tissue hemoglobin content and fractional blood volume in tissue as determined from the isosbestic reflectance and Fig. 8 shows the correlation plot between the tissue hemoglobin saturation vs. arterial and venous saturation. The tissue oxygen saturation in the gut mucosa is much closer to the venous saturation, probably because there is a larger amount of venous volume in tissue. The tissue oxygen saturation is the average of the arterial and venous saturation weighted by the arterial and venous blood volume fraction. Hence, the arterial to venous blood volume distribution can be derived from

$$OS_T = f_v \times OS_v + (1 - f_v) \times OS_a \qquad (8)$$

In the above equation, f_v is the fractional venous blood volume normalized to the total blood volume. The mean f_v value in the gut tissue ranged from 0.85 to 0.60 with the mean value of 0.71, indicating that arterial to venous blood volume fraction in mucosal tissue is 3:7. While this experimental value is low, compared with the value of 0.81 given by Green (1944), this last value is a theoretical value resulting from computation of the average volume of venules in tissue.

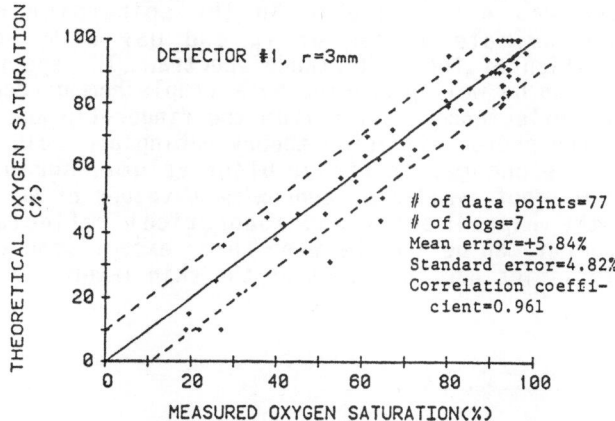

Fig. 7. Correlation plot between the tissue OS obtained by the sensor and meand of arterial and venous blood OS under tissue cooling.

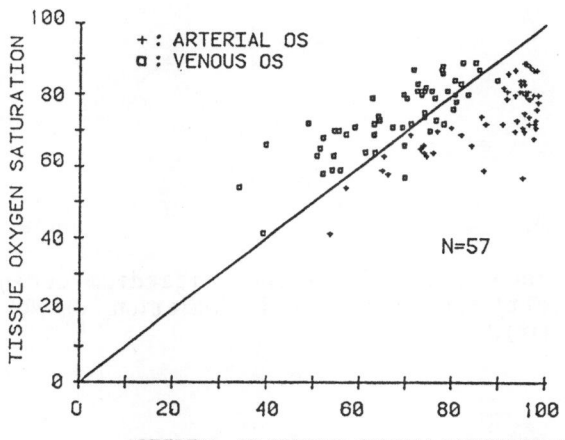

Fig. 8. Correlation plot between the tissue OS obtained by the sensor and arterial and venous blood OS under normal condition.

Table I. Tissue hemoglobin content and derived tissue blood volume fraction.

	Detector #1		Detector #2	
Dog#	Hb_T (gm%)	V_B	Hb_T (gm%)	V_B
1	1.489–0.664	0.087–0.040	1.431–1.150	0.088–0.077
2	1.143–0.598	0.092–0.049	1.192–0.870	0.096–0.071
3	2.099–1.106	0.109–0.058	1.972–1.358	0.102–0.075
4	1.533–0.716	0.090–0.042	2.041–1.543	0.120–0.091
5	1.511–0.658	0.105–0.043	1.296–0.648	0.080–0.046
6	1.287–0.660	0.095–0.063	1.202–1.085	0.086–0.078
7	1.321–0.840	0.090–0.058	1.136–0.733	0.079–0.051
Mean	1.483–0.749	0.095–0.050	1.467–1.055	0.093–0.070
μ	1.161	0.073	1.261	0.082

[a]In canine gut mucosa, V_B is approximately 0.08 or equivalent to 1.2 gm% tissue hemoglobin content
[b]μ = Mean of the maximum and minimum values.

Besides application to gut tissue, the epicardial reflectance has also been measured during coronary artery occlusion experiment. Fig. 9 shows the reflectance spectra from the epicardium and the theoretical prediction using the photon diffusion theory. Since the one-point calibration data was not available in the epicardial reflectance measurement, approximate values of V_b and OS_T were used in the theoretical prediction of the reflectance spectra. The spectral changes in the epicardium can be well predicted by a simple homogeneous analysis. Fig. 10 shows the reflectance spectra from the finger-tip and theoretical prediction using the photon diffusion theory during arm raising maneuver that simulates the changes in tissue blood volume. Again, since the calibration data was not available, approximate values of V_b and OS were used to match the experimental and theoretical reflectances. Good qualitative agreement can be seen between them except some deviation in spectral characteristics due to effect of the skin layer.

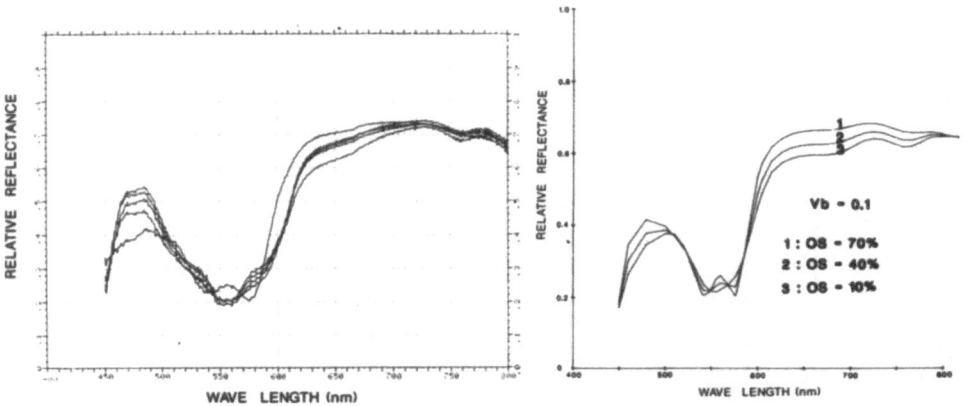

Fig. 9. The reflectance spectra from the epicardium during coronary artery occlusion (left) and theoretical simulation using the photon diffusion theory (right).

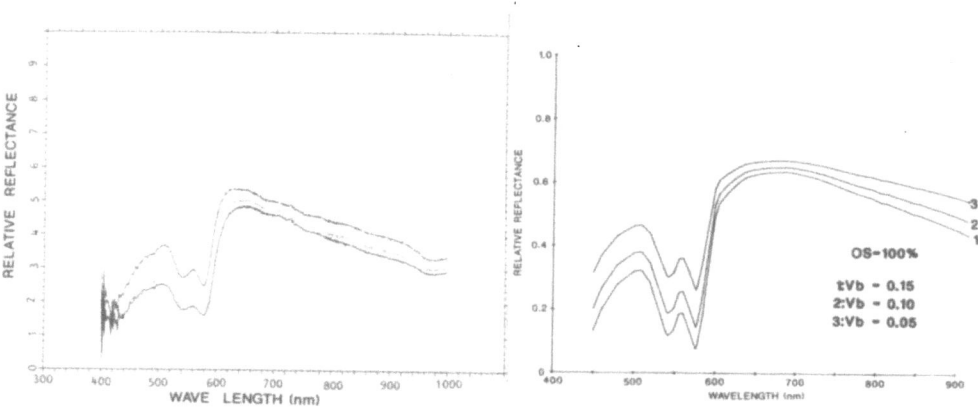

Fig. 10. The reflectance spectra from the finger-tip during arm up and down maneuver (left) and theoretical simulation using the photon diffusion theory (right).

If the experimental medium can be assumed to be homogeneous, then the similar mathematical treatment presented above can be used to obtain theoretical reflectances. However, the tissues are structually inhomogeneous usually consisting of multiple layers with RBC's contained within the capillaries and with smaller and larger blood vessels being present. Thus, the optical propagation in the tissue may not be uniform. In addition, when the skin is illuminated and its reflectance is measured, the measured reflectance from the vascular beds underneath the skin layer would be modified by the skin optical characteristics as shown in Fig. 10. Lubbers and Hoffmann (1980) proposed a transformation method in order to recover true hemoglobin spectra from the measured skin spectra based on the relation between the hemoglobin spectra and skin spectra. In our study, the skin was treated as a two-layer tissue model where absorbing and scattering thin skin layer is placed over a vascular bed which is a homogeneous mixture of the blood and bloodless tissue as shown in Fig. 11. The diffuse reflectance equation for this model was developed earlier and presented elsewhere (Takatani and Graham, 1979). The theoretical study based on this model revealed that through properly designing a transducer the skin layer effect can be minimized. Fig. 12 shows a pictorial representation of the percent reflectance from the second layer and deeper layers as a function of the light source and detector separation distance. As the separation distance between the optical source and receiver (refer to Fig. 11 two-layer tissue model) is increased, the percent reflectance from the second layer increases; this result suggests that by properly designing the transducer geometry (source and detector separation distance with respect to the superficial layer thickness) and by appropriately selecting the illumination wavelength, it is possible to minimize the effects of the superficial layer and to detect the light returning from the specific depth in tissue. In general, if the superficial layer thickness is known, two-layer model can be used to obtain critical source-detector separation distance that can minimize the effect of the superficial layer and to allow detection of the light returning from the specific depth in tissue.

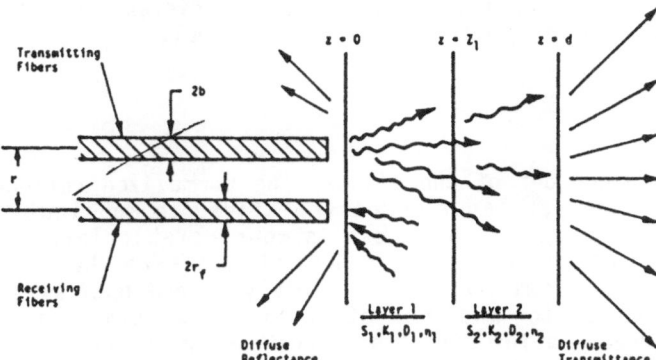

Fig. 11. Geometry of the two-layer tissue model and optical transducer. For the two-layer model, the first and second layers are indicated by the subscripts 1 and 2 on the variables, respectivel. S and K are the scattering and absorption constants from which the diffusion constant D can be calculated as $D = \nu/\eta(K + 2S)$, where ν is the velocity of light in space and η is the refractive index for the medium. The diffusion scattering and absorption constants f_s and f_a are simply $f_s = \nu S$ and $f_a = \nu K$, respectively. For the transducer, the transmitting and receiving fiber-optic bundles are assumed to be normal to the tissue and separated by a distance r. The bundles have radii b and r_f, respectively

SUMMARY

The photon diffusion theory can yield quantitative estimation of tissue hemoglobin saturation, provided that the medium is homogeneous and that one calibration data is available. The error in detection of tissue OS of the gut mucosa ranged from 5 to 10 % in oxygen saturation. In application to skin, the two-layer tissue model suggests that by properly designing the optical sensor and by appropriately selecting the illumination wavelengths, it is possible to capture mainly the light returning from the specific depth in tissue. Since the skin layer thickness is roughly in the order of 1 mm, the source and detector separation distance of approximately 3 mm or larger would ensure that the measured reflectance is truly returning from the deeper layer. When such reflectances are normalized to the blood-free reflectance obtained by

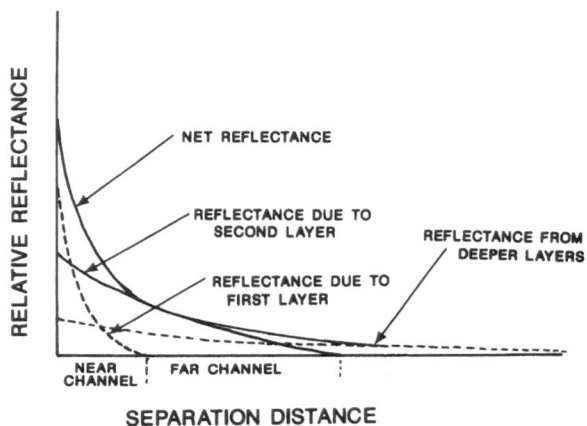

Fig. 12. Total reflectance, reflectance from the second layer, and reflectance from deeper layers vs. separation distance between the light source and detector in a multi-layer tissue model.

squeezing the blood out of the tissue, the normalized reflectance truly represents the deeper layer characteristics. In application to head, since the skin and skull thickness is considerable large, separation distance of 40 mm or greater is required to ensure the reflectance is actually returning from the brain. Closely spaced optical sensor would measure the scattering and absorption characteristics of the skin and skull of the head. As for directional changes in optical propagation due to tissue inhomogeneities, multiple light sources at the equi-distance around the detector can be placed to average out the effect. The resultant reflectance can be analyzed based on the similar mathematical treatment as presented in this study. However, since the absolute reflectance level calculated by the theory and the actual reflectance for a given transducer geometry have some deviation, again one point calibration is required to close the gap between them. This can be accomplished through arterialization of the tissue and ventilating with pure oxygen to yield reflectance from tissue containing 100% saturated blood. As for hemoglobin content, isosbestic reflectance, for example at 805 nm, can be utilized to estimate tissue hemoglobin content. Once one point calibration is accomplished, reflectance changes thereafter due to changes in Hb_T and OS_T can be fairly accurately predicted by the photon diffusion theory in combination with linear analysis.

Concerning separation of arterial and venous blood in tissue, the diastolic and systolic phases of the optical plethysmographic signal can be assumed to relate to venous or DC level, and to arterial or AC component. Since the four components, arterial and venous OS and Hb, are unknowns in the system, four equations or four wavelength measurements are required to sort out each effect. In combination with the photon diffusion theory, sensor design as specified here, the assumptions stated here and development of inhomogeneous tissue model that include effects of both arterial and venous blood distribution, absolute prediction of arterial and venous OS and Hb would become possible. Currently, the methodology is under examination in the tissue model and will be reported in future.

REFERENCES

1. Brinkman, R., Cost, W.S., Koopmans, R.K. and Zylstra, W.G. (1949): Continuous observation on the percentage oxygen saturation of capillary blood in patients. Arch Chir Neerl 1:184-191.
2. Chance, B. (1954): Spectrophotometry of intracellular respiratory pigments'. Science 120:765-775.
3. Cohen, A. and Wadsworth, N. (1972): A light emitting diode skin reflection oximeter. Med Biol Eng 10:385-391.
4. Green, H.D. (1944): Circulatory system; Physical Principles, Medical Physics, In Glasser, O. Editor, Year Book Medical Publishers, Inc., Vol.2, Chicago.
5. Johnson, C.C. (1970): Optical diffusion in blood. IEEE Trans Bio-Med Eng BME-17:129-133.
6. Kramer, K. (1934):Fortlaufende Registrierung der Sauerstroffsattigung im Blute an uneroffneten Blutgefaben. Klin Wschr 13:379-380.
7. Longini, R. L. and Zdrojkowski, R. (1968): A note on the theory of backscattering of light by living tissue. IEEE Trans Bio-Med Eng BME-15:4-10.
8. Lubbers, D.W. and Wodick, R. (1969): The examination of multicomponent systems in biological materials by means of a rapid scanning photometer. Appl Optics 8:1055-1062.
9. Lubbers, D.W. and Hoffmann, J. (1980): Absolute reflection photometry at organ surfaces. Adv Physiol Sci: Cardiovascular Physiology. Heart, Peripheral Circulation and Methodology 8:353-361.
10. Matthes, K. (1934): Über den Einflub der Atmung auf die Sauerstoffsattigung des Arterienblutes. Naunyn-Schmiedeberge Arch. exp Path Pharmk 176:683-696.
11. Mendelson, Y., Cheung, P.W., Neuman, M.R., Fleming, D.G. and Cahn, S.D. (1983): Spectrophotometric investigation of pulsatile blood flow for transcutaneous reflectance oximetry. Oxygen Transport Tissue: 4:93-102.
12. Millikan, G.A. (1942): The oximeter, an instrument for measuring continuously the oxygen saturation of arterial blood in man. Rev Scient Instrum 13:434-442.
13. Reynolds, L.O., Johnson, C.C. and Ishimaru, A. (1976): Diffuse reflectance from a finitie blood medium: Application to the modeling of fiber optic catheters. Appl Opt 15:2050-2067.
14. Takatani, S. (1978): On the theory and development of a noninvasive tissue reflectance oximeter, Ph.D. Dissertation, Dept. Biomed Eng, Case Western Reserve University, Cleveland, OH 44106.
15. Takatani, S. and Graham, M.D. (1979): Theoretical analysis of diffuse reflectance from a two layer tissue model. IEEE Trans Biomed Eng BME-26(12):656-664.
16. Takatani, S., Cheung, P.W. and Ernst, E.A. (1980): Noninvasive tissue

reflectance oximeter: An instrument for measurement of tissue hemoglobin oxygen saturation in vivo. Ann Biomed Eng 8:1-15.

17. Takatani, S., Noda, H., Takano, H. and Akutsu, T. (1988): A miniature hybrid reflection type optical sensor for measurement of hemoglobin content and oxygen saturation of whole blood. IEEE Trans Biome Eng 35(3):187-198.

18. Yoshiya, I., Shimada, Y. and Tanaka, K. (1980): Spectrophotometric monitoring of arterial oxygen saturation in the fingertip. Med Biol Eng 18:27-32.

NEW INSTRUMENT FOR MONITORING HEMOGLOBIN OXYGENATION

Tomomi Tamura, Hideo Eda, Michinosuke Takada
and Toshiya Kubodera

Research and Development, Shimadzu Corporation
Kyoto 604, Japan

INTRODUCTION

Oxygen transport has much relation to the condition of our living tissue. Cerebral hypoxia and ischemia cause many disturbances, and especially in the case of human newborn, it is important for us to monitor the oxygenation state of hemoglobin in cerebrum. But we have no instrument to monitor it non-invasively, directly and simply. In this case we usually use the percutaneous oxygen electrode monitor or the pulse-oximeter. As for the fomer, the response is slow and it is difficult to apply it to the patient of peripheral circulatory incompetence, and for the latter in case of hypotension.

To overcome this, we had been studied the near-infrared spectroscopy. Last year, we developed three-wavelength method and confirmed that the volume changes of oxy-Hb, deoxy-Hb and blood volume in the ventilated and anesthetized rat head were well monitored quantitatively by using it. Next we developed the compact high-performance instrument to apply this method to the oxygenation monitoring of the human newborn. First we applied it to the adult hand and the wrist. We did the brachium compression test and we could get reasonable data.

METHOD

Calculation of Oxy-Hb and Deoxy-Hb Content

Using the perfused rat head in situ, we measured the absorbance changes at the selected three wavelengths with change of the hematocrit values in the flowing perfusate under aerobic and anoxic conditions.

We could find a linear relationship between the absorbance change and hematocrit value in the physiological region. According to the Lambert-Beers Law, we can derive following equations.

$$\Delta O.D._{\lambda_1} = k_1\Delta[HbO_2] + k_1'\Delta[Hb] + S(\lambda_1) \dots\dots\dots\dots\dots\dots (1)$$
$$\Delta O.D._{\lambda_2} = k_2\Delta[HbO_2] + k_2'\Delta[Hb] + S(\lambda_2) \dots\dots\dots\dots\dots\dots (2)$$
$$\Delta O.D._{\lambda_3} = k_3\Delta[HbO_2] + k_3'\Delta[Hb] + S(\lambda_3) \dots\dots\dots\dots\dots\dots (3)$$

where $\Delta OD_{\lambda n}$ means the absorbance change at λn (nm), k_1, k_2, k_3, k_1', k_2', k_3' are absorbance coefficients at each wavelength, and $\Delta[HbO_2]$, $\Delta[Hb]$ indicate the change of the amount of Oxy-Hb and Deoxy-Hb. $S(\lambda n)$ indicate the absorbance change by scattering at λn nm . Assuming $S(\lambda_1) = S(\lambda_2) = S(\lambda_3) = S$ at these wavelengths, we can obtain following equations.

$$\Delta[HbO_2] = \frac{(k_2'-k_3')(\Delta OD_{\lambda_1}-\Delta OD_{\lambda_3})-(k_1'-k_3')(\Delta OD_{\lambda_2}-\Delta OD_{\lambda_3})}{(k_1-k_3)(k_2'-k_3')-(k_2-k_3)(k_1'-k_3')} \quad \cdots\cdots\cdots\cdots (4)$$

$$\Delta[Hb] = \frac{-(k_2-k_3)(\Delta OD_{\lambda_1}-\Delta OD_{\lambda_3})+(k_1-k_3)(\Delta OD_{\lambda_2}-\Delta OD_{\lambda_3})}{(k_1-k_3)(k_2'-k_3')-(k_2-k_3)(k_1'-k_3')} \quad \cdots\cdots\cdots\cdots (5)$$

$$\Delta[IHb] = \frac{(k_2'-k_2-k_3'+k_3)(\Delta OD_{\lambda_1}-\Delta OD_{\lambda_3})-(k_1'-k_1-k_3'+k_3)(\Delta OD_{\lambda_2}-\Delta OD_{\lambda_3})}{(k_1-k_3)(k_2'-k_3')-(k_2-k_3)(k_1'-k_3')} \quad \cdots (6)$$

where $\Delta[IHb] = \Delta[HbO_2] + \Delta[Hb]$ (change of blood volume).

We can obtain these coefficients by measuring the absorbance changes of the completely oxygenated or deoxygenated red blood cell suspension in vitro against changes of hematocrit value. Now we use three wavelengths, 780nm, 805nm and 830nm, then equations (4), (5), (6) become as follows:

$$\Delta[HbO_2] = -3\Delta OD_{805} + 3\Delta OD_{830} \quad \cdots\cdots\cdots\cdots (7)$$
$$\Delta[Hb] = 1.6\Delta OD_{780} - 2.8\Delta OD_{805} + 1.2\Delta OD_{830} \quad \cdots\cdots\cdots\cdots (8)$$
$$\Delta[IHb] = 1.6\Delta OD_{780} - 5.8\Delta OD_{805} + 4.2\Delta OD_{830} \quad \cdots\cdots\cdots\cdots (9)$$

SYSTEM

The block diagram of our system is shown in Fig. 1. The light source consists of three high-power laser diodes, and the light beams being emitted in turn by the time-sharing mode are coupled to one beam through the three-branched optical fiber and illuminates the tissue. The transmitted light is guided to photomultiplier through the receiving optical

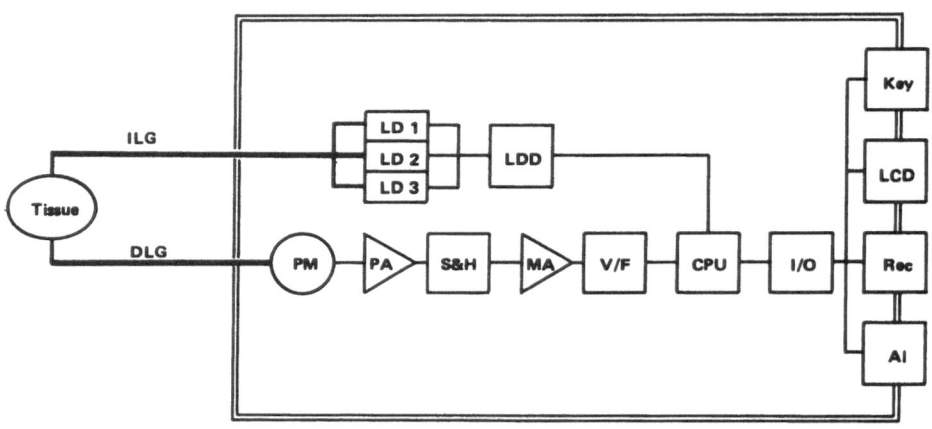

LD 1	Laser Diode 30mW $\lambda_p = 780$nm
LD 2	Laser Diode 30mW $\lambda_p = 805$nm
LD 3	Laser Diode 30mW $\lambda_p = 830$nm
LDD	Laser Diode Driver
PM	Photomultiplier
PA	Pre-Amplifier
S&H	Sample Hold Circuit
MA	Main Amplifier
V/F	Voltage-Frequency converter
Key	Key board
LCD	Display
Rec	Recorder
AI	Alarm
ILG	Illuminating light guide; bundle 5mm dia.
DLG	Detecting light guide; bundle 5mm dia.

Fig. 1 Block Diagram of our System

bundle fiber. The output signal is amplified and processed by the 8 bit micro computer, and finally the time-cource changes of above three hemoglobin values are displayed on the LCD panel and recorded on the recorder in real time.

We can set up the upper and lower limit against the calculated values, and can monitor the condition of the patient by the alarm light and sound.

RESULTS

Measurement in Rat Head

To check our method and our instrument, first we measured the absorbance changes and calculated the parameters, $\Delta[HbO_2]$ and $[Hb]$, with change of the O_2 concentration of the inspired gas in rat head. The calculated values changed by degrees according with the decrease of O_2 concentration as shown in Fig. 2.

Measurement in human hand

We did the brachium compression test. We measured the transmitted light through the hand. We oppressed it by 70mm Hg pressure at which the blood flow of vein was considered to be stopped, with monitoring the above calculated parameters. Ther result is shown in Fig. 3-(a),(b). Calculated Oxy-Hb volume was considered to be constant but that of Deoxy-Hb increased gradually. When the pressure was released, these values returned to the initial level after overshooting.

Next we oppressed it by 160mm Hg pressure at which the blood flow of artery and vein were considered to be stopped. Then the Oxy-Hb volume decreased gradually, but that of Deoxy-Hb increased more.

DISCUSSION

From above results, we can confirm the possibility to monitor the

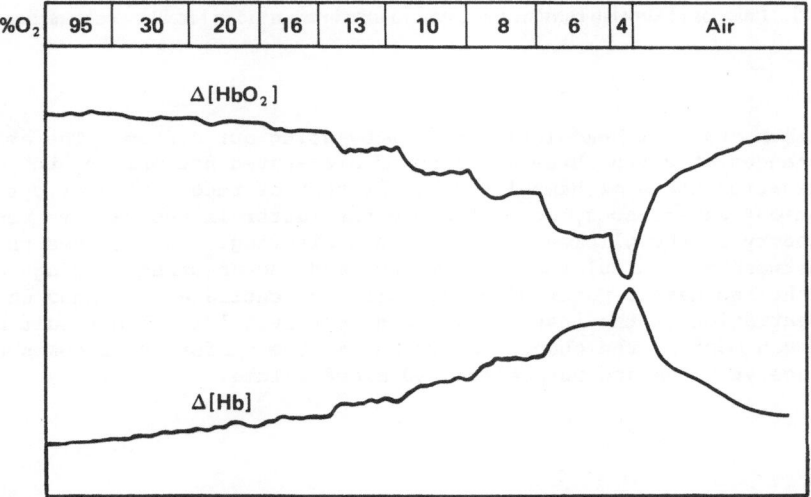

Fig. 2 Changes of Oxy-Hb and Deoxy-Hb content in the rat head according with the aerobic to anoxic changes of the ventilate conditions

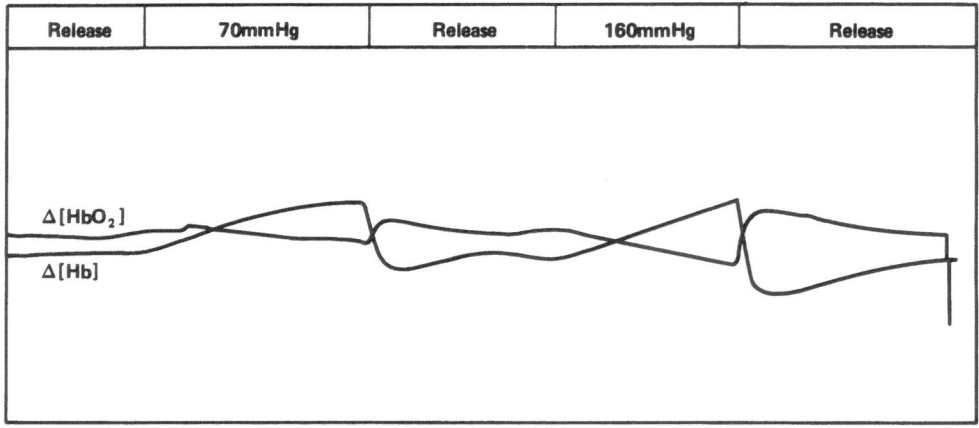

Release	70mmHg	Release	160mmHg	Release

$\Delta[HbO_2]$

$\Delta[Hb]$

Fig. 3-(a)   Changes of Oxy-Hb and Deoxy-Hb content in the hand in brachium Compression Test

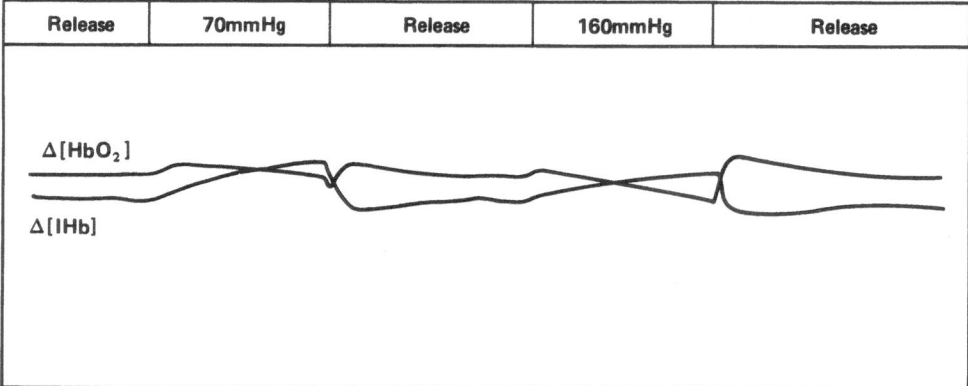

Release	70mmHg	Release	160mmHg	Release

$\Delta[HbO_2]$

$\Delta[IHb]$

Fig. 3-(b)   Changes of Oxy-Hb content and blood volume in the hand in brachium compression test

oxygenation state of hemoglobin in vivo by using our system. The absorbance changes at given three wavelengths are swayed not only by the change of oxygenation state of hemoglobin but by that of redox state of cyt aa_3. But from our experiences, we can ignore the latter in the case we use above theory to the clinical oxygenation monitoring. We obtained the same trace between the calculated blood volume and the absorbance change at 805nm, the isosbestic point of hemoglobin, in ventilated rat head when the O_2 concentration of the inspired gas was kept over 10%. The result shows that we can monitor the change of cyt aa_3 as the difference between the absorbance at 805nm and our calculated blood volume.

SUMMARY

A new compact high-performance instrument for monitoring the oxygenation state of the human newborn are developed. First we applied this instrument to the rat head, and next to the human arm. We could confirm that the volume of Oxy-Hb, Deoxy-Hb and total blood volume were well monitored by using our instrument. Now we started to apply this technique to the human newborn. Furthermore, we started the experiments of the near-infrared tomography for the small animals on the basis of this technique.

REFERENCES

Tamura, T., Hazeki, O., Takada, M., and Tamura, M., 1987, Absorbance profile of red blood cell suspension in vitro and in situ, Adv. Exp. Med. Biol., 222:211-217.
Hazeki, O., and Tamura, M., 1988, The quantitative analysis of hemoglobin oxygenation state of rat brain in situ as monitored by near-infrared spectrophotometry, J. Appl. Physiol., 64:796-802.
Hazeki, O., and Tamura, M., 1988, Near-infrared quadruple wavelength spectrophotometry of rat head, Adv. Exp. Med. Biol., this volume.

OXYGEN DISTRIBUTION IN ISOLATED PERFUSED LIVER OBSERVED BY PHOSPHORESCENCE IMAGING

David F. Wilson, William L. Rumsey and Jane M. Vanderkooi

Department of Biochemistry & Biophysics
University of Pennsylvania School of Medicine
Philadelphia, PA

INTRODUCTION

The measurement of oxygen in intact tissues has proven to be full of technical difficulties. For example, the direct determination of oxygen in tissue using microelectrodes is subject to error in the low range of oxygen pressures (< 1 uM) and is restricted to fixed sites of evaluation (Silver, 1984). Moreover, insertion of the electrode damages the tissue and disturbs the local milieu. Indirect qualitative measurements using surface fluorescence of NADH indicate the presence of anoxia in tissue (Barlow and Chance, 1976) but lacks internal standardization (Koretsky et al., 1987). Spectrophotometric measurements of hemoglobin (Kekonen et al., 1987) and myoglobin (Fabel and Lubbers, 1965; Tamura et al., 1978) are useful for oxygen pressures near their P_{50} but are subject to optical interference from changes in light scattering and from absorbtion by other chromophores. Alternatively, cryomicrospectrophotometry of myoglobin (Gayeski and Honig, 1986) can provide an accurate method for calculating intracellular oxygen pressure in some tissues. Because this method requires freezing of the tissue, it is impractical for monitoring changes in oxygenation during any length of time.

Optical techniques for determining oxygen in tissue have, in general, the advantage that continuous, noninvasive measurements are possible. Moreover, these methods can, in some cases, be used to image a broad area of the organ of interest. An optical method for measuring oxygen based on its ability to quench phosphorescence of selected lumiphores has been developed (Vanderkooi and Wilson, 1987; Vanderkooi et al., 1987). This method has successfully been used to determine the oxygen dependence of respiration in suspensions of isolated mitochondria (Wilson et al., 1988) and cells (Robiolio et al., 1988; this volume). We have recently extended the use of phosphorescence quenching to include the imaging of the distribution of oxygen in perfused tissue (Rumsey et al., 1988; this paper).

METHODS AND MATERIALS

Male Sprague Dawley rats (275-325 g) were fasted for 24 hours prior to the experiment. Sodium pentobarbital (35-40 mg/kg body wt) was administered by intraperitoneal injection. The liver was perfused in situ via the portal vein and the effluent from the vena cava was discarded. The perfusate, a Krebs-Henseleit buffer supplemented with 5 mM glucose, was maintained at 37° C and equilibrated with 95% O_2:5% CO_2.

The liver was excised and placed on a clear plastic petri dish that was mounted on a tripod. From this point on, the perfusate was recirculated and contained 1% bovine serum albumin (Fraction V, ICN Immunobiologicals, Lisle IL) with the oxygen probe, palladium coproporphyrin (1uM). Positioned beneath the tripod were two mirrors: one served to reflect upward the excitation beam whereas the other reflected the phosphorescence (> 665 nm, with a Schott cutoff filter) emitted from the tissue to a silicon intensified target (SIT) video camera (Dage MTI model 66, Michigan City, IN). The illuminating light was a 150 W xenon arc lamp that was filtered with a 1.5 cm thickness of a saturated solution of $CuSO_4$ in 0.5 M H_2SO_4 and a cutoff filter (50% block at 395 nm). The phosphorescence emitted from a lobe of liver was imaged through a 90 mm lens attached to the SIT camera, recorded on a video cassette recorder, digitized, and displayed on a analogue monitor. The images were processed to remove a constant amount of background intensity, about 50 units. The remaining gray levels were linearly stretched to bring the maximal intensity arising from the anaerobic liver to the scale maximum (256 units). A region of the liver lobe, 256 X 232 pixels, was then expanded by two fold, 512 X 465 pixels, for analysis.

The principles of the method have been published previously (Vanderkooi and Wilson, 1987, Vanderkooi et al., 1987). In brief, the illumination of the tissue raises Pd-coproporhyrin to its triplet state from which it returns to the ground state either by emitting light or by energy transfer to other molecules (quenching). Oxygen is the principal quenching agent in biological materials. The concentration of oxygen can be calculated from the intensity of the emitted light or from the lifetime of the triplet state using the Stern-Volmer relation:

$$I_0/I = T_0/T = 1 + k_q T_0 [O_2]$$

where I_0 and T_0 are the phosphorescence intensity and lifetime, respectively, in the absence of oxygen; I and T are the phosphorescence intensity and lifetime, respectively, at a given oxygen concentration $[O_2]$; and k_q is a constant related to the rates of diffusion of oxygen and the selected lumiphor.

RESULTS

Figure 1A shows that a small amount of phosphorescence is emitted from the liver when it is perfused at 30 ml/min, and that the phosphorescence intensity is homogenous throughout the illuminated area. When flow is stopped, phosphorescence intensity increases markedly as the oxygen in the perfusate is exhausted (Figure 1B). The change in intensity rises from about 50 to 250 gray levels.

As an alternative, the image obtained from the well-oxygenated tissue can be subtracted from those where flow has been reduced, thereby providing an image of the change in phosphorescence. This is represented by Figure 2 where it can be seen that the complete stoppage of flow results in an increase of phosphorescence intensity to about 200 gray levels.

When flow was started again, increased from zero to about 10 ml/min, a pattern developed which indicated that a marked heterogeniety occurred during the reoxygenation process. Figure 3A shows an image in which bright irregularly shaped spots, surrounded by areas of lesser intensity, resulted from the increase in flow. This produced a mottled appearance in the image of the liver. When the direction of flow was reversed during reperfusion, i.e., oxygenated buffer was pumped into the vascular system via the vena cava, a pattern (Figure 3B), which was the negative of that in Figure 3A, was obtained. The spots that appeared to be very bright in Figure 3A are dark in Figure 3B whereas the dark areas in Figure 3A are bright in Figure 3B.

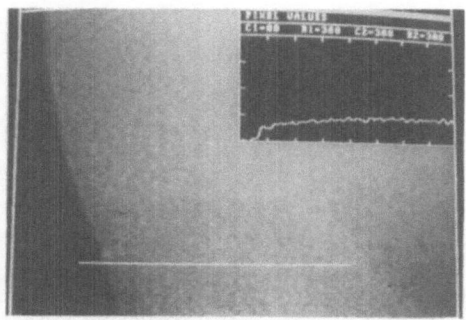

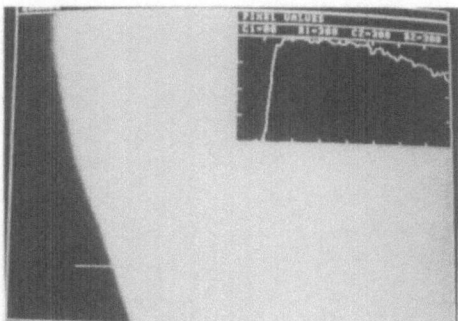

Figure 1. Phosphorescence intensity in response to aerobic perfusion (A) and anoxia (B). The view is from the underside of the liver. The white line (equal to about 10 mm on the liver surface) was arbitrarily scribed to depict the values of phosphorescence appearing in the inset. The line contains 300 pixels (x-axis) with intensity ranging from 0 to 256 gray levels.

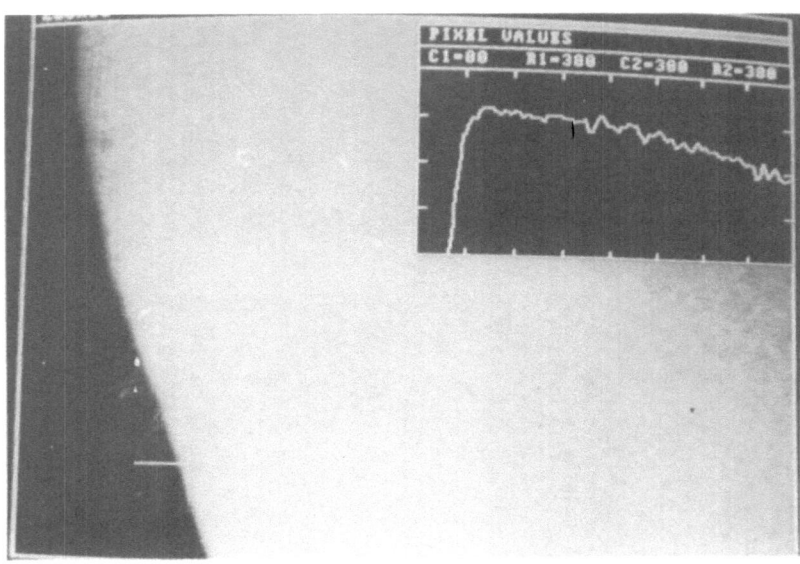

Figure 2. Phosphorescence intensity as a result of anoxia without the contribution of background light. Phosphorescence from the fully oxygenated liver was subtracted from that during anoxia.

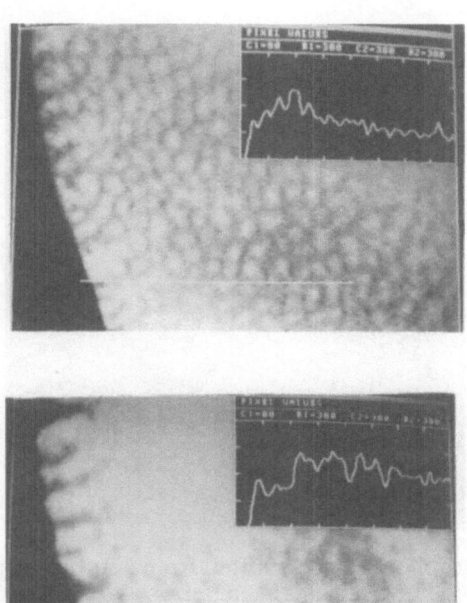

Figure 3. Phosphorescence intensity resulting from reperfusion via the portal vein (A) and the vena cava (B). Following a stoppage in flow, perfusion was reinitiated at low levels of flow, about 10 ml/min. In A, the influent entered the portal vein and exited the vena cava. In B, oxygenated perfusate was supplied through the vena cava and the effluent passed out the portal vein. In both photographs, background intensity was subtracted from the absolute image.

Most experiments were performed over a period of 2-3 hours in which several episodes of no flow-reperfusion occurred. During longer periods of perfusion, it was observed that, in some cases, small regions of the tissue became refractory to reperfusion. This is demonstrated in figure 4, where two hypoxic regions appeared during a time in which the remainder of the tissue was adequately oxygenated.

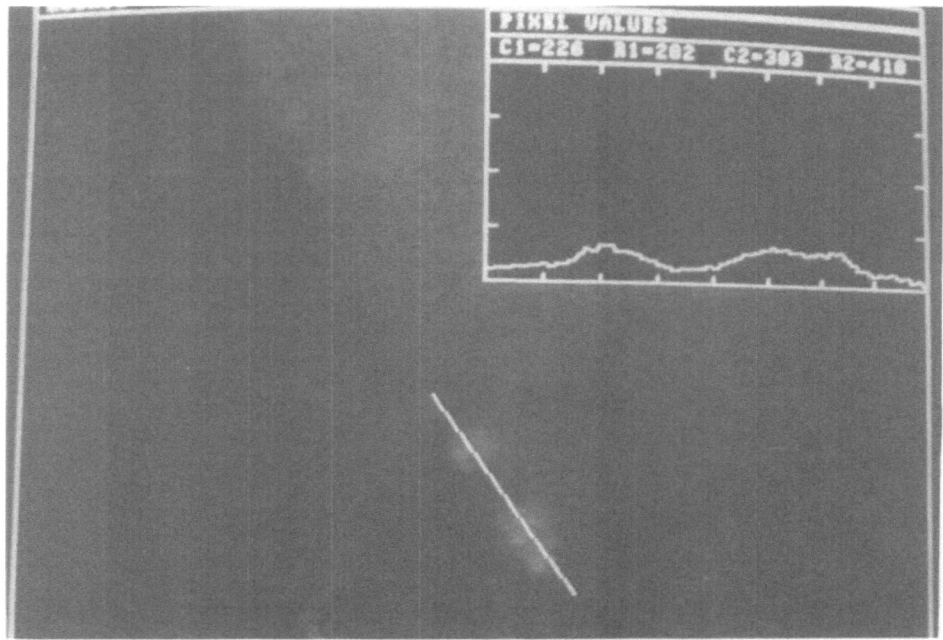

Figure 4. Image of phosphorescence emitted from aerobically perfused liver with hypoxic regions. The liver had been perfused for about 3 hours in which flow was stopped several times, about 10 occurrences.

DISCUSSION

Oxygen dependent quenching of phosphorescence is a powerful new tool for the study of oxygen metabolism in biological systems. In this area of research, it is, perhaps, the imaging of phosphorescence that may prove to be of greatest benefit. Phosphorescence quenching provides accurate serial determinations of oxygen in the physiological range, from 10^{-5} M to less than 10^{-8} M, with a limit of $\pm$ 0.01 uM (Wilson et al., 1987). With phosphorescence imaging, it is possible to define temporal alterations in tissue oxygenation and locate macroscopic areas of tissue hypoxia. The current work also demonstrates that differences in perfusion can be observed, in this case, at the level of the periportal (around the portal vein) and the pericentral (around the central vein) patches of the liver. Moreover, a large area of tissue can be scanned for levels of oxygen content by simply adjusting the location of the excitation light in conjunction with the video camera. Since the oxygen probes do not appear to be toxic, the possible utility of the method spans a wide range of in vivo and in vitro experimental and diagnostic procedures.

In the present study, phosphorescence intensity was not converted into units of oxygen concentration or pressure. It is possible, however, to calculate, pixel by pixel, values of oxygen concentration throughout the image (the development of the appropriate software for these calculations is currently underway).

Acknowledgements: This work was supported by grants GM36393 and GM21524.

REFERENCES

Barlow, C.H. and Chance, B., 1976, Ischemic areas in perfused rat hearts: Measurement by NADH fluorescence photography. Science, 193: 909-910.

Fabel, H. and Lubbers, D.W., 1965, Measurement of reflection spectra of the beating heart. Biochem. Z., 341: 351-356.

Gayeski, T.E.J. and Honig, C.R., 1986, O_2 gradients from sarcolemma to cell interior in red muscle at maximal VO_2. Am. J. Physiol., 251: H789-H799.

Kekonen, E.M., Jauhonen, V.P. and Hassinen, I.E., 1987, Oxygen and substrate dependence of hepatic cellular respiration: Sinusoidal oxygen gradient and effects of ethanol in isolated perfused liver and hepatocytes. J.Cell Physiol., 133: 119-126.

Koretsky, A.P., Katz, L.A. and Balaban R.S., 1987, Determination of pyridine nucleotide fluorescence from the perfused heart using an internal standard. Am. J. Physiol., 253: H856-H862.

Rumsey, W.L., Vanderkooi, J.M., and Wilson, D.F., 1988, Imaging of phosphorescence: A novel method for measuring the oxygen distribution in perfused tissue. Science, in press.

Robiolio, M, Rumsey, W.L. and Wilson, D.F., 1988, Oxygen diffusion and mitochondrial respiration in neuroblastoma cells. Am. J. Physiol., in press.

Silver, I.A. 1984, Polarographic techniques of oxygen measurements. In: "Oxygen: An in-depth study of its Pathophysiology", Gottlieb, S.F., Longmuir, I.S. and Totter, J.R. eds., Undersea Med. Soc. pub No. 62 (ws) 3-1-84, Bethesda, MD. pp 215-238.

Tamura, M. Oshino, N., Chance, B. and Silver, I.A., 1978, Optical measurements of intracellular oxygen concentration of rat heart in vitro. Arch. Biochem. Biophys. 191: 8-22.

Vanderkooi, J.M., Maniara, G., Green, T.J., and Wilson, D.F., 1987, An optical method for measurement of dioxygen based quenching of phosphorescence. J. Biol. Chem., 262: 5476-5482.

Vanderkooi, J.M., and Wilson, D.F., 1986, A new method for measuring oxygen in biological systems. In:"Oxygen transport to tissue VIII, Longmuir, I.A. ed., Plenum Press, New York, p. 189-193.

Wilson, D.F., Rumsey, W.L., Green, T.J., and Vanderkooi, J.M., 1988, The oxygen dependence of mitochondrial oxidative phosphorylation measured by a new optical method for measuring oxygen concentration. J. Biol. Chem., 263: 2712-2718.

Wilson, D.F., Vanderkooi, J.M., Green, T.J., Maniara, G., DeFeo, S.F. and Bloomgarden, D.C., 1987. A Versatile and Sensitive Method for Measuring Oxygen. In: "Oxygen Transport to Tissue IX". Silver, I.A. and Silver, A., ed., Plenum Press, New York and London, (Adv. Exp. Med. Biol. 215, 71-77).

OTHER METHODS AND INSTRUMENTATION

THE USE OF CONDUCTIVE THERMOPLASTIC WIRES AS OXYGEN SENSORS IN MICROWAVE FIELDS

H.I. Bicher, M.D., K. Reesman and D. Moore

Valley Cancer Institute
14427 Chase Street
Panorama City, Los Angeles, CA 91402

Local microwave induced hyperthermia is considered standard treatment in the management of maligant disease, specifically recommended for locally recurrent tumors and for primary cancer where other treatment modalities have a poor history of success Several mechanisms of action of heat on tumors have been determined including, among others, vascular damage which disturbs the already poor tumor blood supply resulting in increased heat retention and deprivation of oxygen and nutrition in tumors (2-10) leading to a decrease in tissue pH and tumor death. In previous publications we have described the possible prognostic value of clinically determining PO_2 levels during hyperthermia treatments to predict the effectiveness of the given therapy (3), however, when measurements are to be made in strong electromagnetic or microwave fields such as occur during hyperthermic treatment of cancerous tumors, the metallic nature of the sensing electrodes cause the formation of electrical eddy currents that induce self heating of the sensors leading to erreous readings and can also induce thermal burns on the treated tissues.

This paper describes the construction, characteristics and preliminary testing of a Polarographic O_2 sensor built using thermoplastic materials devoid of the metallic characteristics causing electromagnetic field disturbances.

MATERIALS AND METHODS

Electrodes were constructed using two different thermoplastic wires encased in a vinyl plastic pippette draw to a small opening exposing a 100 micron tip that was then coated with O_2 pervious water semi pervious membrane.

Electrode construction: A connecting wire was drawn by using an electronic carbon based conductive coating (ECCOCOAT R 258A, Emerson and Cumming, Canton, Mass.) applied over a 5 zero

silk thread, and left to dry in air for 24 hours.   To the tip
end of this wire, a second elastomer coated wire (ECCOCOAT cc-
40-A Emerson and Cumming, Canton, Mass.) is connected.   This
second elastoner is silver filled and constitutes the oxygen
sensing basis.   It is applied on a small 100 micron polivinyl
wire.   Both wires are joined together using a drop of ECCOCOAT
258 A.   They are then pulled through a PTFE pippete 200 micron
i.d. using a small vinyl leader.   The pipette is then sealed
over the sensing tip by gentle pulling over a small Bunsen
flame.   The tip is then cut at a 45 degree angle. (fig 1)

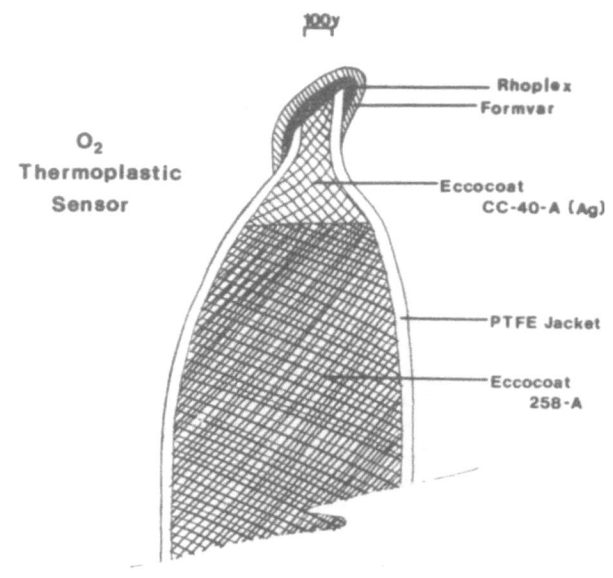

Figure 1

To finish the O2 electrode, a membrane is formed over the
sensing tip by the application of Rhoplex-Formvar, as
previously described (1).

The reference electrode was made of Ag-AgCl as previously
described (1). The basic probe can be used to record changes in
tissue PO , acting as an open electrode. According to Davies
and Brink (8) and Cater (5), ths type of electrode is useful
for continuous mesurements and responds very rapidly to changes
in tissue oxygen tension.   Also as described by Cater and
Silver (6), electrodes of this size are insensitive to
"stirring" (rapid movements of fluid near the tip are), consume
but small amounts of oxygen themselves, and cause little tissue
damage in the area in which they are inserted.   Their main
disadvantage is that they are subject to "poisoning,"
particularly by the electrophoretic deposition of protein on
the active sensing surface.

To overcome this difficulty, the tip of our microelectrode
was coated with a water carrying, highly adhesive oxygen-
pervious material, Rhoplex AC35 (Rohm Haas Co., Philadelphia,
PA).   This coating was achieved by dipping the tip of the
microelectrode in Rhoplex AC35 emulsion, and then drying the
electrode in air for 30 min.   In some experiments as additional

coating with water-repellent substance was deposited, using Formvar (E.R. Fulham, Inc. Schenectady, NY). The formvar coating was not as extensive as the Rhoplex AC35, so that electrical contact can be established between the emulsion and the surrounding tissue, as described by Silver (14).

The electrodes were connected to a Transidyne O_2 amplifier (Transidyne General, Ann Arbor, MI.) and the output recorded using a Beckman S2 Dynograph. (Beckman Instruments, Palo Alto, CA).

Calibration: electrodes were calibrated as described by Silver (13) in saline solutions of known PO_2 values. Polarographic curves were also recorded placing the electrodes in nonrespiring brain tissue, as described by Whalen et al. (16)

The electrodes were "conditioned" by placing them in saline solution and applying 0.8 v current for 2 hrs. After treatment, they were usually very stable. Stirring of the solution by a magnetic rotor did not change the oxygen current. The current reading at zero oxygen tension was very low (residual current) and the response of the microelectrode to changes of oxygen tension was very rapid (95% response time of the order of 0.5 sec)

MICROWAVE INDUCTION AND APPLICATION: Microwaves were generated in the clinical range of 915 or 300 Mhz and delivered using appropriate applicators, with the electrodes tested in muscle phantom material. All these instruments and materials were supplied by HBCI Inc. 14427 Chase St., Panorama City, Ca 91402.

IN VIVO TESTING: The O_2 response was tested in Nembutal Anesthetized Rabbits with the electrodes introduced in the Carotid Artery. Blood PO values were tested using a standard Beckman Gas Analyzer.

RESULTS

The insulation resistance is in the range of 10^{11} Ohms. If the electrode is stored in distilled water the resistance will decrease to 10^{10} or 10^9 Ohms after four weeks.

The polarographic plateau lies in a range of 500 to 900 mV (Fig. 2), other electrodes may have a higher plateau. In order to get the most accurate values, the polarogram of each electrode should be determined before measuring.

The calibration curve (Fig. 3) is linear and shows a very small zero current. The electrodes show a drift of the signal up to 10% /hr.

Exposing the electrodes to microwave radiation caused no change in the Phatom readings between 0 to 300 Watts of incident power at either 915or 300 Mhz. Proper shielding of connecting wires and electroni equipment was necessary.

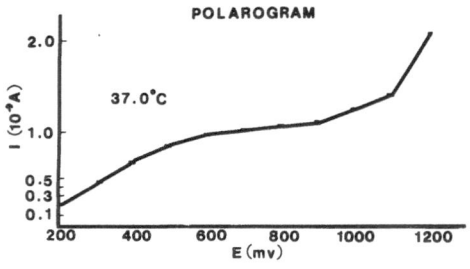

Figure 2

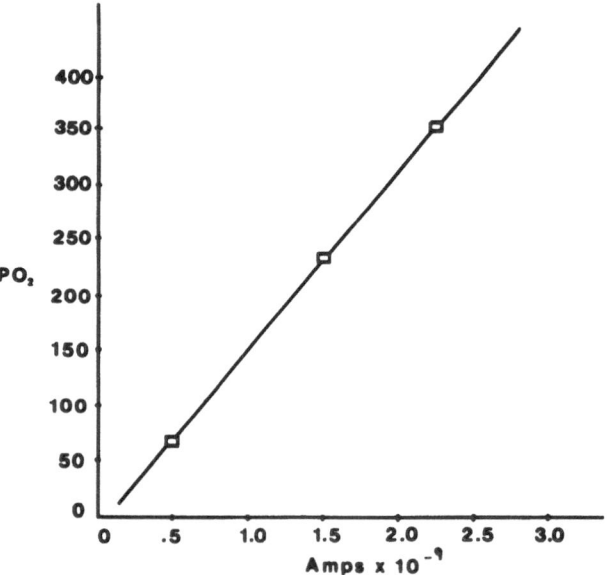

Figure 3

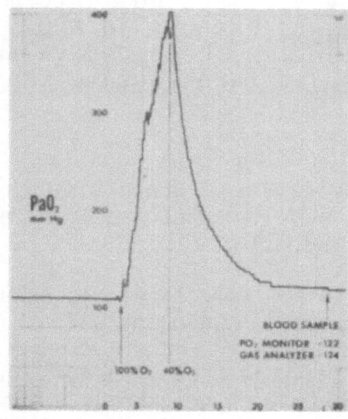

Figure 4

The microelectrodes are linear between 1-300mm Hg,
exhibited a quasi plateau between .5-.9 volts of polarographic
voltage and have minimal storing and pH artifacts. They are
usable in microwave fields with virtually no artifacts when the
recording amplifiers and wires are properly shielded.

DISCUSSION

The newly developed elastomeric oxygen probe shows promise
for oxygen readings in biological tissues subject to microwave
irradiation. Further development and miniaturization are
necessary before the device can become a standard clinical
tool. However our results demonstrate the possibility of
using polarographic elastomer probes, which show
characteristics similar to those of other oxygen micoelectrodes
A major difference seems to be a reduced output per surface area
exposed, since our probe showed a current in the 10 Amperes
range with an exposed area of 100 micron diameter.
However, the lack of perturbation in microwave fields and
linearity of readings makes further research in this probe
promising.

SUMMARY

Determination of PO levels in biological tissues or fluids
using polarography is a well established technique. However,
when measurements are to be made in strong electromagnetic or
microwave fields such as occur during hyperthermia treatment of
cancerous tumors, the metallic nature of the sensing electrodes
cause the formation of electrical eddy currents that induce
self heating of the sensors leading to erroneous readings and
can also induce thermal burns on the treated tissues.

In order to change the metallic nature of the O sensors we
have used conductive polymer materials. These materials are
contructed by the introduction of certain molecules in the
thermoplastic carrier that make it conductive, without causing
it to behave as a metallic antenna in the absorption of
electromagnetic waves. The tested materials in these
experiments included ECCOCOAT doped with carbon or silver
molecules.

An example of in vivo blood PO_2 determination is shown in fig 4. which represents changes in APO_2 upon respiring the rabbit 100% 0 for 2 minutes. Recorded values corresponded with Gas Analyzer blood samples taken during these measurements.

REFERENCES

1. Bicher, H.I. and Knisely, M.H.: Brain tissue reoxygenation time, demonstrated with a new ultramicro oxygen electrode. J of Applied Phys 28: 387-390, 1970

2. Bicher, H.I. and Hetzel F: Effects of hyperthermia on normal and tumor microenvironment. Radiology 137: 523-530, 1980

3. Bicher, H.I. and Mitagvaria N P: Changes in tumor tissue oxygenation during microwave hyperthermia - clinical relevance, in Overgaard J (ed): Hyperthermic Oncology 1984 1: Taylor and Francis, London and Philadelphia, 1984, pp 169-172

4. Bicher, H.I. and Mitagvaria NP: Circulatory responses of malignant tumors during hyperthermia. Microvascular Res 21: 67: 491-501, 1976

5. Cater, D. The measurement of PO2 in tissue. In: Oxygen in Animal Tissue. New York: Macmillan, 1964,·p. 239-246.

6. Cater, D., and I. Silver. Electrodes and microelectrodes used in biology. In: Reference Electrode, edited by J. Janz. New York: Academic, 1961, Chapt. 11.

7. Corry P M, Robinson S, Getz B S: Hyperthermic effects on DNA repair mechanisms. Radiology 123 475-482, 1987

8. Davies, P., and F. Brink. Microelectrodes for measuring local oxygen tension in animal tissue. Rev. Sci. Instr. 13: 524- 533, 1942.

9. Dewhirst M W, Ozimek E J, Gross J, et al: Will hyperthermia conquer the elusive hypoxic cell? Implications of heat effects on tumor and normal tissue microcirculation. Radiology 137: 811-817, 1980

10. Emami B, Nussbaum G H, TenHaken R K, et al: Physiological effects of hyperthermia: response of capillary blood flow and structure to local tumor heating. Radiology 137: 805-809, 1980

11. Li G C, Evans R G, Hahn G M: Modification and inhibition of repair of otentially lethal x-ray damage by hyperthermia. Rad Res 67: 491-501, 1976

12. Mivechi N F, Hofer K E: Evidence for separate modes of action in thermal radiosensitization and direct thermal cell death. Cancer 51: 38-43, 1983

MICROELECTRODE STUDIES OF FACILITATED O_2 TRANSPORT ACROSS HEMOGLOBIN AND MYOGLOBIN LAYERS

Buerk, D.G., *Hoofd, L. and *Turek, Z

Biomedical Engineering and Science Institute, Drexel University
Philadelphia, PA USA
*Department of Physiology, School of Medicine, Catholic University
Nijmegen, The Netherlands

INTRODUCTION

Kreuzer and Hoofd (1987) recently reviewed the experimental evidence, theoretical framework and physiological significance for facilitated O_2 transport by hemoglobin in the red blood cell and by myoglobin in heart and red skeletal muscle. It is now well accepted that these vital biological proteins enhance tissue O_2 delivery by carrier-mediated transport. However, many mechanistic details are not fully understood or their importance *in vivo* have not been completely evaluated. Much of the previous research during the past 30 years has been conducted with flat layers of carrier protein solutions subjected to known O_2 concentration gradients in diffusion chambers. Facilitated O_2 transport theory predicts how PO_2 will vary with distance across the layer. In the present study, we measured PO_2 profiles with recessed cathode microelectrodes (Whalen *et al.*, 1967) across ca. 500 μm layers of aqueous solutions containing either hemoglobin or myoglobin. To our knowledge, there have been no previous attempts to evaluate facilitated O_2 transport theory by actually measuring PO_2 profiles in carrier protein solutions.

THEORY

Experimental PO_2 profiles were compared to profiles predicted from one-dimensional facilitated O_2 transport theory (see APPENDIX for notation). Parameters reported in the literature for carrier-mediated O_2 binding kinetics (association and dissociation rates), diffusion coefficients for dissolved and carrier-bound O_2, and other physical constants for both hemoglobin and myoglobin have been listed by Fletcher (1980) and Jacquez (1984). Myoglobin parameters are also listed by Gonzalez-Fernandez and Atta (1982). Nonequilibrium conditions are possible in some situations, depending on the Damköhler number γ, a ratio of diffusion and chemical reaction rates (Hoofd and Kreuzer, 1979, Kruezer and Hoofd, 1987). When γ is small, nonequilibrium conditions prevail and coupled partial differential equations for 3 chemical species (dissolved O_2, carrier-bound O_2 and deoxygenated carrier) must be solved by numerical methods (Kutchai *et al.*, 1970, Jacquez, 1984). Analytical approximations have also been described (Hoofd and Kreuzer, 1979). When γ is large, chemical equilibrium prevails. The present analysis will be restricted to equilibrium conditions, which is a reasonable assumption with the layer thicknesses and PO_2 gradients for these studies.

When O_2 is in equilibrium with the carrier, the derivative or slope of the O_2 saturation curve (dS/dP) plays an important role in facilitated transport. The O_2 flux is the sum of dissolved and carrier-bound O_2 fluxes

$$J = \{ \alpha D_1 + C_t D_2 \frac{dS}{dP} \} \frac{dP}{dx} \qquad [1]$$

The total flux across a layer is found by integrating eqn. [1], using suitable boundary conditions. The facilitation, or additional O_2 transport by the carrier relative to passive diffusion of dissolved O_2, is

$$F = (C_t D_2 / \alpha D_1) \frac{dS}{dP} \qquad [2]$$

The ratio $(C_t D_2 / \alpha D_1)$, a constant with partial pressure units, is also referred to as the facilitation pressure. F varies with PO_2 and F maximum occurs when the slope is greatest.

For the oxyhemoglobin equilibrium curve, the Easton (1979) double exponential function will be used. As modified by Buerk and Bridges (1986), saturation is a function of PO_2 and 3 parameters (K, P*, S_m), calculated from

$$S = (S_m - S_o) \, \exp^{- \exp \{ KP^* (1 - P / P^*) \}} + S_o \qquad [3]$$

The 2nd scaling factor S_o is related to the other 3 parameters, since S = 0 at P = 0 (Buerk, 1985). Buerk and Bridges (1986) fit eqn. [3] to human whole blood saturation data reported in the literature, and developed a relatively simple algorithm where changes in only 1 parameter (either K or P*) is required for computing O_2 affinity variations with temperature, pH, PCO_2 and 2,3-DPG. The product KP* remains constant. The slope, found by differentiating eqn. [3] with respect to PO_2, has a finite value at P = 0, reaches a maximum at P = P*, then decreases to 0 at high values of P. Therefore, F maximum for hemoglobin occurs at P = P*.

For the oxymyoglobin equilibrium curve, the hyperbolic function for a single step reaction

$$S = P / (P + P_{50}) \qquad [4]$$

will be used. The slope, found by differentiating eqn. [4], has a maximum value = $1/P_{50}$ at P = 0 and decreases to 0 at high values of P. Therefore, F maximum for myoglobin occurs at P = 0.

METHODS

Bovine hemoglobin solutions were prepared from fresh whole blood. Red blood cells were washed and separated, then lysed in distilled water. Salts were removed by deionization with a mixed bed ion exchanger. The hemoglobin solution was divided into 2 batches, with dry salt added to 1 batch for a final concentration of 100 mM KCl. Myoglobin solution was prepared from commercial equine heart myoglobin crystals (Sigma). Protein concentrations of hemoglobin (130 g/l) and myoglobin (30.9 g/l) were measured by a dual beam spectrophotometer (Perkin Elmer Model 124). Both protein solutions were refrigerated at 0 °C (melting ice). Experimental studies were completed within 1 week after preparing myoglobin solutions, and within 2 weeks for hemoglobin.

Discrete data points on bovine hemoglobin equilibrium curves were measured from a different batch of hemoglobin solution, using spectrophotometric methods (Spaan et al., 1980). Layers ca. 50 μm thick were supported on a 3.2 μm thick Teflon membrane in a diffusion chamber fitted with a quartz window. Saturation was determined from relative absorbance at 670 nm wavelength as gas concentration in the diffusion chamber was varied with a mixing pump, allowing sufficient time for equilibration. Data were obtained from 2 hemoglobin solutions and fit to eqn. [3]. Glycerol (200 mM) was added to the salt-free solution to minimize evaporation losses. The 2nd hemoglobin solution contained 100 mM KCl, with Tris-Mops buffer added to adjust the pH from about 7 (salt-free) to 7.4.

Recessed cathode PO_2 microelectrodes (Whalen et al., 1967) with tip diameters < 5 μm were used to measure PO_2 profiles across ca. 500 μm layers of carrier protein solutions placed into a 1 cm diameter well over a 25 μm thick gas permeable membrane (Celgard) in a diffusion chamber. Layers were exposed to a constant O_2 concentration gradient at ambient temperatures (21-22 °C). Gas flowing under the membrane contained 5.3% O_2 (balance N_2) for hemoglobin studies and 2.8% O_2 for myoglobin studies, with 0% O_2 (100% N_2) flowing over the top of the layer in both studies. Gases were humidified by bubbling through water before entering the diffusion chamber to minimize evaporation losses from the layer. Dithionite was added to the humidifier water for the N_2 gas stream to insure that all O_2 was removed. Butyl rubber and stainless steel tubing was used to direct gas flows into the diffusion chamber. Microelectrode measurements were made through a small bore hole in the top chamber. A AgCl reference anode was inserted through a 2nd bore hole and the microelectrode was polarized at 0.7 volt. The microelectrode was advanced with a hydraulic microdrive (Kopf Instruments) to the bottom of the layer, touching the membrane. The microelectrode was then withdrawn slowly (rate < 1 μm/sec) until the current abruptly decreased to the electrical zero of the picoammeter (Keithley Instruments Model 602) when the tip lost contact with solution. PO_2 profiles were analyzed from graphical records using a tablet digitizer. Profile locations were normalized with respect to the measured microdrive distance between 90% and 5% of the total PO_2 gradient, in 10% PO_2 increments above and 5% PO_2 increments below the 30% PO_2 change. Values near the membrane (> 90%) were not analyzed due to possible position errors caused by microelectrode bending and/or membrane stretching. Values near the liquid/gas interface (< 5%) were also eliminated due to possible position errors caused by a fluid meniscus remaining attached to the microelectrode as it was withdrawn from solution.

A computer simulation was developed to fit the experimental data. The O_2 flux was computed from the PO_2 and saturation at the endpoints of the profile, based on parameters for either hemoglobin (eqn. [3]) or myoglobin (eqn. [4]). A shooting technique was employed, calculating dP/dx from eqn. [1] for each increment in x. The PO_2 value at the start of each increment was used for the first calculation, then the average PO_2 in the interval was used to compute the slope for the carrier equilibrium curve and a new value for dP/dx. This step was repeated until an error criteria < 1×10^{-5} was reached, before moving to the next interval. The sum of squares was calculated for the errors between measured and simulated PO_2 values, varying the simulation parameters until the best fit was found.

RESULTS

The Easton (1979) function (eqn. [3]) provided excellent curve fits to the bovine saturation data for the parameters listed in Table I, as shown in Fig. 1. P_{50} values for the 2 equilibrium curves are 0.55 (salt-free) and 1.72 (100 mM KCl) kPa. Variations in the slopes of the 2 equilibrium curves with PO_2 are shown in Fig. 2. Maximum values for dS/dP are 135 (salt-free) and 42 (100 mM KCl) %/kPa as marked in Fig. 2 at the respective P* values (Table I).

TABLE I.   Parameters for least squares fit of oxyhemoglobin equilibrium curve (eqn. [3]) to bovine hemoglobin saturation data (Fig. 1).

	P* (kPa)	KP* (dimensionless)	S_m (%)
Salt-free, 200 mM glycerol	0.430	1.653	94.86
100 mM KCl, Tris-Mops	1.351	1.603	96.10

The mean PO_2 gradients (dP/dx) near the membrane and thicknesses (L) for the PO_2 gradient between 4.71 and 0.26 kPa measured across layers of the bovine hemoglobin solutions are summarized in Table II. Correcting for water vapor, the PO_2 at the membrane was 5.23 kPa. Gradients near the membrane were significantly different (p < 0.05) for the 2 hemoglobin solutions. The parameters in Table I were used for the initial computer simulation to curve fit the experimental PO_2 profiles. In simulations for both hemoglobin solutions, C_t = 0.181 ml O_2/ml solution and D_1 = 1.2×10^{-5} cm²/sec were used, based on the protein concentration (130 g/l). In salt-free hemoglobin solution, α = 0.031 ml O_2/ml solution/atm. In the 100 mM KCl

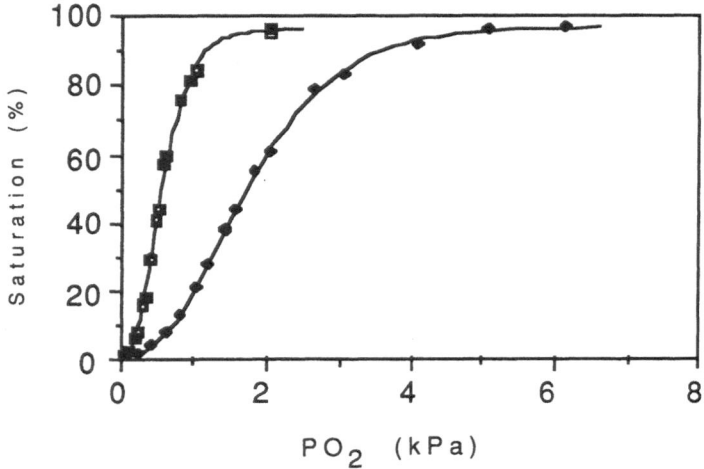

Fig. 1.   Oxyhemoglobin equilibrium curve fits (eqn. [3]) to bovine hemoglobin saturation data (Table I). Hemoglobin solutions are: salt-free with 200 mM glycerol (squares) and with 100 mM KCl and Tris-Mops buffer (diamonds).

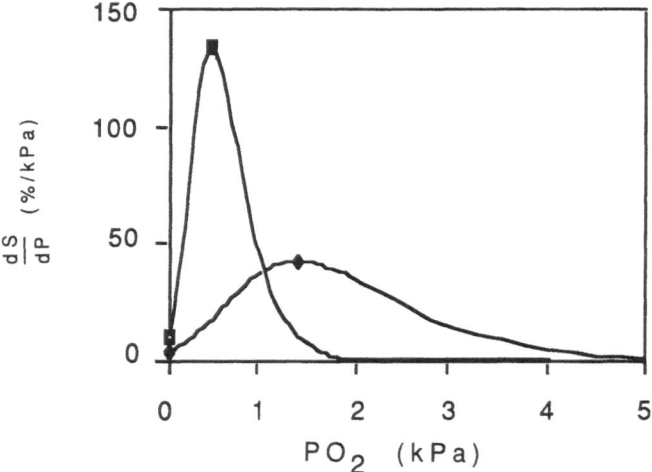

Fig. 2.   Slopes for bovine oxyhemoglobin equilibrium curves from Fig. 1. Maximum values for slopes at $PO_2 = P^*$ (Table I) are marked for salt-free (square) and 100 mM KCl (diamond) hemoglobin solutions.

128

TABLE II.   Summary of experimental PO_2 profile measurements across bovine hemoglobin layers, parameters for fit of equilibrium model computer simulation and calculated fluxes (Fig. 3).

	Measured		Simulation Parameters		Calculated
	$\frac{dP}{dx} \pm SE$	$L \pm SE$	P^*	maximum F	$J \times Area$
	(kPa/cm)	(μm)	(kPa)	(dimensionless)	(μl O_2/min)
Salt-free 23 profiles 5 layers	258 ± 15	427 ± 29	0.90	4.66	0.0450
100 mM KCl 18 profiles 6 layers	175 ± 14	436 ± 32	1.02	1.95	0.0297

and hemoglobin solution, α was assumed to be 3% lower due to the additional salt. The oxygenated carrier diffusion coefficient (D_2) was then varied to match dP/dx measured near the membrane. For salt-free hemoglobin solution, F maximum was 10.89 and D_2 was 73 times smaller than D_1. For 100 mM KCl and hemoglobin solution, F maximum was 1.61 and D_2 was 159 times smaller than D_1. However, simulated profiles using the parameters in Table I did not match the experimental data well, especially at low PO_2.

To investigate whether equilibrium curves with different O_2 affinities would improve the fit, P^* was varied, without changing KP^* and S_m in eqn. [3]. The parameters which provided the best fit to the upper half of the PO_2 profiles (> 2 kPa) are listed in Table II. F maximum was found to be considerably lower than expected for the salt-free hemoglobin solution, with P^* much higher than in Table I. For the 100 mM KCl and hemoglobin solution, P^* was found to be slightly lower and F maximum a little higher than computed with the parameters in Table I. The values for D_2 were slightly lower, 81 times smaller than D_1 for salt-free hemoglobin solution and 174 times smaller for the 100 mM KCl and hemoglobin solution. The O_2 flux through the layer was a little over 1.5 times greater for the salt-free hemoglobin solution. The experimental data, gradients near the membrane and simulated profiles for the 2 hemoglobin solutions with the parameters in Table II are shown in Fig. 3. Simulated profiles provide an excellent match to the experimental data for the higher PO_2 range, but are above the measured profiles at low PO_2. However, the computer simulations for the parameters in Table II did not provide the optimum least squares fits to the data. If forced to find the best fit over the entire PO_2 profile, the simulation deviated systematically, underestimating the data > 2 kPa and overestimating the data < 2 kPa.

The mean PO_2 gradient (dP/dx) near the membrane and layer thickness for the PO_2 gradient between 2.51 and 0.16 kPa measured in equine myoglobin solution are summarized in Table III. Correcting for water vapor, the PO_2 at the membrane was 2.81 kPa. For the computer simulation, $\alpha = 0.03$ ml O_2/ml solution at 1 atmosphere, $D_1 = 1.2 \times 10^{-5}$ cm^2/sec and $C_t = 0.0355$ ml O_2/ml solution were used. The value for C_t was determined experimentally by Scholander analysis. The simulation PO_2 profile based on the equilibrium model (eqn. [4]) provided an excellent fit to the experimental data, with the optimum parameters listed in Table III. For these parameters, the value for D_2 was 209 times smaller than D_1. The resulting curve fit and experimental data are shown in Fig. 4.

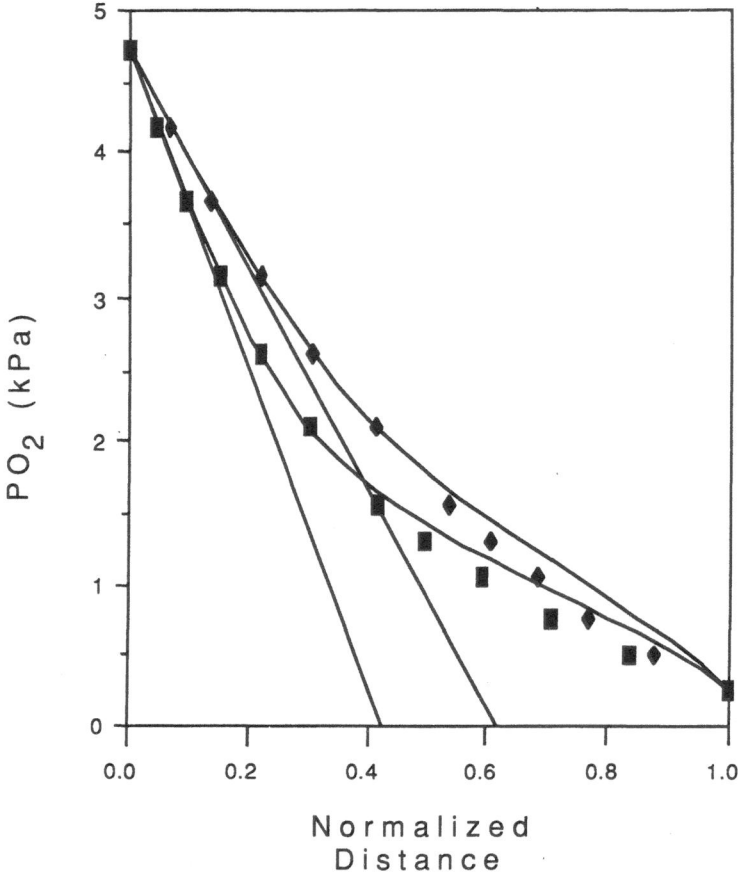

Fig. 3.   Mean values for PO_2 with normalized distances between 90% and 5% of profile measured with recessed cathode PO_2 microelectrodes across layers of salt-free (squares) and 100 mM KCl (diamonds) bovine hemoglobin solutions (130 g /l). PO_2 gradients (dP/dx) measured near membrane (straight lines) and computer simulations for equilibrium model (curved lines) are shown. Experimental data, model parameters and calculated O_2 fluxes across bovine hemoglobin layers are summarized in Table II.

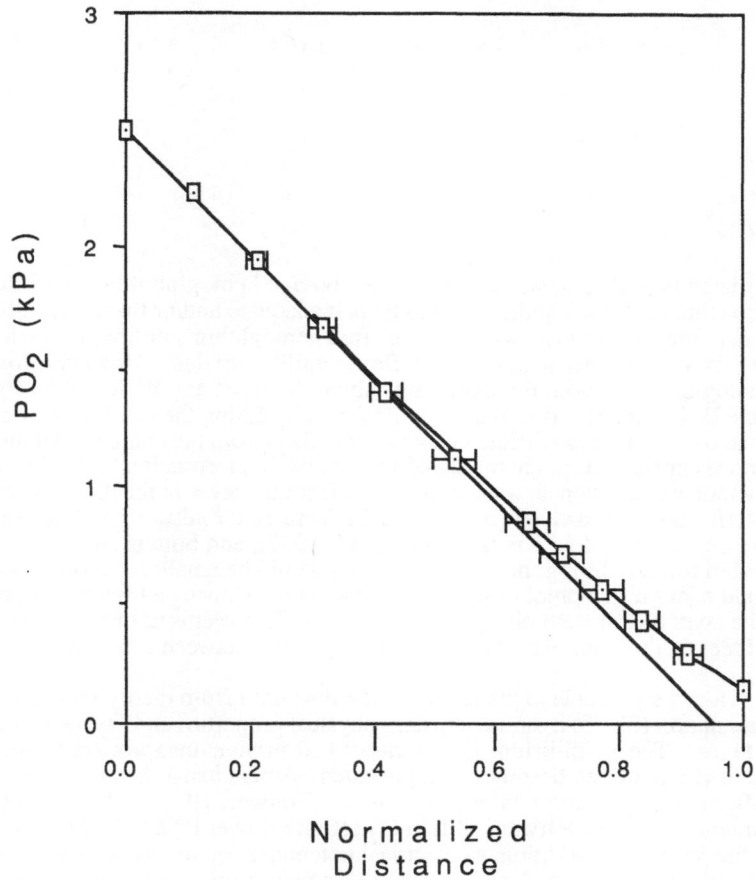

Fig. 4.   Mean values (open squares ± SD) for PO_2 with normalized distances between 90% and 5% of profile measured with recessed cathode PO_2 microelectrodes across layers of equine myoglobin solutions (30.9 g protein/l). PO_2 gradient (dP/dx) measured near membrane (straight line) and computer simulation for equilibrium model (curved line) are shown. Experimental data, model parameters and calculated O_2 flux across myoglobin layers are summarized in Table III.

TABLE III. Summary of experimental PO_2 profile measurements across equine myoglobin layers, parameters for fit of equilibrium model computer simulation and calculated flux (Fig. 4).

	Measured		Simulation Parameters		Calculated
	$\frac{dP}{dx} \pm SE$	$L \pm SE$	P_{50}	maximum F	J x Area
	(kPa/cm)	(μm)	(kPa)	(dimensionless)	(μl O_2/min)
33 profiles 6 layers	62.4 ± 3.6	424 ± 19	0.168	3.42	0.0174

DISCUSSION

From our PO_2 profile measurements across bovine hemoglobin layers (Table II, Fig. 3), one might conclude that the equilibrium model is inadequate and/or that nonequilibrium conditions influence transport at low PO_2. For salt-free hemoglobin solution, F maximum is much lower and P* is much higher than expected from equilibrium data. However, for the 100 mM KCl and hemoglobin solution, F maximum is about the same and P* found by fitting the PO_2 profile (Table II) is not much different from P* found by fitting the equilibrium curve (Table I). Values for the oxyhemoglobin diffusion coefficients (D_2) from the computer simulation fit were smaller than we expected from the measured hemoglobin concentration (130 g/l). This might be caused by hemoconcentration as water evaporated from the layer in the diffusion chamber. The diffusion coefficient for dissolved O_2 (D_1) would decrease (Goldstick and Fatt, 1970), although less than D_2 (eg. see Fig. 5 in Kreuzer and Hoofd, 1987), and both α and C_t will increase. If methemoglobin formed during the experiments, C_t would be smaller. Another possible source of error would occur if hemoglobin settled out of solution, causing a higher concentration at the bottom of the layer near the membrane than at the top. This seems unlikely, since no noticeable settling was seen in the solutions stored under refrigeration between experiments.

There is a more probable explanation for the deviation from theory seen at low PO_2. Our computer simulation (Fig. 3) assumes constant physical properties and O_2 affinity (constant P*) across the layer. The equilibrium theory described earlier does not take into account pH gradients as oxyhemoglobin becomes deoxygenated. A pH change from 7.00 to 7.06 has been measured after deoxygenating bovine hemoglobin (Bouwer, 1987). Consequently, there is a small, continuously varying leftward shift in O_2 affinity (lower P* and higher dS/dP) at the low PO_2 end of the profile. In addition, an electrical potential is set up due to movement of charged species (H^+, OH^- and Cl^-). A 4 mV potential was measured in salt-free bovine hemoglobin solution, which became negligible with KCl > 100 mM (Breepoel et al., 1982). Hoofd et al. (1984) modeled this effect, but found that the facilitation pressure measured at low KCl concentrations was even smaller than predicted from theory. Electrical currents with the movement of charged species may influence D_2, so that facilitation is reduced. Therefore, the assumption that D_2 is constant across the layer may not be valid. This effect should be nearly absent in the 100 mM KCl and hemoglobin solution. Further modeling is under development to test whether the pH gradient and resulting electrical potential explains our experimental results. Until a more complete analysis is done, we are not able to determine whether nonequilibrium conditions were influencing our profile measurements at low PO_2.

Buerk and Goldstick (1982) have measured PO_2 gradients > 1 kPa in blood flowing near walls of large blood vessels in vivo. Similar PO_2 gradients may exist in the microcirculation. Hellums (1977) proposed that there may be a large transport resistance within the red blood cell. Nonequilibrium kinetics may become important in vivo under some conditions, particularly for organs with very high O_2 metabolism and blood flow rates. Impaired O_2 uptake and release has been simulated by Mochizuki and Sagawa (1986), taking into account O_2 association and dissociation kinetics, as well as red blood cell pH and PCO_2. Gutierrez (1986) has simulated the effect of oxyhemoglobin kinetics on O_2 transport to tissue, demonstrating significant differences between PO_2 in the red blood cell and in plasma between cells. Factors that reduce facilitated transport by hemoglobin will result in lower tissue PO_2 values. The existence of a pH gradient and electrical potential within the red blood cell has not been considered in previous simulations. This mechanism may cause a further reduction in facilitated transport under extreme conditions when the outer boundary of the red blood cell is deoxygenated. However, internal mixing as the red blood cell tumbles through the microcirculation may have an opposite effect, enhancing O_2 transport. Further experimental and theoretical work is required before the effects of these and other mechanisms on facilitated O_2 transport by hemoglobin *in vivo* can be evaluated.

Our PO_2 profiles measured across equine myoglobin layers are in excellent agreement with equilibrium theory. Values for F maximum and P_{50} (Table III) are within ranges reported by others. However, as we found for hemoglobin, the oxymyoglobin diffusion coefficient (D_2) was smaller than we anticipated. Perhaps water evaporated from the layer, causing a higher protein concentration and changes in physical properties (α, D_1, C_t), as discussed previously. Ions, pH and PCO_2 do not alter the O_2 affinity of the oxymyoglobin equilibrium curve. Since P_{50} is so small at this temperature (21-22 $^{\circ}C$) and F is not maximum until PO_2 reaches 0, facilitated transport by myoglobin is not noticeable in Fig. 4 until $PO_2 < 1.5$ kPa. Facilitated transport would most likely begin to be significant at a higher tissue PO_2 *in vivo*, since P_{50} for myoglobin increases with temperature.

Gayeski *et al.* (1987) used an experimentally determined $P_{50} = 0.71$ kPa to interpret their cryospectrophotometric measurements of oxymyoglobin saturation in quick-frozen, maximally working dog gracilis muscles. The importance of facilitated transport by myoglobin is readily apparent in this red muscle, which has a maximum metabolic rate around 15 ml O_2/100 g/min at 37-39 $^{\circ}C$. In 7 of 10 muscles, oxymyoglobin was < 77% saturated. In all 10 muscles, the lowest 5% of observations were saturated < 43%. This implies (from eqn. [4]) that nearly all tissue PO_2 values were < 2.4 kPa, with the lowest 5% < 0.5 kPa. Since cellular bioenergetics (ATP and ratio of phosphocreatinine to free creatinine) were unaffected, they concluded that metabolism does not become limited by O_2 availability until tissue PO_2 falls below 0.07 kPa. They also estimated an upper bound for the Michaelis-Menten constant (K_m) around 0.001 kPa, assuming that there is not a large PO_2 gradient between myoglobin and mitochondria. This value is similar to K_m values from studies with isolated mitochondria reported in the literature. It appears that without facilitated transport by myoglobin, working red muscle would not be able to sustain high metabolic rates at such low tissue PO_2 levels.

There may also be a role for facilitated transport by myoglobin in resting muscle as well. Whalen *et al.* (1973a, 1974, 1976) have measured tissue PO_2 with recessed cathode micro-electrodes in resting dog and cat gracilis muscles. Resting metabolic rates measured in isolated, pump-perfused dog gracilis muscles (Whalen *et al.*, 1976) were less than 1/60 of the maximum rate for working muscles (Gayeski *et al.*, 1987). Yet the mean tissue PO_2 was relatively low, 2.8 kPa at a blood flow rate of 7.1 ml/100 g/min, with 27% of tissue PO_2 values < 1.33 kPa and 14% < 0.67 kPa (around P_{50}). After flow was reduced by 50%, the tissue PO_2 distribution shifted to the left, with 45% of tissue $PO_2 < 1.33$ kPa, and 22% $< P_{50}$. Interestingly, the number of lowest PO_2 values (< 0.3 kPa) actually decreased after reducing blood flow, due to autoregulatory adjustments in the microcirculation. Whalen *et al.* (1974) also reported lower tissue PO_2 values in innervated, resting cat gracilis muscle (mean 2.23 kPa, 44% < 1.33 kPa) compared to values after denervation (mean 3.19 kPa, 24% < 1.33 kPa). Myoglobin in red muscle may permit lower blood flow at rest than would be possible for an organ with similar metabolism but no carrier protein.

The heart is clearly another organ where facilitated transport by myoglobin is necessary to insure that O_2 is available to meet high energy demands. Myocardial tissue PO_2 has been measured *in vivo* with microelectrodes in the beating cat heart (Whalen, 1971, Whalen *et al.*, 1973b). Near the surface (depth < 1 mm), mean PO_2 was 1.28 kPa with 65% of values < 1.33 kPa. At deeper locations (> 1 mm), the mean was 0.61 kPa, lower than P_{50} for myoglobin, with 89% of values < 1.33 kPa. Mean values were obtained by electronically damping the PO_2 signal, which fluctuated with the heart beat, reaching or approaching 0 during systole. In this situation, the oxymyoglobin kinetics may buffer PO_2 oscillations as blood flow and metabolism vary during the cardiac cycle. The amount of O_2 facilitation in tissue is difficult to quantify, due to the concentration dependence of O_2 metabolism. This fact made measurements *in vitro* from slices of chicken gizzard difficult to interpret (de Koning *et al.*, 1981). As facilitated transport increases tissue PO_2, metabolism will increase, depending on K_m in the mitochondria, thereby reducing the additional amount of tissue that can be oxygenated. Another situation, where facilitated transport is diminished by the presence of membranes that prevent diffusion of oxymyoglobin, has been simulated (Gonzalez-Fernandez and Atta, 1982). While there is strong evidence for facilitated O_2 transport by myoglobin *in vivo*, further experimental and theoretical work is also required to evaluate its role more completely.

SUMMARY

Experimentally measured PO_2 profiles across layers of hemoglobin and myoglobin solutions were compared with profiles predicted from facilitated transport theory assuming chemical equilibrium. Measurements across myoglobin layers were in excellent agreement with theory, but measurements across hemoglobin layers departed from theory at low PO_2. This departure was greatest for salt-free hemoglobin solution, which may be caused by an electrical potential formed by a pH gradient in the layer as oxyhemoglobin is deoxygenated.

APPENDIX - TABLE OF NOTATION

Symbol	Definition	(units)
α	Solubility coefficient for dissolved O_2	(ml O_2/ml/atm)
C_t	Maximum O_2 bound by carrier protein	(ml O_2/ml)
D_1	Diffusion coefficient for dissolved O_2	(cm^2/sec)
D_2	Diffusion coefficient for bound O_2	(cm^2/sec)
$\frac{dS}{dP}$	Slope of carrier saturation curve	(%/kPa)
F	Facilitation	(dimensionless)
γ	Damköhler number	(dimensionless)
J	Flux of O_2 across layer	(μl O_2/min/cm^2)
KP*	Parameter group for oxyhemoglobin equilibrium curve algorithm	(dimensionless)
L	Thickness of layer	(μm)
P	Partial pressure of O_2	(kPa)
P*	PO_2 where slope of oxyhemoglobin equilibrium curve is maximum	(kPa)
P_{50}	PO_2 where carrier equilibrium curve is 50 % saturated	(kPa)
S	Saturation of carrier protein	(%)
S_m, S_o	Scaling parameters for oxyhemoglobin equilibrium curve algorithm	(%)

ACKNOWLEDGEMENTS

This research project was supported in part by a visiting scientist award to Dr. Buerk from the Dutch government funding agency (Z.W.O.).

REFERENCES

Breepoel, P.M., de Koning, J., Hoofd, L., 1982; Diffusion of oxygen in methemoglobin solutions: dependence on salt concentrations, Biochem. Biophys. Res. Commun., 109:848-850.

Bouwer, S., 1987; Facilitated oxygen diffusion through hemoglobin solutions. Measurement of diffusion and reaction parameters, Ph.D. Thesis, Department of Physiology, Catholic University, Nijmegen, The Netherlands.

Buerk, D.G., Goldstick, T.K., 1982; Arterial wall oxygen consumption rate varies spatially, Am. J. Physiol., 243:H948- H958.

Buerk, D.G., 1985; An evaluation of Easton's paradigm for the oxyhemoglobin dissociation curve, in: "Oxygen Transport to Tissue - VI," Plenum Press, N.Y., Adv. Exp. Med. Biol., 180:333-344.

Buerk, D.G., Bridges, E.W., 1986; A simplified algorithm for computing the variation in oxy-hemoglobin saturation with pH, PCO_2, T and DPG, Chem. Eng. Commun., 47:113-124.

de Koning, J., Hoofd, L.J.C, Kreuzer, F., 1981; Oxygen transport and the function of myoglobin. Theoretical model and experiments in chicken gizzard smooth muscle, Pflügers Arch, 389:211-217.

Easton, D.M., 1979; Oxyhemoglobin dissociation curve as expo-exponential paradigm of asymmetric sigmoid function, J. Theor. Biol., 76:335-349.

Fletcher, J.E., 1980; On facilitated oxygen diffusion in muscle tissues, Biophys. J., 29:437-458.

Gayeski, T.E., Connett, R.J., Honig, C.R., 1987; Minimum intracellular PO_2 for maximum cytochrome turnover in red muscle in situ, Am. J. Physiol., 252:H906-915.

Goldstick, T.K., Fatt, I., 1970; Diffusion of oxygen in solutions of blood proteins, Chem. Eng. Progr. Symp. Ser., 66:101-113.

Gonzalez-Fernandez, J.M., Atta, S.E., 1982; Facilitated transport of oxygen in the presence of membranes in the diffusion path, Biophys. J., 38:133-141.

Gutierrez, G., 1986; The rate of oxygen release and its effect on capillary O_2 tension: a mathematical analysis, Resp. Physiol., 63:79-96.

Hellums, J.D., 1977; The resistance to oxygen transport in the capillaries relative to that in the surrounding tissue, Microvas. Res., 13:131-136.

Hoofd, L., Kreuzer, F., 1979; A new mathematical approach for solving carrier-facilitated diffusion problems, J. Math. Biol., 8:1-13.

Hoofd, L., Breepoel, P., Kreuzer, F., 1984; Facilitated diffusion and electrical potentials in protein solutions with ionic species, in: "Oxygen Transport to Tissue - V," Plenum Press, N.Y., Adv. Exp. Med. Biol., 169:133-143.

Jacquez, J.A., 1984; The physiological role of myoglobin: More than a problem in reaction-diffusion kinetics, Math. Biosci., 68:57-97.

Kreuzer, F., Hoofd, L., 1987; Chapter 6, Facilitated diffusion of oxygen and carbon dioxide, in: "Handbook of Physiology, The Respiratory System, Vol. IV," American Physiological Society, Bethesda, MD., pp. 89-111.

Kutchai, H., Jacquez, J.A., Mather, F.J., 1970; Nonequilibrium facilitated transport in hemoglobin solutions, Biophys. J., 10:38-54.

Mochizuki, M., Kagawa, T., 1986; Numerical solution of partial differential equations describing the simultaneous O_2 and CO_2 diffusions in the red blood cell, Jap. J. Physiol., 36:43-63.

Spaan, J.A.E., Kreuzer, F., van Wely, F.K., 1980; Diffusion coefficients of oxygen and hemoglobin as obtained simultaneously from photometric determination of the oxygenation of layers of hemoglobin solutions, Pflügers Arch, 384:241-251.

Whalen, W.J., Riley, J., Nair, P., 1967; A microelectrode for measuring intracellular PO_2, J. Appl. Physiol., 23:798-801.

Whalen, W.J., 1971; Intracellular PO_2 in heart and skeletal muscle, The Physiologist, 14:69-82.

Whalen, W.J., Buerk, D.G., Thuning, C.A., 1973a; Blood flow-limiting oxygen consumption in resting cat skeletal muscle, Am. J. Physiol., 224:763-768.

Whalen, W.J., Nair, P., Buerk, D., 1973b; Oxygen tension in the beating cat heart in situ, in: "Oxygen Supply - Theoretical and Practical Aspects of Oxygen Supply and Microcirculation of Tissue," Urban and Schwarzenberg, West Germany, pp. 199-201.

Whalen, W.J., Nair, P., Buerk, D., Thuning, C.A., 1974; Tissue PO_2 in normal and denervated cat skeletal muscle, Am. J. Physiol., 227:1221-1225.

Whalen, W.J., Buerk, D.G., Thuning, C.A., Kanoy, Jr., B.D., Duran, W.N., 1976; Tissue PO_2, VO_2, blood flow and perfusion pressure in resting dog gracilis muscle at constant flow, in: "Oxygen Transport to Tissue - II," Plenum Press, N.Y., Adv. Exp. Med. Biol., 75:639-655.

POTENTIOMETRIC POLAROGRAPHIC PO$_2$ ELECTRODE (PO$_2$-PPE)

M. Kessler and J. Höper

Institut für Physiologie und Kardiologie der Universität
Erlangen-Nürnberg, Waldstraße 6, D-8520 Erlangen, FRG

INTRODUCTION

Polarographic electrodes in general have the disadvantage of being "poisoned" by many chemical substances which are dissolved even in only minimal amounts in the electrolyte solution making the contact between the cathode and the anode. This electrode-poisoning causes substantial drift and therefore limits the accuracy of long-term measurements.

New potentiometric-polarographic electrodes (PPE's) developed during recent years (Kessler et al. 1985, Kessler et al. 1986) can overcome this problem. These PPE's are designed and constructed in such a way that the noble metal electrodes are sealed by PVC membranes protecting the metal surface and avoiding electrode poisoning. These sensors can be used for measurements of H$_2$, H$_2$O$_2$, glucose and O$_2$.

Principle

The PVC membrane protecting the noble metal interface contains highly selective, mobile lipophilic ion carrier ligands (n-dodecylamine, Schulthess et al. 1981). These ion selective ligands allow potentiometric measurements of proton gradients.

Figure 1 shows schematically the principle of the oxygen electrode. The oxygen diffusing through the PVC membrane and the hydrated interspace between the membrane and the Pt-interface is reduced at the metal surface, forming oxygen anions. These immediately react with hydrogen ions present in the interspace thus inducing the formation of a H$^+$ gradient. The decline in the proton activity induces a H$^+$ flux across the PVC membrane, the protons are selectively carried by the ion selective ligand forming a further H$^+$ gradient across the PVC membrane. These H$^+$ gradients will cause a potential difference which can be measured.

Long term stabilities can be achieved, which were hitherto impossible, due to the ideal protection of the noble metal interface against any kind of pollution and side reactions occuring in conventional polarographic electrodes.

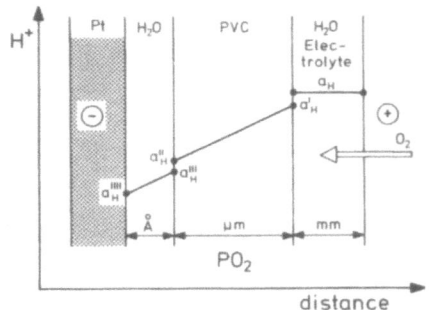

Fig. 1 Schematic drawing of the electrode function. The oxygen molecules diffusing from the outer sample to the noble metal interface with its hydrated interspace are reduced in the presence of protons.

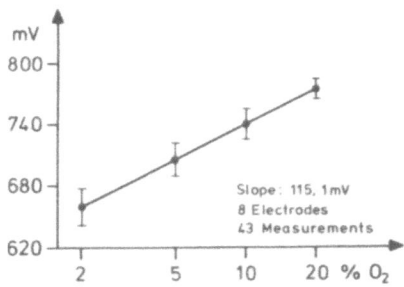

Fig. 2 Mean calibration curve of 8 PO$_2$-PPE`s (mean$\pm$SD).

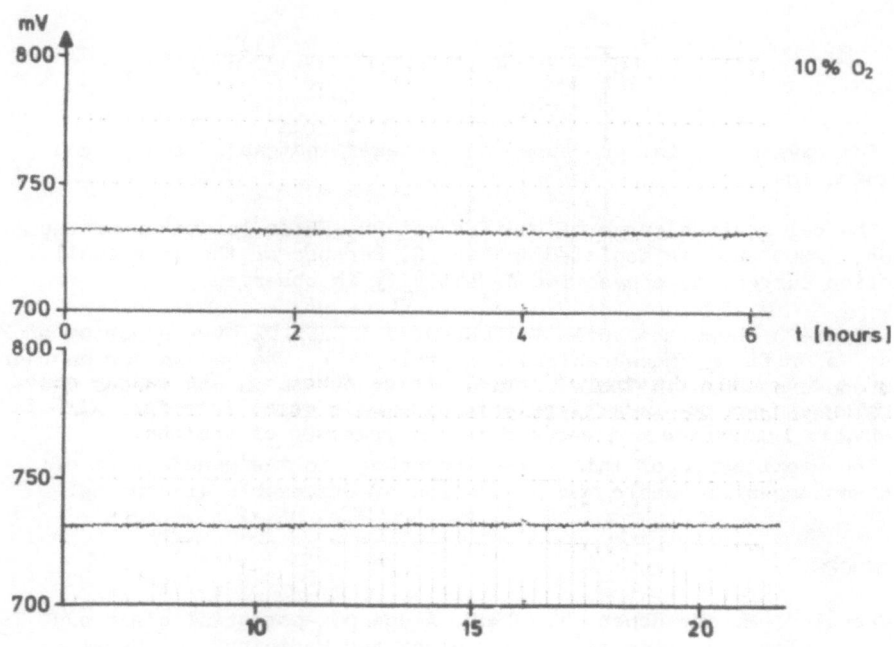

Fig. 3 Continuous measurement in 10% O_2 over a period of 20 hours.

Table 1 Drift characteristics of a single PO_2-PPE.

Time (Days)	0	31
mV	410	410
pO2 (mmHg)	150	150
Drift ($\triangle$ pO2)	< 0.1%	

RESULTS

The proton gradients described in Fig. 1 can be measured potentio-metrically. The relationship between the oxygen concentration and the EMF is given by the equation (1):

$$E = E_o + K * \frac{RT}{F} \log (O_2) \qquad (1)$$

The oxygen partial pressure can be taken instead of the oxygen concentration.

The calibration curve of 8 different PO_2-PPE's (diameter of the cathode: 2000 μm) is depicted in Fig. 2. Because of the very small reduction current no convection sensitivity is observed.

Figure 3 shows the potential measured in 10% O_2 over a period of 20 hours. No drift was measurable during this time. The values for another electrode are shown in Table 1. This electrode was in continuous use over 31 days and showed a drift of less than 0.1%.

The application of thin layer technology to the manufacture of these sensors will enable the production of disposable electrodes.

REFERENCES

Kessler, M. and Höper, J., 1985, A new pO_2-potential electrode. In: "Ion Measurements in Physiology and Medicine", M. Kessler, D.K. Harrison and J. Höper, eds.,Berlin:Springer, 74.

Kessler, M., Harrison, D.K. and Höper, J., 1986, Tissue oxygen measurement techniques. In: "Microcirculatory Technology", C.H. Baker and W.L. Nastuk, eds., New York:Academic Press.

Schulthess, P., Shijo, Y., Pham, H.U., Pretsch, E., Amman, D. and Simon, W., 1981, A hydrogen liquid membrane ISE based on tri-dodecylamine as central carrier. Anal. Chim. Acta, 131:111.

THE HYDROGEN GAS CLEARANCE METHOD FOR LIVER BLOOD FLOW EXAMINATION:

INHALATION OR LOCAL APPLICATION OF HYDROGEN?

Hermann P. Metzger

Arbeitsbereich Biosignalverarbeitung und Kybernetik
Medizinische Hochschule Hannover
3000 Hannover, Postbox 610180, FRG

INTRODUCTION

The liver, the largest gland of the body, with its many important functions such as bile production, carbohydrate storage, ketone body formation, etc., undergoes immediate changes upon lowering of oxygen supply and hepatic blood flow (HBF). So far, reliable data characterizing the hepatic macro- and microcirculation are difficult to obtain because of methodological difficulties. Large HBF differences have been described in the literature which need further clarification especially with regard to ischemia and hemorrhage. Of course, hydrogen gas clearance in response to hydrogen inhalation has been well-established as a method for organ blood flow determination, but in the case of liver blood flow it has certain limitation: during the washout period hydrogen molecules may flow from the gastro-intestinal tract into the portal vein. Furthermore, mean tissue and venous concentration might not always be in equilibrium during the washout process as postulated by the theory.

The intention of this paper is to validate HBF measurements by hydrogen clearance and thereby elucidate the biology.For realising this, a combination method of hydrogen inhalation and local hydrogen production enabled us to study the hepatic blood flow and its temporal as well as spatial heterogeneity within a circumscribed small tissue volume of o.5 mm in diameter located immediately below the liver surface. We studied both, hemorrhagic and control animals.

METHODS AND MATERIALS

1. Electrode production. A multi-wire electrode was developed which consisted of hydrogen producing and hydrogen sensing elements arranged close to each other. The multi-wire electrode was completed by the appropriate reference electrodes necessary for hydrogen production and measurement. Furthermore, a thermistor (K 19, Siemens Corp., München, FRG) was inserted into the electrode body in order to continuously monitor the temperature of the liver surface under test.

Measuring wires of 1oo um in diameter were arranged at 15o um distance from each other. The hydrogen producing platinum wire (2oo um in diameter) was located centrally within less than 15o um distance from the middle of the whole arrangement (Fig. 1). The different wires were soldered on a printed plate which had been prepared as carrier for the different elements. The whole electrode body, 5 mm in diameter and 5 mm in height, was embedded into Hysol (Dexter Corp., New York), a two-component glue, well known from heart pacemaker technique. The weight of the multi-wire surface electrode, together with its connecting wires, was 1-2 g (Fig. 2).

As reference for the hydrogen producing wire, an Ag/AgCl ring was placed around the electrode array. Alternatively, an external calomel electrode served as reference and was found to produce very stable hydrogen pulses.

For hydrogen clearance measurements it was necessary to use a second external calomel electrode. Preliminary tests using the Ag/AgCl reference electrode failed, probably due to the production of chloride anions in the solution. The multi-wire electrode was calibrated in isotonic NaCl-solution equilibrated with gas mixtures of 2% and of 5% hydrogen in nitrogen respectively. The longterm current drift of 5% per hour was not as good as that of the polarographic oxygen electrodes which showed excellent stability of less than 1%/h with gold as the electrode material.

2. Electronic circuit for local hydrogen production and measurement. Both molecules, hydrogen and oxygen, were determined according to the polarographic principle. Hydrogen oxidation or oxygen reduction occurs at the catalytic surface of any noble metal when the surface is at the correct electrical potential for the molecule to be determined by the method. The same electronic equipment originally designed for oxygen measurement was used for hydrogen concentration analysis as well simply by switching the polarisation voltage of the metal electrode from $-8oo$ mV (O_2) to $+15o$ mV (H_2). The influence of any external DC-potentials on the polarisation voltage was eliminated by use of a voltage clamp technique which kept the potential at its prescribed level. The electrons produced by oxygen reduction or hydrogen oxidation were measured with a high input impedance amplifier where $1o^{-12}$ A corresponded to $1o$ mV output voltage.

3. Gas mixing device. Variable mixtures of gas concentrations ($H_2:O_2:N_2$) were supplied by a specially designed gas mixing system. The hydrogen and oxygen contents of the inspired gas mixture was controlled by use of rotameters connected to the laboratory gas supply and completed with an additional gas tank. The different gas streams were conducted through special valves and adjusted according to the rotameter scale and prescribed volume relationship.

4. Experimental protocol. White rats (Wistar-Frömter strain, kept pathogen-free, 18o-23o g bw), N=19 animals, were anesthetized by means of a combination of ketamin-xylazine as described recently (Metzger and Savas, 1988). After cannulation of the carotid artery for MAP registration, the animals were tracheotomized and ventilated artificially (resp. freque. = 7o/min, tidal vol. = 1.5 - 2.2 ml depending upon the body weight). Following laparatomy the different ligaments were carefully dissected. Then the measuring electrode was placed on the surface of lobus sinister. The capsule was left intact. The rest of the liver surface was covered with seran. Before the chest was closed, the liver was rinsed with warmed ringer-lactate in order to keep the surface as moist and homeothermic as possible. Surface temperature was continuously monitored by means of a thermistor inserted into the electrode tip.

Inhalatory hydrogen was applied for 5-8 minutes until a steady-state concentration was achieved while local hydrogen production lasted about 2 minutes before a local plateau was reached. Washout times of 2 and 5 minutes respectively were necessary for an adequate clearance registration.

5. Induction of hemorrhage. Hemorrhagic ischemia was induced by slow withdrawal of blood until 4o mm Hg MAP was obtained. Again, inhalatory and electrochemical hydrogen were applied and the clearance was registered. At the end of the experiment, post mortem, local hydrogen production was continued in order to analyse the spreading of the diffusive front into the tissue. The effect of pure diffusion was evaluated when convection of blood within the sinusoids was zero at death.

Multi-wire surface electrode: scheme and printed plate

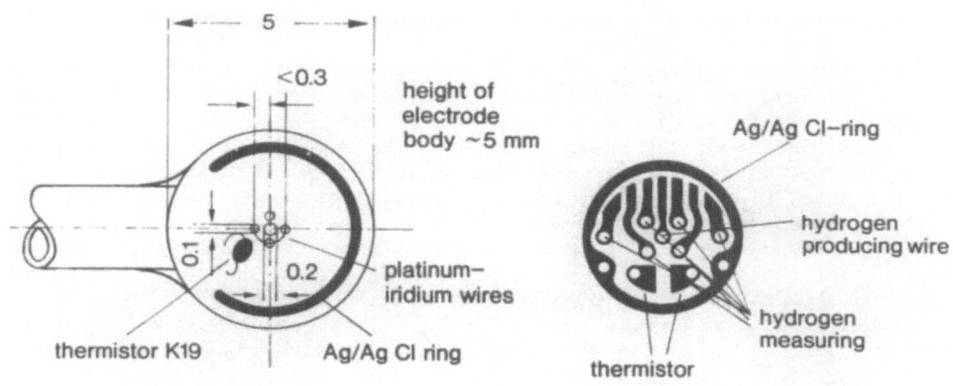

scaling factor 10:1

Fig. 1

6. Calculation of HBF by use of Aukland's equation. Inhalatory and local hydrogen clearance curves were evaluated by use of the equation originally described by Aukland et al. (1964): $HBF = 0.693/t_{1/2}$, with $t_{1/2}$ = time to reach 5o% of the initial hydrogen concentration. The steepest slope of the curve about 1o seconds after returning to air breathing was used for HBF calculation in order to minimize errors due to hydrogen inflow from the gastro-intestinal tract as well as hydrogen recirculation according to the idea of Kakimoto (1982).

As a second step, the pure diffusion process was considered and the time-constant of the diffusive spreading of the hydrogen front into the tissue was subtracted from the combined diffusion-convection process similiar to the procedure derived for heat clearance analysis.

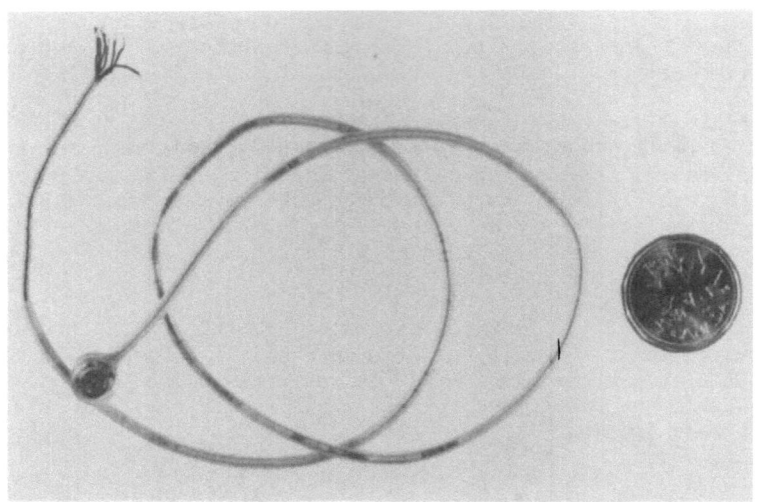

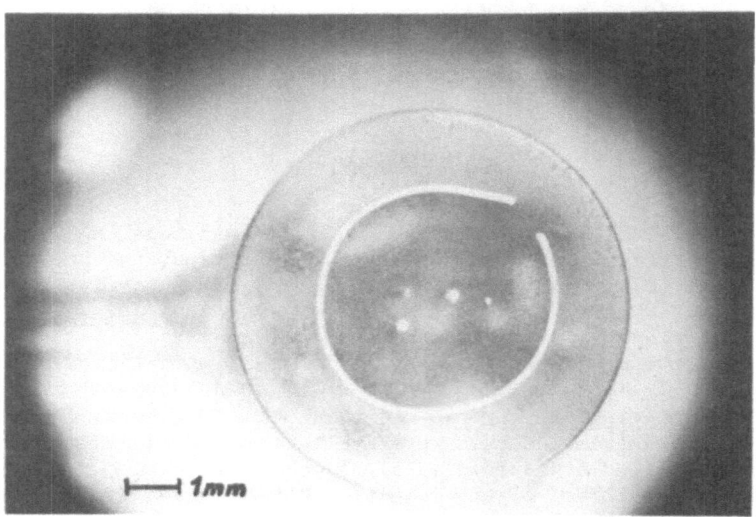

Fig. 2. Multi-wire surface electrode designed for measurement of HBF and LHBF by use of inhalatory and local hydrogen application.

RESULTS

Hepatic blood flow of ketamin-xylazine anesthetized and artificially
ventilated rats was determined from hydrogen washout curves following a
period of about 5 minutes of hydrogen inhalation. The evaluation of the
clearance curves was performed according to the principle described in the
literature (Aukland et al., 1964; Kakimoto, 1982; Gouma et al., 1986). The
resulting HBF data showed a Gaussian-type frequency distribution (MV+SD =
o.5o±o.26 ml/g·min; n = 48, N = 19 animals). Each HBF value was calculated
as an averaged mean value of four hydrogen clearance curves registered at
the same time within a small tissue volume of o.5 mm in diameter.

Surprisingly, a ten-fold higher local hepatic blood flow (LHBF) was
determined from the local clearance curves in response to local hydrogen
production. Again, the same theoretical approach was applied which had
already been used for the analysis of hydrogen clearance curves in response
to hydrogen inhalation. A wide scattering of LHBF values was obtained which,
again, showed a Gaussian-type frequency distribution (MV±SD = 4.66±2.13 ml/
g·min; n = 43, N = 19 animals).

The validity of the combined method of HBF and LHBF determination was
further tested for the conditions of hemorrhagic ischemia. Both parameters,
HBF and LHBF, showed the same trend. In response to blood withdrawal down
to 4o mm Hg MAP the washin and washout periods were enlarged, the slopes
of the decrease and increase were flatter than that for the control animals
(Figs. 3 and 4).

The whole set of HBF and LHBF values including those registered under
normovolemia and hemorrhage were plotted on the same graph (Fig. 5). The
slope m = o.1 of the straight line and the intersection with the y-axis
(y_o = o.oo1) were calculated according to the method of the least squares.
The results demonstrate a reasonable correlation of HBF and LHBF data despite the large scattering of the individual values; the correlation was
characterized by a correlation coefficient of r = o.685.

DISCUSSION

The combined method of hydrogen inhalation and local application has
been used for HBF and LHBF determination. In addition, the frequency distribution of both parameters was evaluated in order to study the flow
heterogeneity statistically. Both parameters were analyzed within a very
small circumscribed tissue volume of about o.5 mm in diameter so that a
direct comparison seems reasonable. Surprisingly, both values were found
to be numerically different by a factor of about 1o despite the fact that
the same test molecules and measuring electrodes were applied and the same
tissue volume considered.

Different diffusion-convection processes might occur if hydrogen is
applied by inhalation or by local production. In response to inhalation, an
equilibrium hydrogen concentration without hydrogen concentration gradients would expected within the whole liver. By contrast, in response to
local hydrogen production a hydrogen concentration profile should exist
with its maximum at the producing electrode tip and the border of the zone
of zero hydrogen concentration within a distance of about 5oo um. As the
hydrogen producing current and its driving electrical potential are not
allowed to exceed 1 uA, corresponding to 1o volt, an enlargement of the
hydrogen production is not possible. Nevertheless, a second hydrogen producing wire might be inserted into the electrode array within 1 mm distance.
Alternatively, the electrode can be moved in random walk across the liver

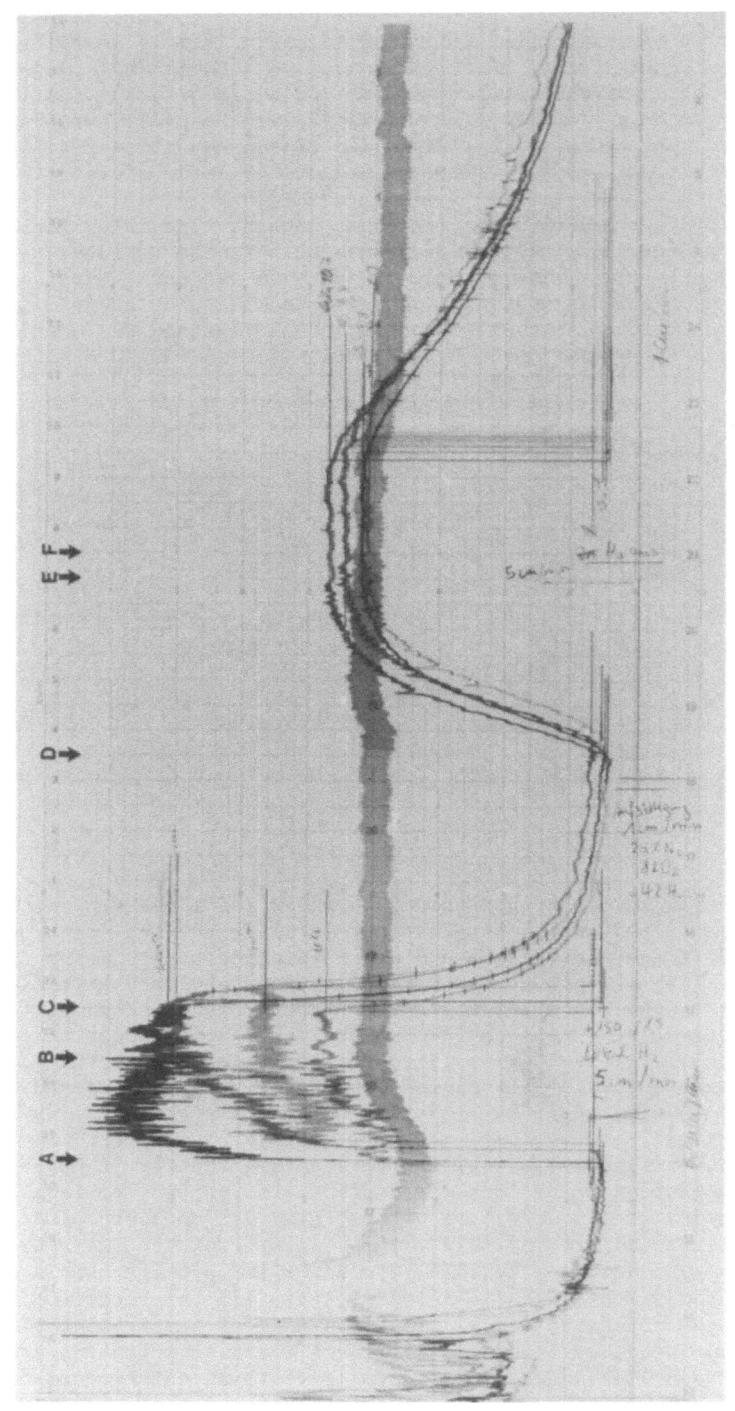

Fig. 3. Original registration of hydrogen clearance curves in response to local hydrogen production (A) and inhalation of hydrogen (D). A: start production, B: time scale 5 cm/min, C: stop production, D: time scale 1 cm/min, E: time scale 5 cm/min, F: stop inhalation. Mean arterial pressure = 1oo mm Hg.

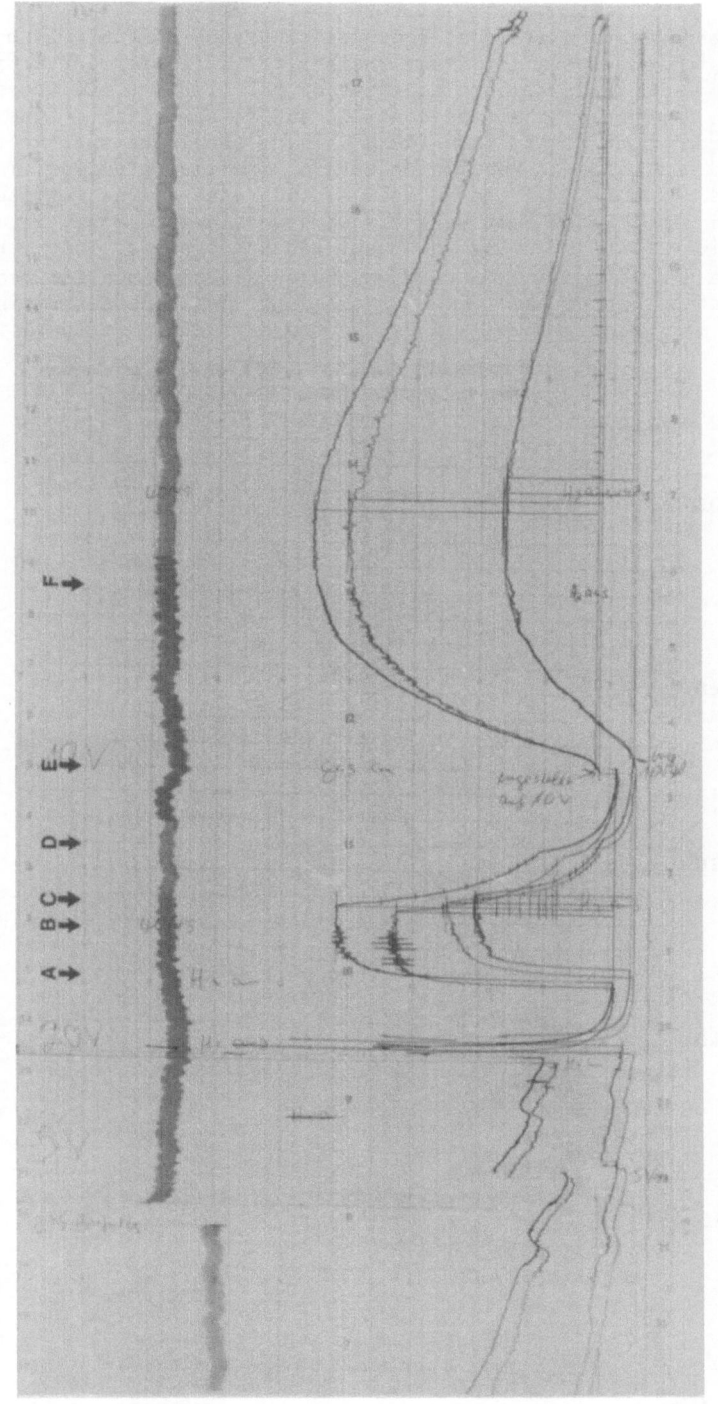

Fig. 4. Original registration of hydrogen clearance curves under hemorrhage: local hydrogen production (A) and inhalation of hydrogen (E). A: start hydrogen production, B: change time scale, C: stop hydrogen production, D: change time scale, E: start inhalation, F: stop inhalation. Mean arterial pressure = 4o mm Hg.

surface. In both cases, the measuring field might be extended over the critical seize of o.5 mm which is typical for the producing/measuring wire arrangement described here.

The interpretation of local hydrogen clearance curves will be a matter of further studies. It is not clear whether the diffusion-convection processes at the electrode tip are additive or not (Lübbers and Stosseck, 197o; Wodick, 1976). Perhaps, the time-constant due to spreading of the hydrogen diffusion front post mortem into "dead" tissue should be substracted from the combined diffusion-convection process of the living animal. This procedure seems reasonable as the individual clearance curves registered surrounding the same tissue spot were the same and, consequently, hydrogen gradients did not exist during the local washout process. Under these conditions, the simplified mathematical approach already used for heat clearance evaluation can be used. However, reliable data characterizing the post mortem diffusion process and enable the extraction of the diffusion

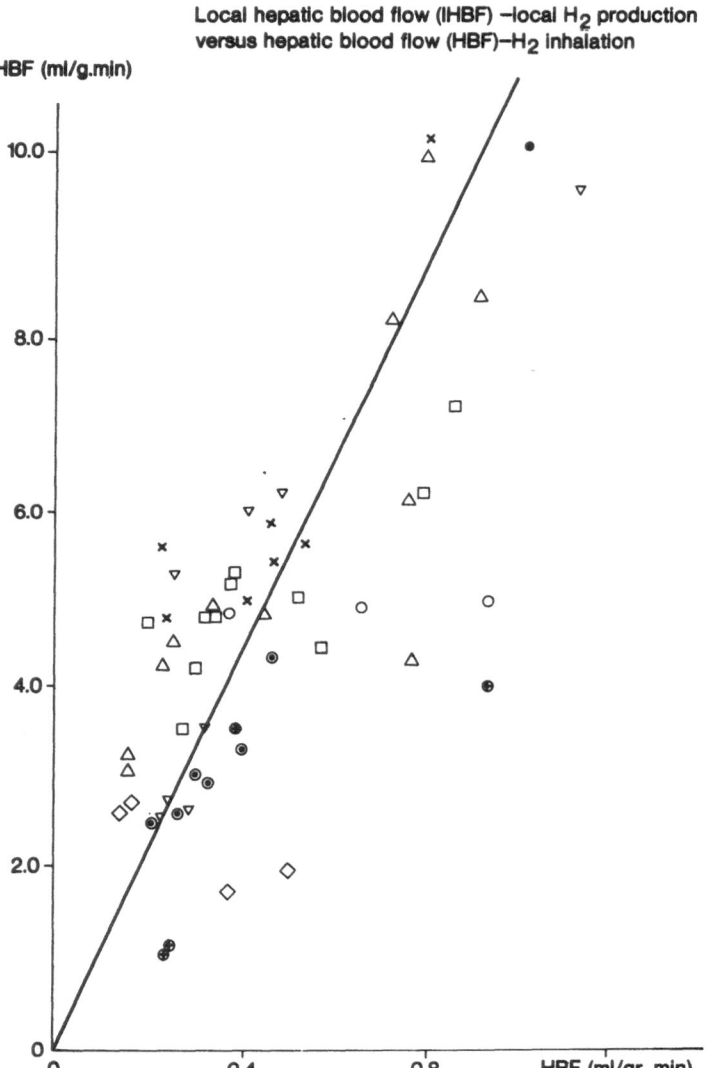

Local hepatic blood flow (lHBF) —local H$_2$ production versus hepatic blood flow (HBF)—H$_2$ inhalation

Fig. 5. Local blood flow versus hepatic blood flow are correlated with each other (r=o.685). The slope of the straight line was calculated according to the method of the least squares.

coefficient are difficult to extract from the registrations because the diffusion kinetics were found to be different at any moment after death.

A number of advantages of the hydrogen clearance registration in response to local production can be summarized:

- The registration of locally induced clearance is less time-consuming (2 min.) than the clearance in response to inhalation (5-8 min.). Therefore experiments can be performed more frequently.

- At t = o, at the beginning of the washout process, no further molecules will be produced and errors due to hydrogen inflow from the gastro-intestinal veins are not to be expected unlike with the hydrogen inhalation method.

- The local hydrogen production is much easier to handle during the experiment because hydrogen is produced electrochemically and does not need to be supplied by an external gas tank and additional mixing device.

SUMMARY

The combined method of hydrogen inhalation and local hydrogen production enable the determination of hepatic blood flow (HBF) and local hepatic blood flow (LHBF). LHBF was registered within a small superficial tissue volume of o.5 mm in diameter by means of a multi-wire electrode having 2oo um producing and 1oo um measuring wires arranged within less than 3oo um distance between the measuring wires. The feeding current for hydrogen production was 1 uA, the potential less than 1o V. The clearance in response to inhalation was registered by means of the same measuring electrodes within the same tissue volume.

Spontaneously breathing rats (Wistar-Frömter strain, 18o-23o g bw, N = 19, ketamin-xylazine anesthesia, artificial respiration) showed the following flow values: HBF+SD = o.5o+o.26 ml/g·min, n = 48 registrations; LHBF+SD = 4.66+2.13 ml/g·min, n = 43. The validity of the combined method is demonstrated in the LHBF/HBF graph which summarizes the data of hemorrhagic and control animals, m = o.1 and y_o = o.oo1. The correlation coefficient of r = o.685 shows a reasonable correlation of the combined data despite the wide scattering of the individual values.

REFERENCES

Aukland, K., Bower, B.F. and Berliner, R.W. (1964): Measurement of local blood flow with hydrogen gas. Circ. Res.14, 164-187

Gouma, D.J., Coelho, J.C.U., Schlegel, J., Fisher, J.D., Li, Y.F. and Moody, F.G. (1986): Estimation of hepatic blood flow by hydrogen gas clearance. Surgery 99 (4), 439-445

Kakimoto, T. (1982): Study of local hepatic blood flow of the rat by hydrogen clearance method. Nippon Shokakibyo Gakkai Zashi 79, 73-82

Lübbers, D.W. and Stosseck, K. (197o): Quantitative Bestimmung der lokalen Durchblutung durch elektrochemisch im Gewebe erzeugten Wasserstoff. Naturwissensch. 57, 311

Metzger, H.P. and Savas, Y. (1988): The influence of the calcium antagonists flunarizine and verapamil on cerebral blood flow and oxygen tension of anesthetized WFS-rats. Adv.Exp.Med.Biol. 222, 411-418

Wodick, R. (1976): Möglichkeiten und Grenzen der Bestimmung der Blutversorgung mit Hilfe der lokalen Wasserstoff-Clearance. Funktionsanalyse biol. Syst. 3, 249-411

THE OVERALL FRACTIONATION EFFECT OF ISOTOPIC OXYGEN MOLECULES DURING OXYGEN TRANSPORT AND UTILIZATION IN HUMANS.

K.-D. Schuster and K.P. Pflug

Institute of Physiology I, University of Bonn, 53 Bonn 1, FRG

INTRODUCTION

Oxygen isotopes have been used various times as a tracer to study phenomena of respiration. The tracer studies are based on the assumption that the labeled as well as the non-labeled molecules behave in a very similar manner. It is assumed that no or only slight fractionations occur between both species of molecules. Few investigations with conflicting results are known concerning fractionation effects of the oxygen molecules $^{16}O^{18}O$ and $^{16}O_2$ during respiration. The fractionations measured indicate that $^{16}O_2$ is consumed in preference to $^{16}O^{18}O$ by 1.8 % (Lane and Dole, 1956) or by 8 % (Muysers et al., 1963), respectively. If the latter value is correct, the effect cannot be neglected in tracer studies. One reason for determining the $^{16}O^{18}O/^{16}O_2$ ratio during respiration was to clarify the above problem. Physiologically more interesting may be the fact that the pathways of oxygen transport to tissues include non-fractionating processes such as ventilation and blood flow and fractionating processes e.g. diffusion and chemical reactions. Measuring fractionation effects could help to differentiate between these processes when investigating respiration under varying conditions. Moreover, if one of the fractionating processes becomes limiting for oxygen transport, its fractionation power presumably determines the overall fractionating effect, and by measuring this effect its origin could be detected.

METHODS

Experiments were performed on 7 healthy humans at rest. The person inspired air and expired into a bag (Fig.1.). Respiratory conditions were quantified by measuring CO_2 and O_2 partial pressures with a mass spectrometer (MS) and ventilatory volume with a pneumotachograph (PT). Samples were drawn from the bag into an evacuated flask (S). In order to apply a high precision technique of isotope ratio analysis as described by Mook and Grootes (1973), the samples drawn were processed as follows: using the system shown in Fig.2, the flask S containing the sample was cooled with

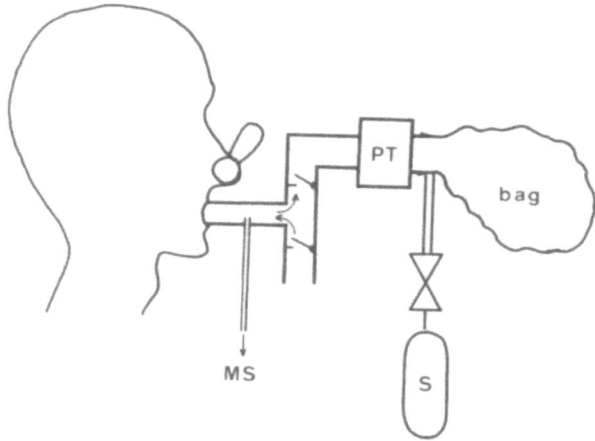

Fig. 1. Scheme of collecting expiratory air. S: flask for drawing samples, MS and PT: respiratory mass spectrometer
and pneumotachograph for quantifying respiratory conditions.

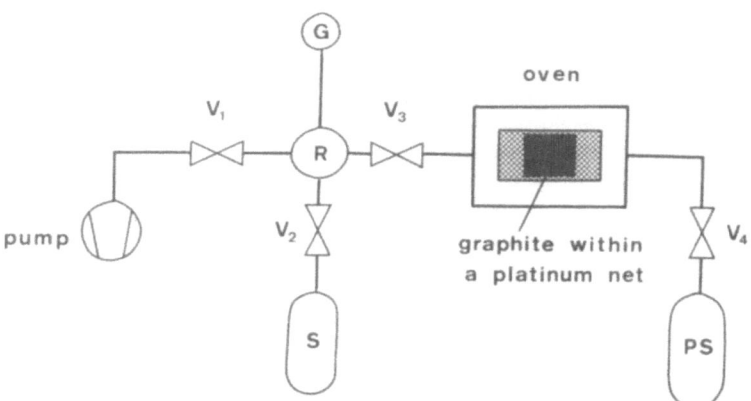

Fig. 2. Scheme of burning oxygen to carbon dioxide. S: sample
to be processed, PS: processed sample R: reservoir,
$V_1...V_4$: valves.

liquid nitrogen to retain expiratory carbon dioxide. Prior to processing, the reservoir R, the oven and flask PS were evacuated (valves V_1, V_3 and V_4 open, V_2 closed). After switching the valves, the oxygen of the sample was burnt inside the oven with graphite to carbon dioxide which was trapped in PS by freezing. Heating the graphite was accomplished by using light energy (Förstel et al., 1976). To achieve complete combustion, platinum served as a catalyst. The process was controlled by measuring the pressure with a gauge (G). The samples of CO_2 generated were analysed with an isotope ratio mass spectrometer (Micromass 602 C, VG Instruments) on their $C^{16}O^{18}O/C^{16}O_2$ ratio by comparing with reference CO_2 of known isotopic composition.

RESULTS

The fractionations are given as δ-values in table 1. With the abbreviation $X=(^{16}O^{18}O)/(^{16}O_2)$, the δ_{IE}-value between inspiratory gas (I) and expiratory gas (E) is defined as

$$\delta_{IE} = \frac{X_I - X_E}{X_I} \quad 100\% \tag{1}$$

Columns 2 and 3 of table 1 contain alveolar partial pressures of O_2 (P_{AO_2}) and CO_2 (P_{ACO_2}) to characterize the respiratory situation. δ_{IE} in column 4 is the percentage deviation of $^{16}O^{18}O$-abundance between inspiratory and expiratory gas according to formula (1).

Table 1. Fractionation effects of isotopic oxygen molecules during respiration.

Subject	P_{AO_2} (mmHg)	P_{ACO_2} (mmHg)	δ_{IE} (%)	δ_{IU} (%)
A	103	41	−0.33	1.40
B	106	36	−0.18	0.97
C	104	37	−0.28	0.95
D	104	39	−0.20	0.85
E	104	38	−0.18	0.80
F	104	36	−0.12	0.73
G	106	37	−0.16	0.69
mean	104.4	37.7	−0.21	0.91
±SD	1.1	1.8	0.07	0.24

δ_{IE}, δ_{IU}: $^{16}O^{18}O$-abundance differences between inspiratory and expiratory oxygen, and between inspiratory oxygen and the oxygen taken up, respectively. P_{AO_2}, P_{ACO_2}: alveolar partial pressures of O_2 and CO_2.

δ_{IU} was calculated from

$$\delta_{IU} = - \frac{\dot{V}EO_2}{\dot{V}O_2} \cdot \delta_{IE} \qquad (2)$$

where $\dot{V}EO_2/\dot{V}O_2$ is the ratio of expiratory oxygen flow to oxygen uptake. δ_{IU} describes the percentage deviation of $^{16}O^{18}O$-abundance between inspiratory and respiratory oxygen.

Its mean value of 0.91 % denotes that uptake, subsequent transport and utilization of oxygen occur with a 0.91 % higher rate for $^{16}O_2$ than for $^{16}O^{18}O$ under stationary conditions at rest. No clear cut relationship becomes apparent between δ_{IU} and the other respiratory parameters.

DISCUSSION

Determinations of $^{16}O^{18}O/^{16}O_2$ ratios by mass spectrometry are known to be influenced by many side effects. To avoid such errors, a well accepted procedure of isotopic chemistry has been applied.

The results of this paper are therefore considered to be more reliable than the 10-fold higher figure of Muysers et al. (1963). This is confirmed by the measurements of Lane and Dole (1956). Their results are only slightly higher than the values of this paper. Several reasons could account for this difference. Lane and Dole used a rebreathing procedure assuming an infinite ventilation rate between the rebreathing system and the lungs (Lane, 1955). There were some problems with leakage of their system. They applied a straightforward correction on their values based on unproved assumptions.

From the δ-value determined for $^{16}O^{18}O$, a similar value for $^{18}O_2$ can be estimated at $\delta_{IU}(^{18}O_2) = 1.7$ %, since the isotope effect of $^{18}O_2$ is approximately twice that of $^{16}O^{18}O$ (Vojta, 1960). Both δ-values, that of $^{16}O^{18}O$ and that of $^{18}O_2$, are small enough to be negligible in most tracer applications with ^{18}O labeled oxygen.

Fig.3 shows the various processes of oxygen transport to tissues and their expected fractionation power. The fractionation power is given as ratio of the rate constants of $^{16}O_2$ and $^{16}O^{18}O$. This figure is known as fractionation factor in the literature concerning isotope effects. A ratio of 1.00 means no fractionation takes place. Fractionations between $^{16}O^{18}O$ and $^{16}O_2$ are expected to be caused by diffusion and reaction processes but not by ventilation or blood flow. For different gases Krogh's diffusion constants K are inversely proportional to the square root of the molecular weights, and proportional to the solubilities β. The ratio of the solubility constants $\beta(^{16}O_2)/\beta(^{16}O^{18}O)$ has been investigated for distilled water within the temperature range 275-300 K by Klots and Benson (1963). Extrapolated to 310 K, it amounts to 0.9995. With the molecular weights 32 for $^{16}O_2$ and 34 for $^{16}O^{18}O$, the ratio of Krogh's diffusion constants $K(^{16}O_2)/K(^{16}O^{18}O)$ can be calculated to $\sqrt{34/32}$ 0.9995 = 1.03. The diffusion rate

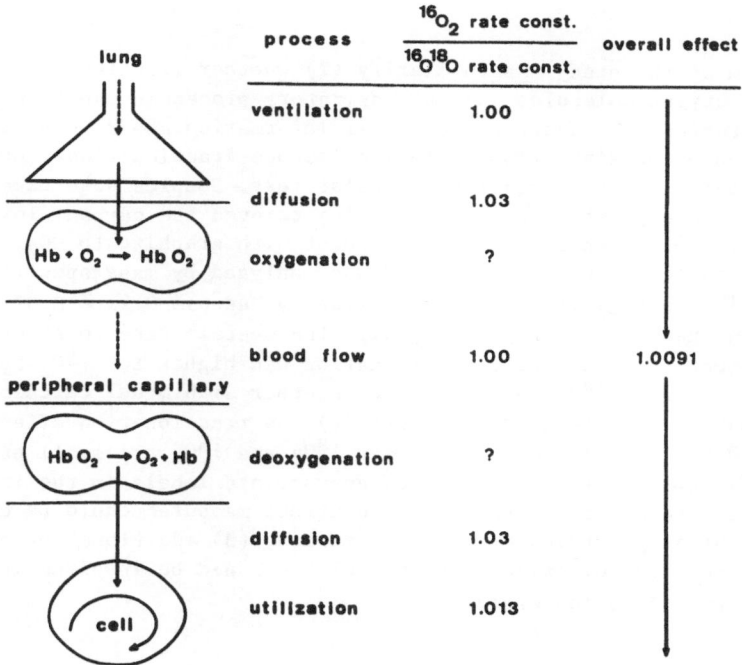

lung	process	$\dfrac{^{16}O_2 \text{ rate const.}}{^{16}O^{18}O \text{ rate const.}}$	overall effect
	ventilation	1.00	
	diffusion	1.03	
Hb + O₂ → Hb O₂	oxygenation	?	
	blood flow	1.00	1.0091
Hb O₂ → O₂ + Hb	deoxygenation	?	
	diffusion	1.03	
cell	utilization	1.013	

Fig. 3. Simplified scheme of the pathways of oxygen transport
to tissues. The fractionation power expected for some
processes with regard to transport of isotopic oxygen
molecules are given as ratio of the rate constants of
$^{16}O_2$ and $^{16}O^{18}O$.

constant is therefore expected to be higher for $^{16}O_2$ than for $^{16}O^{18}O$ by
3 %. Feldman et al. found a fractionation factor of 1.013 for reactions of
the respiratory chain. This means $^{16}O_2$ is metabolized 1.3 % more rapidly
than $^{16}O^{18}O$.

The overall fractionation effect of respiration, δ_{IU}, determined in
this paper, is smaller than the single effects known up to now. Several
reasons could account for this discrepancy. From model analysis for a chain
of fractionating and non-fractionating processes such as shown in fig.3,
the overall fractionation effect δ_{IU} converges against zero if oxygen
transport is supposed to be non-limited by a fractionating process.

The overall effect measured is higher than zero but below 0.013 %, the
fractionation effect found for oxygen utilization by Feldman et al. (1959).
This would be in line with a slight limitation of oxygen transport by
metabolism. However, additional fractionations caused by processes such as
the reaction of oxygen with hemoglobin, could be involved in forming the
overall fractionation effect. More precise interpretations are expected to
become possible if the fractionation factor of every single step of oxygen
transport is known.

SUMMARY

The aim of the study was to clarify (1) whether fractionation effects of isotopic oxygen molecules due to respiratory processes can be neglected in tracer studies, (2) whether additional information about respiratory processes can be obtained from measuring isotope fractionations. Experiments were performed on 7 healthy humans at rest. Samples were taken from inspiratory and expiratory gas. After having removed the carbon dioxide from the samples, the oxygen was completely burnt with graphite to CO_2, and the $^{16}O^{18}O/^{16}O_2$ ratio of the CO_2 generated was analysed by mass spectrometry. The $^{16}O^{18}O/^{16}O_2$ ratio of the expiratory gas was 0.21 ± 0.07 % greater than that of the inspiratory air. The overall rate constant for uptake, subsequent transport and utilization was higher for $^{16}O_2$ by 0.91 ± 0.24 % than for $^{16}O^{18}O$. These results together with model calculations including data from literature suggest: (1) the fractionation effects between $^{16}O^{18}O$ and $^{16}O_2$ as well as between $^{18}O_2$ and $^{16}O_2$ are small enough for both isotopic species to be considered appropriate labels in the investigation of respiratory processes, (2) the effect measured could be due to limitations of oxygen transport by utilization, (3) additional processes such as the reaction of oxygen with hemoglobin could be involved in forming the overall fractionation effect.

REFERENCES

Feldman, D.E., Yost, H.T.Jr., and Benson B.B., 1959, Oxygen isotope fractionation in reactions catalyzed by enzymes, Science, 129:146-147.

Förstel, H., Weiner, B., and Schleser, G., 1976, Preparation of oxygen samples for $^{18}O/^{16}O$ measurements by a combined gas chromatography - burning technique, Appl. Radiat. Isot., 27:211-215.

Klots, C.E., and Benson, B.B., 1963, Isotope effect in the solution of oxygen and nitrogen in distilled water, J. Chem. Phys., 38:890-892.

Lane, G.A., 1955, A study of the fractionation of oxygen isotopes, Dissertation, Northwestern University, 44-63.

Lane, G.A., and Dole, M., 1956, Fractionation of oxygen isotpes during Respiration, Science, 123:574-576.

Mook, W.G., and Grootes, P.M., 1973, The measuring procedure and corrections for the high-precision mass-spectrometric analysis of isotopic abundance ratios, especially referring to carbon, oxygen and nitrogen, Int. J. Mass. Spectrom. Ion. Phys., 12:273-298.

Muysers, K., Siehoff, F. and Worth, G., 1963, Respiratorischer Gasaustausch von Sauerstoffisotopen, Beitr. Silikose-Forsch. S-Bd Grundfragen Silikoseforsch., 5:389-395.

Schuster, K.-D., 1985, Die Kinetik des CO_2-Transfers zwischen Alveolarraum und Blut: Bestimmung der alveolo-kapillären Diffusionskapazität D_M mit Hilfe ^{18}O-markierten Kohlendioxids, in: "Funktionsanalyse biologischer Systeme", 14, G.Thews, ed., Akademie der Wissenschaften und der Literatur Mainz, Franz Steiner Verlag Wiesbaden GmbH, Stuttgart.

Vojta, G., 1960, Grundlagen der statistischen Thermodynamik von Isotopensystemen mit chemischen Reaktionen, Kernenergie, 3:917-927.

EVALUATION OF A HEPARIN-COATED PO$_2$ ELECTRODE FOR CONTINUOUS INTRAVASAL PO$_2$ MONITORING

R.Tenbrinck, W.Schairer, G.J.van Daal,
M.H.Kuypers*,G.F.J.Steeghs*, and B.Lachmann

Dept. of Anesthesiology, Erasmus University
Postbus 1738, 3000 DR, Rotterdam
* PPG Hellige Research Dept., Best, The Netherlands

Intensive care units always require continuous PaO$_2$ monitoring to enable fast and proper ventilator adjustment to patients need. Conventional bloodsampling techniques followed by bloodgas analysis are not able to monitor these sudden changes of the oxygen partial pressure. A better solution would be an intravasal PaO$_2$ electrode. Monitoring the arterial and venous oxygen partial pressure simultanuously will give information about the oxygen extraction rate; threatening perfusion disturbances and acute changes during spontaneous respiration can be detected immediately. The direct effect of extracorporal oxygenators during open heart surgery can be watched; there are also possibilities of closed loop feed back regulation of the respirator for optimal ventilator settings in ARDS-patients.

One of the disadvantages of the few available PaO$_2$ catheters is that the surface of the catheter membrane becomes clotted with blood in a short time, which always leads to a decrease in the intravasal measured PaO$_2$. Therefore, based on semiconductor technology, a heparin-coated intravasal PO$_2$ electrode was developed by PPG Hellige (Best, The Netherlands); this electrode was expected not to have the former disadvantages.

The aim of this study was to investigate the accuracy of this electrode under in vivo conditions in comparison to normal bloodsampling and analysis.

MATERIALS AND METHODS

The PO$_2$-electrode

The oxygen-sensor consists of a chip which can measure the PO$_2$ and the temperature, a metal tip with a window-opening for the chip and a lumen for taking bloodsamples, a flexible polyurethane catheter (F6), a keflar cable for safety and a connector with the calibration resistors for the PO$_2$ and the temperature (figs 1,2). This oxygenchip is based on the Clarkcell-principle (1,2) integrated with the chiptechnology. It is described more detailed by Kimmich (2). Since then some important improvements were made. In the new generation the chloride (Cl$^-$) is extracted from the bloodstream through a semi-permeable hydrogel layer which is impermeable for proteins. In the same layer heparine is immobilized to prevent the blood forming thrombi on the sensing area.

This results in the following characteristics according to the manufactorer: 95% response time is 15 sec; flow dependency (> 8 cm/sec) is 1% ; the lifetime is approximately 7 days. For drifting the following specifications were given: First 30 min 4% ;second 30 min 1.5% ; third 30 min 1% and for every following hour 0.5% provided the catheter is stabilized first.

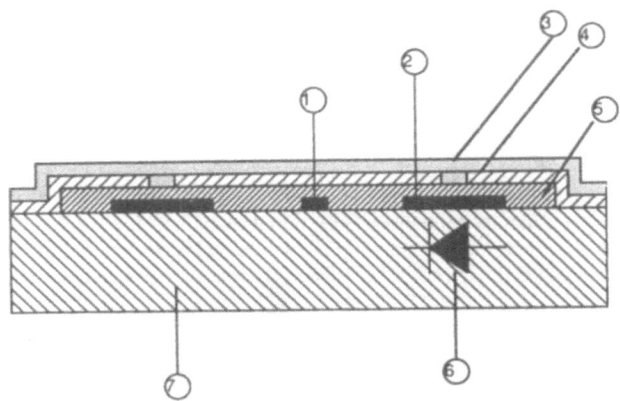

1. GOLD CATHODE, 5 micron width,
                    1200 micron long
2. Ag/AgCl ANODE
3. SEMI-PERMEABLE MEMBRANE
4. HYDRDROPHOBIC MEMBRANE
5. HYDROGEL
6. TEMP. SENSOR
7. CHIP

Fig. 1. Diagram of oxygen sensor.

Method

Fourteen New Zealand white rabbits were anesthetized with pentobarbital (35 mg/kg bw) and tracheotomised; they could breathe spontaneously while they were also prepared to be ventilated. The electrode (tip diameter 1.2 mm) was inserted in the left arteria carotis; the animals were not heparinised. Blood samples were taken from the left arteria femoralis by means of an intravasal catheter. As amplifier and monitor system a Hellige RM 300 (PPG Hellige, Best, The Netherlands) was used. The values of the electrode were recorded continuously on a Honeywell PM 8221 penrecorder (Honeywell, Best, The Netherlands). During 8 hours of experimental time every 30 minutes an arterial blood sample was drawn from the arteria femoralis catheter and analysed by an ABL 300 bloodgas analyser (Radiometer, Copenhagen, Denmark). Throughout the period of monitoring there were no efforts made to adapt the electron values to those obtained by the ABL 300.

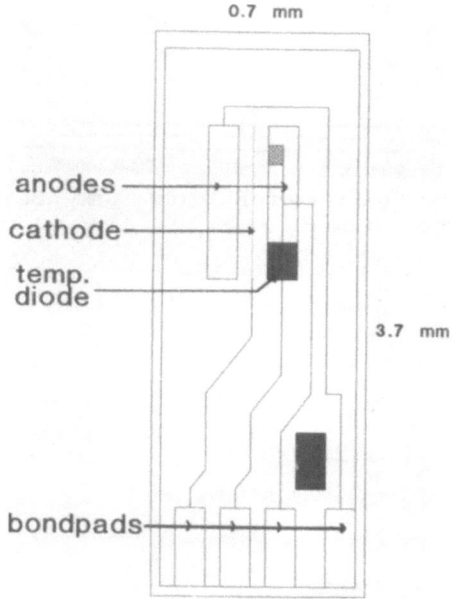

Fig. 2. Drawing of electrode-tip (not to scale).

STATISTICS

For statistical analysis of the data we used the Wilcoxon rank-test for paired measurements and the Student T-test for paired observations. The outcome of both tests was the same in all cases. A P < 0.05 was accepted as statistically significant.

RESULTS

The values monitored by the electrode and ABL 300 were combined. The total number of observations was 83. The data were processed in two ways: first comparison of electrode and ABL 300 value pairs (Fig. 3); secondly the differences of ABL 300 and electrode values were plotted over the time (Fig. 4). After 8 hours the deviation due to drifting had to be less than 10%; however in our investigation this was 18%. During the first 3 hours there was no statistical significant difference between ABL 300 and electrode data (P > > 5%). In the last two hours these differences became significant (P < 1.25%).

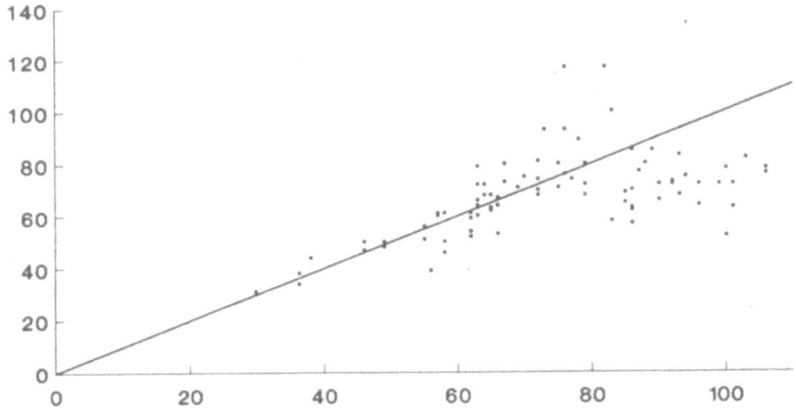

Fig. 3. Comparison of electrode data and ABL 300 data, X-axis ABL 300 data in mm Hg, Y-axis electrode data in mm Hg.

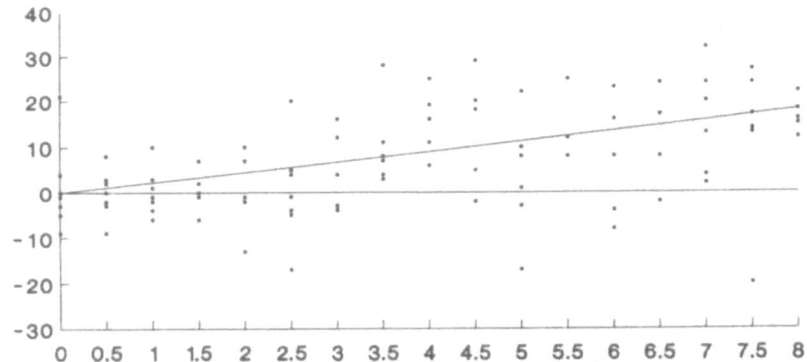

Fig. 4. Differences between ABL 300 data and electrode data over the time, X-axis time in hours, Y-axis ABL 300 data minus electrode data.

DISCUSSION

Similar experiments in beagle dogs and in vitro experiments using saline solutions showed good correlation and a very acceptable drift in time (Fig. 5 ; M.H. Kuypers, unpublished results).

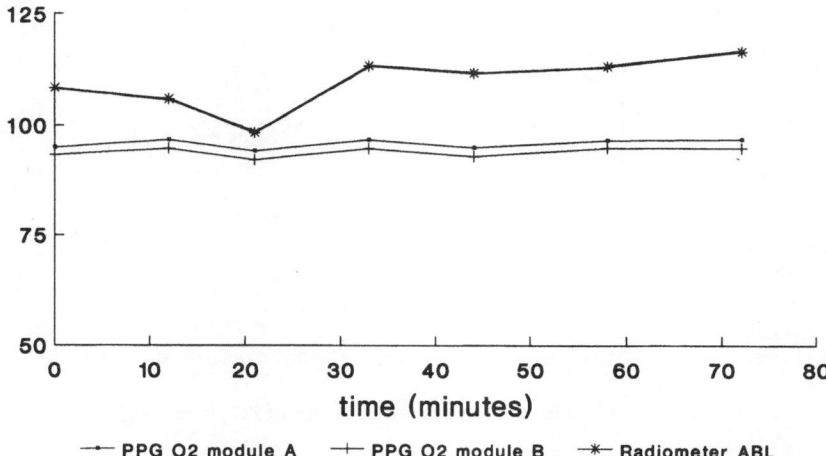

Fig. 6. Electrode PO$_2$ recording in mm Hg, revealing a PO$_2$ dip which could be visualised instantanuously.

An other point of interest is the fact that the PO$_2$ electrode is a very fast acting device while the ABL 300 is relatively slow: 15 seconds for the electrode against 3 minutes for the ABL 300 is the time lapse between two measurements. So the use of the PO$_2$ electrode leads to a visualisation of dips in the PO$_2$-curve that could not be monitored before (Fig. 6). An example: When one takes a bloodsample in the middle of such PO$_2$-dip (without electrode monitoring) this sample would be qualified as bad and a new sample would be taken. When it considers a patient who is being ventilated it could lead to an undesired change in ventilator settings.

CONCLUSION

From these results we conclude that the heparin-coated intravasal PO$_2$ electrode could be routinely used in intensive care units for continuous PO$_2$ monitoring, provided it is placed in a vessel that is large enough to allow the electrode membrane floating freely in the bloodstream. The use of heparin is under these circumstances facultary.

During the experiments we had some problems with the movability of the catheter in the carotic artery. We presumed that the diameter of the rabbit vessels was not large enough to have an optimal functioning catheter. We did not heparinise the animals because we did not want to conceal any possible clotting effect. Due to a changed flow pattern there might be a

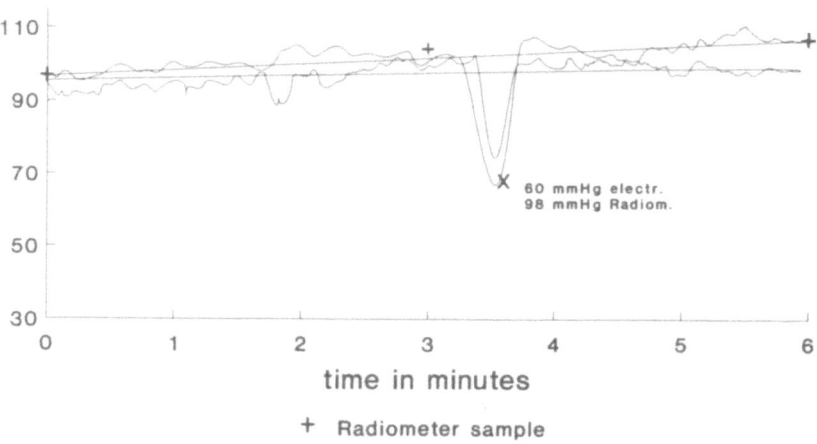

Fig. 5. Results of in-vivo canine measurements of PO$_2$ in mm Hg.

precipitation of clotting factors around the catheter or on the vessel lumen so that the electrode sensor area had a diminished contact with the bloodstream. This finding brought us to the conclusion that the electrode had a limiting factor in the vessel diameter; it also contributed to the deviations in the figures 3 and 4 in comparison with manufactorers data.

REFERENCES

1. Clark LC. (1956) Monitor and control of blood and tissue oxygen tensions. Trans Am Soc Art Int Organs 2: 41-45.

2. Kimmich HP, Kuypers MH, Engels JML, Maas HGR. (1981) Disposable solid state oxygen sensor. Fifth Intern Symp of the Internatinal Soc for Oxygen Transp to Tissue. Detroit: Aug 1981.

MATHEMATICAL MODELS

BUFFERING OF MUSCLE TISSUE PO_2 LEVELS BY THE SUPERPOSITION OF THE OXYGEN FIELD FROM MANY CAPILLARIES

Patricia A. Clark[1], Stephen P. Kennedy[1], and Alfred Clark, Jr.[2]

[1] Mathematics, Rochester Institute of Technology
[2] Mechanical Engineering, University of Rochester
Rochester, N.Y., USA

INTRODUCTION

Characteristic of a Krogh (1919, 1929) analysis is a polygonal array of cells with oxygen supplied by capillaries at each cell vertex. Because of the symmetry of the arrangement, the assignment of tissue PO_2 at any point can be thought of as the consequence of one particular capillary in the array. Thus the PO_2 distribution in an entire muscle tissue sample is reduced to the analysis of a single Krogh cylinder. The Krogh analysis predicts substantial PO_2 variation radially across a muscle as well as longitudinally running parallel to the capillary.

High resolution in vivo measurements by Gayeski and Honig (1986, 1988a) and Honig and Gayeski (1987) show the transverse PO_2 variation across entire cells in multicellular arrays to be very shallow. Such cellular arrays include points both near and far from operating capillaries. They (1988b) found similar small variation in tissue PO_2 from measurements made parallel to capillaries in both skeletal and cardiac muscle. Contrary to a Krogh picture of the oxygen distribution, these measurements appear to require the superposition of the diffusion fields of a larger environment of capillaries. Related measurements have been reported by Wittenberg and Wittenberg (1985) using using other measurement techniques.

The contradiction between measurement and theory has been the subject of a great deal of study. Inclusion of the facilitated diffusion, due to the presence of myoglobin, was the first modification of the Krogh analysis to be examined. Extensive reviews of this literature have been provided by Wittenberg (1970) and Kreuzer (1970). The facilitation can lead to shallow variations only if the PO_2 values in the tissue are down near or below the PO_2 of myoglobin. Hellums (1977), Federspiel and Sarelius (1984), Honig et al. (1984), Federspiel and Popel (1986), and Grobe and Thews (1986) have suggested the possibility of substantial transcapillary resistance to the oxygen transport. Significant precapillary losses as described by Duling and Berne (1970) and Gross (1979) have also been proposed. Recent measurements by Gayeski and Honig (1988) and theoretical studies by Clark et al. (1988) suggest that the precapillary losses are not, however, significant in exercise. Although facilitation can reduce the gradients in tissue PO_2, it can not account for the measured relative lack of correlation between the PO_2 at a point and the proximity of the point relative to the nearest active capillary.

Morphological studies by Myrhage and Erikson (1980) and those reported by Weibel (1984) in both skeletal and cardiac tissue show myocytes typically surrounded by 6-8 capillaries. These observations have led to a variety of models in which a myocyte is subject to diffusion from the capillaries immediately surrounding it. Such models are described by Popel (1978, 1980), Klitzman et al (1983), Federspiel (1986), Grobe and Thews (1987), and

Secomb and Hsu (1988). It is quite straight forward to show that the transcellular PO_2 drops are smaller than the single capillary Krogh model by a factor approximately equal to the inverse of the "effective number of capillaries supplying the myocyte."

The present work seeks to take this reasoning a step further to show that the PO_2 at a point in the tissue depends in an essential way on the global superposition picture. While contributions from local capillaries are certainly a part of the PO_2 at a point, we find that the non-local contributions provide an important buffering effect on the transcapillary PO_2 variations.

MODEL

Presented here is a sequence of steady-state cylindrical models for multicellular muscle tissue samples each of which is penetrated by many capillaries running parallel to the cylinder axis. Reeves and Rakusan (1988), using a colored microsphere tracer in rat myocardium, found that the standard modeling assumption of parallel capillaries with the concurrent flow was characteristic of more than 95% of their observations. Similar models have been studied by Popel (1980).

This sequence of models is characterized by a uniform volume rate of oxygen consumption in which an arbitrary array of cylindrical oxygen sources (model capillaries) is distributed. First the non-myoglobin containing model will be described.

The steady state oxygen pressure is obtained as the solution of

$$\nabla^2 P = \gamma_m/(\pi R_f^2 K_1) - \sum_{j=1}^{N} \Gamma_j \tag{1}$$

where $P(X,Y)$ is the tissue PO_2 in (Torr), γ_m is the consumption rate per unit length in the tissue in (moles/cm-sec) and K_1 is the Krogh diffusivity in (moles/cm-sec-Torr), R_f is the radius of the tissue cylinder and Γ_j is the oxygen production rate (flux) of the jth capillary The boundary condition on the tissue cylinder is

$$dp(R_f, \Theta)/dR = 0. \tag{2}$$

Calculations were also made for $P(R_f, \Theta) = 0$ and the basic conclusions are very similar.

The problem is made dimensionless as follows: $p = P/(\gamma_m/\pi K_1)$, $r = R/d$, $r_c = R_c/d$, $r_f = R_f/d$, and $G_j = \Gamma_j \pi K_1/\gamma_m$ where d is a typical muscle cell diameter. R_c is the capillary radius. In terms of the dimensionless variables, equations (1) and (2) become

$$\nabla^2 p = (1/r_f^2) - \sum_{j=1}^{N} G_j \tag{3}$$

and

$$dp(r_f, \Theta)/dr = 0. \tag{4}$$

Each capillary has a production rate per unit length given by γ_{cj}. Because steady state requires the balance of oxygen production and consumption

$$\gamma_m = \sum_{j=1}^{N} \gamma_{cj}. \tag{5}$$

The dimensionless PO_2 can be written as $p = p_c + p_m$ where p_c is the result of capillary production and p_m is due to mitochondrial consumption. These quantities satisfy

$$\nabla^2 p_m = 1/r_f^2 \tag{6}$$

and

$$\nabla^2 p_c = -\sum_{j=1}^{N} G_j \tag{7}$$

where

$$d(p_m + p_c)/dr = 0 \quad \text{at } r = r_f \tag{8}$$

and

$$p_m + p_c = p_o \quad \text{at } r = 0. \tag{9}$$

p_m can be obtained by twice integrating equation (6) and requiring boundedness at $r = 0$, so that

$$p_m(r) = (r/(2r_f))^2 + c_2. \tag{10}$$

c_2 will be evaluated after p_c has been determined.

To obtain p_c, two regimes must be considered:

(1) $|\mathbf{r} - \mathbf{r}_{oj}| > r_c$ (outside the jth capillary)

(2) $|\mathbf{r} - \mathbf{r}_{oj}| < r_c$ (inside the jth capillary)

using a method of images solution technique. Both the capillary and its image appear as line sources. Thus the solution of equation (7) for the jth capillary-image pair is

$$p_{cj} = -\frac{\gamma_{cj}}{2\gamma_m}\left[\ln|\mathbf{r} - \mathbf{r}_{oj}| + \ln\left|\mathbf{r} - \frac{r_f^2 \mathbf{r}_{oj}}{r_{oj}^2}\right|\right]. \tag{11}$$

To obtain p_c for $|\mathbf{r} - \mathbf{r}_{oj}| < r_c$, the image appears as a line source while the capillary is resolved. Therefore p_{cj} is written

$$p_{cj} = p_{in\ j} + p_{out\ j} \tag{12}$$

where $p_{in\ j}$ is the contribution due to the jth capillary and $p_{out\ j}$ is due to the image of the jth capillary. In a manner similar to that used in obtaining equation (11), it can be shown that

$$p_{out\ j} = -\frac{\gamma_{cj}}{2\gamma_m}\ln\left|\frac{r_{oj}^2 \mathbf{r} - r_f^2 \mathbf{r}_{oj}}{r_{oj}^2}\right|. \tag{13}$$

For the calculation of $p_{in\ j}$, a coordinate system (x^s, y^s) centered on the jth capillary is chosen. In this reference frame the equation to be solved is

$$\nabla^{s2} p_{in\ j} = -\frac{\gamma_{cj}}{r_c^2 \gamma_m} \quad \text{for } r^s < r_c. \tag{14}$$

After twice integrating and requiring $p_{in\ j}$ to be bounded at $r^s = 0$, the solution written in terms of the original r, Θ tissue cylinder centered coordinates is

$$p_{in\ j} = -\frac{\gamma_{cj}(r^2 + r_{oj}^2 - 2r_{oj}r\cos(\Theta - \Theta_{oj}))}{4\gamma_m r_c^2} + c_3. \tag{15}$$

Equations (13) and (15) are substituted into (12). c_3 is determined by making p_c continuous at $|\mathbf{r} - \mathbf{r}_{oj}| = r_c$. Combining these results, it can be shown that

$$\frac{4\gamma_m}{\gamma_{cj}}p_{cj} = \ln(r^2 + r_{oj}^2 - 2r_{oj}r\cos(\Theta - \Theta_{oj}))$$

$$+ \ln(r_{oj}^2 r^2 + r_f^4 - 2r_f^2 r_{oj}r\cos(\Theta - \Theta_{oj})) \tag{16}$$

for $|\mathbf{r} - \mathbf{r}_{oj}| > r_c$ and

167

$$\frac{4\gamma_m}{\gamma_{cj}}p_{cj} = \frac{r^2 + r_{oj}^2 - 2r_{oj}r\cos(\Theta - \Theta_{oj}))}{r_c^2} + \ln \ r_c^2 - 1$$

$$+ \ln\left(r_{oj}^2 r^2 + r_f^4 - 2r_f^2 r_{oj} r \cos(\Theta - \Theta_{oj})\right) \tag{17}$$

for $|\mathbf{r} - \mathbf{r}_{oj}| < r_c$. The total capillary contribution is

$$p_c = \sum_{j=1}^{N} p_{cj}. \tag{18}$$

To evaluate c_2 in equation (10), equations (16), (18), and (10) are substituted into (9). This gives

$$p_m = p_o + \frac{r^2}{4r_f^2} + \ln \ r_f + \frac{1}{4\gamma_m} \sum_{j=1}^{N} \gamma_{cj} \ln \ r_{oj}^2. \tag{19}$$

By requiring that $r_f = (N/\pi)^{\frac{1}{2}}$ in every model, each capillary serves on the average a region of the same cross sectional area.

In order to incorporate the presence of facililated diffusion, the equilibrium approximation as described by Fletcher (1980), is used. This leads to the solution of an equation similar to equation (3) where p is replaced by $p + \alpha Y$. Y, the fraction of myoglobin saturated with oxygen is related to p by

$$Y = \beta p/(1 + \beta p) \tag{20}$$

where $\alpha = \pi c_p D_m/\gamma_m$ and $\beta = k_1 \gamma_m/(\pi k_2 D_o)$. c_p is the concentration of myoglobin in the tissue, D_m is the oxymyoglobin diffusion constant, D_o is the free oxygen diffusion constant, k_1 is the rate of association of myoglobin and oxygen and k_2 is the rate of dissociation of oxymyoglobin. The boundary conditions are

$$p(0, \Theta) = p_o \tag{21}$$

and

$$d(p + \alpha Y)/dr = 0 \quad \text{at } r = r_f, \tag{22}$$

and the matching condition is that p must be continuous at the edge of each capillary. Solution for $p + \alpha Y$ is accomplished in a similar way to that for p in the non-myoglobin case. Y can be eliminated using equation (20) and a quadratic equation solved for p to give

$$p = -\frac{1}{2}(\frac{1}{\beta} + \alpha - M) + \frac{1}{2}\left[(\frac{1}{\beta} + \alpha - M)^2 + \frac{4M}{\beta}\right]^{\frac{1}{2}} \tag{23}$$

where

$$M = r^2/(4r_f^2) + p_o + \alpha\beta p_o/(1 + \beta p_o) + \ln \ r_f$$

$$+(4\gamma_m)^{-1} \sum_{j+1}^{N} \gamma_{cj} \ln \ r_{oj}^2 - (4\gamma_M)^{-1} \sum_{j=1}^{N} \gamma_{cj} \ln \ [r^2 + r_{oj}^2 - 2r_{oj}r\cos(\Theta - \Theta_{oj})]$$

$$-(4\gamma_m)^{-1} \sum_{j=1}^{N} \gamma_{cj} \ln \ [r_{oj}^2 r^2 + r^4 - 2rr_{oj}r_f^2 \cos(\Theta - \Theta_{oj})]. \tag{24}$$

168

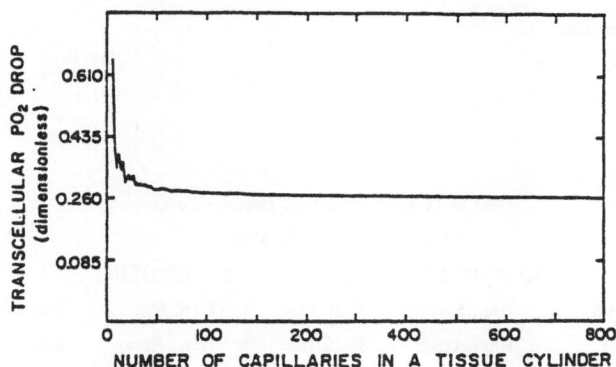

Figure 1. PO$_2$ Drop across the center myocyte from sarcolema to cell center in a tissue cylinder of a regular square array of capillaries as a function of the number, N, of capillaries in the cylinder(no myoglobin)

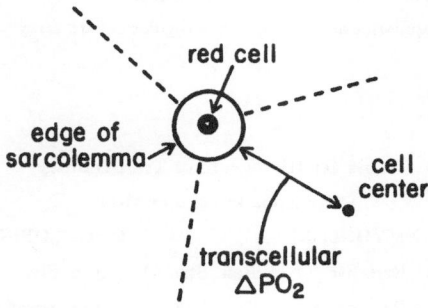

Figure 2. Diagram of transcellular PO$_2$ drop

Table I Average PO$_2$ drop from sarcolema to cell center

Single Krogh Cylinder	Multicellular Tissue Uniform array (no myoglobin)	Multicellular Uniform array (myoglobin)
Skeletal		
39.0 Torr	13.4 Torr	11.4 Torr
Cardiac		
4.8 Torr	1.8 Torr	0.4 Torr

Table II Histogram of Transcellular
PO$_2$ drops in skeletal tissue

Intervals (Torr.)	NO MYOGLOBIN		MYOGLOBIN	
	Ran. Str. Reg. Array	Ran. Str. Ran. Loc.	Ran. Str. Ran. Array	Ran. Str. Ran. Loc.
0-5	33	28	45	48
5-10	13	15	13	13
10-15	7	10	12	8
15-20	15	15	9	13
20-25	14	15	10	8
25-30	18	14	8	8
30-35	0	3	3	2

% mesurements in each interval

Ran. Str. - capillary PO$_2$ values randomly assigned

Ran. Loc. - capillary locations randomly assigned

Reg. Array - capillaries are arranged in a regular square array

Table III Histogram of Transcellular
PO$_2$ drops in cardiac tissue

Intervals (Torr.)	NO MYOGLOBIN		MYOGLOBIN	
	Ran. Str. Reg. Array	Ran. Str. Ran. Loc.	Ran. Str. Ran. Array	Ran. Str. Ran. Loc.
0-1	29	28	75	74
1-2	21	22	13	15
2-3	16	19	6	6
3-4	16	16	2	1
4-5	16	13	4	3
5	2	2	0	1

% mesurements in each interval

Ran. Str. - capillary PO$_2$ values randomly assigned

Ran. Loc. - capillary locations randomly assigned

Reg. Array - capillaries are arranged in a regular square array

CALCULATIONS

The models analyzed were of the following six types of tissue cylinders: regular square arrays of capillaries of equal strengths, γ_m/N, regular square arrays of capillaries with randomly chosen strengths in the range $0 - 2\gamma_m/N$ and randomly located capillaries with randomly chose strengths. Each type of model was calculated both with and without myoglobin. The random assignment of capillary strengths was carried out in such a way as to maintain the

Table IV Regression of transcellular

PO$_2$ drop versus capillary PO$_2$

Transcapillary Δ PO$_2$ = a (Capillary PO$_2$)+b

No myoglobin   (dimensionless)

Regular array; random capillary PO$_2$ values

a	b	Corr.	Remote Cont.[1]
.314	-.0883	.95	42%

Random array; random capillary PO$_2$ values

| .373 | -.0958 | .98 | 40% |

Mycglobin[2]

Random array; random capillary PO$_2$ values

Skeletal

| 15.13 | -4.39 | .95 | 43% |
| 1.02 | -.44 | .82 | 52% |

---

1   Remote cont. -% contribution to PO$_2$ from all other capillaries

(excluding the cell centered one)

2   Because of non-linearity cardiac and skeletal must be considered

separately

steady state requirement in equation (5). It was found that for a regular array of capillaries in the tissue cylinder for $N > 100$, the PO$_2$ at a point near the center of the tissue cylinder changed by less than 2% as more capillaries were included in the model (see Figure 1). The majority of the calculations were performed for tissue cylinders containing 208 capillaries.

The parameter values for models appropriate for the study of skeletal and cardiac tissue require numerical values for $\gamma_m/(\pi K_1) = \dot{V}O_2 \cdot r_f^2 \cdot d^2/K_1$, where $K_1 = 2.13 \cdot 10^{-14}$ moles/cm-sec-Torr (see Clark and Clark (1985)), $r_f = (N/\pi)^{\frac{1}{2}}$ and $\dot{V}O_2$ is the consumption rate per unit volume. In skeletal tissue a typical cell diameter is $d = 60 \cdot 10^{-4}$ cm (Horstman et al. (1976))

and a maximal consumption rate $\dot{V}O_2 = 8.9 \cdot 10^{-8}$ moles/cm^3-sec Horstman et al. (1976) and Connett et al. (1985)) so that $\gamma_m/(\pi K_1)$ (skeletal) $= 47.9$ N Torr. For cardiac tissue with a characteristic cell diameter of $d = 20 \cdot 10^{-4}$ cm in dog and a maximal consumption of $2.7 \cdot 10^{-7}$ moles/cm^3-sec (Gayeski and Honig 1988a), the PO$_2$ scale parameter is $\gamma_m/(\pi K_1)$ (cardiac) $= 16.1$ N Torr. In the presence of myoglobin, the quantities α, β are also required. Using the parameter values provided by Fletcher (1980) $c_p = 5 \cdot 10^{-7}$ moles/cm^3, $D_m = 1.7 \cdot 10^{-6}$ cm^2/sec, $k_1 = 2.4 \cdot 10^{10}$ cm^3/moles-sec, $k_2 = 65$ sec^{-1} and $D_o = 1.5 \cdot 10^{-5}$ cm^2/sec, we obtain: α (skeletal) $= 0.83$/N β (skeletal) $= 25.1$ N, α (cardiac) $= 2.4$/N and β (cardiac) $= 8.6$N.

The calculations described are of two types. First, is the PO$_2$ difference across a myocyte and second, is a regression between capillary flux and transcellular PO$_2$ drops.

For the calculation of the difference across a myocyte, the tissue PO$_2$ is calculated at each vertex of a Krogh size square centered on each capillary. Each of these values is differenced with the PO$_2$ on the diagonal through that point at the near edge of the sarcolema (see Figure 2). The average of these four differences is used as a measure the transcellular PO$_2$ drop. Presented in Table I is a comparison of the results from the Krogh cylinder and regular array multicellular models both with and without myoglobin. Table II presents histograms for the transcellular PO$_2$ drops in the random tissue cylinder models for both skeletal and cardiac tissue contrasting the myoglobin and non-myoglobin cases.

Although transcellular PO$_2$ drops in skeletal tissue average about 13 Torr. for the random models without myoglobin, about 30% of the cells in the model were characterized by PO$_2$ differences of 5 Torr or less. In the case of cardiac tissue (no myoglobin), the average PO$_2$ difference across a cell is about 2 Torr while almost 30% of these cells exhibited drops of less than 1 Torr. The introduction of myoglobin in the models served to enhance the number of cells with low drops. For skeletal tissue 45-48% had drops of 5 Torr of less while 75% of cardiac cells were characterized by differences of 1 Torr or less.

For the random capillary PO$_2$ strength models, PO$_2$ drops were regressed against capillary PO$_2$ strengths. The results of these calculations are presented in Table IV. All correlations are .82 or somewhat higher. It is shown that while there is a strong correlation between a capillary PO$_2$ value and the drop across its surrounding cell, that there exists a significant contribution to the PO$_2$ distribution from the more remote capillaries. For the models considered here, the remote contributions are at least 40% of the total.

CONCLUSIONS

In situations of irregular geometry of capillary distribution and unequal capillary PO$_2$ levels, it is no longer possible to isolate a region of plane completely surrounded by a zero oxygen flux boundary condition and carry out the analysis on it. Such closed zero flux boundaries disappear and it is necessary to consider explicitly the superposition of the diffusional oxygen fields of all the capillaries of the tissue cylinder.

The superposition has three major effects on the tissue PO$_2$ distribution. One is the expected buffering which is the result of adding the diffusional contribution of all the capillaries at their various distances. The effect has been notes earlies in the work of such authors as Popel (1978), Klitzman et al. (1983), Federspiel (1986), Grobe and Thews (1986), and Secomb and Hsu (1987) and is, of course evident in the present models. Note in Tables II and III the large percentage of transcellular PO$_2$ differences of small magnitudes. In myoglobin containing models of skeletal tissue at least 45% of the cells had drops of less than 5 Torr and more than 60% show of 10 Torr or less. For myoglobin containing cardiac tissue the effect was even more dramatic with at least 74% of the drops less than 1 Torr.

The second effect of the superposition is a significant non-local contribution to the PO$_2$ at a point in the tissue. We designate as non-local, the diffusion fields of all but the closest capillary. While these non-local contributions vary from point to point in the tissue, average values are listed in Table IV. For both types of tissues, those non-local contributions are significant averaging 40% or more of the PO$_2$ level at a point. The importance of the non-local effect on tissue PO$_2$ is further evident in Figure 1 where it is shown that the PO$_2$ at a point in the tissue is significantly affected by approximately the nearest 100 capillaries. The size of the non-local effects is not unanticipated from the measurements of Gayeski and

Honig who fail to find any consistent Krogh relation between the PO_2 at a point and the proximity of that point to an active capillary. They also observed significant gradients in myoglobin saturation between contiguous myocytes, leading to PO_2 differences consistent with the present model.

The third, and perhaps most unexpected result of the superposition, may be noted in the regression of transcellular PO_2 drops versus the capillary PO_2 levels. The constant term, representing the non-local contributions to the transcellular PO_2 drop is always negative. That is the PO_2 levels in the tissue are reduced by superposition rather than enhanced. The effect is also noted in Figure 1 where the PO_2 at a point in the tissue is studied as a function of the number of capillaries in the tissue cylinder model. The PO_2 level at the point falls from a Krogh value for $N = 1$ to a value about 1/3 as high as the number of capillaries in the model increases beyond approximately 100 and remains at that level as N continues to increase.

This relative PO_2 deficit, we feel is the result of the relaxation on the characteristic length scale (Krogh diameter). The zero flux boundaries disappear in the absence of symmetry. This means that the distance for PO_2 adjustment is consequently greater so the size of the gradients is correspondingly reduced.

It seems that one of nature's adaptations is that the irregularities in both capillary geometry and capillary PO_2 levels should produce a situation in which the tissue PO_2 is fairly uniform.

The authors wish to thank Drs. Richard Connett, Thomas Gayeski, Karlfriend Grobe, and Carl Honig for many helpful discussions of this work.

SUMMARY

High resolution measurement in both exercised skeletal and cardiac tissue made radially outward from capillaries and longitudinally parallel to capillaries by Gayeski and Honig (1986, 1986a,b) and Honig and Gayeski (1987) indicate shallow variation of tissue PO_2 and the absence of strong causal relation between the PO_2 at a point and the proximity of that point to the nearest active capillary. Proposed as a model for the analysis of this tissue PO_2 distribution, so contrary to the expectations of Krogh type models, are a class of multicellular tissue cylinder models. Each cylinder is penetrated by many parallel capillaries. In order to better represent the natural irregularities of the skeletal and cardiac tissue both with regard to radial placement and the stagger of the capillary inlets, the following types of models both with and without yoglobin are examined: regular square arrays where the capillary PO_2 levels are random, uniform capillary PO_2 levels but random capillary positions, and those with both the capillary PO_2 levels and the positions are random.

The results of the model calculations show that the superposition of the oxygen diffusion fields of all the capillaries produce a tissue PO_2 distribution with the properties:
(1) lower tissue PO_2 levels than those predicted by Krogh theory,
(2) significant non-local contributions to the PO_2 at a point in the tissue which greatly reduces this correlation between PO_2 at a point and its proximity to an active capillary,
(3) shallow transcellular PO_2 variation.

This work was supported in part by NIH Grant HL 37205.

REFERENCES

Clark, A. Jr., and Clark, P.A.A., 1985, Local oxygen gradients near isolated mitochondria, Biophys. J., 48: 931.

Clark, A. Jr., Grobe, K., and Clark, P.A.A., 1988, An upper bound on oxygen loss from arterioles, ISOTT (abstract).

Connett, R.J., Gayeski, T.E.J., and Honig, C.R., 1985, An upper bound on the minimum PO_2 for oxygen consumption in red muscle, Adv. Exper. Med. Biol., 191: 291.

Duling, B.R. and Berne, R.M., 1970, Longitudinal gradients in periarteriolar oxygen tension, Circ. Res., 27: 669.

Federspiel, W.J. and Sarelius, I.H., 1984, An examination of the contribution of red cell spacing to the uniformity of oxygen flux at the capillary walls, Microasc. Res., 27: 273.

Federspiel, W.J., 1986, A model study of intercellular oxygen gradients in a myoglobin-contain ing skeletal muscle fiber, Biophys. J. 49: 857.

Federspiel, W.J., and Popel, A.S., 1986, A theoretical analysis of the effect of the particulate nature of blood on oxygen release in capillaries, Microasc. Res., 32: 164.

Fletcher, J.E., 1980, On facilitated oxygen diffusion in muscle tissue, Biophys. J., 29: 437.

Gayeski, T.E.J., and Honig, C.R., 1986, O_2 gradients from sarcolema to cell interior in red muscle at maximal $\dot{V}O_2$, Am. J. Physiol., 251: 789.

Gayeski, T.E.J., and Honig, C.R., 1988a, Intracellular PO_2 in individual cardiac myocytes in dog, cat, rabbit, ferret, and rat, Am. J. Physiol., in press.

Gayeski, T.E.J., and Honig, C.R., 1988b, Intracellular PO_2 in the long axis of individual fibers in working dog gracilis muscle, Am. J. Physiol., 254: 1179

Grobe, K., and Thews, G., 1986, Theoretical analysis of oxygen supply to contracted skeletal muscle, Adv. Exper. Med. Biol., 200: 495.

Hellums, J.D., 1977, The resistance to oxygen transport in capillaries relative to that in surrounding tissue, Microasc. Res., 13: 131.

Honig, C.R., Gayeski, T.E.J., Connett, R.J., Federspiel, W.J., Clark, A. Jr., and Clark, P.A., 1984, Muscle O_2 gradients from hemoglobin to cytochrome: new concepts, new complexities, Adv. Exper. Med. Biol., 169: 23.

Honig, C.R., and Gayeski, T.E.J., 1987, Comparison of intracellular PO_2 and conditions for blood-tissue O_2 transport in heart and working red skeletal muscle, Adv. Exper. Med. Biol., 215: 309.

Honig, C.R., and Gayeski, T.E.J., 1988, Precapillary O_2 loss and arterio-venous diffusion shunts are below limits of detection in myocardium, ISOTT (abstract).

Horstman, D.H., Gleser, M., Delehunt, J., 1976, Effects of altering O_2 delivery on $\dot{V}O_2$ of isolated working muscle, Am. J. Physiol., 230: 327.

Klitzman, B., Popel, A.S., and Duling, B.R., 1983, Oxygen transport in resting and contract-ing hamster muscles: experimental and theoretical microvascular studies, Microasc. Res., 25: 108.

Kreuzer, F., 1970, Facilitated diffusion of oxygen and its possible significance: a review, Respir. Physiol., 9: 30.

Krogh, A., 1919, The number and distribution of capillaries in muscles with calculations of the oxygen pressure head necessary for supplying the tissue, J. Physiol., 52: 409.

Krogh, A., "The Anatomy and Physiology of Capillaries", Yale U.P., New Haven (1929).

Myrhage, R., and Eriksson, E., 1980, Vascular arrangements in hind muscles of cats, J. Anat., 131: 1.

Popel, A.S., 1978, Analysis of capillary-tissue diffusion in multicapillary systems, Math. Biosci., 39: 187.

Popel. A.S., and Gross, J.F., 1979, Analysis of oxygen diffusion from arteriolar networks, Am. J. Physiol., 237: 681.

Popel, A.S., 1980, Oxygen diffusion from capillary layers with concurrent flow, Math. Biosci., 50: 171.

Reeves, W.J., and Rakusan, K., 1988, Myocardial capillary flow pattern as determined by the method of coloured microspheres, in: "Oxygen Transport to Tissue, X", Mochizuki, M., Honig, C.R., Koyama, T., Goldstick, T.K., and Bruley, D.F., eds., Plenum, New York.

Secomb, T.W., and Hsu, R., 1988, Analysis of oxygen delivery to tissue, in: "Oxygen Trans-port to Tissue, X", Mochizuki, M., Honig, C.R., Koyama, T., Goldstick, T.K., and Bruley, D.F., eds., Plenum, New York.

Weibel, E.R., "The Pathway for Oxygen", Harvard U.P., Cambridge (1984).

Wittenberg, J.B., 1970, Myoglobin-facilitated oxygen diffusion: role of myoglobin in oxygen entry into muscle, Physiol. Rev. 50: 559.

Wittenberg, B.A., and Wittenberg, J.B., 1985, Oxygen gradients in isolated cardiac myocytes, J. Biol. Chem., 260: 6548.

EFFECTS OF RED CELL SPACING AND RED CELL MOVEMENT UPON OXYGEN RELEASE UNDER CONDITIONS OF MAXIMALLY WORKING SKELETAL MUSCLE.

K. GROEBE[1], AND G. THEWS[2]

[1]Dept. of Mechanical Engineering, University of Rochester,
Rochester, NY 14627, USA

[2]Physiologisches Institut der Universität Mainz,
Saarstr. 21, D-6500 Mainz, West Germany

INTRODUCTION

The impacts of the particulate nature of blood upon capillary O_2 release have been studied extensively by FEDERSPIEL and SARELIUS [8] and by FEDERSPIEL and POPEL [9]. The latter authors found that the O_2 flux out of a capillary decreases rapidly as intracapillary red blood cell spacing increases. The O_2 flux out of a single RBC, however, is enhanced as long as the inter-erythrocytic plasma gap does not exceed the "zone-of-influence" of a single RBC, which they determined to be about 1 capillary diameter. In their model, they considered spherical red cells contained in a cylindrical tube filled with plasma, on the lateral surface of which a boundary P_{O_2} was specified. Based on earlier studies by AROESTY and GROSS [2], they neglected the contribution of intracapillary convection of plasma on O_2 transport. By restricting their view to the capillary interior, they disregarded the interactions between moving RBCs inside the capillary on the one hand and the stationary capillary wall and surrounding tissue on the other. This interaction may be characterized by "charging" and "discharging" of the stationary structures with oxygen as an RBC or a plasma gap, respectively, passes by. The importance of this interaction is being addressed in the present study.

Furthermore, in real tissue, P_{O_2} as well as O_2 flux at the capillary wall are not constants but functions of time, in a sense that both of these quantities oscillate at a frequency which is set by the intracapillary passage of RBCs and plasma gaps. The cases of constant boundary P_{O_2} or constant boundary O_2 flux, however, represent two important limiting cases as will be detailed later.

METHODS

Oxygen transport from a capillary is described by the well known system of partial differential equations:

---

[1]Supported by Deutsche Forschungsgemeinschaft grant Gr 887/1–1 and by NIH grant HL37205

$$\left.\begin{array}{rcl}\dfrac{\partial P}{\partial t}+\vec{\mathbf{v}}\cdot\nabla P &=& D_{O_2E}\nabla^2 P+\dfrac{\varrho(P,S)}{\alpha_E}\\[2mm]\dfrac{\partial S}{\partial t}+\vec{\mathbf{v}}\cdot\nabla S &=& D_{Hb}\nabla^2 S-\dfrac{\varrho(P,S)}{[Hb]_T}\end{array}\right\}\qquad\text{in the RBC, }\mathcal{E}\qquad\qquad(1)$$

$$\dfrac{\partial P}{\partial t}+\vec{\mathbf{v}}\cdot\nabla P \;=\; D_{O_2B}\nabla^2 P\qquad\qquad\qquad\text{outside the RBC, }\mathcal{B},$$

where the following conventions apply:

∇	gradient operator
P	oxygen partial pressure
S	hemoglobin O_2 saturation
$\vec{\mathbf{v}}$	local velocity vector of fluid movement
$D_{O_2E},\,D_{Hb},\,D_{O_2B}$	diffusion coefficients of O_2 or Hb inside the erythrocyte, $\mathcal{E}$, or of O_2 inside the rest of the domain of integration, $\mathcal{B}$, respectively
α_E	Bunsen's O_2 solubility coefficient inside the erythrocyte
$[Hb]_T$	total intra-erythrocytic hemoglobin concentration
$\varrho(P,S)$	net exchange rate between free and hemoglobin bound oxygen.

At the interface between the erythrocyte, $\mathcal{E}$, and the rest of the domain of integration, $\mathcal{B}$, P_{O_2} and total O_2 flux are to be continuous, whereas at the outer perimeter of $\mathcal{B}$ a boundary P_{O_2} (or a boundary O_2 flux) is prescribed.

As analytical solutions to this system do not exist, various approaches have been suggested (see e.g. [19,17,4,6]). Unfortunately, the stochastic approach [4] by BRULEY et al. which appears to be best suited to reflect the complex geometry of RBC, capillary, and surrounding tissue space, is impractical. (BRULEY et al. [4] needed 150 cpu minutes for less than 1/20 completion of just one RBC O_2 unloading run, even though they modelled blood as a *homogeneous* medium. This clearly would be an infeasible oversimplification in the case of the questions addressed in the present paper, as it deals with the effects of the particulate nature of blood, i.e. of blood *heterogeneity*.) As a compromise between geometrical exactness and cost efficiency, the system is solved using a previously introduced 2-dimensional finite difference model for RBC oxygen unloading in capillaries [14]. In this model, erythrocyte, plasma and tissue spaces are approximated by axially symmetric cylinders which are contained within one another and the axes of which are aligned. The cylinder representing the RBC is filled with hemoglobin and is shorter than the others thus allowing for an inter-erythrocytic plasma gap.

Following CLARK et al. [5] and AROESTY and GROSS [2], convection inside RBCs and convection in plasma boluses trapped between red blood cells are neglected, and only the net movement of the flowing blood along the capillary axis is considered.

The parameter values employed all refer to the situation in working dog gracilis muscle and are listed in Table 1. The choice of the blood parameters has been discussed by CLARK et al. [6]. The Hb-O_2-dissociation curve used is taken from a nomogram by THEWS [18] and is adjusted to the conditions in working muscle. For dimension and diffusivity of the capillary wall structures, no experimental data are available. Therefore, their thickness is estimated from electronmicroscopic cross sections through muscle capillaries (see e.g. [3]). An estimated average water content of 80 % is used to retrieve O_2 solubility and O_2 diffusion coefficient from the respective relationships established by ZANDER [23] and VAUPEL [20]. A number of measurements of RBC velocities in capillaries of resting muscles has been compiled

Table 1: Parameter values employed in the numerical evaluation (values in parentheses state ranges examined). All data refer to 37^0 C.

MCV	=	RBC volume	= $66\ fl$	[1]
R_E	=	RBC radius	= $2\ \mu m$	
R_C	=	capillary radius	= $2.75\ \mu m$	[7]
$R_{EI}-R_C$	=	thickness of endothelium and interstitial space	= $0.8\ (0.8-3.2)\ \mu m$	[3]
d_e	=	plasma gap length	= $0-4$ RBC lengths	
u_{RBC}	=	RBC velocity	= 2 or $4\ mm/s$	[10]
P_B	=	boundary P_{O_2}	= $3\ mm\,Hg$	[12]
α_E	=	O_2 solubility in the RBC	= $1.56{\cdot}10^{-9}\ \frac{mole}{cm^3{\cdot}mm\,Hg}$	[6]
α_P	=	O_2 solubility in plasma	= $1.3{\cdot}10^{-9}\ \frac{mole}{cm^3{\cdot}mm\,Hg}$	[23]
α_{EI}	=	O_2 solubility in endothelium and interstitial space	= $7.58{\cdot}10^{-10}\ \frac{mole}{cm^3{\cdot}mm\,Hg}$	[23]
D_{O_2E}	=	O_2 diffusion coefficient in the RBC	= $9.5{\cdot}10^{-6}\ cm^2/s$	[6]
D_{O_2P}	=	O_2 diffusion coefficient in plasma	= $2.18{\cdot}10^{-5}\ cm^2/s$	[13]
D_{O_2EI}	=	O_2 diffusion coefficient in endothelium and interstitial space	= $1.3{\cdot}10^{-5}\ cm^2/s$	[20]
D_{Hb}	=	Hb diffusion coefficient in the RBC	= $1.44{\cdot}10^{-7}\ cm^2/s$	[6]
$[Hb]_T$	=	total Hb monomer concentration in the RBC	= $2.03{\cdot}10^{-2}\ mole/l$	[22]
α_M	=	O_2 solubility in the muscle fiber	= $9.4{\cdot}10^{-10}\ \frac{mole}{cm^3{\cdot}mm\,Hg}$	[23]
$[Mb]_T$	=	total Mb concentration in the muscle fiber	= $5.4{\cdot}10^{-4}\ mole/l$	[21]
P_{50}	=	half saturation P_{O_2} of Mb	= $5.3\ mm\,Hg$	[11]
H'	=	slope of the Mb oxygen dissociation curve at $1\ mm\,Hg$	= $0.13\ mm\,Hg^{-1}$	
a	=	O_2 consumption rate of the fiber	= $16\ \frac{ml\,O_2}{100g{\cdot}min}$	[12]

by FRONEK and ZWEIFACH [10]. The ones chosen for this study are meant to be representative for heavily working muscle and therefore range among the greatest values measured in muscle capillaries. The boundary P_{O_2} is chosen in accordance with median Mb-O_2-saturation in maximally working skeletal muscle as measured by GAYESKI and HONIG [12].

RBC desaturation is calculated starting at a saturation of 85 % at the capillary origin and finishing at a saturation of 40 % at the venous capillary ending. These values are typical of dog gracilis muscle working at a twitch frequency of 8 Hz [16,15]. Whenever a desaturation run of the program was repeated with an O_2 flux instead of a P_{O_2} boundary condition, the prescribed O_2 flux was calculated to match the average RBC O_2 flux from the previous run.

The results are being presented in terms of an RBC O_2 conductance k which is defined by

$$F = k(\overline{P} - P_B)$$

where F is the flow of O_2 out of the RBC, $\overline{P}$ is the equilibrium P_{O_2} pertinent to the mean RBC O_2 saturation, and P_B is the boundary P_{O_2}, all of these quantities averaged over the entire O_2 unloading time.

RESULTS

Fig. 1a shows the O_2 conductance of a single RBC unloading O_2 in a capillary of working muscle, as a function of inter-erythrocytic gap length. Each data point involves one full O_2 unloading calculation as detailed above. The solid curve represents a stationary RBC whereas with the dashed curve, RBC movement of 4 mm/s relative to the capillary wall is considered. In both curves, RBC O_2 flux is enhanced by an increase in spacing. This effect saturates, however, for the stationary RBC at gaps of about 1 RBC length and at an enhancement of about 1.4-fold. Conductance in the moving RBC increases to almost 1.7-fold, this increase saturating at gaps of about 2 RBC lengths only. The difference between moving and stationary cases is due to the fact that the stationary capillary structures are "charged" with oxygen as an RBC passes by, which they supply to the tissue during transit of plasma gaps, when O_2 flux out of the capillary is low.

On the scale of an entire capillary, increase in RBC spacing decreases the overall flow of oxygen and hence O_2 conductance, because increasing portions of the capillary wall become non-functional for O_2 exchange. This effect is in part compensated for by an increase in conductance per RBC as shown in Fig. 1b. Again, solid and dashed curves represent the conductances of capillaries filled with stationary or moving erythrocytes, respectively. The dotted curve is the conductance to be expected if enhancement of single RBC conductance were absent. Even though overall capillary conductance drops very rapidly with increasing cell separation in either case, it is, e.g. at gaps of 2 RBC lengths, enhanced by a factor of 1.4 (stationary RBC) or of

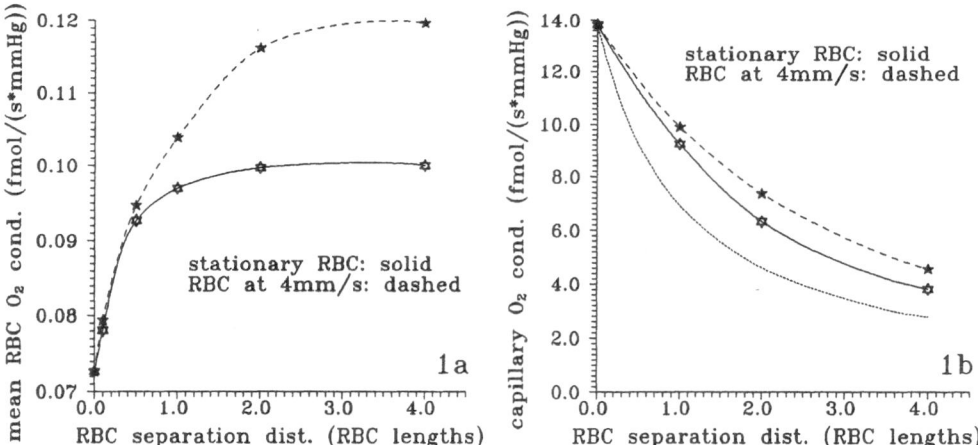

Figure 1. O_2 conductance of a single RBC (Fig. 1a) or of a whole muscle capillary of 1 mm length (Fig. 1b) as a function of inter-erythrocytic plasma gap length. The solid curves represent RBCs stationary inside the capillary whereas the dashed curves consider RBC movement at 4 mm/s. The dotted curve in Fig. 1b gives the conductance to be expected if enhancement of single RBC conductance were absent.

1.6 (RBC at 4 mm/s) over the case of constant conductance per RBC. This in part explains the beneficial effects of hemodilution upon tissue O_2 supply: If systemic hematocrit is reduced such that a capillary RBC spacing of 2 RBC lengths results, the same amount of oxygen can be transferred from the capillaries to the tissue with an RBC flux which is by a factor of 1.6 lower.

Tissues constantly need to be supplied with oxygen. This is particularly true for tissues with high O_2 consumption rates like working red muscle or brain. Therefore, the question of the uniformity of O_2 flux out of a capillary is of importance. The answer to this question furnished by a theoretical model calculation is largely predetermined by the type of boundary condition imposed at the capillary wall. In order to understand this, we ask ourselves which values of P_{O_2} and O_2 flux an observer posted at the capillary wall would see: Each time an erythrocyte passes by, he would sense a raised intracapillary P_{O_2} leading to a burst of oxygen out of the capillary, which, in turn, would increase the local P_{O_2} next to the capillary tissue interface. This would diminish the local O_2 flux out of the capillary and give rise to a longitudinal O_2 flux directed towards locations of lower P_{O_2}. During passage of a plasma gap he would observe the reverse: low intracapillary P_{O_2} and O_2 flux and, therefore, an even lower P_{O_2} at the capillary tissue interface, which successively would intensify O_2 flux from intra- and extracapillary locations into his observation site. As quantitative information on the distributions of both P_{O_2} and O_2 fluxes next to a capillary is lacking, for modelling purposes one generally assumes either a constant P_{O_2} typical of the tissue under consideration or a constant oxygen flux matching tissue O_2 consumption rate and capillarity. It becomes clear from the above that a constant P_{O_2} boundary condition exaggerates the non-uniformity of O_2 flux. On the other hand, a constant flux boundary condition annihilates non-uniformities in O_2 flux at the cost of over-emphasizing variations in boundary P_{O_2}. Thus, the P_{O_2} boundary condition establishes an upper bound on the non-uniformity of O_2 flux across the boundary, whereas the flux boundary condition gives an upper bound on the non-uniformity of boundary P_{O_2}.

The non-uniformity of boundary O_2 flux and boundary P_{O_2} is qualitatively demonstrated in Figs. 2a and 2b, which display longitudinal profiles of boundary O_2 flux density and of boundary P_{O_2}, calculated using a P_{O_2} or an O_2 flux boundary condition, respectively. The solid curves represent the cases of RBCs stationary inside the capillary at spacings of 1, 2, or 4 RBC lengths, respectively, whereas the dashed ones refer to RBCs spaced in the same fashion and moving down the capillary at 4 mm/s. From both Figs. 2a and 2b it becomes intuitively clear that 1) non-uniformity of O_2 flux and of P_{O_2} increases with extending inter-erythrocytic gaps (the ratio of e.g. the maximum over the minimum value of the flux distribution for the stationary case is 2.5 at a spacing of 2 RBC lengths and 120 at a spacing of 4 RBC lengths), and that 2) RBC movement inside the capillary improves uniformity (the ratio of maximum over minimum flux at a spacing of 4 RBC lengths is 120 in the stationary case and 31.5 at 4 mm/s). This may be quantified by considering the coefficients of variation of the boundary O_2 flux distribution (Fig. 3, stars) for the P_{O_2} boundary condition or of the boundary P_{O_2} distribution (Fig. 3, dots) for the O_2 flux boundary condition. Solid or dashed curves represent the cases of stationary or moving RBCs, respectively. The two sets of curves marked by dots or stars both confirm the above statements.

In conclusion, RBC movement not only enhances RBC O_2 release but also renders O_2 flux densities out of the capillary and P_{O_2} at the outer capillary wall more uniform.

DISCUSSION

Relation to work of others

The effects of the discreteness of blood upon oxygen release have been studied before by FEDERSPIEL and SARELIUS [8] and by FEDERSPIEL and POPEL [9]. In these studies, the capillary was modelled as a fluid filled tube containing the erythrocytes, at the inner wall of which an O_2 flux [8] or a P_{O_2} [9] boundary condition was imposed.

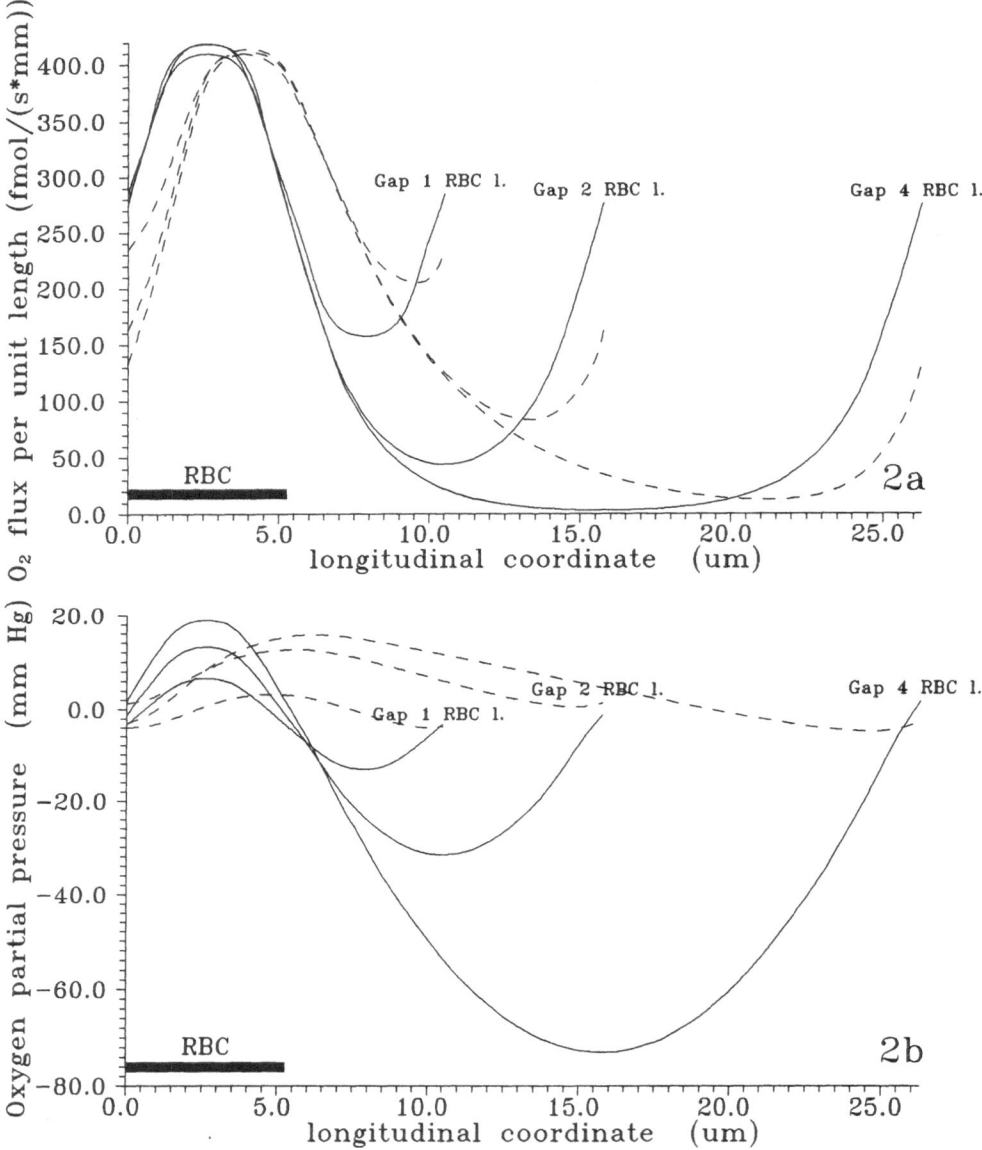

Figure 2. Longitudinal profiles of boundary O_2 flux density (Fig. 2a) and of boundary P_{O_2} (Fig. 2b), calculated using a P_{O_2} (Fig. 2a) or an O_2 flux (Fig. 2b) boundary condition, respectively. The solid curves represent the cases of RBCs stationary inside the capillary at spacings of 1, 2, or 4 RBC lengths (labelled "Gap 1 RBC 1." etc.), respectively, whereas the dashed ones refer to RBCs spaced in the same fashion and moving down the capillary at 4 mm/s.

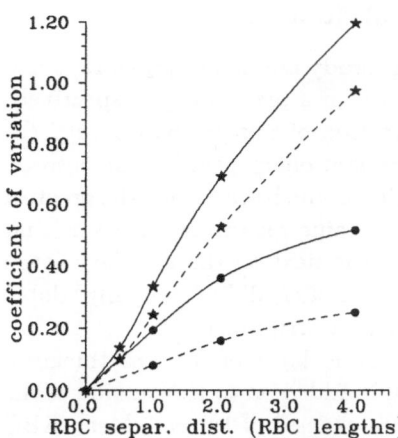

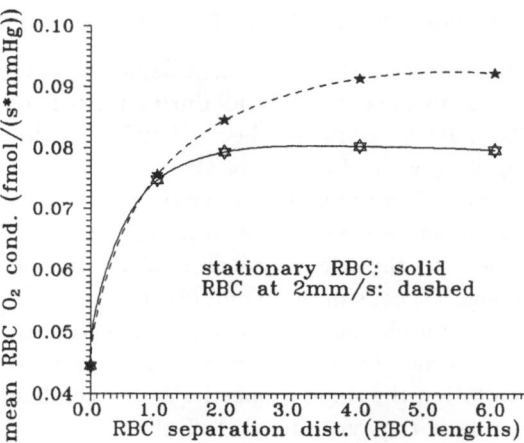

Figure 3. Coefficients of variation of the boundary O_2 flux distribution (stars) for the P_{O_2} boundary condition or of the boundary P_{O_2} distribution (dots) for the O_2 flux boundary condition as functions of inter-erythrocytic gap length. Solid or dashed curves represent the cases of stationary RBCs or RBCs moving at 4 mm/s, respectively.

Figure 4. O_2 conductance of a single RBC as a function of inter-erythrocytic plasma gap length for an "effective layer thickness" of 1.6 μm (about twice the capillary wall thickness). The solid curve represents RBCs stationary inside the capillary whereas the dashed curve considers RBC movement at 2 mm/s. Note that in Figs. 1–3 layer thickness was 0.8 μm and RBC velocity was 4 mm/s.

Therefore, no predictions could be made about the interactions between moving RBCs and stationary capillary wall structures.

FEDERSPIEL and SARELIUS [8] found that a positive P_{O_2} at the inner capillary wall could not be sustained in gaps larger than 1 RBC length, if an O_2 boundary flux typical for maximally working skeletal muscle was assumed. This is in good agreement with our results for stationary RBCs (profiles not shown). The P_{O_2} at the outer capillary wall, however, would clearly have to be negative even at spacings of 1 RBC length (Fig. 2b, solid curves). On the other hand, boundary P_{O_2} drops hardly below zero in the case of moving RBCs, not even at spacings of 4 RBC lengths (Fig. 2b, dashed curves).

FEDERSPIEL and POPEL [9] used a P_{O_2} boundary condition and examined the impacts of RBC spacing and clearance upon average O_2 conductance. Their absolute values of capillary O_2 conductance are greater than ours by a factor of about 2 because their particles are separated from the boundary by a plasma layer, the average thickness of which is only somewhat more than 1/3 of the aggregate thickness of plasma sleeve and capillary wall structures considered here. The enhancement of single RBC O_2 release by inter-erythrocytic plasma gaps they found is in good agreement with our results for the stationary case, in that this effect saturates at gap lengths of about 1 capillary diameter. They conclude from this that the "zone-of-influence" (with respect to O_2 release) of a given particle does not extend substantially beyond the particle. Considering the case of moving RBCs, however, it becomes clear from Fig. 1a that the "zone-of-influence" may extend beyond the RBC by more than 2 RBC lengths, this number being a function of the assumed thickness of the stationary layer (see below).

Extension of results towards the physiological situation

The effects of RBC movement demonstrated in this study are due to O_2 storage and release in the capillary wall during passage of an RBC or a plasma gap, respectively. The authors centered their attention to the interaction of moving RBCs with the *capillary wall*. Because of the small volume of the stationary capillary structures, marked effects could be observed only at RBC velocities, which are among the greatest that have been measured in muscle capillaries. The same reasoning, however, that applies to the capillary wall is also true for the tissue next to the capillary (even though O_2 consumption and O_2 storage and facilitation of O_2 diffusion by myoglobin in red muscle somewhat complicate the mathematical description).

In order to demonstrate this qualitatively, a tissue layer of 0.8 μm thickness exhibiting the same properties as the capillary wall is added on to the outside of the capillary, thus yielding an "effective capillary wall thickness" of 1.6 μm. Mean RBC O_2 conductance as a function of RBC spacing is displayed in Fig. 4 (notations as in Fig. 1a). Note that the RBC velocity considered is only 2 mm/s and thus well in the physiological range to be expected in heavily working muscle. Nevertheless, conductance at a spacing of 4 RBC lengths is enhanced over conductance at zero spacing by a factor of 1.8 (stationary RBC) or of 2.1 (RBC at 2 mm/s) and the enhancement by RBC movement saturates at a spacing of 4 RBC lengths only.

Uniformity of oxygen flux density

Non-uniformity of boundary O_2 flux or of boundary P_{O_2} as illustrated in Fig. 2 is quite dramatic in the case of stationary RBCs. For RBCs travelling at 4 mm/s, the O_2 flux boundary condition yields very well-smoothed boundary P_{O_2} profiles. Boundary O_2 flux calculated using a P_{O_2} boundary condition, however, remains remarkably non-uniform, particularly for wide RBC spacings.

As pointed out in the preceeding paragraph the capillary cannot be studied separately but must be viewed in the context of the surrounding tissue. The importance of the choice of an appropriate "viewpoint" is demonstrated in Fig. 5. Longitudinal profiles of boundary O_2 flux, normalized by the respective mean boundary O_2 flux, are displayed for moving RBCs at a spacing of 4 RBC lengths and for "effective capillary wall thicknesses" of 0.8 μm (solid), 1.6 μm (dashed), and 3.2 μm (dotted). Obviously, addition of tissue to the model drastically decreases boundary O_2 flux heterogeneity. This is reflected in the coefficient of variation of the boundary O_2 flux distributions which drops from 1.0 (solid curve) to 0.4 (dotted curve). Increasing uniformity also suggests that the profiles obtained employing a P_{O_2} or an O_2 flux boundary condition "converge" with an increase in the amount of tissue added on to the model.

Even though there remains some non-uniformity of O_2 flux density out of the capillary even for the thickest layers considered, this is not likely to compromise local O_2 consumption rates. In a capillary in which RBCs travel at a velocity of 2 mm/s, a plasma gap of 4 RBC lengths passes a tissue site next to the capillary within 10.5 ms. If, during this period of time, O_2 flux into this site was stopped completely, this would result in a P_{O_2} drop of no more than 1.5 $mm\,Hg$ (if O_2 storage by Mb is neglected) or of 0.019 $mm\,Hg$ (if O_2 storage by Mb pertinent to a P_{O_2} of 1 $mm\,Hg$ is considered).

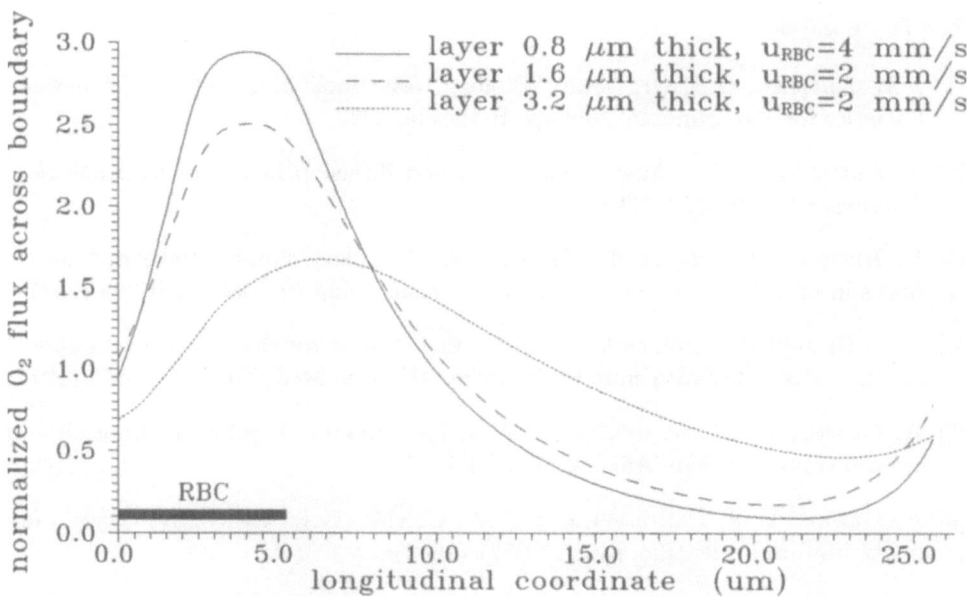

Figure 5. Longitudinal profiles of boundary O_2 flux, normalized by the respective mean boundary O_2 flux, for moving RBCs at a spacing of 4 RBC lengths and for "effective capillary wall thicknesses" of 0.8 μm (solid), 1.6 μm (dashed), and 3.2 μm (dotted). RBC velocity is 4 mm/s (solid) or 2 mm/s (dashed and dotted).

SUMMARY

RBC spacing in capillaries plays an important role in that it determines the total number of RBCs contained in a capillary and, therefore, the total O_2 flux out of the capillary. The detrimental effects of increased RBC spacing upon *capillary O_2 release* are in part compensated for by enhanced O_2 release out of *single RBCs* due to improved diffusion geometry and RBC movement.

Non-uniformity of O_2 flux brought about by the particulate nature of blood is considerably smaller than calculations which do not consider RBC movement indicate. It creates oscillations in the O_2 supply to the tissue, the periodicity of which is fast, however, compared to the time constant of the P_{O_2} decay in a temporarily unsupplied tissue.

We conclude that non-uniformity of O_2 flux out of capillaries due to large intererythrocytic plasma gaps does not play an important role for tissue O_2 supply as long as average RBC spacing is sufficiently small to guarantee an appropriate overall capillary O_2 flux.

Acknowledgements

The helpful comments of Drs. A. CLARK and C. R. HONIG in the preparation of the manuscript are gratefully acknowledged.

References

[1] P.L. ALTMAN, D.S. DITTMER, "Biology Data Book", Federation of American Societies for Experimental Biology, Bethesda, 1972

[2] J. AROESTY, J.F. GROSS, Convection and diffusion in the microcirculation, *Microvasc.Res.* 2:247 (1970)

[3] P. BRODAL, F. INGJER, L. HERMANSEN, Capillary supply of skeletal muscle fibers in untrained and endurancetrained men, *Am.J.Physiol.* 232:H705 (1977)

[4] D.F. BRULEY, L.J. GROOME, H. BICHER, M.H. KNISELY, A stochastic model for the transport of oxygen to brain tissue, *Adv.Exp.Med.Biol.* 75:267-277 (1976)

[5] A. CLARK, JR., G.R. COKELET, W.J. FEDERSPIEL, Erythrocyte motion and oxygen transport, *Bibl.Anat.* 20:385 (1981)

[6] A. CLARK, W.J. FEDERSPIEL, P.A.A. CLARK, G.R. COKELET, Oxygen delivery from red cells, *Biophys.J.* 47:171 (1985)

[7] E. ERIKSSON, M. MYRHAGE, Microvascular dimensions and blood flow in skeletal muscle, *Acta Physiol.Scand.* 86:211 (1972)

[8] W.J. FEDERSPIEL, I.H. SARELIUS, An examination of the contribution of red cell spacing to the uniformity of oxygen flux at the capillary wall, *Microvasc.Res.* 27:273 (1984)

[9] W.J. FEDERSPIEL, A.S. POPEL, A theoretical analysis of the effect of the particulate nature of blood on oxygen release in capillaries, *Microvasc.Res.* 32:164 (1986)

[10] K. FRONEK, B.W. ZWEIFACH, Microvascular blood flow in cat tenuissimus muscle, *Microvasc.Res.* 14:181 (1977)

[11] T.E.J. GAYESKI, C.R. HONIG, O_2 gradients from sarcolemma to cell interior in red muscle at maximal $\dot{V}_{O_2}$, *Am.J.Physiol.* 251:H789 (1986)

[12] T.E.J. GAYESKI, R.J. CONNETT, C.R. HONIG, Minimum intracellular P_{O_2} for maximum cytochrome turnover in red muscle in situ, *Am.J.Physiol.* 252:H906 (1987)

[13] T.K. GOLDSTICK, V.T. CIURYLA, L. ZUCKERMAN, Diffusion of oxygen in plasma and blood, *Adv.Exp.Med.Biol.* 75:183 (1975)

[14] K. GROEBE, G. THEWS, Theoretical analysis of oxygen supply to contracted skeletal muscle, *Adv.Exp.Med.Biol.* 200:495 (1986)

[15] C.R. HONIG, T.E.J. GAYESKI, Precapillary O_2 loss and arteriovenous O_2 diffusion shunt are below limit of detection in myocardium, *Adv.Exp.Med.Biol. (Oxygen Transport to Tissue XI)*, in press

[16] C.R. HONIG, T.E.J. GAYESKI, W. FEDERSPIEL, A. CLARK, P. CLARK, Muscle O_2 gradients from hemoglobin to cytochrome: new concepts, new complexities, *Adv.Exp.Med.Biol.* 169:23 (1984)

[17] W. MOLL, The influence of hemoglobin diffusion on oxygen uptake and release by red cells, *Respir.Physiol.* 6:1 (1968/69)

[18] G. THEWS, "Der Atemgastransport bei körperlicher Arbeit, Funktionsanalyse biologischer Systeme, Bd. 10", Akademie der Wissenschaften und der Literatur, Mainz, 1984

[19] G. THEWS, W. NIESEL Zur Theorie der Sauerstoffdiffusion im Erythrocyten, *Pflügers Arch.* 268:318-333 (1959)

[20] P. VAUPEL, Effect of percentual water content in tissues and liquids on the diffusion coefficients of O_2, CO_2, N_2, and H_2, *Pflügers Arch.* 361:201 (1976)

[21] G.H. WHIPPLE, The hemoglobin of striated muscle. I. Variations due to age and exercise, *Am.J.Physiol.* 76:693-707 (1926)

[22] M. W. WINTROBE, G. R. LEE, D. R. BOGGS, T. C. BITHELLS, J. FOERSTER, J. W. ATHENS, J. N. LUKENS, Clinical Hematology, Lea & Febiger, Philadelphia, 1981

[23] R. ZANDER, Cellular oxygen concentration, *Adv.Exp.Med.Biol.* 75:463 (1975)

CALCULATION OF OXYGEN PRESSURES AND FLUXES IN A FLAT PLANE PERPENDICULAR TO ANY CAPILLARY DISTRIBUTION

Louis Hoofd, Zdenek Turek and Jos Olders

Dept. Physiology
University of Nijmegen
Nijmegen, The Netherlands

INTRODUCTION

Most modelling of oxygen transport in tissue has been based on the classical Krogh-Erlang equation (Krogh, 1919), which describes oxygen diffusion in any circular cross-section of a spherical cylinder around a capillary. Extensions like facilitated diffusion and non-zero-order oxygen consumption can be built in and combined with blood flow and capillary spacing heterogeneity (Turek et al., this volume). Other models are based on computer simulations (Federspiel, 1986; Popel et al., 1986; Groebe, 1987), with the disadvantage of describing only specific situations; thus making application in real heterogeneous tissue impossible.

When the Krogh-Erlang model is compared with other models (e.g., see Rakusan et al., 1984; Hoofd et al., 1987), all these models appear to consist of the same principal form, describing oxygen partial pressure P as:

P = Constant + ConsumptionField + Source(s)

For example, in the Krogh-Erlang equation, ConsumptionField is a term of order r^2, where r is the radial distance from the center, Source is a logarithmic term $\ln(r)$, and Constant is adapted to the boundary conditions: the oxygen partial pressure in the capillary and the size of the Krogh cylinder. In this paper, the above description is generalised, so that it will lead to a mathematical formula describing oxygen partial pressure in a two-dimensional flat plane with arbitrary capillary distribution. The remaining question to be solved is that of the boundary conditions, for which a feasible solution will be presented[1].

The features of the solution are multiple. Firstly, of course, it allows calculation of oxygen partial pressure at any location in the field. Secondly, by following the path of the oxygen flowing out of the capillary, oxygen flux lines can be constructed. These flux lines together form an area around the capillary, that is supplied with O_2 from this capillary. Finally, the actual size of each oxygen supply area can be determined directly from the solution itself, without the need of calculating flux lines.

---

[1]This approach appears to be very similar to that presented by Clark et al. (this volume), except for one of the boundary conditions.

MATHEMATICAL APPROACH

Diffusion of oxygen is described by the mass balance equation:

$$\mathbf{P} \nabla^2 P = M \tag{1}$$

where $\mathbf{P}$ is the permeability for oxygen ("Krogh's diffusion coefficient", equal to $D\alpha$ where D is diffusion coefficient and α is solubility), P is the oxygen partial pressure and M is the oxygen consumption. ∇ is the gradient operator (differentiation with respect to location) and ∇^2 is called the Laplace operator.

In the Krogh-Erlang approach (Krogh, 1919), an essentially two-dimensional solution is derived that is radially symmetric, and consequently is entirely described in terms of the radial distance r from the origin, the center of the capillary:

$$P = P_c - \frac{Mr_c^2}{4\mathbf{P}} + \frac{MR^2}{2\mathbf{P}} \ln(r_c) + \frac{M}{4\mathbf{P}}\left[r^2 - 2R^2 \ln(r)\right] \tag{2}$$

where P_c is capillary pressure, r_c is the capillary radius and R is the size of the circular region where the oxygen is supplied to. The two terms between the brackets assess two different parts of the solution. The term r^2 constitutes a specific solution of the differential equation (1), i.e., the solution $P=Mr^2/4\mathbf{P}$ already satisfies this equation. The logarithmic term $\ln(r)$ is a solution of the harmonic equation, $\nabla^2 P=0$, which can be added to the specific solution. Here, the boundary condition that a circular region with radius R is supplied, is translated into the mathematical condition that the gradient dP/dr is zero at r=R. A term $\ln(r)$ multiplied by $2R^2$ has to be added to account for this condition.

Another solution of the harmonic equation comes up in the equation describing an "asymmetric cylinder", i.e., a non-circular supply region (Hoofd et al., 1987):

$$P = P_c - \ldots + \frac{M}{4\mathbf{P}}\left[r^2 + g_k\{(r/r_c)^k - (r_c/r)^k\}\cos(k\phi) - 2R^2 \ln(r)\right] \tag{3}$$

where the first terms, indicated with dots, are the same as in eq.(2), and g_k and k are constants depending on the form of the region. The term $g_k\{..\}\cos(k\phi)$, which now depends also on the angle ϕ of the polar coordinates, is also a solution of the harmonic equation, but accounts for the different boundary conditions of this situation.

A seemingly different solution was presented for a situation where the capillary is not located at the center, r=0, but two or more capillaries are surrounding a circular region (muscle fiber) where the oxygen now diffuses into:

$$P = P_o + \frac{MR^2}{\mathbf{P}} \ln(R) + \frac{M}{4\mathbf{P}}\left[r^2 - \frac{2R^2}{n} \ln\{r^{2n} - 2r^n R^n \cos(n\phi) + R^{2n}\}\right] \tag{4}$$

where P_o is the pressure at r=0 and n is the number of capillaries, located at distance r=R, which is also the size of the region, at angles $\phi=0$, $\phi=2\pi/n$, $\phi=4\pi/n$,.. etc. At first sight, this equation looks different from the foregoing ones, but it can be mathematically rewritten as:

$$P = P_o' + \frac{Mr^2}{4\mathbf{P}} - \frac{M}{4\mathbf{P}} \sum_{j=1}^{n} \frac{2R^2}{n} \ln(|\underline{r} - \underline{R_j}|^2) \tag{5}$$

where $\underline{r}$ is the vector (r,ϕ) and $\underline{R_j}$ is the place where the jth capillary is located and is equal to $(R, 2\pi j/n)$. Now, the harmonic solution consists of a set of ln()-functions, one for each capillary, in the same way as for the central capillary of the previous solutions.

Equation (5) already shows the features of the general solution. P_0' is a constant depending on the general level of the oxygen partial pressure. The term $Mr^2/4P$ is the simplest solution of eq.(1) for diffusion and consumption in a circular field. Each capillary is associated with an oxygen source term $\ln(|\underline{r}-\underline{R}_j|^2)$; this is analogous to, e.g., an electric charge in electromagnetic theory. The total amount of oxygen coming from a source $C\cdot\ln(|\underline{r}-\underline{R}_j|^2)$ can be found by drawing a (small) circle around $\underline{r}=\underline{R}_j$, calculating the radial flux, across this circle, defined as $J=P\partial P/\partial n$ where $\underline{n}$ is the vector perpendicular to the circle at the specific location, and integrating this flux over its circumference. Defining this delivered amount as O_{2Del}, the described mathematical process leads to:

$$O_{2Del} = 4\pi \; \mathbf{P} \; C \tag{6}$$

With a tissue oxygen consumption M, this oxygen is to supply an area A_j (for the jth capillary) equal to O_{2Del}/M, so that C can be expressed in terms of this oxygen supply area A_j as:

$$C = \frac{M}{4\mathbf{P}} \frac{A_j}{\pi} \tag{7}$$

Then, a general solution of differential equation (1) for n capillaries in a circular field is constructed:

$$P = C_P + \frac{Mr^2}{4\mathbf{P}} - \frac{M}{4\mathbf{P}} \sum_{j=1}^{n} \frac{A_j}{\pi} \ln(|\underline{r}-\underline{R}_j|^2) \tag{8}$$

where, again, C_P is a constant determined by the boundary conditions. Together with the n values of A_j, there is a total of n+1 constants to be determined.

A straightforward set of boundary conditions is, that the oxygen partial pressure at (the rim of) each capillary is given. However, around the ith capillary with capillary radius $r_{c,i}$, P can be different in each direction. Therefore, we adapt this boundary condition slightly to state that the average value over the capillary contour will be equal to this pressure $P_{rc,i}$:

$$P_{rc,i} = \frac{1}{2\pi r_{c,i}} \int_{-\pi}^{\pi} P(\underline{r}+(r_{c,i},\phi')) \; r_{c,i} \; d\phi' \tag{9}$$

where P(...) is according to eq.(8), evaluated at location $\underline{r}+(r_{c,i},\phi')$. This equation can be rewritten as:

$$P_{rc,i} = C_P + \frac{M}{4\mathbf{P}}\Big[R_i^2+r_{c,i}^2 - \frac{2A_i}{\pi}\ln(r_{c,i}) - \sum_{j\neq i}^{n} \frac{A_j}{\pi}\ln(|\underline{R}_i-\underline{R}_j|^2)\Big] \tag{10}$$

for each capillary i=1,.. n.

For the last boundary condition, the $(n+1)^{th}$, it has been chosen that all the n capillaries together supply an area that is as large as the total area A, or:

$$A = \sum_{j=1}^{n} A_j \tag{11}$$

For a circular area with radius R, A is equal to πR^2.

Two more extensions of the basic model of eq.(8) have been made:
- Facilitation of oxygen diffusion by myoglobin (Mb) can be incorporated, according to Hoofd et al. (1987), replacing P by $Po_2 + P_F S$, where Po_2 is the oxygen partial pressure and S is the myoglobin saturation with oxygen. P_F is called Facilitation Pressure (de Koning at al., 1981) and is equal to $D_{Mb} C_{Mb}/P$, where D_{Mb} is myoglobin diffusion coefficient and C_{Mb} is myoglobin concentration. In case of chemical equilibrium between O_2 and Mb, which is a reasonable approximation here (Hoofd, 1987), S is determined by the oxygen saturation curve of the myoglobin:

$$S = \frac{Po_2}{Po_2 + P_{so}} \tag{12}$$

where P_{so} is the half-saturation pressure. Then, also the left side of the boundary condition eq.(10) has to be rewritten as $P_{rc,i} + P_F S_i$, where S_i is calculated by substituting $Po_2 = P_{rc,i}$ in eq.(12).
- The definition of the capillary pressure $P_{rc,i}$ might not reflect what is generally called the capillary pressure. Here, it is defined as the average pressure just outside the capillary, where diffusion into the tissue commences. This value is likely to be lower than $P_{c,i}$, the (mean) pressure <u>within</u> the capillary, due to several resistances (e.g., diffusion in erythrocyte, plasma and capillary wall; nonequilibrium between hemoglobin and oxygen, etc.). Here, this topic is handled very generally: the additional resistance will result in a pressure difference, which is called "Capillary Barrier", CB (see Turek et al., this volume):

$$CB_i = P_{c,i} - P_{rc,i} \tag{13}$$

As resulting from transport resistances, the value of CB_i depends on the actual oxygen delivery of the capillary and here will be assumed to be proportional to the capillary flux and thus to the supply area A_i. "The" CB for a whole field containing several capillaries will be indicated as **CB** and is defined as the mean capillary barrier, i.e., for a mean area A/n. Then, according to eq.(13), the boundary condition pressure $P_{rc,i}$ is related to the capillary pressure $P_{c,i}$ as:

$$P_{rc,i} = P_{c,i} - \mathbf{CB}\,\frac{nA_i}{A} \tag{14}$$

Now, oxygen partial pressures can be calculated from eq.(8), after having solved the values of C_P and A_j from the set of eqs.(10) and (11); for facilitated diffusion, the substitution $P=Po_2+P_F S$ must be made in eqs.(8) and (10) with S according to eq.(12); for a capillary barrier, also eq.(14) has to be incorporated.

RESULTS

The above derived set of equations was applied to an exemplary region, a cross section of the subendocardial region of a normal heart of a Wistar rat. Capillary locations in a rectangular photomicrograph of this section were read in into a computer as described by Hoofd et al. (1985), and a circular region of radius 60 μm was selected (see fig.1). For all capillaries, a capillary radius of 2.3 μm was assumed. Data for M and **P** were 0.4 $ml \cdot g^{-1} \cdot min^{-1}$ and $3.53\ 10^{-10}\ ml \cdot s^{-1} \cdot cm^{-1} \cdot mmHg^{-1}$ respectively, taken from Turek and Rakusan (1981). All calculations were done including Mb facilitated diffusion with a P_F of 14 mmHg (Hoofd et al., 1987) and P_{so} of 5.3 mmHg (Gayeski, 1982). Although the latter value seems to be unusually high, it is adopted here for better comparison with literature calculations.

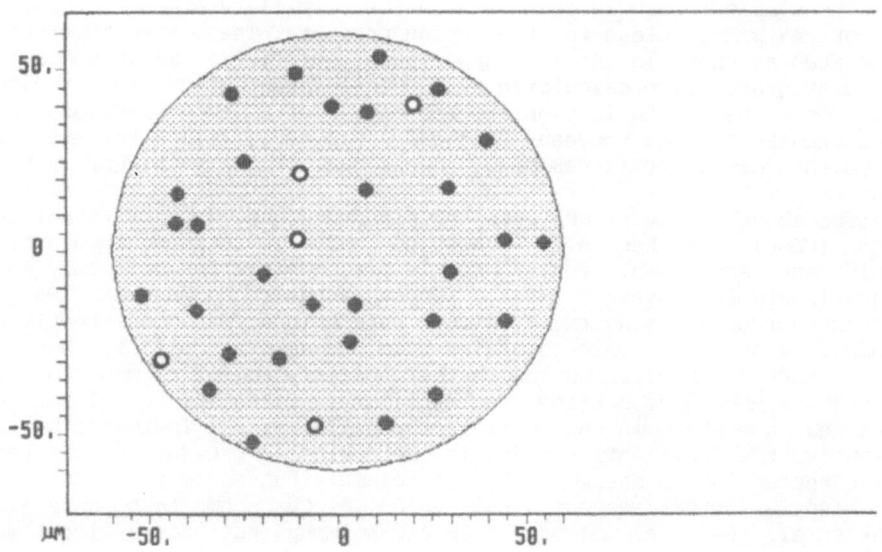

Fig. 1. Computer representation of the exemplary circular region, with the 35 capillaries indicated as small circles. Those 5 indicated with open circles, randomly selected, were given a high capillary pressure in some calculations (see text).

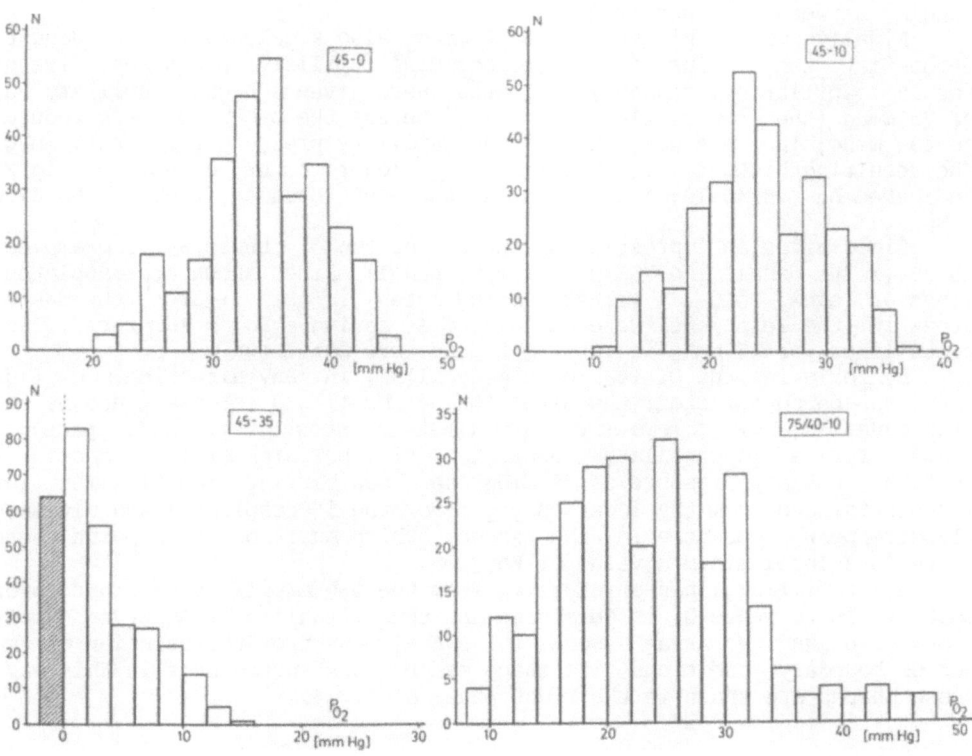

Fig. 2. Po_2-histograms, number N of Po_2's in each class of width 2 mmHg, calculated for different conditions of capillary pressure(s) – capillary barrier, as indicated in the panels.

The simplest case is that of a uniform capillary pressure, for which we took 45 mmHg, close to the venous pressure. Then, a Po_2 histogram is calculated as shown in the left upper panel of fig.2, indicated by 45-0. (All histograms were calculated at 300 equidistant points in the circular area). Note that this is a histogram only for a two-dimensional plane! Calculated Po_2's are between 21.4 and 45.4 mmHg; since the average Po_2 around the · capillary is 45 mmHg, occasional slightly higher values are found.

The above calculations were obtained wihtout any additional resistance (**CB**=0). It has been suggested that in canine myocardium a significant additional resistance is responsible for uniformly low Po_2 values (Honig and Gayeski, 1987). These authors found that Po_2, when calculated from the measured myoglobin saturations, was consistently below 10 mmHg. In order to mimic this situation, we need a capillary barrier **CB** of at least 35 mmHg (which lowers the capillary Po_2 of 45 mmHg to a value of 45-35=10 mmHg just outside the capillary). Calculations with this set of values are shown in the lower left panel of fig.2, indicated by 45-35. Obviously, the facilitation by Mb is not able to prevent anoxic regions (all gathered in the dashed histogram column situated left from 0).

Also for other reasons, this value of **CB** seems to be unrealistic. Turek et al. (this volume) used a **CB** of 35 mmHg, but for skeletal muscle with a mean capillary supply area of 1225 μm^2. Here, with rat heart muscle tissue, the mean area is 323 μm^2; accordingly, we might choose a value of **CB** of 35(323/1225)=10 mmHg. Calculations with this barrier value are presented in the upper right panel of fig.2, indicated by 45-10. Po_2 values now are between 11.9 and 36.9 mmHg; here, the values above 35 mmHg are mainly due to the variation in the individual CB_i for each capillary, ranging between 7.75 and 12.9 mmHg.

Finally, with this value, **CB**=10 mmHg, also a calculation was done to demonstrate the influence of heterogeneous capillary pressures. Five of the 35 capillaries, randomly selected, were given a higher capillary Po_2 of 75 mmHg (the open circles in fig.1), whereas the other 30 were reduced to 40 mmHg; in this way, the average capillary pressure remains 45 mmHg. The resulting histogram is shown in the lower right corner of fig.2, indicated by 75/40-10; Po_2 range is now much broader, from 9.7 to 47.0 mmHg.

Fig.3 gives an impression of the Po_2 of two of these above examples, 45-35 (left panel) and 75/40-10 (right panel). Lines shown are isopleths, lines of equal Po_2, at values as indicated in the figures; the dashed areas in the left panel were calculated as having a Po_2 below zero. (Isopleth lines for 45 mmHg in the right panel are not shown.)

By following the O_2 leaving the capillary in any direction, O_2 flux lines can be constructed; the direction of the flux $\underline{J}$ is determined by the flux equation $\underline{J}=\nabla P$. Examples of flux lines are shown in the left panel of fig.4, for a 13-capillary subsection of the area of fig.1 and for a uniform capillary pressure of 45 mmHg and no barrier (**CB**=0). The flux lines originate from the location $\underline{r}=\underline{R}i$ (for the ith capilary) and ultimately disappear somewhere in the tissue, which must be at a point where there is a local minimum value of Po_2.

All the flux lines originating from the ith capillary together cover that region to where O_2 is delivered and thus visualize the oxygen supply area. Although the area size A_i follows already from the solution of the set of boundary conditions, its shape can be constructed only in this way. These shapes are shown in the right panel of fig.4.

DISCUSSION

Although the above results were calculated only for one exemplary region, some conclusions can already be drawn.

First of all, even for identical values of capillary Po_2, there is a dispersion in tissue Po_2, as shown in fig.2, larger than predicted by other models. The Krogh model, eq.(2), applied to a mean area of 323 μm^2 (radius 10.1 μm) leads to a minimum Po_2 of 35.2 mmHg; when facilitated diffusion is included, this value is slightly higher, 35.3 mmHg; when also capillary barrier is included, we find 25.7 (for $CB=10$) and 3.6 (for $CB=35$) respectively (all in mmHg). In the latter case, most of the Krogh area, 85%, is between 3.6 and 6 mmHg. This by far does not resemble the calculated histograms in fig.2; in particular, these come down to much lower values. So, such predictions, based on only one typical example area, are incorrect.

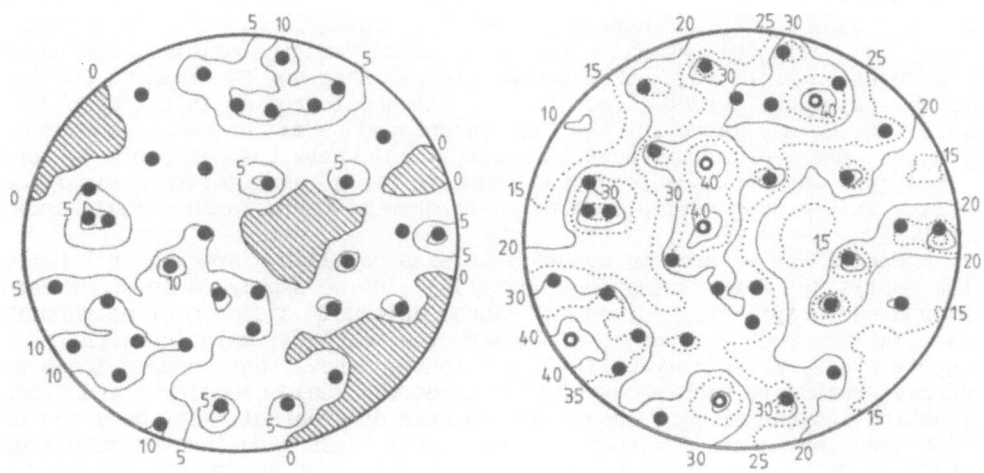

Fig. 3. Po_2 isopleths, lines of equal Po_2, for two choices of capillary pressure(s) - capillary barrier: 45-35 (left panel) and 75/40-10 (right panel). The areas where Po_2 is calculated negative, for the 45-35 case, are dashed.

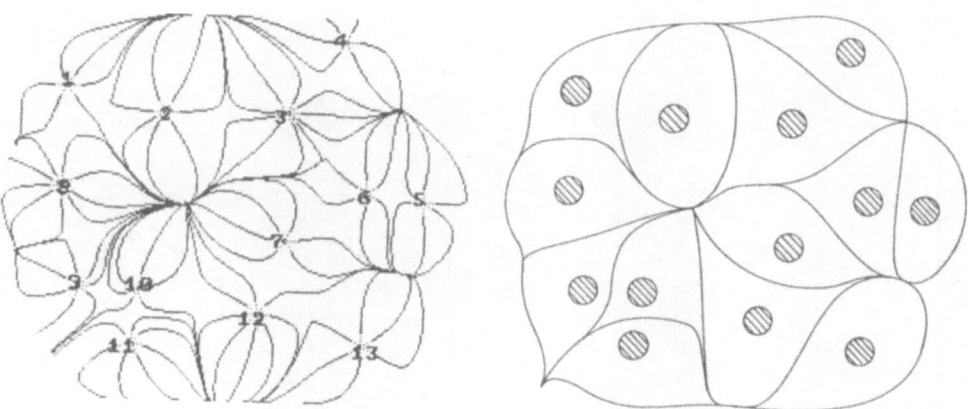

Fig. 4. Subsection of the original circular area of fig. 1, showing (left panel) flux lines emanating from the capillaries, indicated with numbers 1-13, and the areas where these lines are enclosed in (right panel), being the oxygen supply areas of the capillaries, indicated as circles of 2.3 μm radius.

Due to the heterogeneity in capillary spacing, the individual O_2 supply areas will be different. For the case 45-0 ($P_{c,1}$-**CB**), calculated area size varies between 220 and 436 μm^2, for 45-10 between 250 and 416 μm^2 (and for 45-35 between 274 and 375 μm^2, but see below for this case). Again, if taken as the circular region of a Krogh cylinder, the minimum Po_2's calculated for the largest areas will be 30.5, 19.1 and -0.4 mmHg respectively. Even this does not reflect the actual drop in Po_2. The reason must be that the regions deviate from circular, as can be seen on fig.4, in accordance with previous findings (Hoofd et al., 1987).

A capillary barrier of 35 mmHg leads to a considerable area with negative Po_2's calculated using eq.(8). This, of course is not possible. What will happen is that, since there is not enough O_2 to supply the whole area, some of it will remain devoid of O_2 and consequently cannot consume O_2; therefore, no O_2 flux goes to these areas and can be redirected to other areas. In the Krogh-Erlang model, for comparison, this is reflected in a new radius of the outer border R, in between the original radius and the location where Po_2 would become zero, so that now Po_2 remains positive up to this new border. Accordingly, the anoxic region in the case 45-35 will not be as large as the 21% calculated here; however, some anoxic regions must occur, since the situation where the available O_2 would supply the whole tissue leads to negative Po_2's and therefore is impossible. This argumentation remains valid even if there is one single anoxic spot.

One of the methods to quantify tissue capillary spacing is that of the capillary domains (Hoofd et al, 1985), where a polygon is constructed around each capillary, formed of lines at equal distance of neighboring capillaries. Such polygons, if constructed in fig.4, coincide with the O_2 supply areas remarkably well. Fig.5 indeed shows that domain areas and supply areas are highly correlated inside the whole circular field (those reaching its border line should be omitted of course), not only for the 45-0 case but also for the 45-10 case; the 45-35 case is not shown since the O_2 supply areas are not correct, as pointed out above (but also led to a high correlation). Without barrier, both domain and supply areas around the same capillary are almost the same; for **CB**=10 mmHg, the minimum O_2 supply area is larger and the maximum is lower, but the relationship remains linear with a high correlation. The same was calculated for the 45-20 case, not shown here, where Po_2 just remains above zero everywhere. Potentially there may be a way to predict O_2 supply areas from a modified domain method also for non-uniform capillary Po_2's.

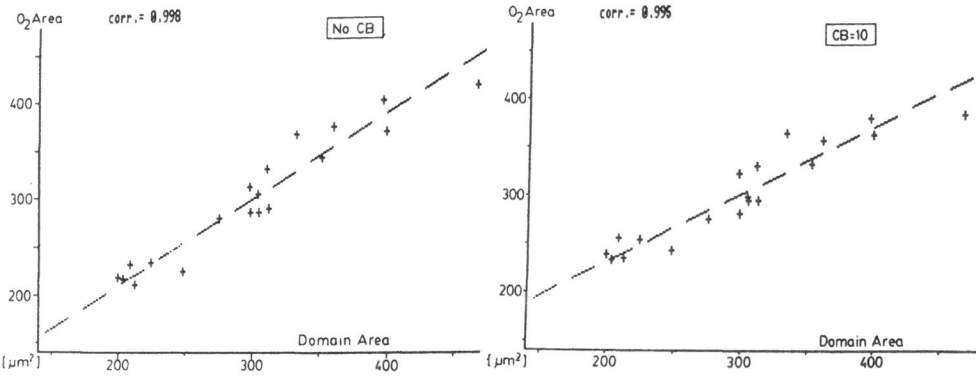

Fig. 5. Correlation between domain areas and the 19 inner O_2 supply areas for **CB**=0 (left panel) and **CB**=10 mmHg (right panel).

Domain distribution can be taken as lognormal (van Haelst et al., 1985). Then, the distribution predicts that 93% of all domain areas will be smaller than 440 μm^2, a log(area) of median + 1.5 times log(SD). If that is taken as an exemplary large value for the Krogh cylinder, minimum Po_2's are calculated for the cases 45-0, 45-10 and 45-35 of 30.3, 17.3 and -3.6 mmHg respectively. This still is not according to fig.2 in the first two cases; for 45-35, it now is lower than the actual minimum of -2.5 mmHg calculated, but it is higher for the other two cases (45-0 and 45-10). So, even this method of accounting for heterogeneity in capillary spacing does underestimate the resulting spread in Po_2 for a not too large capillary barrier.

If also capillary Po_2 is heterogeneous, the O_2 supply areas are different. In the case 75/40-10, the mean A_i for the 5 capillaries at 75 mmHg is 641 μm^2, 2.37 times larger than the mean A_i of the other 30 capillaries at 40 mmHg. Consequently, their individual barrier CB_i is much larger, about 20 mmHg, and also the flux and thus the gradient close to the capillary are larger; therefore, few values above 36 mmHg were found (see fig.2). So, the oxygen-rich capillaries "assist" the other ones in supplying more oxygen, resulting in a larger Po_2 drop from the high values and a less heterogeneous tissue Po_2 than expected from the capillary Po_2 distribution.

However, Po_2 heterogeneity is not at all abolished. There can be supply of more distant capillaries to some areas, reflected by long flux lines, as some can be seen emanating from the capillaries 1 and 6 in fig.4, and reducing Po_2 variations on this scale. The same is seen from fig.3. But that figure also shows that now a "large-scale" heterogeneity in Po_2 arises, over areas involving several capillaries, even for uniform capillary pressure.

It is not possible at the moment, to draw more conclusions from the model. As presented here, it is two-dimensional; it should be extended into a three-dimensional tissue model yet. Furthermore, topics like anoxic zones and Po_2 or locally dependent oxygen consumption could extend the range of applicability. But the model seems to be already much more promising than those available hithero and a better tool for calculating oxygen distribution in heterogeneous tissue.

REFERENCES

Clark, P.A., Kennedy, S.P., Clark Jr., A.C., Buffering of muscle Po_2 levels by the superposition of the oxygen field from many capillaries, this volume.

De Koning, J., Hoofd, L.J.C., Kreuzer, F., 1981, Oxygen transport and the function of myoglobin: theoretical model and experiments in chicken gizzard smooth muscle, Pflügers Arch., 389: 211-217.

Federspiel, W.J., 1986, A model study of intgracellular oxygen gradients in a myoglobin-containing skeletal muscle fiber, Biophys. J., 49: 857-868.

Gayeski, T.E.J., 1982, "A cryogenic microspectrophotometric method for measuring myoglobin saturation in subcellular volumes: Application to resting dog gracilis muscle", Dissertation Thesis, University of Rochester, Rochester, NY, USA.

Groebe, K., 1987, "Theoretische Untersuchung des Sauerstofftransportes in der Skeletmuskulatur bei maximaler Sauerstoffaufnahme", Dissertation Thesis, University of Mainz, Mainz, FRG.

Honig, C.R., Gayeski, T.E.J., 1987, Comparison of intracellular Po_2 and conditions for blood-tissue O_2 transport in heart and working red skeletal muscle, in: "Oxygen Transport to Tissue IX (Adv. Exp. Med. Biol. 215)," I. A. Silver and A. Silver, eds., Plenum Press, New York & London, pp. 309-321.

Hoofd, L., 1987, "Facilitated diffusion of oxygen in tissue and model systems", **Dissertation Thesis**, University of Nijmegen, Nijmegen, the Netherlands.

Hoofd, L., Turek, Z., Kubat, K., Ringnalda, B.E.M., Kazda, S., 1985, Variability of intercapillary distance estimated on histological sections of rat heart, in: "Oxygen Transport to Tissue VII (Adv. Exp. Med. Biol. 191)," F. Kreuzer, S. M. Cain, Z. Turek and T. K. Goldstick, eds., Plenum Press, New York & London, pp. 239-247.

Hoofd, L., Turek, Z., Rakusan, K., 1987, Diffusion pathways in oxygen supply of cardiac muscle, in: "Oxygen Transport to Tissue IX (Adv. Exp. Med. Biol. 215)," I. A. Silver and A. Silver, eds., Plenum Press, New York & London, pp. 171-177.

Krogh, A., 1919, The number and distribution of capillaries in muscles with calculations of the oxygen pressure head necessary for supplying the tissue. J. Physiol., 52: 409-415.

Popel, A.S., Charny, C.K., Dvinsky, A.S., 1986, Effect of heterogeneous oxygen delivery on the oxygen distribution in skeletal muscle, Math. Biosci. 81: 91-113.

Rakusan, K., Hoofd, L., Turek, Z. 1984, The effect of cell size and capillary spacing on myocardial oxygen supply, in: "Oxygen Transport to Tissue VI (Adv. Exp. Med. Biol. 180)." D. Bruley, H. I. Bicher and D. Reneau, eds., Plenum Press, New York & London, pp. 463-477.

Turek, Z., Olders, J., Hoofd, L., Egginton, S., Kreuzer, F., Rakusan, K., PO2 histograms in various models of tissue oxygenation in skeletal muscle, this volume.

Turek, Z., Rakusan, K., 1981, Lognormal distribution of intercapillary distance in normal and hypertrophic rat heart as estimated by the method of concentric circles: its effect on tissue oxygenation, Pflügers Arch., 391: 17-21.

Van Haelst, A.C.T.A., Hoofd, L., Turek, Z., 1985, Lognormal distribution of capillary domains in the rat myocardium. J. Physiol. (London), 366: 114P.

STRUCTURED MODELING AND SIMULATION OF OXYGEN TRANSPORT TO HYBRIDOMA CELL IN A SUSPENSION CULTURE: THEORETICAL ANALYSIS

Kyung A. Kang and Dewey D.Y. Ryu

Department of Chemical Engineering
University of California
Davis, CA 95616, USA.

INTRODUCTION

As the demand for monoclonal antibodies increases rapidly because of their diverse use for clinical and analytical application, and protein separation, monoclonal antibody production by hybridoma cell culture in large scale became necessary.

Among many variables which control the cell viability and production rate of monoclonal antibodies, the dissolved oxygen tension (DOT) in the media is one of the most important factors to be considered. Too low a DOT causes cell-death because the cells can not metabolize the energy source and, at the same time, too high a DOT could be cytotoxic. For that reason, many researchers have studied obtain the optimum oxygen tension for their hybridoma cell lines. However, different cell lines give different optimum values. Even for the two hybridoma cell lines from the same myeloma and spleen cell origins, there is no guarantee that these two will give the same optimum oxygen demand values because there is no possibility to make the two identical hybridoma cell lines during the fusion. For these reasons, the optimum dissolved oxygen value is reported to vary widely from 0.5 - 100 % of air saturation. (Boraston, et al., 1982, Reuveny, et al., 1986a, Miller, et al., 1987, and Phillips, et al., 1987). As a consequence, one of the difficulties of hybridoma culture is that the previous research results obtained by others may not be applicable to the new study if they use different hybridoma cell lines. This means that we need to find the optimum value of DOT for each and new hybridoma cell lines. Most of animal cell culture research takes long time due to the slower growth rate as compared to the microbial fermentation. At the same time, the culture media used for hybridoma cell culture are very expensive. Therefore, it is necessary to look for ways to predict the DOT for optimum growth and maximum monoclonal antibody production, as a rational basis.

Most of the biotechnologists in the field of hybridoma cell culture are mainly interested in the result from the culture without considering the physiological phenomena that govern the outcome. Nevertheless, it is necessary to understand the cell physiology and cytology if we were to reduce research time and expenses.

In the cell, the organells that may be involved in oxygen transport and utilization are membranes and mitochondria. Although it is very important to consider both the membrane and mitochondrion for this study, we mainly focused on the the mitochondria as a first attempt to analyze the problem more systematically. When DOT is not high, it is assumed that the resistance of the cell membrane against the oxygen transport into the cell is almost negligible and that oxygen could be transported passively.

In the process of hybridoma cell culture, the difference from other normal cells is a high concentration of lactic acid. The reason for high lactic acid concentration is that hybridoma cell metabolize the carbon source by anaerobic pathway. Hybridoma cell is originated from two cells, usually, one of which is myeloma cell and the other is a spleen cell. Myeloma cell is a tumor cell and the usage of this cell is to give immortality of hybridoma cell. Most of the tumor cells, has much lower mitochondrial content in the cell compared with normal cells. According to Pedersen (1978), with few exceptions, the mitochondrial content of most tumors is less than 50 % of normal cells. Especially, fast growing and highly glycolitic tumor cells show markedly lower mitochondrial content compared to the origin. The respitory capacity of the tumor decreases and the glycolytic capacity of the tumor increases because of marked reduction of mytochondria. Therefore, it seems that there is a correlation between mitochondiral content and cell oxygen consumption, i.e. respiration.

There is a scarcity of quantitative cytological data for the myeloma cells that are used for the hybridoma cell fusion. However, we expect that the mitochondrial content might be much lower than the origin since the myeloma cells used for the hybridoma production grow fast and a large amount of lactic acid formation from the hybridoma cell culture. Therefore, it is reasonable to assume that the hybridoma cells contain significantly less mitochondria and the have the unique combination of cellular content of the mitochondria inside the new resulting cell. This might be the reason for each hybridoma cell line having the different optimal oxygen demand.

Thus, the study of correlation between the content of the mitochondria in the hybridoma cell and the oxygen demand is considered to be very important for the hybridoma cell culture.

The main objective of this paper is to find correlation between the physiological factor, specifically mitochondrial content in the cell to the optimum dissolved oxygen tension in the media by analyzing the oxygen tension inside the hybridoma cell for varying content of the mitochondria.

THE GEOMETRY, PARAMETERS, AND SIMULATION CONDITIONS

In our biochemical engineering laboratory at the University of California, Davis, we have been studying in hybridoma cell culture system for monoclonal antibody production and affinity separation of beta-lactam antibiotic producing enzyme. We have completed basic batch experiments for the perfusion suspension culture. Perfusion suspension culture culture gives better performance in hybridoma cell and monoclonal antibody production. Suspension culture gives very good control of DOT and purfusion enables high productivity of monoclonal antibody (Reuveny, et al., 1986b)

At the present time, we do not as yet have enough cytological data for our hybridoma cell line. Most of the quantitative cytological data available are from the hybridoma cell originated from the myeloma cell, SP2/0, from the research of Renau-Piqueras, et al. (1983).

For our simulation studies, a single hybridoma cell line was chosen as a model system. With changing content of mitochondria in the cell, the oxygen tension in the cell was examined in order to find how mitochondrial content could actually affect the oxygen requirement of the cell.

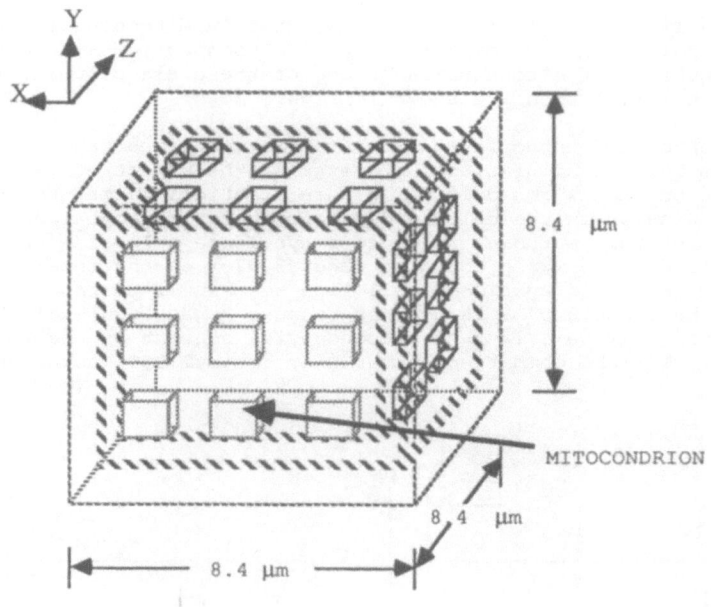

Figure 1. The Model System Simulated. A Hybridoma Cell with
       Mitochondria.

Followings are are simulation model descriptions.

(1)   Total Cell Volume: 8.4 x 8.4 x 8.4 μm cube (as an assumed shape).

For more realistic simulation, the shape should be spheroid or
ovoid. However, for the purpose of the simulation of oxygen
diffusion and oxygen consumption by mitochondria, the distance from
the cell membrane to mitochondria is the same order of magnitude.
The assumed cell shape should be reasonable and cubic shape should
also make the simulation much easier (Figure 1). From the research
of Renau-Piqueras, et al.(1983), the averaged hybridoma cell from
the myeloma cell of SP2/O has 598.43 μm^3· In our simulation, 592.7
μm^3 cube was chosen.

(2) Mitochondria

A single Mitochondrion: 0.6 x 0.6 x 0.6 μm cube (assumed).
The shape is chosen for the same reason that used for the cell model
shape. In Renau-Piqueras, et al. found, from their SP2/O originated
hybridoma cells, total averaged mitochondrial volume was 24.63 μm^3
and the averaged number of the mitochondria were 112.39, which gives
a single mitochondrion volume about 0.22 μm^3. Therefore, our single
mitochondrion volume was chosen to be 0.216 μm^3. For the simulation
we assumed that each mitochondrion has same volume and same shape
(Figure 2a, 2b, 2c).

Distribution of Mitochondria: In the Renau-Piqueras' paper, the
nucleus occupies the volume of 266.40 μm^3, which is almost half of
the total cell volume. Therefore, we assume that mitochondria are
surrounded around the nucleus which locates center of the cell. In
other words, mitochondria are evenly located on six plane of the
rectangle in one layer between 1.5 μm and 0.9 μm from cell membrane

(Figure 1). Each of six planes that contain mitochondria has the
same total volume and distribution of mitochondria. The detailed
distributions of mitochondria on one of these six planes for three
different simulation are shown in figure 2.

Contents of Mitochondria in a Cell: The most important aspect to
observe in this simulation is to examine the effect of mitochondrial
content on the oxygen demand, in more practical terms, the optimum
oxygen tension in the culture media for a certain hybridoma cell
line. For this purpose, we chose a hybridoma cell line which has
110 mitochondria which occupies about 3.97 % of the total cell
volume. For this hybridoma cell line we assumed to know its optimum
DOT. The value 3.97 % is decided because Renau-Piquiera's hybridoma
cells had average 4.12 % of mitochondrial content of the total cell
volume. For the comparison with this, optimum DOT known hybridoma

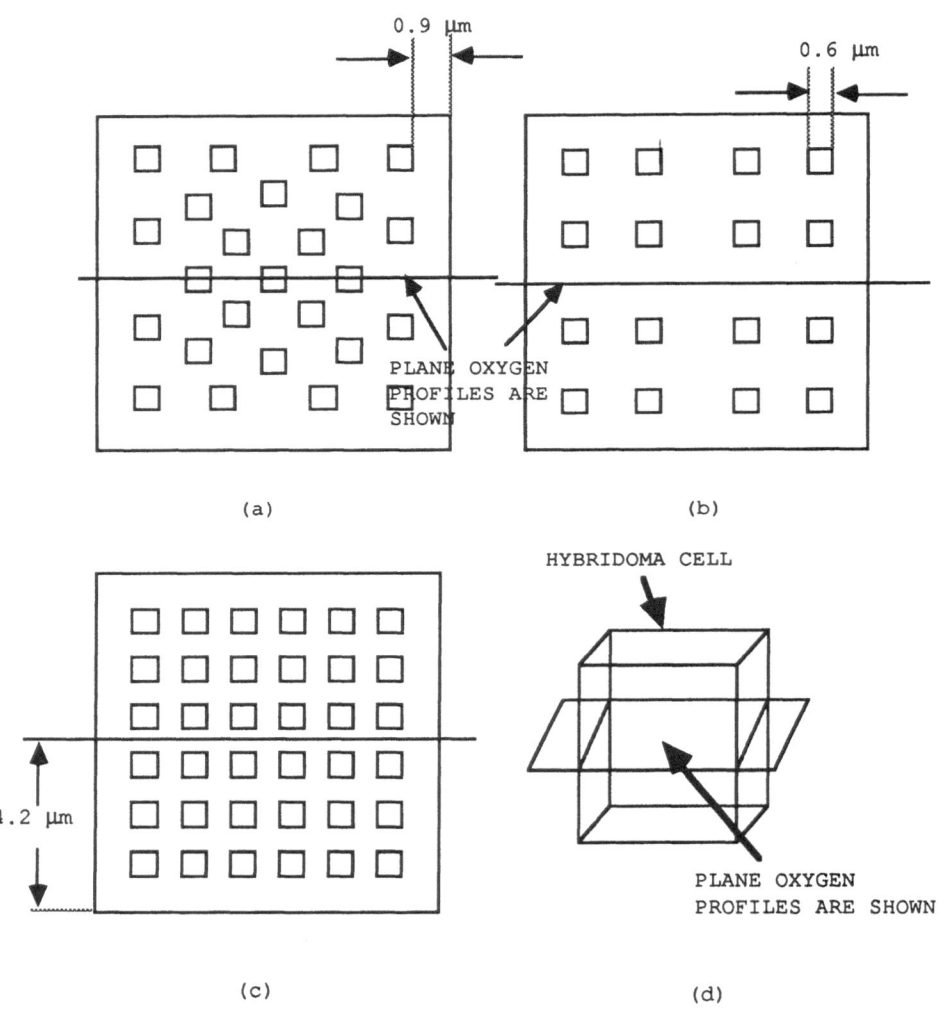

Figure 2. Frontal View of Mitochondrial Distributions in the cell.
(a) Simulation I, (b) Simulation II, (c) Simulation III,
(d) Plane on that oxygen profiles are shown in results.

cell line, there were two more arbitrarily chosen hybridoma cell lines. An attempt was made to distribute these mitochondria uniformly around the cell nucleus as much as possible (Figure 2).

Simulation I: Hybridoma cell with 110 mitochondria, 3.97 % of the total cell volume. It was assumed that we know its optimum culture DOT (Figure 2a).

Simulation II: Hybridoma cell with 56 mitochondria, 2.02 % of the total cell volume. For this, we would like to predict its optimum culture DOT (Figure 2b).
Simulation III: Hybridoma cell with 152 mitochondria, 5.48 % of the total cell volume. For which the optimum culture DOT was predicted (Figure 2c).

(3) Oxygen Diffusion Coefficient: 1750 μm^2/sec.

Because of scarity of data, in this simulation, the oxygen diffusion content of tumor tissue (DS-Carcinosarcoma) at 37 °C (Grote, et al, 1977) was used. This value is also reasonably close to the value (1960 μm^2/sec) that Clark's oxygen gradient simulation of a single mitochondrion (1985).

(4) Oxygen Consumption Rate of the Mitochondria: 1.64×10^{-6} mole-O_2 /cm^3-mitochondrion/sec.

This value was induced from research of Miller and his coworkers (1987), who used same myeloma cell origin, SP2/O . In their research, they found the oxygen consumption rate per a hybridoma cell to be 0.14×10^{-9} m mole/cell/hr. Assuming that this hybridoma cell has same volume and the same quantity of mitochondria as our model cell line, the above oxygen consumption rate of mitochondria was obtained.

(5) Solubility of Oxygen in Tissue: 1.11×10^{-9} mole-O_2/cm^3-cell mm Hg at 37 °C (Clark and Clark, 1985).

(6) Technique used for the simulation:

The technique used for this simulation is a deterministic stochastic numerical technique, the Williford-Bruley (W-B) method. This technique calculates the time dependent solution by using transient density function on the three dimensional space divided by grids, which saves computation time as compared with actual random walks using the Monte Carlo technique (Williford, 1972 and Kang, 1984). This method also has advantages in that the computation is rather simple such that three dimensional time dependent conduction, convection, and reaction problem can be solved readily.

(7) Simulation Time: 5.0×10^4 microseconds.

(8) Grid size: 0.3 μm.

(9) Time step size: 16.4 microseconds. For this grid size chosen the optimum time step size which give more accurate solutions is determined.

(8) Computer used and CPU time

The computer used for this simulation was the VAX 11/785 at University of California Davis, California. The computation time for each simulation is about 39 minutes for 5.0×10^4 microseconds simulation.

GOVERNING EQUATION FOR THE SIMULATION

(1) Equation.

In the cell, oxygen is transferred by diffusion and reaction takes place. Therefore the governing equation becomes,

$$\partial^2 P/\partial x^2 + \partial^2 P/\partial y^2 + \partial^2 P/\partial z^2 - Rx/Sb = (1/D)\,(\partial P/\partial t) \qquad (1)$$

where P is oxygen partial pressure in the cell, Rx is the oxygen consumption rate of the mitochondrion, Sb is the oxygen solubility constant of oxygen in the cell at 37 $^{\circ}$C, and D is oxygen diffusion coefficient in the hybridoma cell. For the cytoplasm except mitochondria, Rx=0.0 because oxygen is consumed by mitochondria only.

(2) Initial Condition.

$$P_o(x,y,z,t=0.0) = 0.83 \quad mm\ Hg \qquad (2)$$

where P_o is the initial oxygen partial pressure in the cell.

(3) Boundary Conditions.

$$P(x=0,\ y,\ z) = P_b \qquad (3)$$

$$P(x=L,\ y,\ z) = P_b \qquad (4)$$

$$P(x,\ y=0,\ z) = P_b \qquad (5)$$

$$P(x,\ y=L,\ z) = P_b \qquad (6)$$

$$P(x,\ y,\ z=0) = P_b \qquad (7)$$

$$P(x,\ y,\ Z=L) = P_b \qquad (8)$$

where P_b is 0.83 mm Hg and L=8.4 μm.

The boundary oxygen tension, 0.83 mm Hg, is from the research of Miller and his coworkers (1987). In their research, the steady state concentration of viable cells showed the highest value when the dissolved oxygen concentration was 0.5 %. We chose this value because their hybridoma cell is derived from the myeloma cell line SP2/O, which that we took cytological data from (Renau-Piqueras, et al, 1983). At the same time the oxygen tension is low enough to ignore the membrane resistance to oxygen transport through it. 0.5 % oxygen concentration was converted to 0.83 mm Hg.

ASSUMPTIONS FOR THE SIMULATION

Since the initial oxygen tension is not known inside of the cell, we decided to see the effect of the different contents of the mitochondria in some different ways. At first, we assume that the oxygen concentration in side of the cell is the same as that of medium outside. Then, after the mitochondria consumes the oxygen transport for a certain period of time, the oxygen profiles in each different mitochondrial contents are examined to see the difference.

For this simulation, following assumptions were used.

(1) Considering hybridoma cell fraction to the total suspension reactor is small and the oxygen is mixed well in the reactor, the liquid film outside of the cell is assumed to be negligible.

(2)   Supension hybridoma cell reactor perform well enough to control the DOT accurately.

(3)   The resistance to oxygen transport through membranes (cytoplasmic and mitochondrial) is assumed to be negligible and the oxygen in the culture media diffuses passively.

(4)   Oxygen is consumed only by mitochondria.

(5)   Each mitochondrion has the same volume and shape and has same oxygen consumption rate.

(6)   Oxygen consumption rate is zeroth order and constant.

(7)   The oxygen diffusion property is constant inside of the cell.

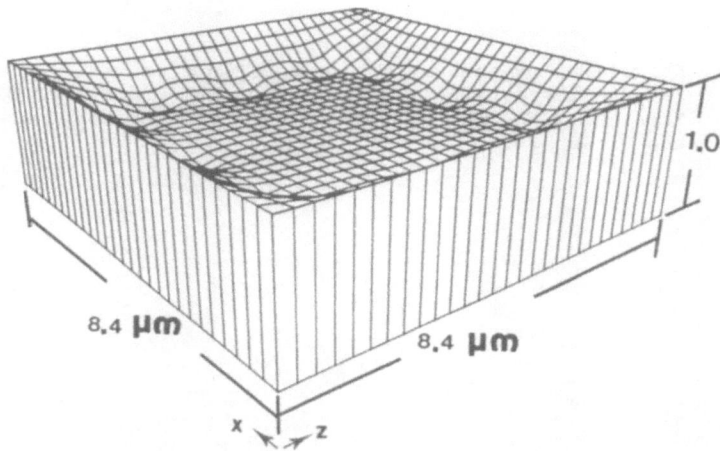

Figure 3. Normalized Oxygen Profile on the Plane of y=4.2 μm (Figure 2d). The Cell has Mitochondrial Content, 3.97 % of Total Cell. Volume. Mitochondria are distributed as in the Figure 2a.

RESULTS AND DISCUSSION

Results from the simulation studies is presented in terms of normalized value of the oxygen tension, which is P/P_o. The surface that shows its oxygen profile is the plane, y = 4.2 μm. This plane passes through the middle of any x-z plane (Figure 2d).

Figure 3, 4, and 5 show that the oxygen profile on the plane through the center of the cell (Figure 2d) of the simulation I, II, and III, respectively. Since we assumed that the culture media dissolved oxygen tention, Pb is the known optimum oxygen tension of the media for the hybridoma cells whose mitochondrial content is 3.98 % of the total cell volume, Figure 1 depicts the oxygen profile of the optimized DOT.

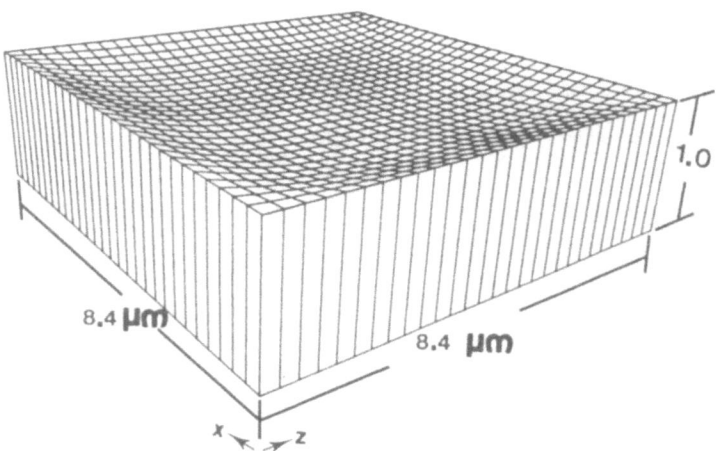

Figure 4. Normalized Oxygen Profile on the Plane of y=4.2 μm
        (Figure 2d). The Cell as Mitochondrial Content, 2.02 % of
        Total Cell Volume. Mitochondria are distributed as in
        the Figure 2b.

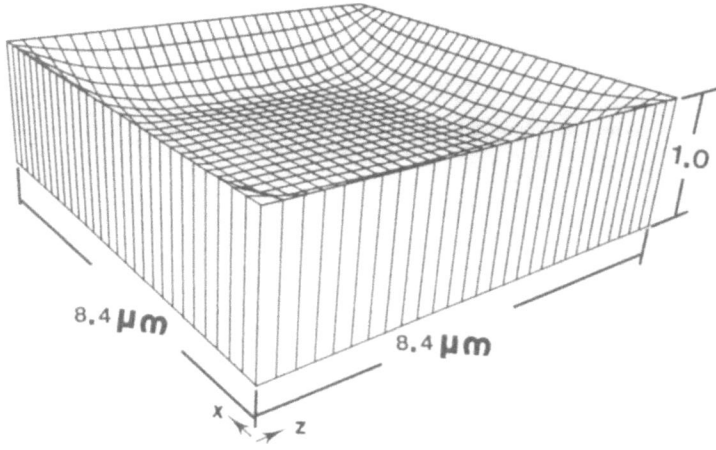

Figure 5. Normalized Oxygen Profile on the Plane of y=4.2 μm
        (Figure 2d). The Cell has Mitochondrial Content, 5.48 %
        of Total Cell Volume. Mitochondria are distributed as in
        the Figure 2c.

Compared to Figure 3, Figure 4, which is the simulation of oxygen tension profile for 2 % mitochondrial content, shows very little oxygen tension difference between inside oxygen concentration and the media outside of the cell after 50000 microseconds. Actual simulated value for the lowest oxygen tension in normalized form for Figure 4 is 0.54, while that of Figure 5 is 0.8. This could be interpreted that if we use the same DOT for this 2.02 % mitochondrial hybridoma cell culture as for 3.98 %, the oxygen tension in the cell may be too high. This may result in an unnecessary power consumption for oxygen transfer in the fermenters or in an citotoxicity of the cell.

Figure 5 shows the oxygen profile in the cell with 5.48 % of mitochondrial content in the cell. The differencee between the simulation I and III are not as much as that between I and II. However, in the simulation III, the lowest normalized oxygen tension on the plane shown is 0.45. This could be interpreted that cells need more oxygen to maximize their growth or cells may metabolize their carbon source more in anaerobic pathway, which could produce more lactic acid and lower media PH.

Although the mitochondrial content in a cell is normally less than 10 % of the total cell volume, as we can see in the simulation, the content of mitochondria in hybridoma cells are directly related to the intra cellular oxygen tension, thereby, uptake rate from outside of the cell. Knowing the mitochondrial content of the newly developed hybridoma cell line and comparing it with known experimental data with other hybridoma cell line may help designing better reactor system.

There may be many factors related to the oxygen consumption rate or to oxygen transport to the mitochondira in hybridoma cells, and some of the factors may be too difficult to find their correlation between them in a quantitative way. However, instead of doing hybridoma cell culture research only by repeated trial and error to find the optimal condition for cell growth and product, it will be beneficial, for economic reasons of time and expense and for future researchers, to adapt engineering approach as demonstrated in this paper

CONCLUSIONS

It can be concluded from the computer simulation that,

(1) For the better prediction of optimum conditions in hybridoma cell culture, it is necessary to understand physiological and cytological nature of hybridoma cells.

(2) It may be worthwhile to study the correlation of the content of mitochondria in hybridoma cells and the optimum DOT (dissolved oxygen tension) in the culture media.

ABSTRACT

The importance of the consideration of micro-anatomical and cytological observation for the oxygen utilization by hybridoma cells are emphasized.

Tumor cells have considerably lower content of mitochondria and that cells metabolize much of their carbon source by anaerobic pathway, therefore, consumes much less oxygen. Because of the fact that hybridoma cells are derived from a myeloma cell, which is a type of tumor cell, there may be possibility that the oxygen demand could be closely related to mitochondrial content of the resulting cell. Possible correlation between the optimum DOT and mitochondrial content of the hybridoma cells is hypothesized. By simulating oxygen transport inside of a single hybridoma cells for the cases of three different mitochondrial contents, possible differences in the optimum DOT according to the mitochondrial contents were studied and the results reported.

REFERENCES

Boraston, R, Thompson, P.W., Garland, S., and Birch, J., 1984, Growth and Oxygen Requirements of Antibody Producing Mouse Hybridoma Cells in Suspension Culture, Develop. biol. Standard., 55, 103-111.

Clark Jr., A. and Clark, P., Local Oxygen Gradients near Isolated Mitochondria, 1985, Biophys. J. 48, 931-938.

Grote, J, Susskind, R., and Vauel, P, Oxygen Diffusivity in Tumor (DS-Carcinosarcoma) under Temperature Conditions within the Range of 20-40 °C, 1977, Pflugers Arch., 372, 37-42.

Kang, K., 1984, An Analysis of the Williford Bruley Technique, A Simulation of Oxygen Transport in Brain Tissue, MS Thesis, Biomedical Engineering Department, Louisiana Tech University.

Matsuno, T.,1987, Bioenergetics of Tumor Cells: glutamine Metabolism in Tumor Cell Mitochondria, Int.J.biochem., 19:4, 303-307.

Miller, W., Wilke, C., and Blanch, H., 1987, Effects of Dissolved Oxygen Concentration on hybridoma Growth and Metabolism in Continuous Culture, Journal of Cellular Physiology, 132, 524-530.

Mizrahi, A., Vosseller, G., Yagi, Y., and Moore, G.,1972, The Effect of Dissolved Oxygen Partial Pressure on Growth, Metabolism and Immunoglobulin Production in a Permanent Human Lymphocyte Cell Line Culture (36092), Proc. Soc. Exp. Biol. Med., 139, 118-122.

Pedersen, P., Tumor Mitochondria and the Bioenergetics of Cancer, Prog. exp. Tumor Res., vol. 22, 190-274.

Phillips, H., Scharer, J., Bols, N., and Moo-Young, M.,1987, Effect of Oxygen on Antibody Productivity in Hybridoma Culture, Biotechnology Letters, 9:11, 745-750.

Randerson, D., Large-Scale Cultivation of Hybridoma Cells, 1985, Journal of Biotechnology, 2, 241-255.

Reith, A. and Boysen, M., 1986, Light and Electron Microscopic Results of a Morphometric/Stereologic Analysis of Precancerous Lesions, in the book, of Quantitative Image Analysis in Cancer Cytology and Histology, edited by Mary, J. and Rigaut, J., 197-204, Elsevier Science Publishers.

Reuveny, S., Velex, D., Macmillan, J., and Miller, M., 1986a, Factors Affecting Cell Growth and Monoclonal Antibody Production in Stirred Reactors, Journal of Immunological Methods, 86, 53-59.

Reuveny, S., Velez, D., Miller, L., and Macmillan, J., 1986b, Comparison of Cell Propagation Methods of their Effect on Monoclonal Antibody Yield in Fermenters, Journal of Immunological Methods, 86, 61-69.

Renau-Piqueras, J., Perez-Serrano, M., and Martinez-Ramon, A., 1983, Stereological Study of Murine Myeloma and Hybridoma Cells in Vitro and in Vivo, J. Submicrosc. Cytol., 10 (3), 607-618.

Simon, L., Robin, E., ;and Theodore, J., 1981, Differences in Oxygen Dependent Regulation of Enzyme Between Tumor and Normal Cell Systems in Culture, Journal of Cellular Physiology, 108:393-400.

Williford, Jr., C., Bruley, D., and Artigue, R., 1974, Probabilistic Modelling of Oxygen Transport in Brain Tissue, NeuroResearch 2, 153-170.

RELATIONSHIP BETWEEN THE TRANSFER RATE AND THE DIFFERENCE IN PARTIAL PRESSURE OF GAS MOLECULES AT A HETEROGENEOUS INTERFACE

Masaji Mochizuki

Geriatric Respiratory Research Centre
Nishimaruyama Hospital
064 Sapporo/Chuo-Ku, Japan

INTRODUCTION

When a gas passes across a gas-liquid interface from the gas to liquid side, the gas molecule undergoes an increase of partial pressure at the liquid side because of its low solubility. Thus, in order for a molecule to cross the interface, it is necessary that the gas concentration at the liquid side is reduced below the equilibrium level. When the concentration falls, a discontinuity in the partial pressure inevitably occurs at the border because of difference in solubility. The relationship between the gap in partial pressure at the interface and the transfer rate is theoretically derived from the solubility difference, where the gap size is evaluated by dividing the transfer rate by the transfer coefficient, η. Comparing the theoretical and experimental η values obtained at the gas-liquid interface and red blood cell (RBC) boundary, the transfer rate across the heterogeneous interface is found to be proportional to the difference in partial pressure from the equilibrium level. In addition, conductivity, k, in aqueous solution at the interface is calculated to be 6.6×10^{-3} cm·sec^{-1}·Atm^{-1}, regardless of gas type. Furthermore, for O_2, CO_2 and CO solubility in the RBC membrane is estimated to be about 22% of that in water, regardless of gas type.

THEORETICAL ANALYSIS OF THE GAP IN PARTIAL PRESSURE

The gas-liquid interface

In the passage of gas across a gas-liquid interface, the partial pressure on the liquid side cannot exceed that on the gas side. Thus, to initiate the passage of a gas molecule across the interface the gas concentration on the liquid side must be lowered from the equilibrium level. When the gas concentration at the interface is decreased by ε from the equilibrium level, as shown in Fig. 1,A, the decreases in partial pressure on the liquid and gas sides are expressed by ε/α and ε, where α is gas solubility in liquid. Because the partial pressure in the gas phase is given by the same value as fractional concentration, the change in partial pressure on the gas side is given by ε (Atm). In Fig. 1, B-1, two values for partial pressure are shown on the two sides of an imaginary boundary layer, while the concentration profile (A) is shown as a single

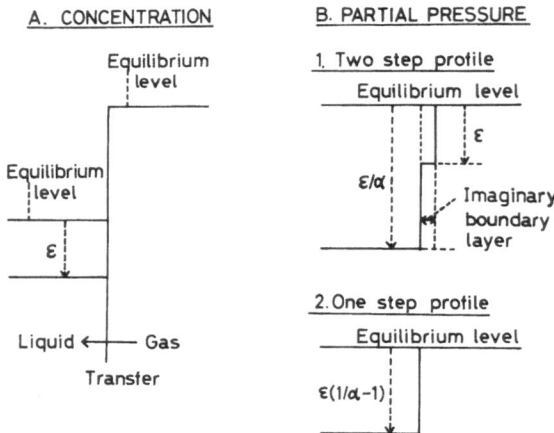

Fig. 1. Schematic profiles of concentration and partial pressure at a gas-
liquid interface during a gas transfer from the gas to liquid side.
The two-step profile of partial pressure (B-1) is approximated by
the one-step pattern (B-2) in solving the differential equation for
diffusion, where the gap size is given by $\varepsilon(1/\alpha - 1)$.

step. For obtaining the relation between the gap in partial pressure and
the transfer rate, a simulation technique has hitherto been adopted, where
the differential equation for diffusion is solved by approximating the
partial pressure at the interface to a single step profile as shown in
Fig. 1, B-2. That is, the diffusion rate in the gas phase is assumed to
be so fast that the gap in partial pressure on the gas side, ε, is instan-
taneously countered by the arrival of more molecules, and therefore, the
partial pressure on the liquid side, also, rises by ε as a result of an
increase in concentration, $\alpha \cdot \varepsilon$. Consequently, as shown in Fig. 1, B-2, the
gap in partial pressure (ΔP) is numerically expressed as:

$$\Delta P = \varepsilon(1/\alpha - 1). \tag{1}$$

The conductivity across the interface is thought to be proportional
to the product of the frequency of collisions with the interface and the
momentum on collision. According to the kinetic theory of gases, the
above product, namely, the momentum on collisions per second becomes
proportional to the kinetic energy of gas molecules in a unit volume, and
subsequently to the partial pressure. Thus, the conductivity across the
interface per unit partial pressure difference becomes a constant, regardless of gas type. When the partial pressure on the liquid side decreases,
gas transfer occurs from the gas to liquid side. Designating the conductivity per unit partial pressure on the liquid side by k, the transfer
rate, J, across the interface is given by

$$J = k \cdot \varepsilon/\alpha. \tag{2}$$

Setting the relation of Eq. (2) into Eq.(1) and eliminating ε, ΔP is given
by

$$\Delta P = (1 - \alpha)J/k. \tag{3}$$

Hitherto we have expressed the relation between J and ΔP by using
the transfer coefficient (η : $cm \cdot sec^{-1} \cdot Torr^{-1}$)(Mochizuki and Fukuoka,
1958; Mochizuki, 1966, 1975; Fukui and Mochizuki, 1972; Kagawa and
Mochizuki, 1982, 1984; Uchida et al., 1983, 1986; Niizeki et al., 1983,

1984), as follows:

$$J = \eta \cdot P_B \cdot \Delta P, \tag{4}$$

where P_B is the barometric pressure in Torr. Thus, η is given from Eqs. (3) and (4) by

$$\eta = k/\{P_B(1 - \alpha)\}. \tag{5}$$

The RBC membrane

Since gas solubility in the RBC membrane is lower than that in aqueous solution, the concentration in the membrane decreases from the equilibrium level, when a gas passes across the interface from the liquid to membrane side. Let α_m be the gas solubility in the membrane. The changes in partial pressure on the liquid and membrane sides will be described by ε/α and ε/α_m, respectively, for the change in concentration ε. Thus, the gap in partial pressure between the two sides will be given by

$$\Delta P = \varepsilon \, (1/\alpha_m - 1/\alpha). \tag{6}$$

Let k_m be the transfer rate per unit partial pressure difference, or conductivity in the membrane. Then, for a change in partial pressure of ε/α_m from the equilibrium level, the inward gas transfer across the interface is given by $k_m \cdot \varepsilon/\alpha_m$. In a membrane such as that of the RBC, where gas transfer occurs through an aqueous channel in the pore, the ratios, k_m/k and α_m/α, are taken to be equal to water content in the membrane, yielding, $k_m/\alpha_m = k/\alpha$. Thus, the transfer rate J across the liquid–membrane interface is also given by Eq. (2). Inserting ε obtained from Eq. (2) into Eq. (6), ΔP is rewritten as:

$$\Delta P = J(S - 1)/k, \tag{7}$$

where $S = \alpha/\alpha_m$. Eliminating J from Eqs. (4) and (7), η is expressed as

$$\eta = k/\{P_B(S - 1)\}. \tag{8}$$

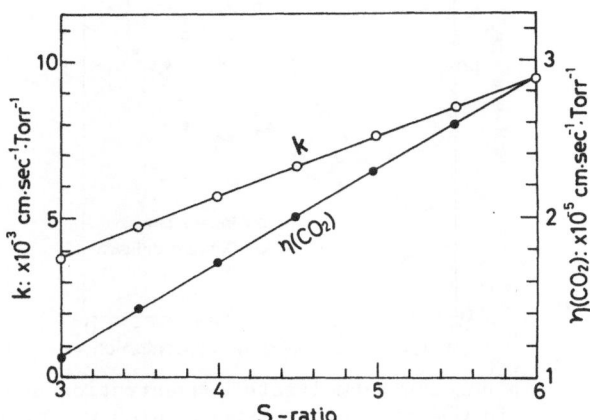

Fig. 2. Relationship between k and S obtained from Eq. (8) using $\eta = 2.5 \times 10^{-6}$ cm·sec^{-1}·Torr^{-1}, and that between S and $\eta(CO_2)$ at the gas-liquid interface computed by setting these k values into Eq. (5).

The conductivity k is an important parameter for estimating the diffusivity across the heterogeneous interface, and S is also important to confirm the validity of the k value. It has hitherto been shown experimentally that the η value at the RBC membrane is 2.5×10^{-6} cm·sec^{-1}·Torr^{-1} and is the same for O_2, CO_2 and CO (Mochizuki, 1966, 1975; Fukui and Mochizuki, 1972; Kagawa and Mochizuki, 1982, 1984; Niizeki et al., 1983, 1984). Since k is constant and independent of gas type, the constancy of η indicates that the ratio α/α_m is also independent of gas type, suggesting that the gas transfer occurs mainly through the aqueous channel. Figure 2 shows k and $\eta(CO_2)$ at the gas-liquid interface plotted against S values ranging from 3 to 6. The k values were obtained from Eq. (9) at 6 different S values, where $\eta(CO_2)$ at the RBC boundary is taken to be 2.5×10^{-6} cm·sec^{-1}·Torr^{-1}. Then, the $\eta(CO_2)$ values at the gas-liquid interface were obtained by setting these values for k into Eq. (5). Both k and $\eta(CO_2)$ are linearly related to S. For S between 4 and 5, k ranges between 5.7×10^{-3} and 7.6×10^{-3} cm·sec^{-1}·Atm^{-1} and $\eta(CO_2)$, between 1.7×10^{-5} and 2.3×10^{-5} cm·sec^{-1}·Torr^{-1}.

To obtain the experimental η value, we first measured the diffusion rate of gases in a simple system, and then solved the partial differential equation for diffusion by varying η under boundary conditions similar to those in the experimental situation. Finally, by comparing the computed parameter values such as concentration and partial pressure with experimental values, the value of η was determined. We measured the CO_2 diffusion in a thin layer of haemolysate at various concentrations using pH sensitive fluorescence (Uchida et al., 1983). Hydration and dehydration reactions of CO_2 proceed rapidly in haemolysates because of the abundance of carbonic anhydrase, and the rate of CO_2 diffusion could therefore be measured from the pH change. However, in haemolysate-free solution, it could not be measured in this way because of the absence of carbonic anhydrase. Thus, the diffusion rate in haemolysate-free solution was estimated by extrapolation from the relationship between the diffusion rate and haemolysate concentration. Figure 3 illustrates the relationship be-

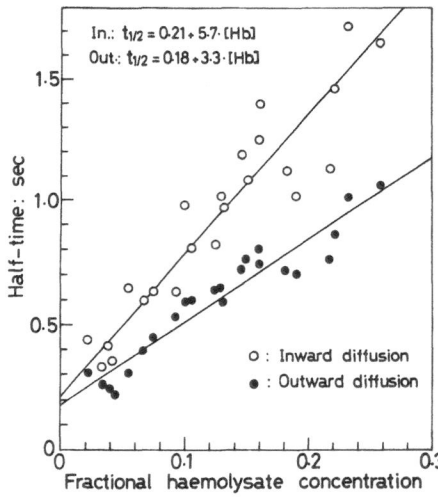

Fig. 3. Relationship between haemolysate concentration and the half-time of the change in CO_2 partial pressure, which was measured in a thin haemolysate layer during CO_2 diffusion (Uchida et al., 1983). The open and closed circles refer to the inward and outward directions, respectively. The half-time in haemolysate-free solution was estimated from the regression lines to be about 0.2 sec.

tween the half-time of the change in CO_2 partial pressure and haemolysate concentration. The half-time linearly increases with an increase in haemolysate concentration, because the diffusion coefficient and solubility of CO_2 are reduced with increasing concentration, and in addition, hydration and dehydration reactions are increased by an increase in haemoglobin concentration. The open and closed circles show the half-times for the inward and outward directions, respectively. The half-time for the inward direction is longer than that for the outward direction. This is because the slope of the CO_2 disociation curve for the inward direction is reduced by increasing CO_2 partial pressure at the gas-liquid interface. Both the regression lines, however, coincide with the ordinate for zero haemolysate concentration at about 0.2 sec, suggesting this to be the half-time in haemolysate-free solution.

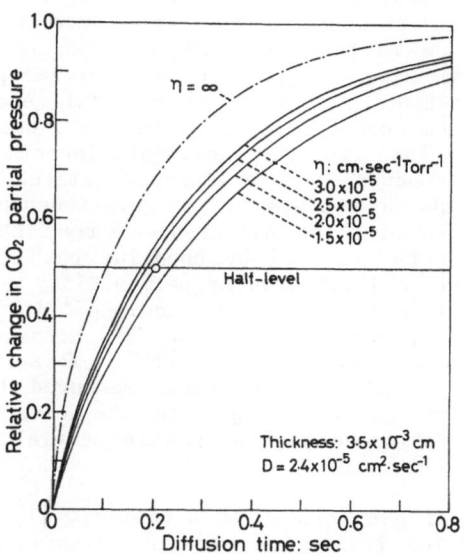

Fig. 4. The relative change in CO_2 partial pressure in a water layer of 3.5×10^{-3} cm in thickness computed from a one-dimensional diffusion equation. $\eta(CO_2)$ was varied from 1.5×10^{-5} cm·sec^{-1}·Torr^{-1} to infinity, whereas diffusion coefficient (D) is 2.4×10^{-5} cm^2·sec^{-1} and $\alpha(CO_2) = 0.567$ Atm^{-1}. The open circle shows the average of measured half-times, suggesting $\eta(CO_2) = 2 \times 10^{-5}$ cm·sec^{-1}·Torr^{-1}.

The change in CO_2 partial pressure in a one-dimensional model is obtained by solving the differential equation analytically (Mochizuki, 1975). Figure 4 shows the average CO_2 partial pressure obtained in a layer by varying η in a range of 1.5×10^{-5} to 2.5×10^{-5} cm·sec^{-1}·Torr^{-1}, where the thickness of the layer is kept constant at 3.5×10^{-3} cm, $\alpha(CO_2)$ = 0.567 Atm^{-1} (Opitz and Bartels, 1955), and the diffusion coefficient (D) is 2.4×10^{-5} cm^2·sec^{-1} (Gros and Moll, 1971). The broken line illustrates the change in CO_2 partial pressure computed by taking $\eta(CO_2)$ to be infinity, where the half-time is 0.1 sec. The half-times of the solid curves are in a range of 0.18 to 0.23 sec, being fairly close to the experimental value. Table 1 shows the average values for k, S and η, which are expected from the experimental data so far obtained.

DISCUSSION

To date there has been no theoretical treatment of the relationship between conductivity, k, in water and the magnitude of the momentum on

Table 1. The average k, S and η values estimated from the diffusion rate so far measured across the gas-liquid interface and RBC boundary.

Parameters	Average values
k	6.6×10^{-3} cm·sec^{-1}·Atm^{-1}
S	4.5
RBC-mem η	2.5×10^{-5} cm·sec^{-1}·Torr^{-1}
Gas-liq η(CO_2)	2.0×10^{-5} cm·sec^{-1}·Torr^{-1}
Gas-liq η(O_2)	8.9×10^{-6} cm·sec^{-1}·Torr^{-1}
Gas-liq η(CO)	8.8×10^{-6} cm·sec^{-1}·Torr^{-1}

collision with the interface, or the impulses per sec, whose dimension is equal to the partial pressure. However, since the transfer rate is linearly related to the difference in impulses per sec between two phases of the interface, the transfer rate for unit partial pressure, or conductivity becomes a constant, k. Since no chemical reaction occurs between the gas and water, the conductivity in water is independent of gas type. Thus, the k value obtained from the gas-liquid interface can be applied to the membrane, whose conductivity is linearly related to solubility. The permeability of a membrane represents the relationship between the transfer rate and the concentration difference across the membrane. Since membrane permeability is measured by changing gas concentration outside the membrane, it will be influenced by permeability across the interface, k_m/α_m, or k/α. Since $k = 6.6 \times 10^{-3}$ cm·sec^{-1}·Atm^{-1} from Table 1 and $\alpha(CO_2)$ is 0.567 Atm^{-1}, permeability for CO_2 across the usual liquid-membrane interface will become 0.011 cm·sec^{-1}. This value is fairly close to CO_2 permeability of the RBC membrane measured by Silverman et al. (1976), 0.009 cm·sec^{-1}, suggesting that the permeability of the RBC membrane is largely influenced by the gap in partial pressure across the heterogeneous interface.

In the passage of gas across the alveolar wall, gas molecules first enter into the solution lining the alveolar membrane, then pass through the capillary wall to reach the plasma. Thus, the partial pressure drops at the gas-liquid and liquid-membrane interfaces. Because the total gap in partial pressure is the sum of the individual gaps, the following relationship is derived for the total η value:

$$1/\text{total } \eta = 1/ \text{ gas-liq } \eta + 1/\text{liq-mem } \eta. \tag{9}$$

Since the energy of gas in liquid is increased by an external force or convection, the η value at the interface between flowing plasma and the inner wall of the capillary is thought to exceed the value expected in a stationary state. Thus, η at the inner wall of the capillary is disregarded in Eq. (10). According to Grote (1967), O_2 solubility in the pulmonary tissue (α_t) at 37° C is 1.8×10^{-2} Atm^{-1}. Assuming that O_2 solubility in the lining solution around the alveolar membrane be the same as that in water, 2.4×10^{-2} Atm^{-1} (Opitz and Bartels, 1955), the ratio of solubility, α/α_t, is calculated to be 1.32. From an equation similar to Eq. (8) this gives $\eta(O_2) = 2.7 \times 10^{-5}$ cm·sec^{-1}·Torr^{-1} for the liquid-membrane interface. As shown in Table 1, $\eta(O_2)$ at the gas-liquid interface is 8.9×10^{-6} cm·sec^{-1}·Torr^{-1}. Thus, the total $\eta(O_2)$ for the alveolar membrane is estimated to be 6.7×10^{-6} cm·sec^{-1}·Torr^{-1}. Similarly, η(CO) is estimated to be 6.6×10^{-6} cm·sec^{-1}·Torr^{-1}. Uchida et al. (1986) measured the pulmonary diffusing capacity for CO in normoxia and hyperoxia together with the cardiac output, and estimated η(CO) to be 5.3×10^{-6} cm·sec^{-1}·Torr^{-1}. In this estimation, η(CO) for the RBC membrane was taken to be 2.5×10^{-6} cm·sec^{-1}·Torr^{-1}. However, in plasma α and k will somewhat be reduced by the albumin content of ca. 60 gr·1^{-1} (Yamaguchi et al.,

1985). Assuming $\eta(CO)$ for the RBC membrane in blood plasma is 2.3 x 10^{-6} cm·sec^{-1}·Torr^{-1}, the total $\eta(CO)$ value for the RBC in the pulmonary capillary is estimated to be 1.7 x 10^{-6} cm·sec^{-1}·Torr^{-1} from the above $\eta(CO)$ value of the alveolar membrane. This value agrees fairly well with the experimental value of (1.7 ± 0.4) x 10^{-6} cm·sec^{-1}·Torr^{-1}. Therefore, the data shown in Table 1 appear justifiable.

SUMMARY

When a gas passes across the interface, a gap in partial pressure appears between the two sides. On the other hand, the transfer rate is linearly related to the change in partial pressure at the side with lower solubility from an equilibrium level. From the relationship between the gap in partial pressure and the transfer rate measured at a gas-liquid interface and the RBC boundary, conductivity in aqueous solution is estimated to be 6.6 x 10^{-3} cm·sec^{-1}·Atm^{-1}, regardless of O_2, CO_2 and CO. Water content in the RBC membrane is then computed to be about 22%.

ACKNOWLEDGEMENT

The author would like to express his sincere thanks to Dr Ann Silver, Cambridge, GB. for her painstaking work in revising the manuscript.

REFERENCES

Fukui, K. and Mochizuki, M. (1972) Some basic problems on the pulmonary diffusing capacity for carbon monoxide. I. The reaction rate of CO with oxygenated hemoglobin in the red blood cell. Monograph Series of Research Institute of Applied Electricity. 20, 69-78.

Gros, G. and Moll, W. (1971) The diffusion of carbon dioxide in erythrocytes and hemoglobin solutions. Pflügers Arch. 324, 249-266.

Grote, J. (1967) Die Sauerstoffdiffusionskonstanten im Lungengewebe und Wasser und ihre Temperaturabhängigkeit. Pflügers Arch. 295, 245-254.

Kagawa, T. and Mochizuki, M. (1982) Numerical solution of partial differential equation describing oxygenation rate of the red blood cell. Jpn. J. Physiol. 32, 197-218.

Kagawa, T. and Mochizuki, M. (1984) Numerical solution of partial differential equations for CO_2 diffusion accompanying HCO_3^- shift in red blood cells. Jpn. J. Physiol. 34, 1029-1047.

Mochizuki, M. and Fukuoka, J. (1958) The diffusion of oxygen inside the red cell. Jpn. J. Physiol. 8, 206-224.

Mochizuki, M. (1966) Study on the oxygenation velocity of human red cell. Jpn. J. Physiol. 16, 635-648.

Mochizuki, M. (1975) Graphical analysis of oxygenation and CO-combination rates of the red cells in the lung. Hirokawa Publ. Co., Tokyo. pp. 1-72.

Niizeki, K., Mochizuki, M. and Uchida, K. (1983) Rate of CO_2 diffusion in the human red blood cell measured with pH-sensitive fluorescence. Jpn. J. Physiol. 33, 635-650.

Niizeki, K., Mochizuki, M. and Kagawa, T. (1984) Secondary CO_2 diffusion following the HCO_3^- shift across the red blood cell membrane. Jpn. J. Physiol. 34, 1003-1013.

Opitz, E. and Bartels, H. (1955) Gasanalyse. Hoppe-Seyler/Thierfelder, Handbuch der physiologisch- und pathologisch-chemischen Analyse, 10. Aufl. Bd. 11, Springer Verlag, Berlin, Göttingen und Heidelberg. pp. 305-308.

Silverman, D. N., Tu, C. and Wynns, G. C. (1976) Depletion of ^{18}O from $C^{18}O_2$ in erythrocyte suspensions. J. Biol. Chem. 251, 4428-4435.

Uchida, K., Mochizuki, M. and Niizeki, K. (1983) Diffusion coefficients of CO_2 molecule and bicarbonate ion in hemoglobin solution measured by fluorescence technique. Jpn. J. Physiol. <u>33</u>, 619–634.

Uchida, K., Sibuya, I. and Mochizuki, M. (1986) Simultaneous measurement of cardiac output and pulmonary diffusing capacity for CO by a re-breathing method. Jpn. J. Physiol. <u>36</u>, 657–670.

Yamaguchi, K., Nguyen-phu, D., Scheid, P. and Piiper, J. (1985) Kinetics of O_2 uptake and release by human erythrocytes by a stopped-flow technique. J. Appl. Physiol. <u>58</u>, 1215–1224.

MEASUREMENTS OF OXYGEN FLUX FROM ARTERIOLES IMPLY HIGH PERMEABILITY OF PERFUSED TISSUE TO OXYGEN

A.S. Popel, R.N. Pittman[*], M.L. Ellsworth[*], and D.P.V. Weerappuli

Department of Biomedical Engineering, The Johns Hopkins University School of Medicine, Baltimore, MD 21205

[*]Department of Physiology, Medical College of Virginia, Virginia Commonwealth University, Richmond, VA 23298

A mathematical model developed for the analysis of oxygen flux from an arteriole surrounded by perfused tissue was used to analyze experimental data in the resting hamster cheek pouch retractor muscle. The flux predicted by the model, with the commonly accepted values of tissue permeability to oxygen (the Krogh diffusion coefficient), was an order of magnitude smaller than the average value of experimentally observed oxygen flux. The values of permeability required by the model to equate the predicted and observed oxygen flux are one to two orders of magnitude higher than the accepted values. Also, the values of the oxygen tension gradient in the arteriolar wall predicted by Fick's law are an order of magnitude greater than the measured values reported in the literature. Since the accepted values of permeability are based on experiments with unperfused tissue and the values predicted by the model are for blood-perfused tissue, we conjecture that tissue permeability is a function of the perfusion conditions. Hence, there is a need for re-examination of the distribution of resistance to oxygen transport along the pathway between red blood cells and mitochondria. Theoretical estimates based on accepted values of tissue permeability to oxygen show resistances to oxygen transport inside and outside the capillaries to be of similar magnitude. However, if the tissue component of the resistance is significantly reduced because of greater permeability, the intracapillary resistance becomes dominant and is responsible for most of the drop in the oxygen tension between the red blood cells and the tissue.

INTRODUCTION

Since the classic studies of August Krogh (1919), it has been assumed that permeability to oxygen is the same in both perfused and unperfused tissue. The diffusion (D) and solubility (α) coefficients, whose product is equal to tissue permeability or the Krogh diffusion coefficient (K), have been measured only in unperfused tissue samples. Reported values of tissue permeability to oxygen are typically between 40 and 90% of the permeability in water (Altman and Dittmer, 1971; Evans et al., 1981; Ellsworth and Pittman, 1984). According to Krogh's mathematical model of oxygen diffusion from blood capillaries, the difference between oxygen tension (PO_2) in capillaries and the minimum PO_2 in surrounding tissue is inversely proportional to the permeability of that tissue. Thus, tissue permeability is a major determinant of tissue oxygenation.

Experiments on different tissues (hamster cheek pouch, striated muscle, brain) have shown that a substantial fraction of oxygen is exchanged in the precapillary circulation (Duling and Berne, 1970; Pittman and Duling, 1977; Duling et al., 1982; Ivanov et al., 1982). It has also been reported that the PO_2 difference across the arteriolar wall is small, only a few torr.

In the present work we analyze recent experimental data on oxygen flux from arterioles in the resting hamster cheek pouch retractor muscle (Kuo and Pittman, 1988) using a new mathematical model based on mass-balance arguments (Weerappuli and Popel, 1989a,b).

EXPERIMENTAL METHODS AND RESULTS

The experimental methods used and results presented by Kuo and Pittman (1988) are summarized here.

Retractor Muscle Preparation

Experiments were carried out on 23 male golden hamsters weighing between 68 and 85 g (75±1.4 (SE), age = 33±2 (SE) days). Animals were initially anesthetized with sodium pentobarbital (65 mg/kg body weight, i.p.) and supplemental anesthetic was administered by continuous infusion (0.15 mg/min) through the left femoral vein. The trachea was cannulated to ensure a patent airway and animals breathed room air spontaneously. The left femoral artery was cannulated with PE-10 tubing to monitor systemic arterial blood pressure continuously. Data were collected only if mean arterial blood pressure exceeded 80 torr.

The right cheek pouch retractor muscle was prepared for in vivo microscopy as described by Sullivan and Pittman (1982). Briefly, the retractor muscle was exposed by an incision in the skin along the upper middle back, the spinal end of the retractor muscle was separated from underlying back muscles and a hemoclip (Edward Weck) was used to clamp two ligatures to the muscle. The spinal end of the muscle was then severed and the muscle was placed ventral side up on a clear Plexiglas platform. Several ligatures were then attached to the edge of the muscle to flatten it and maintain it in its in situ dimensions. The muscle was covered with a thin transparent plastic film (Saran, Dow Corning) to prevent desiccation and to minimize gas exchange between the muscle and the atmosphere. Deep esophageal and muscle temperatures were monitored and maintained at $37\pm1^{o}C$ by separate heat exchangers within the animal platform. A small amount (100 μl) of arterial blood was taken to determine systemic blood gases and pH using a blood gas analyzer (Radiometer NS, Copenhagen, Denmark), and systemic arterial hematocrit (Hct) was determined by microcentrifugation.

Measurements of Hemodynamic Variables and Hemoglobin Oxygen Saturation

The retractor muscle was transilluminated with a xenon lamp and the microcirculation was observed with a Leitz Labolux II microscope equipped with a long working distance objective (Leitz UMK 50/0.60). The objective was used dry to give an effective magnification/numerical aperture of 32x/0.40. The microscope image was displayed on a television monitor, and the vessel luminal diameter (d) was measured from the monitor using a video analyzer (Colorado Video Model 321). Centerline dual-sensor RBC velocity

(v_{ds}) was determined with the dual sensor cross-correlation method of Wayland and Johnson (1967) as modified by Intaglietta et al. (1970). The hemoglobin-oxygen saturation (S) in single microvessels was determined by the three-wavelength spectrophotometric method developed by Pittman and Duling (1975). The microvessel Hct (H) was determined by the method of Lipowsky et al. (1982). These measurements were performed simultaneously with the on-line microcomputer system described by Duling et al. (1983).

Arterioles of the retractor muscle were classified by branching order. The large input arterioles that originate outside of the retractor muscle were designated as first order (1°). Branches of first order arterioles were designated as second order (2°) and so on. Usually, either capillaries or small terminal arterioles branched directly from a given 4° arteriole. Clearly visible vessel segments were selected for this study. In four branching orders of the arteriolar network, simultaneous determinations of d, v_{ds}, H, and S were made at the same upstream and downstream sites of the longest unbranched segment of that vessel order. Unbranched segments were used to ensure that any differences in oxygen saturation at the two sites were due to diffusion across the intervening arteriolar wall rather than the result of a differential distribution of convective oxygen flow at bifurcations. The length of each segment (Δz) was also measured. The change in saturation (ΔS) for each segment was expressed as the difference between the upstream and downstream saturation values and the longitudinal saturation gradient ($\Delta S/\Delta z$) was computed for each segment.

Calculation of Blood Flow and Oxygen Flow

To quantify the rate at which oxygen is delivered by convection to a particular location within the microcirculation, it is necessary to estimate both blood flow and oxygen content. Oxygen content, $[O_2]$, is determined primarily by the Hct and the saturation as expressed by the following equation:

$$[O_2] = C_{Hb}[Hb]S \tag{1}$$

where C_{Hb} is the oxygen-binding capacity of hemoglobin (1.34 ml O_2/g Hb) and [Hb] is the hemoglobin concentration (g Hb/100 ml of arteriolar blood), which is the product of Hct and mean corpuscular hemoglobin concentration (31.55 g Hb/100 ml RBC; Meyerstein and Cassuto, 1970). The physically dissolved oxygen was neglected in this calculation due to the low solubility of oxygen in plasma.

Blood flow was estimated as the product of mean blood velocity and vessel luminal cross-sectional area assuming a circular cross section. Mean blood velocity ($\bar{v}$) was determined from the centerline dual sensor velocity with the method described by Pittman and Ellsworth (1986) that takes into account the spatial averaging inherent in the method, as well as the bluntness of the velocity distribution. Blood flow (Q) and oxygen flow (QO_2) in a single microvessel were computed as:

$$Q = (\pi d^2/4)\bar{v} \tag{2}$$

$$QO_2 = [O_2]Q \tag{3}$$

Summary of Experimental Results

Average microvessel Hct in four branching orders of arterioles was 42.5±0.4%, which is significantly lower than the corresponding systemic value (51.8±0.4%). There were no statistical differences in microvessel Hct among the four branching orders. Mean vessel diameter gradually fell from 62.1±4.0 μm in 1° arterioles to 23.4±0.9 μm in 4° arterioles and mean

RBC velocity significantly decreased from 18.0±1.2 mm/sec in 1° arterioles to 3.7±0.3 mm/sec in 4° arterioles. Hemoglobin oxygen saturation progressively declined from 78.1±3.1% in systemic arterial blood to 66.8±2.4% in 1° arterioles and then gradually fell to 51.2±1.7% in 4° arterioles. The longitudinal gradient in S ($\Delta S/\Delta z$) increased with increasing branching order from an average of 1.1±0.2%/mm in 1° arterioles to 13.1±1.0%/mm in 4° arterioles. The change in $\Delta S/\Delta z$ from 1° to 4° arterioles was statistically significant ($p < 0.05$).

ANALYSIS OF EXPERIMENTAL DATA

Analysis of Oxygen Transport through the Arteriolar Wall

Consider a cylindrical blood vessel with luminal diameter d, and wall thickness w. Oxygen mass balance in the vessel is expressed by:

$$QC(\Delta S/\Delta z) = -\pi d j_m \qquad (4)$$

where Q is the volumetric blood flow rate, C is the oxygen-binding capacity of blood, S is the fractional hemoglobin-oxygen saturation, $\Delta S/\Delta z$ is the saturation gradient along the vessel, and j_m is the oxygen flux per unit area at the vessel wall averaged over the vessel circumference. The local oxygen flux at the blood-vessel wall interface can be expressed by Fick's law

$$j = -K_w(\partial P/\partial r) \qquad (5)$$

where P is the oxygen tension, r is the radial coordinate, $K_w = D_w \alpha_w$ is the wall permeability. Using the data presented above for each arteriolar branching order under control conditions, we can calculate the values of oxygen flux, j_m, from Equation 4. These results are presented in Table 1 for 1° and 4° arterioles; each value is the average for 16 vessels. There are no data on permeability of the arteriolar wall; thus, we assume the range of values reported for other tissues (Evans et al., 1981): $K_w = (4.2-6.8) \times 10^{-10}$ cm^2/s/torr. Using the highest permeability value, we find from Equation 2 a minimum estimate of the PO_2 gradient at the wall to be approximately 15 torr/μm. If the wall thickness is significantly smaller than the vessel diameter, then the PO_2 gradient should be approximately constant across the wall. The effect of oxygen consumption within the wall is negligible (Popel and Gross, 1979). Thus, for a wall thickness larger than 4 μm, we would estimate a transmural PO_2 difference that exceeds 60 torr.

TABLE 1

Branching order	Diameter, d (μm)	O_2 flux, j_m (ml O_2/cm^2/s)	P_ℓ (torr)
1°	62.1	$59.0 \cdot 10^{-6}$	37.7
4°	23.4	$56.9 \cdot 10^{-6}$	31.6

Arteriolar diameter, d, oxygen flux from the arteriole, j_m, and oxygen tension in the lumen, P_ℓ, for vessels of two branching orders.

However, these predicted values are not realistic since the luminal PO_2 corresponding to the experimental oxygen saturation values was only between 30 and 40 torr (Table 1), i.e., smaller than the predicted transmural PO_2 difference. In addition, small transmural PO_2 values of only a few torr have been measured with oxygen microelectrodes in hamster cheek pouch, hamster and rat cremaster muscle, and rabbit and cat pial vessels (Duling and Berne, 1970; Pittman and Duling, 1977; Duling et al., 1982; Ivanov et al., 1982). Duling et al. (1982) reported a radial PO_2 gradient in the wall of pial vessels of approximately 1 torr/μm. Thus, the values of the experimentally measured transmural PO_2 gradient are an order of magnitude smaller than the values predicted by Fick's law (Equation 5). Nevertheless, the predicted and measured values can be made consistent with each other if we assume that the permeability of the arteriolar wall is an order of magnitude (about 15 times) higher than the value we used in the calculations. Next, we consider oxygen transport outside the arteriolar wall.

Analysis of Oxygen Transport outside the Arteriolar Wall

The mathematical model presented by Weerappuli and Popel (1989a) can be summarized as follows. Consider oxygen diffusion from an arteriole surrounded by an infinite array of capillaries carrying blood at a velocity v_c normal to the flow in the arteriole. The problem of oxygen exchange between the arteriole and an array of discrete capillaries is very complex, and, in addition, the detailed information on the arteriole-capillary geometry is not available at present. As a first approximation, we adopted a "continuum" model, formulated by Salathe (1982), in which the discrete capillary flow was replaced by a distributed flow of hemoglobin (Figure 1). Instead of two values of oxygen tension for tissue and capillaries, a single value P is introduced that can be considered as a local volume-averaged oxygen tension.

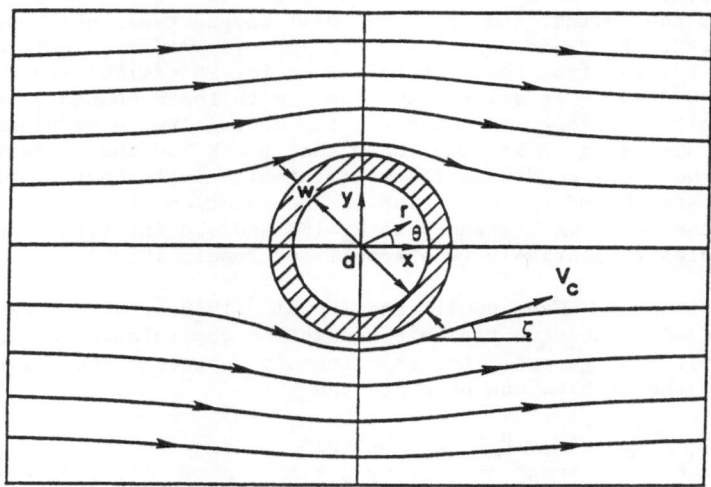

Fig. 1. An arteriole of luminal diameter d with an avascular wall of thickness w is surrounded by a distributed tissue-capillary structure. The hemoglobin flows (from left to right) normal to the arteriolar longitudinal axis along paths closely representing the actual capillary alignment.

In the steady state, the oxygen tension P satisfies a mass balance equation

$$K_t \nabla^2 P - C\psi v_c \cdot \nabla S = M_t(1-\psi) \qquad (6)$$

where ∇^2 is the two-dimensional Laplacian operator in the plane perpendicular to the arteriolar axis, K_t is the tissue permeability, C is the oxygen-binding capacity of blood, ψ is the fractional volume occupied by the capillaries, and M_t is the oxygen consumption rate in the perfused region. Diffusion in the longitudinal direction of the arteriole is neglected. The hemoglobin paths in the distributed capillary-tissue structure were assumed to be given by

$$f(r,\theta) = r \sin\theta \, [\, 1 - (d_0/2r)^2] = constant \qquad (7)$$

where $d_0 = d+w$. This representation coincides with the streamline pattern for potential fluid flow around an infinite cylinder.

The oxygen tension P at a point was expressed as the sum of P_0 and P', where P_0 is the tissue oxygen tension in the absence of the arteriole. Representing the oxygen dissociation curve by the Hill equation, and defining a local Peclet number Pe_c as

$$Pe_c = \frac{C\psi v_c d}{K_t} \left. \frac{dS}{dP}\right|_P \qquad (8)$$

the governing equation was expressed as

$$\nabla^2 P' - \frac{1}{d} Pe_c \cos\varsigma \frac{\partial P_0}{\partial x} - \frac{1}{d} Pe_c \left[G \frac{\partial P'}{\partial r} + \frac{1}{r} F \frac{\partial P'}{\partial \theta} \right] = 0 \qquad (9)$$

where F and G are the directional cosines of the capillary velocity v_c given by $\sin(\varsigma-\theta)$ and $\cos(\varsigma-\theta)$, respectively, with ς the inclination of v_c to the x-axis.

The following boundary conditions were imposed: (1) at $r=d/2$, $P_w=P_\ell$, where P_ℓ is the luminal and P_w the wall oxygen tensions, respectively; (2) at $r=d_0/2$, $P_w=P$ and $K_w \partial P_w/\partial r = K_t \partial P/\partial r$; (3) at a fictitious outer boundary $r=R_\infty$, far from the arteriole $P'=0$ for $90°<\theta<180°$; (4) at $r=R_\infty$, $\partial P'/\partial x=0$ for $0<\theta<90°$. We solved Equation 9 with these boundary conditions using a finite difference scheme with successive overrelaxation and appropriate upwinding. It was found that for $R_\infty>5d$ the effect of the "downstream" boundary condition (4) was minimal. To further eliminate the effect of this boundary condition on the solution for flux from the arteriole, the mesh was extended to $R_\infty=16d$ and $33d$ for first and fourth order arterioles respectively (Weerappuli and Popel, 1989b).

Calculations by Weerappuli and Popel (1989a) were based on an assumption $K_w=K_t$; subsequently this assumption was relaxed (Weerappuli and Popel, 1989b). In general, the relationship between the dimensionless parameters of the problem can be expressed as:

$$Sh = \Psi\left\{ Pe, \frac{P_\infty}{P_{50}}, \frac{(P_\ell-P_\infty)}{P_{50}}, \frac{2w}{d}, \frac{M(1-\psi)d^2}{K_t P_{50}}, \frac{K_t}{K_w}, \alpha P_{50}, n, C \right\} \qquad (10)$$

Here the Sherwood number Sh can be considered as the dimensionless oxygen flux at the lumen-wall interface

$$Sh = \frac{J/\pi}{K_t(P_\ell - P_\infty)} \qquad (11)$$

where J is the oxygen flux per unit length of the arteriole, $J = \pi d j_m$, and P_∞ is the "background" PO_2, far from the arteriole, on the y axis. The Peclet number Pe is defined as

$$Pe = \frac{C\psi v_c d}{K_t} \left. \frac{dS}{dP} \right|_{P_{50}} \qquad (12)$$

where n is the dimensionless exponent in the Hill equation, and P_{50} is the oxygen tension at 50% hemoglobin oxygen saturation. Relationship (10) was systematically investigated by Weerappuli and Popel (1989a,b).

The following parameter values were used in the analysis of oxygen transport outside the arteriolar wall: the diffusion coefficient $D_t = D_w = 1.39 \times 10^{-5}$ cm^2/s, the solubility coefficient $\alpha_t = \alpha_w = 2.84 \times 10^{-5}$ ml O_2/ml/torr; $\psi = 0.0146$, corresponding to a capillary density of 1435 mm^{-2} and capillary diameter 3.6 μm; capillary velocity $v_c = 0.01$ cm/s; oxygen-binding capacity of blood $C = 0.25$ ml O_2/ml corresponding to a hematocrit of 50%; and $n = 2.52$ and $P_{50} = 29.1$ torr in the Hill equation (Ellsworth and Pittman, 1984; Ellsworth et al., 1988). Note that the values of the oxygen diffusion and solubility coefficients were obtained experimentally in unperfused muscles; these values are similar to those reported for other unperfused tissues (Altman and Dittmer, 1971; Evans et al., 1981). We denote the corresponding "standard" permeability K. The values for d and P_ℓ in the standard case are taken from Table 1. Wall thicknesses of 10.6 and 4μm for 1° and 4° arterioles, respectively, were assumed, such that the ratio w/d was the same in both cases. P_∞ was assumed to be 20 torr. The oxygen consumption rate for the wall and perfused tissue was 1.57×10^{-4} ml O_2/ml/s.

We obtained the oxygen flux from the arterioles, J_{theor}, using the values of parameters given above. Comparison of J_{theor} with the experimentally determined flux, J_{exp}, gives: $J_{exp}/J_{theor} = 16.2$ and 15.7 for 1° and 4° vessels, respectively. For $P_\infty = 10$ torr these ratios become 9.9 and 7.9. Again, as in the case of the arteriolar wall considered above, the experimental values of oxygen flux are an order of magnitude higher than the predicted values.

These flux ratios can be made equal to unity if the tissue permeability used in the calculations is allowed to increase. According to the foregoing calculations for the arteriolar wall, the wall permeability should be increased by a factor of 15, thus we chose $K_w = 15K$. We then calculated the flux, J_{theor}, as a function of tissue permeability K_t. Figure 2 shows the variation of J_{exp}/J_{theor} with K_t/K for 1° and 4° arterioles. The values of tissue permeability predicted by the model that yield the values of oxygen flux equal to those experimentally measured, are up to two orders of magnitude higher (144 times from the data on 1° vessels and 75 times from the data on 4° vessels) than currently accepted values.

DISCUSSION AND SUMMARY

If, as our analysis suggests, the permeability of tissue to oxygen is much greater than has been thought, we must reassess the distribution of resistance to oxygen transport along the pathway between red blood cells and mitochondria. Theoretical estimates (Hellums, 1977) based on accepted values of tissue permeability to oxygen showed resistances to oxygen transport inside and outside the capillaries to be of similar magnitude; these estimates were subsequently confirmed by more detailed calculations (Federspiel and Popel, 1986). However, if the tissue component of the resistance is significantly reduced because of greater permeability, the intracapillary resistance becomes dominant and is responsible for most of the drop in PO_2 between the red blood cells and the tissue. This picture is consistent with the arguments presented by Honig et al. (1984) and Gayeski et al. (1988) based on the findings that PO_2 gradients in myoglobin-rich muscle fibers are small. However, the higher permeability predicted in this study is not caused by the presence of myoglobin.

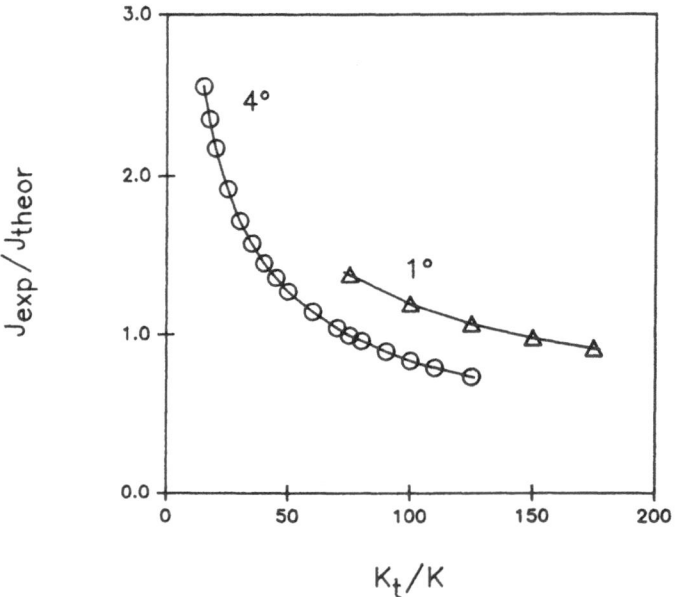

Fig. 2. The variation in the ratio J_{exp}/J_{theor} with K_t/K for 1° and 4° arterioles. The points generated by the model are shown by symbols.

In the present calculations we have not considered intra-arteriolar PO_2 gradients. It is clear, however, that for the values of oxygen flux involved, the plasma layer adjacent to the arteriolar endothelium should represent a major resistance to oxygen transport, just like the plasma layer in capillaries around red blood cells at high oxygen flux (Gayeski at al., 1988). A theory of oxygen transport in the arteriolar lumen has yet to be developed (Popel, 1989), but taking this transport into account would not change qualitative conclusions reached in the present work.

The conjecture that the arteriolar wall permeability is an order of magnitude higher than commonly accepted values follows directly from (a) Fick's law and (b) experimental data on transmural PO_2 differences reported in the literature; it is not based on the mathematical model. To confirm this conjecture, simultaneous measurements of the longitudinal gradients of saturation and the transmural differences of PO_2 are necessary.

On the other hand, the conjecture that the tissue permeability outside the arteriolar wall has to be one to two orders of magnitude higher than commonly accepted values is less definitive since it is based on the mathematical model; and the model may have a limitation in representing capillary flow as a continuous flow of hemoglobin. A possible error can be assessed by solving the problem for discrete capillaries. This is a problem for future studies.

In summary, using experimental data reported by Kuo and Pittman (1988) for control conditions, we have shown that the observed oxygen flux from arterioles in the resting hamster retractor muscle is inconsistent with commonly accepted values of tissue permeability measured in unperfused tissues. Analysis of the data on hemodilution reported by Kuo and Pittman (1988) leads to the same conclusion. The mathematical model predicts values of tissue permeability that are one to two orders of magnitude higher for both the vascular wall and extravascular tissue. Since all previously reported measurements were made in unperfused tissues, this conclusion is not contradictory to the accepted values of tissue permeability. Thus, tissue permeability appears to be a function of the perfusion conditions. Taking this conjecture even further, it might be hypothesized that tissue permeability to oxygen is a regulated variable.

The mechanism of increased permeability is not known. The existence of preferential intracellular oxygen channels has been proposed (Longmuir, 1980). Further experimental investigation of the dependence of tissue permeability on blood perfusion conditions and, if such dependence is confirmed, identification of the molecular mechanisms of increased permeability are subjects for future studies.

ACKNOWLEDGMENT

Supported by grant HL 18292 from the Heart, Lung, and Blood Institute.

REFERENCES

Altman, P.S., and D.S. Dittmer (1971). Respiration and Circulation, Fed. Amer. Soc. Exp. Biol., Bethesda, Md., pp. 21-22.

Duling, B.R., and R.M. Berne (1970). Longitudinal gradients in periarteriolar oxygen tension: a possible mechanism for the participation of oxygen in local regulation of blood flow. Circ. Res. 27: 669-678.

Duling, B.R., W. Kuschinsky, and M. Wahl (1982). Measurements of the perivascular PO_2 in the vicinity of the pial vessels of the cat. Pflugers Arch. 383: 29-34.

Duling, B.R., D.N. Damon, S.R. Donaldson, and R.N. Pittman (1983). A computerized system for densitometric analysis of the microcirculation. J. Appl. Physiol. 55: 642-651.

Ellsworth, M.L., and R.N. Pittman (1984). Heterogeneity of oxygen diffusion through hamster striated muscle. Am. J. Physiol. 246 (Heart Circ. Physiol. 15): H161-H167.

Ellsworth, M.L., A.S. Popel, and R.N. Pittman (1988). Assessment and impact of heterogeneities of convective oxygen transport parameters in capillaries of striated muscle: experimental and theoretical. Microvasc. Res. 35: 341-362.

Evans, N.T.S., P.F.D. Naylor, and T.H. Quinton (1981). The diffusion coefficient of oxygen in respiring kidney and tumor tissue. Respir. Physiol. 43: 179-188.

Federspiel, W.J., and A.S. Popel (1986). A theoretical analysis of the effect of the particulate nature of blood on oxygen release in capillaries. Microvasc. Res. 32: 164-189.

Gayeski, T.E.J., W.J. Federspiel, and C.R. Honig (1988). A graphical analysis of the influence of red cell transit time, carrier-free layer thickness, and intracellular PO_2 on blood-tissue O_2 transport. Adv. Exp. Med. Biol. 222: 25-35.

Hellums, J.D. (1977). The resistance to oxygen transport in the capillaries relative to that in the surrounding tissue. Microvasc. Res. 13: 131-136.

Honig, C.R., T.E.J. Gayeski, W. Federspiel, A. Clark, and P. Clark (1984). Muscle O_2 gradients from hemoglobin to cytochrome: new concepts, new complexities. Adv. Exp. Med. Biol. 169: 23-38.

Krogh, A. (1919). The number and distribution of capillaries in muscles with calculations of the oxygen pressure head necessary for supplying the tissue. J. Physiol. (London) 52: 409-415.

Kuo, L., and R.N. Pittman (1988). Effect of hemodilution on oxygen transport in arteriolar networks of hamster striated muscle. Am. J. Physiol. 254 (Heart Circ. Physiol. 23): H331-H339.

Intaglietta, M., W.R. Thompkins, and D.R. Richardson (1970). Velocity measurements in the microvasculature of the cat omentum by on-line method. Microvasc. Res. 2: 462-473.

Ivanov, K.P., A.N. Derry, E.P. Vovenko, M.O. Samoilov, and D.G. Semionov (1982). Direct measurements of oxygen tension at the surface of arterioles, capillaries and venules of the cerebral cortex. Pflugers Arch. 393: 118-120.

Lipowsky, H.H., S. Usami, S. Chien, and R.N. Pittman (1982). Hematocrit determination in small bore tubes by differential spectrophotometry. Microvasc. Res. 24: 42-55.

Longmuir, I.S. (1980). Channels of oxygen transport from blood to mitochondria. Adv. Physiol. Sci. 25: 19-22.

Meyerstein, N., and Y. Cassuto (1970). Haematological changes in heat-acclimated Golden hamster. Br. J. Haematol. 18: 417-423.

Pittman, R.N., and B.R. Duling (1975). Measurement of percent oxyhemoglobin in the microvasculature. J. Appl. Physiol. 38: 321-327.

Pittman, R.N., and B.R. Duling (1977). Effects of altered carbon dioxide tension on oxygenation in hamster cheek pouch microvessels. Microvasc. Res. 13: 211-224.

Pittman, R.N., and M.L. Ellsworth (1986). Estimation of red cell flow in microvessels: consequences of the Baker-Wayland spatial averaging model. Microvasc. Res. 32: 371-388.

Popel, A.S., and J.F. Gross (1979). Analysis of oxygen diffusion from arteriolar networks. Am. J. Physiol. 237 (Heart Circ. Physiol. 6): H681-H689.

Popel, A.S. (1989). Theory of oxygen transport to tissue. CRC Crit. Rev. Biomed. Eng., in press.

Salathe, E.P. (1982). Mathematical modeling of oxygen transport in skeletal muscle. Math. Biosci. 58: 171-184.

Sullivan, S.M., and R.N. Pittman (1982). Hamster retractor muscle: A new preparation for intravital microscopy. Microvasc. Res. 23: 329-335.

Wayland, H., and P.C. Johnson (1967). Erythrocyte velocity measurement in microvessels by a two-slit photometric method. J. Appl. Physiol. 22: 333-337.

Weerappuli, D.P.V., and A.S. Popel (1989a). A model of oxygen exchange between an arteriole or venule and the surrounding tissue. J. Biomech. Eng., submitted for publication.

Weerappuli, D.P.V., and A.S. Popel (1989b). Calculation of oxygen flux from an arteriole surrounded by tissue of high permeability. Submitted for publication.

Po2 HISTOGRAMS IN VARIOUS MODELS OF TISSUE OXYGENATION IN SKELETAL MUSCLE

Zdenek Turek, Jos Olders, Louis Hoofd, Stuart Egginton, Ferdinand Kreuzer and Karel Rakusan

Depts. of Physiology
Catholic University of Nijmegen
Nijmegen, The Netherlands
University of Birmingham
Birmingham, U.K.
and University of Ottawa
Ottawa, Canada

INTRODUCTION

A natural way of testing mathematical models is to compare them with experimental results. Until recently, the most common experimental results on tissue oxygenation were in the form of Po2 histograms obtained with surface or needle Po2 electrodes. In recent years Po2 histograms derived from myoglobin cryospectrophotometry were presented for skeletal (Honig, 1984; Gayeski et al., 1985; Gayeski and Honig, 1986; Gayeski and Honig, 1988) and cardiac muscle (Honig and Gayeski, 1987). It came as a great surprise that the Po2 histograms derived from myoglobin saturation measured by myoglobin cryospectrophotometry were different from those obtained with the electrodes, having a very high percentage of low Po2 values even at rest. In stimulated skeletal muscle or in the beating heart a great majority of the Po2 values were lower than 10 mm Hg. An additional resistance operating near the capillary wall was suggested as an explanation for these results. Similar additional resistance in the capillary wall has been also proposed by Rose and Goresky (1985). This would be expected to result in hypoxia of a high percentage of tissue if only a diffusion of physically dissolved O2 is considered. Honig and Gayeski proposed that the diffusion of O2 facilitated by myoglobin might be the mechanism that compensates for this danger so efficiently that the Po2 profiles become very low and flat, but higher than zero. This was supported by theoretical analysis indicating that the facilitation of O2 diffusion by myoglobin can compensate for the effect of an additional resistance when average size of a muscle fiber was considered (Federspiel, 1986; Gayeski and Honig, 1986; Groebe and Thews, 1986).

Several models of skeletal or cardiac muscle oxygenation have been developed that could produce Po2 histograms resembling at least roughly those obtained with the Po2 electrodes. These models included either the spatial heterogeneity in capillary spacing (Rakusan et al., 1984), the flow heterogeneity (Popel et al., 1986), or both (Rakusan and Turek, 1985). The model of Popel et al. (1986) used the Michaelis-Menten kinetics of O2 consumption, whereas in our previous papers we used zero-order

227

kinetics. In our original communication we used the method of concentric circles for determination of a value of the mean radius of Kroghian tissue cylinders and a logarithmic standard deviation (log SD), as a heterogeneity index (Turek and Rakusan, 1981). In this communication, the method of capillary domains (Hoofd et al., 1985), was used for the evaluation of the heterogeneity in capillary spacing. This method yields an almost identical value of the mean radius of the tissue cylinders but a smaller log SD than the method of concentric circles, used previously (Turek et al., 1986; 1987).

The effect of O_2 diffusion facilitated by myoglobin was assessed in our recent publication (Hoofd et al. 1987). We found that facilitated diffusion had a minor effect in the beating heart during rest, becoming more important with increased cardiac work but not efficient enough to prevent the occurrence of anoxic tissue even without any additional barrier to O_2 transport. Our subsequent calculations (unpublished) indicated that when such a barrier was assumed, a large percentage of anoxic tissue occurred in the beating heart, even at rest.

In the present communication we extend our model so that both the effect of the myoglobin-facilitated diffusion of O_2 as well as the Po_2 dependent O_2 consumption could be evaluated. We assume that this Po_2 dependency operates according to Michaelis-Menten kinetics which may be representative for other O_2-dependent consumption models (Hoofd, 1987). Our program also allows for consideration of the effect of an additional barrier. Morphometric data obtained from the perfused capillaries of an oxidative skeletal muscle were used.

METHODS

Experiments were performed on male Sprague-Dawley rats (450-550 g body weight) under pentobarbitone anesthesia (Sagatal; 6mg/100 g body weight intraperitoneally). Tibialis anterior (TA) muscles were dissected free of fascia and underlying muscles, and attached to a strain gauge via cut tendons. Muscles were held at optimal lenth, determined by maximal tension developed in response to indirect stimulation of the motor nerves. Three levels of muscle activity were studied: rest, stimulation at low voltage and high frequency, and high voltage with low frequency. The first protocol recruits predominantly fast glycolytic fibers and produces less than 20 percent of maximal tension, while the second protocol elicits maximal hyperaemia. Thioflavin-S conjugated with BSA was injected via a carotid arterial cannula, and blood flow to the TA stopped by ligation of the common iliac artery at 7.5 seconds (Egginton et al., 1987). Muscles were rapidly excised and frozen; unstained cryostat sections were examined by epifluorescent microscopy and photographs taken from 6 sample sites chosen at random from the oxidative core. The sections were subsequently stained for alkaline phosphatase activity in order to locate the total capillary bed, and photographs of the same fields were taken. Values of the mean tissue cylinder radius and log SD were obtained experimentally by the method of capillary domains (Hoofd et al., 1985).

The model consisted of a series of parallel tissue cylinders with varying radii following a lognormal distribution (see Turek and Rakusan, 1981; Turek et al., 1986). A capillary radius of 2.3 μm was taken. The whole distribution of the tissue cylinder radii was integrated from a radius slightly larger than that of the capillary to a radius equal to the median plus four log SD, on a logarithmic scale.

The O_2 and CO_2 contents at the beginning of each cylinder were calculated from the arterial Po_2 (100 mm Hg), Pco_2 (40 mm Hg), pH (7.4) and O_2 capacity (20 ml/dl), using combined O_2 and CO_2 dissociation curves and correcting for Bohr and Haldane effects. The Adair model of the O_2 dissociation curve (Adair, 1925) was used with constants of human blood (Roughton and Severinghaus, 1973), corrected for the rat P50 (Turek et al., 1973). Algorithms of Kelman (1967) for CO_2 blood dissociation curve, and the equations for the acid-base equilibrium as published by Maas et al. (1972) were used.

Each tissue cylinder was divided longitudinally into one hundred slabs, for each, radial Po_2 profiles were calculated from the capillary wall to the periphery, using an iterative procedure. It was assumed that the slabs consisted of layers of concentric rings ("peels") and an O_2 pressure gradient across each layer was computed using the Fick equation. The flux of O_2 out of each layer had to be equal to the flux into the layer minus the O_2 consumed, which was assumed to follow either the zero-order kinetics or the Michaelis-Menten kinetics (Buerk and Saidel, 1978; Hoofd, 1987). The value of q_{50} (Po_2 at which O_2 consumption equals half of its maximal value, Qmax) was assumed to be 3 mm Hg (De Koning et al., 1981; Jones, 1986; Kennedy and Jones, 1986). The flux through the capillary wall had to be the same as the O_2 consumption in each particular slab and the flux at the periphery of the cylinder had to equal zero. The procedure used in this program was that a value for the flux through the capillary wall was chosen and repeatedly adjusted until it resulted in a zero flux at the periphery.

The effect of the diffusion of O_2 facilitated by myoglobin was expressed as an additional pressure gradient above the pressure gradient of the physically dissolved O_2 (Hoofd et al. 1987). The maximal value of this additional gradient is called facilitation pressure and depends on myoglobin concentration (De Koning et al., 1981; Hoofd et al., this volume). The advantage of this approach is that the additive force of the facilitated diffusion can be computed analytically, saving computer time. This procedure assumes an equilibrium between Po_2 and myoglobin saturation and thus represents an upper limit of the facilitation. A facilitation pressure of 14 mm Hg was used (Hoofd et al., 1987).

Using this procedure, not only a Po_2 profile but also the actual O_2 consumption in each slab was derived. CO_2 production was taken the same as O_2 consumption (RQ=1). This allowed us to calculate O_2 and CO_2 blood contents for the following slab, using values of blood flow and hematocrit (45 %). By trial and error, Po_2, Pco_2 and pH were calculated from the O_2 and CO_2 contents, using the respective dissociation curves and acid-base equations, assuming no change in Base Excess.

Two flow arrangements were considered. Flow type A ("homogeneous flow") assumed that blood flow per volume of tissue was always the same. Thus flow in the capillaries of thicker cylinders had to be proportionally higher than that in thin cylinders. Flow type B ("heterogeneous flow") assumed that flow in each capillary was the same, i.e. the flow per volume tissue was smaller in thick and larger in thin cylinders.

With Michaelis-Menten kinetics of O_2 consumption the computer calculated an O_2 consumption from Qmax and other input data. In order to obtain a computed O_2 consumption close to the measured one, it was necessary to adjust Qmax accordingly. Without the additional resistance the difference between Qmax and the computed oxygen consumption was rather small and easily adjustable. However, when an additional resistance was operational, this involved several repetitions of the calculation of the complete histogram. As calculation of one histogram took a long time, we were satisfied when the computed O_2 consumption was within 5 per cent of the measured one. This inaccuracy did not affect the histograms appreciably. In zero-order O_2 consumption the experimentally obtained and computed O_2 consumptions were identical, unless anoxic tissue, not consuming O_2, was present.

The additional resistance, when applied, had an Ohmic character. This means that Po_2 gradients due to this additional resistance depend on the O_2 flux which is equal to the O_2 consumption in the slab. Consequently, at the same consumption per volume of tissue, the Po_2 gradients in thick cylinders had to be larger that in the thin ones. This resistance may be located near or within the capillary. The effect of this capillary

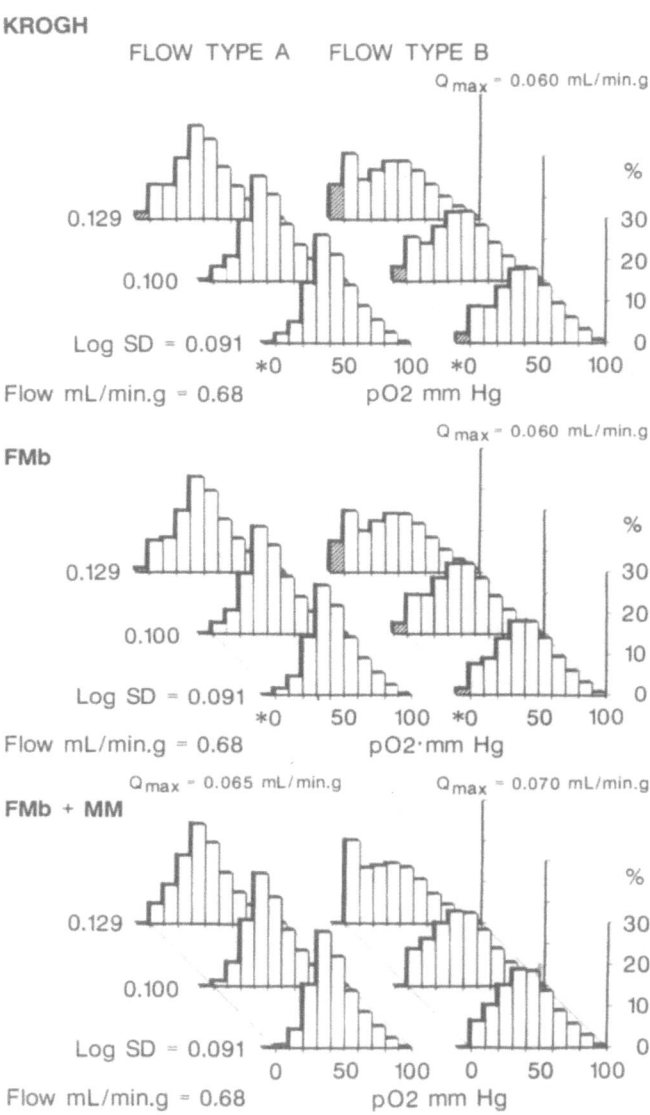

Fig. 1.   Po₂ histograms calculated for situations with the diffusion
of physically dissolved O_2 (Krogh), with the diffusion of O_2
facilitated by myoglobin (FMb) and FMb together with Michaelis-Menten kinetics of O_2 consumption (FMb+MM). For heterogeneities in flow (A and B) and heterogeneity in capillary
spacing (log SD), see text, * indicates the anoxic tissue
($P < 0.005$ mm Hg).

barrier was expressed quantitatively as a pressure drop in an average cylinder, at average tissue consumption.

Po_2 histograms were calculated for each slab, summed and finally expressed as a percentage of the total "tissue" volume.

In order to demonstrate the effect of the spatial heterogeneity on Po_2 histograms in models with and without capillary barrier, we used a single value of the radius of tissue cylinders (23.1 μm), the average of all measurements. The values of log SD were: value at rest (0.129), during muscle stimulation (0.100; it was the same for both stimulation protocols) and the lowest log SD (0.091), as found when all histochemically stained capillaries were considered. The value of capillary barrier, expressed as the equivalent pressure difference (35 mm Hg), was chosen after preliminary calculations as to obtain about 80 per cent of Po_2 values between 0 and 10 mm Hg, as is typical for results derived from myoglobin cryospectrophotometry, mentioned in the Introduction. Such histograms could only be derived when Michaelis-Menten kinetics of O_2 consumption was applied. Consequently, also the value of Qmax had to be assumed considerably higher (0.100 ml/min.g) than the actually measured, or computed, O_2 consumption (0.060 ml/min.g). In histograms where the effect of a gradual decrease of blood flow was simulated, we kept Qmax at the same level (0.065 ml/min.g). The remaining input data are in the corresponding Figures.

Our program allows also a calculation of histograms of O_2 saturation in capillary blood, which is, however, not part of this communication.

RESULTS

Figures 1 to 4 depict the effect of the heterogeneity in capillary spacing and of the heterogeneity of blood flow on the calculated Po_2 histograms in various models of oxygenation. Fig. 1 contains the Po_2 histograms calculated with diffusion of the physically dissolved O_2 (Krogh), with the diffusion of O_2 facilitated by myoglobin (FMb) alone or combined with Michaelis-Menten kinetics of O_2 consumption (MM). In the first situation (Krogh), a small percentage of anoxic tissue (Po_2 < 0.005 mm Hg) occurs, which increases with a larger heterogeneity in capillary spacing and is even larger when also flow becomes heterogeneous (flow type B vs. flow type A). This is not changed considerably with FMb. On the other hand, addition of MM eliminates the anoxic foci but the effect of both spatial and flow heterogeneity remains apparent.

A tissue model containing a combination of FMb and capillary barrier (CB, Fig. 2) results in a large percentage of Po_2 values between 0 and 10 mm Hg as well as a considerable percentage of anoxic tissue, preserving the sensitivity to both types of heterogeneity. However, if the above situation is combined with MM (Fig. 3), the anoxic tissue disappears and the histograms become uniform, with most values being located in the interval 0-10 mm Hg. They also become insensitive to both changes in spatial and flow heterogeneities, thus resembling results derived from myoglobin cryospectrophotometry. Corresponding effect can be also achieved by a combination of only CB and MM (Fig. 4).

Fig. 5 depicts the effect of gradually decreasing blood flow in a model with FMb and MM, without CB. Average R as well as a middle value of log SD were taken as input data. Decrease of perfusion shifts the histograms gradually to the low values, until a very skewed histogram is obtained.

DISCUSSION

Many of the assumptions of our model are also inherent to the original Krogh model. As it is beyond the scope of this paper to discuss them in detail, we refer to our previously published papers (Kreuzer, 1982;

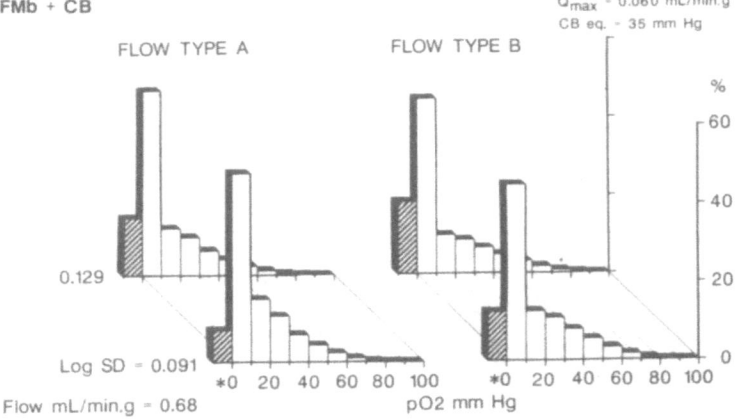

Fig. 2. Po₂ histograms calculated for situations with capillary barrier (CB) and diffusion of O_2 facilitated by myoglobin (FMb). The rest as in Fig. 1.

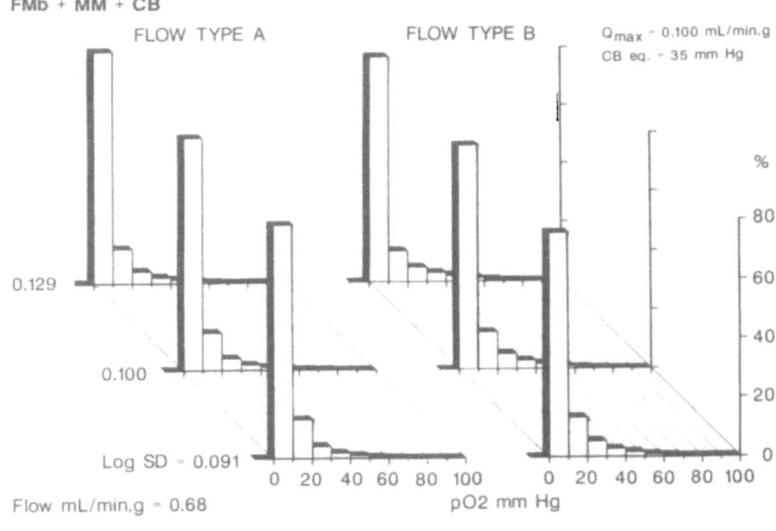

Fig. 3. Po₂ histograms calculated for situations in which FMb, MM and CB are combined.

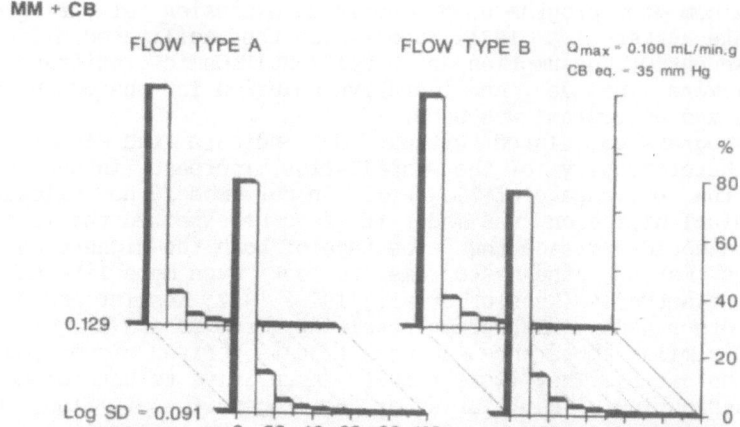

Fig. 4. Po_2 histograms calculated for situations, in which only MM and CB are combined.

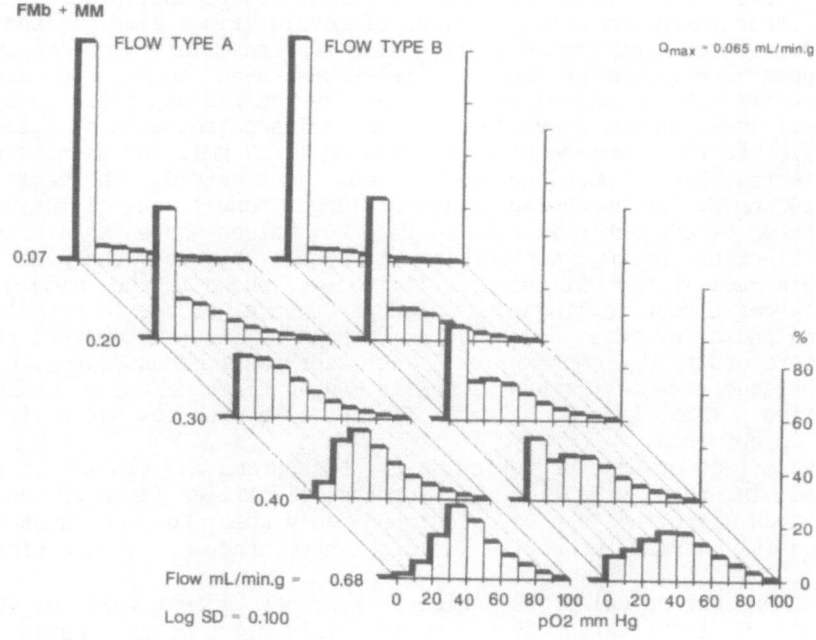

Fig. 5. The effect of underperfusion on Po_2 histograms calculated for situations in which FMb and MM are combined. Muscle blood flow decreases gradually from normal values of 0.68 to 0.07 ml/min.g.

Hoofd et al., 1987). The theories of facilitated diffusion and of Po_2-dependent O_2 consumption have been presented elsewhere (Hoofd, 1987; Jacquez, 1984; Kreuzer and Hoofd, 1987; Kreuzer and Cain, 1985; Kessler and Höper, 1985).

The computer program used here is an extension of our previous program (Rakusan et al., 1984) into which the facilitated diffusion of O_2, Po_2-dependent O_2 consumption and additional Ohmic resistance (capillary barrier) were included. An iterative solution for the calculation of Po_2 profiles and histograms was used.

Histograms calculated without CB indicate the importance of the spatial heterogeneity of the capillaries. Increase in heterogeneity increases the percentage of low Po_2. In our model, no allowance for the longitudinal diffusion was made. It is to be expected that a longitudinal diffusion would decrease the percentage of both the highest and the lowest Po_2, thus making the histograms to look even more like those obtained with Po_2 electrodes (Kessler et al., 1976; 1982; Kessler and Höper, 1985). On the other hand, histograms resulting from CB and MM, with or without FMb, had similar appearance as the results derived from myoglobin cryospectrophotometry. They were rather insensitive to heterogeneity in spatial distribution of capillaries or heterogeneity of flow. However, to obtain histograms so much shifted to the low Po_2 values required presence of a high CB and a high Qmax. Without the inclusion of MM, we invariably obtained a notable percentage of anoxic tissue, which was dependent on log SD (Fig. 2).

Our finding of anoxic tissue at a combination of CB and FMb with zero-order kinetics of O_2 consumption is at variance with conclusions of other authors (Federspiel, 1986; Gayeski and Honig, 1986; Groebe and Thews, 1988). The reason for this apparent discrepancy probably is that these authors used the average values of muscle fiber size in their calculations, which implies an average size intercapillary distance and thus a homogeneous capillarization. This is demonstrated in Fig. 6, where Po_2 profiles are shown, calculated for an average radius of tissue cylinder (23.1 µm) and also for a cylinder, which radius corresponds on a logarithmic scale to the average plus 1.5 log SD (32.7 µm). The magnitude of CB was selected high enough to yield a Po_2 just outside the capillary of about 10 mm Hg. In an average cylinder (upper panel), the diffusion of O_2 facilitated by myoglobin is able to keep Po_2 values above zero when zero-order kinetics of O_2 consumption is used. However, the power of the myoglobin-facilitated diffusion is limited, as shown in the middle panel: in a larger tissue cylinder, a high percentage of anoxic tissue occurs (between radius of 26.9 and 32.7 µm), when CB and FMb operate together with zero-order O_2 consumption. The lower panel demonstrates, that the anoxic tissue disappears when Michaelis-Menten kinetics of O_2 consumption is applied. Then, indeed, a very flat profile with low but positive Po_2 values is obtained.

The effect of underperfusion on Po_2 histograms is similar to results obtained in this situation by surface electrodes (Kessler and Höper, 1985). Such histograms can be calculated only when Po_2-dependent O_2 consumption is assumed; otherwise a large percentage of anoxic tissue results.

Our present calculations indicate a rather modest role of the diffusion of O_2 facilitated by myoglobin. This might be the reason, why so many experiments on the facilitated diffusion by myoglobin yield contradictory results (for a review see Kreuzer and Hoofd, 1987). It is quite conceivable that when the facilitation is only moderately effective, it escapes detection by present experimental procedures.

The Po_2 histograms computed with our program can resemble those obtained by Po_2 electrodes or those derived from myoglobin cryospectrophotometry, depending on the input data used. This cannot, however, help to solve the apparent discrepancy between results of these different methods, as no theoretical paper can.

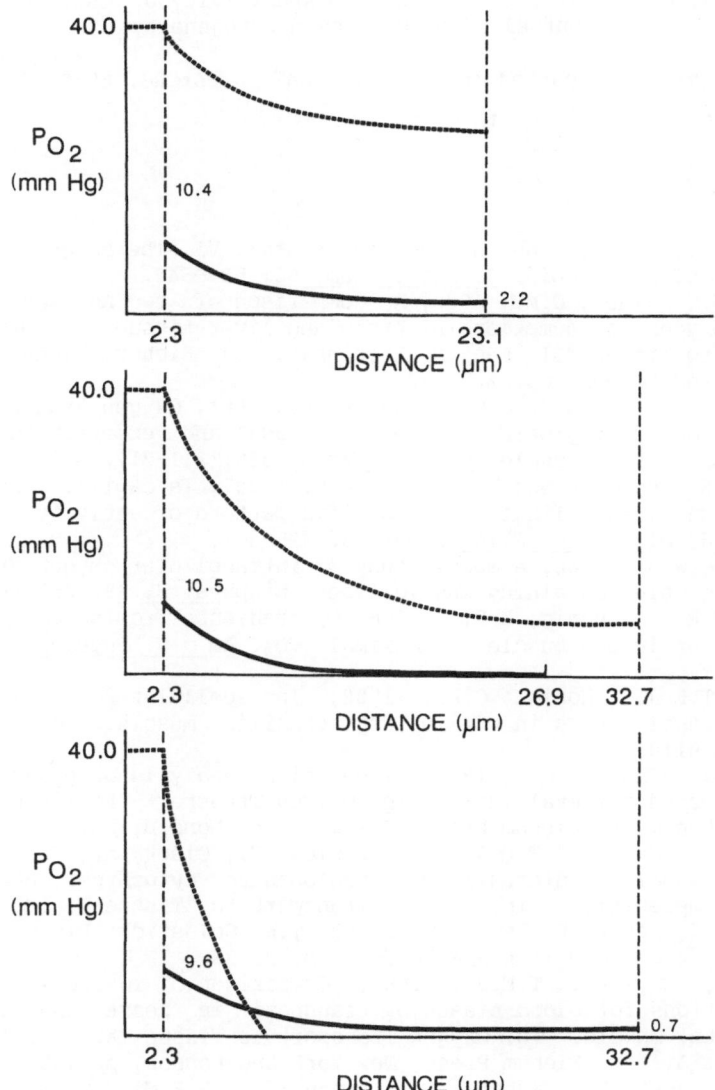

Fig. 6. Calculated radial Po_2 profiles (thick lines). Upper panel:
CB and FMb in an average-sized cylinder (R=23.1 µm). Middle
panel: CB and FMb in a larger cylinder (R=32.7 µm). Note
that the area between R=32.7 µm and R=26.9 µm becomes anoxic. Lower panel: CB, FMb and MM in a cylinder of the same
size (32.7 µm). Dotted lines depicting Po_2 profiles calculated with the Krogh equation are shown for comparison. For
details see text.

In conclusion: our computed results are dependent upon the type of model and its entry data. In models without additional resistance there is a strong dependence of the Po_2 histograms on spatial and flow heterogeneity. On the other hand, an inclusion of an additional barrier to O_2 transport from the capillary to tissue, combined with Po_2-dependent O_2 consumption, essentially eliminates the sensitivity of Po_2 histograms not only to the spatial but also to the flow heterogeneity.

Acknowledgement: Supported in part by a NATO grant RG. 86/0073 and by MRC of Canada

REFERENCES

Adair, C.S., 1925, The hemoglobin system. VI. The oxygen dissociation curve of hemoglobin. J. Biol. Chem. 63: 529-545.

Buerk, D.G., Saidel, G.M., 1978, A comparison of two nonclassical models for oxygen consumption in brain and liver tissue. in: Oxygen Transport to Tissue-III. Eds. M. Erecinska, H.J. Bicher, Plenum Press, New York and London, pp. 225-232.

De Koning, J., Hoofd, L.J.C., Kreuzer, F., 1981, Oxygen transport and the function of myoglobin: theoretical model and experiments in chicken gizzard smooth muscle. Pflügers Arch. 389: 211-217.

Egginton, S., Hargreaves, D., Hudlicka O., 1987, Is capillary perfusion in skeletal muscle linked to a specific pattern of activity? Int. J. Microcirc. Clin. Exper. 8: 295-301.

Federspiel, W.J., 1986, A model study of intracellular oxygen gradients in a myoglobin-containing muscle fiber. Biophys. J. 49: 857-868.

Gayeski, T.E.J., Honig, C.R., 1986, O_2 gradients from sarcolemma to cell interior in red muscle at maximal VO_2. Am. J. Physiol. 251: H789-H799.

Gayeski, T.E.J., Honig, C.R., 1988, Intracellular Po_2 in long axis of individual fibers in working dog gracilis muscle. Am. J. Physiol. H1179-H1186.

Groebe, K., Thews, G., 1986, Theoretical analysis of oxygen supply to contracted skeletal muscle. in: Oxygen Transport to Tissue-VIII. Ed. I.S. Longmuir, Plenum Press, New York and London, pp.495-514.

Honig, C.R., Gayeski, T.E.J., Federspiel, W., Clark, A., Jr., Clark, P., 1984, Muscle gradients from hemoglobin to cytochrome: new concepts, new complexities. in: Oxy en Transport to Tissue-V. Eds. D.W. Lübbers, H. Acker, E. Leniger-Follert, T.K. Goldstick, Plenum Publishing Corporation, New York and London, pp. 23-38.

Honig, C., Gayeski, T.E.J., 1987, Comparison of intracellular Po_2 and conditions for blood-tissue O_2 transport in heart and working red skeletal muscle. in: Oxygen Transport to Tissue-IX. Eds. I.A. Silver and A. Silver, Plenum Press, New York and London, pp. 309-321.

Hoofd, L., Turek, Z., Kubat, K., Ringnalda, B.E.M., Kazda, S., 1985, Variability of intercapillary distance estimated on histological sections of rat heart. in: Oxygen Transport to Tissue-VII. Eds. F. Kreuzer, S.M. Cain, Z. Turek, T.K. Goldstick, Plenum Press, New York and London, pp. 239-247.

Hoofd, L., 1987, Facilitated diffusion of oxygen in tissue and model systems. Dissertation Thesis, Catholic University, Nijmegen, The Netherlands.

Hoofd, L., Turek, Z., Rakusan, K., 1987, Diffusion pathways in O_2 supply of cardiac muscle. in: Oxygen Transport to Tissue-IX. Eds. I.A. Silver and A. Silver, Plenum Press, New York and London, pp. 171-177.

Hoofd, L., Turek, Z., Olders, J., Calculation of oxygen pressures and fluxes in a flat plane perpendicular to any capillary distribution, this volume.

Jacquez, J.A., 1984, The physiological role of myoglobin: more than a problem in reaction-diffusion kinetics. Math. Biosci. 68: 57-97.

Jones, D.P., 1986, Intracellular diffusion gradients of O_2 and ATP. Am. J. Physiol. 250: C663-C675.

Kelman, G.R., 1967, Digital computer procedure for the conversion of Pco_2 into CO_2 content. Respir. Physiol. 3: 111-115.

Kennedy, F.G., Jones, D.P., 1986, Oxygen dependence of mitochondrial function in isolated rat cardiac myocytes. Am. J. Physiol. 250: C374-C383.

Kessler, M., Höper, J., Krumme, B.A., 1976, Monitoring of tissue perfusion and cellular function. Anesthesiology 45: 184-197.

Kessler, M., Höper, J., Pohl, U., 1982, Monitoring of local Po_2 in skeletal muscle of critically ill patients. in: Handbook of Critical Care. Ed. J.L. Berk, 2nd edition, Little Brown, Boston, pp. 599-609.

Kessler, M., Höper, J., 1985, Signaloxidasen, Signalketten in Leber, Niere und Myocard. in: Festschrift aus Anlass der Emeritierung von Prof. Dr. med. D.W. Lübbers. Eds. R. Kinne, H. Acker, E. Leniger-Follerst, Max-Planck-Institut für Systemphysiologie, Dortmund, pp.121-155.

Kreuzer, F., 1982, Oxygen supply to tissues: the Krogh model and its assumptions. Experientia 38: 1415-1426.

Kreuzer, F., Cain, S.M., 1985, Regulation of the peripheral vasculature and tissue oxygenation in health and disease. Critical Care Clinics 1: 453-470.

Kreuzer, F., Hoofd, L., 1987, Facilitated diffusion of oxygen and carbon dioxide. in: Handbook of Physiology, Section 3: The Respiratory System, Volume IV: Gas Exchange, eds. L.E. Farhi and S.M. Tenney, American Physiological Society, Bethesda, Maryland.

Maas, A.H.J., Kreuger, J.A., Hoelen, A.J., Visser, B.F., 1972, A computer program for calculating the acid-base parameters in samples of blood using a mini-computer. Pflügers Arch. 334: 264-275.

Popel, A.S., Charny, C.K., Dvinsky A.S., 1986, Effect of heterogeneous oxygen delivery on the oxygen distribution in skeletal muscle. Math. Biosci. 81: 91-113.

Rakusan, K., Hoofd, L., Turek, Z., 1984, The effect of cell size and capillary spacing on myocardial oxygen supply. in: Oxygen Transport to Tissue-VI. Eds. D. Bruley, H.I. Bicher, D. Reneau, Plenum Publishing Corporation, London and New York, pp. 463-475.

Rakusan, K., Turek, Z., 1985, The effect of heterogeneity of capillary spacing and O_2 consumption-blood flow mismatching on myocardial oxygenation. in: Oxygen Transport to Tissue-VII. Eds. F. Kreuzer, S.M. Cain, Z. Turek, T.K. Goldstick, Plenum Press, New York and London, pp. 257-261.

Rose, C.P., Goresky, C.A., 1985, Limitations of tracer oxygen uptake in the canine coronary circulation. Circ. Res. 56: 57-71.

Roughton, F.J.W., Severinghaus, J.W., 1973, Accurate determination of O_2 dissociation curve of human blood above 98.7% saturation with data on O_2 solubility in unmodified human blood from 0oC to 37oC. J. Appl. Physiol. 35: 861-863.

Turek, Z., Kreuzer, F., Hoofd, L.J.C., 1973, Advantage or disadvantage of a decrease of blood oxygen affinity for tissue oxygen supply at hypoxia. Pflügers Arch. 342: 185-197.

Turek, Z., Rakusan, K., 1981, Lognormal distribution of intercapillary distances in normal and hypertrophic rat heart as estimated by the method of concentric circles. Pflügers Arch. 391: 17-21.

Turek, Z., Hoofd, L., Rakusan, K., 1986, Myocardial capillaries and tissue oxygenation. Can. J. Cardiol. 2: 98-103.

Turek, Z., Hoofd, L., Rakusan, K., 1987, A comparison of the methods for assessment of the heterogeneity of myocardial capillary spacing. in: Oxygen Transport to Tissue-IX. Eds. I.A. Silver and A. Silver, Plenum Press, New York and London, pp.13-19.

TISSUE GEOMETRY

SPATIAL DISTRIBUTION OF CORONARY

CAPILLARIES: A-V SEGMENT STAGGERING

S. Batra, C. Kuo, and K. Rakusan

Department of Physiology
University of Ottawa
Ottawa, Ontario Canada K1H 8M5

INTRODUCTION

The modelling of oxygen transport to tissue necessitates a concerted effort in linking structural and functional data. In our laboratory, we are interested in the geometrical distribution of coronary capillaries. Traditionally, capillary supply has been characterized only by measures of capillary density, from which it is possible to calculate average inter-capillary distance (ICD). The deficiency in such a calculation is the assumption of a uniform distribution of capillaries. The heterogeneity of inter-capillary spacing is clearly an important factor in myocardial oxygenation, over and above average ICD. Methods for assessing the heterogeneity of capillary spacing, and it's effect on myocardial oxygenation have been recently analyzed (Rakusan and Turek, 1985 and Turek et. al., 1987). Another important parameter for modelling oxygen transport is the knowledge of the direction of blood flow in adjacent capillaries. Our recent application of coloured microspheres, for the analysis of myocardial flow pattern, revealed a predominance of concurrent flow in neighboring capillaries (Reeves and Rakusan, 1987). Nonetheless, a uniformity in flow direction does not ensure that the spatial position and PO2 values of neighboring capillaries are synchronous. One may envision a situation where the transverse arteriole furnishes capillaries at staggered levels in the tissue.

The histochemical technique applied in this study enabled us to distinguish between arteriolar and venular portions of individual capillaries. And as such, offered a new approach in assessing capillary geometry: as a function of the relative positions of the capillary arterial and venous portions, i.e. A-V segment staggering. Applying a combination of histological methods, that demonstrated dual enzyme activity along the length of individual capillaries, the distribution of capillaries as a function of histochemical type was studied. The staining technique employed, enables one to distinguish between the arterial and venous portions of individual capillaries. Alkaline Phosphatase (AP) activity has been demonstrated by the azo-coupling method. This treatment stains the arterial portion of capillaries blue. Dipeptidyl Peptidase IV (DPP IV) activity has also been demonstrated, giving rise to the venous portion of capillaries staining red (Lojda,

In tissue cross-sections, a statistical test, sensitive to departures from a random distribution in a two dimensional space was used. The test criterion is based upon the number of adjacent capillaries of differing histochemical type (blue/red), which serves as an index of capillary grouping. Most recently, this test has been used for the detection of type grouping in muscle fibre patterns (Venema, 1988). This is the first time that this approach has been applied to the study of the distribution of coronary capillaries. In selected longitudinal sections, the dual staining technique has been used to follow individual capillaries from terminal arteriole to collecting venule. Preliminary data is presented, as a function of histochemical type, for: minimal capillary length, capillary set length and capillary segment number and length.

METHODS

15 male rats (Sprague-Dawley; 250-300 g) were used in this study. Following anesthesia with sodium pentobarbital, the heart was quickly excised and frozen in liquid nitrogen. Cryostat sections (16 μm) were prepared of the mid-myocardium of the anterior left ventricle. The sections were prefixed for 5 minutes in chloroform and acetone (1:1); after rinsing in distilled water, the sections were incubated for 100 minutes in a medium sensitive to DPP IV in the endothelial wall. After this incubation, the sections were rinsed and transferred to an AP sensitive medium for 25 minutes. The constituents of the incubating solution are listed in Table 1.

Table 1.   Incubating media for the demonstration of enzyme activites in the endothelium of arterial and venous segments of the capillary bed (Lojda, 1979; Mrazkova et al., 1986). [Sigma Chemical Co., St. Louis MO.]

Alkaline Phosphatase (AP)	Dipeptidyl Peptidase IV (DPP IV)
1. 40 mg Naphthol AS-MX Phosphate	1. 16 mg Glycy-L-proline-4-methoxy
2. 2 ml N,N-dimethylformamide	B-naphthylamide
3. 80 mg Variance Blue Salt Rt	2. 2 ml N,N-dimethylformamide
4. 40 ml 0.1M tris-HCl buffer pH9.2	3. 40 mg Fast Blue B salt
	4. 40 mg 0.1M acetate buffer pH5.4

The spatial position of capillaries in selected fields were digitized with the aid of a graphic tablet linked into a software program (Bioquant, R&M Biometrics, Inc.). Line segments connecting adjacent capillaries were then drawn. By convention, no two line segments could cross. In the case where more than one choice existed, the shortest segment prevailed. In this manner, the pattern of capillary neighbors was unambiguous. To determine whether the distribution of red and blue capillary profiles was random, a statistical test was employed. A field consisted of N capillaries, with **Nr** red and **Nb** blue capillaries (N = Nr + Nb). With respect to the colour of neighboring pairs, **Nt** was the total number of inter-connecting segments. The number of neighbors of the same colour were denoted by either **N-rr** (both red) or **N-bb** (both blue). Neighboring capillaries of differing colour were denoted by **N-br** (blue-red). In this manner, **Nt** = **N-rr** + **N-bb** + **N-br**. In general, it may be said that when capillary type grouping exists, the number of similar neighbors will be greater than what would be expected in a randomly distributed pattern. The test criterion that was employed was based upon the number of unlike neighboring capillaries. The mean and variance of the expected number of unlike neighbors (EN-br) for a random pattern of dualistic data has been previously derived (Cliff and Ord, 1973):

$$EN\text{-}br = \frac{2NrNb}{N(N-1)} \; Nt \qquad (1)$$

$$Var \; EN\text{-}br = \frac{(EN\text{-}br)^2}{Nt} \qquad (2)$$

For each field, the difference between the actual and expected number of unlike neighbors was appraised by the Z statistic:

$$Z = \frac{N\text{-}br - EN\text{-}br \pm 1/2}{SD \; (EN\text{-}br)} \qquad (3)$$

From the digitized data on capillary cross-sections, inter-capillary distances, as a function of histochemical type were also calculated; that is, the distance between neighboring capillaries of the same (red-red or blue-blue) and different (blue-red) type. These values were compared using a One-Way Analysis of Variance.

In longitudinal sections, preliminary data has been collected for individual capillaries that could be directly followed from terminal arteriole to collecting venule. From the data, the following working definitions have been proposed:

Minimal Capillary Length (CL)
- the shortest, direct pathway leading from terminal arteriole to collecting venule.
- the proportions of CL that represent arterial (blue) and venous (red) segments is also noted.

Capillary Set Length (CSL)
- the total path length of all micro-vessels connected to the same terminal arteriole and collecting venule
- the proportions of CL that represent arterial (blue) and venous (red) segments is also noted.

Capillary Segment Number and length
- the number of segments and their lengths the distance between two successive branch points) is also recorded as a function of segment colour: red, blue and mixed.

RESULTS

Data for the number of red and blue capillaries on cross-section, and the test for a non-random distribution pattern is presented in table 2. One field was selected from each heart. As all the data originate from the same population of rat hearts, pooled results are also given.

Table 2.   Results of a test for a random distribution based upon the expected and actual number of neighboring capillaries of different histochemical type.

Heart	N	Nr	Nb	Nt	N-br	EN-br	SD EN-br	Z	
1	274	47	227	784	159	223.64	7.99	- 8.03	*
2	193	37	156	511	117	159.19	7.04	- 5.92	*
3	249	48	201	697	139	217.80	8.25	- 9.49	*
4	174	43	131	481	148	180.02	8.21	- 3.84	*
5	246	63	183	689	198	263.60	10.04	- 6.48	*
6	256	78	178	712	241	302.86	11.35	- 5.41	*
7	221	47	174	615	153	206.89	8.34	- 6.40	*
8	278	43	235	691	143	181.35	6.90	- 5.49	*
9	271	54	217	743	190	237.98	8.73	- 5.44	*
10	211	72	139	578	195	261.10	10.86	- 6.04	*
11	514	43	471	1412	104	216.91	5.77	-19.48	*
12	427	92	335	1233	174	417.82	11.90	-20.45	*
13	362	56	306	1112	117	291.63	8.75	-19.90	*
14	413	79	334	1181	161	366.27	10.66	-19.21	*
15	511	97	414	1465	277	451.49	11.80	-14.74	*
pooled	4600	899	3701	12904	2516	4058.93	35.73	-43.17	*

* denotes significance at $\alpha = 0.01$

Data for inter-capillary distances as a function of histochemical type are presented in the frequency histograms of Figure 1. Results of the One-Way ANOVA show that there is a significant difference ($\alpha = 0.05$) between the inter-capillary distance (x), of group A (red-red) and group B (blue-red), as well as between group A and group C (blue-blue).

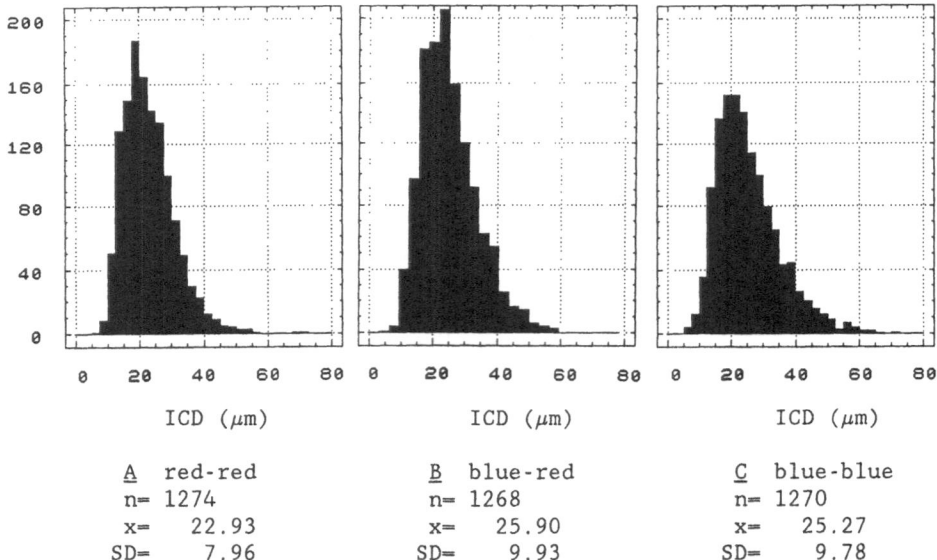

A red-red	B blue-red	C blue-blue
n= 1274	n= 1268	n= 1270
x= 22.93	x= 25.90	x= 25.27
SD= 7.96	SD= 9.93	SD= 9.78

Figure 1.   Illustrates histogram of inter-capillary distances for: A red-red, B blue-red, and C blue-blue neighboring capillaries.

The preliminary data presented here, for capillary lengths on longitudinal section, represent only 20 fields for which a capillary could clearly be traced from terminal arteriole to collecting venule. More data, with the appropriate analysis is being collected at present, and will be presented in a separate paper. Our preliminary data suggest that minimal capillary length is approximately 366 ± 35 μm; capillary unit length is 747 ± 103 μm; and capillary segment length is 63 ± 7 for arterial and 51 ± 15 μm for venous segments. Preliminary analysis indicates that there are more segments on the venous side of the capillary bed, although more data will be needed to corroborate this observation.

DISCUSSION

The Z score for the pooled data was an astonishing -43.17. All sections showed a highly significant Z value under the null hypothesis: Z displays a standard normal distribution. For a two tailed test at α = 0.01, the critical Z value is ± 2.576. The greatly negative values indicate capillary grouping as a function of histochemical type (red/blue). It is readily apparent that as the number of capillaries in a field increase, so too, does the observable degree of capillary grouping; that is, the Z score value is more negative with a larger field size. The next step in this area of work will be to look at the size and shape of the capillary groups (clusters). Using a program written in SAS macro-language, this is already possible. Of specific interest will be the patterns of clusters in serial sections. If for example, the cluster size is very large for venous portion capillaries, this might represent a potentially hypoxic region of tissue. These patterns across the heart muscle should prove to be very informative.

The data for inter-capillary distances also proved to be quite remarkable. Red to Red inter-capillary distance was significantly lower than for Blue to Red or Blue to Blue neighboring capillaries. In light of the fact that PO2 values are lower on the venous side, it follows that the tissue space that the venous portion of a capillary could supply would also be smaller. Along the length of a given capillary, the volume of tissue supplied would also be reduced. The fall in ICD from arterial to venous end, would be more pronounced in terms of it's effect on the volume of tissue supplied. In addition, at the venous side, the variability of inter-capillary distance was also much lower than at the arterial or mixed region (7.96 vs 9.93 and 9.78 μm, ±1 S.D). From this, it may be conjectured that at the venous side, where hypoxic regions of tissue may be more likely, it is critical that O2 distribution be uniform, and hence, the smaller variability of inter-capillary distance.

There appears to be a level of capillary organization for myocardium that has not been previously elucidated. Capillaries are not randomly distributed in a 3-dimensional space as a function of their arterial and venous portions. The model being proposed, favours the following schema: pre-terminal arterioles run at an oblique angle to the axis of anisotropy. Thus, the furnished capillaries may also be staggered in this model. Accordingly, the A-V portions of individual capillaries would not be in register with one another. Depending upon the angle of the terminal arteriole, neighboring capillaries could be grouped together, and as such, form a bundle of capillaries in register. Figure 2 illustrates this model. Further insight with respect to the three

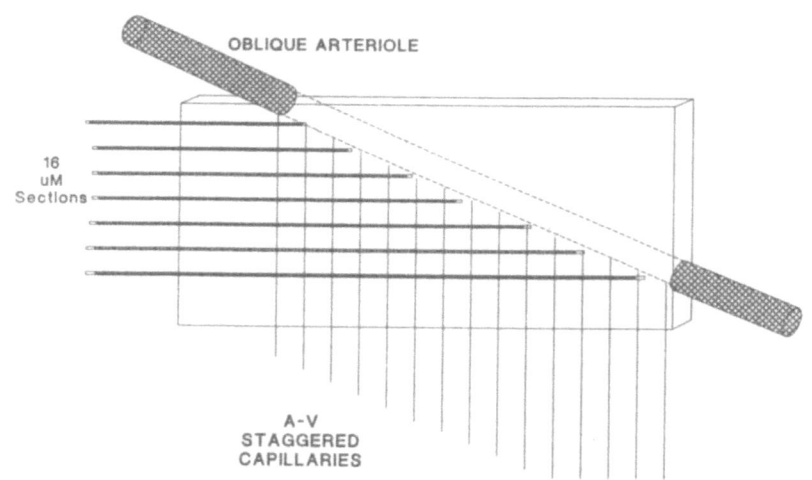

A. MODEL OF CAPILLARY STAGGERING

OBLIQUE ARTERIOLE

16
uM
Sections

A-V
STAGGERED
CAPILLARIES

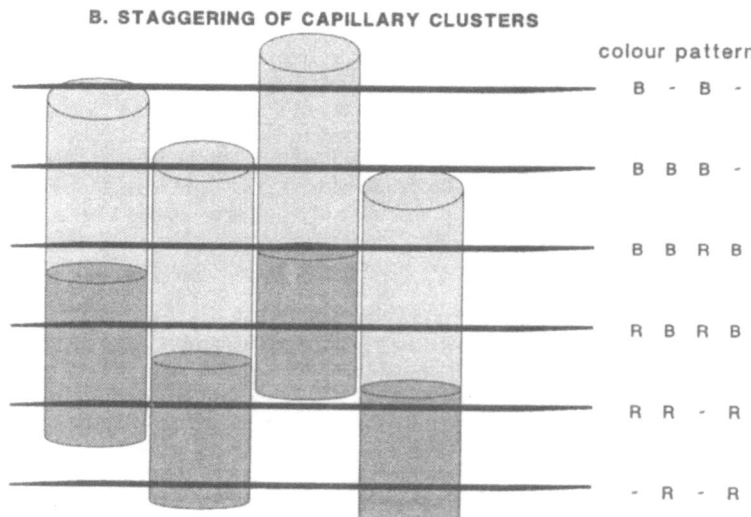

B. STAGGERING OF CAPILLARY CLUSTERS

colour pattern

B - B -

B B B -

B B R B

R B R B

R R - R

- R - R

Figure 2.   Colour Pattern, B=Blue, R=Red.

dimensional organization of coronary capillaries will be possible when our cross-sectional and longitudinal measurements are consolidated in the same tissue block.

The essence of this study, then, was to assess the three dimensional organization of coronary capillaries using both transverse and longitudinal sections. The histological technique employed was inaugural in that it enabled the 3-dimensional modelling of coronary capillaries in a new fashion: as a function of their arterial and venous portions.

REFERENCES

Cliff, A.D., Ord, J.K., 1973, Spatial Autocorrelation. Pion., London.

Lojda, Z., 1979, Studies on Dipeptidyl(Amino)Peptidase IV (Glycyl-Proline Naphthylamidase. Histochemistry. 59:153-166.

Mrazkova, O., Grim, M., Carlson, B.M., 1986, Enzymatic Heterogeneity of the Capillary Bed of Rat Skeletal Muscles. Amer. J. Anat. 177:141-148.

Rakusan, K., Turek, Z., 1985, The effect of heterogeneity of capillary spacing and O2 consumption-blood flow mismatching on myocardial oxygenation of cardiac muscle. in: Oxygen Transport to Tissue-VII. Eds. F. Kreuzer, S.M. Cain, Z. Turek, T.K. Goldstick, Plenum Press, New York and London, pp. 257-261.

Reeves, W.J., Rakusan, K., 1988, Myocardial Capillary Flow Pattern as Determined by the method of Coloured Microspheres. in: Oxygen Transport to Tissue-X. Eds. M. Mochizuki, C.R. Honig, T. Koyama, T.K. Goldstick, and D.F. Bruley, Plenum Press, New York and London, pp. 447-453.

Turek, Z., Hoofd, L., Rakusan, K., 1987, A comparison of the methods for assessment of the heterogeneity of myocardial capillary spacing. in: Oxygen Transport to Tissue-IX. Eds. I.A. Silver and A. Silver, Plenum Press, New York and London, pp.13-19.

Venema, H.W., 1988, Spatial Distribution of Fiber Types in Skeletal Muscle: Test for a Random Distribution. Muscle & Nerve II:301-311.

Acknowledgement: Supported by the Medical Research Council of Canada. The authors wish to thank Jimmy Gao for his expert technical assistance.

FRACTAL NETWORKS EXPLAIN

REGIONAL MYOCARDIAL FLOW HETEROGENEITY

Johannes H.G.M. van Beek, James B. Bassingthwaighte and
Stephen A. Roger

Center for Bioengineering WD-12, University of Washington,
Seattle WA 98195 and Laboratory for Physiology, Free University
van der Boechorststraat 7, 1081 BT Amsterdam, the Netherlands

INTRODUCTION

Regional myocardial blood flow is very heterogeneous. This has been found by injection of radioactively labeled microspheres, or the "molecular microsphere" iododesmethylimipramine, and measuring the deposition of these flow indicators in small sample pieces into which the heart has been cut (King et al., 1985; Bassingthwaighte et al., 1987; Bassingthwaighte et al., 1988). It was also found that the spread of the flow distribution increases with the spatial resolution of the measurement. This could be expressed (Bassingthwaighte et al., 1988; Bassingthwaighte, 1988) via a mathematical relation between the relative dispersion RD, defined as the standard deviation of the flow distribution divided by its mean, and the average mass, m, of the sample pieces into which the heart was divided:

$$RD(m) = RD(m_0) \cdot \left[\frac{m}{m_0} \right]^{1-D} \tag{1}$$

where m_0 is an arbitrary reference mass. Since the number of sample pieces equals the total mass M divided by the average mass of the sample piece it follows that

$$RD(N) = RN(N_0) \cdot \left[\frac{N}{N_0} \right]^{D-1} \tag{2}$$

The power laws fit the measurements very well for all except the larger sample pieces (see Figure 1). Such a relation between a measure and the spatial resolution of the measurement has been found for the geometrical features of certain types of mathematical sets that are called fractals. Equations 1 and 2 are therefore often called fractal relationships and parameter D is called the fractal dimension.

WHAT ARE FRACTALS?

Fractals are mathematical geometrical constructs that resemble the geometrical patterns found in nature (Mandelbrot, 1983). The latter are called natural fractals. The concept of fractals has been invented and propagated by B. Mandelbrot (1983) and is explained and visualized in an esthetically pleasing way by Peitgen and Richter (1986). Mathematical fractals exhibit an infinite amount of detail, which they have acquired via an endless recurrent iteration defining the $(n+1)^{th}$ generation from the nth.

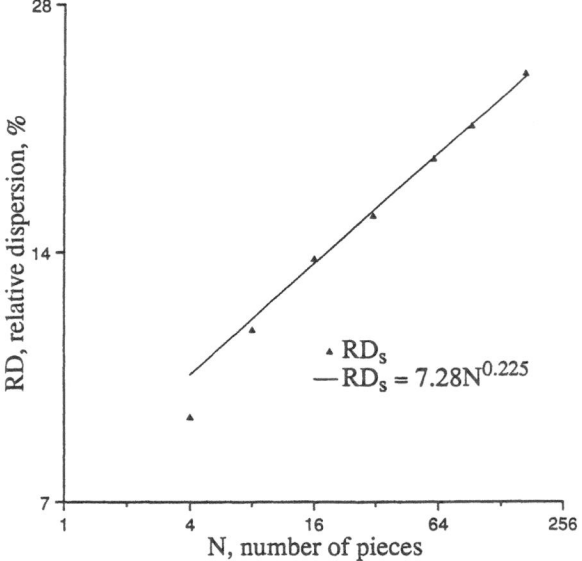

Heterogeneity of Regional Blood Flows as a
Fractal Function of Region Size
(Heart of an Awake Baboon, Microsphere Data)
Baboon #7 ; $RD_s = 7.28N^{0.225}$, r = 0.999

Figure 1. Relative distribution of myocardial blood flow as a function of the number of sample pieces into which the heart is divided. The relative dispersion of flow, RD, has been corrected for measurement noise. Results for one baboon.

Recursive Extension of a Line ... a Simple Fractal System

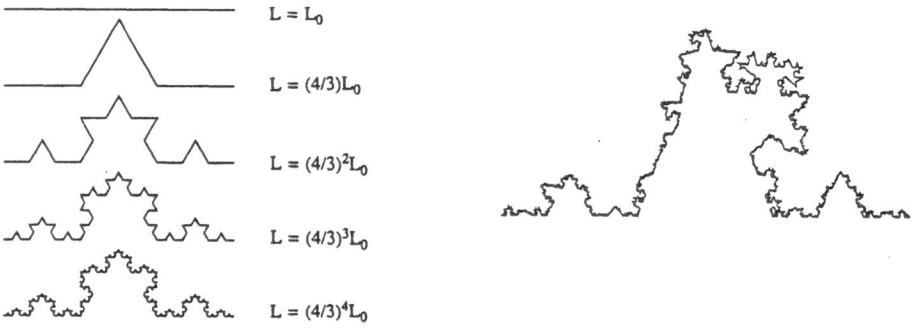

Figure 2. The Koch snowflake. The left panel shows a deterministic recursive process. The right panel shows the result of a recursion with a random process.

An example is given in Figure 2, where a side of a snowflake is generated by replacing the middle third of all straight lines repeatedly by two line segments of similar length. The resulting figure forms one side of Koch's snowflake. It can be shown that the length L of the resulting figure depends on the length of the line segment ε in that particular iteration

$$L(\varepsilon) = L(\varepsilon_0) \cdot \left[\frac{\varepsilon}{\varepsilon_0} \right]^{1-D} \tag{3}$$

where ε_0 is an arbitrary reference length (Mandelbrot, 1983). The parameter D turns out to have the properties of a dimension and is called the fractal dimension. By repeating the iteration a figure of "infinite" complexity can be generated, which has "infinite" length. It is also clear that a part of Koch's snowflake is a small replica of the whole. It shares this property of self-similarity with many other fractal structures. The pattern of the figure resembles that of a coastline. Indeed, Equation 1, which is generally valid for fractals, has been applied to real coastlines. Now ε represents the scale of the details on a geographical map that are taken into account when determining the length of the coastline. In practice one can do that by setting a caliper to width ε and walking it along the coastline on a map. Equation 2 turns out to be valid for the coastal length found in this way: the coast is a natural fractal and the Koch snowflake can serve as a crude mathematical model for it. Introducing an element of chance in the iterations for the generation of the Koch snowflake makes it look more like a natural coastline (Figure 2, right panel). The relative dispersion of blood flows in the heart, given by Equation 1, seems to behave according to the fractal law, Equation 3. However, it is not immediately obvious how the relative dispersion of flow can be compared with the length of an object. We will explore this in the next section.

THE FRACTAL RELATIONSHIP FOR DISPERSION OF FLOW CAN BE DERIVED FROM CORRELATION BETWEEN NEIGHBORS

Imagine that an organ is divided into voxels (volume elements) and that the flow to the voxels has a spread with a certain standard deviation. The flow in neighboring voxels also has a certain correlation coefficient r. We have shown (van Beek et al., submitted). that when r is constant, irrespective of the size of the voxels into which the organ is divided, the relation between voxel size and relative dispersion of flow is given by Equations 1 and 2 We found that the relation between r and fractal dimension D is given by

$$r = 2^{3-2D} - 1 \tag{4}$$

This derivation showed that the fractal relationship holds as long as the correlation of the flows in neighboring voxels is constant. This idea of a property being constant over many magnitudes of scale, self-similarity, is a feature of fractals (Mandelbrot, 1983). When the local flows are completely random, r = 0, the fractal dimension D is 1.5; when they are highly correlated with r = 1, it follows that D is 1.0.

The fractal law of Equation 1 seems to describe the experimental findings well. It therefore resolves the question of how to compare results on flow heterogeneity obtained by various investigators who used different spatial resolutions in their measurements. The fractal D found for Equation 1 is around 1.15 for myocardial blood flow, which means that the correlation between flows in adjacent voxels is around 0.63, according to Equation 4, no matter the voxel size, so long as equal sizes are compared.

FRACTAL NETWORKS

Blood vessel networks tend to show the same pattern which repeats at various levels of scale. The fact that a fractal relation described the measurements very well inspired us to investigate whether the distribution of flow might be explained by a network of blood vessels having a fractal geometry. We started to look at one of the simplest geometries: a repeatedly branching bifurcating network (see Figure 3). The radius r_0 of the parent vessel in this network is reduced by a factor a in the longer daughter branch at the bifurcation, and by a factor b in the shorter one.

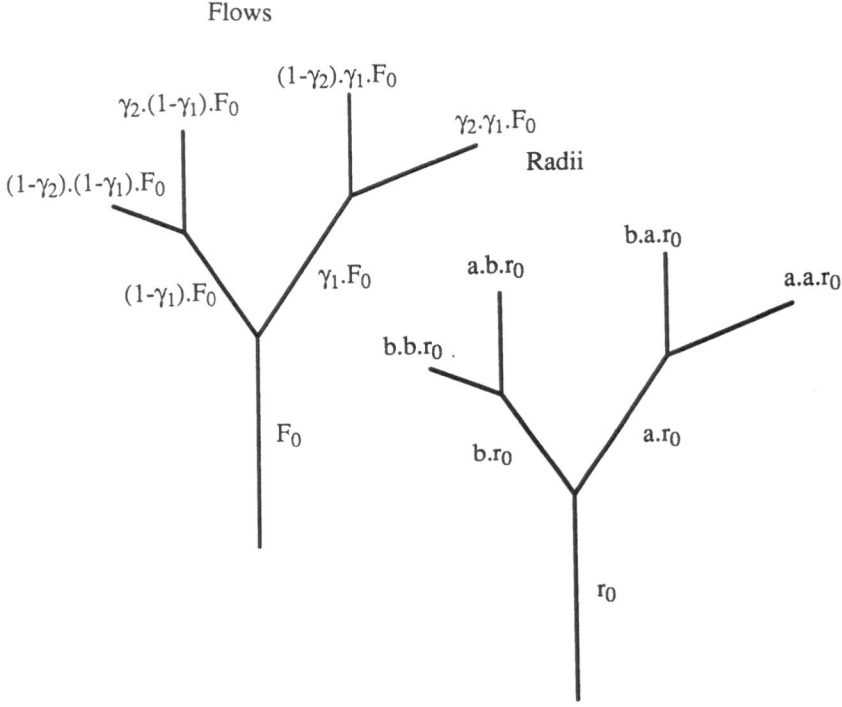

Figure 3. Bifurcation of blood vessels. In the right panel the recursion for the vessel radius is shown. A similar recursion applies to the length of the vessel segments. The left hand panel shows the distribution of flow. A fraction γ of the blood flows into one of the branches, $1 - \gamma$ flows into the other. This fraction is the same for all bifurcations of one generation, but may vary across the generations.

The radius r_0 of the parent vessel in this network is reduced by a factor a in the longer daughter branch at the bifurcation, and by a factor b in the shorter one. The length L_0 is decreased by a factor f_L and f_S for the longer and shorter branches. This pattern of branching is repeated and an arterial tree is generated. It is connected to a similarly constructed venous tree with a different r_0 and L_0, but with the same factors a, b, f_L and f_S. Using Poiseuille's law, we deduced from this geometry that the flow distribution in such a network can be represented by an asymmetry parameter γ which equals $f_S \cdot a^4/(f_L \cdot b^4 + f_S \cdot a^4)$. A fraction γ of the flow enters one branch; a fraction $1-\gamma$ enters the other. When the geometrical branching parameters are constant all through the network (Model I), the asymmetry parameter γ is independent of the generation. For this case we derived the following relation between the relative dispersion of flow found in the branches after a certain number of bifurcations and the number of branches N concerned (van Beek et al., submitted):

$$RD = \sqrt{N(\gamma^2 + (1-\gamma)^2)^{\log N/\log 2}_{-1}} \tag{5}$$

The number of symmetrical branches, with $\gamma = 0.5$, is another free parameter in this model. In order to fit the experimental data we assumed a one to one correspondence between blood vessels and sample pieces. The fit to data on one baboon is given in Figure 4.

Heterogeneity of Regional Blood Flows as a

Fractal Function of Region Size

(Heart of an Awake Baboon, Microsphere Data)

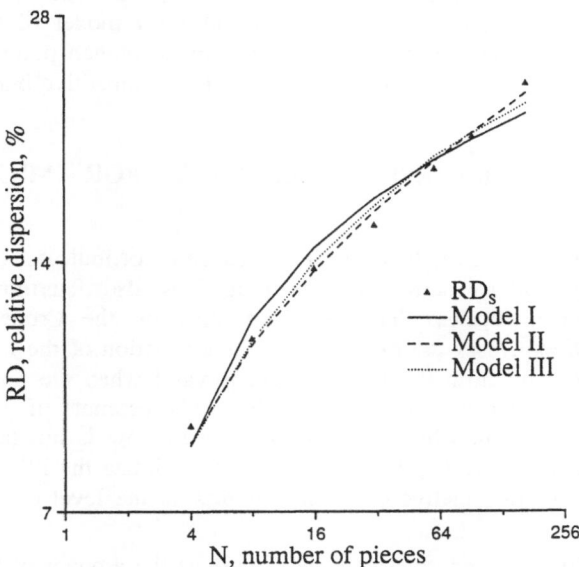

Figure 4. Fit of the models to the RD obtained experimentally in one baboon. The highest number of pieces corresponds to the number of sample pieces into which the heart was actually divided. The data for lower numbers of sample pieces (corresponding to bigger pieces) were obtained by aggregating a number of the smallest pieces.

When the geometrical parameters depend on the generation of branching, so does γ. Model II is defined by a geometric progression for γ:

$$\gamma_i = \gamma_1 \cdot f^{i-1} \tag{6}$$

where i is the index for the generation of branching and γ_1 is the asymmetry parameter for the first generation. The relation derived (Van Beek et al., submitted) for this model is:

$$RD = \sqrt{[2^n \cdot \prod_{i=1}^{n}(2 \cdot f^{2i-2} \cdot \gamma_1^2 - 2 \cdot f^{i-1} \cdot \gamma_1 + 1)] - 1} \tag{7}$$

The resulting fit to the data is given in Figure 4.

Model III was generated by drawing γ for each branch point randomly from the normal distribution with mean 0.5 and variance σ^2. To do this 10 trees were generated on the computer for variance 1. They were then scaled to have variance σ^2. Parameter γ_1 for the first generation was an independently chosen deterministic parameter. The fit of model III to the data is also shown in Figure 4.

The goodness of fit of the models to the data was assessed by using the coefficient of variation given by:

$$CV = \frac{\sqrt{\sum(y_{measured} - y_{model})^2 / (n-2)}}{\sum y_{measured}/n} \tag{8}$$

where $y_{measured}$ are the observed values, y_{model} are the estimated values given by the model, and n is the number of measured points from which the degrees of freedom of the model are subtracted. The mean coefficient of variation for model fits to measurements on 10 baboons was 0.0647, 0.0336 and 0.0411 for models I, II and III respectively. For measurements on 11 sheep this was 0.0720, 0.0438 and 0.0499 respectively. Model II fitted the best in general, while the stochastic model III was almost as good. The value of gamma was around 0.45 and σ for model III was around 0.05. This indicates that a moderate degree of asymmetry at branch points can lead to the marked heterogeneity of flow in the myocardium, due to repetitive branching.

THE RELATIVE DISPERSION RELEVANT FOR MICROVASCULAR EXCHANGE

The local flow determines how much of a substrate or indicator is delivered to the tissue. When the capillary permeability is finite, flow also determines the extraction fraction E of the molecules. This is exemplified by the Crone-Renkin formula $E = 1 - e^{-PS/F}$ which expresses the extraction, E, as a function of the local permeability-surface product PS and flow F. This formula is valid when the reflux of molecules from tissue back into the capillary is negligible. The estimate of permeability in an organ will be quite wrong when the heterogeneity in flow is not taken into account (Bassingthwaighte and Goresky, 1984). In order to estimate the PS correctly we must obtain an estimate of the relative dispersion of flow at the level of the microvascular exchange region.

We have defined the microvascular unit mass as the amount of tissue to which a single arteriole delivers blood and nutrients. Flow heterogeneity within the microvascular unit might exist but is of less importance since diffusion tends to equalize concentration gradients.

The first indication that every heart might approach a similar maximum relative dispersion was reported by Bassingthwaighte et al. (submitted). For 21 animals studied (10 baboons and 11 sheep) it was found that hearts with a larger RD at the reference sample size of 1 gram had a smaller fractal dimension D, when Equation 1 was fit to the data. In other words, the best fit lines to each animal on a log RD versus log m plot tended to have a shallower slope when the RD(m_0) was large, and tended to have a steeper slope when the RD(m_0) was small.

However, the power law might fail when the maximum RD is approached asymptotically. For this reason, we believe that our best fitting branching model (model II), which can account for the curvature seen in the experimental data, would be the favorable tool for estimating the RD at the exchange level.

With an analytical solution available to model II (Equation 7), the model function can be extrapolated beyond the range of the experimental data. The idea was to introduce a common point at a very high spatial resolution (large N) into the data set of each animal. After being forced through this common point, the best model fits are found for each animal. Figure 5 shows the solution to the data for 10 baboons when the common point introduced was at an RD of 55%, and the hearts were theoretically sliced into 524288 pieces, average mass 52 μg which corresponds to a cube with a side of about 400 μm. The overall correlation coefficient between the model and the data from ten animals was > 0.99.

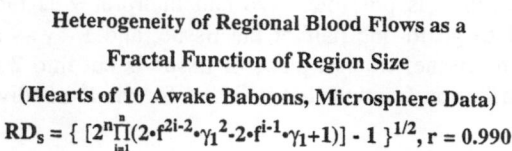

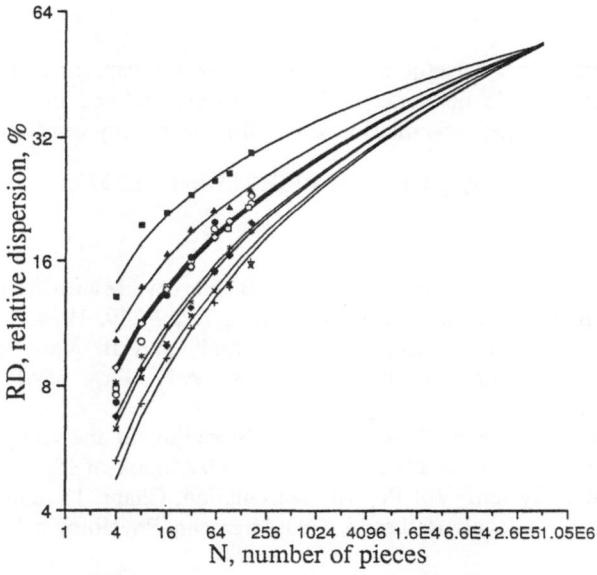

Figure 5. Relative dispersion of myocardial perfusion as a function of the number of sample pieces into which the heart is cut. The common point introduced is at N equals 524288, and an RD of 55%. The data are for 10 baboons.

Is this the microvascular unit point? This question cannot yet be answered. Although a trend was seen in the solutions, no clear cut "winner" was obtained after numerous different common points were introduced into the data sets. The next step will be to use these results in accordance with anatomical studies and autoradiographic flow heterogeneity studies to gain further insight into the branching and flow characteristics of the myocardium at the microscopic scale. However, it did become very clear that the branching networks were much more accurate in assessing the spatial heterogeneity of regional myocardial perfusion than the power law equation.

DISCUSSION

Although the fractal networks described here do not resemble the coronary vasculature in detail (Bassingthwaighte et al., 1974), they seem to describe the measured relative dispersion of flow rather well. The network is similar to the one used by Pelosi et al. (1987) to analyze total resistance. Zamir and Chee (1987) analyzed vessel lengths and diameters in the arterial part of the human coronary network, using a labeling scheme compatible with our bifurcating network. They found that length and diameter of vessel segments decreased steadily as a function of generation number.

An alternative interpretation of the mathematics used for deriving the distribution of flow in the networks is possible. We can interpret γ as indicating the fraction of flow that is found to go to one half of the tissue, and $1 - \gamma$ as the flow which goes to the other half of the tissue, when a piece of tissue is cut into 2 sample pieces. Then γ_i is the average distribution fraction that is found for the ith subdivision of the tissue.

CONCLUSION

Fractals form a very useful tool for modeling the capricious form and behavior found in Nature. Simple networks of blood vessels, defined in an economic fractal way, explain the heterogeneous distribution of flow in the myocardium very well.

This work was supported by NIH grants RR01243 and HL19139.

REFERENCES

Bassingthwaighte, J. B., T. Yipintsoi, and R. B. Harvey. Microvasculature of the dog left ventricular myocardium. Microvasc. Res. 7:229-249, 1974.

Bassingthwaighte, J.B., R.B. King, J.E. Sambrook, and B. Van Steenwyk. Fractal analysis of blood-tissue exchange kinetics. Adv. Exp. Med. Biol. 222:15-23, 1988.

Bassingthwaighte, J. B., and C. A. Goresky. Modeling in the analysis of solute and water exchange in the microvasculature. In: Handbook of Physiology, Sect. 2 The Cardiovascular System, Vol IV, Microcirculation, Chapt. 13, edited by E. M. Renkin, and C. C. Michel. Bethesda, MD:American Physiological Society, 1984, p. 549-626.

Bassingthwaighte, J. B. Physiological heterogeneity: Fractals link determinism and randomness in structures and functions. *News in Physiol. Sci.* 3:5-10, 1988.

Bassingthwaighte, J. B., M. A. Malone, T. C. Moffett, R. B. King, S. E. Little, J. M. Link, and K. A. Krohn. Validity of microsphere depositions for regional myocardial flows. Am. J. Physiol. 253 (Heart. Circ. Physiol. 22):H184-H193, 1987.

Bassingthwaighte, J.B., R.B. King, and S.A. Roger. Fractal nature of regional myocardial blood flow heterogeneity. (submitted)

King, R. B., J. B. Bassingthwaighte, J. R. S. Hales, and L. B. Rowell. Stability of heterogeneity of myocardial blood flow in normal awake baboons. Circ. Res. 57:285-295, 1985.

Mandelbrot, B. B. *The fractal geometry of nature.* San Francisco: W.H. Freeman and Co., 1983.

Peitgen, H. O., and P. H. Richter. *The beauty of fractals: images of complex dynamical systems.* Berlin/Heidelberg: Springer-Verlag, 1986.

Pelosi, G., G. Sarossi, M.G. Trivella, and A. L'Abbate. Small artery occlusion: a theoretical approach to the definition of coronary architecture and resistance by a branching tree model. Microvasc. Res. 34: 318-335, 1987.

Van Beek, J.H.G.M., S.A. Roger, and J.B. Bassingthwaighte. Regional myocardial flow heterogeneity explained with fractal networks. (submitted)

Zamir, M. and H. Chee. Segment analysis of human coronary arteries. Blood Vessels 24:76-84, 1987.

CAPILLARY SPATIAL PATTERN AND MUSCLE FIBER GEOMETRY IN THREE

HAMSTER STRIATED MUSCLES

R. A. O. Bennett, R. N. Pittman, and S. M. Sullivan

Department of Physiology
Box 551
Medical College of Virginia
Virginia Commonwealth University
Richmond, Virginia 23298 0551

INTRODUCTION

The maintenance of an adequate tissue oxygen tension is a principal function of the respiratory and circulatory systems. In 1919, Krogh presented a mathematical model that described the diffusion of oxygen from capillaries to the surrounding tissue (Krogh, 1919). This was the first model to relate oxygen diffusion and capillary spacing within muscles and the first to quantify the relationship between the rate of delivery of the oxygen to various sites in the tissue to the rate of consumption of oxygen at those sites. Determinants of tissue oxygen tension include the PO_2 of blood, blood flow, red blood cell spacing, diameter of microvessels, hemoglobin oxygen saturation, as well as the spatial pattern of capillaries within the tissue. Krogh's model assumes homogeneity in the composition of blood and muscle tissue, uniform consumption of oxygen at every point in the tissue, and a uniform and radial movement of oxygen outward from the center of a capillary into the surrounding tissue independent of oxygen diffusion from all other capillaries within the tissue. Since the model neglects the intrinsic heterogeneities of capillary spacing and shape of the muscle fibers, it fails to characterize completely the movement of oxygen from the hemoglobin in the red blood cells to the sites of oxygen utilization in muscle tissue.

Heterogeneities in the transit of oxygen in striated muscle tissue are due to the composition of the medium through which oxygen must travel before reaching its destination. Histologically, muscles vary in size, capillarity, and fiber type composition. Variations also exist in the proportions of aqueous cytosol and lipidic membranes and droplets, all of which have unique oxygen solubilities (Ellsworth and Pittman, 1984).

An examination of histological cross sections of striated muscle tissue, both light and electron microscopic, suggests that capillaries are not arranged in a regular pattern parallel to muscle fibers. Since tissue PO_2 from Krogh's equation depends roughly on the square of the distance between capillary and tissue site, deviations from uniformity of capillary placement around muscle fibers could have a large impact on tissue oxygenation, leading to hypoxia or anoxia at distant sites.

Three general types of spatial pattern of capillaries in cross section are possible. The capillary pattern is described as being a regular, random, or aggregated (clustered, contagious) array. A regularly spaced capillary pattern provides a uniform distribution of oxygen in the tissue provided there are no heterogeneities in the composition of the tissue or in the source of oxygen serving the tissue. Each capillary then serves an equal volume of tissue. The opposite extreme of the regular array is that of an aggregated pattern of capillaries. Compared with the regular pattern the likelihood that the tissue is adequately oxygenated in this case is less, and this pattern could give rise to anoxic regions of tissue. Between these two extremes lies the case of randomness. This situation may or may not provide adequate tissue oxygenation.

The three methods employed to analyze capillary spacing utilize the theory of stochastic processes and have been developed by plant ecologists (Cottam and Curtis, 1956; Moore, 1954), geographers (Bartels et al., 1979; Cliff and Ord, 1981; Getis and Boots, 1978), an animal ecologist (Southwood, 1978), and mining engineers (Guarascio et al., 1976; Journel and Huijbregts, 1978). Kayar et al. (1982b) have previously applied one of the methods (i.e., closest individual method) to the analysis of capillary spacing in muscle tissue.

Capillary spatial patterns of three hamster muscles were analyzed by computing the empirical statistical distributions of distances between all pairs of capillaries, distances between nearest neighbor capillaries, and distances between computer-generated sample points in the tissue to closest capillaries. These methods for the analysis of mapped point patterns are easily adapted to the study of capillarity, and provide information not previously gathered from earlier studies. The results of these analyses are used to formulate a mathematical model for capillary placement in striated muscle.

MATERIALS AND METHODS

Tissue Preparation and Histochemistry

Male golden hamsters were prepared according to the methods of Sullivan and Pittman (1984). Magnified images (×200-450) of stained histological sections were projected onto an image analysis platform, from which the border of each muscle fiber was traced. Muscle fiber geometric parameters included (1) muscle fiber type, (2) cross-sectional area of fiber, (3) perimeter of fiber, (4) number of fibers of a particular type surrounding a fiber, (5) number of capillaries at the junction of three or more fibers derived from the position and number of capillaries surrounding each fiber, and (6) number of vertex regions in a random sample of the section.

Spatial pattern analysis

Pattern analysis of capillaries in striated muscle and the detection of non-randomness in pattern is a straightforward application of stochastic point process theory. For purposes of this study, capillary locations were classified as one of two types. Capillaries that were at the junction of three or more fibers were called muscle fiber *vertex capillaries*. The reason for choosing such a definition comes from the observation that muscle fibers in general take on a polygonal shape having pseudo-edges and pseudo-vertices. If a capillary were to reside along the edge of one muscle fiber, but yet at the corner of another fiber, or if it were to reside along an edge of one fiber that was adjacent to an open space, then that capillary was also defined to be a vertex capillary. When the capillary did not fit into one of these situations, it was sandwiched between the sides of two fibers and termed a muscle fiber *edge capillary*. Subsets of the data from the histochemically-prepared cross sections of muscle were further investigated for number of vertices per unit area (vertex density) and for number of capillaries associated with vertex regions. Values that were calculated from these parameters included: (1) fractional area of muscle occupied by a given fiber type, (2) muscle fiber area per capillary, (3) muscle fiber perimeter per number of edges, (4) muscle fiber perimeter per capillary, (5) number of edges for each fiber, (6) average edge length for each fiber, (7) fraction of capillaries located at vertex regions, (8) number of vertices per unit area (vertex density), and (9) fraction of vertices containing capillaries (vertex occupation).

The fractional area of muscle occupied by each fiber type was calculated from the number of fibers of a given type and the mean fiber area of that fiber type. The number of edges that a fiber possessed was determined by the number of fibers that surrounded it, and the average edge length of a fiber was calculated from the perimeter of that fiber divided by the number of fibers that surrounded it.

The spatial pattern of capillaries was analyzed by the *mapped pattern* technique, which is outlined by Diggle (1983b). The basis of this analysis rests on test statistics formulated from three variations on the measures of distances between capillaries and the distributions of distances.

A preliminary analysis of any mapped pattern begins at least with a test of complete spatial randomness (CSR). Several test statistics are formulated on the distributions of inter-capillary distances between all pairs of capillaries, first nearest neighbor capillary distances, and random tissue sample point to closest capillary distances. The data that were investigated were normalized so that all capillaries occurred in a square of unit side length. The distance was divided into 100 bins of equal size and the number of distances that fell into each bin gives the probability density function which was then numerically integrated over the 100 points to give the cumulative distribution function.

The empirical cumulative distribution functions (EDFs) are then plotted against the theoretical cumulative distribution of distances between two events which are independently and uniformly distributed on the region A (Diggle, 1983b). If the data are compatible with CSR, the plot looks roughly linear with intercept = 0 and slope = 1. Exact Monte Carlo tests were constructed to aid in the interpretation of the EDF plots.

The empirical distribution functions (EDFs) are:
(1) inter-capillary distances, t_{ij}, for the region A, a square of unit side,

$$\hat{H}_1(t) = \{ n(n - 1)/2 \}^{-1} \times \#(t_{ij} \leq t), \ 0 \leq t \leq \sqrt{2};$$

(2) first nearest neighbor capillary distances, y_i, and,

$$\hat{G}_1(y) = n^{-1} \times \#(y_i \leq y), \ 0 \leq y \leq \sqrt{2};$$

(3) point to closest capillary distances, x_i,

$$\hat{F}_1(x) = m^{-1} \times \#(x_i \leq x), \ 0 \leq x \leq \sqrt{2},$$

where # means "the number of" and n is the number of events in A. And, the theoretical distribution functions are, respectively,

$$H(t) = \begin{cases} \pi t^2 - 8t^3/3 + t^4/2, & 0 \leq t \leq 1 \\ 1/3 - 2t^2 - t^4/2 + 4/3(2t + 1)\sqrt{(t^2 - 1)} \\ \quad + 2t^2 \sin^{-1}(2t^{-1} - 1), & 0 \leq t \leq \sqrt{2}; \end{cases}$$

$$G(y) = 1 - \exp(-\lambda \pi y^2), \qquad y \geq 0;$$

$$F(x) = 1 - \exp(-\lambda \pi x^2), \qquad x \geq 0,$$

where $\lambda = n|A|^{-1}$ and $|A|$ denotes the area of the region A. The theoretical distribution functions $G(y)$ and $F(x)$ are only approximations for large n. The reason for these approximations is that true distributions of distances depend not on A, but rather on n and thus are not expressible in closed form (Diggle, 1983b).

Plots were made of $\hat{H}_j(t)$ as ordinate against $H(t)$ as abscissa. A somewhat linear plot would not suggest rejection of the null hypothesis of CSR, which is defined to be a realization of a two-dimensional, Poisson point process; that is, events do not interact with each other, multiple occurrences of events are not allowed at the same location, and the expected number of events per unit area [i.e., intensity (λ)] is constant. To assess any departure from linearity we computed the EDFs of 99 computer simulations of n

events independently and uniformly distributed on A and defined an upper simulation envelope and a lower simulation envelope as such

$$U(.) \quad = \quad \max_{i=2,...,100} \hat{H}_i(.); \qquad\qquad L(.) \quad = \quad \min_{i=2,...,100} \hat{H}_i(.).$$

Similar definitions are made for the cumulative distribution functions $\hat{G}_i$ and $\hat{F}_i$, $i = 2$, . . ., 100. These simulation envelopes have the property that, for each t, y, or x, the probability for each cumulative distribution function to be greater than or less than U or L is equal to 0.01.

To further aid in the interpretation of the EDF plots, exact Monte Carlo tests[*] based on the integrated squared distance and 99 computer simulations mentioned above can be constructed. Define h_i to be a discrepancy between $\hat{H}_i(t)$ and $H(t)$ over the whole range of t,

$$h_i = \int \{ \hat{H}_i(t) - H_i(t) \}^2 \, dt.$$

The test is based on the rank of h_1 against the other h_i. If h_1 is ranked in the top five (from largest to smallest), then the null hypothesis of CSR can be rejected at the 5% level. Similarly,

$$g_i = \int \{ \hat{G}_i(y) - \overline{G}_i(y) \}^2 \, dy,$$

where $\overline{G}_i(y) \quad = \quad (s - 1)^{-1} \sum_{j \neq i} \hat{G}_j(y),$

and,

$$f_i = \int \{ \hat{F}_i(x) - \overline{F}_i(x) \}^2 \, dx,$$

where $\overline{F}_i(x)$ is defined analogously to $\overline{G}_i(y)$. Computer-generated, pseudo-random numbers for Monte Carlo testing were obtained from subroutines by Numerical Algorithms Group (NAG) and linked to FORTRAN programs of our own design.

Statistics

Results of muscle fiber and capillary geometric data are expressed as means ± SE. Data from the different muscles and data from fiber types were compared using Student's t test. Statistical analyses were carried out with the Statistical Analysis System software (SAS Institute). Monte Carlo testing was performed as described previously.

RESULTS

EDF Plots

Ten muscle fields were analyzed by this method using inter-capillary distances, first nearest neighbor capillary distances, and random sample point to closest capillary distances. Figure 1 illustrates these methods. Inspection of the EDF plots of the inter-capillary distances shows that only one field supported a regular alternative. First nearest neighbor capillary distances of the muscle fields and their corresponding EDF plots provided more interesting results than the analysis using the inter-capillary distance

---

[*] The Monte Carlo tests used in this investigation are derived from simulations of the two-dimensional, Poisson point process by n independently and uniformly distributed, computer-generated, pseudo-random points in the region A. In other words, 99 EDFs, $H_i(t)$, $G_i(y)$, and $F_i(x)$, $i = 2, \ldots, 100$, were computed from $99n$ independently and uniformly distributed events.

method. Using this method one field in the sartorius, one field in the retractor, and all five fields of the soleus muscle supported the regular alternative. Finally, an inspection of the EDF plots generated by the random sample point to closest capillary distance method showed that one field in the sartorius and three fields in the soleus were suggestive of a regular alternative to the null hypothesis of CSR. Figure 2 is an example of two of these plots: one showing evidence for regularity and the other indicating a random pattern.

Test Statistics h_1, g_1, and f_1

Exact one-sided Monte Carlo tests were developed based on $s - 1$ simulations ($s = 100$) and the integrated squared difference of the observed and predicted cumulative frequency distributions over the whole range of distances. The rank of the test statistic gives a test of size k/s, where k is the rank of the observed statistic among all s_i sorted

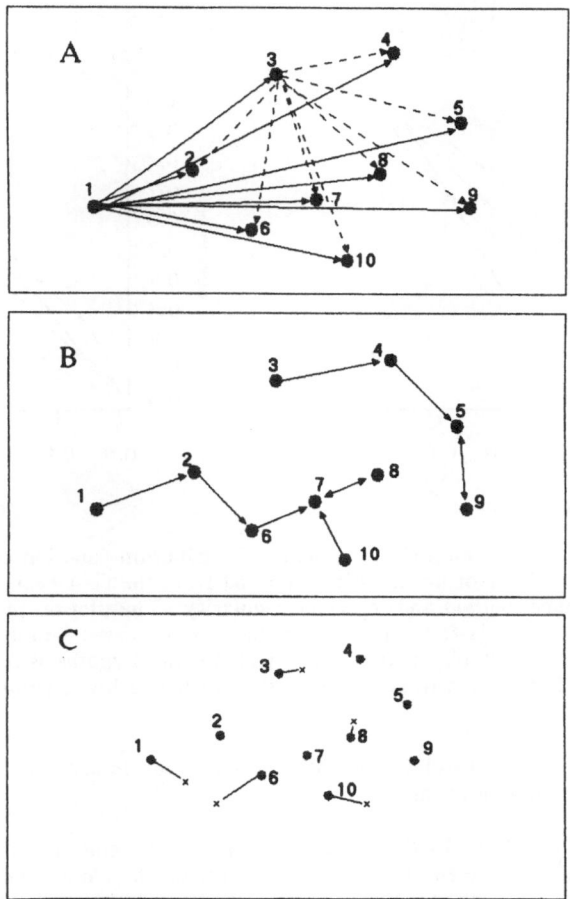

Figure 1. Three distance methods used in the analysis of capillary spatial pattern according to the method of Diggle (1983b). Panel A depicts the inter-capillary distance method where distances between all pairs of capillaries are recorded. Panels B and C depict the distances between a capillary and its first nearest neighboring capillary and the distances between random tissue sample points and the closest capillary, respectively.

from largest to smallest (Diggle, 1983c). Thus, if the rank of the test statistic for the observed data is 5 or less, then the null hypothesis is rejected at the 5% level. These test statistics are given in Table 1 for all of the muscle fields analyzed. The three methods of analysis—inter-capillary, first nearest neighbor capillary, and random sample point to closest capillary distances—are subtly different with respect to their sensitivity in detecting one type of pattern over another. The first nearest neighbor capillary distance method has the greatest power to detect regularity within a pattern. Therefore, eight out of ten cases were found to be suggestive of a regular pattern in the first nearest neighbor capillary distance analysis. Only one out of ten showed evidence for regularity by the inter-capillary distance method and three out of ten by the random sample point to closest capillary method.

DISCUSSION

Analysis of a pattern of n events in a specified region A by statistical techniques has its origins in stochastic point process theory. (One point in A is considered one event.) Stochastic processes are statistical phenomena that create a whole range of patterns of points in some specified region through random variables of space and/or time. For example, the stationary, homogeneous, Poisson point process places events in space

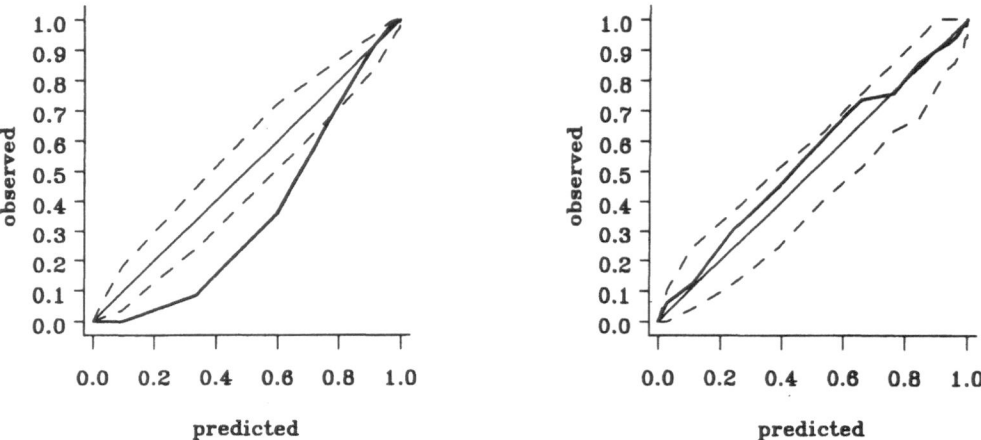

Figure 2. Examples of two empirical cumulative distribution function plots of hamster soleus muscle. The plot on the left is derived from the first nearest neighbor capillary distance method and indicates regularity. The plot on the right is a different soleus muscle field analyzed by the random tissue sample point to closest capillary distance method. It does not reject the null hypothesis of complete spatial randomness. The broken lines represent the upper and lower simulation envelopes.

such that all events are independent and that every location in space has an equal probability of containing an event.

Cottam and Curtis (1956) were plant ecologists who became the first to apply these methods to their field. They employed what they termed the *closest individual* method and the *nearest neighbor* method. The former method considers randomly-placed sampling points across a study region and the measure of the distance from each to the closest event in the region. The latter method involves the measure of the distance from each event inside the region to its first nearest neighbor whether it be inside or outside the region. Cottam and Curtis used these methods to determine the density of the population they were sampling. More than just density of the population can be extracted from the information these techniques provide. Skellam (1952) was the first to apply these methods

Table 1. Test statistics h_1, g_1, and f_1 for empirical and modelled data

Field	h_1	g_1	f_1
empirical			
RET060481_FIELD1.CAP, sector 0	0.56	0.88	0.20
RET060481_FIELD1.CAP, sector 1	0.03*	0.05*	0.18
RET060481_FIELD2.CAP, sector 0	0.20	0.01*	0.84*
SAR060981.CAP, sector 0	0.77	0.01*	0.04*
SAR062381.CAP, sector 0	0.33	0.22	0.14
SOL070781_SLIDE5.CAP, sector 0	0.50	0.02*	0.24
SOL070781_SLIDE5.CAP, sector 1	0.70	0.01*	0.45
SOL070781_SLIDE7.CAP, sector 0	0.37	0.01*	0.01*
SOL070781_SLIDE7.CAP, sector 1	0.45	0.02*	0.03*
SOL070781_SLIDE7.CAP, sector 2	0.19	0.01*	0.08
modelled			
HEX031988_RETPOP.MOD, sector 0	0.43	0.03*	0.71
HEX031988_SARPOP.MOD, sector 1	0.32	0.05*	0.47
HEX031988_SOLPOP.MOD, sector 0	0.11	0.04*	0.50

* significant at $p \leq 0.05$

to analyze quantitatively non-randomness in spatial patterns of plants. Test statistics are then based on departures from the null hypothesis that $\mu = \sigma^2$. Testing spatial point patterns, assuming that in the random case the events are a realization of a Poisson point process, has become the accepted standard for testing the null hypothesis of complete spatial randomness (CSR) (Diggle, 1979).

> Although the Poisson process provides a standard of comparison for more complex processes, we must remember the mildly worded caution of (Cox and Lewis, 1966) that "The Poisson process is a mathematical concept and no real phenomenon can be expected to be exactly in accord with it" (Cliff and Ord, 1981).

None of the fields analyzed by the mapped pattern techniques elicited evidence for the aggregated alternative to CSR. There were, however, several cases where regularity was detected. Each test has a certain power and the test based on first nearest neighbor capillary distances is more powerful in detecting regular alternatives than is the test based on random tissue sample point to closest capillary distances. The opposite is true for aggregated alternatives.

The test based on $\hat{G}_1(y)$, which is known to be powerful in the detection of regularity, detected this pattern at least two-thirds of the time. The reason so many fields suggest a regular alternative is not surprising when one looks at the cross sections of tissue. Upon first glance at the sections one might not notice any sort of regular pattern with respect to the capillaries; however, that the muscle fibers are reminiscent of polygonal shapes is quite apparent. This interesting observation led to the formulation of a mathematical model that would provide a good approximation to the actual situation occurring in the muscle.

DEVELOPMENT OF MODEL

The model places capillaries around hexagonal muscle fibers according to the follow-ing prescription. The capillaries are randomly assigned (x, y) coordinates along the edges of fibers or at the vertices of fibers which are represented by space-filling regular hexagons of side length l. Figure 3 displays the orientation of the regular hexagons. The model requires (1) number of hexagons r along one dimension to generate an $r \times r$ array of space-filling hexagons, (2) number of (x, y) coordinates to specify capillary locations, (3) edge length of hexagons in microns, (4) fraction of capillaries at vertices, (5) name of file to contain (x, y) coordinates of the capillaries, (6) number of test points, and (7) bin size in microns for frequency distribution of distances. Given these parameters the model appropriately places "capillaries" either at the vertex of a hexagon or along the edge of a hexagon using pseudo-random numbers such that the fraction of capillaries placed at vertices has a mean equal to the number of capillaries at vertices input to the program. Output from the computer simulation includes (1) total number of hexagons in the array, (2) number of vertices in the array, (3) (x, y) coordinates of the vertices, (4) total area of space including the area of the hexagons and the area exterior to the hexagons on the right and left sides (see Figure 3), (5) total number of capillaries placed at vertices, (6) file containing capillary (x, y) coordinates, number of capillaries, and date, (7) number of test points placed in a $k \times k$ grid, and, (8) frequency distribution, cumulative frequency distribution, and normalized cumulative frequency distribution of distances.

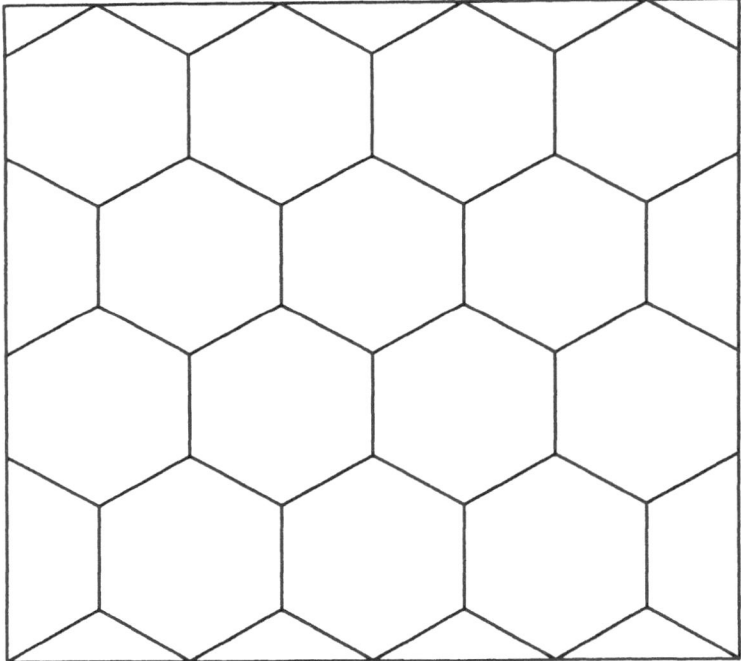

Figure 3. Orientation of the hexagons that represent the muscle fibers in the mathe-matical model. The solid lines represent the interstitial space.

RESULTS OF MODELLING

Sullivan and Pittman (1984) have shown previously that mean muscle fiber area was greatest for FG fibers in the sartorius and smallest for SO fibers of the retractor muscle.

Mean areas, perimeters, and perimeter/# edges for the three fiber types in the retractor were statistically different from each other. In the soleus muscle all parameters for SO fibers were statistically different from those of FOG fibers (p < 0.01). The retractor muscle was observed to be of mixed fiber type, having approximately 70% glycolytic and 30% oxidative fibers. Ninety-eight percent of the fibers in the sartorius were glycolytic, and greater than 99% of the fibers in the soleus were oxidative.

The number of capillaries around a fiber was found to be significantly greater (p < 0.01) for oxidative fibers than for glycolytic fibers in the retractor muscle; however, no significant difference was found between these fibers in the sartorius muscle. The area/capillary values suggest that capillaries around FG fibers must supply oxygen and nutrients to a greater volume of muscle tissue than capillaries serving oxidative fibers (Sullivan and Pittman, 1984).

The greatest number of capillaries per unit area (capillary density) was observed in the soleus, the most oxidative muscle studied, and the 98%-glycolytic sartorius muscle displayed significantly lower values of capillary density and vertex density from the other muscles. It is interesting to note that though the fiber density is greater for the retractor than for the soleus, the vertex density is less for the retractor than for the soleus. This fact indicates there are more fibers surrounding a fiber in the soleus muscle than in the retractor.

Figure 4 shows that the fraction of capillaries associated with a vertex region is greater than 50%, and in the case of the retractor muscle, as many as 75% of the capillaries are associated with a vertex region. This figure also shows the vertex occupancy for each muscle. Only eight muscle fields for each muscle were analyzed for the number of vertices per unit area (vertex density). The fraction of vertices occupied in the sartorius was not significantly different from that of the retractor, but the soleus muscle's vertex occupancy was significantly greater than either the retractor or the sartorius muscle's vertex occupancy.

EDF Plots

Three modelled muscle fields were analyzed using the inter-capillary, first nearest neighbor, and random sample point to closest capillary distance methods. Each muscle field corresponded to a retractor, sartorius, or soleus muscle having characteristics of the mean values for the parameters given to the model.

At the very smallest inter-capillary distances the empirical distribution functions drift outside the simulation envelopes in all three modelled muscles providing evidence to reject CSR. First nearest neighbor capillary distances of the modelled data and their corresponding EDF plots reject CSR in two fields, namely, modelled sartorius and modelled soleus. Lastly, CSR was not rejected in the modelled muscle fields analyzed by the random sample point to closest capillary distance method.

Test Statistics h_1, g_1, *and* f_1

The first nearest neighbor capillary distance method again exhibited its power in detecting regularity. First, however, test statistic h_1 found no evidence for rejection of CSR, whereas for test statistic g_1, all three modelled muscle fields rejected CSR. The test statistic f_1 from the random sample point to closest capillary distance method gave results similar to h_1's results in that there is no evidence for a rejection of CSR. See Table 1 for these results.

DISCUSSION OF MODELLING

The results of the EDF plots by the inter-capillary distance method for modelled data gave results that were not consistent with what was found in the analysis of the empirical data by the same method. If one were to examine only the plots given by this method, one would conclude that the null hypothesis of CSR is rejected for some other alternative. Consistency was found when the modelled data were subjected to the same test statistics as the real muscle data. Again, the inter-capillary and random sample point to closest individual distance methods were unable to detect any deviation from CSR. Only the test statistic g_1 (first nearest neighbor capillary distance method) detected deviation from CSR; it did so in all three modelled muscle fields.

The deficiencies in the model are due in part to assumptions made in its development. The assumptions that were made are: (1) muscle fibers are regular hexagons of identical area; (2) no provision is made to classify hexagons according to fiber type; (3) hexagons fill region of interest completely leaving no interstitial space; (4) vertices and capillaries are dimensionless; (5) capillaries are located at vertices or along the edges of the hexagons; (6) capillaries not occurring at a vertex are randomly distributed (uniform distribution) along the edges of the hexagons; and, (7) the placement of an edge capillary is independent of the placement of any previously placed capillary. Some of the more imp rtant assumptions that might make a difference in the analysis are (1), (3), and (7) above. All muscle fibers are not six-sided and do not have identical cross-sectional area. A modification to the model could allow for a variable number of edges for the polygons. This would create interstitial space that might more closely depict the actual situation where sometimes one observes the growth of a capillary against one muscle fiber but open to interstitial space on its opposite side. The last assumption was made because there was no information available that would indicate any preferential location of a capillary along an edge. However, as mentioned earlier, 60-75% of all capillaries in the muscle fields studied were associated with a vertex region. These values for the location of capillaries at vertex regions are approximately three times what one would expect if capillaries were just randomly placed within the interstitial space. This should be considered a significant finding. Why should such a significant fraction of the capillaries be located at these regions? One reason might be

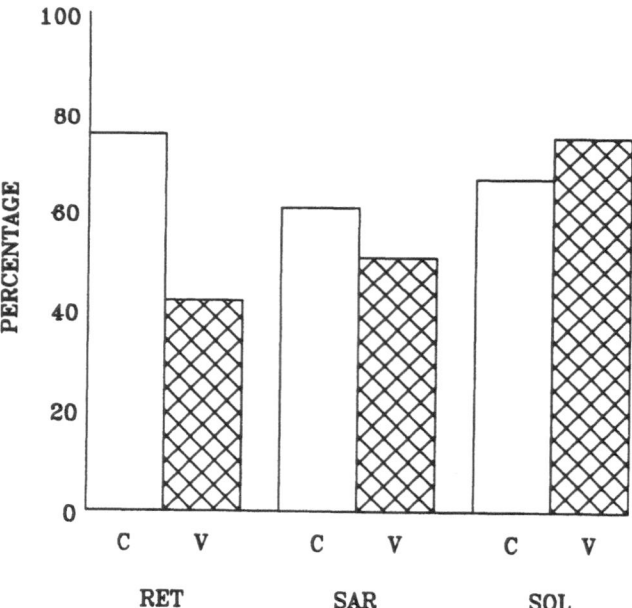

Figure 4. Percentage of capillaries located at vertices and vertex occupancies for three muscles. C, (number of capillaries at vertices)/(total number of capillaries) and V, (number of capillaries at vertices)/(total number of vertices).

purely mechanical: There is more room for a capillary to grow in the vertex region than between muscle fibers.

Modelling capillary spatial pattern is not new. In fact Snyder (1987a,b) has been working on this same topic. Snyder (1987a) has addressed the estimation of diffusion distances in muscle by developing a model of his own design which is based on models by Kayar et al. (1982b) and Plyley and Groom (1975). Uniform sizes of fibers that are space filling were used (square and hexagonal) and capillaries were placed at the vertices of the fibers in an ordered fashion as determined by a capillary to fiber ratio (C/F) ranging from 0.5 to 4.0. The positioning of the capillaries follows the scheme developed by Plyley and Groom (1975). The analysis he performed is similar to that performed by Kayar et al. (1982b) in that diffusion distances were measured as the percent cumulative frequency of fiber area within a given distance of a capillary, which is similar to one method in this investigation termed random sample point to closest capillary distance method. His approaches improve on Kayar et al.'s (1982b) by considering fiber cross-sectional areas of fibers (FCSA) and C/F for two geometric arrangements of fibers and capillaries that Kayar et al. did not. They previously developed models based on capillary density (CD) which are affected by both capillary growth and fiber growth. Also, they only considered one condition of C/F and FCSA for each arrangement of capillaries. Snyder suggests that though Kayar et al.'s analysis is not invalid, it is less flexible than his analysis.

These investigators have been interested in the more traditional descriptors of capillarity, that is, capillary density, capillary to fiber ratio, and number of capillaries surrounding a fiber. Others suggest that a better assessment of potential oxygen supply to tissue is the percent of muscle fiber perimeter in contact with the capillary membrane (capillary-fiber contact) (Sullivan and Pittman, 1987). Their results indicate significantly higher percent capillary-fiber contact in oxidative fibers as compared to glycolytic fibers. They also found no significant difference in the number of capillaries surrounding oxidative and glycolytic fibers within a given muscle. Measures of capillarity such as these give valuable information about numbers and relationships between capillaries and muscle fibers. However, information about any sort of spatial pattern (two-dimensional cross sections) is not available. The methods employed here yield a statistical picture of the spatial pattern of capillaries in a muscle field providing a more complete description of capillarity. These statistical methods in conjunction with measures of capillarity mentioned earlier ought to give modellers of oxygen distribution a better understanding of the role played by microvascular architecture in providing an adequate oxygen supply to tissue.

SUMMARY

The primary goal of this project was to elucidate the spatial pattern of capillaries in three hamster striated muscles according to statistical techniques of pattern analysis. The spatial pattern of capillaries and traditional measures of capillarity are important to understanding the supply and distribution of oxygen and nutrients in a tissue. Statistical tests based on the distance between nearest neighbor capillaries are the most sensitive for detecting regularity in a pattern.

A mathematical model was created to simulate the observed muscle fields. The same statistical tests that were performed on the empirical data were performed on the modelled data. The results of the analysis of the modelled data agree sufficiently with those of the empirical data to justify overall confidence in the assumptions.

Conclusions that may be drawn from this investigation are (1) the spatial pattern of capillaries tends to be more regular than random and never was there evidence for aggregation using the test statistics; (2) as many as 60-75% of capillaries are located at the corners of muscle fibers indicating that there is some preferential placement for capillaries, and (3) the model developed is a good first approximation to the real situation.

ACKNOWLEDGEMENT

This study was supported by Grant HL-18292 from the National Heart, Lung, and Blood Institute.

REFERENCES

Bartels, C.P.A. and R.H. Ketellapper, eds. *Exploratory and explanatory statistical analysis of spatial data*. Boston: Martinus Nijhoff Publishing, 1979.

Cliff, A.D. and J.K. Ord. *Spatial Processes Models & Applications*. London: Pion Limited, 1981, pp. 86-117.

Cottam, G. and J.T. Curtis. The use of distance measures in phytosociological sampling. *Ecology* 37, 1956, 451-460.

Cox, D.R. and P.A.W. Lewis. *The Statistical Analysis of Series of Events*. London: Methuen, 1966.

Diggle, P.J. On parameter estimation and goodness-of-fit testing for spatial point patterns. *Biometrics* 35, 1979, 87-101.

Diggle, P.J. Preliminary testing for mapped patterns. In *Statistical Analysis of Spatial Point Patterns*, London: Academic Press, 1983b, pp. 10-30.

Diggle, P.J. Introduction. In *Statistical Analysis of Spatial Point Patterns*, London: Academic Press, 1983c, pp. 1-9.

Ellsworth, M.L. and R.N. Pittman. Heterogeneity of oxygen diffusion through hamster striated muscles. *American Journal of Physiology* 246(15), 1984, H161-H167.

Getis, A. and B. Boots. *Models of Spatial Processes: An approach to the study of point, line and area patterns*. Cambridge: Cambridge University Press, 1978, pp. 1-85.

Guarascio, M., C.J. Huijbregts and M. David. *Advanced Geostatistics*. Dordrecht: Reidel, 1976.

Journel, A.G. and C.J. Huijbregts. *Mining Geostatistics*. London: Academic Press, 1978.

Kayar, S.R., P.G. Archer, A.J. Lechner and N. Banchero. The closest-individual method in the analysis of the distribution of capillaries. *Microvascular Research* 24, 1982b, 326-341.

Krogh, A. The number and distribution of capillaries in muscles with calculations of the oxygen pressure head necessary for supplying the tissue. *Journal of Physiology (London)* 52, 1919, 409-415.

Moore, P.G. Spacing in plant populations. *Ecology* 35, 1954, 222-227.

Plyley, M.J. and A.C. Groom. Geometrical distribution of capillaries in mammalian striated muscle. *American Journal of Physiology* 228, 1975, 1376-1383.

Skellam, J.G. Studies in statistical ecology. I. Spatial pattern. *Biometrika* 39, 1952, 346-362.

Snyder, G.K. Estimating diffusion distances in muscle. *Journal of Applied Physiology* 63(5), 1987a, 2154-2158.

Snyder, G.K. Capillary growth and diffusion distances in muscle. *Comp. Biochem. Physiol.* 87A, No. 4, 1987b, 859-861.

Southwood, T.R.E. *Ecological Methods*. London: Chapman and Hall, 1978.

Sullivan, S.M. and R.N. Pittman. In vitro O_2 uptake and histochemical fiber type of resting hamster muscles. *Journal of Applied Physiology* 57, 1984, 246-253.

Sullivan, S.M. and R.N. Pittman. Relationship between mitochondrial volume density and capillarity in hamster muscles. *American Journal of Physiology* 252, 1987, H149-H155.

QUANTIFYING CAPILLARY DISTRIBUTION IN FOUR DIMENSIONS

S. Egginton and H.F. Ross

Department of Physiology, The Medical School
University of Birmingham, Birmingham B15 2TJ, UK.

INTRODUCTION

Capillary supply to tissue has for decades been quantified simply using capillary number, usually capillary density (CD), on the assumption that the intensity of point sources of oxygen in a section through any tissue will be a limiting factor for, or accurately reflect, the level of oxidative metabolism. This approach has met with limited success, being mainly applicable for highly aerobic tissue with a high CD and fairly regular geometry of the capillary bed. Many tissues, perhaps the majority, do not fall into this category, necessitating alternative approaches in order to provide an adequate description of capillary distribution as a basis for models of oxygen transport. However limitations are still evident. For example, estimates of overall capillary volume are available which exclude information at higher spatial resolutions, while powerful statistical treatments of point distributions often require unrealistic physiological assumptions.

In order to maximise the utility of available mathematical models of transport processes it is important to utilise analytical methods capable of accounting for heterogeneity of the vascular oxygen conduits in both time and space.

ESSENTIAL REQUIREMENTS

The following discussion assumes unambiguous identification of the total (anatomical) or functional (patent/perfused) capillary bed from histological sections, and unbiased data collection. These are not trivial problems, but are beyond the scope of this article (see Egginton, 1989). Given these conditions, analysis of spatial distribution may involve measured parameters of 0, 1 or 2 dimensions, which may then be used in analyses incorporating 3 or 4 dimensions. Striated muscle is used as an example, although the principles involved are generally applicable.

ZERO DIMENSION

Numerical (0-D) indices based on counts of capillaries and fibres within a given field are obviously the most direct form of analysis, commonly expressed as the ratio of capillaries to fibres (C:F) or sample area (CD). Even these simple indices may produce quite biased estimates if an inappropriate sampling protocol is followed. Apart from the ambiguity of reference phase (some workers choose to express CD per unit area of tissue; we prefer unit area of muscle fibres) other major considerations include sufficiency of sample size for a stable estimate of population means (Egginton and Johnston, 1982). Adopting an unbiased counting rule is also important as simply counting all capillaries within a field of view is not sufficiently robust, particularly when comparative counts of different objects are required (see Egginton and Ross, this volume).

Such indices tend to be restricted to regional or global estimates, providing information about inter muscular and inter animal ('biological') variation, although experimental designs are available whereby subdivision of a sample area into smaller quadrats may permit an index of intra muscular heterogeneity to be computed by comparing quadrat and sample means. Attempts have also been made to improve the analytical resolution by weighting data according to the relative proximity of different fibre types (see Egginton and Ross, this volume). Despite many problems, these indices are useful in describing gross changes in capillary supply during physiological adaptation. C:F is an adequate index of capillarisation where neovascularisation is not accompanied by cellular hypertrophy or atrophy, and has the advantage of being relatively insensitive to section orientation or dimensional artefacts associated with tissue processing. The use of apparent CD of gelatin-filled vessels in muscles at rest and during contractions led August Krogh to develop the concept of capillary recruitment during hyperaemia, and the cylinder model still widely used as a basis for oxygen transport studies.

ONE DIMENSION

Linear (1-D) separation of capillaries is a popular index with modellers of transport processes as the mean intercapillary distance (ICD) is generally assumed to be a limiting factor in muscle performance, although we feel there is sufficient evidence to dispute this (Hudlická et al, 1988). This may in part be due to problems with sampling and/or analysis. As ICD is often derived rather than measured, errors may be compounded, although unbiased sampling rules for linear analyses may be obtained from both the number and associated area distributions. Four basic approaches may be used.

Spatial point process statistics

This uses the assumption, implicit though not stated in many studies, that capillaries may be treated as point sources and their distribution modelled in order to derive information about their overall spatial rela-tionships. Perhaps the most appropriate is the K-function (Diggle, 1983) whereby all pairwise distances are measured and cumulative probability plots used to describe the mean number of points within a given distance of an arbitrary point. These may then be used to test for aggregations/repulsions (equivalent to local concentrations of angiogenesis or zones of capillary growth inhibition), which are particularly useful in comparing observed with known ('benchmark') distributions. However the form of dis-tribution may be misleading. The basic statistical comparison relies on the concept of complete spatial randomness (CSR), where any point is equally likely to occur anywhere in the plane as any other, and is not excluded from any area. Point process statistics may be very sensitive to changes, and useful in a comparative manner at the tissue level of organisation but, as a 'random' distribution of capillaries cannot physically equate with CSR due to the presence of muscle fibres, this approach is inappropriate for more detailed analyses such as oxygen transport calculations at a cellular or individual capillary level.

Nearest neighbour analysis

This is a specific application of point process statistics, and hence subject to similar limitations, whereby the distance from an object to its nearest neighbour is calculated. The original protocol used an annular sampling frame centred over a capillary to estimate the minimal ICD (Loats et al, 1978). Although estimates of minimal ICD are relatively insensitive to section orientation, the functional capacity of the capillary bed is more sensitive to maximal diffusion distances and heterogeneity of spacing. Kayar et al (1982) introduced a statistical refinement estimating the distance from a point over the tissue, chosen in a systematic random manner, to the nearest capillary. This is useful in describing both the minimal and maximal intra muscular diffusion distances. From the distribution of such values Krogh's radius, R_k, can be derived which is assumed to be equal to ICD/2.

In using nearest neighbour statistics one needs to make certain assumptions regarding the probability of point distribution which may be physiologically unrealistic. Cumulative probability plots are sensitive tools with which to detect changes, but open to false interpretation in equating calculated and observed patterns. Whether an averaged relationship actually reflects the in situ distribution needs careful consideration: again, that a distribution of spacing on average follows a curve computed for a random distribution may be of comparative interest, but is of little further use as one is applying different definitions of 'random'. Although this approach may adequately describe a given geometric arrangement and subsequent perturbations, it is unlikely to provide much information regarding the underlying stimuli or functional consequences.

Regular geometric arrays

The mean ICD is often estimated from capillary density and the maximal diffusion distance in the Krogh's cylinder, R_k, assumed to be half:

$$R_k = \frac{1}{\sqrt{\Pi . CD}} = \tfrac{1}{2} \text{ ICD}$$

However, values can be derived from global parameters only if a regular geometric array of capillaries is assumed. Rather than assuming tissue is supplied by overlapping cylinders as in the above example a common alternative is the square array where $R_k = \frac{1}{2\sqrt{CD}}$. The question of nomenclature may cause some confusion in the case of an hexagonal array as it is often not clear whether R_k is being measured to the centre of the circle containing or contained by the hexagon, or along its sides. This particular arrangement is popular with modellers of O_2 transport as it encloses the largest area of tissue and maximises ICD estimates, while triangular and square-ordered arrays are assumed to equate with an average of 3 or 4 capillaries around a fibre. However such a direct relationship between capillary spacing and CD is strictly valid only with a normal (Gaussian) distribution of values, whereas ICD's have been shown by many authors to follow a right-skewed (lognormal) distribution. Therefore if muscles share a similar constant of proportionality in the relationship $R_k \propto \frac{1}{\sqrt{CD}}$ this only <u>suggests</u> a similar distribution, rather than <u>demon</u>strates it.

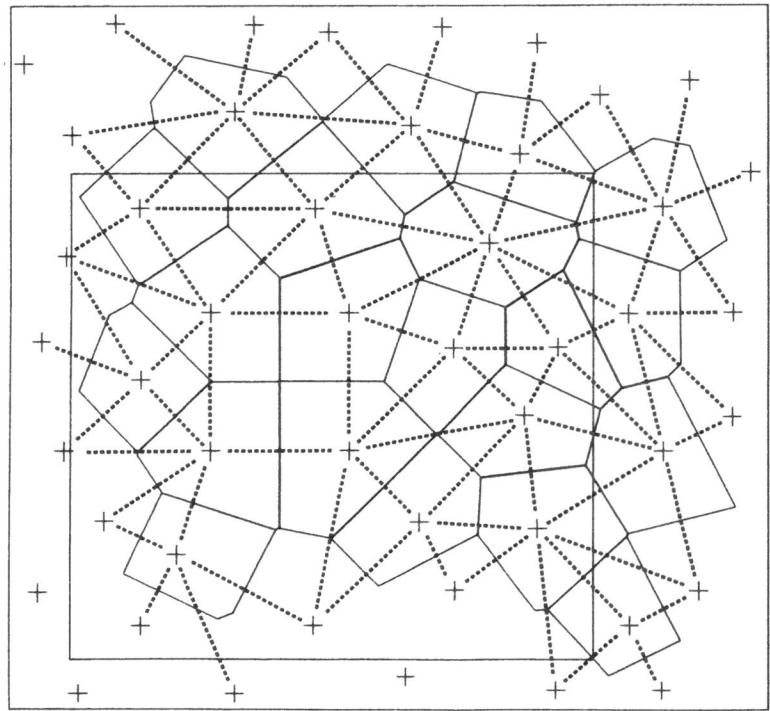

Fig 1   The Voronoi tessellation (solid lines), using capillaries (crosses)
as generating points, is useful in describing planar neighbourhood conditions.   There is one derivative which permits a more objective, and
possibly more efficient, estimate of ICD than by other linear methods.
Previously described methods of calculating ICD suffer from an inherent
uncertainty as to which capillaries are contiguous with (adjacent to) an
individual vessel.   Nearest neighbours are strictly valid only for estimating minimum diffusion distances, while an objective scheme is necessary
to account for non-nearest neighbours.   Lines bounding capillaries whose
domains are contiguous, i.e. share a common edge, and meet at a vertex form
a secondary (Delaunay) tessellation (broken lines) to the Voronoi.   This
also covers the plane in a space-filling manner with lines whose length
distribution is clearly linked to that of the primary (Voronoi) tessellation.   Given the coordinates of points (capillaries) and domain boundaries we can therefore unambiguously identify the interpoint distance (ICD)
between adjacent vessels.   This method has the advantages that distances
are measured not derived, limits to the algorithm are natural not imposed,
and analysis is not limited to nearest or k-nearest neighbours although
such information may be readily obtained.   The resulting planar graph
resembles the non-overlapping triangulation method of Renkin et al (1981),
although the form of construction is not imposed.   For example, in tissue
where capillaries are thought to be ordered in a square array, the
separation of contiguous capillaries will form a square of ICD's.

Direct mensuration

A direct, unbiased method of determining the mean ICD and type of distribution may be provided by adopting the inter point distance (IPD) procedure of Ripley (1981). In this case distances would be measured between adjacent capillaries within a sampling frame, and to those capillaries lying in a surrounding 'guard' area, but not between those within the guard area. A similar, though less rigorous approach was adopted by Renkin et al (1981) who used a series of non-overlapping triangles to connect adjacent capillaries, such that half the sides = R_k.

We have examined a complementary (Delaunay) tessellation to that used as a 2-D index of capillary spacing for this purpose. This provides a more objective sampling protocol, and a self-limiting algorithm which avoids ambiguity of which capillaries are classified as 'adjacent' (Fig 1). Plots of ICD's from three muscles of differing composition show similar, right-skewed distributions which appear to scale inversely with oxidative capacity (Fig 2A).

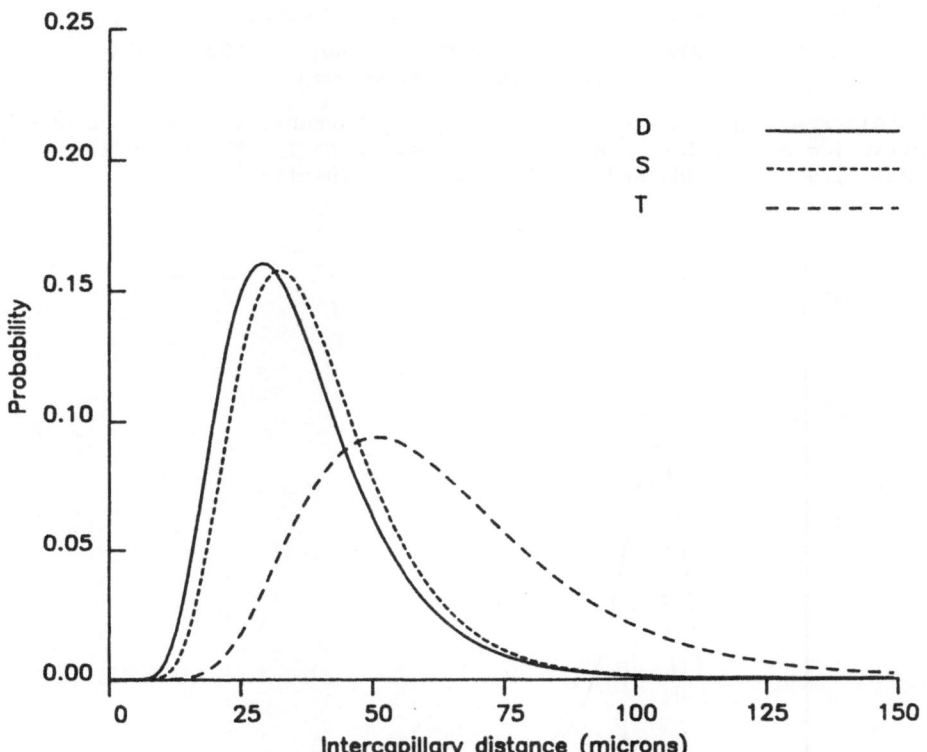

Fig 2 [A] Inter point distance (ICD) distributions for three rat muscles, determined by Delaunay tessellation from the capillary domain data sets of Egginton and Ross (this volume). The magnitude of ICD is seen to scale inversely with muscle oxidative capacity, although the broad distribution of fibre size in the mixed diaphragm leads to difficulty in resolving this curve from that of the purely oxidative soleus with larger fibres. S = soleus (oxidative), D = diaphragm (mixed), T = cortex of TA (glycolytic). Mean $\log_{10}$ ICD ($\pm$ SEM) = 1.588 (0.0022), 1.532 (0.0025) and 1.771 (0.0031), respectively.

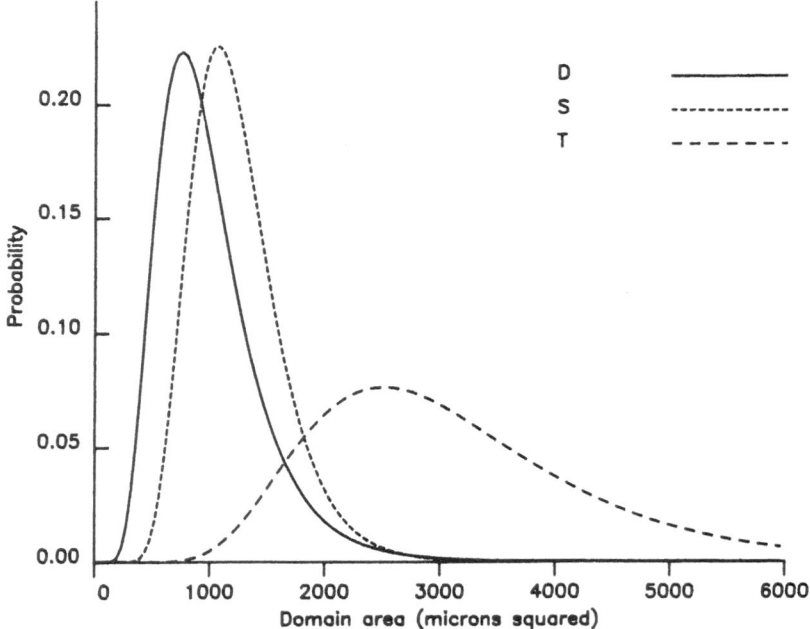

Fig 2 [B] Area distributions of the primary, Voronoi, tessellation showing a similar log normal form. Mean $\log_{10}$ domain area ($\pm$ SEM) = 3.071 (0.0028), 2.959 (0.0038) and 3.464 (0.0056), respectively.

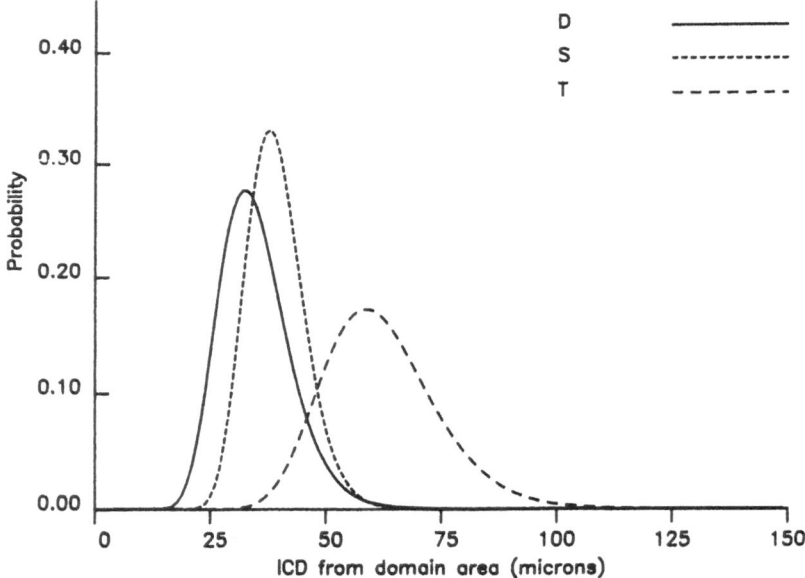

Fig 2 [C] ICD distributions derived from the Voronoi polygons, with ICD $2.\sqrt{(A/\Pi)}$. Note the similar mean values but narrower distribution to those in [A], resulting from the use of domains directly which averages local heterogeneities in capillary separations. Mean $\log_{10}$ ICD ($\pm$ SEM) = 1.588 (0.0008), 1.532 (0.0019) and 1.785 (0.0028), respectively.

TWO DIMENSIONS

As functional resistance to intra muscular diffusion is an integration of all linear separations between an individual capillary and its neighbours, it follows that analysis of the associated area, or domain, of influence for a population of capillaries should provide a more realistic estimate of capillary distribution. 2-D space can be divided into bounded and connected subsets (or tiles), with a tessellation being an aggregate of tiles that cover the plane without overlapping. A Voronoi tessellation is generated by a finite set of distinct points, each of which acquires a tile comprising that part of the plane closer to it than any other. While many authors have documented the limitations to Krogh's cylinder model it still forms the basis of many analyses of oxygen supply, and may provide a reasonable approximation of global oxygen delivery in aerobic tissue. For use at the local level, however, it requires an irregular space-filling, non-overlapping representation of the zone of capillary influence (Fig 1). The convex polygons fulfill these requirements and therefore provide a realisation of Krogh's tissue 'cylinder' for individual vessels, providing an estimate of local diffusion distances, R_k, given as the radius of a circle with equivalent area to the domain.

This approach was first used by Hoofd et al (1985) to study heterogeneity of capillary spacing in myocardium, although dealing with area per se, rather than derived linear quantities, has a number of advantages. If we consider what is meant in the present context by 'heterogeneity' the advantages of a 2-D index should become clear. A plot of ICD's in a regular array will have a tight distribution, and will be broader for a hexagonal than an equivalent square array. In one sense the former is therefore more heterogeneous than the latter. When modelling intra muscular oxygen tensions it is clear that the longest diffusion distances have the greatest effect, and so in situ variability (which by definition is minimised in assumed regular arrays of whatever type) is potentially of great importance. The use of a 2-D index of separation must therefore provide more information about the geometry of the capillary bed as it describes not only the separation of individual vessels, but also the influence of non-nearest neighbours. This index is, however, sensitive to inappropriate sampling protocols and clearly illustrates an inverse relationship between sampling bias and data variance. For example, in rat EHP (a mixed fast muscle) sampling all domains entirely contained within a counting frame produced both the smallest mean domain area and SD ($853\pm413\mu m^2$) as a result of sample bias towards smallest domains, compared to an unbiased sample ($992\pm507\mu m^2$) which correctly includes larger domains in proportion to their numerical distribution. Sampling domains associated with all capillaries within the counting frame, or deletion of peripheral domains intersecting the frame boundaries, constitute an intermediate level of bias.

Perhaps the greatest advantage of an analysis in 2-D is the potential to investigate the interaction or causal relationship between heterogeneity in capillary spacing and other forms of heterogeneity within the tissue. The most clear example of this, the influence of muscle phenotype on local capillary supply, is described elsewhere (Egginton and Ross, this volume) and it may also be useful in describing other functional differences. For example, domain area may be weighted in proportion to the relative or absolute level of perfusion of individual vessels, or the complementary differential oxygen demand of the enclosed tissue.

THREE DIMENSIONS

The mean volume of tissue supplied by individual capillaries may be derived from global estimates of ICD or CD, by integrating the calculated area of tissue around a capillary over its pathlength. However, this is valid for strictly linear networks, and only if a regular geometric arrangement can be assumed. Extrapolation of capillary domains into 3-D may present a way of circumventing these restrictions for irregular networks.

A domain of area A and section thickness δT has volume $A.\delta T$ associated with a capillary of length L. In a linear network the domain volume would then be $A.L$, although in practice we wish to analyse non-parallel networks. Length density (J_v), the length of capillaries associated with unit volume of tissue, is a global index that accounts for both changes in CD and tortuosity. Its reciprocal therefore represents $A.\delta T$ associated with the fraction of capillary length equal to δT, and $1/J_v$ = mean domain area. In this way a discrepancy between these two parameters may be used to determine regional differences in tortuosity of capillary pathlengths, while also setting a margin of error within which mean domain area may be extrapolated to domain volume in partially anisotropic systems.

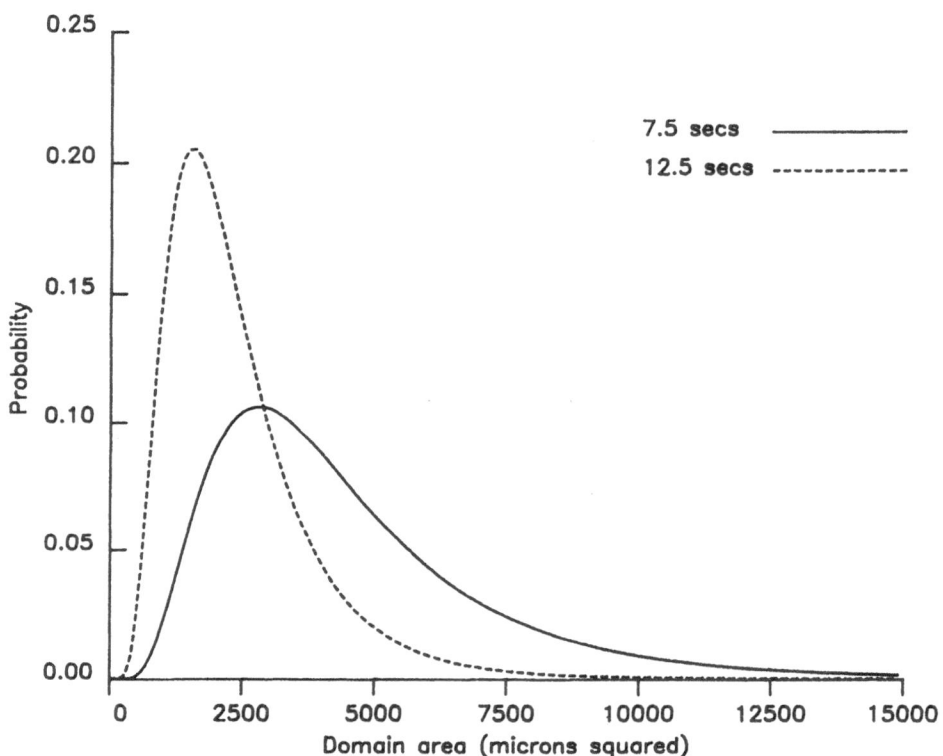

Fig 3 Frequency distribution of domain area for perfused capillaries identified with timed infusion of a fluorochrome (Thioflavin S, conjugated with 6% BSA in saline). Times shown represent the period between start of infusion and cessation of flow by ligation of right common iliac artery. Samples taken from glycolytic cortex of tibialis anterior, Sprague-Dawley rats, using 14μm cryostat sections and epifluorescence microscopy. Given the approximately logarithmic normal distribution, parametric statistics may be performed on log-transformed variates: mean $\log_{10}$ domain area ($\pm$ SD) = 3.5885 $\pm$ 0.2480 and 3.3175 $\pm$ 0.2349, with a sample size of 1014 and 1228 for 7.5 and 12.5 sec samples, respectively.

FOUR DIMENSIONS

Temporal heterogeneity may be assessed in a similar manner to spatial heterogeneity, using 0 - 2-D indices, by quantifying the extent of vascular perfusion at known time increments. This is similar to the approach of Krogh, effectively compressing the time base of hyperaemia in contracting muscle by single point determinations. Infused fluorochromes have much in their favour over established particulate markers, providing an adequate resolution for the study of temporal variation in capillary supply (Fig 3).

Certain advantages may accrue from using simple counts of perfused capillaries, as bivariate spatial point process statistics may be used to assess the degree of cooperativity between sequentially perfused populations of capillaries, or capillary beds. Likewise, as domains are not an additive index, being the result of cumulative divisions with time, apparent C:F or CD of fluorescently labelled capillaries may be useful indicators of change in relative vascular perfusion. However, as a basis for modelling oxygen transport capacity it must be advantageous to utilise an index of higher dimensionality, hence analysis of temporal heterogeneity combined with 3-D capillary domains will provide a direct 4-D index of functional capillary spacing.

Supported by the Wellcome Trust.

SUMMARY

Analysis of spatial distribution using numerical (0-D) distribution is limited to global estimates, while linear (1-D) separation of capillaries may be used to describe unrealistic spatial patterns. Intramuscular diffusion is best viewed as an integration of all distances between surrounding capillaries, or area (2-D) of influence for individual vessels. True planar analysis also accommodates other forms of heterogeneity, and may be extrapolated to give the volume (3-D) of tissue supplied by capillaries. Temporal (4-D) heterogeneity in functional spacing may then be quantified.

REFERENCES

Diggle, P.J., 1983, Statistical analysis of spatial point patterns, Academic Press, London.

Egginton, S., 1989, Morphometric analysis of tissue capillary supply, in: "Gas exchange from environment to cell", Boutilier, R.G., ed., Springer Verlag (in press).

Egginton, S., Johnston, I.A., 1982, Suitability of measured parameters and minimal sample sizes required to quantify capillary supply to fish muscle, Acta Stereologica, 1: 309-319.

Hoofd, L., Turek, Z., Kubat, K., Ringnalda, B.E.M., Kazda, S., 1985, Variability of intercapillary distance estimated on histological sections of rat heart, Adv. Exp. Med. Biol., 191: 239-247.

Hudlická, O., Egginton, S., Brown, M.D., 1988, Capillary diffusion distances - their importance for cardiac and skeletal muscle performance, N.I.P.S. (in press).

Kayar, S.R., Archer, P.G., Lechner, A.J., Banchero, N., 1982, The closest-individual method in the analysis of the distribution of capillaries, Microvasc. Res., 24: 326-341.

Loats, J.T., Sillau, A.H., Banchero, N., 1978, How to quantify skeletal muscle capillarity, Adv. Exp. Med. Biol., 94: 41-48.

Renkin, E.M., Gray, S.D., Dodd, L.R., Lia, B.D., 1981, Heterogeneity of
    capillary distribution and capillary circulation in mammalian skeletal
    muscles, in: "Underwater Physiology VII", Bachrach, A.J., Matzen, M.M.,
    eds., Undersea Med. Soc., Bethesda, pp465-474.
Ripley, B.D., 1981, "Spatial Statistics", John Wiley, NY.

INFLUENCE OF MUSCLE PHENOTYPE ON LOCAL CAPILLARY SUPPLY

S. Egginton and H.F. Ross

Department of Physiology, The Medical School
University of Birmingham, Birmingham B15 2TJ, UK.

INTRODUCTION

Numerous attempts have been made to quantify the capillary supply to muscles of varying metabolic character, e.g. oxidative or glycolytic, using indices based on simple counts. Such studies have mainly been concerned with global estimates, i.e. mean values for individual muscles, using the numerical capillary to fibre ratio (C:F) or capillary density (CD). While one may infer the relative extent of capillarisation to different fibre types from the variation among muscles, this can only be in rather general terms as information about intramuscular heterogeneity is missing. One of the first studies to address the specificity of capillary supply with respect to individual fibres was that of Plyley and Groom (1975), who quantified the number of capillaries around a fibre (CAF) for individual categories or fibre types. However, the limited range of values found among mammalian muscles essentially limits its usefulness to mono-phenotypic muscle such as found in lower vertebrates, or where neovascularisation occurs without any change in the fibre population such as in response to chronic hyperaemia. These authors also attempted to view the microvascular supply from the point of individual capillaries, by quantifying the number of fibres surrounding each capillary or sharing factor (SF). It is unfortunate that variations in such indices are often assumed to directly reflect oxygen demand, attaching undue importance to such data. This ignores the conceptual limitations inherent in these indices; in particular the influence of adjacent fibres, fibre size, and proximity of neighbouring capillaries are not taken into account.

Perhaps the best attempt at partitioning capillary supply on the basis of muscle composition was that of Gray and Renkin (1978), which addressed many of these important considerations. This study again analysed individual capillaries, essentially using the SF to weight supply according to the relative number of a given fibre type around each capillary. The mean fraction of SF then provides a specific C:F (SCF), which when normalised for mean fibre area gives a specific capillary density (SCD), for each category of fibre. Although this was a significant improvement over other methods based on capillary counts it is unable to scale the microvascular supply according to local intercapillary distances (ICD's). In addition, integer-based indices produce a discontinuous, rather coarse distribution of values which limits their descriptive power.

Many of these analytical limitations may be resolved by a 2-D realisation of capillary distributions. The area of tissue supplied by individual capillaries, or domain of influence, was introduced by Hoofd et al (1985) as a method of analysing capillary spacing in the plane of transverse sections, and provides an integration of linear separation between all neighbouring capillaries. The interaction of capillary domains and individual fibres may be quantified as the sum of partial domains overlapping a fibre profile, being the areal equivalent of Gray and Renkin's numerical fraction of SF. This local capillary to fibre ratio (LCFR) provides a non-integer index of potential supply, essentially a continuous distribution of 'capillary equivalents', which enables the geometry of the capillary bed to be analysed with respect to both fibre area and oxidative capacity, while implicitly taking the proximity of other capillaries into account. We feel this approach provides the best anatomical data for modelling the functional capacity of the microcirculation in mixed striated muscles.

MATERIAL AND METHODS

Three muscles were chosen in order to cover a range of fibre composition and metabolic character: oxidative soleus, mixed diaphragm, and the predominantly glycolytic cortex of tibialis anterior. Tissue was excised from a male Sprague-Dawley rat of 250g BW following excess pentobarbitone anaesthesia, mounted in OCT compound and rapidly frozen in isopentane (2-methylbutane) cooled in liquid nitrogen. Serial cryostat sections (12µm) were cut and stained for myosin ATPase and SDH by standard methods, to indicate the phenotype and relative oxidative capacity of individual fibres. Three major fibre types were identified as SO (β), FO (αR) and FG (αW). Alkaline phosphatase activity was used to localise individual capillaries. A large sample area was chosen for study in each of the muscles where the least common fibre type was adequately represented, covering 0.359, 0.284 and 0.540mm^2 for soleus, diaphragm and TA, respectively. Camera lucida tracings of histochemical sections were made and the data was digitised using a Summagraphics MM1812 A3 digitising tablet interfaced via an RS232 line to a PDP-11/23 computer. During the digitising procedure results were displayed on a colour VDU screen (Microvitec CUB) driven by the TERMULATOR program in a BBC model B computer. Different buttons on the digitising cursor were used to identify the fibre types which were then displayed in three different colours. Fibre areas and the centres of gravity of fibre outlines were computed as the data was digitised. Those parameters together with the coordinates of each capillary centre and the outline coordinates of each of the fibres were written to disk. Fig. 2 shows a partial data set in which only one fibre type (FG) has been digitised. Normally, all three types are digitised together, but the resulting display is confusing in black and white. A full data set occupies about 250kb of disk space. Two methods were used to characterise the distribution of capillaries to different fibre types.

Integer (GR) method

This analysis followed that of Gray and Renkin (1978). Fibres of each type were counted, with peripheral fibres that intersected the rectangular counting frame scored as 0.5, in order to compute the numerical density (or number fraction)

$$N_N(i) = \frac{n_i}{(n_i + n_j + n_k)}$$

where n_i, n_j and n_k are the number of fibres of type i, j and k, respec-

tively.   The areal density (or area fraction) of individual fibre types
were computed from the total number and mean fibre area

$$A_A(i) = \frac{n_i \bar{a}_i}{(n_i \bar{a}_i + n_j \bar{a}_j + n_k \bar{a}_k)}$$

where $\bar{a}_i$, $\bar{a}_j$ and $\bar{a}_k$ are the mean fibre areas of type i, j and k, respectively. In this study, however, we used the measured value for mean fibre
area of all fibres rather than calculated mean value from axes (diameters)
of a small subsample of fibres. This avoids the necessity for normalisation

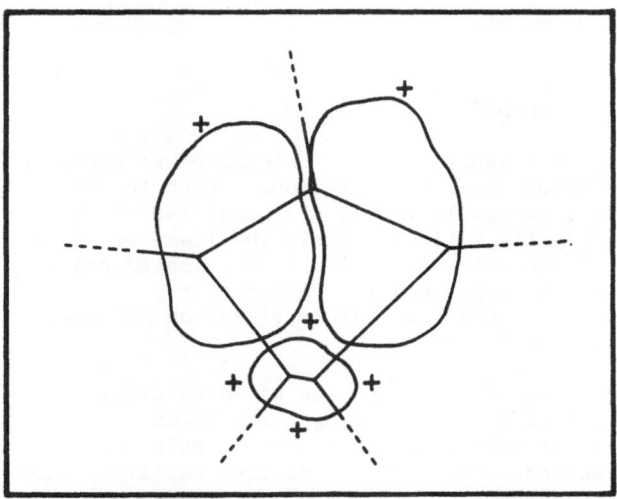

Fig 1 Comparison of scoring methods based on numerical (GR method) or areal
(Domain method) fraction of a capillary to be ascribed to different fibre
types. In this example the central capillary serves two large glycolytic
fibres and a smaller oxidative fibre. The numerical fraction of this
capillary serving glycolytic fibres is therefore 0.66 while the areal
fraction is much greater, around 0.80. In addition to this discrepancy, the
GR method only allows for a distribution analysis on the basis of capillary
fractions per fibre type, while the Domain method provides an additional
level of resolution by summing capillary fractions per individual fibre.

if total sample area (A) differs from the quantity $(n_i \bar{a}_i + n_j \bar{a}_j + n_k \bar{a}_k)$ by
more than 10%, as adopted in the original study. The necessity of such a
correction arises from the assumption that total sample area is the desired
reference for calculation of capillary density, and hence the need to over
estimate actual fibre area in proportion to the cumulative fraction of
interstitial space (nerves, vascular elements, connective tissue, etc). We
prefer a more objective approach, defining capillary density as the number
of capillaries per unit area of muscle fibre within a given reference space,
$N_A(c,f)$.

A specific capillary fibre ratio (SCF) was computed for each fibre type in the following manner. Each capillary serves, i.e. is adjacent to, at least 2 and typically up to 4 fibres of varying metabolic type (the SF). For each capillary within the sample field, the fraction 1/SF is added to a running total for the 3 categories of fibres for each fibre encountered (see Fig. 1), providing a mean value of C:F for specific fibre types

$$\overline{SCF} = \sum_{s=1}^{ncap} (1/SF_s)/ncap$$

where ncap is the total number of capillaries, and SF_s is the sharing factor for that fibre type and capillary s. This quantity on an individual capillary basis can also be used to describe in general terms the characteristics of surrounding tissue (see Table 2). SCF is then normalised to give a capillary density for specific fibre types by division with mean fibre area

$$SCD = \overline{SCF}/\bar{a}(f_i)$$

Non-integer (Domain) method

Numerical and areal densities were calculated as above, except for the adoption of an unbiased sampling criterion. Using the same sampling frame as before (the inner rectangle in Fig. 2), profiles of whole fibres are included only if they lie entirely within the frame or if they exclusively intersect either of the pair of previously designated adjacent inclusion edges of the frame (the upper and left edges in Fig. 2). Counts of capillaries made in the same manner therefore yields an unbiased estimate of global (i.e. overall) C:F and CD.

The planar analysis of capillary bed geometry utilised the division of 2-D space into a set of non-overlapping tiles known as a Voronoi tessellation (Voronoi, 1908; also referred to as Dirichlet or Theissen polygons) in the following manner. Construction of each capillary domain was by a process of reduction starting from a rectangle large enough to contain all the capillaries (the outer rectangle in Fig. 2). That polygon was then reduced by removing from it the area which consisted of points which were nearer to other capillaries than to the one under consideration. These other capillaries were examined in order of proximity to the current one, thus the reduction procedure needed to be continued only so long as the distance from the capillary to the most remote vertex of the domain was greater than twice the distance from the capillary to the nearest capillary which has not yet contributed to the slicing process. At each step of the reduction procedure the area sliced off the polygon was defined by the points where the perpendicular bisector of a line joining the two capillaries intersected the polygon. The part of the polygon on the side of that line opposite the capillary was removed and replaced by part of the line. Each domain has on average only about 6 neighbours, so the reduction procedure needed to be applied only a similar number of times if the neighbouring capillaries are examined in order of increasing distance from the current one. At the beginning of the computation of the domain for each capillary, the squares of the distances between it and each of the others was computed and stored. At each reduction step the smallest unused item on this list determined the next capillary to be used in the reduction. This algorithm appears to be less efficient than the one of Green and Sibson (1978) but it used less memory and seemed easier to implement. Only the size of individual domains (the distance from the capillary to the furthest vertex) was saved for use in the next part of the analysis.

To calculate the local capillary fibre ratio for each fibre which satisfies the sampling criterion defined at the beginning of this section the coordinates of its boundary and its centre of gravity were read from the data file. The radius of the fibre was computed as the largest distance between the centre of gravity and any point on the profile. Each capillary was examined in turn, if the distance between the capillary and the centre of gravity of the fibre was greater than the sum of the size of the domain (computed in the first phase of the analysis) and the radius of the capillary (as defined above) then there was no possibility of any intersection between the domain and the capillary, and the capillary could be ignored. Otherwise, the domain for the capillary had to be computed and then the polygon defining the intersection between the domain and the fibre calculated. That was done by finding the intersections of edges of the domain and fibre polygons, then cutting and stitching the vertex lists appropriately to form the intersection polygon. Since the fibre polygon had, typically, between 100 and 200 edges that was the most time consuming part of the whole procedure. The area of the intersection polygon was computed and expressed as a fraction of the whole domain area and that number was used as the contribution of that capillary to the local capillary fibre ratio for that fibre. If the capillary was closer to any point of the fibre than a certain (definable) critical distance it was labelled as adjacent to the fibre. The procedure was repeated for all capillaries. For each fibre the following parameters were recorded; fibre area ($\bar{a}$), the number of capillaries around a fibre (CAF, the number of adjacent capillaries), the capillary density (FCD = CAF/$\bar{a}$), local capillary/fibre ratio (LCFR) and the local capillary fibre density (CFD = LCFR/$\bar{a}$).

The program consists of almost 1000 lines of C code, excluding the graphics routines. It was run on a Digital Equipment PDP 11/23 computer with the KEF11-AA floating point option and RLO2 cartridge disks. The data set in Fig. 2 took about 35 minutes to process (if all 3 fibre types were handled at once the time was increased to 90 minutes) of which 15 minutes was required to complete the domains.

RESULTS

Muscle composition

Normalising data for $\bar{a}(f)$ in order to fill the sample area (as per original GR protocol) introduced an error of 10% (soleus), 2.8% (diaphragm) and 9.4% (TA) relative to measured fibre area. Likewise, a small error in N_N was evident using the GR vs unbiased sampling rule. For example, in soleus FO = 0.40 vs 0.41 and SO = 0.60 vs 0.59, respectively. Clearly, if N_N is slightly different this will be magnified by any difference in estimated vs measured fibre area, and be reflected in values of A_A which in diaphragm were 0.14 vs 0.14 (SO), 0.336 vs 0.375 (FO) and 0.525 vs 0.485 (FG), respectively. These are likely to represent maximal differences, however, as values for other tissue differed by only 1-2%.

Fibre type composition of the muscle subsamples broadly reflect the known oxidative capacity determined biochemically. The large differential in size between glycolytic and oxidative fibres produces a significant difference between muscle composition based on number (N_N) and area (A_A).

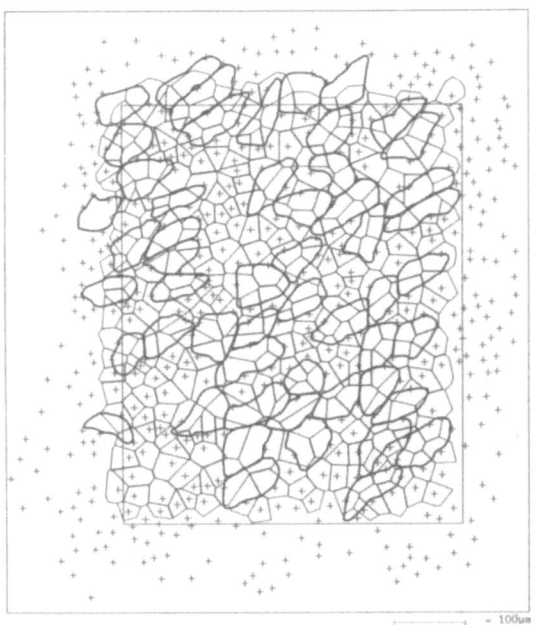

Fig 2 Illustration of the application of the domain method of computing the local capillary fibre ratio. The data set shown is from diaphragm muscle. The outer rectangle defines the limits of the working region for the computer. The inner rectangle is the sampling frame. The crosses indicate the location of capillaries, and the irregular, thick-lined, profiles outline the fast glycolytic fibres in this specimen. All of those items are supplied as data via the digitiser. The thin-lined polygons are the domains associated with each of the capillaries which are computed, as described in the text. Though the domains for all 511 capillaries are computed only the ones accepted by the unbiased sampling rule are shown here. Likewise for the fibre profiles; all fibres of the desired type which intersected the sampling frame were digitised but only those which satisfied the sampling criterion (see text), as applied by the computer, were retained. In computing the local capillary fibre ratio all domains which intersected any of the sampled fibres were used. Note that it is necessary to have a rather wide guard zone of capillaries around the sampling frame.

<table>
<thead>
<tr><th colspan="5">Table 1 Muscle composition</th></tr>
<tr><th></th><th></th><th colspan="3">Fibre type</th></tr>
<tr><th>Muscle</th><th>Index</th><th>SO</th><th>FO</th><th>FG</th></tr>
</thead>
<tbody>
<tr><td>Soleus</td><td></td><td></td><td></td><td></td></tr>
<tr><td></td><td>N_N</td><td>0.590</td><td>0.410</td><td>–</td></tr>
<tr><td></td><td>A_A</td><td>0.643</td><td>0.357</td><td>–</td></tr>
<tr><td></td><td>$\bar{a}(f)$</td><td>2723±79</td><td>2190±55</td><td>–</td></tr>
<tr><td>Diaphragm</td><td></td><td></td><td></td><td></td></tr>
<tr><td></td><td>N_N</td><td>0.222</td><td>0.497</td><td>0.280</td></tr>
<tr><td></td><td>A_A</td><td>0.140</td><td>0.375</td><td>0.485</td></tr>
<tr><td></td><td>$\bar{a}(f)$</td><td>918±37</td><td>981±34</td><td>2729±70</td></tr>
<tr><td>TA, cortex</td><td></td><td></td><td></td><td></td></tr>
<tr><td></td><td>N_N</td><td>–</td><td>0.130</td><td>0.870</td></tr>
<tr><td></td><td>A_A</td><td>–</td><td>0.081</td><td>0.919</td></tr>
<tr><td></td><td>$\bar{a}(f)$</td><td>–</td><td>1550±79</td><td>2655±52</td></tr>
</tbody>
</table>

Analysis of muscle fibre composition using the unbiased counting rule. Number of fibres analysed was 130, 189 and 184, respectively. N_N, numerical fraction; A_A, areal fraction; $\bar{a}(f)$, mean fibre cross section (μm^2). Note the influence of fibre size in producing a discrepancy between fibre composition expressed as N_N or A_A.

Classification of capillary neighbourhood

Table 2 lists the fraction of total capillaries which served particular combinations of muscle fibre types. In the mixed diaphragm 50% were surrounded by predominantly oxidative fibres, while in the cortex of TA over 50% were surrounded by purely glycolytic fibres.

Table 2 Neighbourhood specificity of local capillary supply

Muscle	Number of capillaries	Fraction of total capillaries					
		AW	PW	AR	Aβ	ARβ	PRβ
Soleus	295	–	–	0.051	0.227	0.722	–
Diaphragm	279	0.029	0.029	0.100	0.151	0.204	0.495
TA, cortex	179	0.553	0.374	0.006	–	–	0.067

Weighting of capillary supply according to fibre type composition of the surrounding tissue for the following categories: AW, surrounded by all white fibres; PW, greater than 50% white; AR, only αR fibres; Aβ only β; ARB, all red (i.e. β + αR); PRB, greater than 50% red (after Gray & Renkin, 1978). An equivalent analysis using true planar, domain, data would simply involve substituting number with relative area for individual fibre types. Given the differential in size between fibre types, neighbourhood specificity of apparent capillary supply based on relative number or area will be quite different in mixed muscles.

Specificity of capillary supply

Indices based on simple capillary counts (SCF or CAF) varied directly with the proportion of oxidative fibres. However, the different size of a given fibre type between muscles resulted in a variable local capillary density (SCD or FCD). Within a muscle there is a greater density of supply to oxidative fibres which is not as apparent from C:F ratios, and presumably reflects their smaller size. In contrast, the index based on area (LCFR) produced lower estimates of local capillary supply, while normalising for fibre area resulted in only small differences in apparent capillary density between fibre types within a given muscle (CFD) (Tables 3 and 4).

DISCUSSION

Attempts at quantifying capillary supply to different fibre types have usually been restricted to counting the number of adjacent capillaries. However as a capillary usually supplies 2 or more fibres which may be quite distinct in terms of their metabolic oxygen demand, this information is insufficient to determine the specificity of supply to individual fibres. Given that few muscles are of homogeneous composition, analysis of heterogeneity within the capillary bed requires an objective protocol for characterising the distribution of capillaries within mixed muscles with respect to different fibre types. In addition, realistic modelling of the oxygen transport capacity of the microcirculation requires a classification of capillaries according to the metabolic characteristic of surrounding fibres which, if a reasonable sample size is to be obtained, essentially requires this to be based on histological data.

Scaling capillarisation on the basis of fibre type is possible by dividing integer values by the SF to give an approximation of local capillary supply. This clearly accounts only for global supply, and ignores the differential metabolic capacity of different fibre types which must, for example, weight oxygen efflux from a perfused capillary in proportion to the level of cellular oxygen demand in the surrounding tissue. If an estimate of the relative $\bar{V}O_2$ is available this could be used to scale this index; e.g. if $\bar{V}O_2$ of FG fibres is 1/4 that of SO fibres then in the example given in Fig. 1, the fraction of capillary supply to each FG fibre would be $0.33 \times 0.25 = 0.08$. The effect of fibre area on C:F ratios is only partially accounted for by re-calculation on the basis of $\bar{a}(f)$, as this ignores both local variation and overall heterogeneity of capillary spacing.

Table 3 Numeric analysis of local capillary supply (GR method)

Fibre type

Index	Muscle	SO	FO	FG
SCF				
	Soleus	2.343	2.159	–
	Diaphragm	1.419	1.467	1.518
	TA	–	1.123	0.902
SCD				
	Soleus	782	896	–
	Diaphragm	1506	1449	541
	TA	–	644	303

Partitioning of relative capillary supply on the basis of capillary counts indicates a differential supply among muscles, but not at the level of individual fibre types. SCF, specific capillary to fibre ratio; SCD, specific capillary density (μm^{-2})

Both within and between muscles the number of capillaries surrounding oxidative fibres may be less than around glycolytic fibres where they are significantly smaller. Where fibre size is similar, e.g. diaphragm FG and soleus SO, this integer-based index is greater around the oxidative fibres. However, we suggest that interpretation of this data in terms of a direct correlation between anatomical capillary supply and oxidative capacity may be erroneous as LCFR shows no significant difference between these fibre types. In contrast, the direct scaling of LCFR with fibre area suggests that capillarisation is primarily determined by fibre size, although some small variation in CFD (LCFR normalised for fibre area) for different fibre types shows that this basic relationship may be modulated by the absolute level of oxidative metabolism. Furthermore, it is clear that the local

environment within which a fibre is operating will affect the intensity of
capillary supply; for example FG fibres in the TA receive on average less

Table 4 Planar analysis of local capillary supply (Domain method)

Index	Muscle	Fibre type SO	FO	FG
CAF				
	Soleus	5.2 ± 0.19	4.9 ± 16	–
	Diaphragm	3.5 ± 0.18	3.4 ± 0.12	3.9 ± 0.21
	TA	–	2.6 ± 0.20	2.2 ± 0.26
FCD				
	Soleus	1998 ± 85	2285 ± 80	–
	Diaphragm	4065 ± 255	3666 ± 149	1451 ± 80
	TA	–	1710 ± 183	760 ± 38
LCFR				
	Soleus	2.31 ± 0.07	1.93 ± 0.05	–
	Diaphragm	1.05 ± 0.06	1.07 ± 0.04	2.24 ± 0.08
	TA	–	0.63 ± 0.04	0.84 ± 0.02
CFD				
	Soleus	856 ± 16	894 ± 24	–
	Diaphragm	1145 ± 47	1100 ± 27	825 ± 24
	TA	–	403 ± 28	315 ± 8

Focusing on the object of neighbourhood specificity, i.e. adopting an
unbiased sample of muscle fibres rather than capillaries, provides a more
direct comparison between numerical and planar indices of local capillary
supply. Where all three fibre types are found in close proximity, e.g. in
the mixed diaphragm, the number of capillaries surrounding individual
fibres (CAF) shows little difference among fibre types. When corrected for
fibre area to give an estimate of local capillary density (FCD) there is a
2.8-fold range in values. However, this results from multiplying an index
of poor spatial resolution by another parameter which differs greatly among
fibre types. That such results appear to be an analytical artefact can be
seen when local capillary supply is estimated from the relative area of
different fibre types supplied by individual capillaries (LCFR). This
gives a 2-fold range of values which is almost abolished when normalised
per unit area of fibre (CFD). Scaling of capillary supply therefore
appears to be largely on the basis of fibre size, with only a relatively
small effect of metabolic character evident. CAF, number of capillaries
around a fibre; FCD, fibre capillary density (μm^{-2}); LCFR, local capillary
to fibre ratio; CFD, capillary fibre density (μm^{-2}).

than one capillary-equivalent compared to FG fibres in the diaphragm where this is more than doubled by virtue of the capillary supply to adjacent oxidative fibres. At this level of resolution both the GR and Domain methods produce qualitatively similar conclusions: Gray and Renkin (1978) showed that SCF for *oW* fibres was inversely related to the proportion of *oW* fibres in a sample, in other words that the capillary supply depends to a large extent on the proximity of oxidative fibres.

Supported by the Wellcome Trust.

SUMMARY

The general method used to compute a local capillary fibre ratio for each fibre is as follows. The selected area of the section is covered by a tessellation of domains. For each fibre, the contribution of each capillary whose domain intersects the fibre is computed as the proportion of the domain area which overlaps the fibre. The sum of all the contributions from overlapping domains is taken to be the effective number of capillaries contributing to that fibre, and is called here the local capillary fibre ratio (LCFR). This parameter may be normalised by dividing by fibre area to give an index, independent of fibre size, which is capable of identifying the differential capillary supply to fibre types within a mixed muscle.

REFERENCES

Gray, S.D., and Renkin, E.M., 1978, Microvascular supply in relation to fiber metabolic type in mixed skeletal muscles of rabbits, Microvasc. Res., 16: 406-425.
Green, P.J., and Sibson, R., 1978, Computing dirichlet tessellations in the plane, The Computer Journal, 21: 168-173.
Hoofd, L., Turek, Z., Kubat, K., Ringnalda, B.E.M., and Kazda, S., 1985, Variability of intercapillary distance estimated on histological sections of rat heart. Adv. Exp. Med. Biol., 191: 239-247.
Plyley, M.J., and Groom, A.C., 1975, Geometrical distribution of capillaries in mammalian striated muscle, Am. J. Physiol., 228: 1376-1388.
Voronoi, G., 1908, Nouvelles applications des paramètres continus á la théorie des formes quadratiques, deuxième memoire, recherches sur les parallelloèdres primitifs, Jnl. für die Reine und Angewandte Mathematik, 134: 198-287.

COMPUTER SIMULATION OF CEREBRAL MICROHEMODYNAMICS

Antal G. Hudetz, James G. Spaulding, and Mohammad F. Kiani

Departments of Biomedical Engineering and Zoology, Louisiana
Tech University, Ruston, LA, USA; and Experimental Research
Department, Semmelweis Medical University, Budapest, Hungary

INTRODUCTION

Microvascular path length and transit time are important factors which
influence the extraction of oxygen from red blood cells in cerebral tissue.
The direct measurement of these variables is hindered by the difficulty in
observing the deep cortical capillary circulation in a noninvasive manner
(Pawlik et al, 1981). An alternative approach to estimate microvascular path
lengths and transit times is to simulate the red cell flow distribution in
the microvessel networks by computer.

This paper presents results of a computer simulation of path length and
transit time in the cerebrocortical microcirculation. The calculation was
based on the true geometry and topology of a cortical microvascular network
of the rat, which was measured in an anatomical sample and reconstructed by
computer. The non-newtonian characteristics of blood and the flow properties
of red cells at arteriolar bifurcations were accounted for in the
simulation. The effect of arterial hematocrit on the distribution of path
length and transit time was investigated.

METHODS

Experimental procedures

Vascular casts of the brain were made using a modification of the
techniques of Yasuji and Fusahiro (1984) and Motti et al (1986). Rats of
approximately 250g were anesthetized using Sodium Pentobarbital (50mg/kg
body wt i.p.). The common carotid arteries and the external jugular veins
were exposed. Both common carotids were cannulated using PE 50 tubing. The
external jugular veins were opened to provide drainage. The first perfusate
was 0.9% NaCl containing 10 IU/ml heparin and 1×10^{-7}g/ml papaverine, at 37 C
and perfused from a height of 1m above the animal. The second solution of 4%
buffered formalin was perfused for 10 minutes. This was followed by the
perfusion of modified Batson's No. 17 casting solution made to the following
proportions: 8ml of Batson's 17 monomer, 4ml Sevriton (DeTrey Dentsply),
2.5ml catalyst, 0.17ml promoter. This was injected using a 250ml syringe
attached to the three way valve. When the cast material appeared in the
external carotid veins perfusion was stopped. Following perfusion the head
of the rat was put on ice for 2-3 hours. Approximately 0.5 mm sections were

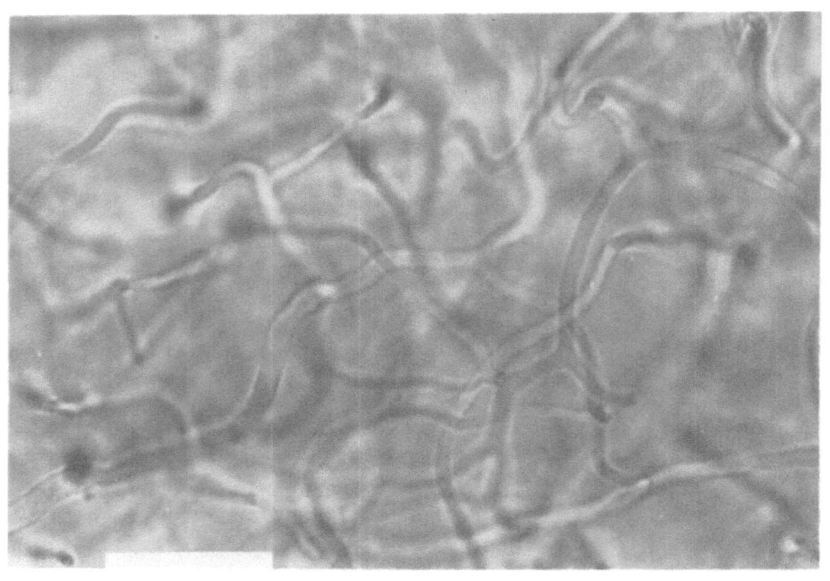

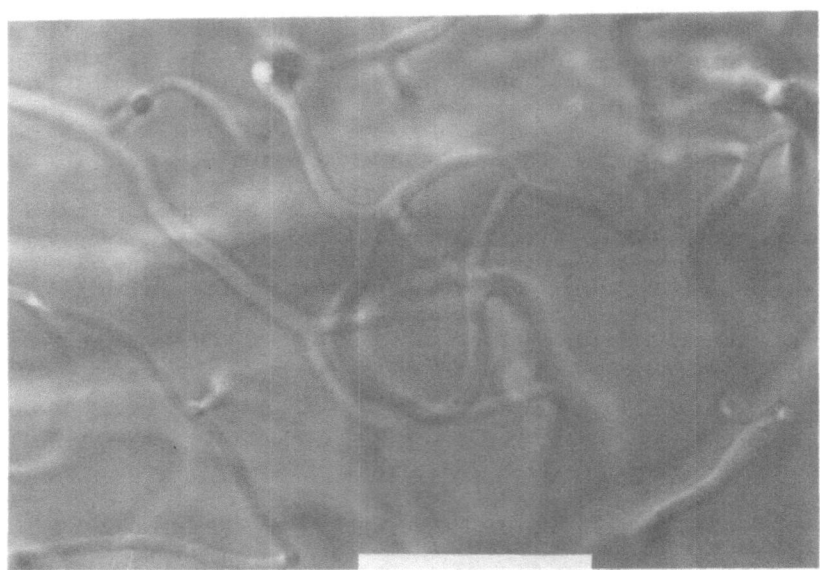

Fig. 1. Microphotographs of the cerebrocortical microvascular cast used in
the present study. A major bifurcation of the terminal artery is seen
in the center and several venules somewhat deeper (above). Typical
cortical capillary loop supplied/drained by four vessels (below).
Photographed at 20X and 25X, respectively. Bar = 50 um.

taken coronally from the mid area of the cerebral hemispheres. The tissue was removed by soaking the section in 20% NaOH for 12 to 24 hours followed by soaking in 5% NaClO with gentle agitation and solution changes for 1 to 3 weeks. When clean, the cast was dehydrated in a graded series of ethanol, transferred to propylene oxide and finally embedded in LX-112 epoxy (Ladd) in a deep depression slide (Figure 1).

Measurement of vascular geometry

The three-dimensional geometry of the vessels were measured by optical sectioning microscopy using a custom-modified version of the Bioquant IV image analysis system (R and M Biometrics). The vessels were traced in the video image of the vascular cast using a mouse as pointing device (Figure 2). The movement of the mouse was fed in the computer (IBM AT) as a series of X and Y coordinates along the vessel. In order to assess dimensions along the third axis, the focusing plane was moved to keep the observed object at the current X, Y location in sharp focus. The position of the focusing knob on the microscope was fed into the computer as the Z coordinate. In addition, the diameter of each segment and branch point was measured interactively. Departing from a small artery, all branches were traced until a small vein was reached. Arteries and veins were identified by the difference in the characteristics of their arborization as observed at low magnification (Duvernoy et al., 1981).

Computer simulation

The 3-dimensional topological connections of the microvascular network were reconstructed from the vascular coordinates by computer. A first approximation of blood flow distribution in the network was calculated assuming constant blood viscosity. The hydraulic resistance of each capillary was estimated from its mean diameter and length based on

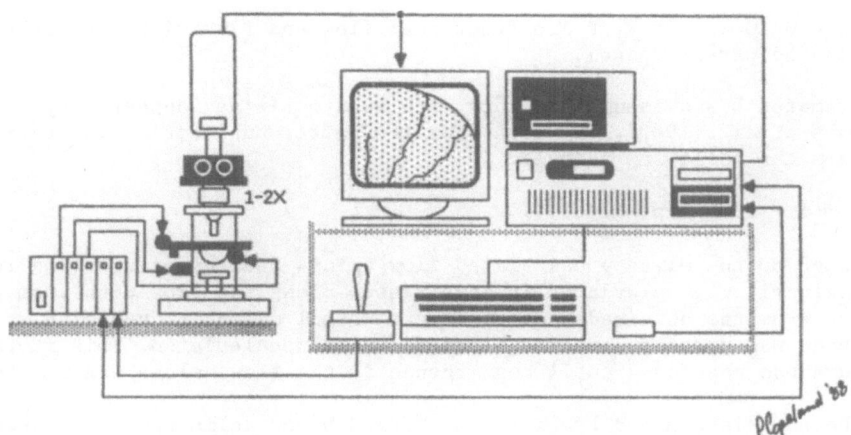

Fig. 2. Schematic diagram of the video microscope system for three-dimensional measurement and reconstruction of microvascular networks. The video image of the specimen under the microscope and the computer graphics are mixed and displayed on a high resolution color monitor. The course of microvessels is digitized by tracing them on the screen. The X, Y, Z coordinates are fed into the computer which also controls the microscope stage through a stage controller interface.

Poiseuille flow. Equations for flow were written for each capillary segment and at each branch point the flow balance was written as a second equation. The resulting system of equations was solved by standard computer methods using matrix inversion using a Microvax II computer.

Next, the effect of local variations in apparent blood viscosity were considered. A mathematical model was developed to describe the dependence of apparent relative viscosity on vessel diameter and vessel discharge hematocrit. The model was to approximate the experimental data of Jay et al. (1972) and Reinke et al. (1986) as close as possible for diameters between 3 and 100 um, and also to behave in accordance with the Fahraeus effect. By combining models presented by Haynes (1960) and Whitmore (1967) a satisfactory solution in the following form was obtained (Figure 3):

$$u_{app} = [1 - (1 - u_p/u_c)(1 - 2w/D)^4]^{-1} [1 - (D_m/D)^4]^{-1}$$

u_p = viscosity of plasma (= 1.7cp)
u_c = core viscosity in a large tube
w = width of marginal layer
D = vessel diameter
D_m = minimum vessel diameter red cell can traverse (= 2.7um)
Ht = discharge hematocrit

The core viscosity was obtained from the following fit to the viscosity data of Reinke et al. (1986):

$$u_c = \exp(0.48 + 2.35\ Ht)$$

The dependence of marginal layer on hematocrit was taken as:

$$w = 2.03 - 2\ Ht$$

In addition, the partitioning of red cells at bifurcations was described by the following formula (Klitzman and Johnson, 1982):

$$[(1 - f)/f] = [(1-q)/q]^B$$

where $q = Q_1/Q$ and $f = F_1/F$ are fractional flow and flux with respect to those in the parent vessel.

The parameter B was assumed to vary with hematocrit as suggested by Dellimore et al. (1983). The data from the latter publication was refitted to obtain the following formula:

$$B = 1 + 2.24\ \exp(-Ht/0.15)$$

Based on the already calculated flow values and the bifurcation rule, the hematocrit was calculated in each vessel along the flow path from artery to vein. From the obtained hematocrit and vessel diameter, vascular resistance was updated, and the blood flow was recalculated. This iterative procedure was continued until convergence in the flow values was reached.

The path length and transit time of red blood cells was calculated by a probabilistic simulation. Individual transits from arteriole to venule were simulated by the Monte Carlo method. At each bifurcation the probability of selecting one of the channels by the cell was determined in proportion to the previously calculated cell flux distribution. The total path length and transit time were given as the sum of selected segment lengths/transit times. In order to estimate capillary path length, segment lengths were also summed while weighing them inversely by the segment blood flow. For each parameter combination, 10000 transits were simulated.

RESULTS

Red cell flux and vessel hematocrit

Figure 4 displays the calculated red cell flux distribution in the reconstructed network at feed hematocrit of 40 percent. The obtained cell fluxes vary over at least 3 orders of magnitude indicating a significant heterogeneity in red cell perfusion. Vessels with zero cell flux correspond to those with very low blood flow.

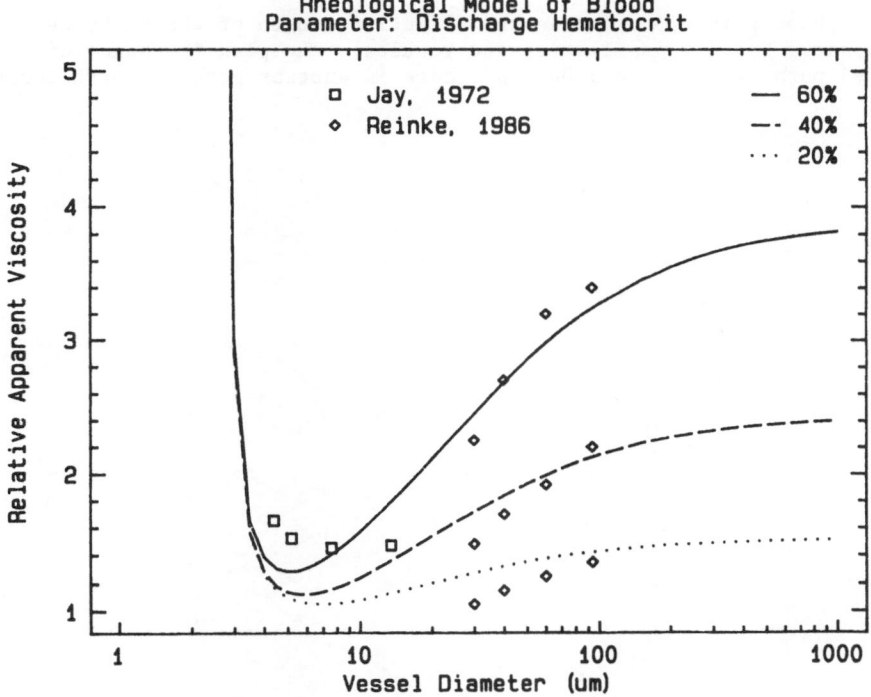

Fig. 3. Variation of apparent relative viscosity with vessel diameter and discharge hematocrit in the model developed. The model was fitted to the experimental data at a discharge hematocrit of 40% and adjusted for the dependence of marginal layer width and core viscosity on hematocrit. The fit for hematocrits of 40% and 60% was closer than that of 20%. The reversal of the Fahraeus effect in vessels smaller than 8 um and the leveling-off of the apparent relative viscosity at large diameters are clearly shown.

To analyze heterogeneity further the frequency distribution of local discharge hematocrit was evaluated. Figure 5 indicates that this distribution is bimodal at various feed hematocrit values. One mode represents the low hematocrit "plasma" channels. In the other mode, discharge hematocrit is distributed around the arterial feed hematocrit. Thus, in some vessels discharge hematocrit exceeds the feed hematocrit value.

Path length and transit time

Path length and transit time in the network were simulated at different arterial feed hematocrit values (5-40%). The total path lengths through the network had a peak frequency at about 1.7mm at all feed hematocrit values (Figure 6). The height of the peak decreased with feed hematocrit, implying that cell perfusion is more homogeneous at higher hematocrit. There was another group of paths at 3.6mm, which may represent some long collateral routes.

Weighted path approximates the effective length of the exchange microvessels (mainly capillaries and venules). The peak frequency of weighted path is close to 300um and there is another group of path lengths

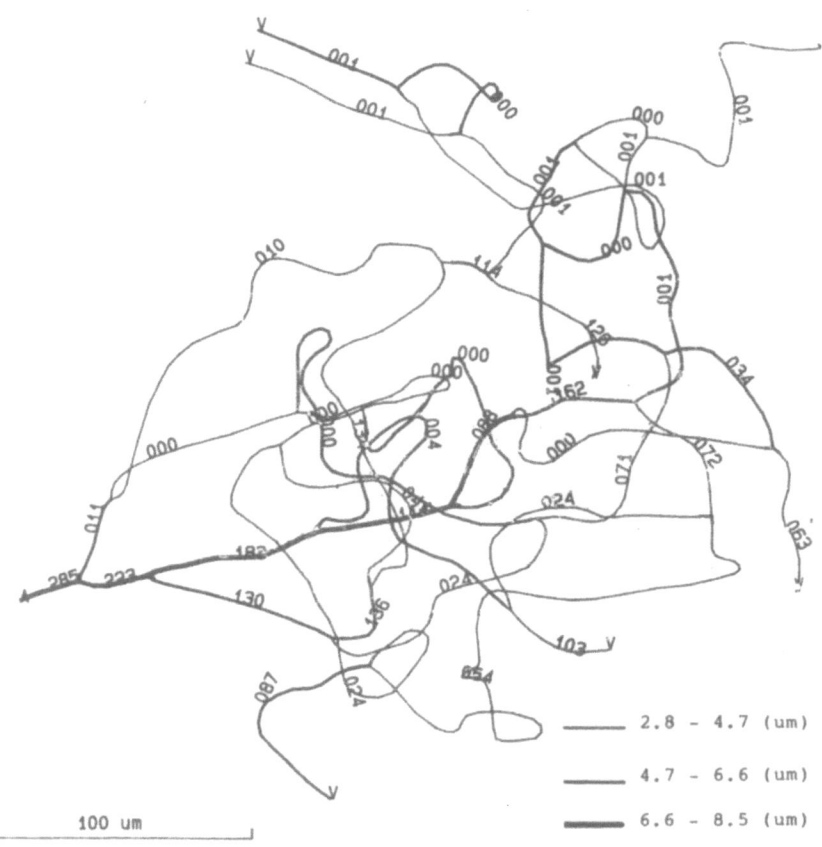

Fig. 4. Distribution of calculated red cell flux in the reconstructed cerebrocortical microvascular network. Cell flux is shown in um/s units. Different line thicknesses indicate vessel diameters in three ranges.

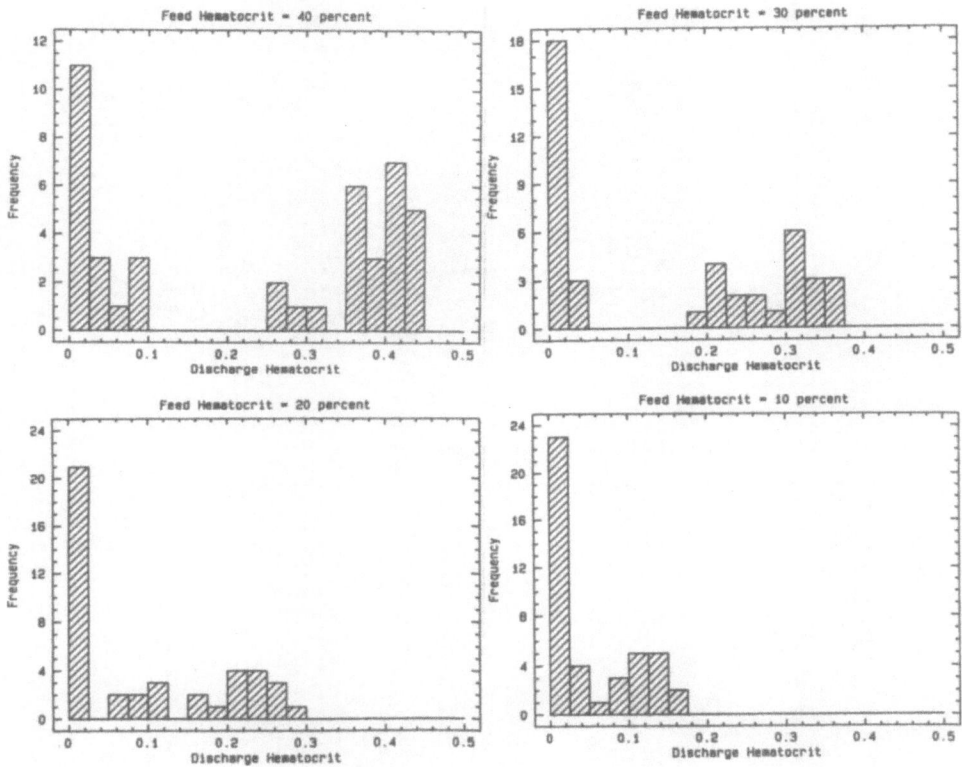

Fig. 5. Frequency histogram of microvessel discharge hematocrit calculated
at different feed hematocrits. The distribution is bimodal in all
cases, representing the two groups of normally perfused and plasma-
perfused vessels. In the former group, the hematocrit values are
distributed around the feed hematocrit value.

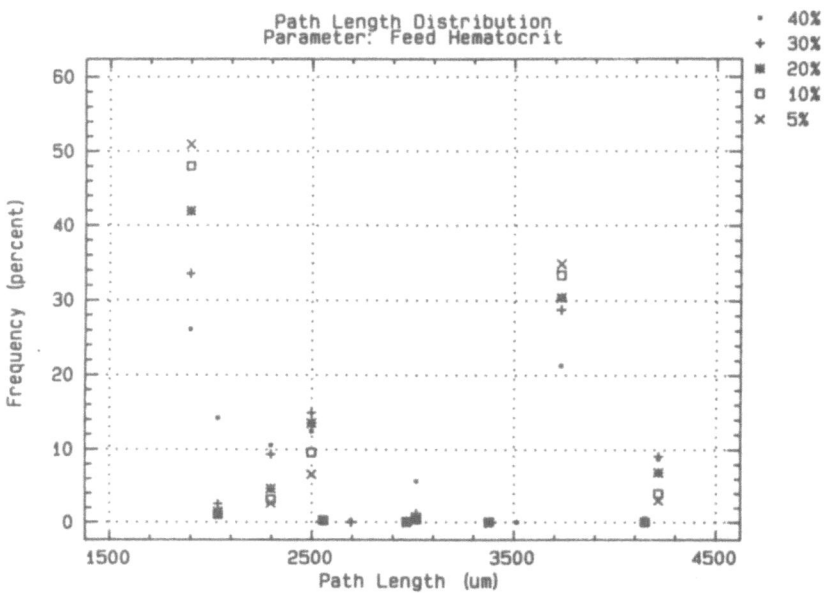

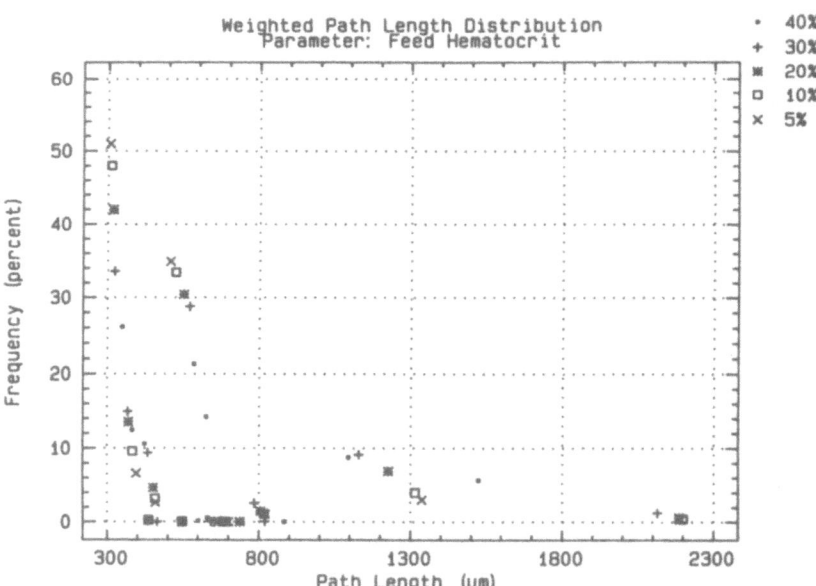

Fig. 6. Frequency distribution of microvascular path lengths simulated by computer based on the calculated cell flux. Total arterio-venous path lengths are shown above. Weighed path lengths (below) represent mainly capillary path lengths and, therefore, they are shorter than the total paths. Although the distributions vary with feed hematocrit, paths with the highest frequency seem to be independent of the feed hematocrit.

around 5-600 um (<u>Figure 6</u>). These values are much lower than the non-weighted ones, but they are still higher than anatomical capillary lengths (Motti et al, 1986).

The distribution of transit times is very similar to the distribution of weighted path lengths, except that the units are different (<u>Figure 7</u>). The peak transit time is at 4-5s, which agrees with experimental findings (Tomita et al, 1983).

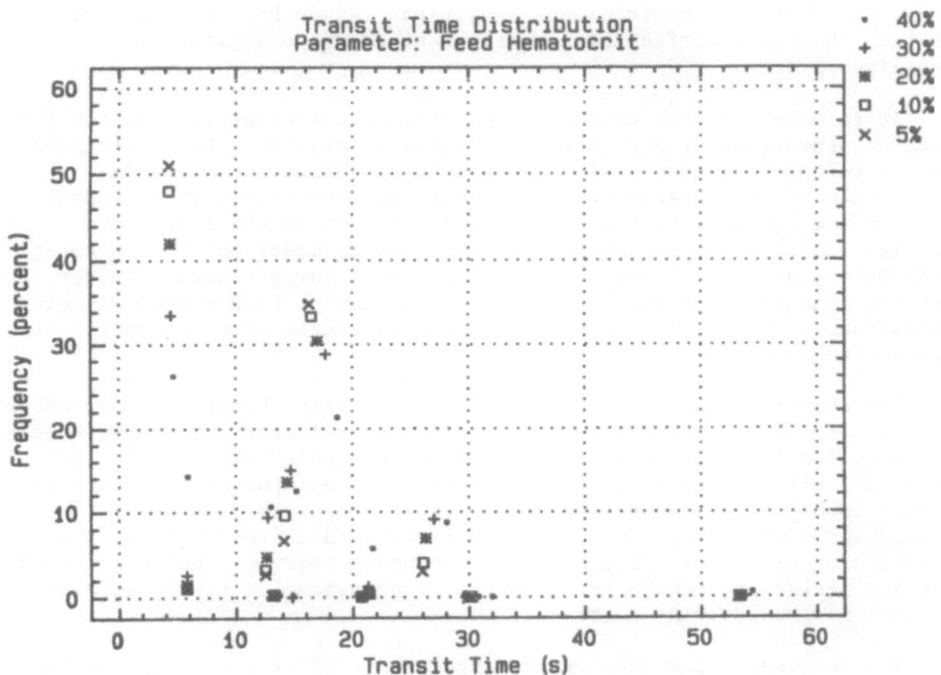

Fig. 7. Frequency distribution of transit times in the microvascular network. Although the distribution seems to be wider than expected from a gamma distribution, the highest frequency is found at 4s, which is in agreement with experimental values.

DISCUSSION

The purpose of the present work was to illustrate the application of computer simulation methods to estimate path length and cell transit time distributions in the cerebrocortical microvascular network. It is essential, that such a simulation is based on the true geometrical and topological properties of the vascular network.

A fundamental characteristic of the cerebrocortical capillaries is their irregular pattern (Wiederhold et al., 1976; Duvernoy et al., 1981; Pawlik et al., 1981). The organization of the microvascular bed is significantly different in various layers of the cerebral cortex (Duvernoy et al., 1981). Because the cerebral vasculature is essentially 3-dimensional, the analysis of the branching pattern of deep cortical vasculature requires special techniques.

In this work, a novel method was applied to digitize, store and reconstruct the 3-dimensional geometry of microvascular networks, including the complete spatial course, diameter and branching pattern of the vessels contained by a vascular cast. The geometry of the vasculature is well preserved by the vascular cast, thus, most of the topological connections of the network within the sample could be mapped. This approach eliminates the need for reconstruction from serial sections (Wiederhold et al, 1976) which has been the major difficulty in earlier attempts to digitize large tree-like objects.

It is also critical that the microrheological properties of the flowing blood be appropriately accounted for in the simulation. The rheological model developed and used in the present study offers a reasonable description of the dependence of apparent relative viscosity on vessel diameter and hematocrit. Compared to other models in the literature (Cokelet, 1976; Papenfuss and Gross, 1981), this model has the advantage that the viscosity does not tend to infinity at large vessel diameter, and that the transition of viscosity values from low to high vessel diameter is realistically smooth. The model is in good agreement with several experimental data.

The computer simulation predicts significantly heterogeneous red cell flux in the microvessel network at normal arterial feed hematocrit. Vessels with very low flow are converted to "plasma channels" as a result of nonlinear cell partitioning at their entrance. An important question is whether such low flow vessels exist in vivo. Nonperfused or near-nonperfused capillaries have indeed been observed in the brain cortex (Yamakawa et al, 1986). It can not be excluded, however, that some of the very low flows were due to the presence of occasionally truncated side branches in the vascular cast.

The frequency distribution of calculated transit times was similar to those obtained by indirect measurements (Tomita et al, 1983). Therefore, it is likely that the computer predictions of path length are also correct. Weighted and total path lengths are estimations of functional capillary and arterio-venous lengths traversed by the red cells. Since the measurement of full capillary paths has not been possible in vivo (Pawlik et al, 1981) a comparison of simulated to experimental data cannot be made at this time.

SUMMARY

Microvascular network hemodynamics was simulated by computer in an anatomically reconstructed cerebral microvascular network. A video microscope system was used for three-dimensional mapping of the vessel network in the rat brain cortex. The complete topology, length and mean diameter of the microvessels were determined. The distribution of blood flow and red cell flux in the network was calculated based on vessel resistance estimated from geometrical data and a rheological model of blood. This model described apparent relative blood viscosity as a function of vessel diameter and local discharge hematocrit. The calculations predicted highly heterogeneous cell flux distribution at any feed hematocrit between 10 and 40 percent. The frequency distribution of microvessel hematocrit was bimodal and included values exceeding the feed hematocrit value. A probabilistic simulation of cell transit resulted in transit time distributions which agree with experimental findings. The most probable transit time and capillary path length and 4s and 300um, respectively.

ACKNOWLEDGEMENT

The authors express their appreciation to Dr. Charles R. Horton for his advice and invaluable support to the progress of this work. They express gratitude to Mr. Paul Copland for his photographic work.

This work was supported by the State of Louisiana, Division of Rehabilitation Services.

REFERENCES

Cokelet, G., 1976, Blood rheology interpreted through the flow properties of the red cell. Microcirculation, 1:9.

Duvernoy, H. M., Delon, S., and Vannson, J. L., 1981, Cortical blood vessels of the human brain, Brain Res. Bull., 7:519.

Dellimore, J.W.; Dunlop, M.J.; and Canham P.B., 1983, Ratio of Cells and Plasma in Blood Flowing Past Branches in Small Plastic Channels. Am. J. Physiol., 244:H635.

Haynes, R.H., 1960, Physical basis of the dependence of blood viscosity on tube radius. Am. J. Physiol., 198:1193.

Jay, A.; Rowlands, S. and Skibo, L., 1972, The resistance to blood flow in the capillaries. Canad. J. Phys. Phar., 50:1007.

Klitzman, B., and Johnson, P.C., 1982, Capillary network geometry and red cell distribution in hamster creamaster muscle. Am. J. Physiol., 242:H211.

Motti, E. D. F., Imhof, H.-G., and Yasargil, M. G., 1986, The terminal vascular bed in the superficial cortex of the rat, J. Neurosurg., 65:834.

Papenfuss H.-D., and Gross, J. F., 1981, Microhemodynamics of capillary networks, Biorheology, 18:673.

Pawlik, G., Rakl, A., and Bing, R.J., 1981, Quantitative capillary topography and blood flow in the cerebral cortex of cats: an in vivo microscopic study, Brain Res., 208:35.

Reinke, W.; Johnson, P.C. and Gaehtgens, P., 1986, Effect of shear rate variation on apparent viscosity of human blood in tubes of 29 to 94 um diameter. Circulation Res., 59:124.

Tomita, M., Gotoh, F., Amano, T., Tanahashi, N., Kobari, M., Shinohara, T., and Mihara, B., 1983, Transfer function through regional cerebral cortex evaluated by photoelectric method. Am. J. Physiol., 245:H385.

Whitmore, R.L., 1967, A theory of blood flow in small vessels. J. Appl. Physiol., 22:767.

Wiederhold, K.-H., Bielser, W., Schultz, U., Jr., Veteau, M.-J., and
    Hunziker, O., 1976, Three dimensional reconstruction of brain
    capillaries from frozen serial sections, <u>Microvasc. Res.</u>, 11:175.

Yamakawa, T., Niimi, H., Sugiayama, I., and Yamaguchi, S., 1986, Red blood
    cell flow distribution and capillary hematocrit in the cerebral cortex
    microcirculation of cat: intravital microscopic study, <u>Proc. IUPS.</u>,
    16:222.

Yoshida, Y. and Fusahiro I., 1984, Three-dimensional architecture of
    cerebral microvessels with a scanning electron microscope: a
    cerebrovascular casting method for fetal and adult rats. <u>J. Cereb.</u>
    <u>Blood Flow Metabol.</u> 4:290.

MUSCLE FIBER SIZE AND CHRONIC EXPOSURE TO HYPOXIA

Odile Mathieu-Costello, David C. Poole and Richard B. Logemann

Department of Medicine, M-023A, University of California, San Diego, La Jolla, CA 92093-0623

Fiber size is an important variable to consider when estimating several aspects of the geometry of peripheral tissue gas exchange. It directly affects diffusion and intercapillary distances, capillary density (often estimated as capillary counts per fiber cross-sectional area, capillary/ mm^2 of fiber), and capillary-to-fiber perimeter ratio.

There is experimental evidence that muscle fiber volume remains constant during shortening (Dulhunty and Franzini-Armstrong, 1975). Consequently, within a given muscle, fiber cross-sectional area, $\bar{a}(f)$, is a direct function of sarcomere length, l_o. The product $\bar{a}(f) * l_o$ is constant, and $\bar{a}(f)$ is an hyperbolic function of l_o. Proportional changes in $\bar{a}(f)$ with fiber shortening have been found in muscles frozen without fixation of length (Gray and Renkin, 1978), as well as in muscles perfusion-fixed *in situ* at different sarcomere lengths from extension to tetanic contraction (l_o range, 1.62-2.85 μm; Mathieu-Costello, 1987).

Considering the effect of muscle shortening on fiber cross-sectional area, it is surprising that sarcomere length is practically never reported when estimates of fiber size and/or related variables (e.g. capillary/ mm^2 of fiber in transverse sections) are compared between muscles and/or experimental conditions. In perfusion-fixed muscles, the range in l_o can be very large (e.g. 1.74-2.85 μm; Mathieu-Costello, 1987) depending on joint position during fixation. Even in muscle biopsies, where fibers are generally shortened after removal from the intact muscle, substantial differences are found in l_o between samples (i.e. 1.6-2.3 μm; Conley et al, 1987). Over these ranges of sarcomere length, fiber cross-sectional area is expected to vary by as much as 30 to 64%.

To date, studies of the effect of chronic exposure to hypoxia on fiber size have been contradictory, possibly due to differences in sarcomere length between samples. The purpose of this study was to examine the effect of chronic exposure to hypoxia on fiber cross-sectional area at measured sarcomere length. Specifically, we addressed the question of whether or not there are differences in muscle fiber cross-sectional area with chronic exposure to an altitude of 3800 m, when the confounding effect of sarcomere length variability between samples is removed.

MATERIAL AND METHODS

Twenty four perfusion-fixed muscles from a total of seventeen animals were used in this study. Twelve 9 week-old female Sprague-Dawley rats (body weight, 180-200 g) were randomly separated into two groups. Six rats (high altitude, HA) were kept at the UC Station at White Mountain, Barcroft, 3800 m (inspired PO_2, 91 mmHg), for 5 months, and 6 rats (sea level controls, SL) were kept at UCSD. Three HA deer mice (Peromyscus maniculatus) were native to Barcroft (3800 m), and two SL deer mice were taken from a population kept at sea-level (UC Riverside) for several generations. There was no difference in body weight between HA and SL animals of either species. All animals were anesthetized by intravenous injection of pentobarbital (3 mg/ 100 g). Vascular perfusion-fixation of the muscles was performed as described previously (Mathieu-Costello, 1987), using a 6.25% solution of glutaraldehyde in 0.1 M sodium cacodylate buffer (total osmolarity of the fixative, 1100 mOsm; pH 7.4). All HA animals were perfusion-fixed at the UC Station at White Mountain (Barcroft). After the muscles were excised, samples (approx. 1 cm x 4 mm x 1 mm) taken from the midbelly portion of M. Soleus (rats) and deep portions of the calf and thigh muscles (mice) were processed for electron microscopy using standard procedures (Mathieu-Costello, 1987).

One micrometer thick sections were cut on a LKB Ultrotome III and stained with 0.1% aqueous toluidine blue solution. Eight blocks were cut from each muscle, yielding 4 transverse and 4 longitudinal sections. The angle of sectioning was carefully controlled following a procedure described elsewhere in detail (Mathieu-Costello, 1987). Briefly, a minimum of three sections were taken parallel to the fiber axis at consecutive angles, α, approximately 1° apart. Sarcomere length, l_o, was measured on each section at magnification 630x (10 measurements of 10 consecutive sarcomeres/ section). We considered a section to be longitudinal when changing α in either direction produced an increase in l_o. For the transverse sections, the blocks were cut perpendicular to the fiber axis, at consecutive angles approximately 5° apart. Sections were considered transverse when changing α in either direction produced a decrease in l_o.

Fiber cross-sectional area, $\bar{a}(f)$, was obtained by one of two methods: standard point counting, using 140 to 200 fibers per muscle, measured at magnification 2060x (Mathieu-Costello, 1987), or tracing the contours of a total of 70 to 180 fibers per muscle at magnification 1050x, with a Videometric 150 image analyzer (American Innovision Inc.). The estimates of $\bar{a}(f)$ obtained in the same tissue with both methods were not significantly different and showed standard errors typically below 10%.

The standard error of the estimates of l_o and $\bar{a}(f)$ represents the variability between sections and micrographs, respectively. Differences between groups were identified by means of t-test for unpaired samples ($p < 0.05$).

RESULTS AND DISCUSSION

Fiber cross-sectional area demonstrated a considerable variability in both mice (range, 372 - 890 μm^2; Table 1) and rat (1759 - 2457 μm^2; Table 2). In mice, there were no systematic differences in the value of $\bar{a}(f)$ in calf compared to thigh muscles in HA and SL groups (Table 1). Banchero and colleagues (rev., Banchero, 1987) were the first to recognize the importance of considering body weight when analyzing the effect of hypoxia on muscle fiber cross-sectional area. However, because of the narrow range of age and/or body weight (21-27 g in mice; 244 - 305 g in rats) examined in the present investigation, differences in fiber cross-sectional area could not be explained by differences in body size between animals (Fig. 1).

TABLE 1. Sarcomere length, l_o, mean cross-sectional area of fibers, $\bar{a}(f)$, and fiber cross-sectional area normalized to $l_o = 2.1$ μm, $\bar{a}(f)_{2.1}$, in mice.

Muscle #	site	l_o (μm)	$\bar{a}(f)$ (μm^2)		$\bar{a}(f)_{2.1}$ (μm^2)
High altitude					
1	calf	2.42 ± 0.04	372 ± 26		429 ± 30
2	calf	2.35 ± 0.18	489 ± 55		547 ± 62
3	calf	1.99 ± 0.04	755 ± 51		715 ± 48
4	calf	1.89 ± 0.03	565 ± 24		508 ± 22
				$\bar{x}$	550 ± 60
5	thigh	2.28 ± 0.05	454 ± 16		493 ± 17
6	thigh	2.03 ± 0.06	672 ± 34		650 ± 33
				$\bar{x}$	572 ± 79
				$\bar{x}$ (N=6)	557 ± 43
Sea-level					
7	calf	2.43 ± 0.14	539 ± 30		624 ± 35
8	calf	1.93 ± 0.08	575 ± 28		528 ± 26
9	calf	1.87 ± 0.04	890 ± 57		793 ± 51
				$\bar{x}$	648 ± 77
10	thigh	2.23 ± 0.04	666 ± 19		707 ± 20
11	thigh	2.14 ± 0.14	578 ± 27		589 ± 28
12	thigh	2.13 ± 0.07	779 ± 55		790 ± 56
				$\bar{x}$	695 ± 58
				$\bar{x}$ (N=6)	672 ± 45

TABLE 2. Sarcomere length, l_o, mean cross-sectional area of fibers, $\bar{a}(f)$, and fiber cross-sectional area normalized to $l_o = 2.1$ μm, $\bar{a}(f)_{2.1}$, in rat M. Soleus.

Muscle #	l_o (μm)	$\bar{a}(f)$ (μm^2)		$\bar{a}(f)_{2.1}$ (μm^2)
High altitude				
1	2.00 ± 0.04	2087 ± 75		1988 ± 71
2	1.95 ± 0.03	2278 ± 110		2115 ± 102
3	2.13 ± 0.03	1939 ± 76		1967 ± 77
4	2.41 ± 0.03	2001 ± 87		2296 ± 100
5	1.98 ± 0.02	2305 ± 83		2173 ± 78
6	2.04 ± 0.02	1805 ± 79		1753 ± 77
			$\bar{x}$	2049 ± 77
Sea-level				
7	1.93 ± 0.02	2213 ± 98		2034 ± 90
8	1.76 ± 0.05	2457 ± 117		2059 ± 98
9	1.99 ± 0.04	2096 ± 112		1986 ± 106
10	2.07 ± 0.03	1999 ± 119		1970 ± 117
11	2.30 ± 0.02	1759 ± 98		1927 ± 107
12	2.35 ± 0.11	1771 ± 114		1981 ± 127
			$\bar{x}$	1993 ± 19

Sarcomere length, l_o, ranged from 1.87 to 2.43 μm in mice and 1.76 to 2.41 μm in rats (Tables 1 and 2), suggesting that as much as 23 to 37% variability in fiber cross-sectional area between muscles could be related to differences in l_o, rather than inter-animal variations and/or experimental conditions. As expected, ā(f) was inversely related to l_o (Fig. 2). The correlation of the linear transform, ā(f) vs $1/l_o$, was r= 0.65 (p<0.05) in mice, and r=0.80 (p<0.01) in rats.

There was no difference in fiber cross-sectional area normalized to sarcomere length, ā(f)$_{2.1}$, between high altitude and control animals in either species (Fig. 2; Tables 1 and 2), confirming previous findings that hypoxia alone, especially at moderate altitude, may not be severe enough to induce changes in muscle fiber cross-sectional area (rev. Banchero, 1987). In contrast, there is evidence of muscle wasting at a much higher altitude (above 5400 m) in man (Pugh, 1964; Boyer and Blume, 1984). Whether these contradictory findings are related to species, degree of hypoxia, concomitant exposure to cold, animal activity and/or muscle aerobic capacity remains to be determined. The HA mice in this investigation were free-ranging and therefore

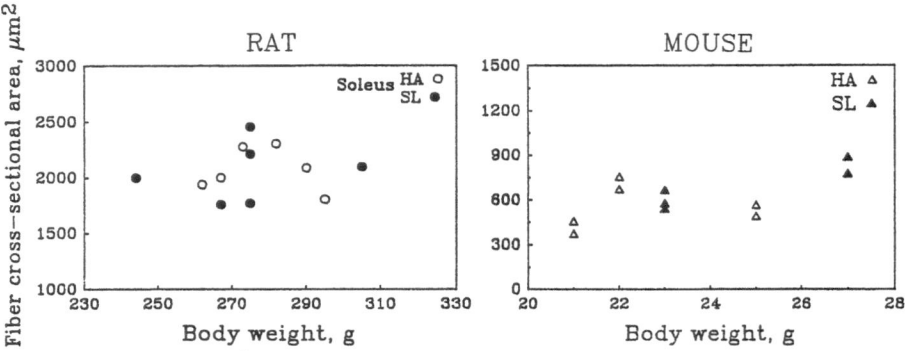

Fig. 1. Relationship between fiber cross-sectional area and body weight.
Left Panel: rat; r=0.10; ns
Right Panel: mouse; r=0.61; ns
Note narrow range in body weight for both species.

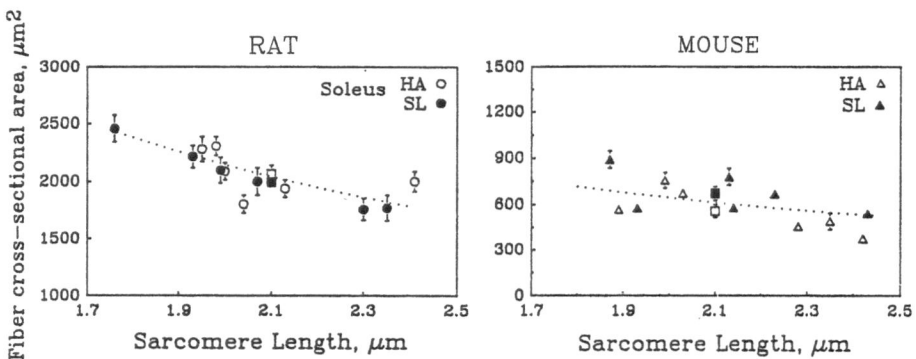

Fig. 2. Relationship between fiber cross-sectional area and sarcomere length.
Left Panel: rat; r=0.80; p<0.01
Right Panel: mouse; r=0.65; p<0.05
Broken line: predicted change in fiber cross-sectional area with sarcomere length, assuming constant fiber volume.

subjected to ambient temperatures in the region of White Mountain. These mice were therefore likely to be cold- as well as altitude-acclimatized, and yet no change in fiber size was evident. Activity levels were not monitored in any of the groups: all groups except for the HA mice were caged. It is possible that differences between man and animal studies relate to the metabolic potential and capillary bed of the muscles.

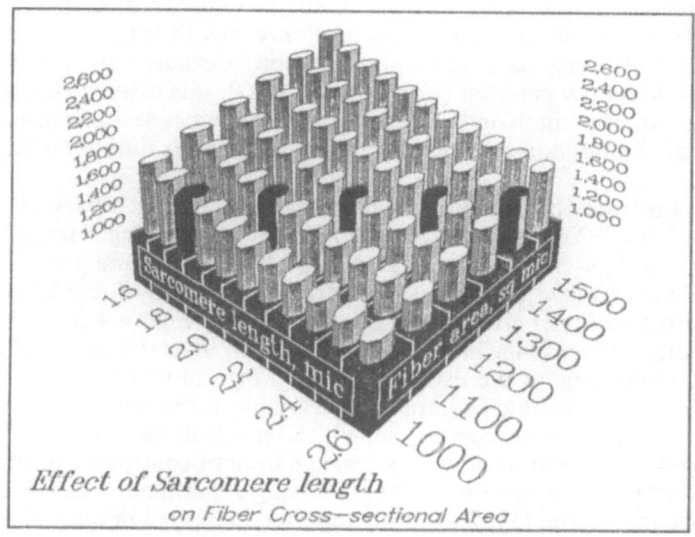

Fig. 3. Three-dimensional schematic of the effect of changing sarcomere length on fiber cross-sectional area. A given fiber cross-sectional area, e.g. 1400 μm^2 (dark fibers), can be produced by fibers of very different size (1000 to 1400 μm^2 at l_o=2.6 μm), depending on sarcomere length.

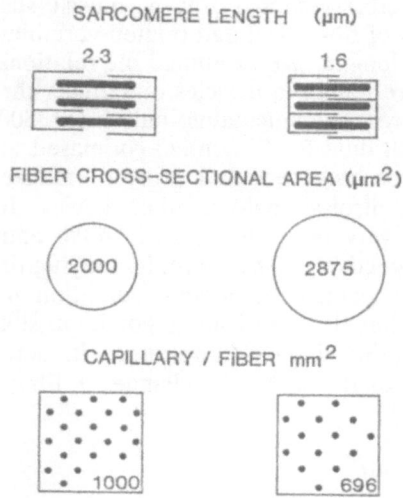

Fig. 4. Schematic illustrating the effect of changed sarcomere length on fiber cross-sectional area and capillary/ mm^2 of fiber.

However, considering the differences in metabolism between the two species (mice and rats), and the 2-3 fold differences in the muscle capillary supply (Mathieu-Costello, 1988), it is pertinent that no systematic difference in muscle fiber size was found after chronic exposure to hypoxia in mice or rats. In the experimental conditions under study, no reduction in diffusion distance from capillary to fiber mitochondria was achieved via a decrease of fiber size in either species.

Although the importance of normalizing fiber cross-sectional area for sarcomere length has been recognized previously (Gray and Renkin, 1978; Lieber et al, 1986), to our knowledge, this is the first study where differences in sarcomere length between experimental groups have been taken into account. Because sarcomere length varies by at least 30% in biopsy material (Conley et al, 1987), it is evident that demonstrating relatively small experimentally-induced changes in fiber cross-sectional area (i.e. < 30%) in such material without normalizing for sarcomere length is difficult if not impossible.

Figure 3 illustrates the change in cross-sectional area of fibers of different size (1000-1500 μm^2 at l_o =2.6 μm), with alteration in sarcomere length from 1.6 to 2.6 μm. A given measured fiber cross-sectional area, e.g. 1400 μm^2, can be produced by a range of fibers (dark fibers, Fig. 3). Unless sarcomere length is known, the unique identity of a given fiber cross-sectional area cannot be determined. Figure 4 is a schematic which illustrates the effect of a change in sarcomere length on fiber size and capillary number/fiber cross-sectional area, at the extremes of the range of sarcomere lengths found in biopsy material. Besides reducing intra-group variability, the normalization of fiber size and related variables to sarcomere length is particularly important when some experimentally-induced perturbation (e.g. change in fiber contractile properties) results in systematic differences in sarcomere length among experimental and control muscles. Indeed, Goubel and Marini (1987) recently reported increased muscle stiffness in rat M. soleus after endurance training.

SUMMARY

Sarcomere length is practically never considered when fiber size and dependent variables are compared between muscles or experimental conditions. Because of the direct dependence of fiber cross-sectional area on muscle shortening, it is imperative to normalize measurements of fiber size, and related variables (e.g. capillary number/mm^2 of fiber) to sarcomere length. We examined the relationship between fiber cross-sectional area and sarcomere length in muscles of animals chronically exposed to high altitude (deer mice, Peromyscus maniculatus, native to 3800 m, and rats, Sprague-Dawley, kept at the same altitude for 5 months) compared to sea-level controls. We found no difference in fiber cross-sectional area, normalized to sarcomere length, between high altitude and control animals in either species. It has been demonstrated that sarcomere length can vary by as much as 30-44% and 43-76% in biopsy and perfusion-fixed muscles, respectively. Therefore, identifying relatively small changes in fiber size in response to a given experimental condition in such material without normalizing for sarcomere length is difficult if not impossible. Furthermore, if the conditions of the investigation induce differences in sarcomere length between experimental and control animals, artifactual changes in fiber cross-sectional area will be produced.

ACKNOWLEDGEMENTS

We are grateful to Dr. Mark Chappell (Dept. of Biology, University of California, Riverside) for generously donating the mice used in this study, and Ms Carlita Durand for technical assistance. This work was supported by grants from the National Institutes of Health (5P01 HL17731 and HL40015) and the American Lung Association of California.

REFERENCES

Banchero, N., 1987, Cardiovascular responses to chronic hypoxia. Ann. Rev. Physiol., 49:465-476.

Boyer, S.J., and Blume, F.D., 1984, Weight loss and changes in body composition at high altitude. J. Appl. Physiol., 57:1580-1585.

Conley, K.E., Kayar, S.R., Rösler, K., Hoppeler, H., Weibel, E.R., and Taylor, C.R., 1987, Adaptive variation in the mammalian respiratory system in relation to energetic demand: IV. Capillaries and their relationship to oxidative capacity. Resp. Physiol., 69:47-64.

Dulhunty, A.F., and Franzini-Armstrong, C., 1975, The relative contributions of the folds and caveolae to the surface membrane of frog skeletal muscle fibers at different sarcomere lengths. J. Physiol. (London), 250:513-539.

Goubel, F., and Marini, J.F., 1987, Fibre type transition and stiffness modification of soleus muscle of trained rats. Pflügers Arch.,410:321-325.

Gray, S.D, and Renkin, E.M., 1978, Microvascular supply in relation to fiber metabolic types in mixed skeletal muscles of rabbits. Microvasc. Res., 16:406-425.

Lieber, R.L., Boakes, J.L., Kitabayashi, L.R., Hargens, A.R., Roy, R.R., and Edgerton, V.R., 1986. Normalization of fiber area to sarcomere length by optical diffraction of freeze-substituted whole skeletal muscle. Biophys. J., 49:473a.

Mathieu-Costello, O., 1987, Capillary tortuosity and degree of contraction or extension of skeletal muscles. Microvasc. Res., 33:98-117.

Mathieu-Costello, O., 1988, Capillary configuration in contracted muscles: comparative aspects. In: "Oxygen Transfer from Atmosphere to Tissues", edited by N.C. Gonzalez and M.R. Fedde, pp. 229-236. Plenum Press, New York.

Pugh, L.G.C.E., 1964, Animals in high altitudes: man above 5,000 meters-mountain exploration. In: "Handbook of Physiology, Section 4: Adaptation to the environment", edited by D.B. Dill, E.F. Adolph and C.G. Wilber, pp. 861-868, American Physiological Society, Washington D.C., Williams & Wilkins, Baltimore.

CAPILLARY LENGTHS AND ANASTOMOSES IN RAT HINDLIMB MUSCLES, STUDIED BY AQUABLAK PERFUSION DURING REST VERSUS EXERCISE

R. F. Potter, S. Houghton and A.C. Groom

Department of Medical Biophysics, University of Western Ontario and The J.P. Robarts Research Institute London, Ontario, Canada. N6A 5C1

An understanding of oxygen transport to tissue demands a knowledge of the three-dimensional arrangement of capillaries. In skeletal muscle the distribution of capillaries within the plane normal to the fiber axes has been studied extensively (see Plyley and Groom (1975), for review), whereas the network arrangement of capillaries in a plane parallel to the fiber axis has received relatively little attention. Plyley et al. (1976) and Honig et al. (1977) presented frequency distributions of the dimensions and interconnections of capillaries in frog sartorius and rat gracilis, respectively. They measured capillary pathlengths from terminal arteriole to collecting venule, segment lengths (i.e. distances between successive branch points) and the number of segments per path. Plyley's measurements were made from muscle vasodilated with papaverine and perfused with a silicone elastomer (Microfil), whereas those of Honig were carried out in vivo under resting conditions and also following 2 min electrical stimulation. From a study of rat spinotrapezius muscles injected with India ink, Skalak and Schmid-Schonbein (1986) reported the presence of discrete microvascular units ('capillary bundles'), and similar results have been obtained by Lund et al. (1987) from tibialis anterior of the hamster, using intravascular injection of fluorescein-labelled albumin to outline the capillary network.

Using methods similar to those described above, it should be possible to make comparisons of capillary network geometry in different striated muscles, in a search for the general concepts underlying microvascular architecture in this particular tissue. Furthermore, from a comparison of measurements made in resting versus hyperemic muscles, it should be possible to determine to what extent unperfused capillaries may be present at any particular instant in resting muscles. This report addresses these two issues, based on studies following Aquablak perfusion of three different muscles of rat hindlimb: (1) medial head of the gastrocnemius, a muscle of mixed fiber type, (2) gracilis, also a muscle of mixed fiber type, and (3) soleus, a muscle composed predominantly of slow oxidative fibers. Our observations were made both on resting muscle and on muscle rendered hyperemic by electrical stimulation. The results showed that the capillary network in the soleus exhibited a much greater degree of branching than that in gastrocnemius or gracilis, although neither capillary pathlengths nor arteriolar-to-venular distances differed dramatically in the three muscles. The results are also interpreted to mean that when the blood flow per g muscle is initially low, as in a basal resting state, arterioles as well as capillaries may be recruited during exercise.

MATERIALS AND METHODS

Eight male Wistar rats weighing 50 to 100g were anesthetized with sodium pentobarbital (6.5mg/100g, i.p.). Following induction of anesthesia, all animals were heparinized by injection of heparin sodium (1000 I.U./kg) subcutaneously. Four animals were used to study the gastrocnemius and soleus muscles, while the remaining four were used to study the gracilis. In each group, two animals were used for measurements on resting muscle, and two others for measurements during hyperemia. In all experiments a gracilis muscle, or the medial head of a gastrocnemius muscle was exposed by reflection of the overlying skin, and covered with a thin film of transparent plastic material (Saran) impermeable to oxygen and water vapor. The limb was then positioned so that the muscle was at full extension, and a period of one hour was allowed for resting conditions to become fully established.

The method employed was the percutaneous intracardiac injection of a small bolus of 0.1-0.2 ml Aquablak (Borden Chemicals: a colloid carbon suspension) which became distributed throughout the circulation under normal perfusion pressure. Views of the microcirculation at the muscle surface, obtained using epi-illumination (Schott fiber optic light guide) were videorecorded continuously (Panasonic NV 9240 XD recorder) prior to and during the injection of Aquablak, using a microscope (Nikon Optiphot) equipped with a video camera (Panasonic Model WV 1550). Immediately prior to the injection a hyperemic response was produced in four animals; for gastrocnemius and soleus the sciatic nerve was stimulated electrically (2-3v, pulses of 5ms duration, 1 Hz) for one minute, using platinum electrodes, whereas for gracilis direct stimulation of the muscle itself was used. Filling of the capillary network with Aquablak was seen to be complete (i.e. no further capillaries were becoming filled) at approximately 30s after injection of the material. Between 1 and 2 min after the injection, the animal was killed rapidly by intracardiac injection of an overdose of sodium pentobarbital. The hindlimbs were then pinned to rigid wooden supports (to ensure that the muscles remained in the extended position), removed from the body and placed immediately in 10% neutral buffered formalin at 4°C for 48 hours. Following this initial fixation the muscle of interest was dissected free, placed between glass slides, and stored in formalin for subsequent analysis.

Composite photomicrographs of the microvasculature at the surface of each muscle were prepared (final magnification 600x or 1100x),and from these the lengths of the individual capillary segments were measured by means of a map wheel. Each path that could be taken by a red cell travelling from arteriole to venule was characterized in terms of the particular segments involved (Fig 1), the sum of these segment lengths yielding the total pathlength. In this way, measurements of pathlengths, numbers of segments per path, and segment lengths were obtained from each muscle preparation. In addition, the straight line distances between each successive arteriole and venule were measured at a magnification of 100x from the muscle itself. Statistical evaluations of differences between mean values were carried out using Student's t-test; since the distributions of segment lengths were exponential, these data were transformed to a logarithmic scale before application of the t-test.

RESULTS

Mean values for all measured parameters in resting gastrocnemius, gracilis, and soleus muscles are presented in Table 1. No significant differences were found, between muscles, in terms of mean capillary pathlength or mean arteriolar-to-venular distance. However, significant differences ($p < 0.05$) were found between gastrocnemius and gracilis, with regard to both mean segment length and the mean number of segments per path. In gastrocnemius the mean segment length was 19% greater than in gracilis, while the mean number of segments per path was 27% smaller.

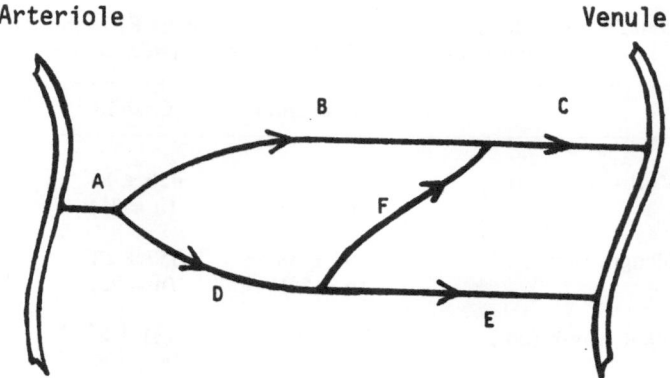

Fig. 1. Schematic illustrating measurement of capillary pathlengths. Each path that could be taken by a red cell travelling from arteriole to venule was characterized in terms of the capillary segments involved. The sum of the lengths of these segments yielded the capillary pathlength. Thus, Pathlength #1 = A + B + C; #2 = A + D + F + C; #3 = A + D + E.

During the hyperemia which followed electric stimulation of the muscle, changes in magnitude of several of the parameters occurred (Table 2). Significant reductions in mean path length and in mean arteriolar-to-venular distance were found in gracilis and soleus, whereas in gastrocnemius the values remained unchanged. In all three muscles the mean segment lengths were reduced significantly (though by different amounts), and in gastrocnemius and soleus significant increases were found in the numbers of segments per path.

Table 3 shows the percentage changes found in mean values for all parameters in the hyperemic as compared to the resting state. In gastrocnemius there was a 22% reduction in mean segment length in conjunction with a 27% increase in mean number of segments per path, while mean pathlength stayed constant. In gracilis the

Table 1. Capillary network parameters (mean ± SE) in resting muscle.

	Gastrocnemius	Gracilis	Soleus
Mean A-V distance (μm)	575 ± 37 (n = 19)	554 ± 39 (n = 19)	499 ± 26 (n = 15)
Mean Pathlength (μm)	616 ± 29 (n = 54)	576 ± 27 (n = 86)	572 ± 28 (n = 56)
Mean Segment length (μm)	225 ± 16[†] (n = 109)	182 ± 12[†] (n = 160)	205 ± 15 (n = 109)
No. Segments per path	2.9 ± 0.2[*] (n = 54)	3.8 ± 0.2[*] (n = 86)	3.3 ± 0.2 (n = 56)

[†], [*] significantly different at p<0.05.

Table 2. Capillary network parameters (mean ± SE) in muscle, during hyperemia following electrical stimulation.

	Gastrocnemius	Gracilis	Soleus
Mean A-V distance (μm)	510 ± 44 (n = 15)	420 ± 13[*] (n = 35)	399 ± 19[*] (n = 24)
Mean Pathlength (μm)	602 ± 38 (n = 58)	357 ± 13[*] (n = 95)	490 ± 17[*] (n = 71)
Mean Segment length (μm)	176 ± 11[†] (n = 147)	141 ± 8[†] (n = 196)	73 ± 4[*] (n = 367)
No. Segments per path	3.7 ± 0.2[†] (n = 58)	3.4 ± 0.3 (n = 95)	6.8 ± 0.3[*] (n = 71)

[†] significantly different at p<0.05 from resting muscle.
[*] significantly different at p<0.01 from resting muscle.

Table 3. Changes (%) in mean capillary network parameters during hyperemia following electrical stimulation, with respect to the values in resting muscle.

	Gastrocnemius	Gracilis	Soleus
Mean A-V distance (μm)	N/C	-24%	-20%
Mean Pathlength (μm)	N/C	-38%	-14%
Mean Segment length (μm)	-22%	-23%	-64%
No. Segments per path	+27%	N/C	+106%

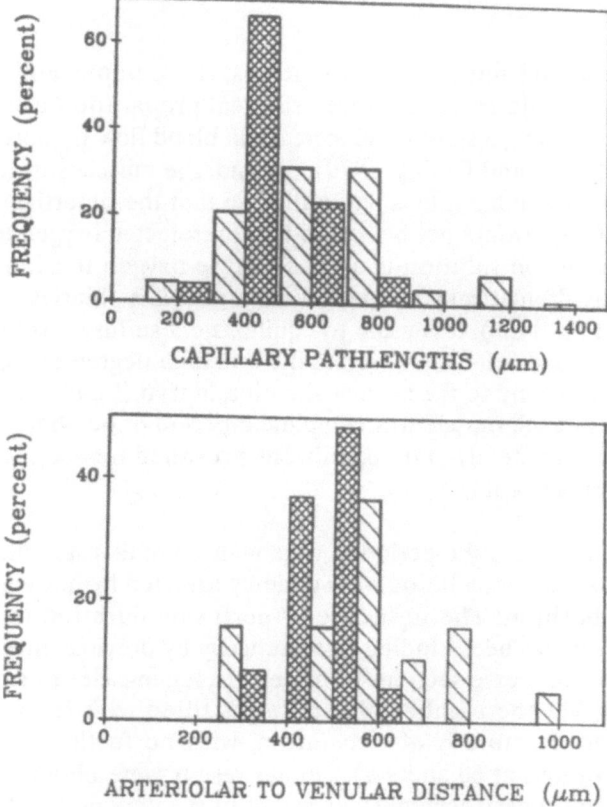

Fig. 2. Frequency distributions of capillary pathlengths *(upper)* and arteriolar-to-venular distances *(lower)* from rat gracilis muscle. Lightly shaded bars denote resting muscle; darkly shaded bars denote hyperemic muscle. Light and dark bars have been staggered for ease of viewing. Note loss of higher values of pathlengths and A-V distances in hyperemia.

mean number of segments per path remained unchanged, but substantial reductions in mean pathlength (38%), mean arteriolar-to-venular distance (24%) and mean segment length (23%) were evident. In soleus the mean number of segments per path actually doubled and mean segment length fell by 64%, while modest reductions in the mean arteriolar-to-venular distance (20%) and mean pathlength (14%) occurred. Frequency distributions of mean pathlength and mean arteriolar-to-venular distance for gracilis muscle, under resting versus hyperemic conditions, are shown in Figure 2. For both parameters a loss of the higher values, preferentially, was found during hyperemia.

DISCUSSION

In order to compare our results with others, three important features of the preparation used should be noted. First, surgical preparation of the muscle was minimal, thereby avoiding a substantial increase in blood flow to more than twice the resting level (Klitzman and Duling, 1979). Second, the muscle surface was isolated from the gaseous environment by a Saran film, so that the superficial vessels of the microvascular network would not be forced to a preselected oxygen tension, such as occurs when a suffusion solution is used. Because oxygen tension at the muscle surface can modify dramatically the density of perfused capillaries (Lindbom et al., 1980; Klitzman et al., 1982), it is usual to equilibrate a suffusion solution with a gas mixture of 95% N_2 and 5% CO_2. This results in a high degree of capillary patency which may not correspond to the normal situation in vivo (Lund et al., 1987). Third, after preparation of each muscle was complete a period of one hour was allowed to elapse, to ensure that steady-state conditions prevailed before the experimental procedures were carried out.

Of crucial importance is the period of time allowed to elapse, following injection of Aquablak, before the circulation was suddenly arrested by intracardiac injection of sodium pentobarbital. The influence of perfusion duration on the extent of microvascular filling has been studied systematically by previous investigators, from measurements on transverse sections of muscle. In leg muscles of rabbit Renkin et al. (1981) found that the number of capillaries filled with India ink increased progressively from 3.5 to 30s of circulation, with no further filling occurring thereafter (observations at 60 and 90s). Similar results were obtained by Kayar and Banchero (1985) in rat gastrocnemius, by use of time intervals 10, 15 and 30s; their data indicated that at 30s after the injection of a fluorescent dye into the aorta, the material had passed through and labelled 98 ± 5% (SE) of all capillaries. In light of these findings, in the present study an interval of 60 to 120s was allowed to elapse before arrest of the circulation was induced. In gracilis and gastrocnemius the entry of Aquablak into the superficial vessels of the microvascular network was videorecorded, and in no instance was the filling of additional capillaries with Aquablak discernible visually after 30s had elapsed. On no occasion was any aggregation and clumping of the Aquablak in microvessels observed. We conclude that under the conditions of these experiments all perfused capillaries should have become labelled with Aquablak.

The values we obtained for the mean capillary pathlengths in resting gastrocnemius, gracilis and soleus muscles of rat (Table 1) are not significantly different from each other and they show good agreement with values reported for rat cremaster (615 ± 85 μm (SD): Smaje et al., 1970) and rat extensor hallicus proprius (535 ± 25 μm (SD): Myrhage and Hudlicka, 1976). However, the pathlengths we found differ greatly from those for rat gracilis given in a previous report (1012 ± 484 μm (SD): Honig et al., 1977). Part of the discrepancy may be associated with differences in both strain and body weight of the animals used in the two sets of experiments (Sprague-Dawley, 160-170g: Honig et al., versus Wistar, 50-100g: present report). Our results are also in reasonable agreement with values reported recently for hamster tibialis anterior muscle (805 ± 330 μm (SD): Lund et al., 1987).

In resting muscle, the values we obtained for mean segment length (Table 1) are in good agreement with those reported for rat cremaster (210 μm: Smaje et al., 1970), cat tenuissimus (200 μm: Eriksson and Myrhage, 1972), hamster tibialis anterior (227 ± 201 μm (SD): Lund et al., 1987), and rat soleus (205 ± 17 μm (SE): Dawson et al.,1987). However, all the mean segment lengths cited above, including our own values for resting muscle, are smaller by a factor of two than those reported for rat gracilis by Honig et al. (1977); it would appear that in that particular muscle preparation many capillary segments were unperfused, for reasons that are unclear. Taken together, our data from resting muscle appear to indicate a very similar capillary network geometry in all three muscles studied. However, the significant changes in magnitude of capillary network parameters found in hyperemic as opposed to resting muscles (Tables 2,3) suggest that during the resting state there were present (i) unperfused capillaries (in all muscles, especially soleus), and (ii) unperfused arterioles (in gracilis and soleus).

The concept of capillary recruitment during exercise (Krogh, 1919) has been the subject of considerable scrutiny during the past decade or so. Some investigations have concluded that in resting muscle a fraction of the total number of capillaries present does indeed remain unperfused (Gorczynski et al., 1978; Honig et al., 1980; Renkin et al., 1981; Klitzman et al., 1982; Gray et al., 1983), whereas other investigations have supported the opposite view (Smaje et al., 1970; Burton and Johnson, 1972; Hudlicka et al., 1982; Kayar and Banchero, 1985). According to this latter group of studies, the concept that capillaries unperfused in resting muscle are recruited during exercise represents an oversimplification, since capillary perfusion is not an "all-or-none" phenomenon. The view has emerged that at low values of total blood flow (as in resting muscle) the distribution of perfusion rates among capillaries is very heterogeneous, leading to long transit times in some capillaries. In contrast, at higher values of total blood flow (as in exercise) "flow recruitment" occurs in all capillaries; this gives rise to a more uniform distribution of perfusion rates and reduces dramatically the long capillary transit times.

Recent studies in which red cell supply rate parameters have been measured for superficial capillaries in muscle, from intravital videomicroscopic recordings (Tyml, 1986; Ellis et al., 1987), confirm that capillaries with a very low mean red cell velocity (many of which are observed in resting muscle) show periods of zero flow. Secomb (1987) has proposed that the elastic deformational energy associated with red blood cells passing through capillaries having irregularities in lumenal cross-section (Ellis et al., 1987) may be sufficient, at low velocities, to account for zero flow in some capillaries while flow is maintained in other vessels. On this basis, the results obtained from investigations such as our own and those listed in the previous paragraph will depend critically on the magnitude of the prevailing total blood flow (ml/min.100g). In any experiment where indicator material is injected intravascularly, results indicative of unperfused capillaries in muscle will be obtained if either (i) the time allowed for perfusion is too short (i.e. results are due to artefact: see above), or (ii) the time for perfusion is adequate and muscle blood flow (ml/min.100g) is at its lowest, as in a basal metabolic state. The experimental protocol in the present investigation was designed so as to ensure that the conditions of (ii) above were met. Therefore, we interpret the results obtained as firm evidence for the existence, in resting muscle, of a proportion of capillaries having zero or near-zero flow throughout the period in which Aquablak was in transit through the muscle.

It was somewhat surprising to find, in gracilis and soleus, significant reductions in mean capillary pathlength in hyperemic versus resting muscle. The substantial loss of longer pathlengths, evident from the frequency distributions (Fig. 2: upper), implied that in resting muscle the average distance between (perfused) terminal arterioles and venules must be greater than in hyperemic muscle. This conclusion was tested experimentally, and verified, by measuring A-V distances over a large area of muscle; the changes in frequency distributions of pathlengths and A-V distances during hyperemia (Fig. 2) showed marked similarities. On the basis of indirect evidence, Honig et al. (1980) concluded that active vasomotion of small arterioles must be the process controlling flow into groups of capillaries in resting muscle, and that within 5s after the onset of exercise relaxation of these vessels occurs, making all capillaries accessible to blood flow. By means of intravital videomicroscopy in resting muscle the reverse process has been observed directly, on raising the oxygen tension of a suffusion solution to 40 mm Hg (Lund et al., 1987). Further confirmation of the concept of arteriolar recruitment during hyperemia has come from a recent study using fluorescently-labelled red cells as markers of flow paths and perfusion rates (Sweeney and Sarelius, 1988). The results of this study showed that in resting muscle unperfused vessels feeding terminal arterioles are present in a proportion which reflects the capillary recruitment found during hyperemia.

From the present investigation it is clear that valid comparisons of capillary network geometry can be made only in fully hyperemic muscle. Under these conditions the mean arteriolar-to-venular distances and mean capillary pathlengths were not dramatically different in all three muscles examined. What was striking, however, was the fact that the capillary pathlengths in soleus were divided into twice the number of segments that were found in medial gastrocnemius and gracilis. This suggests the possibility that in oxidative muscles (in general) the capillary network may exhibit a much higher degree of branching than that found in glycolytic muscles. Recent studies in pigeon pectoralis, a highly oxidative muscle, lead to a similar conclusion (Mathieu-Costello, 1987). An increase in the number of anastomotic channels between capillaries would augment the area for diffusional exchange of oxygen and solutes between blood and tissue.

SUMMARY

This investigation shows that provided an adequate perfusion time of the capillary network is allowed following injection of Aquablak, the presence of arterioles and capillaries having zero or near-zero flow rates can be demonstrated in resting muscle. During hyperemia, "flow recruitment" occurs in these vessels, as indicated by their perfusion with Aquablak.

Our observations of Aquablak perfusion in hyperemic muscles show that in medial gastrocnemius, gracilis, and soleus the mean arteriolar-to-venular distances, and also the mean capillary pathlengths, were not dramatically different. What was striking, however, was the fact that capillary pathlengths in soleus were divided into twice the number of segments found in gastrocnemius and gracilis. This suggests the possibility that in oxidative muscles the capillary network may exhibit a much higher degree of branching than in glycolytic muscles. This would increase the area for diffusional exchange between blood and tissue in oxidative compared to glycolytic muscle.

ACKNOWLEDGEMENTS

We wish to thank Mrs. Barbara Anderson for assistance in the preparation of this manuscript. S. Houghton was supported by Summer Research Studentships (1986, 1987) and this research by an operating grant to A.C. Groom, from the Heart and Stroke Foundation of Ontario.

REFERENCES

Burton, K.S. and Johnson, P.C., 1972, Reactive hyperemia in individual capillaries of skeletal muscle. Amer. J. Physiol. 223:517.

Dawson, J.M., Tyler, K.R. and Hudlicka, O., 1987, A comparison of the microcirculation in rat fast glycolytic and slow oxidative muscles at rest and during contractions. Microvasc. Res. 33:167.

Ellis, C.G., Tyml, K. and Burgess, W., 1987, Quantification of red cell movement in microvessels: a new application of interactive computer graphics. Microvasc. Res. 33:428.

Ellis, C.G., Wrigley, S.M., Potter, R.F. and Groom, A.C., 1988, Temporal distributions of red cell supply rate to individual capillaries of resting skeletal muscle, in frog and rat. Int. J. Microcirc.: Clin. Exp. (submitted).

Eriksson, E. and Myrhage, R., 1972, Microvascular dimensions and blood flow in skeletal muscle. Acta Physiol. Scand., 86:211.

Gorczynski, R.J., Klitzman, B. and Duling, B.R., 1978, Interrelations between contracting striated muscle and precapillary microvessels. Amer. J. Physiol. 235:H494.

Gray, S.D., McDonagh, R.F. and Gore, R.W., 1983, Comparison of functional and total capillary densities in fast and slow muscles of the chicken. Pflügers Arch. 397:209.

Honig, C.R., Feldstein, M.L. and Frierson, J.L., 1977, Capillary lengths, anastomoses, and estimated capillary transit times in skeletal muscle. Amer. J. Physiol. 233:H122.

Honig, C.R., Odoroff, C.L. and Frierson, J.L., 1980, Capillary recruitment in exercise: rate, extent, uniformity, and relation to blood flow. Amer. J. Physiol. 238:H31.

Hudlicka, O., Zweifach, B.W. and Tyler, K.R., 1982, Capillary recruitment and flow velocity in skeletal muscle after contractions. Microvasc. Res. 23:201.

Kayar, S.R. and Banchero, N., 1985, Sequential perfusion of skeletal muscle capillaries. Microvasc. Res. 30:298.

Klitzman, B. and Duling, B.R., 1979, Microvascular hematocrit and red cell flow in resting and contracting striated muscle. Amer. J. Physiol. 237:H481.

Klitzman, B., Damon, D., Gorczynski, R. and Duling, B., 1982, Augmented tissue oxygen supply during striated muscle contraction in the hamster: Relative contributions of capillary recruitment, functional dilation, and reduced tissue P_{O_2}. Circ. Res. 51:711.

Krogh, A., 1919, The supply of oxygen to the tissues and the regulation of the capillary circulation. J. Physiol. (Lond.) 52:457.

Lindbom, L., Tuma, R. and Arfors, K., 1980, Influence of oxygen on perfused capillary density and capillary red cell velocity in rabbit skeletal muscle. Microvasc. Res. 19:197.

Lund, N., Damon, D.H., Damon, D.N. and Duling, B.R., 1987, Capillary grouping in hamster tibialis anterior muscles: flow patterns, and physiological significance. Int. J. Microcirc.: Clin. Exp. 5:359.

Mathieu-Costello, O., 1987, Capillary geometry and blood-tissue exchange in bird muscles under normoxic and hypoxic conditions. Physiologist 30:202. Abstract 61.0.

Myrhage, R.O. and Hudlicka, O., 1976, The microvascular bed and capillary surface area in rat extensor hallicus proprius muscle (EHP). Microvasc. Res. 11:315.

Plyley, M.J. and Groom, A.C., 1975, Geometrical distribution of capillaries in mammalian striated muscle. Amer. J. Physiol. 228:1376.

Plyley, M.J., Sutherland, G.J. and Groom, A.C., 1976, Geometry of the capillary network in skeletal muscle. Microvasc. Res. 11:161.

Renkin, E.M., Gray, S.D. and Dodd, L.R., 1981, Filling of microcirculation in skeletal muscles during timed India ink perfusion. Amer. J. Physiol. 241:H174.

Secomb, T.W., 1987, Flow-dependent rheological properties of blood in capillaries. Microvasc. Res. 34:46.

Skalak, T.C. and Schmid-Schönbein, G.W., 1986, The microvasculature in skeletal muscle. IV. A model of the capillary network. Microvasc. Res. 32:333.

Smaje, L., Zweifach, B.W. and Intaglietta, M., 1970, Micropressures and capillary filtration coefficients in single vessels of the cremaster muscle of the rat. Microvasc. Res. 2:96.

Sweeney, T.E. and Sarelius, I.H., 1988, Arteriolar control of capillary cell flow. Faseb J. 2:A1880, M150 (Abstract).

Tyml, K., 1986, Capillary recruitment and heterogeneity of microvascular flow in skeletal muscle before and after contraction. Microvasc. Res. 32:84.

BLOOD AND BLOOD SUBSTITUTES

MYOCARDIAL OXYGEN TRANSPORT DURING LEFTWARD SHIFTS OF THE OXYGEN DISSOCIATION CURVE BY CARBAMYLATION OR HYPOTHERMIA

Robert W. Baer

Kirksville College of Osteopathic Medicine
800 W. Jefferson St., Kirksville, Missouri 63501 USA

INTRODUCTION

The heart is a highly aerobic organ whose ability to adjust to shifts in the O_2 dissociation curve is of considerable interest (Harkin, 1977). For example, hypothermia is known to increase hemoglobin O_2 affinity and is often used by surgeons in conjunction with other forms of cardioplegia during cardiac surgery (Hearse, 1981). In evaluating the usefulness of hypothermia for myocardial protection it is important to understand how the coronary circulation adapts to the potential ill effects of an increase in hemoglobin-O_2 affinity.

The movement of O_2 to the tissues depends first on its convective transport to the capillaries and then on its transcapillary diffusion to the cells where it is to be utilized. These two transport processes occur in series and are coupled by the O_2 dissociation curve, since convective transport is related to the arterial O_2 content, whereas transcapillary diffusion is related to the mean capillary partial pressure of O_2. Physiological conditions which increase hemoglobin-O_2 affinity and left-shift the O_2 dissociation curve interfere with the normal flux of O_2 from the red cells to the tissues. Under steady state conditons this O_2 flux must equal tissue O_2 consumption. O_2 consumption will decrease when hemoglobin O_2 affinity is increased unless: 1) capillary P_{O_2} decreases enough to maintain a constant arterial-venous O_2 content difference; or 2) convective O_2 transport increases enough to compensate for a reduced O_2 extraction.

Recent evidence suggests that metabolic vasodilation may differ from autoregulatory vasodilation, ie. vasodilation in response to a decrease in perfusion pressure (Vergroessen et al, 1987). To date there have been no studies of the interaction between increased hemoglobin-O_2 affinity and coronary autoregulation. The experiments to be described were undertaken to determine the adaptive response of the coronary circulation to an increase in O_2 affinity across the normal autoregulatory range of perfusion pressures. An isolated dog heart preparation was used to study increased O_2 affinity induced either by hypothermia to 30° C or by carbamylation of hemoglobin with sodium cyanate. The results are consistent with the idea that coronary compensation for increased O_2 affinity has little effect on the autoregulatory process and occurs primarily by a decrease in coronary sinus P_{O_2} rather than an increase in O_2 transport.

METHODS

Isolated Dog Heart Preparation

The perfusion circuit used for the isolated dog heart preparation consisted of 2 parallel sets of tubing connected to a common perfusion nipple. The nipple was fastened into the aorta for retrograde perfusion into the coronary circulation. Coronary venous drainage was collected in a 34 cm funnel placed below the heart. Each of the parallel sets of tubing began at the venous funnel and drained passively into its own Bos Spiroflo 5s oxygenator and heat exchanger. Pumps were used in conjunction with level detectors to supply the oxygenated blood to two separate pressure-controlled, plexiglass reservoirs. Blood from each side of the circuit passed through its own filter (Pall, SQ40S). The 2 sides of the circuit joined to pass through a common electromagnetic flow transducer (6mm) adjacent to the aortic perfusion nipple (13mm, o.d.). An air trap and bleed line attached above the perfusion nipple prevented air embolism of the coronary bed. Clamps were used to perfuse from and return blood to only a single side of the circuit at any one time. In the hypothermia protocol only one side of the perfusion circuit was used. In the carbamylation protocol one side of the circuit contained control blood, and the other side of the circuit contained carbamylated blood.

Mongrel dogs of either sex were anesthetized with pentobarbital, sodium (25-35 mg/kg) and ventilated on a Harvard constant volume respirator. These dogs served as both heart and control blood-prime donors. Control blood-prime was obtained by slowly draining blood from a polyethylene cannula in the jugular vein. Arterial pressure was monitored to prevent it from falling below 50 mmHg. Periodically the dog's volume was supplemented with 6% dextran warmed to 38° C. Total dextran volume did not exceed 500 ml. The total blood-prime obtained by this procedure ranged between 500-1000 ml. Up to 350 ml of additional blood was recovered from the chest after removing the heart. This blood was filtered before adding it to the perfusate. Blood gases and pH were maintained in the normal range during the bleeding procedure by adjusting the ventilator and administering bicarbonate as necessary.

Final blood gas and pH adjustments of the initial blood perfusate were made, and the heart was mounted as follows. The left subclavian and bracheocephalic arteries were tied off, and the aorta was cross-clamped. The heart was electrically fibrillated to reduce O_2 consumption and rapidly excised from the chest. A short section of aorta distal to the cross clamp was pulled over the perfusion nipple and tied. The perfusion line was opened to allow filling to the aortic cross clamp and temporarily closed again. The aortic cross clamp was removed so that any remaining air could be removed from the proximal aorta by gentle massage. The perfusion line was reopened and perfusion was begun at a low pressure. The perfusion pressure was gradually increased to 80 mmHg. A vent was inserted through the left ventricular apex. The mounting procedure was generally completed in 30 sec. Blood gases, pH, perfusion pressure and flow were adjusted as necessary during a 20 min equilibration period.

Carbamylated Blood-Prime

Blood was carbamylated using a procedure similar to that described by Ross and Hlastala (1981). Blood obtained from donor dogs was centrifuged at 3000 rpm for 15 min. The plasma layer was removed, given supplemental heparin (5000 U), and refrigerated overnight. The packed red cells were incubated in an equal volume of wash buffer containing 24 mM sodium cyanate (NaOCN). The wash buffer was composed of 130 mM sodium chloride, 5 mM potassium chloride, 10 mM glucose, and 20 mM HEPES buffer brought to pH = 7.55 at 20° C. This suspension was incubated in the refrigerator overnight to carbamylate the red cells. The carbamylated red cells were recovered from the incubation solution by centrifugation and rinsed three times in wash buffer containing all the ingredients except the NaOCN. The carbamylated red cells were remixed with the original plasma and placed in the perfusion circuit. Blood gases and pH were corrected as necessary.

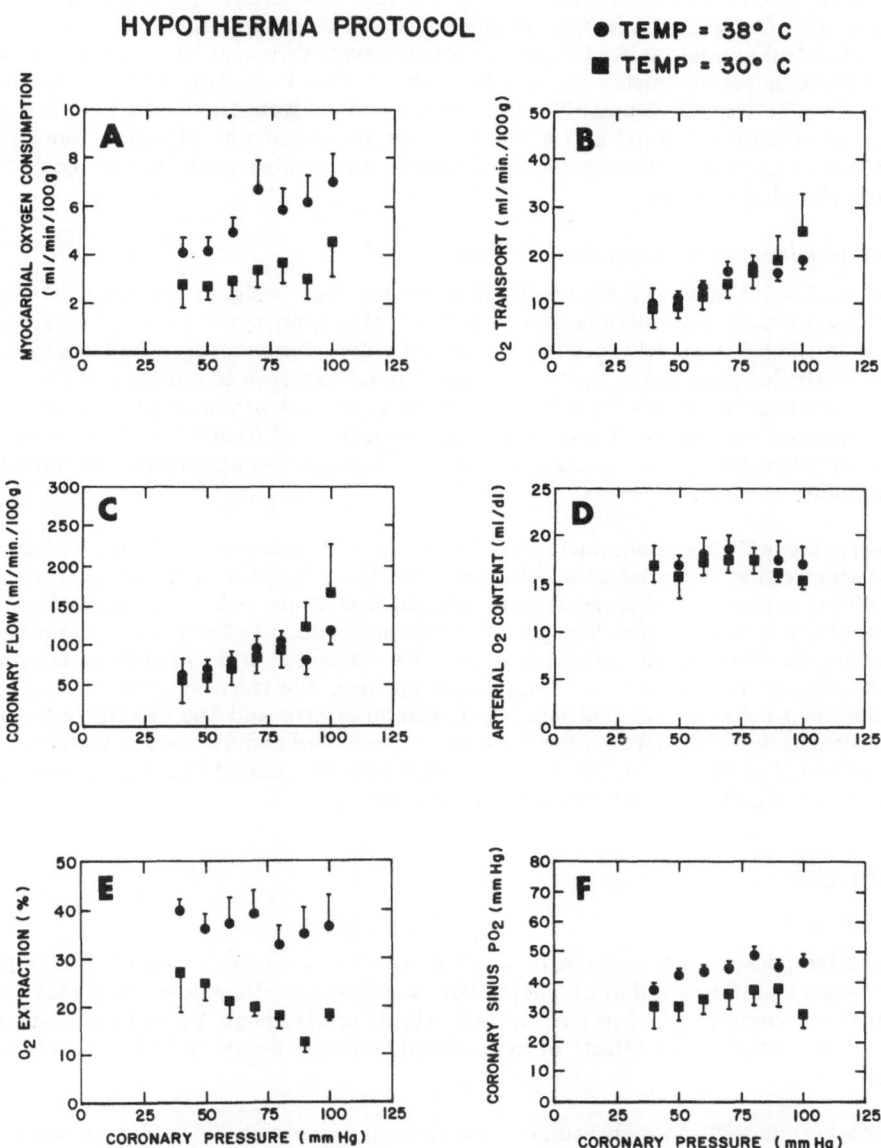

Figure 1. Effect of hypothermia. Panel A = myocardial oxygen consumption; Panel B = oxygen transport; Panel C = coronary flow; Panel D = arterial oxygen content; Panel E = percent oxygen extraction; Panel F = coronary sinus PO₂. Hypothermia had a significant effect on: myocardial oxygen consumption (p < 0.001); oxygen extraction (p < 0.001); and coronary sinus PO₂ (p < 0.001). Perfusion pressure had a significant effect on oxygen transport (p < 0.005) and coronary flow (p < 0.05).

Blood samples were collected in syringes and analyzed for blood gases and pH (IL213, Instrumentation Laboratories) and hemoglobin and O_2 saturation (OSM2 Hemoxymeter, Radiometer). Oxygen content was calculated as 1.36 x hemoglobin x saturation + 0.003 x P_{O_2}. Control and left-shifted O_2 dissociation curves were generated at the end of the experiment using tonometry (IL237, Instrumentation Laboratories) to obtain 6-8 O_2 saturations between 20% and 80%. P_{50} was determined from Hill Plots. Pco_2 was maintained at 40mmHg and pH at 7.4. Curves were generated with electrode temperature set at each experimental temperature. Separate curves were generated for control and carbamylated perfusates.

Hypothermia and Carbamylation Protocols

Hearts from 6 dogs weighing 9-19 Kg and with heart weights averaging 175 ± 52 g were used to study hypothermia. Hearts beat isovolumically against a saline-filled balloon placed in the left ventricle through the mitral valve. Temperature was initially set to 38° C. Perfusion pressure was set to a range of values between 40 mmHg and 100 mmHg while recording flow and left ventricular pressure. At each pressure arterial and venous blood samples were taken. Temperature was then reduced to 30° C and corresponding data were collected. Pressure order was varied. Pressure was returned to 80 mmHg between points.

Hearts from 12 dogs weighing 14-24 Kg were used to compare perfusion with control and carbamylated blood. Heart weights averaged 185 ± 35 g. Six dogs were perfused with control blood first; six with carbamylated blood first. Since order made no difference to the resuls, data were pooled. Temperature was maintained at 38° C throughout the experiment. To avoid the alterations in LV work encountered in the hypothermia protocol no LV balloon was used in the carbamylation protocol. For the first perfusate, perfusion pressure was set to a range of values between 30 mmHg and 100 mmHg returning it to 80 mmHg between points. Flow data and arterial and coronary sinus blood samples were collected at each point. Perfusion was shifted to the second circuit and corresponding data were collected with the second perfusate.

RESULTS

Hypothermia

When temperature was reduced from 38° C to 30° C heart rate slowed ($p < 0.001$), and there was a significant fall ($p < 0.001$) in left ventricular systolic pressure (Table 1). Occasionally, cooling resulted in a slight arrhythmia or bigemeny. P_{50} fell from 32.5 ± 3.8 to 18.0 ± 3.5 mmHg. The effects of hypothermia on O_2 delivery variables are shown in figure 1.

Hypothermia reduced myocardial O_2 consumption to about 60% of its control value. Neither coronary flow nor arterial O_2 content changed significantly with hypothermia, and thus O_2 transport (their product) was unchanged. Coronary sinus P_{O_2} decreased by about 10 mmHg ($p < 0.001$) with hypothermia suggesting a similar decrease in capillary P_{O_2}. Fractional myocardial O_2 extraction is considerably lower in this isolated dog heart preparation (Figure 1 and 2) than is normally observed in intact dogs (Baer et al, 1987). With hypothermia myocardial O_2 extraction fell significantly ($p < 0.001$).

Autoregulatory adjustments in vasomotor tone did not maintain flow absolutely constant as perfusion pressure was changed, but flow changes were attenuated relative to those normally observed during maximal vasodilation (Baer et al, 1987). Autoregulation tended to be better at control temperatures than during hypothermia. The influence of

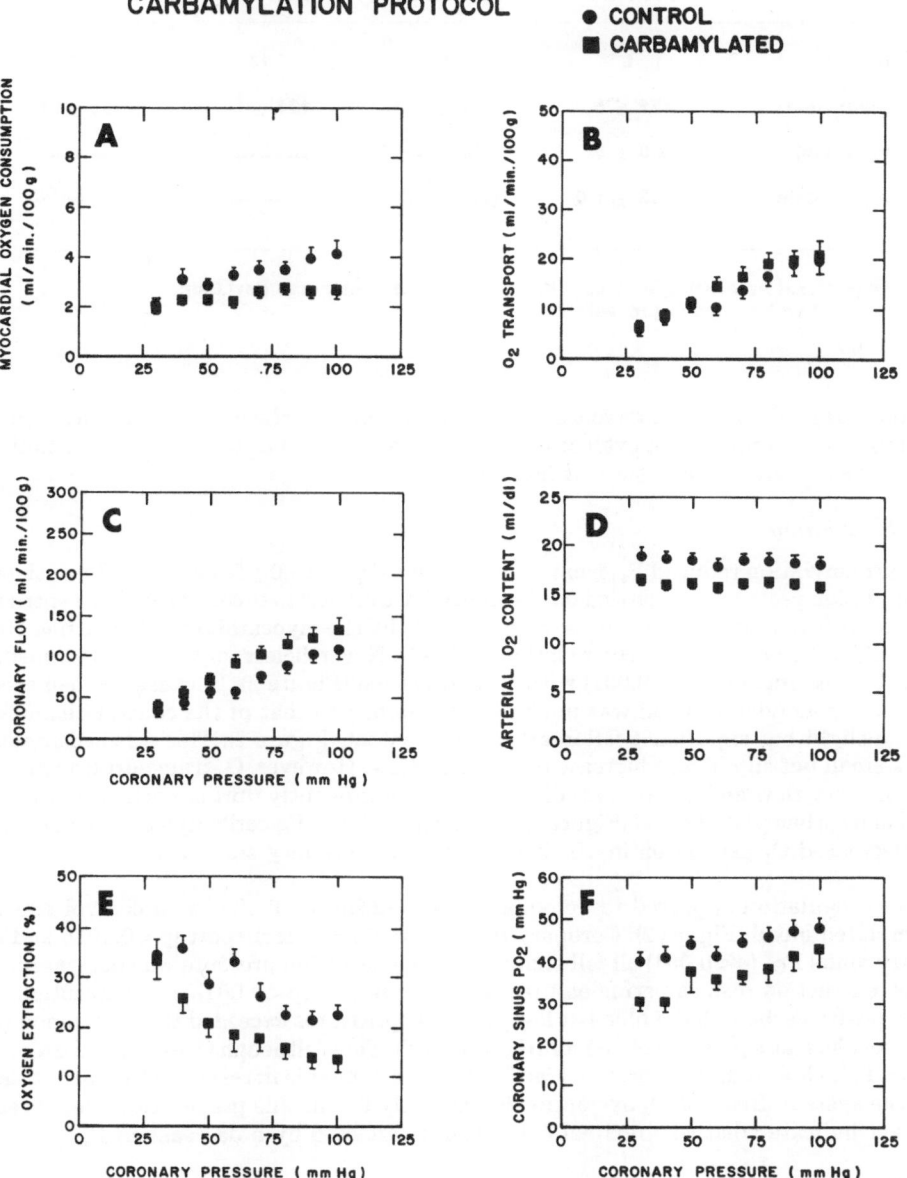

Figure 2. Effect of perfusion with carbamylated blood. Panel A = myocardial oxygen consumption; Panel B = oxygen transport; Panel C = coronary flow; Panel D = arterial oxygen content; Panel E = percent oxygen extraction; Panel F = coronary sinus PO_2. Carbamylation had a significant effect on: myocardial oxygen consumption ($p < 0.001$); coronary flow ($p < 0.001$); arterial oxygen content ($p < 0.001$); oxygen extraction ($p < 0.001$) and coronary sinus PO_2 ($p < 0.001$). Perfusion pressure had a significant effect on: myocardial oxygen consumption ($p < 0.05$); oxygen transport ($p < 0.001$); coronary flow ($p < 0.001$); oxygen extraction ($p < 0.001$); and coronary sinus PO_2 ($p < 0.02$).

TABLE 1. HEMODYNAMIC VARIABLES

TABLE 1. HEMODYNAMIC VARIABLES

	CONTROL	HYPOTHERMIA	CONTROL	CARBAMYLATION
n	6	6	12	12
Heart Rate	131 ± 8	102 ± 2	126 ± 8	120 ± 11
P_{LV} systolic	82.8 ± 5	68 ± 10*	———	———
P_{LV} diastolic	6.5 ± 1.0	10.7 ± 2.0	———	———

*Significant by ANOVA, $p < 0.001$. Values are mean ± standard deviation.
P_{LV} = Left Ventricular pressure.

hypothermia did not change markedly as a function of perfusion pressure although the fall in O_2 consumption, O_2 extraction, and coronary sinus P_{O_2} was somewhat smaller at the lowest perfusion pressures (Figure 1).

Carbamylation

Carbamylation reduced P_{50} from 30.8 ± 2.0 mmHg to 19.0 ± 2.5 mmHg. Since the carbamylation protocol was carried out in empty beating hearts there was no left ventricular pressure development. The only work performed by the myocardium was internal. Heart rate did not change with carbamylation (Table 1). Nevertheless, myocardial O_2 consumption fell significantly ($p < 0.001$) with carbamylation (Figure 2). Because the hematocrit of the carbamylated blood was not exactly matched to that of the control blood there was a slight but significant fall in arterial O_2 content (Figure 2). This was accompanied by a small but significant increase in coronary flow. However, O_2 transport (the product of coronary flow and O_2 content) did not differ significantly during perfusion with control and carbamylated blood (Figure 2). Perfusion with low P_{50} carbamylated blood resulted in decreased O_2 extraction ($p < 0.001$) and a lower coronary sinus P_{O_2}.

Autoregulation appeared to be equally good during perfusion with control and carbamylated blood (Figure 2). Coronary flow ($p < 0.001$), O_2 transport ($p < 0.001$), and coronary sinus P_{O_2} ($p < 0.001$) all fell significantly as perfusion pressure was decreased. Oxygen extraction rose as perfusion pressure was lowered ($p < 0.001$). The extractions obtained with carbamylated blood at low perfusion pressures exceeded those obtained with control blood at a pressure of 100 mmHg (Figure 2). Thus, although O_2 extraction decreases when P_{50} is lowered, there is an extraction reserve available in response to other stresses on the system. Because O_2 extraction is relatively low in this preparation these results do not indicate whether maximal extraction is affected by a decrease in P_{50}.

DISCUSSION

The two methods of increasing hemoglobin-O_2 affinity used in the present study, hypothermia and hemoglobin carbamylation, caused similar P_{50} decreases and yielded qualitatively similar results. Both decreased myocardial O_2 consumption. This decrease was accompanied by a decrease in O_2 extraction and coronary sinus P_{O_2}. However, there was no compensatory increase in O_2 transport. Hypothermia decreased myocardial O_2 consumption at least in part by slowing myocardial metabolism. This decrease in O_2 consumption was accompanied by a decrease in heart rate and systolic pressure development, two of the main determinants of myocardial O_2 consumption (Braunwald, 1971). In pilot experiments an attempt was made to match left ventricular systolic pressure and heart rates to control values but frequent arrhythmias developed. It was therefore decided to treat pressure and heart rate as dependent variables. This approach does not allow one to decide whether O_2 consumption decreased because of decreased function or whether function decreased because of decreased O_2 consumption.

Changes in performance-based determinates of myocardial O_2 consumption were avoided in the carbamylation protocol by using an empty beating heart, and heart rate was not affected by carbamylation. This makes it more likely that the decrease in O_2 consumption in this protocol was due to O_2 supply limitation. Allison et al (1976) have reported biochemical changes consistent with an hypoxia-induced activation of myocardial glycolysis in cyanate-fed rats. Sodium cyanate decreases P_{50} by carbamylating the NH_2 terminal valine on hemoglobin, and may also affect other proteins (Cerami et al, 1973). Therefore, the possibility of toxic effects is an important alternative to consider (Teisseire et al, 1986). Because triply-washed carbamylated red cells were used in the present study, it does not seem likely that cyanate had direct toxic effects on the myocardial metabolic machinery. However, final conclusions about possible toxic effects must await more knowledge of cyanate cellular biochemistry.

Left-shifting the O_2 dissociation curve by hypothermia or carbamylation did not cause a compensatory increase in O_2 transport for the range of perfusion pressures examined in the present study. Turek et al (1978) have previously reported no difference in flow between control and sodium cyanate fed rats. Similarly, Martin et al (1979) found no significant change in the absolute coronary flow to isolated rat hearts when the O_2 dissociation curve was left-shifted by either carbon monoxide or perfusion with human red cells containing hemoglobin Creteil. When flows for the hemoglobin Creteil group were expressed as percent change there was a small significant increase in flow. In contrast, Woodson and Auerbach (1982) have observed a large increase in coronary flow in anesthetized rats subjected to whole body exchange transfusion with red cells incubated with both sodium cyanate and sodium metabisulfate. This procedure reduced P_{50} by 23 mmHg and depleted red cell 2,3-DPG. Stucker et al (1985) reported a 36% fall in coronary flow when the O_2 dissociation curve was right-shifted by 28 mmHg. They found a significant negative correlation between P_{50} and coronary flow. However, the scatter in flow at a single P_{50} (20 mmHg) was 2.5 times as large as the flow increase predicted for the P_{50} change used in the present experiment. Together these studies would tend to indicate that O_2 transport can increase in compensation for increased hemoglobin O_2 affinity, but that such increases are not large for moderate shifts in the dissociation curve such as those studied here.

For hemoglobin to release the same amount of O_2 after P_{50} is lowered, there must be a fall in capillary P_{O_2}. Capillary P_{O_2} decreased during hypothermia and carbamylation as reflected by a decrease in coronary sinus P_{O_2}. However, the fall in capillary P_{O_2} was not sufficiently large to maintain O_2 unloading at its control level, so O_2 extraction by the myocardium fell. This is consistent with the previous studies of increased hemoglobin O_2 affinity in the systemic circulation which have found that although mixed venous P_{O_2} decreases, the magnitude of this fall is insufficient to maintain O_2 extraction (Schumaker et al, 1987).

Woodson (1988) has predicted a 25% decrease in O_2 consumption for an 8 mmHg decrease in P_{50} under conditions where coronary flow does not change. In the present study where O_2 transport showed no change with P_{50} reduction, we observed a decreased O_2 consumption of approximately this magnitude. Several types of experiments support the idea that the capillary wall represents a permeability barrier for the movement of O_2 to the tissues (Rose and Goresky, 1985; Honig and Gayeski, 1987). If this is so, then the fall in capillary P_{O_2} necessary to remove O_2 from hemoglobin must also reduce transcapillary O_2 flux unless there is a compensatory reduction in tissue P_{O_2} or a compensatory increase in the transcapillary O_2 conductance. In the present experiments coronary sinus P_{O_2} fell by 8-10 mmHg with both hypothermia and carbamylation. Honig and Gayeski (1987) have found that tissue P_{O_2} based on myoglobin saturation is less than 10 mmHg in cat hearts. Therefore, tissue P_{O_2} would have had to reach very low levels in the present experiments to maintain transcapillary O_2 flux.

Duvelleroy et al (1980) have proposed a model suggesting the potential importance of capillary recruitment and increased transcapillary O_2 conductance in response to increased hemoglobin-O_2 affinity. Data from rat hearts perfused with human red cells containing high O_2 affinity hemoglobin Creteil supported their model. Capillary recruitment of sufficient magnitude to maintian O_2 consumption did not occur in the present experiments. It is possible that control of precapillary sphincter tone is abnormal in our isolated dog heart preparation. However, we also have observed slightly lower O_2 consumptions in intact dogs following chronic carbamylation (Baer, 1987). The results do not imply that no adaptation occurs at the capillary level, only that adaptation is not sufficient in magnitude to maintain O_2 consumption constant under the conditions of the present experiment. An important area for additional work is to independently determine how much adapatation can occur at the capillary level and what factors optimize capillary responsiveness to interventions such as increased hemoglobin-O_2 affinity.

A comment on the isolated heart preparation used in this study is in order. These hearts had relatively normal coronary flows. However, O_2 extraction was lower and coronary sinus P_{O_2} was higher than normally observed in intact dogs. A similar pattern has been seen by previous investigators using isolated heart preparations of other species (Martin et al, 1979; Apstein et al, 1985). Such a pattern is indicative of relative over-perfusion with no compensatory increase in vasomotor tone. Whether such a patttern is a function of altered intramyocardial compressive forces or some other property intrinsic to isolated heart preparations is unclear. Nevertheless, it is important to recognize such differences as we study control of coronary flow and myocardial O_2 delivery. Studies done in intact preparations in our laboratory and others support most of the major findings of the present study. Nevertheless, differences between intact preparations and isolated heart preparations may in themselves contain hidden insights into the control of myocardial O_2 transport.

SUMMARY

An isolated dog heart preparation was used to study the effect of left-shifting the O_2 dissociation curve by carbamylation or hypothermia. The two interventions had a similar effect on the variables of O_2 delivery. There were significant decreases in myocardial O_2 consumption, coronary sinus P_{O_2}, and O_2 extraction. There was no compensatory increase in O_2 transport. Coronary flow autoregulation was somewhat blunted by hypothermia but not by carbamylation. We conclude that an increase in hemoglobin-O_2 affinity is capable of limiting myocardial O_2 delivery and that increases in convective O_2 transport play a minor role at best in the coronary adaptation to small decreases in P_{50}.

ACKNOWLEDGEMENTS

This work was generously supported by a Grant-in-Aid from the American Heart Association, Missouri Affiliate, and a grant from the Warner Fund.

REFERENCES

Allison, T. B. G., Pieper, G. M., Clayton, F. C., and Eliot, R. S., 1976, Reduced high enery phosphate levels in rat hearts. *Am. J. Physiol.* 230: 1751-1754.

Apstein, C. S., Dennis, R. C., Briggs, L., Vogel, W. M., Frazer, J., and Valeri, C. R., 1985, Effect of erythrocyte storage and oxyhemoglobin affinity changes on cardiac function. *Am. J. Physiol.* 248 (Heart Circ. Physiol. 17): H508-H515.

Baer, R. W., 1987, Myocardial oxygenation after increasing oxyhemoglobin affinity by carbamylation. *Fed. Proc.* 46: 1113 (Abstr.).

Baer, R. W., Vlahakes, G. J., Uhlig, P. N., and Hoffman, J. I. E., 1987, Maximum myocardial oxygen transport during anemia and polycythemia in dogs. *Am. J. Physiol.* 252 (Heart Circ. Physiol. 21): H1086-H1095.

Braunwald, E., 1971, Control of myocardial oxygen consumption. Physiologic and clinical considerations. *Am. J. Cardiol.* 27: 416-432.

Cerami, A., Allen, T.A., Graziano, J. H., DeFuria, F. G., Manning, J. M., and Gillete, P. N., 1973, Pharmacology of Cyanate. I. General effects on experimental animals. *J. Pharm. Exper. Ther.* 185: 653-666.

Duvelleroy, M. A., Martin, J. L., Teisseire, B., Gauduel, Y., and Durable, M., 1980, Abnormal hemoglobin oxygen affinity and the coronary circulation. *Biblthca. Haemat.* 46: 70-77.

Harkin, A. H., 1977, The surgical significance of the oxyhemoglobin dissociation curve. *Surg. Gyn. Obst.* 144: 935-955.

Hearse, D. J., Braimbridge, M. V., and Jynge, P., 1981, Protection of the ischemic myocardium: cardioplegia. Raven Press. New York. pp167-208.

Honig, C. R. and Gayeski, T. E. J., 1987, Comparison of intracellular P_{O_2} and conditions for blood-tissue transport in heart and working red skeletal muscle. *Adv. Exper. Med. Biol.* 215: 309-321.

Martin, J. L., Duvelleroy, M., Teisseire, B., and Duruble, M., 1979, Effect of increased HbO_2 affinity on the calculated capillary recruitment of an isolated rat heart. *Pfleugers Arch.* 382: 57-61.

Rose, C. P., and C. A. Goresky, 1985, Limitations of tracer oxygen uptake in the canine coronary circulation. *Circ. Res.* 56: 57-71.

Ross, B. K., and Hlastala, M. P., 1981, Increased hemoglobin oxygen affinity does not decrease skeletal muscle oxygen consumption. *J. Appl. Physiol.: Respirat. Environ. Exercise Physiol.* 51: 864-870.

Schumacker, P. T., Long, G. R., and Wood, L. D. H., 1987, Tissue oxygen extraction during hypovolemia: role of hemoglobin P_{50}. *J. Appl. Physiol.* 62: 1801-1807.

Stucker, O., Vicaut, E., Villereal, M., Ropars, C., Teisseire, B. P., and Duvelleroy, M. A., 1985, Coronary response to large decreases of hemoglobin-O_2 affinity in isolated rat heart. *Am. J. Physiol.* 249 (Heart Circ. Physiol. 18): H1224-H1227.

Teisseire, B. P., Vieilledent, C. C., Teisseire, L. J., Vallez, M. O., Herigault, R. A., and Laurent, D. N., 1986, Chronic sodium cyanate treatment induces "hypoxia-like" effects in rats. *J. Appl. Physiol.* 60: 1145-1149.

Turek, Z., Kreuzer, F., Turek-Maischeider, M., and Ringnalda, B. E. M., 1978, Blood oxygen content, cardiac output, and flow to organs at several levels of oxygenation in rats with a left-shifted blood oxygen dissociation curve. *Pfleugers Arch.* 376: 201-207.

Vergroesen, I., Noble, M. I. M., Wieringa, P. A., and Spaan, J. A. E., 1987, Quantification of O_2 consumption and arterial pressure as independent determinants of coronary flow. *Am. J. Physiol.* 252 (*Heart Circ. Physiol.* 21): H545-H553.

Woodson, R. D., 1988, Evidence that changes in blood oxygen affinity modulate oxygen delivery: implications for control of tissue P_{O_2} gradients. *Adv. Exper. Med. Biol.* 222: 309-313.

Woodson, R. D., and Auerbach, S., 1982, Effect of increased oxygen affinity and anemia on cardiac output and its distribution. *J. Appl. Physiol.: Respirat. Environ. Exercise Physiol.* 53: 1299-1306.

LOW VISCOSITY OF DENSELY AND HIGHLY POLYMERIZED HUMAN HEMO-GLOBIN IN AQUEOUS SOLUTION - THE PROBLEM OF STABILITY

W.K.R. Barnikol and O. Burkhard

Institut für Physiologie der Universität
Saarstraße 21
D-6500 Mainz (Bundesrepublik Deutschland)

INTRODUCTION

In case of chronic and acute tissue oxygen deficit it is of advantage to have an artificial oxygen carrying blood substitute in order to support a least temporarily blood function. From a physico-chemical point of view an artificial oxygen carrying blood substitute must meet 4 main requirements at the desired concentration.

(1) The oxygen carrying function must be decoupled from the oncotic pressure of blood plasma and therewith from the intravasal volume.

(2) The transporting molecules must have a sufficiently high residence time.

(3) The pressure of half saturation (P_{50}) must be high enough.

(4) The viscosity must be low enough.

In nature the demands are achieved by 2 principles of dispersing hemoglobin in the plasma: (1) by packing hemoglobin (Hb) into cells (higher animals), (2) by synthesis of huge compact soluble oxygen carrying molecules with approximately 200 binding sites (low animals: earthworm "hemoglobin").

In developing an artificial oxygen carrying blood substitute we imitate the second principle.

In earlier investigations we showed that human hemoglobin can be highly polymerized with the aid of glutardialdehyde (GDA) as a crosslinker [1]. The product has a pressure of half saturation of 13 mmHg, its molecular weight (weight average Mw) is at least 6×10^6 dalton [2]. Crosslinking with 1.4-cyclo-hexylendiisocyanate (DIC) is also possible [3], giving a Mw of 3×10^6 dalton. Polymerisation raises the residence time to the desired values (see for instance Venuto, Zegna [4]). The polymers have a broad distribution with a high nonuniformity as indicated by the high U-values (see below). After fractionation by ultrafiltration (cutoff at 300 000 dalton) the DIC-polymer has at 4.5 g/dl a colloid osmotic pressure of 3.5 mmHg, which is only 13% of the normal value in blood plasma.

So within the concept of soluble huge polymers three of four aforementioned requirements are met to a sufficient extent. The fourth requirement remains unsettled. Therefore we have studied the viscous properties of huge compact polymers in aqueous solution compared with native human hemoglobin.

A problem with respect to viscosity of huge molecules is that the form influences the viscosity of their solutions greatly. Especially rods (e.g. fibrinogen) or freely drained coiled thread molecules, like synthetic polymers in a good solvent, generate a high viscosity. Compact globular molecules cause the lowest possible viscosity. The earthworm "hemoglobin" is a compact tube approximately equal in height and diameter as shown by electron microscope. With our special procedure we produce huge molecules by crosslinking the preformed associates occuring in very high concentration (35 g/dl). So we expect that compact polymers increase the viscosity only to a minor degree.

METHODS

Human hemoglobin was polymerized within the erythrocytes by GDA and DIC in a special procedure. Details at the procedure are given elsewhere [5],[6]. The polymers were solved in an electrolyte solution (mmol/l): NaCl 125, KCl 4.5, $NaHCO_3$ 20). Before use the solutions have passed 0.22 µm filters. Determinations of hemoglobin were done with the Cyan-Met-Hb- method. The kinematic viscosity was determined with an micro-Ubbelohde-viscosimeter at 37 °C (type 10, Schott, D-6500 Mainz, West Germany). Molecular weight determinations (Mw and Mn: number average) were done by gel permeation chromatography with Sephacryl 400 (Pharmacia, D-7800 Freiburg, West Germany), column diameter 1 cm, height 80 cm, flow 5 ml/h, temperature 22 °C. For calibration in the high molecular weight range earthworm hemoglobin was used (molecular weight = 3.4×10^6 dalton). Mn-values were also measured with a membrane-osmometer (Knauer KG, D-1000 Berlin, West Germany). Ultrafiltrations were done in a Minitan-system (Millipore, D-6236 Eschborn, West Germany. The nonuniformity (U) is calculated as Mw/Mn-1.

RESULTS WITH DISCUSSION

Figure 1 shows all measurements in a comprehensive manner.

As is seen from fig. 1 the viscosity of native hemoglobin increases only slightly and linearly up to 10 g Hb/dl.

Fig. 2 shows this behaviour on an extended scale.

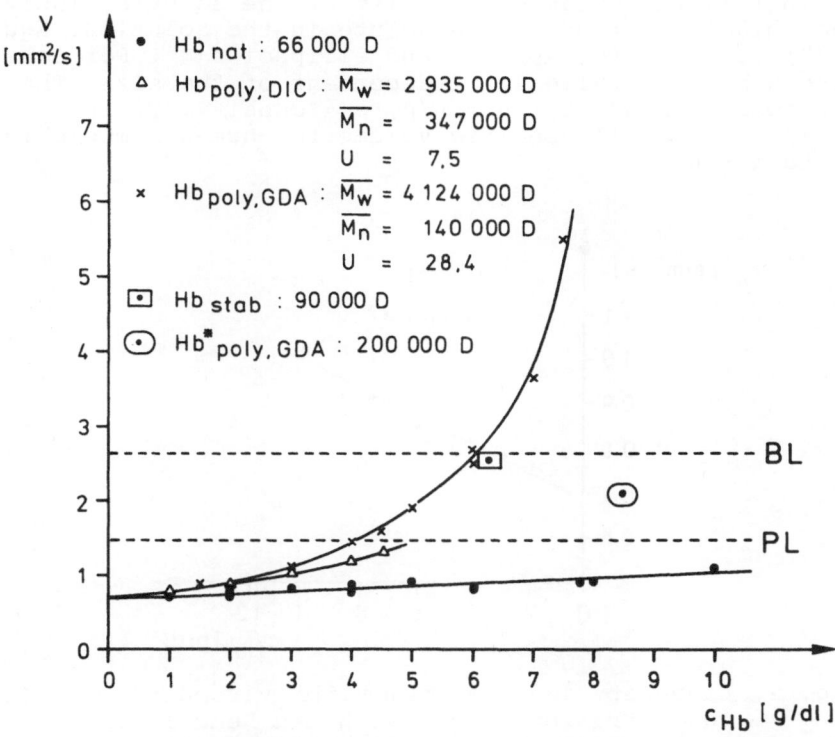

Fig. 1. Kinematic viscosity (ν) in dependence on concentration (c_{Hb}) for native hemoglobin (●), GDA-polymer (×) and cutoff (300 000 dalton) DIC-polymer (Δ). Normal viscosity of plasma (PL) and whole blood (BL) is indicated for comparison. Single value form other preparations are also given: ▣ stabilized hemoglobin 90 000 dalton [7]; ◉ polymerized in diluted solution with GDA, 200 000 dalton [8].

The values fitted by a straight line give

$$\nu = 0.031 \ c_{Hb} + 0.685 \tag{1}$$

Such behaviour is predicted by Einstein [9,10]:

$$\eta = \eta_0 \ (1 + a\varphi) \tag{2}$$

η (η_0) is the dynamic viscosity of the solution (solvent), φ is the volume fraction of the solute in the solution. Equation (2) is valid for spheres and ellipsoids [11]. For equal spheres a has the value 2.5 independent of the size. The relation between ν and η is: $\nu = \eta/\rho$, ρ = density. Taking 0.75 ml/g as partial specific volume for human hemoglobin [12] leads to a = 6.03.

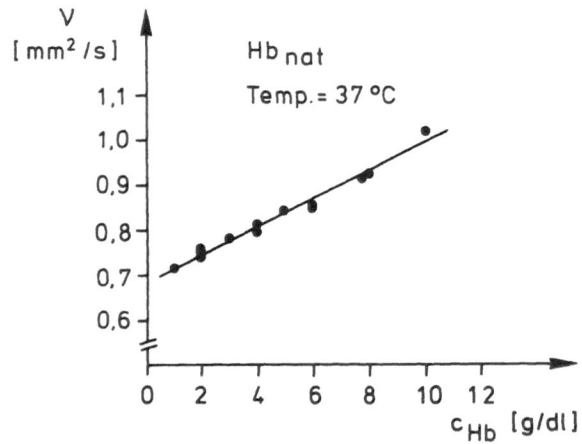

Fig. 2. Linearity between kinematic viscosity (ν) and concentration of native human hemoglobin.

So, hemoglobin fulfills the prediction of Einstein as concerns linearity, but not as concern the factor a: the increase of viscosity with concentration is 2.41 times higher. The last is not expected, since the hemoglobin molecule has not the shape of a sphere, but the sphere-shaped 4 subunits are arranged like a tetragon.

In contrast the viscosity of the GDA-polymer increases non-linearly with concentration. But for 5 g/dl it is still well below the viscosity of whole blood. The so called stabilized hemogobin [7] having 6 polyoxyethylene side chains possesses at 6 g/dl the same viscosity as the huge and broad distributed GDA-polymer. This demonstrates the great influence of the form of the molecule: As can be calculated from data of fig. 1 and 2 the relatively short polyoxyethylene side chains increase the specific viscosity (ν/ν_0-1) of hemoglobin by a factor 10.2. Compatible with our measurements are those with low molecular weight polymer of Kothe and Bonhard [8].

With respect to viscosity our GDA-polymer is suitable for exchange transfusion.

Einstein's theory of viscosity and the measurements shown here suggest to synthesize compact and uniform sphere-shaped O_2-binding macromolecules on order to achieve minimal viscosity independent from the size of the macromolecules. This is a very important aspect of the strategy for designing an artificial support of the oxygen carrying function. Molecular uniformity can be effected by ultrafiltration, as shown here, or by a special polymerisation procedure. Another possibility is a gentechnical approach.

But in case of GDA-polymer it is impossible to cutoff the small molecules. This indicates strongly that the GDA-polymer is unstable.

As our polymer has a very high molecular weight only a minor degree of degradation reduces the mean molecular weight greatly. Because of this big effect already a small degradation of the polymer can be detected easily by ultrafiltration combined with gelchromatography.

The binding points of the bifunctional glutardialdehyde are Schiff's bases which are known to decay slowly. So GDA-Hb-polymer when applied to organism releases presumably GDA which is toxic [13]. This explains the observed mild toxicity of thoroughly purified GDA-Hb-polymer in rabbits [14].

The instability of the GDA-polymer was the reason why we have polymerized human hemoglobin with DIC to huge molecules. The binding points in this case are urea derivatives which are known to be stable. A cutoff ultrafiltration (300 000 dalton) increases the molecular weight and reduces the nonuniformity greatly (see fig. 1). The kinematic viscosity of the fractionated polymer is definitely lower than that of the GDA-polymer, the viscosity equals that of plasma at concentration of 5 g/dl.

SUMMARY

The viscosity of highly and densely polymerized human hemoglobin is low enough in reasonable concentrations to serve as an effective oxygen carrying blood substitute. But polymerisation with glutardialdehyde leads to instable polymers. In contrast polymerisation with diisocyanate gives stable polymers which can be fractionated. Fractionation reduces viscosity. The viscosity of the fractionated polymers is at 5 g/dl as low as the viscosity of plasma.

We thank Miss B. Krumm and Mr. H. Pötzschke for having done carefully measurements.

REFERENCES
1. W.K.R. Barnikol and O. Burkhard, Highly Polymerized Human Haemoglobin for Oxygen Carrying Blood Substitue, <u>Advances in Biology and Medicine</u> 215:129 (1987).

2. W.K.R. Barnikol and O. Burkhard, Huge Compact Soluble Molecules: A New Old Concept to Develop an Oxygen Carrying Blood Substitute, J. Biomaterial, Artificial Cells and Artificial Organs, 1988, in press.

3. W.K.R. Barnikol and H. Pötzschke, Ein stabiles Polymere aus menschlichem Hämoglobin mit niedrigem kolloidosmotischem Druck als Kandidat eines sauerstofftransportierenden Blutersatzes, Hoppe-Seylers Physiologische Chemie 369: 793 (1988).

4. F. De Venuto and A. Zegna, Blood exchange with pyridoxolated and polymerized human hemoglobin solution, Surg., Gynecol., Obstetrics 155:342 (1982).

5. W.K.R. Barnikol and O. Burkhard, Verfahren zur Polymerisation von Hämoglobin mittels verknüpfenden Reagenzien, European Patent 85106057.4 (1985).

6. W.K.R. Barnikol and O. Burkhard, Verfahren zur Modifikation, insbesondere zur Polymerisation von Hämoglobin in vitro, German Patent P 3714351.4 (1987).

7. K. Iwasaki and Y. Iwashita, Preparation and evaluation of hemoglobin-polyethylen glycol conjugate (pyridoxalated polyethylen glycol hemoglobin) as an oxygen-carrying resuscitation fluid, Artif. Organs 10:411 (1986).

8. N. Kothe and K. Bonhard, Characterization of a modified stroma-free hemoglobin solution as an oxygen-carrying plasma substitute, Surg., Gynecol., Obstetrics 161:563 (1985).

9. A. Einstein, Eine neue Bestimmung der Moleküldimensionen, Ann. Physik 12:289 (1906).

10. A. Einstein, Berichtigung zu meiner Arbeit: "Eine neue Bestimmung der Moleküldimensionen, Ann. Physik 34:591 (1911).

11. Ch. Tanford, Physical Chemistry of macromolecules, John Wiley & Sons, 1967.

12. E. Antonini and M. Brunori, Hemoglobin and myoglobin in their reactions with ligands, Frontiers of Biology, Vol. 21, North-Holland Publishing Company, Amsterdamm, London, 1971.

13. E. Gendler, S. Gendler, and M.E. Nimni, Toxic reactions evoked by glutardialdehyde-fixed pericardium and cardiac valve tissue bioprosthesis, J. Biomed. Mater. Res. 18:727 (1984).

14. M. Feola, J. Simoni, P.C. Canizaro, R. Tran, G. Raschbaum, F.J. Behal, Toxicity of polymerized hemoglobin solutions, Surg., Gynecol, Obstetrics 166:211 (1988).

THE NATURE OF FLUOROCARBON ENHANCED CEREBRAL OXYGEN TRANSPORT

Leland C. Clark, Jr., Robert B. Spokane, Richard E. Hoffmann, and Ranjan Sudan*

Division of Neurophysiology, Children's Hospital Research Foundation, Cincinnati, Ohio 45229-2899 and *Department of Psychiatry, Wright State University School of Medicine, Dayton, Ohio 45401

INTRODUCTION

This brief communication is intended to epitomize our on-going research relating blood and brain fluorocarbon levels to polarographic oxygen availability (aO_2) as recorded from healed platinum electrodes in the brain of conscious rabbits. The current from such cathodes becomes stable in about ten days and remains steady for months, or even years, with mean currents of about 600 nanoamperes (nA) and oscillations at 10 per minute with an amplitude of up to 300nA. When oxygen is administered the aO_2 current increases to about 900nA. Following the administration of a variety of fluorocarbon emulsions and oxygen breathing the aO_2 current is further increased even though cerebral venous (sagittal sinus) pO_2 is not increased. The results presented here, based upon several hundred hours of continuous recording of aO_2 currents, were selected to emphasize this apparent paradox.

The reader is referred to excellent recent reviews on fluorocarbon physiology by Biro and Blais (1987) and by Faithfull (1987). Oxygen sensors are reviewed by Fatt (1976) and Hitchman (1978). There is a book on substrate specific (enzyme) electrodes by Turner (1987). Potentiometric hydrogen sensors for physiological use were first described by Clark and Misrahy (1956), by amperometry by Clark (1960), for intracardiac shunt detection by Clark and Bargeron (1959) and blood flow measurements have been reviewed by Young (1980). Harabin and Fahri (1987) have reported on blood gas transport in awake rabbits exposed to normobaric oxygen. Moore and Haenel (1988) have recently summarized the state of the art in oxygen monitoring in trauma.

METHODS

Most of the procedures used here have been described in previous publications from this laboratory (for example, Clark et al., 1958; Travis and Clark, 1965; Clark et al., 1988a; Clark et al., 1988b). For an interesting and authoritative historical account of blood gases, acids and bases consult Astrup and Severinghaus (1987).

Animals. Pasteurella-free white New Zealand rabbits, weighing between 2 and 5 kg, were obtained from Hazelton Laboratories, Pennsylvania.

TABLE 1. BRAIN SAMPLES

PERFLUOROCARBON CONTENT (MICROLITERS/100 GRAMS)

BRAIN SECTION	RABBIT 143 PP5/PP9			RABBIT 215 FC-43		RABBIT 219 FC-43		RABBIT 218 PP5/PP9	
	BEFORE HEAT	FIRST HEAT	SECOND HEAT	FIRST HEAT	SECOND HEAT	FIRST HEAT	SECOND HEAT	FIRST HEAT	SECOND HEAT
1	2.81	19.7	35.2	48.4	295	158	68.2	13.1	14.4
2	1.44	31.5	49.7	75.4	213	177	175	14.6	14.8
3	2.09	33.5	44.9	33.9	200	135	130	15.4	17.1
4	2.00	21.4	35.6	54.4	228	167	124	10.3	14.7
5	1.69	17.1	27.0	54.7	258	176	43.9	13.7	15.0
6	4.33	35.8	46.5	82.9	405	186	35.9	7.05	14.3
7	1.97	37.0	56.3	68.4	312	196	173	4.63	16.0
Mean	2.33	28.0	42.2	59.7	273	171	107	11.3	15.2
SD	0.98	8.3	10.0	16.8	71.4	20.0	58.2	4.09	1.01

Brains were sliced anterior to posterior in six slices. Section 7 is the cerebellum.
Electron capture gas chromatography of microwave heated samples in sealed bottles.

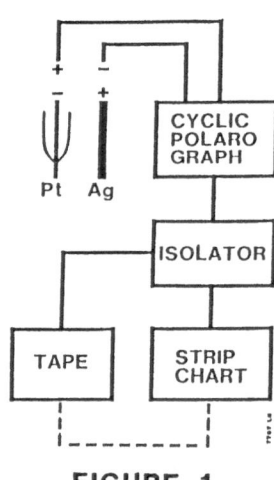

FIGURE 1

Electrodes were installed under sterile conditions using sodium pentobarbital anesthesia and oxygen breathing. Animals were fully recovered shortly after surgery; regular 6-8 hour sessions of monitoring were begun after a 10 day period. They were transported in special zippered canvas bags made here and immobilized in standard plastic restrainers during monitoring. Chlorpromazine, given subcutaneously, was used throughout as a tranquillizer. Antibiotics were used sparingly to avoid possible biochemical and electrochemical interferences from such potentially electro-active drugs. Animals were maintained in an airconditioned room, fed Purina rabbit chow ad lib, and had automatic access to water.

Instrumentation. The polarographic circuits used for this research were specially designed for 2- and 3-electrode steady state and cyclic polarography in living animals. Applied potential ranges and current densities are limited to avoid undue electrical, chemical, or physiological perturbations on or near implanted electrodes. The circuits have proven to be flawless, about a dozen are in constant use, and are remarkably stable and reproducible. They were manufactured by Dr. Jeff Huntington, Yellow Springs, Ohio. Outputs from the polarographs were led to optical-type isolators and then to either Model MR-10 or HR-10G tape decks manufactured by TEAC. Continuous recording of all voltammographic currents was also conducted using conventional strip chart (paper) recorders (Figure 1).

Electrodes. All electrodes were made from 28 B&S gauge platinum wire by sealing flame-cleaned wire in glass tubing having a matching coefficient of expansion. Glass-to-platinum seals must be perfect to function for long periods of time. The inner bore of the glass tubing was partly filled with indium to lessen the flexion strain on the wire from repeated fastening connections. The subcutaneous reference electrodes were fashioned from 18 gauge pure silver wire. Stainless steel screws and quick setting acrylate were used for anchoring the electrodes to the skull. Several hundred electrodes, all made in our laboratory, have been implanted, and kept operable for months, with but rare breakage. Problems from infection or from loosened mounts seldom arose.

Assembly. The system used for 2 cerebrocortical platinum and two subcutaneous silver electrodes is shown diagrammatically in Figure 2.

Blood sampling. Blood was collected in heparinized syringes or capillary tubes for measurement of blood gases and pH (Radiometer ABL30). Hematocrit and fluorocrit were measured with a clinical type microhematocrit (Guest-Siler) centrifuge. Lactate and glucose were measured with Yellow Springs Instrument Company analyzers. Samples were collected from arteries and veins in the ear and from the sagittal sinus using procedures published by Scremin et al, (1982) and shown in figure 3. Saggital sinus blood is the effluent from the same area of brain as the two aO_2 electrodes are normally placed.

Gas chromatography. Samples of blood and brain were analyzed by gas chromatography using an HP 5880A and a column generally at 120°C. 95% argon/5% methane was used as the carrier gas. The output from the electron capture detector was stored on an HP 9122 disc memory and printed out on an HP 3393A integrator. Suitable calibration standards were made by dilution of the vapor phase of stock standards of 2uL of fluorocarbon evaporated in sealed 120 mL serum bottles. Blood and brain samples were volatilized by heating in a microwave oven at 540 watts for 4 minutes, or until a stable reading was obtained.

Hydrogen. Hydrogen wash -in and wash -out curves were recorded from the same platinum electrodes (figure 1) used for the aO_2 measurements. For hydrogen, the platinum was maintained at plus 0.65 volts versus the

TABLE 2. FLUOROCARBON IN BLOOD AND BRAIN GC PEAK
HEIGHTS AND RATIOS OF F-DECALIN(PP5)/F-METHYLDECALIN
(PP5)/F-METHYLDECALIN(PP9) RABBITS

SAMPLE RAB#	TYPE	PP5 PEAK HEIGHTS @4.05	@4.28	PP9 PK HT @5.20	RATIO @4.05 @4.28	RATIO PP5 PP9
	standard 659	392.56	209.72	225.91	1.87	2.67
	vapor 659	1517.45	766.04	259.58	1.98	8.80
143	blood	108676.00	69361.60	66110.60	1.57	2.69
143	brain sec. 1	2989.54	1761.76	1471.53	1.69	3.23
143	brain sec. 2	13613.30	8502.75	6877.11	1.60	3.22
143	brain sec. 3	22473.40	14398.10	10575.30	1.56	3.49
143	brain sec. 4	28070.90	18175.40	13440.20	1.54	3.44
143	brain sec. 5	16344.00	10415.30	7447.89	1.57	3.59
143	brain sec. 6	9435.94	5864.06	4704.73	1.61	3.25
143	cerebellum	19253.90	12222.30	9659.15	1.58	3.26
	standard 5/9	574.54	300.61	169.45	1.91	5.16
	vapor 5/9	2932.39	1567.46	236.29	1.87	19.0
218	blood	119.61	62.53	26.12	1.91	6.97
218	brain sec. 1	4988.00	3182.19	1392.03	1.57	5.87
218	brain sec. 2	8390.36	5494.49	2237.11	1.53	6.21
218	brain sec. 3	14324.20	9513.06	3690.53	1.51	6.46
218	brain sec. 4	15109.40	10038.30	3680.17	1.51	6.83
218	brain sec. 5	8644.55	5642.77	2114.35	1.53	6.76
218	brain sec. 6	2904.45	1779.27	809.25	1.63	5.79
218	cerebellum	12397.30	8141.38	3156.71	1.52	6.51

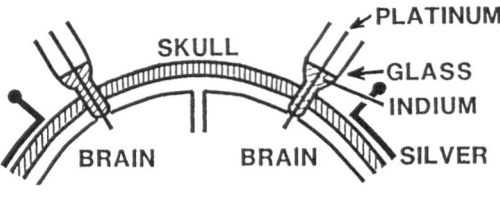

FIGURE 2

subcutaneous silver wire. The hydrogen curves were initiated by having
the awake rabbit breathe pure hydrogen gas for ten seconds. Hence, the
dissolved hydrogen was measured in the exact same brain space as the
dissolved oxygen. Local blood flow by the method Metzger (1988) presented
in a poster here would be better because the hydrogen could be generated
in the microcirculation near the measuring electrode.

Fluorocarbon emulsions. Fluorocarbon emulsions were made using mechanical
procedures (Microfluidics Model 110 TC, Newton, MA) and Pluronic F68
only as the emulsifer. Isotonic salts were added just before use in
the animal. Oxypherol was purchased from Alpha Therapeutics. Cultures
of our emulsions were taken and contaminated preparations set aside.
The F-tributylamine neat liquid was obtained from 3M as FC-43 or FC-47;
it is a mixture of many closely related substances. Pluronic F68 was
obtained from BASF Wyandotte. Fluorocarbons were purified when necessary
using diethylamine wash, reflux with sodium hydroxide, spinning band
column distillation or computer-assisted preparative gas chromatography
(Varex). Quality control was done by a color test developed here (Clark
and Moore, 1979; Clark, unpublished 1988) and by IR and UV spectroscopy
and other means.

RESULTS AND DISCUSSION

Figures 4,5,6 and 7 are playbacks of analog signals recorded on tape;
the time scale is greatly compressed. Oxygen (O_2) and carbogen (C) responses to 5 minutes of breathing the gas through a cup-mask are shown.
All the animals were conscious and appeared completely normal throughout
these experiments. Calibration signals (standard resistor) are not shown.
The width of the line is not noise but is the amplitude of normal physiological fluctuations in aO_2 current.

In figures 4 and 5 the remarkable increase in oxygen induced aO_2 currents
as a result of giving an emulsion of F-tributylamine is shown. This
enhanced current is still clearly present after the first two days even
though the blood FC level has fallen considerably by this time. On day
3,4, and 5 the air-breathng aO_2 and the oxygen and carbogen responses
are back to normal.

Blood glucose and lactate values were well within normal during the experiments shown in figures 4-7. Arterial blood gases and pH were within
the range of values to be expected. Over many years we have consistently
observed (thousands of analyses) an increase in arterial pO_2 as a result
of giving fluorocarbon emulsion, such increases were observed here.

Our work now involves the testing of a large number of pharmacological
agents, including free radical trapping agents, ion channel blockers,
and antiinflammatory agents. We are particularly trying to learn what
controls the "oxygen waves," (oscillation in aO_2 current at 10 per minute
and an amplitude of plus or minus 30 per cent of the mean) as well as
the oxygen enhancement of the aO_2 current with fluorocarbon. The aO_2
waves and the FC enhancement may both represent metabolic interactions
at the microcirculatory level. Both at present are not understood.
It is possible that fluorocarbons can travel between bilayer membranes,
where they may enhance oxygen transport, or that they are distributed
in the brain, via microglia where they enhance oxygen diffusion.

Oxygen diffusion through fluorocarbon is faster than through blood.
Polarographic techniques have been used to study oxygen diffusion in
cells growing in fermentation media (Ju et al., 1988; Linek and Vacek,
1988; Yang et al.,1988) but have not yet been developed to measure oxygen

TABLE 3. FLUOROCARBON IN BLOOD AND BRAIN GC
PEAK HEIGHTS AND RATIOS OF FC-43 RABBITS

SAMPLE RAB#	TYPE	FC-43 PEAK HEIGHTS @3.70	@4.13	@4.45	@5.80
	standard 478	189.24	191.71	443.12	100.72
215	blood	2962.69	4807.52	14963.90	4955.29
215	brain sec. 1	769.64	994.43	3094.24	823.80
215	brain sec. 2	1672.79	2542.35	8198.32	2774.16
215	brain sec. 3	3061.50	4617.18	15149.30	5478.57
215	brain sec. 4	4012.89	6165.69	20484.30	7453.04
215	brain sec. 5	4156.69	6264.62	20908.50	7606.27
215	brain sec. 6	3369.75	5217.13	17266.00	6267.40
215	cerebellum	3406.01	5459.14	18054.90	6834.44
219	blood	2275.52	3976.51	11984.30	4142.36
219	brain sec. 1	502.29	633.33	1898.67	498.36
219	brain sec. 2	2055.57	3420.65	11003.80	3974.67
219	brain sec. 3	2764.69	4821.41	15361.30	5739.90
219	brain sec. 4	3034.19	5122.09	16545.90	6255.48
219	brain sec. 5	1426.50	2360.76	7198.10	2416.78
219	brain sec. 6	953.16	1469.33	4389.68	1289.22
219	cerebellum	2289.18	3937.35	12484.50	4451.91

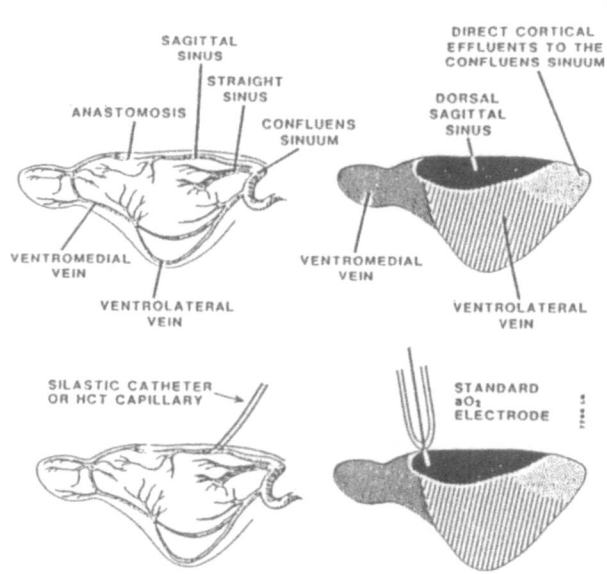

FIGURE 3

diffusion in brain in vivo, a method really needed. Vorob'ev et al. (1988) have proposed that modification of myocardial membranes by fluorocarbon explains its beneficial effect in small doses in ischemia.

Figures 6 and 7 show how the fluorocarbon enhancement is suppressed by the administration of methylprednisolone, followed by recovery of the enhancement when this antiinflammatory agent was discontinued. This tends to show that the enhancement involves the tissue, the brain parenchyma, and not the blood level. We showed some time ago that corticosteroids prolong the FC blood dwell time. Fluorocrits for 5, 10, 20, and 30 minutes centrifugation were percentage wise, for rabbit 71: 57, 37, 19 and 18 for rabbit 91: 56, 36, 18 and 18. This shows that the lack of FC ˙aO$_2$ enhancement in the methylprednisolone rabbit must be due to something other than a difference in circulating fluorocarbon level, or even a difference in circulating FC particle size or particle charge.

Osterholm (1983,1984) has published on the extravascular oxygenation of brain in hypoxia and ischemia and Bose et al. (1986) have described ventriculocisternal perfusion with fluorocarbon emulsions.

It is hoped that further understanding of the means by which relatively low doses of FC are beneficial to cerebral oxygenation may provide insight into their value in the treatment of stroke, ischemia, and trauma.

We have begun the analysis of brain tissue from fluorocarbon treated animals using gas chromatographic techniques and, because of its great sensitivity, electron capture detection. These procedures have been used for thousands of measurements of blood and liver tissue in our laboratory. We are also taking a closer look at the ultrastructure using EM.

To illustrate what such analyses can reveal we have selected data shown in figures 8, 9, and 10 and Tables 1 and 2. The letter A indicates the air peaks and here is a "proof" of the instrument sensitivity. The chromatograph in figure 8 is that from a standard prepared by complete volatilization of a 2uL liquid sample to 120cc and suitable dilution with air. The liquid is a mixture of 1/3 F-methyldecalin and 2/3 F-decalin. The vapor phase above this mixture and in the presence of an excess of liquid, was diluted with air to give the chromatogram shown in figure 9. Note the relatively large double peak, the F-cis/trans decalin pair, compared with the later peak complex, that of the F-methyldecalin isomers. This is due to the greater vapor pressure of the F-decalin. Totally volatilized fluorocarbons follow the gas laws, while fluorocarbons in contact with the liquid follow the vapor laws. Under certain steady-state conditions such information can be used to determine the presence of neat liquid via the composition of the vapor phase. We are in the process of an intense scrutiny of the interchanges of FC vapors and liquids in animals in steady states and various dynamic conditions. The curve shown in figure 10 was selected to illustrate the way in which tissue levels can be related to vapor phase in vivo. The brain analyzed here (Figure 10) came from a rabbit which had received 20cc/kg of a 10%v/v emulsion of F-methyldecalin followed by a 40cc/kg injection of a similar emulsion of F-decalin. The sample was obtained at autopsy thirteen days after the infusion. It can be seen that the F-methyldecalin is much higher than from a vapor phase standard (Figure 9). This is due, in this case, to the fact that the more volatile F-decalin is leaving the animal much faster than the F-methyldecalin. Prediction of vapor pressure conditions in alveolar and other microcirculation is not yet possible because of these dynamic, non-equilibria phenomena. (Clark et al., 1988c)

TABLE 4. LACK OF EFFECT OF PFC ON BRAIN VENOUS EFFLUENT

CONDI-TION	pO_2		pCO_2		pH		Lactate		Glucose		HCT		FCT 30 MIN	
	A	V	A	V	A	V	A	V	A	V	A	V	A	V
	mmHg		mmHg				mM		mg%		%		%	
Before Infusion														
air	75.1		38.5		7.33									
air	74.3		36.3		7.38		3.0		120		33			
O_2	550	32.6	36.1	48.7	7.41	7.33	1.7	1.7	107	101				
O_2	530	45.4	44.8	55.3	7.34	7.26	2.1	1.8	152	137	31	37		
60cc/kg. 10% v/v FC43/Pluronic														
O_2	691	36.0	33.9	46.7	7.36	7.27	1.6	1.4	91	82	28		16	
O_2	511	41.6	34.9	51.4	7.36	7.25	1.6		141		30	31	18*	17*
1 hour post infusion														
O_2	668	43.0	33.5	47.3	7.35	7.25	2.7		91		21	29	19	12
O_2		31.4		51.4		7.25		2.4		93				

Averages of rabbits 215 and 219. A=arterial, V=sagittal sinus.
*Fluorocrits at 5, 10, 20, and 30 minutes were 59, 49, 24, 18 and 57, 43, 20, 17.

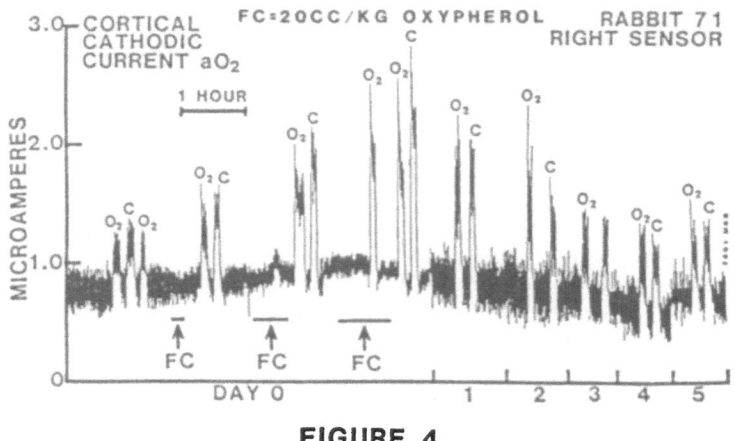

FIGURE 4

Table 1 illustrates the importance of, and difficulty of, vaporizing fluorocarbons from tissue samples. Allowing a small sample, e.g. 0.1 gm, to stand in a large sealed flask (500cc) for days, or even, weeks is not adequate to volatilize F-tributylamine isomers or even the more volatile F-decalin type compounds, as can be seen in the first 3 columns. Enzymatic, alkaline, and other wet digestion methods, as well as infrared beam heating, have not been as reproducible and as accurate as microwave heating.

Tables 2 and 3 are presented to illustrate the kind of information that one can obtain by gas chromatography. There are many possibilities for examining the relative fates of various isomers and analogs following their intravenous infusion and distribution in the body. We are currently compounding known mixtures of Varex separated fluorocarbons to learn more of the biophysical chemistry of these substances.

Hydrogen curves. In a standard 2 platinum electrode rabbit we recorded anodic hydrogen wash-out curves before and after the intravenous administration of 60cc/kg of commercial Oxypherol. Before infusion the calculated blood flow was 55cc/minute/100gm of brain for the right and 49 for the left. Shortly after FC-infusion, the blood flow was 78 for the right and 76 for the left. These calculations are based upon a single compartment model but may well be misleading or inconclusive because of complications introduced by the high solubility of hydrogen gas in the fluorocarbon phase and possibly also because the circulating blood may have to be regarded as a two compartment system due to the presence of fluorocarbon particles. We are investigating the effect of hydrogen solubility in fluorocarbons upon the hydrogen washout determinations done in infused rabbits. Oscillations in anodic hydrogen currents are not present.

Cerebrovenous pO_2

Table 4 summarizes data obtained by analysis of arterial, indicated by "A" and cerebrocortical venous (sagittal sinus), indicated by "V", blood samples before, directly following and one hour after infusion of 60cc/kg of a 10%v/v emulsion of F-tributylamine (FC43) made by mechanical homogenization. Two rabbits, anesthetized throughout with sodium pentobarbital, unlike the other experiments reported here on an aO_2 recording done wth conscious animals, were used and the data averaged. We endeavored to keep the arterial pCO_2 at 35 by the use of a Harvard respirator and endotracheal cannulation. Because oxygen breathing so greatly increases the aO_2 in fluorocarbon treated rabbits we expected to see a great increase in sagittal sinus pO_2. After all, it is common to observe mixed venous pO_2's over 100mm, often as high as 300. We were surprised to find little, if any, increase in cerebral venous pO_2. The brain was well oxygenated as judged by the lactate values and there was adequate glucose to glycolyze. Before our results are accepted as completely comparable, experiments must be extended to include measurements of sagittal sinus blood made concurrently with continuous aO_2 recording in unanesthetized animals. At present, though, it appears that aO_2 can be greatly increased without a concomitant increase in the pO_2 of the venous effluent from the same region of the brain.

GENERAL DISCUSSION

Transport of oxygen to the brain is measured here by recording the current from chronic platinum oxygen cathodes (0.60V vs. Sub Q Ag) implanted in the cerebral cortex of conscious rabbits. The in vivo currents from such healed cathodes are characterized by cyclic variations of about

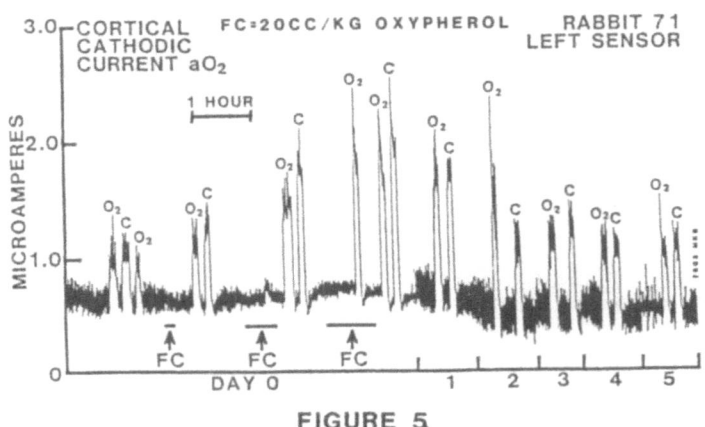

FIGURE 5

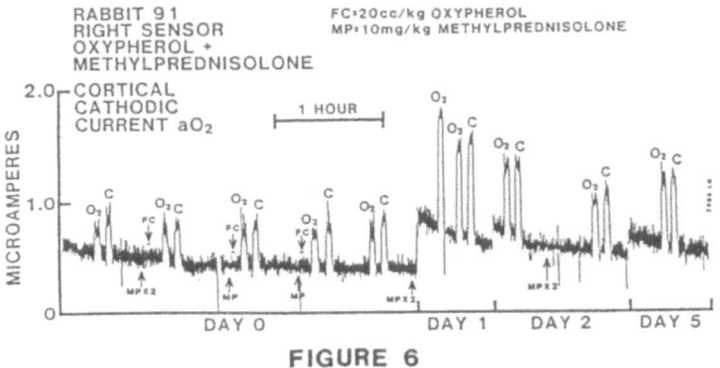

FIGURE 6

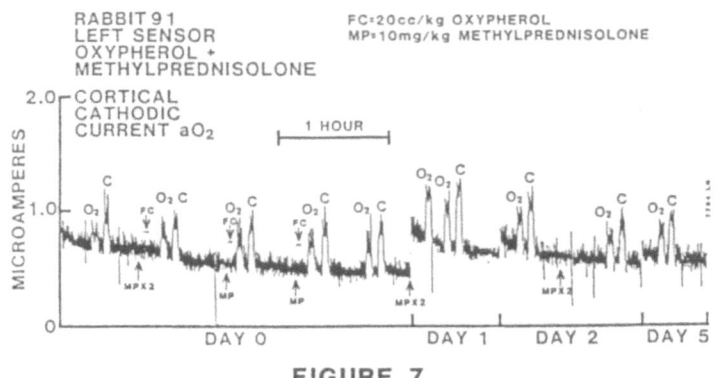

FIGURE 7

10 cycles per minute and an amplitude of up to plus or minus 30% of the mean. These cycles are not related to heart rate, fluctuations in arterial pressure, or respiration and, after a pause of thirty years since their discovery, are under intense study at present. This current represents the oxygen available (aO_2) to the cathode surface and hence it reflects transports by both flow and diffusion. The mean current increases by 50% on oxygen breathing and somewhat more on carbogen breathing. Stable but cyclic aO_2 currents are being routinely recorded from electrodes which have been in place over two years. As part of this study over 200 electrodes have been implanted in 100 rabbits and approximately 500 hours of continuous recording have been accomplished. Upon administration of fluorocarbon emulsions the most prominent observation is an increase, often double or triple that normally observed, in aO_2 current during oxygen or carbogen breathing. This bilateral fluorocarbon/oxygen enhanced aO_2 response is usually found after infusion of 6cc of fluorocarbon per kilogram of body weight, persists for two days, and is almost absent by the third day. There is no fluorocarbon-enhanced increase in aO_2 in the air-breathing phase of fluorocarbon administration. The oxygen enhanced aO_2 response is greatly decreased by large doses of methylprednisolone. For the first 5 days after fluorocarbon infusion the blood level of fluorocarbon is much higher than the brain. Later the brain and blood levels are nearly the same and finally it appears that the brain level is higher than the blood. For example in a rabbit which received 20cc/kg of a 10%v/v emulsion of F-methyldecalin followed by 40cc of a 10% v/v emulsion of F-decalin the blood level was 1.71uL/100 gm while they average brain content was 15.2uL/100gm (N=7).

Again, we have observed great variations in the rate at which fluorocarbon particles are packed during microcentrifugation, depending mainly upon the type of emulsion. A considerable overestimation of the fluorocrit can result, introducing great errors in the calculation of oxygen transport in vivo. The role of heparin and other anticoagulants in this process remains to be studied.

The clinical use of fluorocarbon emulsions in angioplasty and cancer chemotherapy and its probable widespread use as a blood substitute make it desirable to have rapid, accurate methods for its determination in blood and tissue and a better understanding of its function in oxygen transport, particularly in the microcirculation.

CONCLUSIONS

1. CHRONIC BRAIN-IMPLANTED PLATINUM ELECTRODES ARE VALUABLE TO MEASURE OXYGEN AND HYDROGEN TRANSPORT TO BRAIN.
2. OXYGEN-CONSUMING POLAROGRAPHIC CATHODES MIMIC METABOLIZING BRAIN CELLS AND INDICATE OXYGEN AVAILABLE (aO_2) BOTH BY MASS TRANSPORT AND BY DIFFUSION. HYDROGEN-CONSUMING ANODES INDICATE HYDROGEN AVAILABILITY (aH_2).
3. CIRCULATING FLUOROCARBON EMULSIONS DO NOT INCREASE BRAIN aO_2 BUT MAY INCREASE HYDROGEN TRANSPORT AT $FIO_2 = 0.2$.
4. OXYGEN BREATHING ($FIO_2 = 1.0$) INCREASES aO_2 BY 50%.
5. FLUOROCARBON EMULSIONS (6ML PFC/KG) INCREASE aO_2 BY OVER 100% AT $FIO_2 = 1.0$, A RESPONSE WHICH LASTS AT LEAST 24 HOURS.
6. THE OXYGEN-FLUOROCARBON aO_2 RESPONSE IS GREATLY INHIBITED BY METHYLPREDNISOLONE.
7. THE FLUOROCARBON CONTENT OF THE BRAIN PARENCHYMA IS INCREASED.
8. SAGITTAL SINUS PO_2 DOES NOT INCREASE WITH PFC EMULSION AND $FIO_2 = 1.0$.
9. AT PRESENT, THE MECHANISM OF THE FLUOROCARBON-ENHANCED OXYGEN AVAILABILITY INCREASE IN BRAIN IS NOT KNOWN. USE OF METHODS FOR MEASURING

ELECTRON CAPTURE GAS CHROMATOGRAMS

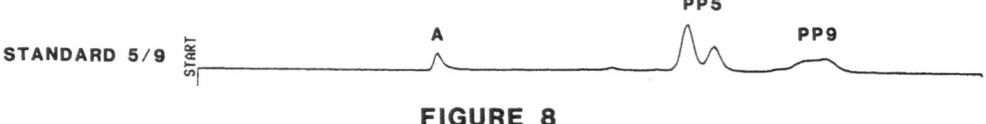

STANDARD 5/9

FIGURE 8

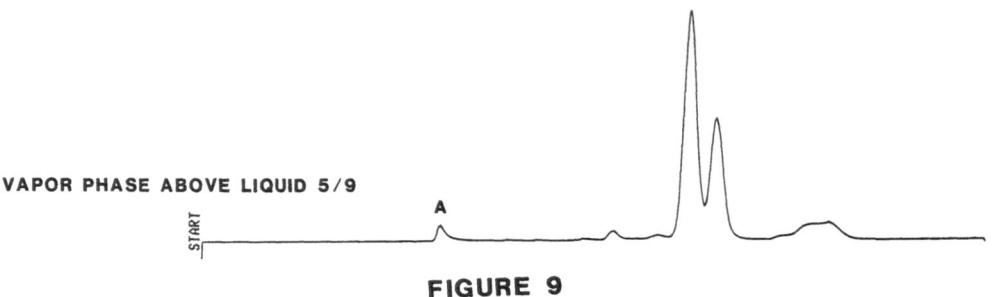

VAPOR PHASE ABOVE LIQUID 5/9

FIGURE 9

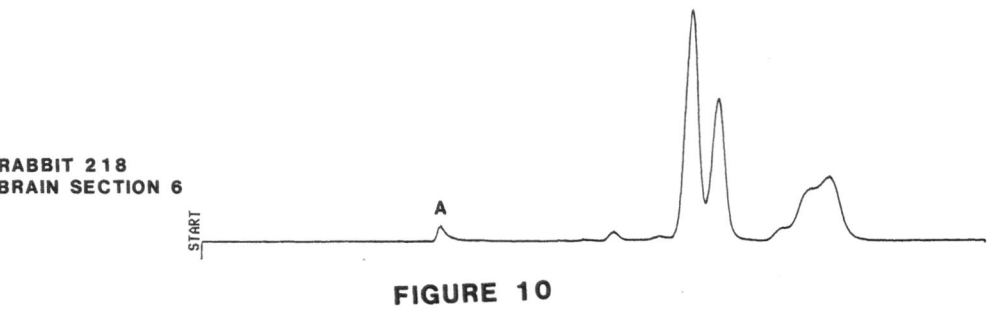

RABBIT 218
BRAIN SECTION 6

FIGURE 10

SPECIFIC SUBSTRATES IN REGIONAL METABOLISM NEAR THE ELECTRODE, WHICH
ARE BEING DEVELOPED, MAY BE NECESSARY TO UNDERSTAND THE EFFECTS
DESCRIBED HERE.

ACKNOWLEDGEMENTS

This work is supported by NIH Grants: HL39874; NS17975; DK31054. The
authors are indebted to Eleanor Clark, Linda Noyes and Estelle Riley
for their assistance in the preparation of this manuscript. Steve Jacobs,
Timothy Stroup, Mary Kay Burden, Stephanie Fouraker, and Daryl Franks
provided valuable laboratory assistance. We are grateful to Dr. Patricia
Tornheim for valuable guidance in neuroanatomical problems. We thank
Malte von Matthiessen, Yellow Springs Instrument Co. for the loan of
a Model 27 Analyzer. The cyclic bipolarographs, isolators, and interfacing
electronics were built by Dr. Jeff Huntington, Analytical Research, Yellow
Springs, Ohio. This research was greatly impeded by the procedures imposed
by the local IACUC.

REFERENCES

Astrup, P., and Severinghaus, J.W., 1987, "The History of Blood Gases,
    Acids and Bases," Munksgaard International Publishers, Copenhagen,
    Denmark.

Biro, G.P. and Blais, P. 1987, Perfluorocarbon blood substitutes, CRC
    Crit Rev in Onc/Hemat, 6:311.

Bose, B., Osterholm, J.L., Payne, J.B., and Chambers, K., 1986, Preser-
    vation of neuronal function during prolonged focal cerebral ischemia
    by ventriculocisternal perfusion with oxygenated fluorocarbon emul-
    sion, Neurosurgery, 18:270.

Clark, L.C., Jr., Spokane, R. B., Stroup, T. L., and Sudan, R., 1988a,
    Polarographic servo control of brain ascorbate in conscious rabbits
    as a model for glucose regulation. ASAIO Abstracts, 17:23.

Clark, L.C., Jr., Spokane, R. B., Hoffmann, R.E., Sudan, R., Homan,
    M.M., Maloney, A.C., Jacobs, S.J., Stroup, T.L., and Winston, P.E.,
    1988b, Polarographic cerebral oxygen availability, fluorocarbon
    blood levels and efficacy of oxygen transport by emulsions, Biomat.
    Art. Cells Art. Organs, 16:375.

Clark, L.C., Jr., Hoffmann, R.E., Spokane, R. B., and Winston, P.E.,
    1988c, Physiological evaluation of fluorocarbon emulsions with notes
    on F-decalin and pulmonary inflation in rabbits, Mater. Res. Soc.
    Symp. Proc., In press.

Clark, L.C., Jr., Misrahy, G., Fox, R.P., 1958, Chronically implanted polarographic electrodes, Journal of Applied Physiology, 13:85.

Clark, L.C., Jr., and Bargeron, L.M., Jr., 1959, Left-to-right shunt detection by an intravascular electrode with hydrogen as an indicator, Science, 46:797.

Clark, L.C., Jr., and Misrahy, G.A., 1956, An electrochemical method to measure hydrogen gas in biological tissues, Abstract, The XXth International Physiological Congress, Brussels, Belgium, July 30 - August 4.

Clark, L.C., Jr., Wolf, R., Granger, D., and Taylor, Z., 1953, Continuous recording of blood oxygen tensions by polarography, J. Appl. Physiol., 6:189.

Clark, L.C., Jr., 1960, Intravascular polarographic and potentiometric electrodes for the study of circulation, Trans. Amer. Soc. for Artif. Internal Org. 6:348.

Clark, L.C., Jr., and Moore, R.E., 1979, Diethylamine purification of perfluororchemicals for biological use. (Abstract) The American Chemical Society Fourth Winter Fluorine Conference, Daytona Beach, Florida, 1/28-2/2.

Faithfull, N.S., 1987, Fluorocarbons. Current status and future applications, Anaesthesia, 42:234.

Fatt, I., 1976, "Polarographic Oxygen Sensors," Robert E. Krieger Publishing Co.

Harabin, A.L., and Fahri, L.E., 1987, Blood-gas transport in awake rabbits exposed to normobaric hyperoxia, Undersea Biomed. Res., 14:133.

Hitchman, M.L., 1978, Measurement of dissolved oxygen, In: Chemical Analysis: A series of monographs on analytical chemistry and its applications, (eds. P.J. Elving, J.D. Winefordner), John Wiley & Sons, New York, Vol. 49.

Ju, L.K., Ho, C.S., and Baddour, R.F., 1988, Simultaneous measurements of oxygen diffusion coefficients and solubilities in fermentation media with polarographic oxygen electrodes, Biotechnol. Bioeng., 31:995.

Linek, V., and Vacek, V., 1988, Comments on validity of measuring oygen diffusion coefficients with polarographic oxygen electrodes, Biotechnol. Bioeng., 31:100

Metzger, H.P., 1988, The hydrogen gas clearance method for liver blood flow examination: inhalation or local application of hydrogen? Book of Abstracts, ISOTT, Ottawa, Quebec, Canada 8/7-8/11.

Moore, F.A., and Haenel, J.B., 1988, Advances in Oxygen Monitoring of trauma patients, Med. Instrumentation, 22:135.

Osterholm, J.L., Extravascular circulation of oxygenated synthetic nutrients to treat tissue hypoxic and ischemic disorders, Patent No.
4,393,863, Issued July 19, 1983.

Osterholm, J.L., Stroke treatment utilizing extravascular circulation
of oxygenated synthetic nutrients to treat tissue hypoxic and ischemic
disorders, Patent No. 4,445,886, Issued May 1, 1984.

Scremin, O.U., Sonnenschein, R.R., and Rubinstein, E.H., 1982, Cerebrovascular anatomy and blood flow mesurements in the rabbit, J Cerebr
Blood Flow Metab., 2:55.

Travis, R.P., Jr., and Clark, L.C. Jr., 1965, Changes in evoked brain
oxygen during sensory stimulation and conditioning. Electroencepha-
lography and Clinical Neurophysiology 19:484.

Turner, A.P.F., 1987, Biosensors. Fundamentals and Applications, APF
Turner, I Karube, GS Wilson (eds). Oxford University Press.

Vorob'ev, S.I., Iuanitskii, G.R., Ladilov, I.V., Obrazrsov, V.V., and
Skilfas, A.N., 1988, Modification of membranes with perfluorocarbons
as a possible mechanism of reduced ischemic damage of the myocardium,
Dokl. Akad. Nauk. USSR, 299:228.

Yang, X-M, Mao, Z-X, and Mao, W-Y, 1988, An improved method for determination of the volumetric oxygen transfer coefficient in fermentation
processes, Biotechnol. Bioeng., 31:1006.

Young, W., 1980, H_2 clearance measurement of blood flow: a review of
technique and polarographic principles. Stroke, 11:552.

THE EFFECT OF FLUOROCARBON EMULSION ON PLACENTAL INSUFFICIENCY

N.S. Faithfull and H.W. Marshall

Department of Anaesthesia, University of Manchester and
Northwest Injury Research Group, Manchester, UK

Oxygen-carrying plasma substitutes based on emulsified perfluorocarbons (PFCs) have undergone extensive experimental and clinical evaluation over the last 10 years. Though they were originally introduced as "artificial blood", it is clear that their use will probably be largely in more specialised areas such as improvement of microcirculatory oxygenation in, for instance, the treatment of myocardial and cerebral ischaemia (Bose et al. 1986; Faithfull et al. 1986) or improvement of oxygenation, and hence radiosensitivity, of malignant tumours (Teicher and Rose 1984).

PFC emulsions can enhance microcirculatory distribution probably because of their low viscosity, particularly at low shear rates, and the extremely small size, and hence good penetrability, of the emulsion particles (Faithfull et al. 1985). Recent work (Faithfull and Cain 1987) has suggested that they may improve oxygen diffusion through endothelial membranes. This is supported by the fact that haemodilution with PFC's causes increases in myocardial oxygen tensions (Rude et al. 1984) and improvement in cerebral oxygen diffusion (Erdmann 1982).

The clinical management of the growth retarded foetus is an important obstetrical problem. Maintenance of pregnancy under conditions of placental insufficiency is essential to ensure that a sufficiently mature foetus is delivered to lessen the risks of pulmonary hyaline membrane disease and respiratory distress syndrome. Experimental placental insufficiency obtained by severe restriction of uterine blood flow results in foetal hypoxia and contributes to growth retardation (Moll 1973). Maternal supplemental oxygen inhalation improves survival and growth of such foetuses (Vileisis 1985), almost certainly due to increases in maternal PaO_2 and increased pressure gradients for transplacental diffusion of oxygen. If PFC emulsions improve oxygen diffusion, they should improve foetal growth in the presence of placental insufficiency.

METHODS

Three groups of female rats, allowed free access to stock diets, were allowed to mate between 1900 and 1100 hours. On day 17 of gestation the uterus was exposed via a laparotomy incision performed under ketamine anaesthesia (20-30 mg ip) and the uterine artery supplying one horn of the uterus tied to produce unilateral placental insufficiency by the Wigglesworth method (Wigglesworth 1964) see Fig. 1. The number of foetuses in each horn were recorded. After closure of the abdomen the animals were kept in individual cages.

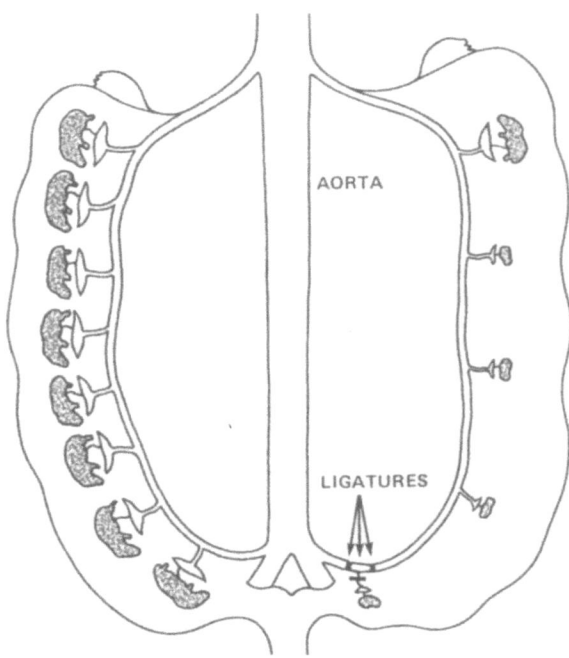

Fig. 1. Diagram to illustrate the placement of ligatures to produce placental insufficiency. Typical results at the conclusion of an experiment are given. Redrawn from Wigglesworth (1964).

One group of animals received daily injections of 10 ml/kg of FC-43 - a commercially available fluorocarbon containing plasma substitute. A second group received 10 ml/kg injections of 6% Hydroxyethyl-starch solution (HES) - the oncotic component of FC-43, while a third group received no further treatment. The injections were given intravenously via the tail vein and commenced on the day of operation. The pregnancies were allowed to progress to day 21 when the animals were sacrificed. Surviving foetuses were rapidly delivered by caesarian section and individually weighed.

Statistical analyses were performed using the paired Student t test and analyses of variance as appropriate. The nul hypothesis was rejected at a 'p' value of less than 0.05.

RESULTS

There were 15 animals in the no treatment (control) group, 14 in the FC-43 treated group and 16 in the group receiving HES. No animals died as a result of the operative procedures. Fig. 1 illustrates the results of a typical experiment and indicates the position of the vascular ligations performed.

The weights of the maternal rats are given in Table I. There were no significant weight differences between the groups at the beginning of the experiments (17 days). Though the animals in the control and HES groups showed statistically significant weight gains between gestation days 17 and 21, the animals in the FC-43 treated group had significant loss of weight over the same period.

Table I. Maternal weights (grams) at 17 and 21 days of gestation. Figures given are means and standard error of the mean. The number of observations is given in brackets. Statistical significance on paired t tests is indicated by asterisks - *** = p < 0.001; * = p < 0.05.

GROUP	17 DAYS	21 DAYS
NIL	304 5.1 (15)	313 * 6.1 (15)
HES	303 6.1 (16)	312 *** 5.8 (16)
FC-43	315 10.1 (14)	303 * 7.8 (14)

Foetal weights in the three treatment groups are shown in Fig.2 from both the operated and unoperated uterine horns. In all groups uterine artery ligation produced highly significant decreases in weight of the foetuses. In the unoperated horn, treatment with FC-43 resulted in significantly smaller foetuses than those obtained in the HES or no treatment groups. In the operated horn, FC-43 resulted in foetuses that were significantly smaller than in the no treatment group.

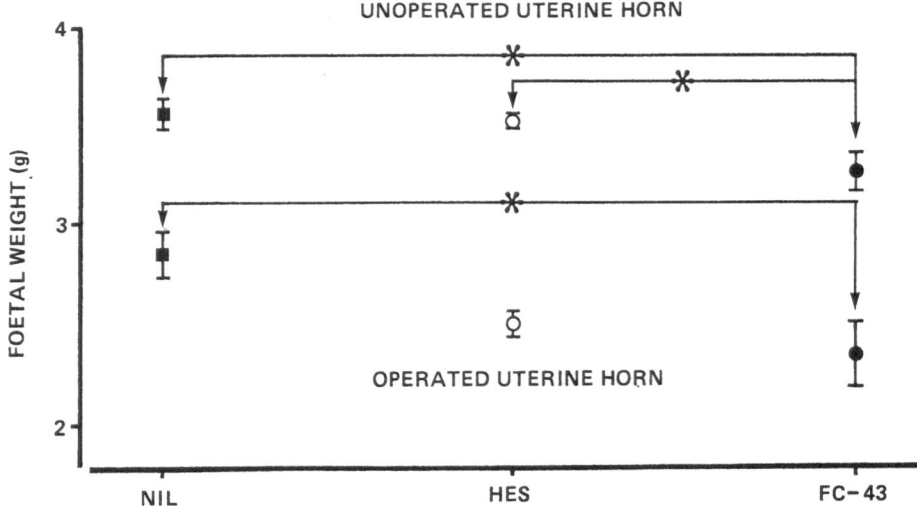

Fig. 2. Foetal weights from unoperated (top) and operated (bottom) uterine horns for the no treatment group (nil) and the animals receiving 10 ml/kg of hydroxyethyl starch solution (HES) or FC-43. Significance is given by the asterisk. * = p <0.05.

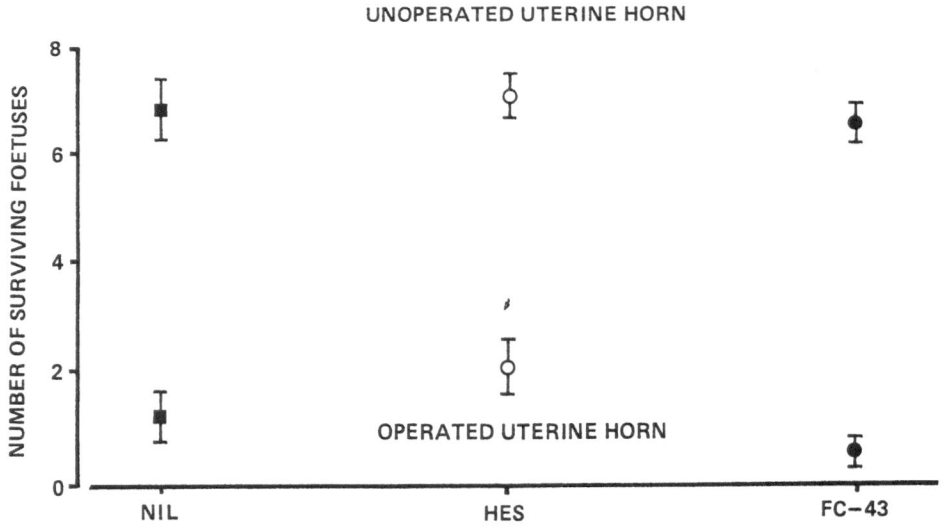

Fig. 3. Foetal survival from unoperated (top) and operated (bottom) uterine horns for the no treatment group (nil) and the animals receiving 10 ml/kg of hydroxyethyl starch solution (HES) or FC-43.

The number of foetuses surviving in each group in both operated and unoperated uterine horns is presented in Fig. 3. In all groups uterine artery ligation produced highly significant decreases in survival of the foetuses. No differences in survival in the unoperated horn were seen between the three treatment groups. However in the operated horns survival in the HES group was more than in the other two groups - this reached borderline significance (p = 0.06).

DISCUSSION

The results presented above would indicate that FC-43 was producing some form of toxicity in that it caused decrease in maternal weights and decrease in foetal weights in both the operated and unoperated uterine horns. Why should this be?

Little has been published on the long term effects of chronic PFC administration to pregnant animals and their foetuses. Naito et al. (1978) reported the results of administration to pregnant rats of a daily intravenous dose of 10 ml/kg of Fluosol-DA (FDA) - a fluorocarbon emulsion containing perfluoro-decalin and perfluoro-tripropylamine. The emulsion was given on gestation days 7 to 17 inclusive; the animals were sacrificed and the foetuses were examined on day 20. A control group of animals received injections of 0.9% saline.

The investigators reported that "During the experimental period of 20 days of gestation the weight of the animals treated with Fluosol-DA increased well but slightly inhibited as compared to that of the saline control group." They did not however comment on the apparently very markedly smaller weight gain between 17 and 20 days in the FDA group. The mean weights of the foetuses in the FDA group was less than that in the saline group and it was stated that this was not significant. Naito and colleagues did not state if they were presenting standard deviations of the means or standard errors of the mean. The figures given are in accord with our standard deviations for foetal weight. Our own recalculation of their figures based on this assumption reveal that the foetal weights were indeed significantly different in the two groups.

Acute haemodilution with FDA to a haematocrit of 5 percent was performed by Cefalo et al. (1985) in pregnant sheep respiring 100 percent O2. Increases in maternal PaO2 occurred in both FDA and control groups - though more so in the former. Foetal PaO2 increased in the FDA group; this decreased in the control animals - an unsurprising result in view of the lack of oxygen transport afforded by the ringers lactate and plasma that was administered. Based on foetal fluorocrits, (percentage of fluorocarbons layered in a spun haemtocrit tube), there appeared to be no gross placental transfer of the fluorocarbon particles. At similar haematocrits, similar maternal oxygen contents and similar cardiac outputs, the foetal arterial oxygen contents appeared significantly higher in the FDA group and this could be interpreted as confirming improved diffusion of oxygen accross the placental barrier.

Why then should chronic administration of FC-43, a very similar product to FDS, be so apparently toxic to both mother and foetus? Previous biological and clinical studies have indicated that the currently commercially available fluorocarbon emulsions FDA and FC-43 can produce acute reactions when injected intravenously. Vercellotti et al. (1982)

showed that FDA 20% activated complement in human plasma in vitro via the alternate pathway and that this was caused by the emulsifier Pluronic F68. This complement activation caused aggregation of granulocytes and they were able to induce in vivo granulocytopenia in rabbits. Clinical studies with FDA (Tremper et al. 1984) have revealed transient but very marked decreases in white cell and platelet counts which have largely recovered at 12 to 15 minutes after administration of the FDA test dose. These effects may or may not be accompanied by increases in pulmonary and systemic vascular resistances and may be attributed to release of C5a causing leucoembolism or leucostasis due to increased stickiness of polymorpho-nuclear neutrophils.

Reactions have been observed in various animal species and are accompanied by varying degrees of haemodynamic upset. For instance in the dog these reactions are accompanied by acute systemic hypotension; they are usually short lasting and were origionally thought to be of little consequence (Clark et al. 1970). However, microvascular reactivity may be impaired long after the reaction is apparently over, the degree of impairment being closely correlated with the severity of the reaction (Faithfull et al. 1987). In the dog FC-43 appears to be more potent in initiating reactions and these are in some respects more severe than those caused by FDA (Faithfull and Cain 1988).

It should be stressed that further administration of PFC emulsion after recovery from the initial reaction does not usually result in further changes. However it is not known how long impaired microcirculatory function remains. We also do not know how long it is before the animal will produce a further reaction when rechallenged by the origional emulsion. Our results could be explained on the basis of complement activation following each administration of FC-43.

Though haemodynamic and microvascular changes are usually attributed to leucoembolic phenomena, O'Brodovitch et al. (1985) have concluded that pulmonary hypertension following FDA in sheep is caused by vasoconstriction and postulated that release of prostanoids might be occuring. However, Shakir and Williams (1982) demonstrated that FC-43 can inhibit phospholipase A2 activity in vivo in rats and might interfere with release of arachadonic acid and subsequent reduction of prostanoids. Saeed et al. (1987) have shown that FC-43 inhibited contractile responses of isolated rabbit aortic strips to norepinephrine, seratonin and histamine; the agents responsible were Pluronic F68 and hydroxyethyl starch.

In rats and guinea pigs Steinberg et al. (1979) showed that FC-80 in albumen caused accelerated removal of platelets in isolated lung perfusion models. Colman et al. (1980) demonstrated decreases in platelet counts after injection of 2 to 8 ml of a 30-35% (v/v) emulsion of FC-80 emulsified in bovine albumin (these are in the range of volumes used in our experiments but contains about four times as much fluorocarbon). Two hours after administration platelets were 34% to 74% of initial levels. No falls were seen when similar volumes and strength of a cholesterol lecithin based emulsion was used.

In conclusion we have demonstrated that FC-43 is toxic to both mother and foetus when administered to pregnant rats between days 17 and 20. It is proposed that this toxicity is most likely caused by Pluronic F68 and is associated with chronic complement activation and microvascular dysfunction. Much work is underway in many centres to produce stable second generation emulsions with "cleaned up" emulsifiers; it would seem that the model used in these experiments would be useful for initial screening of these products.

REFERENCES

Bose, B., Osterholm, J.L., Payne, J.B. and Chambers, K., 1986. Preservation of neuronal function during prolonged focal cerebral ischemia by ventriculocisternal perfusion with oxygenated fluorocarbon emulsion. Neurosurgery 18:270.

Colman, R.W., Chang, L.K., Mukherji, B. and Sloviter, H.A., 1980. Effects of a perfluoro erythrocyte substitute on platelets in vitro and in vivo. J. Lab. Clin. Med. 95:553.

Clark , L.C., Kaplan, S. and Becattini, F., 1970. The physiology of synthetic blood. J. Thorac. Cardiovasc. Surg. 60: 757.

Erdmann, W., 1982. O2 diffusion coefficients. In:Oxygen Carrying Colloidal Blood Substitutes, Eds. Frey R, Beisbarth H, Stossek K. W Zuckschwerdt Verlag, Munich. p143.

Faithfull, N.S., Fennema, M., Erdmann, W., Lapin, R., Smith, A.R., van Alphen, W., Essed, C.E. and Trouwborst, A., 1985. Tissue oxygenation by fluorocarbons. Adv. Exp. Med. Biol. 180:569.

Faithfull, N.S., Erdmann, W., Fennema, M. and Kok, A., 1986. The effects of haemodilution with fluorocarbons or dextran on oxygen tensions in the acutely ischaemic myocardium. Brit.. J. Anaesth. 58:1031.

Faithfull, N.S., King, C.E. and Cain, S.M., 1987. Peripheral vascular responses to fluorocarbon administration. Microvasc. Res. 33:183.

Faithfull, N.S. and Cain, S.M., 1988. Critical oxygen delivery following haemodilution with dextran or Fluosol-DA. J. Crit. Care 3:14.

Faithfull, N.S. and Cain, S.M., 1988. Cardiorespiratory consequences of fluorocarbon rections in dogs. Biomat. Art. Cells and Art. Organs 16:463.

Moll, W., 1973. Placental Function and oxygenation in the foetus. Adv. Exp. Med. Biol. 37:1017.

Naito, R., Fujita, Y. and Suyama, T., 1978. Studies on teratogenicity of Fluosol-DA in rats. In: Proceedings of the Symposiun on Research on Perfluorochemicals in Medicine and Biology. V. Novakova and L-O. Plantin, eds. GOTAB, Stockholm.

O'Brodovich, H., Belbeck, L., Andrew, M. and coates, G., 1985. Fluosol-DA causes pulmonary hypertension and increased lymph lung flow in un anesthetised sheep. Clin. Invest. Med. 8:15.

Rude, R.E., Bush, L.R. and Tilton, G.D., 1984. Effects of fluorocarbons with and without oxygen supplementation on cardiac hemodynamics and Energetics. Am. J. Cardiol. 54:80.

Saeed, M., Hartmann, A. and Bing, R., 1987. Inhibition of vasoactive agents by perfluorochemical emulsion. Life Sci. 40:1971.

Shakir, K.M.M., and Williams, T.J., 1982. Inhibition of Phospholipase A2 acticity by Fluosol, and artificial blood substitute. Prostoglandins 23:919.

Teicher, B.A. and Rose, C.M, 1984. Oxygen-carrying perfluorochemical emulsion as an adjuvent to radiation therapy in mice. Cancer Res. 44:4285.

Tremper, K.K., Vercellotti, G.M. and Hammerschmidt, D.E., 1984. Hemodynamic profile of adverse clinical reactions to Fluosol- DA 20%. Crit. Care Med. 12:428.

Vercellotti, G.M., Hammerschmidt, D.E., Craddock, P.R., and Jacob, H.S., 1982. Activation of plasma complement by perfluorocarbon artificial blood: probable mechanism of adverse pulmonary reactions in treated patients and rationale for corticosteroid prophylaxis. Blood 59:1299.

Vileisis, R.A., 1985. Effect of Maternal Oxygen Inhalation on the fetus with growth retardation. Ped. Res. 19:324.

Wigglesworth, J.S., 1964. Experimental growth retardation in the foetal rat. Pathol. Bacteriol. 88:1.

ADENOSINE DEAMINASE IN STROMA-FREE HEMOGLOBIN SOLUTION IS NOT RESPONSIBLE FOR CORONARY VASOCONSTRICTION

K. Lawless, P. J. Anderson and G.P. Biro

Departments of Biochemistry and Physiology, University of

Ottawa 451 Smyth Road, Ottawa, Ontario, Canada K1H 8M5

INTRODUCTION

One of us (Biro, 1982), using an unmodified preparation of stroma-free hemoglobin solution (SFHS) to hemodilute dogs, observed that coronary blood flow was inadequate to maintain normal coronary sinus pO_2. In view of the likely involvement of adenosine in flow autoregulation of the coronary circulation (Berne, 1980) and of the presence of significant concentrations of adenosine deaminase (ADA) in human red blood cells (Lerner et al., 1970), the suggestion was made (Biro, 1982) that ADA activity in SFHS may have interfered with endogenous adenosine and thereby prevented adequate coronary vasodilation.

This suggestion seemed all the more plausible, since extraneous ADA has been used to test the "adenosine hypothesis" and has been found to blunt vasodilator responses presumably mediated by endogenously released adenosine (Saito et al., 1981; Dole et al., 1985). More recently, we measured ADA in SFHS and found it to be 64-70 U/ml (Biro, 1988). This activity is about 20-40 times that in perfusate and in cardiac interstitial fluid which significantly blunted the reactive hyperemia-responses (Saito et al., 1981; Dole et al., 1985).

While there is evidence to suggest that hemoglobin per se may be a potent vasoconstrictor, (Biro, 1988),the above associations suggest the need to test directly whether ADA present in SFHS may be responsible for the coronary vasoconstriction observed in isolated hearts perfused with SFHS-containing media (Vogel et al., 1986; Biro et al., 1988). Accordingly, we tested for coronary vasoconstrictor activity SFHS preparations from which we succeeded to remove significant proportions of ADA activity and tested directly perfusion medium containing extraneous ADA.

METHODS

Male Sprague-Dawley rats (220-280 g) were used to prepare Langendorff-perfused isolated hearts. The hearts were paced at 274/minute by electrodes attached to the atria. The Krebs-Ringer-Bicarbonate (KRB) perfusate was oxygenated in a temperature controlled glass reservoir kept at 37°C. The aortic cannula was attached, through a bubble-trap, to a Cole-Palmer roller pump which perfused the heart at a constant flow of 10-13 ml/minute. This flow rate was adjusted to maintain perfusion pressure between 50-55 mm Hg, measured through a side-arm in the cannula. A small saline-filled balloon was inserted into the left ventricular cavity and end-diastolic pressure was set at 0-5 mm Hg. A small syringe-pump was connected to the trap, so that the solutions to be tested could be added to the KRB. The test solutions were added to the perfusate flow at 1 ml/minute. In all cases, a thirty-second period of the addition of KRB by the syringe-pump was used to establish the effect of the added volume alone upon coronary perfusion pressure. Since this represented an 8-10% increment in volume-flow without coronary vasoactivity, it resulted in an 8-10% rise in perfusion pressure. The effect of test-solutions on perfusion pressure was "read" referenced to this baseline.

SFHS was prepared from the same batch of human blood, by phosphate precipitation (DeVenuto et al., 1977) and ultrafiltration methods (Sehgal et al., 1983). The initial preparation involved washing with 0.9% NaCl three times, followed by lysis in one volume of distilled water, centrifugation at 2000g and filtration through sterile gauze.

SFHS prepared by the phosphate precipitation method was produced by dialysis against 2.8 M potassium phosphate, followed by dissolving the crystals in distilled water. This was dialysed repeatedly, first against distilled water, then against KRB. The final concentration was adjusted to 5-6 g/dl. This product is referred to as SFHS1.

SFHS prepared by the ultra-filtration method was produced by filtration through a 0.22 μm Amicon filter. This was followed by filtration through a 100 KD molecular weight cut-off filter, and was concentrated using a 10 KD molecular-weight cut-off filter and dialysed against distilled water and finally against KRB. This product is referred to as SFHS2.

The supernatant resulting from lysis of red cells was also filtered through 0.22 μm ultrafilter and dialysed against distilled water and KRB. This product is referred to as lysate.

Methemoglobin was determined spectrophotometrically (Evelyn and Malloy, 1938). ADA activity was determined by the method of Hopkinson et al. (1969) whereby one unit of activity equals one mole of adenosine consumed per minute at 25°C. Removal of ADA was accomplished by phosphocellulose chromatography. Hemoglobin was bound to the phosphocellulose column (Sigma) under low ionic strength conditions (10 mM PIPES at pH=7.4) and was eluted with 100 mM PIPES at pH=7.4, yielding Fraction II, of reduced ADA activity.

We also tested the effect of the addition of KRB "spiked" with ADA (Sigma, prepared from calf intestinal mucosa) in concentrations of 25, 50 and 100 U/ml, bracketing the concentration of ADA present in SFHS1 and SFHS2.

RESULTS

Phosphocellulose chromatography reduced ADA activity in SFHS1, from 60.1 ± 2.2 to 7.8 ± 0.7 U/ml, achieving 87% purification. The average change in perfusion pressure was 27 ± 4% with SFHS1 and 28 ± 3% with Fraction II of SFHS1. Purification of SFHS2 reduced ADA activity from 36.0 ± 5.2 U/ml to 5.7 ± 0.9 U/ml, representing an 84% reduction of this activity. There was no difference in the change in perfusion pressure induced by SFHS II or its purified fraction. The findings are summarized in the scatter plot of Figure 1 showing no evident correlation between ADA activity and apparent coronary constrictor activity.

The addition of KRB "spiked" with ADA (25, 50, 100 U/ml) produced no change in coronary perfusion pressure, indicating that it did not cause coronary vasoconstriction directly.

In addition to the change in perfusion pressure, the presence of SFHS in the perfusate also caused a depression of contractile activity of the left ventricle which in some cases assumed dramatic proportions. Particularly, during perfusion with unpurified SFHS1, left ventricular systolic pressure and its maximal rate of development were drastically depressed. These findings are detailed in Table I.

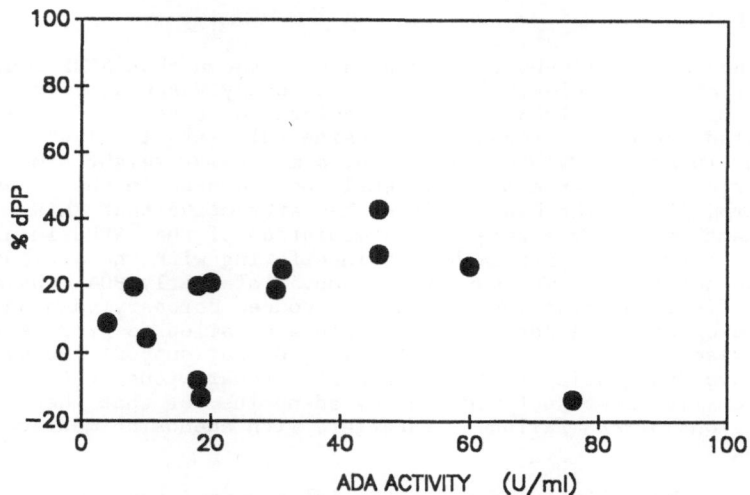

Figure 1. Scatter plot of the change in perfusion pressure against ADA activity of measured SFHS preparations. No significant correlation was found.

TABLE I. ADA activity and changes in left ventricular pressure and its derivative (dP/dt) during the addition of SFHS preparations to the perfusate, expressed as % change from control:

	ADA (U/ml)	Δ% LVSP	Δ% LV(dP/dt)
SFHS1	60.1±2.2	-88.5±10.6✓ *	-62.0±3.9✓ *
SFHS1/fr.II	7.8±0.7	-10.0±6.7	-12.0±4.6
SFHS2	36.0±5.2	-60.0±9.6✓	-40.0±6.9✓
SFHS2/fr.II	5.7±0.9	-43.8±6.0✓	-29.6±7.2✓

Abbreviations:
Δ% LVSP: percentage change from control in left ventricular systolic pressure;
Δ% LV(dP/dt): percentage change from control in rate of left ventricular pressure development;
fr. II: the eluate from the phosphocellulose column;
* indicates a difference in the effect of fraction II and starting material (p<0.05);
✓ indicates significant change from the control value.

It is clear that both SFHS1 and SFHS induced significant contractile depression; although quantitatively different, the degree of depression induced by the two preparations is not significantly different. Fraction II prepared from SFHS1 exhibited a highly significantly reduced depressant activity when compared to the starting material. In the case of SFHS2, however, the purification did not reduce the contractility depressant activity to a statistically significant extent.

DISCUSSION

A variety of experimental techniques have shown that SFHS induces a vasoconstriction in relatively isolated coronary vascular beds (Vogel et al, 1986; Biro et al, 1988). Such vascular beds also fail to respond in the expected manner to exogenous adenosine (Biro et al., 1988). ADA which is present in human erythrocytes, is of a molecular weight similar to that of a dimeric hemoglobin and is evidently not removed in the preparative process from SFHS. The hypothesis seemed attractive that this "contaminant" may play a role in the mediation of the SFHS-induced coronary constriction, presumably by interfering with the actions of endogenous adenosine. We found that removal of nearly 90% ADA activity does not affect quantitatively the SFHS-induced coronary vasoconstriction and that exogenous ADA added to the perfusate failed to produce any coronary vasoconstriction. These findings do not support the hypothesis stated above, suggesting that the "basal" coronary tone in these preparations is either not mediated by adenosine, or that these concentrations of ADA failed to interfere with adenosine-mediated processes.

We have thus eliminated one potential "contaminant" in SFHS which may be responsible for the observed constrictor activity. The alternative candidate thus appears to be hemoglobin <u>per se</u> or one of its spontaneous breakdown products. The neurosurgical literature (Wellum et al., 1982; Okwuasaba et al., 1981; Tonishima, 1980; Wellum et al., 1980; Toda et al., 1980; Cheung et al., 1980) contains strong evidence indicative of the vasoconstrictor potency of hemoglobin upon cerebral blood vessels. This constrictor potency is evident at low concentrations (10^{-6} M) which are roughly 10^{-3} times those present in SFHS preparations and 10^{-2} those present in the final mixture delivered to the perfused hearts.

Evidence is now available which suggests a mechanism whereby hemoglobin may affect vascular tone. Vascular endothelium will release, in response to a variety of exogenous and endogenous stimuli, a potent but exquisitely short-lived vascular smooth muscle-relaxing factor (Furchgott, 1984; Vanhoutte et al, 1986). The release of this "endothelium-derived relaxing factor" (EDRF) is inhibited by hemoglobin in concentration as low as 10^{-6} M (Martin et al., 1985, 1986; Fujiwara et al., 1986). It is not yet clear, however, whether EDRF is the mediator involved in physiologically (rather than pharmacologically) relevant vasodilator responses. Therefore, it is not clear whether the mechanism of hemoglobin's apparent vasoconstrictor-action involves its effect on EDRF-release. Such a possibility deserves further investigation.

Our experiments also revealed that, SFHS preparations contain, in addition to the coronary constrictor activity, a contractility-depressant principle. A substantial proportion of this depressant activity appears to be removed from SFHS1 by phosphocellulose chromatography. It seems reasonable that this activity is not directly related to ADA or to hemoglobin and would therefore appear that phosphocellulose chromatography may achieve the removal of other potentially deleterious constituents from SFHS. The identity of this contractility-depressant activity is not evident but a concentrated effort aimed at the identification of this activity is of paramount importance.

SUMMARY

Erythrocytes contain a high concentration of cytosolic adenosine deaminase and this enzyme activity is present in preparations of stroma-free hemoglobin (SFHS). The documented vasoconstrictor activity of SFHS preparations does not appear to be due to interference with endogenous adenosine-mediated mechanisms, because removal of >80% of adenosine deaminase activity failed to affect the vasoconstrictor potency of SFHS preparations. We have also demonstrated the presence of a contractility-depressant activity in SFHS when the latter was added to aqueous buffer perfusing rat Langendorff-hearts. This activity is diminished in phosphocellulose-purified preparations with reduced adenosine deaminase activity but is not necessarily causally related to this enzyme.

ACKNOWLEDGEMENTS

Financial support for these investigations was provided by the Defence and Civil Institute of Environmental Medicine of Canada.

We are grateful to Mrs. M. Masika for technical assistance.

REFERENCES

Berne, R. M., 1980, The role of adenosine in the regulation of coronary blood flow, Circ. Res., 47(6):807.

Biro, G. P., 1982, Comparison of acute cardiovascular effects and oxygen supply following haemodilution with dextran, stroma-free haemoglobin solution and fluorocarbon suspension, Cardiovasc. Res., 16: 194.

Biro, G. P., 1988, Blood substitutes and the cardiovascular system, Biomat., Art. Cells, Art. Org., 16:595.

Biro, G. P., Taichman, G. C., Lada, B., Keon, W. J., Rosen, A. L., and Sehgal, L. R., 1988, Coronary vascular actions of stroma-free hemoglobin preparations, Artif. Organs, 12:40.

Cheung, S., T., McIlhany, M., P., Lim, R., and Mullan, S., 1980, Preliminary characterization of vasocontractive activities in erythrocytes J. Neurosurg., 53:37.

DeVenuto, F., Moores, W., Y., Zegna, A., I., and Zuck, T., F., 1977, Total and partial blood exchange in the rat with hemoglobin prepared by crystallization, Transfusion, 17:555.

Dcle W. D., Yamada, N., Bishop, V. S., and Olsson, R. A., 1985, Role of adenosine in coronary blood flow regulation after reductions in perfusion pressure, Circ. Res., 56:517.

Evelyn, K., A., and Malloy, H., T., 1938, MIcrodetermination of oxhemoglobin, methemoglobin, and sulfhemoglobin in single sample blood, J. Biol. Chem., 126:655.

Fujiwara, S., Kassell, N. F., Sasaki, T., Nakagomi, T., and Lehman, R. M., 1986, Selective hemoglobin inhibition of endothelium-dependent vasodilation of rabbit basilar artery, J. Neurosurg., 64:445.

Furchgott, R. F., 1984, The role of endothelium in the responses of vascular smooth muscle to drugs, Ann. Rev. Pharmacol. Toxicol., 24:175

Hopkinson, D., A., Cook, P., J., L., and Harris, H., 1969, Further data on the adenosine deaminase (ADA) polymorphism and a report of a new phenotype, Ann. Hum. Genet., 32:361.

Lerner, M., H., and Rubenstein, D., 1970, The role of adenine and adenosine as precursors for adenine nucleotide synthesis by fresh and preserved human erythrocytes, Biochim. Biophys. Acta, 224:301.

369

Martin, W., Villani, G. M., Jothianandan, D,, and Furchgott, 1985, Blockade of endothelium-dependant and glyceryl trinitrate-induced relaxation of rabbit aorta by certain ferrous hemoproteins, <u>J. Pharmacol. Exp. Ther.</u>, 233(3):679.

Martin, M., Smith, J. A., and White, D. G., 1986, The mechanisms by which haemoglobin inhibits the relaxation of rabbit aorta induced by nitrovasodilators, nitric oxide, or bovine retractor penis inhibitory factor, <u>Br. J. Pharmac.</u>, 89:563.

Okwuasaba, F., Cook, D., and Weir, B., 1981, Changes in vasoactive properties of blood products with time and attempted identification of the spasmogens, <u>Stroke</u>, 12:775.

Saito, D., Steinhart, C. R., Nixon, D. G., and Olsson, R. A., 1981, Intracoronary adenosine deaminase reduces canine myocardial reactive hyperemia, <u>Circ. Res.</u>, 49:1262.

Sehgal, L. R., Rosen, A. L.,Gould, S. A., Sehgal, H. L., and Moss, G. S., 1983, Preparation and in vitro characteristics of polymerized pyridoxylated hemoglobin, <u>Transfusion</u>, 23:158.

Tanishima, T., 1980, Cerebral vasospasm: Contractile activity of hemoglobin in isolated canine basilar arteries, <u>J. Neurosurg.</u> 53:787.

Toda, N., Shimuzu, K., and Ohta, T., 1980, Mechanism of cerebral arterial contraction induced by blood constituents, <u>J. Neurosurg.</u>, 53:312.

Vanhoutte, P. M., Rubanyi, G. M., Miller, V. M., and Houston, D.S., 1986, Modulation of vascular smooth muscle contraction by the endothelium, <u>Ann. Rev. Physiol.</u>, 48:307.

Vogel W. M., Dennis, R. C., Cassidy, G., Apstein, C. S., and Valeri, C. R., 1986, Coronary constrictor effect of stroma-free hemoglobin solutions, <u>Am. J. Physiol.</u>, 251:H413.

Wellum G., R., Irvine, T., W. jr, and Zervas, N., T., 1980, Dose response of cerebral arteries of the dog, rabbit, and man to human hemoglobin, <u>J.Neurosurg.</u>, 53:486.

Wellum G. R., Irvine, T. W., and Zervas, N. T., 1982, Cerebral vasoactivity of heme proteins in vitro, <u>J. Neurosurg.</u>, 56:777.

OXYGEN BINDING OF MODIFIED HEMOGLOBINS IN SOLUTION

D. Mauldin, R.L. Dalpé, P.J. Anderson, and G.P. Biro

Departments of Physiology and Biochemistry
University of Ottawa, Canada, K1H 8M5

INTRODUCTION

The need for an oxygen-carrying blood substitute has been recognized for many years, and while various solutions and fluorocarbons have been tried, these have not met with the desired success. Though hemoglobin itself would seem a natural choice for this task, several undesirable properties must be overcome to render it suitable for large volume replacement applications. Hemoglobin, in concentrations appropriate for sufficient oxygen delivery to tissues, exhibits an undesirably high oncotic pressure, and is rapidly eliminated from the circulation. Additionally, in human red blood cells, the oxygen affinity of hemoglobin is partially regulated by 2,3-bisphosphoglycerate (BPG), (P_{50} = 24-30mm Hg), whereas hemoglobin, free in solution and devoid of BPG, has a much higher affinity for oxygen (P_{50} = 12-16mm Hg); its ability to unload oxygen to tissues thus is substantially impaired.

To decrease the oxygen affinity of hemoglobin in an extracellular environment, pyridoxal-5'-phosphate (PLP), which binds to the BPG binding site of hemoglobin, has been linked to hemoglobin by a stable bond by reduction using sodium borohydride[1]. We have altered the procedure along the lines suggested by Jentoft[2] by substituting sodium cyanoborohydride as the reducing agent. In the present work, we compare the properties of the modified hemoglobins using the two reducing agents.

METHODS

Stroma-free human hemoglobin solution prepared by ultrafiltration, and free of BPG, was diluted to 10 mg/mL with 20mM MOPS, pH 6.8; isoamyl alcohol (1μL/mL) was added to prevent foaming, and the mixture was deoxygenated by bubbling nitrogen for 30' in vented amber vials at 10°C. Pyridoxal-5'-PO_4 was added via the vent with a syringe at ten times the molar concentration of hemoglobin. SFHS was allowed to react with the aldehyde for 10 hr before addition of the reducing agent. Both reducing agents were added to a final concentration of twice that of the aldehyde. Nitrogen was bubbled continuously during all phases of the procedure, including the one hour reduction stage. After dialysis against Krebs Ringer Phosphate buffer (pH 7.4), the samples were concentrated in an

Amicon concentration unit to the original volume of SFHS and collected under nitrogen. Hemoglobin was determined by conversion to cyanmethemoglobin, and methemoglobin by the method of Evelyn and Malloy[3]. Oxygen affinities were measured using a Hemoscan (Aminco Instruments), and oxygen saturation with a Co-oximeter (Corning Instrumentation Laboratories), which also determined hemoglobin and methemoglobin in good agreement with the other methods used. Phosphate analyses were performed according to Ames and Dubin[4].

RESULTS

Figure 1 shows the oxygen saturation curves of various preparations of stroma-free hemoglobin solution in the presence and absence of 4 mM BPG. The control (SFHS) and cyanoborohydride ($SFHS_c$) curves both exhibit a plateau of saturation at 100 mm Hg O_2, while the borohydride curves ($SFHS_B$) indicate that saturation is incomplete at this pressure. The control is responsive to the addition of BPG, displaying both a significant shift to lower oxygen affinity and a decrease in cooperativity. This latter would effect significantly better oxygen unloading at PO_2 values around 30 mm Hg whereinsert such unloading would normally be required. Curve B in Figure 1

										+4 mM BPG	
Table 1 SUMMARY OF MEAUREMENT PARAMETERS											
Sample	P_{50}	Hill slope	[BPG]	PO_4 *	[Hb] g/dL	MetHb %	O_2 sat. %	O_2 content mL/L	O_2 capa- city	P_{50}	Hill slope
Fresh, whole blood	22.6	2.74	2.32	-	177.0	0.0	96.8	238	1.34	30.3	2.48
Red +, 9 days old	20.3	2.98	0.06	-	-	-	-	-	-	27.1	2.30
SFHS untreated	16.0	3.14	0.00	0.0	-	-	-	-	-	-	-
SFHS control	16.1	2.85	0.00	-	30.0	0-2.9	94.0	39	1.30	21.5	2.73
$SFHS-P_B$	16.1	2.06	0.00	1.8	39.0	7.0-22.5	81.1	45	1.15	17.0	2.06
$SFHS-P_C$	23.3	2.46	0.00	3.2	39.0	0-1.5	92.0	50	1.28	25.4	2.34

* mol/mol Hb

shows that $SFHS-P_C$ has similar characteristics to control SFHS in the presence of BPG; this contrasts with the oxygen saturation curve for $SFHS-P_B$ (C) which does not resemble either control or control+BPG curves. Figure 2 is a replot of published data[1] reporting the effect of pyridoxalation and subsequent reduction with sodium borohydride on oxygen binding, and is included for comparative purposes. Table 1 shows that controls incubated along with pyridoxalated samples manifest a somewhat lowered Hill slope compared with untreated controls; this result may be due in part to a tendency for hemoglobin to undergo spontaneous dissociation to dimers with time. No significant sensitivity to the addition of BPG is exhibited by SFHS-P produced by either the cyanoborohydride or borohydride methods. The reason for BPG insensitivity in the case of $SFHS-P_B$ is not known, but it is presumed that $SFHS-P_C$ is not sensitive to BPG because nearly all the BPG binding sites are occupied by PLP. This conclusion is borne out by the extent of phosphate incorporation, which is nearly equimolar with the hemoglobin concentration. These results are not complicated by the presence of endogenous BPG, as the ultrafiltered starting product is completely devoid of BPG. (See Table 1).

Oxygen saturation curves determined using the Hemoscan provide only a relative measure of oxygen saturation, since the 0 and 100% oxyhemoglobin values are set using potentiometers. We investigated the absolute values of oxygen saturation in our stroma free preparations using a Co-Oximeter, which provided the additional measures of O_2 content and methemoglobin levels. While SFHS controls and $SFHS-P_C$ samples showed oxygen saturation values between 92 and 96% at ambient PO_2's, $SFHS-P_B$ samples achieved only 81% saturation under the same conditions.

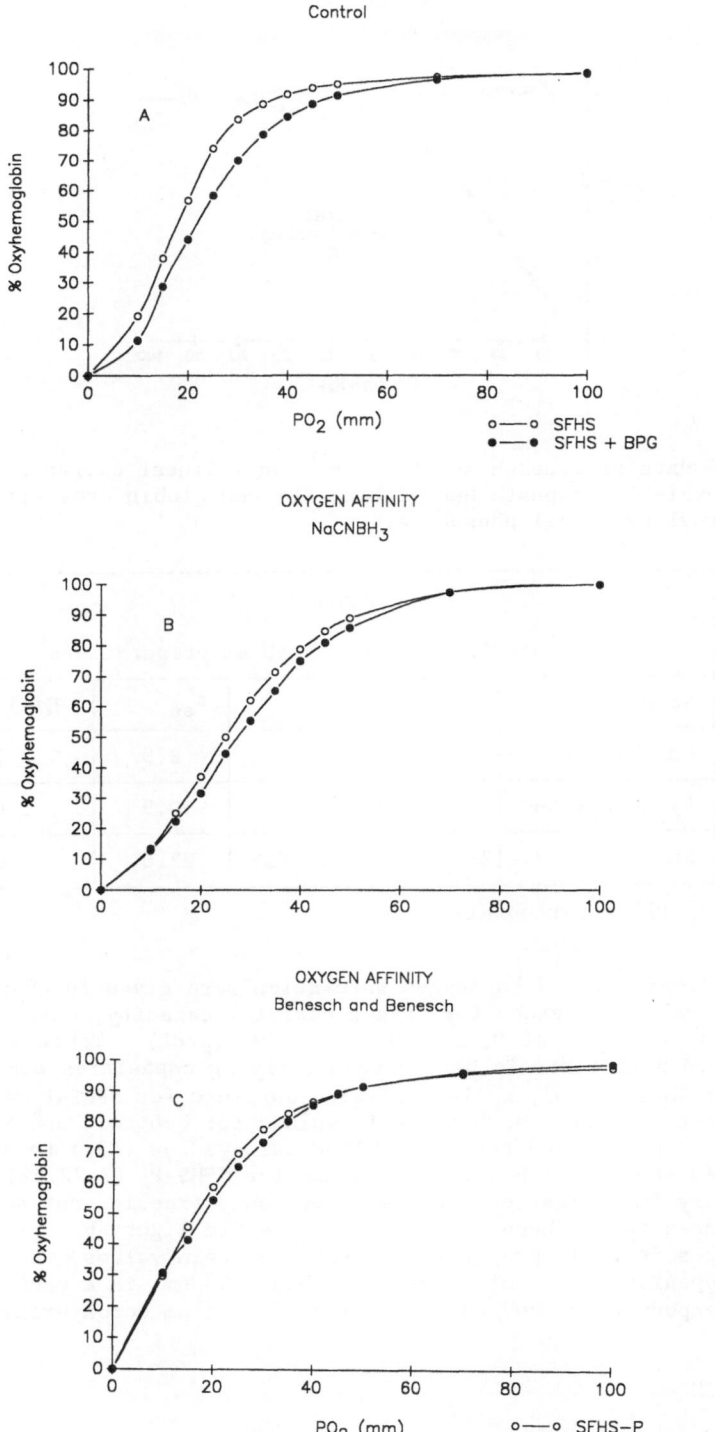

Figure 1

Oxygen affinities of various preparations of stroma free hemoglobin solutions in the absence and presence of 2,3-BPG. A) Unpyridoxylated. B) NaCNBH$_3$ reduction. C)NaBH$_4$ according to the method of Benesch and Benesch[1]. Note distinct lack of sigmoidal character of the curves in C.

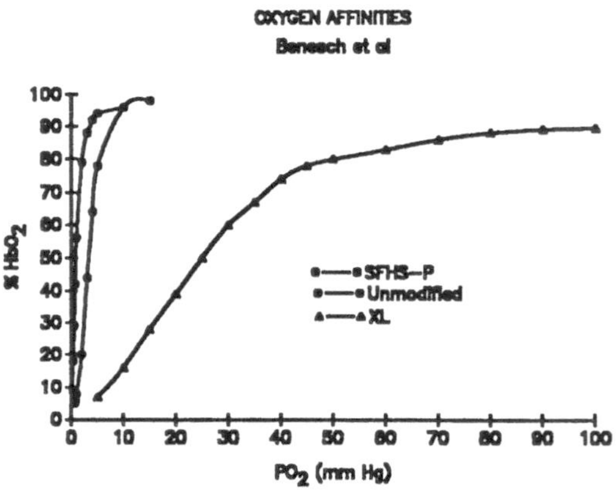

OXYGEN AFFINITIES

Benesch et al

%HbO$_2$

PO$_2$ (mm Hg)

●——● SFHS—P
■——■ Unmodified
▲——▲ XL

Figure 2
Replot of data of Benesch and Benesch[1] on a linear oxygen scale. SFHS-P: di-pyridoxal-5'-phosphate hemoglobin; KL: hemoglobin cross-linked with 2-nor-2-formyl-pyridoxal-phosphate.

Table 2			
Properties of Benesch et al preparations			
Sample	Ligand	P$_{50}$	Hill slope
Unmodified	-	2.9	2.24
Pyridoxalated	PLP	0.9	1.65
XL	2-nor-2-formyl-PLP	23.6	2.00

PLP — pyridoxal-5'-PO$_4$

The observations on oxygen saturation were given further substance from the values obtained for oxygen carrying capacity, calculated from the ratio of O$_2$ content (mL O$_2$/L solution) to Hb (g/dL). Table 1 shows that SFHS-P$_c$ and both controls have oxygen carrying capacities similar to that found for whole blood, while the value obtained for SFHS-P$_B$ was significantly lower. Methemoglobin values for controls and SFHS-P$_c$ were similar to those found for whole blood, always low (<3%) to absent, and contrasted sharply with values obtained for SFHS-P$_B$ (7-22.5%). In a preliminary investigation, we also have found (results not shown) that disturbances to the heme and porphyrin visible light absorption profiles, and changes in their proportion to total protein values, were accompanied by the appearance of a minor peak at about 500 nm, in a manner dependent on time of exposure to, and concentration of, sodium borohydride.

DISCUSSION

In this paper we compare the effects of reduction following pyridoxalation using two different reducing agents on P$_{50}$, Hill slope, oxygen carrying capacity and methemoglobin content of stroma-free hemoglobin preparations. Pyridoxalation has been widely adopted as a preliminary chemical modification in the production of hemoglobin solutions of potential use as blood substitutes[1, 5-10]. Our results indicated that on the basis of the properties measured, the mild reducing agent sodium cyanoborohydride produces a product that is superior to that produced when the commly used, less specific, reducing agent sodium borohydride is used. The original data on the effect of pyridoxalation on the oxygen carrying properties of hemoglobin [1], when replotted on the more common linear scale for oxygen (Fig. 2) indicate that the P_{50} of this product was 0.9 torr and the Hill slope of the binding curve was 1.65. These values indicate that this preparation, made with sodium borohydride as the reducing agent, would be of little use as a blood substitute capable of supplying oxygen to the tissues under physiological conditions. The product we have obtained with sodium borohydride appears to be superior to the modified hemoglobin obtained by Benesch et al, but even this degree of modification appears to be inadequate when compared to the properties of intraerythrocytic hemoglobin. It has been pointed out that the lack of specificity and danger of breaking peptide bonds render sodium borohydride a potentially unsatisfactory reagent in the chemical modification of proteins [2, 11, 12]. With respect to oxygen dissociation curves, cooperativity of binding and oxygen carrying capacity, reduction using cyanoborohydride yields a product which appears to offer more potential as a starting material for further chemical modification to produce a blood substitute of optimal physiological and biochemical characteristics.

CONCLUSION

Modification of the oxygen affinity of stroma-free hemoglobin solutions by covalent linkage with pyridoxal-5'-PO$_4$ is markedly decreased over older methods when NaCNBH$_3$ is used as the reducing agent. The oxygen-carrying capacity of the product modified by this method is not adversely affected compared with unmodified SFHS, and methemoglobin levels are very low. Oxygen affinities and subunit cooperativity values are very similar to those found in whole blood. SFHS-P prepared by this method appears to yield an acceptable starting product for further modification by crosslinking, and may be a better candidate for eventual use as an oxygen carrying blood replacement product.

SUMMARY

SFHS modified by covalent incorporation of pyridoxal phosphate using NaCNBH$_3$ as reducing agent had P_{50} values of 21-25mm Hg, and was insensitive to the addition of BPG. The Hill slope of the modified product was 2.34, demonstrating good functional cooperativity of the subunits. Oxygen carrying capacity, calculated as the ratio of O_2 content to the concentration of functional hemoglobin, was found to be about 1.3, very close to the "textbook" ratio of 1.34 for whole blood. These values were achieved with oxygen saturation values over 90% and methemoglobin levels of less than 3%. The results obtained from pyridoxalation using NaCNBH$_3$ contrasted sharply with those obtained using NaBH$_4$, which yielded a product having high oxygen affinity and high methemoglobin values. This latter product was poorly saturated with oxygen at ambient pressures and demonstrated a significantly lowered oxygen carrying capacity. The NaBH$_4$ product was also insensitive to the addition of BPG; the reason for this was not ascertained. SFHS-P by the borohydride method thus exhibited marked deficiencies relative to that prepared with cyanoborohydride, as defined by the characterizations employed for this study.

ACKNOWLEDGEMENTS

Financial support for these experiments was provided by a contract
from the Defence and Civil Institute of Environmental Medicine of Canada.

REFERENCES

1. R. Benesch and R. E. Benesch, Preparation and Properties of hemoglobin
   modified with derivatives of pyridoxal. Methods in Enzymology
   76:147-59 (1981).
2. N. Jentoft and D. G. Dearborn, Protein labeling by reductive alkylation,
   J. Biol. Chem. 254:4359-65 (1979).
3. K. A. Evelyn and H. T. Malloy, Microdetermination of oxyhemoglobin,
   methemoglobin and sulfhemoglobin in a single sample of blood. J.
   Biol. Chem. 126:655-62 (1938).
4. B. N. Ames and D. T. Dubin, The role of polyamines in the neutralization
   of bacteriophage deoxyribonucleic acid. J. Biol. Chem. 235:769-75
   (1960).
5. L. J. J. Hronoski, Review of modified hemoglobin solutions for use as
   temporary blood substitutes. DCIEM No 85-R-51.
6. W. Lieberthal, E. F. Wolf, E. M. Merrill, N. G. Levinski and C. R.
   Valeri, Hemodynamic effects of different preparations of stroma free
   hemolysates in the isolated perfused rat kidney. Life Sciences
   142(5):265-271 (1987).
7. D. H. Marks, D. L. Moore, F. Medina, G. Boswell, L. R. Zieske, R. B.
   Bolin and A. I. Zegna, Optimization of synthesis of pyridoxalated
   polymerized strom-free hemoglobin solution. Military Medicine
   153(1):44-9 (1988).
8. L. R. Sehgal, S. A. Gould, A. I. Rosen, H. L. Sehgal and G. S. Moss,
   Polymerized pyridoxalated hemoglobin: a red cell substitute with
   normal oxygen capacity. Surgery 94(4):433-438 (1984).
9. T. Tanishima, Cerebral vasospasm: contractile activity of hemoglobin in
   isolated canine basilar arteries. J. Neurosurg. 53:787-93 (1980)
10. W. M. Vogel, R. C. Dennis, G. Cassidy, C. S. Apstein, and R. Valeri,
    Coronary constrictor effect of stroma-free hemoglobin solutions. Am.
    J. Physiol. (Heart Circ. Physiol. 20) H413-H420 (1986).
11. G. E. Means and R. E. Feeney, Reductive alkylation of amino groups in
    proteins. Biochemistry 7:2191-201 (1968).
12. A. M. Crestfield, S. Moore and W. H.Stein, The preparation and
    enzymatic hydrolysis of reduced and S-carboxymethylated proteins. J.
    Biol. Chem. 238:622-7 (1963).

INTRAUTERINE GROWTH RETARDATION RELATED TO MATERNAL ERYTHROCYTE OXYGEN TRANSPORT

Glenn J.B. Mendoza
Department of Pediatrics
Beth Israel Medical Center
New York, N.Y.

Jaclyn Calem-Grunat
Department of Pediatrics
Mt. Sinai Medical Center
New York, N.Y.
(Student Summer Fellowship)

Bernard Z. Karmel
Department of Pediatrics
Mt. Sinai Medical Center
New York, N.Y.

Michael H. LeBlanc,
Department of Pediatrics
University of Mississippi
Medical Center
Jackson, MS

Edwin G. Brown
Department of Pediatrics
University of Mississippi
Medical Center
Jackson, MS

Frank Chervenak
Department of Maternal-Fetal
Medicine
Cornell Medical Center-NY
Hospital, New York, N.Y.

Richard W. Krouskop
Department of Pediatrics
LSU Medical Center
Shreveport, LA

Robert M. Winslow
Division of Blood Research
Letterman Army Instituteof Research
Research
San Francisco, CA

Intrauterine growth retardation (IUGR), is a term applied to fetuses who, at any week of gestation, are undersized for the duration of pregnancy. IUGR is a common problem contributing to 3.5% to 43% of all births (Lin and Evans, 1984). These infants frequently have asphyxia, meconium aspiration, persistent fetal circulation, polycythemia, hypoglycemia and congenital malformations. The causes of IUGR are diverse and can be attributed to maternal, fetal, placental or other unknown factors. Included in the latter group is a subset with inadequate oxygenation whose etiology remains elusive.

During pregnancy, the components of the maternal oxygen transport system must compensate for increased fetal oxygen demands in order to ensure adequate growth. Consequently, any decrease in oxygen delivery to the fetus could contribute to IUGR. Oxygen transport during pregnancy is partly dependent on hemoglobin-oxygen affinity (Sacks and Papadopoulos, 1984) as indicated by the P_{50} (the partial pressure of oxygen at half saturation) of the oxygen dissociation curve (ODC). Shifts in the glycolytic intermediate, red blood cell 2,3-Diphosphoglycerate (2,3-DPG), alter tissue oxygenation as the result of a change in the position or the shape of the ODC. A rise in maternal 2,3-DPG favors a rightward shift in the ODC, a decreased hemoglobin-oxygen affinity and therefore increased oxygen release to the tissues. Conversely, when the 2,3-DPG concentration is reduced, a leftward shift in the ODC occurs and less oxygen is released.

Several investigators have studied oxygen transport during pregnancy. Bauer et al (1969) compared 2,3-DPG levels in pregnant and non-pregnant women. They noted 2,3-DPG to increase 10% above controls at approximately two hours before delivery. Other researchers also have observed an increase in the concentration of 2,3-DPG during pregnancy (Rorth and Bill-Brahe, 1971; MacLennan et al, 1976; Weiss et al, 1982; Tsai and Leeuw, 1982; Madsen and Ditzel, 1984). Rorth and Bill-Brahe (1971) reported a 30% increase in 2,3-DPG in healthy pregnant women at term. MacLennan (1976) reported a fall in 2,3-DPG levels at the beginning of pregnancy, a significant rise during pregnancy and a rapid fall postpartum. Weiss et al (1976) noted a significant increase in 2,3-DPG concentration early in pregnancy, followed by a steady increase until delivery. More recently, Madsen and Ditzel (1984) looked at 2,3-DPG concentration and hemoglobin-oxygen affinity in normal pregnancies. They demonstrated a significant increase in 2,3-DPG levels from the first to the third trimester and a significant decrease postpartum.

The physiological significance of an increased 2,3-DPG during pregnancy has been discussed by Rorth and Bill-Brahe (1971). In uncomplicated pregnancies, hyperventilation leads to a respiratory alkalosis which is only partly compensated by renal bicarbonate excretion (Lucius et al 1970). The decreased arterial PCO_2 and increased pH lead to an increase in the oxygen affinity of maternal hemoglobin as a result of the Bohr Effect. Consequently, the oxygen-releasing capacity of maternal blood is decreased. Although the effect of hyperventilation facilitates transport of carbon dioxide from the fetus to the mother, it impairs oxygen release from the mother's blood to the fetus. However, the persistent respiratory alkalosis of pregnancy is thought to result in a compensatory increase in red blood cell 2,3-DPG ultimately facilitating oxygen transport to the fetus via a rightward shift in the ODC. It has been suggested that estriol may be a factor in stimulating the formation of 2,3-DPG (Madsen and Ditzel, 1984).

Few investigators have examined 2,3-DPG concentrations in women with "high-risk pregnancies". Weiss et al (1976) looked at the 2,3-DPG concentration in normal and hypertensive pregnant women and their newborn infants. At similar gestational ages, maternal 2,3-DPG was significantly higher among the hypertensive group compared to the controls. No difference in 2,3-DPG values was noted between infants of hypertensive mothers and those of normal pregnancies. Two infants of hypertensive mothers were noted to have IUGR and one of the mothers had a low 2,3-DPG level. MacLennan et al (1976) measured 2,3-DPG in normal and abnormal pregnancies and they noted no significant difference between 2,3-DPG values of pre-eclamptic and diabetic patients compared to normal third trimester subjects. There were, however, significantly lower 2,3-DPG values in mothers whose pregnancies resulted in IUGR and stillbirths.

It is the purpose of the present study to further investigate the relationship between fetal growth and maternal oxygen transport parameters. By considering red cell 2,3-DPG, hemoglobin-oxygen affinity (P_{50}), hemoglobin concentration, pH and PCO_2 further insights into the physiology of pregnancies complicated by IUGR might be obtained.

PATIENTS AND METHODS

Twenty-one pregnant women suspected of having IUGR by clinical and ultrasonographic assessment were studied during the third trimester. The antenatal diagnosis of IUGR was based on the estimated fetal weight derived by serial ultrasonographic measurement of the biparietal diameter. The course of each pregnancy was followed until delivery. Patients with evidence of diabetes mellitus, hemoglobinopathy, the TORCH group of infections or a chromosomally abnormal fetus were excluded from the study.

At the time IUGR was diagnosed, informed consent was obtained from the mother. The subject's hand was heated with a commercially available heating pad for fifteen minutes to 106 degrees Fahrenheit (Forster et al, 1972; Morgan et al, 1979) to arterialize the blood. Blood was then drawn anaerobically from the dorsal hand vein into a heparinized syringe. The arterialized venous blood was analyzed for pH and PCO_2 in a Corning 175 blood gas system. Hemoglobin concentration was determined by a Coulter Counter and electrophoresis was done on each sample. The red blood cell 2,3-DPG values were measured at 340 nm in a Bausch and Lomb Spectrophotometric 100 spectrophotometer according to the coupled enzyme system of Sigma Chemical Company (Sigma Tech Bullentin 35, 1982). The red cell 2,3-DPG values were expressed in mole/mole of hemoglobin. The 2,3-DPG:Hemoglobin (2,3-DPG:Hgb) molar ratio was calculated and the P_{50} was determined from a nomogram described by Samaja et al (1981). (Figure 1)

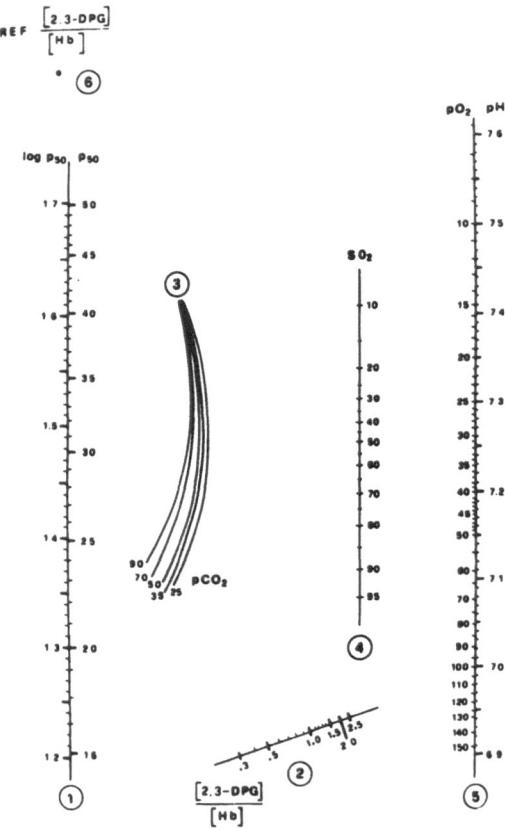

Figure 1. Nomogram relating PO_2, SO_2, pH, PCO_2, P_{50}, and the (2,3-DPG-Hgb) molar ratio in human blood at 37.0 C

M. Samaja, A. Mosca, M. Luzzana, L. Rossi-Bernardi and R. M. Winslow, Equations and nomogram for the relationship of human blood P_{50} to 2,3 Diphosphoglycerate CO_2 and H+. Clin. Chem. 27:1856, 1981 (with permission)

Umbilical artery velocimetry was done prenatally and the placentas were examined histologically postnatally whenever possible. The newborn infant's gestational age was assessed shortly following birth using the Ballard scoring system (Ballard, 1979). To assess the extent of IUGR, the growth curves constructed by Brenner et al (1976) were used as the standard. Deviation of the infant's weight at the estimated gestational age was calculated as a Z-score normalized to the growth curve.

The data were analyzed using the Pearson product moment correlation matrix, regression analysis and analysis of varience.

RESULTS

Infant Characteristics

The mean birth weight of the twenty-one infants was 2368 ± 572 gm ranging from 1040 gm to 3240 gm. The mean gestational age was 37 ± 2 wk and ranged from 31 to 40 wk. There were nine infants whose weight relative to gestational age Z-score was 1.295 standard deviations below the mean. These infants were at the tenth percentile or below according to the Brenner growth curve and were classified as small for gestational age (SGA). The mean birth weight of these infants was 1876 ± 481 gm and the mean estimated gestational age was 36 ± 2.3 wk. The mean gestational age when IUGR was diagnosed was 34.9 wk. The remaining twelve infants were appropriate for gestational age (AGA) and the mean birth weight of these infants was 2737 ± 283 gm. The mean estimated gestational age for the AGA group was 38.0 ± 1.3 wk.

Maternal Characteristics

Table 1 compares maternal 2,3-DPG:Hgb molar ratio, P_{50}, hemoglobin concentration, pH, and PCO_2 between the SGA and AGA groups. Maternal hemoglobin concentration, pH, and PCO_2 showed no significant relationship to the deviations from the Z-score estimate of weight relative to gestational age. The mean 2,3-DPG:Hgb molar ratio and P_{50} are significantly decreased in the SGA group.

Figure 2 shows a significant linear relationship of the maternal 2,3-DPG:Hgb molar ratio to the Z-score estimate of weight relative to gestational age.

Table 1. Maternal Oxygen Transport Parameters

	IUGR/SGA (N = 9)		CONTROL/AGA (N = 12)	
	Mean	SD	Mean	SD
2,3-DPG:Hemoglobin	0.86	$\pm$0.32	1.13	$\pm$0.25
P_{50} (mmHg)	27.20	$\pm$3.97	30.20	$\pm$2.50
Hemoglobin (gm/dl)	12.32	$\pm$2.28	11.19	$\pm$0.84
pH	7.42	$\pm$0.02	7.42	$\pm$0.03
PCO_2 (mmHg)	36.62	$\pm$3.56	31.55	$\pm$5.50

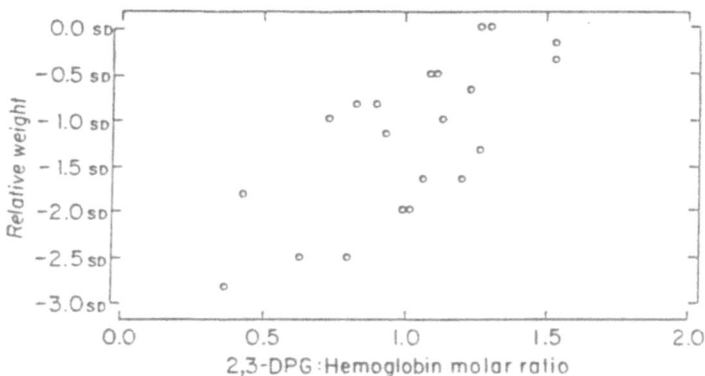

Figure 2. Relationship of maternal red cell 2,3 diphosphoglycerate:
hemoglobin molar ratio to birth weight and gestational age
normalized to the Brenner growth curve.

The correlation coefficient of this relationship is shown in Table 2.

Table 2. Correlation of maternal oxygen transport parameters
relative to birth weight normalized to the Brenner
Growth Cruve

	r	F	p
Maternal DPG:Hemoglobin molar ratio	0.72	20.61	0.001
Maternal P_{50}	0.69	16.90	0.001
Maternal DPG	0.65	13.59	0.01
Maternal Hemoglobin	−0.33	2.30	NS

The smaller the weight of the infant relative to gestational age the lower
the 2,3-DPG:Hgb molar ratio.

Figure 3 shows the relationship of the P_{50} to the estimate of weight
relative to gestational age.

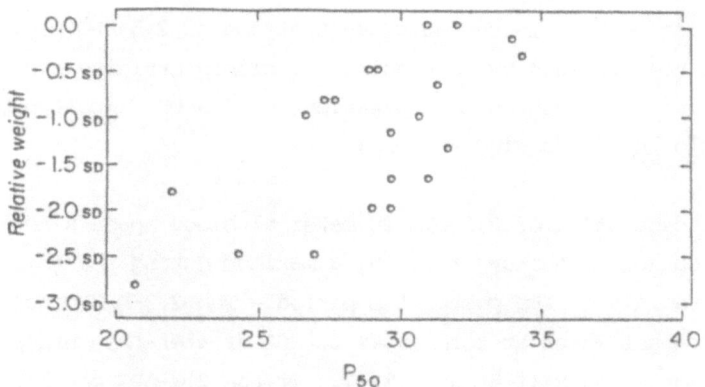

Figure 3. Relationship of maternal P_{50} to birth weight
normalized to the Brenner growth curve.

The P_{50} was also linearly correlated to the relative growth for
gestational age (Table 2). Lower P_{50} values were found among women whose
infants were of low birth weight relative to gestational age. That P_{50}

was strongly dependent on the 2,3-DPG concentration was verified by a very high correlation (R=0.91, p < 0.001) between these two oxygen transport variables.

Placental Characteristics

Histologic examination of six placentas in the IUGR group revealed evidence of vascular insufficiency and perivillous fibrin deposits in five of the six placentas (83.3%).

Umbilical Artery Velocimetry

Doppler velocimetry of the umbilical artery in eight fetuses revealed a higher than normal systolic/diastolic ratio with a mean of 3.8, (normal 3.0) in four of them who were IUGR. The other four fetuses had a normal ratio and were found to be appropriate for gestational age.

DISCUSSION

Red cell 2,3-DPG is known to be an important regulator of hemoglobin-oxygen affinity. Its role in tissue oxygenation is to shift the position of the oxygen dissociation curve whenever changes in its concentration occur. Increased concentrations of 2,3-DPG promote oxygen release to the tissues by decreasing hemoglobin affinity (increasing the P_{50}). Conversely, reduced concentrations of 2,3-DPG impair oxygen release from hemoglobin (decreasing the P_{50}).

The concentration of 2,3-DPG in maternal blood appears to increase gradually during pregnancy, reaching a maximum during the third trimester and falls rapidly in the postpartum period. The 2,3-DPG concentration of healthy pregnant women at term is 7% to 30% greater than non-pregnant controls (Rorth and Bill-Brahe, 1971). If the 2,3-DPG concentration should fail to rise during pregnancy the effect would be to increase the hemoglobin oxygen affinity by a left shift of the P_{50}. This could impair fetal oxygenation by the reduced net diffusion of oxygen across the placenta.

Cordocentesis, a method of sampling umbilical venous blood under ultrasound guidance, is being used to study blood gases and acid base parameters on the fetal side of the placenta in high risk pregnancies.

Preliminary work in 38 IUGR pregnancies has shown PO_2 values less than the mean for gestation in 87% and less than the fifth percentile in 36% of the cases. (Soothill and Nicolaides, 1987). There is clear evidence that IUGR is associated with diminished fetal oxygen.

It is not clear why the 2,3-DPG/Hbg fails to rise in women with an IUGR fetus. Madzen and Ditzel, 1984 found a direct correlation between 2,3-DPG and serum estriol in pregnancies complicated by diabetes. They noted these observations were consistent with the findings that patients with placental insufficiency who have low levels of serum unconjugated estriol tend to have a reduced concentration of 2,3-DPG. Serum unconjugated estriol might participate in mechanisms responsible for oxygen release to the tissues. Logically, any interference with fetal, placental and maternal estrogen metabolism could adversely alter maternal oxygen transport, thereby increasing the risk for intrauterine growth retardation.

The present investigation suggests that maternal oxygen transport abnormalities may result in inadequate oxygenation of the fetus and thereby play a role in the pathogenesis of IUGR. Our results suggest that the degree of IUGR in our study population is related, in part, to a reduced maternal 2,3-DPG:Hgb molar ratio and the resultant lowering of the P_{50}. We were unable to demonstrate a significant difference in maternal pH and PCO_2 between the SGA and AGA groups. These parameters, therefore, do not appear to be responsible for the decreased concentration of 2,3-DPG in the IUGR pregnancies and apparently have no effect on the position of the ODC. Any effect on P_{50} is primarily due to 2,3-DPG. Variations in certain maternal oxygen transport parameters do seem to affect fetal growth even at sea level. Alternatively, it is possible that those factors causing fetal growth retardation are responsible for the changes in maternal oxgyen transport parameters.

REFERENCES

Ballard, J. L., 1977, Simple assignment of gestational age, Pediatrics, 11:274.
Bauer, C., Ludwig, M., Ludwig, I., Bartels, H., 1969, Factors governing the oxygen affinity of human adult and fetal blood, Respir Physiol, 7:271-277.

Brenner, W. E., Edelman, D. A. and Hendricks, C. H., 1976, A standard of fetal growth for the United States of America, Am J Obstet Gyn, 126(5):555.

Forster, H. V., Dempsey, J. A., Thomson, J., et al, 1972, Estimation of arterial PO_2, PCO_2, pH and lactate from arterialized venous blood, J Appl Physiol, 32(1):134.

Lin, C. C. and Evans, M. I, 1984, "Intrauterine growth retardation: pathophysiology and clinical management", McGraw Hill, New York.

Lucius, H., Gahlenbeck, H., Lleine, H. O., et al, 1970, Respiratory functions buffer system and electrolyte concentrations of blood during human pregnancy, Respir Physiol, 9:311.

MacLennan, A. H., Emerson, P. M., Hunter, D. J., et al, 1976, Tissue oxygenation and red cell 2,3-DPG in normal and abnormal pregnancy, Br J Obstet and Gyn, 83:378.

Madsen, H. and Ditzel, J., 1984, Red cell 2,3-DPG and hemoglobin-oxygen affinity during normal pregnancy, Acta Obstet Scand, 63:399.

Madsen, H. and Ditzel, J, 1984, Red cell 2,3-DPG and hemoglobin-oxygen affinity during diabetic pregnancy, Acta Obstet Scand, 63:403.

Moore, L. G., Rounds, S. S., Jahnigen, D., et al, 1982, Infants birth weight is related to maternal arterial oxygenation at high altitude, J Appl Physiol, 52:695.

Morgan, E. J., Baidwan, B., Petty, T. L., et al, 1979, The effects of unanaesthetized arterial puncture on PCO_2 and pH, Amer Rev Resp Dis, 120:795.

Rorth, M. and Bill-Brahe, N. E., 1971, 2,3-DPG and creatine in the red cell during human pregnancy, Scand J Clin Lab Invest, 28, 271.

Sacks, L. M. and Delivoria-Papadopouos, M., 1984, Hemoglobin-oxygen interactions, Semin Perinatol, 8(3)168.

Samaja, M., Mosca, A., Luzzana, M., et al, 1981, Equations and nomogram for the relationship of human blood P_{50} with 2,3-DPG, CO_2 and H+, Clin Chem, 27:1856.

Soothill, P. W. and Nicolaides, K. H., 1987, Blood gases and blood flow in IUGR pregnancies. Soc of Perinatal Obstet, 7th Annual Mtg, (Unpublished communication).

Tsai, C. H. and Leeuw, N. K., 1982, Changes in 2,3-DPG during pregnancy and puerperium in normal women and in B-Thalassemia heterozygous women, Am J Obstet Gyn, 142:520.

Weiss, R. R., Roginsky, M. S., Melner, A., et al, 1976, Erythrocyte 2,3-DPG in normal and hypertensive gravid women and their newborn infants, Am J Obstet Gyn, 124(7):692.

TISSUE OXYGENATION AFTER PROLONGED ISCHEMIA IN SKELETAL MUSCLE:

THERAPEUTIC EFFECT OF PROPHYLACTIC ISOVOLEMIC HEMODILUTION

Michael D. Menger, Falk-Udo Sack*, Frithjof Hammersen⁺, and Konrad Messmer*

Dept. of General Surgery, University of Saarland, Homburg, Saar/FRG, *Dept. of Experimental Surgery, University of Heidelberg, Heidelberg/FRG, ⁺Dept. of Anatomy, University of Munich, Munich/FRG

INTRODUCTION

Tissue damage in skeletal muscle caused by prolonged ischemia is due to energy failure as consequence of inadequate oxygen supply of the cells. It is well known that not only ischemia per se but particularly reperfusion of the ischemic tissue induces deterioration of the nutritive blood supply (McCord, 1985; Schmid-Schönbein, 1987). Although the manifestations of reperfusion injury after prolonged ischemia, such as impaired oxygen supply to the tissue, edema and finally tissue necrosis are well known, the mechanisms, causing microcirculatory failure during reperfusion, are still a matter of controversy.

Recent studies on postischemic reperfusion injury have focussed on the microvasculature (Arfors and Smedegard, 1987; Schmid-Schönbein, 1987; Suval et al., 1987; Menger et al., 1988a). Microcirculatory failure during reperfusion within the microvasculature of postischemic skeletal muscle was first described by Harman in 1948. Lateron, Ames and coworkers (1968) demonstrated that hindrance of microvascular perfusion after ischemia in the rabbit brain leads to a 'point of no return'; they coined therefore the term "no reflow phenomenon". Subsequently postischemic microvascular failure was observed in the adrenal gland (Kovacs et al., 1966), kidneys (Summers and Jamison, 1971; Flores et al., 1972; Johnston and Latta 1977), myocardium (Krug et al., 1966; Poche et al., 1969; Kloner et al. 1974; Darsee and Kloner, 1980) and striated muscle (Strock and Majno, 1969; Romanus et al., 1977; Gidlöf et al. 1988).

Recently, we have demonstrated in an intravital microscopic study that reperfusion injury after 4 hours of tourniquet ischemia in skeletal muscle is characterized by a decrease of functional capillary density, a reduction of capillary RBC-velocity and an impairment of blood flow in postcapillary venules (Menger et al., 1988b).

Supported by a grant from the 'Deutsche Forschungsgemeinschaft', Me 900/1-1

Various factors have been considered instrumental in postischemic reperfusion failure, such as endothelial cell swelling (Poche et al., 1969; Gidlöf et al., 1988), leukocyte-endothelium interaction (Schmid-Schönbein, 1987; Vedder et al., 1988), impaired blood fluidity (Schmid-Schönbein and Rieger, 1981; Lewis, 1984; Menger et al., 1988a) and oxygen-derived free radicals (Parks et al., 1982; McCord, 1985; Granger et al., 1986).

Studies from our laboratory have demonstrated that prophylactic isovolemic hemodilution with dextran to a hematocrit of 30% has the potential to prevent microcirculatory deterioration in postischemic skeletal muscle as result of improved flow properties and flow conditions of the blood (Menger et al., 1988a). Mirashemi and coworkers (1987) suggested on the basis of a theoretical model that hemodilution to a hematocrit of 30-33% should be particularly effective in increasing the oxygenation of ischemic tissue, while it achieves comparatively small effects in nonischemic tissues.

The aim of the present study was therefore to analyze the effect of normovolemic hemodilution with Dextran 60 on the oxygenation of ischemic skeletal muscle tissue.

METHODS

For our studies we have used the hamster dorsal skinfold chamber, which contains skeletal muscle and skin tissue. The preparation allows for intravital microscopic observations and determination of tissue PO2 in the awake animal. In contrast to the hamster cheek pouch, mesenterium or the tenuissimus muscle preparation, in our model microcirculatory changes can be assessed for a prolonged period of time, avoiding the effects of anesthesia and surgical trauma (Endrich and Messmer, 1984).

The chamber and the implantation procedure have been described in detail (Endrich et al., 1980). In brief, under Nembutal anesthesia (50 mg/kg bw) two symmetrical teflon-coated aluminium frames were positioned on the dorsal skinfold, sandwiching the extended double layer of skin. One layer was completely removed in a circular area of approximately 15 mm, and the remaining layer, containing skeletal muscle and subcutaneous tissue, was covered with a removable cover slip, incorporated in one of the aluminium frames. In addition, two permanent catheters were passed from the dorsal to the ventral side of the neck and placed into the carotid artery and the jugular vein. A recovery period of 48 hours after the implantation procedure was allowed prior to the first measurements.

Ischemia was induced on the lower part of the tissue within the observation window by means of a transparent stamp, fixed to the skinfold chamber (Sack et al., 1987). An external pressure of approximately 40 - 50 mmHg was applied to the tissue by means of an adjustable screw clamp, just sufficient to empty the microvessels within the tissue. The non-compressed tissue in the upper part of the observation window served for investigation of changes in the non-ischemic tissue.

In order to achieve a hematocrit of 30%, isovolemic hemodilution was performed by withdrawl of a total of 1.4 ml blood from the carotid artery in three steps, and infusion of 1.4 ml Dextran 60 (Macrodex 6%, Schiwa, Glandorf/FRG). Mean arterial blood pressure and heart rate were monitored during the exchange procedure.

Tissue oxygenation in skeletal muscle was determined by means of a multiwire platinum electrode as described by Kessler and Grunewald (1969) and Lübbers (1969).

EXPERIMENTAL DESIGN

In group A (control) 9 Syrian golden hamsters (bw 60 - 80 g) were fitted with a dorsal skinfold chamber and two permanent catheters were implanted. After a recovery period of 48 hours tissue PO_2 was determined within the observation window of the chamber. Subsequently, pressure-induced ischemia was applied for 4 hours to the lower half of the window, followed by repeated measurements of tissue PO_2 at 15 minutes, 2 and 24 hours after release of ischemia.

In the 8 animals of group B, isovolemic hemodilution with Dextran 60 to a hematocrit of 30% was performed prior to induction of 4 hours of ischemia. Tissue PO_2 was analyzed prior to and 30 minutes after hemo-dilution as well as 15 minutes, 2 and 24 hours after release of ischemia.

Each determination of tissue PO_2 included measurements in the lower half of the observation window (ischemic area) as well as in the upper half (non-ischemic area). Histological analysis of the skinfold preparation was performed using semi-thin plastic sections and electron-microscopy after in vivo fixation of the animal with 3% glutaraldehyde.

STATISTICS

Data were tested for normal distribution and either an analysis of variance and t-test (normal distribution) or the Mann Whitney U-test (non normal distribution) was used to test significance between the groups. A paired t-test accompanied by Bonferroni probabilities (normal distribution) or Wilcoxon signed rank test (non normal distribution) followed by comparisons within the groups. Differences were considered significant at a $p < 0.05$ level.

RESULTS

In the control animals (group A) mean aterial hematocrit was 43.1 ± 2.3 % (n=9). Within the lower part of the observation window (ischemic area) tissue PO_2 in skeletal muscle had decreased significantly ($p < 0.01$) from 20.7 ± 2.4 mmHg (n=916) prior to ischemia to 8.8 ± 3.1 mmHg (n=928) upon 4 hours of ischemia followed by 15 minutes of reperfusion. After 24 hours tissue PO_2 was 15.6 ± 6.1 mmHg (n=900).

In the non-ischemic area (upper half of the observation window) tissue PO_2 decreased also from 20.2 ± 3.4 mmHg (n=852) to 14.6 ± 2.7 mmHg (n=928, $p < 0.05$) in the early reperfusion phase; 24 hours after induction of ischemia tissue PO_2 was 18.5 ± 6.1 mmHg (n=892).

In group B mean systemic hematocrit was reduced from 42.3 ± 2.1 % (n=8) to 29.4 ± 2.5 % by isovolemic hemodilution with Dextran 60. Alterations of mean arterial blood pressure and heart rate were not encountered. Tissue PO_2 in skeletal muscle increased significantly ($p < 0.05$) in response to hemodilution from 20.9 ± 1.6 mmHg (n=936) to 23.5 ± 2.5 mmHg (n=944). In contrast to control animals, upon 4 hours of ischemia and 15 minutes of reperfusion tissue PO2 amounted to 19.8 $\pm$ 6.8 mmHg (n=872) within the ischemic area and remained unchanged for 24 hours (20.0 $\pm$ 2.5 mmHg (n= 776)).

In correspondance with the results within the lower half of the observation window, tissue PO2 increased (p<0.05) from 20.6 $\pm$ 1.7 mmHg (n=960) to 23.5 $\pm$ 4.0 mmHg (n=912) within the upper half of the observation window (non-ischemic area) due to hemodilution. In this area no changes of tissue oxygenation were observed in the early reperfusion phase (23.0 $\pm$ 4.6 mmHg (n=864)) nor 24 hours after release of ischemia (19.7 $\pm$ 1.2 mmHg (n=768)).

Histological examinations of control animals after 4 hours of ischemia and 15 minutes reperfusion revealed swelling of capillary endothelium and accumulation of polymorphonuclear leukocytes (PMNs) within the microvasculature. In addition, adherence of PMNs to the endothelial wall of postcapillary venules was prominent. However, when isovolemic hemodilution with Dextran 60 was performed prior to ischemia, accumulation of PMNs was absent during the postischemic reperfusion phase.

DISCUSSION

Ischemia-Reperfusion

Reperfusion injury following prolonged ischemia is thought to be caused primarily by microvascular failure. Romanus and coworkers (1977) demonstrated in postischemic skeletal muscle impairment of capillary perfusion, associated with diminished blood flow in postcapillary venules. In parallel, quantitative analysis of microcirculatory changes in the hamster dorsal skinfold preparation revealed a decrease of functional capillary density to approximately 30% after 4 hours of ischemia and 15 minutes of reperfusion and only 50% of the initially perfused capillaries were reperfused 24 hours later (Sack et al., 1987).

Since maintainance of tissue oxygenation requires an adequate nutritional capillary perfusion, the decrease of functional capillary density during the early reperfusion phase is associated with a poor tissue oxygenation as demonstrated in the present study. However, a slight recovery of tissue PO2 was observed at 24 hours after release of ischemia.

Different pathophysiological mechanisms are discussed to be responsible for reperfusion injury. Gidlöf and coworkers (1988) postulated that in skeletal muscle swelling of only a few percent of capillary endothelial cells after ischemia would have the potential to alter the microvascular hydraulic conductance to reperfusion. Diminished vascular conductance appears as reduction or abolition of peak reactive reflow, delaying reoxygenation despite tissue needs.

In contrast, various authors have focussed their efforts to elucidate the mechanisms of ischemia-reperfusion injury in terms of leukocyte-endothelium interaction (Arfors et al. 1986; Schmid-Schönbein, 1987; Vedder et al., 1988) and thereby on the release of oxygen-derived free radicals (Del Maestro et al., 1980; McCord, 1985; Granger et al., 1986). Furthermore, impaired blood fluidity in low flow states could play an important role in the pathogenesis of the no reflow phenomenon following prolonged ischemia (Schmid-Schönbein and Rieger, 1981; Lewis, 1984; Menger et al., 1988a).

Hemodilution

Hemodilution as a therapeutic procedure originates from the experimental studies of Messmer et al. in 1969. These authors have lateron demonstrated that local tissue hypoxia did not develop despite a significant decrease in oxygen capacity of the blood as proved by measurements of spatial PO_2 in liver, kidney, pancreas, small intestine and skeletal muscle (Messmer et al., 1973). Further studies showed that isovolemic exchange of whole blood for colloids, such as Dextran 60/70, yields an increase in oxygen transport and delivery to the tissue (Sunder-Plassmann et al., 1975; Mirashemi et al., 1987). Hint (1968) and Duruble et al. (1979) had initially forwarded this idea based on theoretical considerations and computer models, respectively.

In the present study isovolemic hemodilution to a hematocrit of approximately 30% was associated with an increase in mean tissue PO_2 of skeletal muscle by 14%. This is in accordance with the data of Mirashemi et al. (1987), who reported that at a hematocrit of 30-33% the amount of oxygen delivered to the tissue should increase by 5-15%.

Hemodilution and Ischemia

From recent studies it was suggested that in ischemic tissues isovolemic hemodilution should be particularly effective in preventing microcirculatory disorders (Stucker et al., 1983; Hellberg and Källskog, 1986; Menger et al., 1988a) and increase tissue oxygenation (Sunder-Plassmann et al., 1981; Mirashemi et al., 1987). Stucker et al. (1983) demonstrated a better cardiac performance after ischemia when the hearts were reperfused at a hematocrit of 30%. These authors suggested that hemodilution results in a better distribution of oxygen. Measurements of tissue oxygenation in experimental studies (Sunder-Plassmann et al. 1981; Messmer et al., 1982) have shown that the average oxygen tension in ischemic skeletal muscle tissue can be shifted from 16 mmHg at normal systemic hematocrit to 32 mmHg when the blood is diluted to 29% hematocrit. Using a hydraulic analogue for simulation of the cardio-vascular system, it was demonstrated that in the case of severe ischemia the total increase in oxygen delivery at 30-33% hematocrit can reach 66%, due to less arteriolar oxygen diffusion and less arterio-venous shunting of oxygen (Mirashemi et al., 1987).

In the present study prophylactic isovolemic hemodilution to a hematocrit of approximately 30% had the potential to prevent tissue hypoxia during the entire postischemic reperfusion phase. This finding is in agreement with intravital microscopic studies, performed in the identical model in our laboratory. While 4 hours of ischemia in skeletal muscle revealed a marked reduction of functional capillary density accompanied by heterogeneous distribution of capillary perfusion, prophylactic hemodilution (hct 30%) was associated with only a slight impairment of capillary perfusion, and 24 hours after ischemia 90% of the capillaries initially perfused were permanently reperfused (Menger et al., 1988a). In addition, a homogeneous distribution of the perfused capillaries was observed.

Fischer and Ames (1972) attributed the absence of no reflow after prolonged ischemia to the higher perfusion pressure established with intentional hemodilution. However, while PMN accumulation within the microvasculature and PMN interaction with the endothelial wall of the microvessels was prominent after ischemia-reperfusion in non-diluted animals (hct 43%), this phenomenon was prevented by prophylactic hemodilution with Dextran 60 (hct 29%). Therefore, further studies have to elucidate whether prevention of postischemic reperfusion injury by hemodilution with Dextran 60 is a dilution dependent effect or due to pharmacological properties of Dextran per se.

SUMMARY

Prolonged ischemia is known to cause severe damage in skeletal muscle and skin as a result of reperfusion failure. Isovolemic hemodilution has been suggested as a modality to reverse microcirculatory disorders by improving flow properties and flow conditions of the blood. The aim of the present study was to investigate whether prophylactic isovolemic hemodilution could improve tissue oxygenation after 4h of pressure induced ischemia in skeletal muscle.

In 17 Syrian golden hamsters a dorsal skin fold chamber and two permanent arterial and venous catheters were implanted. Following a recovery period of 48h ischemia was induced for 4h by means of a transparent stamp compressing the tissue within the chamber. In 9 animals (control, hct 43%) measurements of tissue PO_2 (platinum multiwire electrode) were performed prior to and 15 min, 2h and 24h after release of ischemia. In 8 animals isovolemic hemodilution with Dextran 60 (hct 29%) was carried out prior to ischemia and measurements of local tissue PO_2 were performed as reported for the control group with an additional measurement 30 min after hemodilution.

In control animals tissue PO_2 decreased significantly ($p<0.01$) from 20.7 ± 2.4 mmHg prior to ischemia to 8.8 ± 3.1 mmHg after 15 min of reperfusion; after 24h tissue PO_2 was 15.6 ± 6.1 mmHg. In hemodiluted animals tissue PO_2 increased due to hemodilution from 20.9 ± 1.6 mmHg to 23.5 ± 2.5 mmHg ($p<0.05$); after 15 min of reperfusion tissue PO_2 was 19.8 ± 6.8 mmHg and remained unchanged for 24h (20.0 ± 2.5 mmHg). In control animals, histological examinations revealed swelling of capillary endothelium and accumulation of PMNs with adherence to the endothelial wall, while prophylactic hemodilution prevented these phenomena.

Four hours of pressure induced ischemia in skeletal muscle is associated with a decrease of nutritional blood flow and tissue PO_2. Prophylactic isovolemic hemodilution has the potential to reduce microvascular reperfusion failure and to ensure oxygen supply during postischemic reperfusion. There is good evidence that one of the mechanisms of action of hemodilution with Dextran 60 is the prevention of accumulation of PMNs and their interaction with the endothelial wall.

REFERENCES

Ames, A. III, Wright, R.L., Kowada, M., Thurston, J.M., and Majno, G., 1968, Cerebral ischemia. II. The no-reflow phenomenon, Am. J. Pathol., 52:437.

Arfors, K.-E., Lundberg, C., Lindbom, L., Lundberg, L., Beatty, P.G., and Harlan, J.M., 1986, A monoclonal antibody to the membrane glycoprotein complex CDw 18 (LFA), inhibits PMN accumulation and plasma leakage in vivo, Blood, 69:338.

Arfors, K.-E., and Smedegard, G., 1987, Permeability of macromolecules as affected by inflammatory cells, Prog. appl. Microcirc., 12:90.

Endrich, B., Asaishi, K., Götz, A., and Messmer, K., 1980, Technical report - A new chamber technique for microvascular studies in unanesthetized hamsters, Res. Exp. Med., 177:125.

Endrich, B., and Messmer, K., 1984, Quantitative analysis of the microcirculation in the awake animal, in: "Handbook of Microsurgery", W. Olszewski, ed. CRC press, Miami.

Darsee, J.E., and Kloner, R.A., 1980, The no-reflow phenomenon: A time-limiting factor for reperfusion after coronary occlusion? Am. J. Cardiol., 46:800.

Del Maestro, R.F., Thaw, H., Björk, J., Planker, M., and Arfors, K.-E., 1980, Free radicals as mediators of tissue injury, Acta physiol. Scand. (Suppl.), 492:91.

Duruble, M., Martin, J.L., and Duvelleroy, M., 1979, Effects theoriques, experimentaux, et cliniques des variations de l'hematocrite au cours de l'hemodilution, Ann. anesthesiol. Fr., 9:805.

Fischer, E.G., and Ames, A. III, 1972, Studies on mechanisms of impairment of cerebral circulation following ischemia: Effect of hemodilution and perfusion pressure, Stroke, 3:538.

Flores, J., DiBona, D.R., Beck, C.H., and Leaf, A., 1972, The role of cell swelling in ischemic renal damage and the protective effect of hypertonic solute, J. Clin. Invest., 51:118.

Gidlöf, A., Lewis, D.H., and Hammersen F., 1988, The effect of prolonged total ischemia on the ultrastructure of human skeletal muscle capillaries. A morphometric analysis, Int. J. Microcirc: Clin. exp., 7:67.

Granger, D.N., Höllwarth, M.E., and Parks, D.A., 1986, Ischemia-reperfusion injury: Role of oxygen-derived free radicals, Acta physiol. Scand. (Suppl.), 548:47.

Harman, J.W., 1948, The significance of local vascular phenomena in production of ischemic necrosis in skeletal muscle, Am. J. Pathol., 24:625.

Hellberg, O., and Kallskög, Ö., 1986, Influence of hematocrit in post-ischemic kidney damage, Int. J. Microcirc: Clin. exp., 5:279.

Hint, H., 1968, The pharmacology of dextran and the physiological background for the clinical use of rheomacrodex and macrodex, Acta anaesthesiol. Belg., 19:119.

Johnston, W.H., and Latta, H., 1977, Glomerular mesangial and endothelial cell swelling following temporary renal ischemia and its role in the no-reflow phenomenon, Am. J. Pathol., 89:153.

Kessler, M., and Grunewald, W.A., 1969, Possibilities of measuring oxygen pressure fields in tissue by multiwire platinum electrodes, Prog. Resp. Res., 3:147.

Kloner, R.A., Ganote, C.E., and Jennings, R.B., 1974, The "no-reflow" phenomenon after temporary coronary occlusion in the dog, J. Clin. Invest., 54:1496.

Kovacs, K., Caroll, R., and Tapp, E., 1966, Temporary ischemia of the adrenal gland, J. Pathol. Bacteriol., 91:235.

Krug, A., du Mesnil de Rochemont, W., and Korb, G., 1966, Blood supply of the myocardium after temporary coronary occlusion, Circ. Res., 19:57.

Lewis, D.H., 1984, The response of the microvasculature in skeletal muscle to hemorrhage, trauma, and ischemia, Prog. appl. Microcirc., 5:127.

Lübbers, D.W., 1969, The meaning of the tissue oxygen distribution curve and its measurement by means of Pt electrodes, Prog. Resp. Res., 3:112.

McCord, J.M., 1985, Oxygen-derived free radicals in postischemic tissue injury, N. Engl. J. Med., 312:159.

Menger, M.D., Sack, F.-U., Barker, J.H., Feifel, G., and Messmer, K., 1988a, Quantitative analysis of microcirculatory disorders after prolonged ischemia in skeletal muscle. Therapeutic effects of prophylactic isovolemic hemodilution, Res. Exp. Med., 188:151.

Menger, M.D., Hammersen, F., Barker, J.H., Feifel, G., and Messmer, K., 1988b, Ischemia and reperfusion in skeletal muscle, Prog. appl. Microcirc., 13:in press.

Messmer, K., Brendel, W., Sunder-Plassmann, L., and Holper, K., 1969, The use of colloidal solutions for extreme hemodilution, Bibl. haematol., 33:261.

Messmer, K., Sunder-Plassmann, L., Jesch, F., Görnandt, L., Sinagowitz, E., and Kessler, M., 1973, Oxygen supply to the tissues after limited normovolemic hemodilution, Res. Exp. Med., 159:152.

Messmer, K., Sunder-Plassmann, L., v. Hessler, F., and Endrich, B., 1982, Hemodilution in peripheral occlusive disease: A hemorheological approach, Clinical Hemorheology, 2:721.

Mirashemi, S., Ertefai, S., Messmer, K., and Intaglietta, M., 1987, Model analysis of the enhancement of tissue oxygenation by hemodilution due to increased microvascular flow velocity, Microvasc. Res., 34:290.

Parks, D.A., Bulkley, G.B., Granger, D.N., Hamilton, S.R., and McCord, J.M., 1982, Ischemic injury in the cat small intestine: Role of superoxide radicals, Gastroenterology, 82:9.

Poche, R., Arnold, G., and Nier, H., 1969, Die Ultrastruktur der Muskelzellen und der Blutkapillaren des isolierten Rattenherzens nach diffuser Ischämie und Hyperkapnie, Virchows Arch. (Pathol. Anat.), 346:249.

Romanus, M., Stenqvist, O., Haljamäe H.,0 and Seifert, F., 1977, Pressure-induced ischemia. I. An experimental model for intravital microscopic studies in hamster cheek pouch, Eur. Surg. Res., 9:444.

Sack, F.-U., Funk, W., Hammersen, F., and Messmer, K., 1987, Microvascular injury of skeletal muscle and skin after different periods of pressure induced ischemia, Prog. appl. Microcirc., 12:282.

Schmid-Schönbein, G.W., 1987, Capillary plugging by granulocytes and the no-reflow phenomenon in the microcirculation, Fed. Proc., 7:2397.

Schmid-Schönbein, H., and Rieger, H., 1981, Why hemodilution in low flow states? Bibl. haematol., 47:99.

Strock, P.E., and Majno, G., 1969, Microvascular changes in acutely ischemic rat muscle, Surg. Gynecol. Obstet., 129:1213.

Stucker, O., Trouve, R., Vicaut, E., Charansonney, O., Teisseire, B., Duruble, M., and Duvelleroy, M., 1983, Effects of different hematocrits on the isolated working rabbit heart reperfused after ischemia, Int. J. Microcirc: Clin. exp., 2:235.

Sunder-Plassmann, L., Kessler, M., Jesch, F., Dieterle, R., and Messmer, K., 1975, Acute normovolemic hemodilution changes in tissue oxygen supply and hemodilution oxygen affinity, Bibl. haematol., 41:44.

Sunder-Plassmann, L., von Hesler, F., Endrich, B., and Messmer, K., 1981, Improvement of collateral circulation in chronic vascular occlusive disease of the lower extremity, Bibl. haematol., 47:43.

Suval, W.D., Duran, W.N., Boric, M.P., Hobson, R.W. II, Berendsen, P.B., and Ritter, A.B., 1987, Microvascular transport and endothelial cell alterations preceding skeletal muscle damage in ischemia and reperfusion injury, Am. J. Surg., 154:211.

Vedder, N.B., Winn, R.K., Rice, C.L., Chi, E.Y., Arfors, K.-E., and Harlan, J.M., 1988, A monoclonal antibody to the adherence-promoting leukocyte glycoprotein, CD 18, reduces organ injury and improves survival from hemorrhagic shock and resuscitation in rabbits, <u>J. Clin. Invest.</u>, 81:939.

HYPEROXIC INTUBATION APNOEA: AN IN VIVO MODEL FOR THE PROOF OF THE CHRISTIANSEN-DOUGLAS-HALDANE EFFECT

Fritz Mertzlufft and Ludwig Brandt

Klinik für Anaesthesiologie,
Johannes Gutenberg-Universität
D-6500 Mainz, F.R.G.

INTRODUCTION

The Christiansen-Douglas-Haldane effect (HALDANE effect) describes the different CO_2 binding capacity of haemoglobin on its degree of oxygenation and was first demonstrated in vitro in 1914 (Christiansen, Douglas and Haldane, 1914).

In vivo it applies to both processes, the CO_2 uptake in the tissues as well as the CO_2 excretion by the lungs, that the CO_2 exchange via diffusion is supported by the Christiansen-Douglas-Haldane effect (Roughton, 1965). The content of chemically bound CO_2 in the blood is substantially dependent on the partial pressure of CO_2 (pCO_2; mmHg), which is determined by the CO_2 production in the tissue and its excretion by the lungs. For the same CO_2 concentration (cCO_2; ml/dl) the pCO_2 will be higher in oxygenated haemoglobin (O_2Hb) than in desoxygenated haemoglobin (Hb). Stated vice versa, for the same pCO_2, less CO_2 is bound by O_2Hb.

The fact, that oxygen itself has no effect on the CO_2 bound to blood while on the other hand the binding capacity for CO_2 decreases with an increasing oxygen saturation, has been shown among other things by Christiansen, Douglas and Haldane at pO_2 values ranging between 25.8 mmHg (sO_2 40%) and 660 mmHg (sO_2 98%) (Christiansen, Douglas and Haldane, 1914).

As CO_2 is removed from the alveoli by respiration, the amount of CO_2 in the blood during its passage through the lungs is reduced so far that the partial pressure of arterial CO_2 ($paCO_2$; mmHg) falls below the value of pCO_2 in mixed venous blood ($p\bar{v}CO_2$; mmHg).

Since $paCO_2$ is lower than $p\bar{v}CO_2$ in physiological respect it follows that, inevitably, the pH value of arterial blood (pHa) has to range above the mixed venous value ($pH\bar{v}$), (a smaller concentration of carbonic acid in arterial than in venous blood). A fall in acidity of blood due to CO_2 excretion predominates well over the increase in acidity due to oxygen uptake during the lung passage.

Therefore, the fact that the Haldane effect cannot be seen under physiologically normal circumstances in vivo, is due to the elimination of CO_2 by the lungs.

Should the CO_2 diffusing from the blood into the alveoli not be excreted during its passage through the lungs, it would have to be expected that, as a consequence of the Haldane effect, at a constant CO_2 concentration - with a maintained oxygenation of blood in the lung (high alveolar pO_2; pAO_2) - not only would the arterial carbon dioxide partial pressure ($paCO_2$) become identical with the mixed-venous value ($p\bar{v}CO_2$), but would even exceed it by the amount induced by the Haldane effect.

Because of the tremendous physiological significance of the HALDANE effect as described above, one has long tried to verify its mechanism in vivo. For its proof mainly animal experiments have been used (e.g. CO_2 rebreathing) (Rudlof, 1988).

For the correct interpretation of any in vivo proof, a distinction in principle between the physiologically so-called "open" and the unphysiologically "closed" system must be made.

During the unphysiological situation of a "closed" system such as for example CO_2 rebreathing and hyperoxic apnea, a continuous uptake of O_2 with a lack of CO_2 delivery should be expected which will consequently lead to the following development: initially $paCO_2$ and pHa align with the values of mixed venous blood (CO_2 production continues) while, at the same time, $p\bar{v}CO_2$ as well as $paCO_2$ climb constantly whereas pHa and $pH\bar{v}$ keep falling. Only now the different acidity of Hb and O_2Hb - usually covered up by the influence of $c\bar{v}CO_2$ or $caCO_2$ on the pH value - can be measured: it results in a drop of pHa below $pH\bar{v}$. As a consequence, the CO_2 binding capacity of arterial blood falls. The value of $paCO_2$ climbs above that of $p\bar{v}CO_2$ and the difference between pHa and $pH\bar{v}$ increases even further.

Under these circumstances (continuing oxygen uptake with lacking CO_2 delivery), it is therefore possible to substantiate the Haldane effect in vivo. Nevertheless, no previous investigations have yet described the influence of the Haldane effect on man under normal clinical conditions.

It therefore only suggested itself to examine the theoretical considerations in an appropriate survey which would possibly corroborate the Haldane effect in vivo under clinical circumstances using the "closed" system during hyperoxic apnea.

METHODS

Patient Population

Eighty patients (men and women) attending the clinic for cardiothoracic and vascular surgery of Mainz Medical School for an elective coronary surgical treatment were included in the examination. Valid exclusion criteria were:
* concomitant pulmonary diseases
* concomitant shunt defects
* concomitant valvular abnormality
* the need for the administration of circulatory drug support during induction of anaesthesia

Preparations

The patients were informed as regards content and risks of the study and their written consent was obtained during the premedication round. On the evening before operation, all patients received 2.0 mg flunitrazepam orally. A further 2.0 mg of flunitrazepam were administered orally as a premedication on the morning of their surgery.

On the arrival in the induction room and following instrumentation with a peripheral venous line, a further 0.01 mg/kgBW of flunitrazepam were given intravenously. Afterwards, under regional anaesthesia, an arterial line was secured in the left radial artery and a Swan-Ganz catheter was inserted via the external or internal jugular vein using Seldinger's technique. Previous to induction of anaesthesia all patients were preoxygenated for three minutes with a flow of 6-10 l O_2/min, using a close fitting mask and breathing spontaneously in the semiclosed system (Duda et al., 1988).

Induction of Anesthesia

Fentanyl 25 yg/kgBW, etomidate 0.1 mg/kgBW and pancuronium 0.1 mg/kgBW were administered with further oxygenation. As soon as spontaneous breathing had ceased, patients were normoventilated with a minute volume of 80-100 ml/kgBW and 6 l O_2/min fresh gas flow using a circle system (Normocap , Datex). Five minutes after pancuronium had been administered, blood was drawn simultaneously for an arterial and a mixed venous blood gas analysis which was measured immediately. Then artificial respiration was stopped, laryngoscopy performed, a stomach tube and an esophageal temperature probe were inserted and the patient was then intubated orotracheally. Before recommencing ventilation, further blood was drawn for a simultaneous arterial and mixed venous gas analysis, which was assessed without any loss of time, too.

Measurements

Intubation apnoea times were recorded. All arterial and mixed venous blood samples were analyzed for blood gas and oxygen status (Corning: pH/170 blood gas analyzer and CO-oximeter 2.500). From all parameters obtained in this way (pH, BE, HCO_3^-, pO_2, sO_2, cHb, COHb, MetHb, Hb, cO_2), the following were evaluated:

pHa and $pH\overline{v}$, saO_2 and $s\overline{v}O_2$ (%), $paCO_2$ and $p\overline{v}CO_2$ (mmHg).

Evaluation

Statistical analysis of the data was performed with the NWA-STAT-PAK program. After an examination of values on normal distribution, the paired and unpaired Student t-test was used for the evaluation of statistical significance. An error probability of 1% was assumed as the level of significance.

RESULTS

According to the different apnoea times of 60-180 seconds, the data were assigned to three groups:

    Group   I: 60-100 sec of apnoea,
    Group II: 101-140 sec of apnoea,
    Group III: 141-180 sec of apnoea.

Table 1 shows the development of parameters investigated. Values marked with an asterix (*) describe the significance in proportion to the initial value ($p <= 0.05$). The numbers in brackets represent the standard deviation from the mean.

The so-called "pCO_2 reversal", i.e. $paCO_2$ exceeding the value for $p\overline{v}CO_2$, can already be seen after one minute. For an apnea lasting more than 140 seconds, the $p\overline{v}CO_2$ lies on average 2.8 mmHg blow the value for $paCO_2$. The corresponding pH values react in the opposite way:

Before apnoea the pHa was 0.03 units below the $pH\overline{v}$, but then exceeded pHa by 0.02 units after one minute. This relationship remained constant in all three apnoea groups.

Moreover, during the entire period of observation the measurement results for arterial (saO_2) and mixed venous ($s\overline{v}O_2$) oxygen saturation show stable values, which correspond to a continuous oxygenation of haemoglobin from 80% towards 100% during the lung passage.

Table 1.   Development of the parameters $paCO_2$, $p\overline{v}CO_2$
           and $dpCO_2$ in mmHg, saO_2 and $s\overline{v}O_2$ in % as well
           as pHa and $pH\overline{v}$, as measured before apnoea up
           to the different apnoea times (60 - 180 sec)
           in the above groups I - III.
           Mean values with standard deviations in
           brackets.
           * denotes $p <= .05$ compared to values prior
           apnoea.
           Used symbols and definitions according to
           Zander and Mertzlufft (1988).

    {see following page}

400

Table 1.

	prior apnoea	I	II	III
pHa	7.39 (.05)	7.32 (.04*)	7.31 (.03*)	7.29 (.04*)
pH$\bar{v}$	7.36 (.03)	7.34 (.04*)	7.33 (.03*)	7.31 (.04*)
saO$_2$	97.6 (.7)	97.5 (.5)	97.5 (1.0)	97.4 (.4)
s$\bar{v}$O$_2$	80.5 (4.1)	81.5 (4.4)	81.7 (2.9)	80.4 (2.0)
paCO$_2$	40.9 (6.1)	50.5 (5.3*)	54.0 (5.4*)	58.6 (7.2*)
p$\bar{v}$CO$_2$	46.7 (5.3)	49.0 (5,4*)	52.7 (5.1*)	55.8 (7.6*)
dpCO$_2$	-5.8 (2.5)	+1.5 (2.3*)	+1.3 (2.5*)	+2.8 (1.8*)
n	80	35	26	19

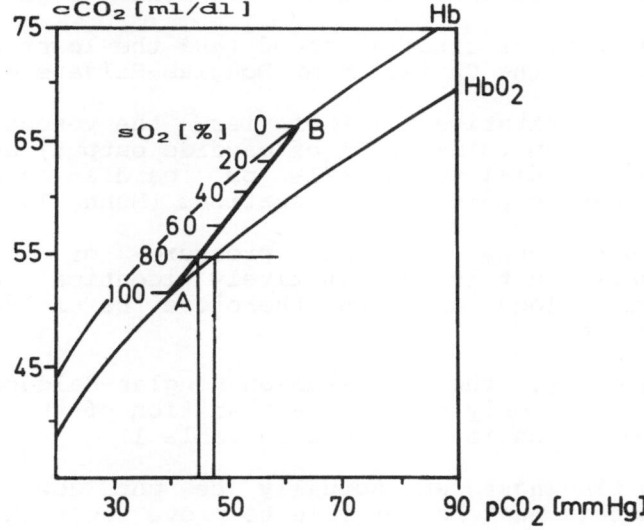

Fig. 1. Section of the CO$_2$ binding curves of oxyhaemoglobin (O$_2$Hb; sO$_2$ = 100%) and desoxyhaemoglobin (Hb; sO$_2$= 0%) (modified from Christiansen et al., 1914). Line AB represents the nexus of cCO$_2$ and pCO$_2$ ("CO$_2$ absorption curve" by Christiansen et al., 1914) in dependence on the degree of oxygenation (sO$_2$; %). Both vertical lines mark the range of pCO$_2$ increase of blood under the circumstances of a "closed" system and an increase in saturation from 80 to 100%. The increase in pCO$_2$ is 2.758 mmHg (distance between both vertical lines) (According to Brandt et al., 1988).

DISCUSSION

Within the "closed" system of hyperoxic apnoea and - owing to a lack of CO_2 delivery alongside with a maintained uptake of O_2 - one would expect a rapid rapprochement between $paCO_2$ and $p\bar{v}CO_2$.

The findings described here indicate not only a rapid alignment of $paCO_2$ and $p\bar{v}CO_2$ but a time dependent further increase of both parameters where, ultimately, the $paCO_2$ is greater than the $p\bar{v}CO_2$ (cf. table 1).

Consequently a similar - though reverse - development shows for the values of arterial and mixed venous pH: the pHa falls below the value of $pH\bar{v}$.

Theoretically, several possible explanations offer to elucidate this paradoxical phenomenon:

- the higher partial pressure of CO_2 in arterial blood (and lower pHa) is a consequence of venous shunting via the Thebesian veins,
- the higher pCO_2 of arterial blood (and the lower pHa) is a consequence of venous shunting via the bronchial veins,
- the higher pCO_2 of arterial blood (and the lower pHa) is a consequence of an increasing intrapulmonary right-left shunt,
- the higher pCO_2 of arterial blood (and the lower pHa) is a consequence of the Christiansen-Douglas-Haldane effect.

From a quantitative point of view, the venous shunting over the Thebesian veins (0.3% of cardiac output) as well as over the bronchial veins (< 1% of cardiac output) have to be ruled out as possible explanations (Nunn, 1977).

Concerning the partial pressure of CO_2, the intrapulmonary shunt is qualitatively identical with the mixed venous blood and can therefore be excluded as an explanation, too.

Consequently, the Christiansen-Douglas-Haldane effect remains as the only possible explanation of the parameter constellation which is described in table 1.

If the findings can actually be put down to this effect, then it must be possible to prove their theoretical considerations, which can be done with figure 1. It shows the physiological part of the CO_2 binding curves of haemoglobin.

If the oxygen saturation of haemoglobin is reduced from 100% to 0% in vivo - a CO_2 partial pressure of 40 mmHg given - the CO_2 concentration (cCO_2), alongside an increasing CO_2 binding capacity of Hb and dependent on the metabolic rate, would rise by 15 ml/dl (Christiansen et al., 1914)). This results in a pCO_2 increase of 22 mmHg (from 40 mmHg to 62 mmHg), which is indicated by the line between AB in the figure.

During anaesthesia, the desaturation in peripheral tissues and the saturation in the lungs amount to a dsO_2 of about 20% only (Brandt, 1988). However, with a maintained CO_2 delivery in the lungs, a mixed venous-arterial pCO_2 difference of -4.4 mmHg will result according to the figure.

If however, as in our case the elimination of CO_2 via the lung is ruled out, the CO_2 concentration during the lung passage (i.e. from mixed venous to arterial) remains constant.

Therefore, because of the unaltered CO_2 binding capacity of blood, a mixed venous-arterial pCO_2 difference of 0 mmHg would have to be expected.
But, since oxygenation of blood during the lung passage continues simultaneously (hyperoxic preoxygenation) the CO_2 binding curve changes. According to Christiansen et al. (1914), the CO_2 binding capacity is reduced as a result of the oxygenation of haemoglobin. Due to the constant CO_2 concentration, the pCO_2 has to rise and shift the binding curve to the right.

The theoretically expected increase in pCO_2 with an oxygen saturation rising from 80% to 100% O_2Hb amounts to 2.758 mmHg (the interval on abscissa between both vertical lines).

In this investigation, with an increase in oxygen saturation of 80.4% $s\bar{v}O_2$ to 97.4% saO_2 (dsO_2 = 17%), we measured a corresponding pCO_2 increase of 2.8 +/- 1.8 mmHg in Group III for example (cf. table 1).

This implies that the theoretically expected value as well as the actually measured value - at least in Group III who sustained the longest apnea - lie within the confidence limits maximally to be achieved in a clinical investigation.

The observed pH changes are also in agreement with the theoretically expected changes.

According to Nunn (1977), the combination of a pCO_2 increase of 6 mmHg and a 25% fall in sO_2 will result in a decrease of pH by 0.033. In comparison, a drop of sO_2 alone from 100% to 0% at a constant pCO_2 of 40 mmHg will produce a pH increase of 0.03. It therefore follows that with a constant pCO_2 and an increase of sO_2 by 17% the pH will fall 0.004 units only.

In this investigation, an increase in pCO_2 of 2.8 mmHg caused a fall in pH of 0.018 which has to be added.

The actually measured pH drop of 0.02 units differs from the theoretically expected fall of 0.022 by 0.002 pH units only. We therefore believe to have reached an acceptable degree of accuracy for the clinical methods chosen.

The described results and derivations therefore prove the Christiansen-Douglas-Haldane effect under clinical circumstances for the first time: Adaequate oxygenation of the blood during apnoea causes the arterial pCO_2 to exceed the value of the mixed venous pCO_2.

CONCLUSIONS

   *   In contrast to the physiological "open" system ($paCO_2$ < $p\overline{v}CO_2$ and pHa > $pH\overline{v}$) a supposedly "paradoxical reversal" of proportions is demonstrated in the unphysiological "closed" system with the example of hyperoxic intubation apnoea: $paCO_2$ > $p\overline{v}CO_2$ and pHa < $pH\overline{v}$. The results presented prove for the first time that the so-called Christiansen-Douglas-Haldane effect can also be shown under clinical circumstances.

   *   The results and their interpretation make clear that the position of $paCO_2$ > $p\overline{v}CO_2$ and pHa < $pH\overline{v}$ is not an error in measurement and equipment or caused by misinterpretations but a consequence of the Christiansen-Douglas-Haldane effect.

SUMMARY

   The CHRISTIANSEN-DOUGLAS-HALDANE effect, commonly known as the HALDANE effect, describes the dependence of the CO_2 binding of blood on the degree of oxygenation of haemoglobin. Under the physiological conditions of an "open" system between blood and alveoli the partial pressure of arterial CO_2, after CO_2 delivery to the alveoli for example, can only range below the value of mixed venous blood.

   However, during the unphysiological circumstances of a "closed" system, e.g. the state of hyperoxic apnoea, i.e. oxygen uptake and lacking CO_2 delivery, the $paCO_2$ cannot only approximate the mixed venous value but must even exceed it.

   Without the HALDANE effect coming into force, a rapid adjustment of arterial to mixed venous pCO_2 would have to be expected during apnoea, due to the lacking CO_2 delivery.

   If however, as a consequence of adaequate preoxygenation (a high alveolar pO_2) and failure to eliminate CO_2 (i.e. the CO_2 concentration remains constant) a sufficient oxygenation of blood takes place during the passage through the lung capillaries, then this leads to a rightwards shift of the CO_2 binding curve - the HALDANE effect.

   The resulting increase in pCO_2 as shown here, actually leads to an arterial-mixed venous CO_2 partial pressure difference of 2.8 +/- 1.8 mmHg, where $pvCO_2$ decreases $paCO_2$.

   The results described substantiate for the first time the existence of the HALDANE effect under clinical conditions, too.

LITERATURE

   Brandt, L., 1988, Bedeutung des gemischtvenösen O_2-Status als Ergänzung zum arteriellen O_2-Status, in: "Der Sauerstoff-Status des arteriellen Blutes", R. Zander and F.O. Mertzlufft, eds., Karger, Basel.

Brandt, L., Mertzlufft, F., Rudlof, B. and Dick, W., 1988, In-vivo-Nachweis des Christiansen-Douglas-Haldane Effektes unter klinischen Bedingungen, <u>Anaesthesist</u>, 37: (in print).

Christiansen, J., Douglas, C.G. and Haldane, J.S., 1914, The absorption and dissociation of carbon dioxide by human blood, <u>J Physiol</u>., 48:244.

Duda, D., Brandt, L., Rudlof, B., Mertzlufft, F. and Dick, W., 1988, Der Einfluß unterschiedlicher Präoxygenierungsverfahren auf den arteriellen Sauerstoff-Status, <u>Anaesthesist</u>, 37:408

Nunn, J.F., 1977, "Applied respiratory physiology", Butterworth & Co Ltd., London.

Roughton, F.J.W., 1965, Transport of oxygen and carbon dioxide, <u>in</u>: "Handbook of Physiology, Respiration I", Amer Physiol Soc, ed., Washington.

Rudlof, B., 1988, Untersuchungen über Veränderungen des arteriellen und gemischtvenösen Blutstatus unter hyperoxischer Intubationsapnoe, <u>Dissertationsarbeit</u>, Mainz.

Zander, R. and Mertzlufft, F.O., 1988, "Der Sauerstoff-Status des arteriellen Blutes", Karger, Basel.

FRACTIONATION OF OXYGEN ISOTOPES DUE TO EQUILIBRATION OF OXYGEN WITH

HEMOGLOBIN

K.P. Pflug, and K.-D. Schuster

Institute of Physiology I, University of Bonn, 53 Bonn 1, FRG

INTRODUCTION

Due to different molecular weights, fractionations occur between the isotopic oxygen molecules $^{16}O^{18}O$ and $^{16}O_2$ when passing through the pathways of oxygen transport and metabolism. In man, the rate constant of $^{16}O_2$ uptake was found to be 0.91 % above that of the heavier molecule (Schuster and Pflug, 1988). Several constituent processes are known to show higher fractionations. The diffusion rate constant should be greater for $^{16}O_2$ than for $^{16}O^{18}O$ by 3 %. Feldman et al. (1959) found for reactions of the respiratory chain that the rate constant of oxidation was higher for $^{16}O_2$ than for $^{16}O^{18}O$ by 1.3 %.

The overall effect measured is far below the fractionation effect expected for diffusion and slightly below the effect found for oxygen utilization. This could be due to a limitation of oxygen transport by metabolism. However, additional fractionations caused by processes such as the reaction of oxygen with hemoglobin, could also contribute to the overall fractionation effect, so that no reliable conclusions can be drawn up to now. The aim of the present study is to answer the question as to whether the chemical reaction of oxygen with hemoglobin exhibits an additional source of isotopic fractionation, which could be of significance in forming the overall effect.

METHODS

Investigations were performed on bovine hemoglobin solutions prepared according to Rossi-Fanelli et al. (1961). Hb concentration and pH of the solution were adjusted to normal values. Fig. 1 exhibits the equilibration set up. After degassing in vacuo within the reaction vessel, the hemoglobin solution was equilibrated at 37°C with oxygen of suitable pressure, so that oxygen saturation levels of 30 %, 50 % and 100 % were achieved. The process of equilibration was controlled by the pressure gauge G and the oxygen electrode O_2E. After complete equilibration, samples of Hb-solution were drawn into the Hb-vessel Hb-V, and samples of the gas phase into flask S.

For deoxygenating the Hb-sample, a second system similar to that of Fig.1 was used. The Hb-sample was drawn via the inlet into the pre-evacuated reaction vessel and the oxygen liberated from Hb was pushed into flask S by the Toepler pump. The oxygen sample was completely burnt to CO_2 which was analysed on its $({}^{16}O^{18}O)/({}^{16}O_2)$ ratio by mass spectrometry as described (Schuster and Pflug, 1988).

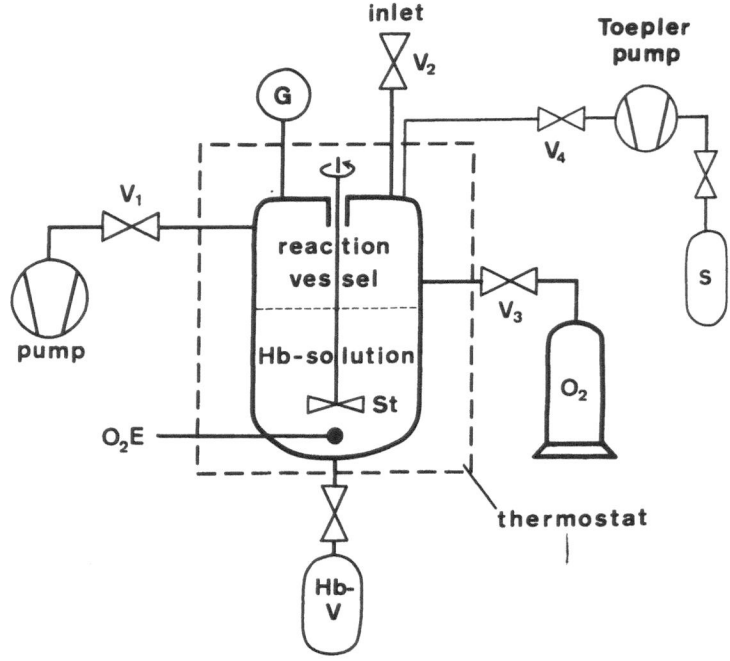

Fig. 1. Simplified scheme of the equilibration unit. V1, V2, V3, V4: valves, G: pressure gauge, O_2E: oxygen electrode, St: Stirrer, Hb-V: vessel for drawing samples of equilibrated Hb-solution. S: flask into which the oxygen of the gas phase was collected via the Toepler pump.

Quantifying Fractionation Effects

The fractionations are given as δ-values. With the abbreviation $X = ({}^{16}O^{18}O)/({}^{16}O_2)$, the δ_{ab}-value between two compartments a and b is defined as

$$\delta_{ab} = \frac{X_a - X_b}{X_a} \cdot 100\% \qquad (1)$$

where δ_{ab} describes the percentage deviation of ${}^{16}O^{18}O$-abundance between both compartments a and b. With respect to hemoglobin, compartment a represents the gas phase (g) and compartment b the oxygen bound to hemoglobin (Hb).

RESULTS

The results are compiled in table 1. Oxygen saturation (SO_2) is given as parameter of the measurements in column 1. The third column contains the fractionation effect δ_{gHb} which represents the percentage deviation between the $^{16}O^{18}O$-abundance of the gas phase and of the oxygen bound to hemoglobin according to formula (1).

δ_{gHb} is independent of SO_2. An overall mean was therefore calculated amounting to δ_{gHb} = 0.35 $\pm$ 0.02 %. This value means that the affinity to combine with hemoglobin is 0.35 % higher for $^{16}O_2$ than for $^{16}O^{18}O$.

Table 1. Fractionation effects (δ_{gHb}) of isotopic oxygen molecules between gas phase and the oxygen bound to hemoglobin.

SO_2 (%)	n	δ_{gHb} (%)	SD (%)
100	5	0.352	0.015
50	3	0.350	0.025
30	2	0.350	0.016
overall mean		0.351	
overall SD		0.023	

SO_2: oxygen saturation of hemoglobin, n: number of independent measurements, δ_{gHb}: $^{16}O^{18}O$-abundance difference between gas phase and the oxygen liberated from hemoglobin according to equ. (1), SD: standard deviation.

DISCUSSION

The value of δ_{gHb} was measured during the situation of complete equilibration. It is therefore not known whether its value is brought about by differences in the association constants of the reaction between hemoglobin and $^{16}O_2$ or $^{16}O^{18}O$, in the dissociation constants or in a combination of both. The fact that δ_{gHb} of bovine hemoglobin was found to be independent of SO_2 implies that δ_{gHb} of human hemoglobin should be of similar size.

It can be extrapolated that the δ_{gHb} value of doubly ^{18}O labeled oxygen, $^{18}O_2$, should be about twice that of $^{16}O^{18}O$: $\delta_{gHb}(C^{18}O_2)$=0.7 %. Both values, that of $^{16}O^{18}O$ and that of $^{18}O_2$, are low enough to be negligible in most applications of ^{18}O as a tracer.

With the δ_{gHb}-value of $^{16}O^{18}O$, model calculations were performed to explain the overall fractionation effect of isotopic oxygen molecules occurring during respiration. Fig. 2 shows the oxygenation of blood taking place along a pulmonary capillary. The right ordinate exhibits PO_2, the left one the percentage deviation of $^{16}O^{18}O$-abundance δ_{AC} between alveolar gas (A) and capillary blood (C). δ_{AC}, δ_{AV} and δ_{Aa} are defined according to formula (1). Since the alveolar-venous difference of $^{16}O^{18}O$-abundance, δ_{AV}, is not known, calculations have been carried out using the two limits $\delta_{AV} = 0$ and $\delta_{AV} = 3\ \%$ as initial conditions. As can be seen from fig. 2, both curves converge against δ_{gHb}. Real δ_{AV}-values are expected to range between the limits presumed, so that arterial blood should leave the lungs with an alveolar-arterial difference δ_{Aa} of $^{16}O^{18}O$-abundance of about 0.35 %. It can be shown that Fick's Principle holds in the following form

$$\delta_{Aa}C_a\dot{Q} = \delta_{Av}C_v\dot{Q} + \delta_{AU}\dot{V}O_2 \qquad (2)$$

where $\dot{Q}$ and $\dot{V}O_2$ denote cardiac output and oxygen uptake, C_a and C_v arterial and venous oxygen concentration and δ_{AU} the difference of $^{16}O^{18}O$-abundance between alveolar oxygen and the oxygen taken up, respectively. δ_{AU} can be calculated from the overall effect at 1.13 %. With $\delta_{Aa} = 0.35\ \%$ and normal values for cardiac output, oxygen uptake and arterial and venous oxygen concentration, δ_{AV} is obtained to 0.03 % from equation (2), confirming the above mentioned supposition with regard to δ_{AV}. Model calculations performed for oxygen transport from peripheral capillary to cell yielded similar values when supposing the oxygen transfer rate to be limited by the

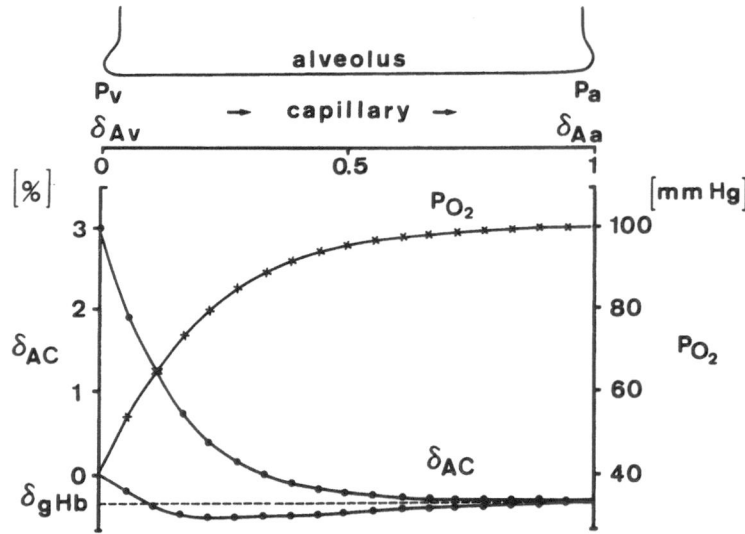

Fig. 2. Oxygenation of blood when passing through a pulmonary capillary. Abscissa: length of capillary (relative units), right ordinate: oxygen partial pressure of blood, left ordinate: percentage deviation δ_{AC} of the $^{18}O^{16}O$-abundance between alveolar gas and capillary blood analogue to formula (1). δ_{Av} and δ_{Aa}: initial and final values of δ_{AC}, respectively. δ_{gHb}: fractionation value of hemoglobin, P_v, P_a: oxygen concentration of venous and arterial blood.

rate of utilization and not by diffusion or by delivering the oxygen from hemoglobin. From these calculations it can be predicted that $^{16}O^{18}O$ enrichment of venous blood and also the overall fractionation effect of respiration should increase if the limitation of oxygen transfer by metabolism diminishes.

SUMMARY

The aim of the study was to answer the question as to whether the chemical reaction of oxygen with hemoglobin exhibits a source of isotopic fractionation, which could be of significance in forming the overall fractionation effect of respiration recently determined in man. Investigations were performed on bovine hemoglobin solutions adjusted to normal values of Hb concentration and pH. After degassing in vacuo, the hemoglobin solution was equilibrated in a closed system with pure oxygen of suitable pressure so that oxygen saturation levels of 30, 50 and 100 % were achieved. After complete equilibration, isotope analysis of oxygen by mass spectrometry resulted in $^{16}O^{18}O/^{16}O_2$ ratios which were 0.35 ± 0.02 % lower in the oxygen bound to hemoglobin than that of the gas phase during all levels of oxygenation. Model calculations suggest the following: (1) the fractionation during oxygen uptake in the lung at rest is primarily determined by the reaction with hemoglobin, (2) the overall fractionation effect of respiration can be explained as due to single effects of the constituent processes when assuming the oxygen transport to be limited by utilization.

REFERENCES

Feldman, D.E., Yost, H.T.Jr., and Benson B.B., 1959, Oxygen isotope fractionation in reactions catalyzed by enzymes, Science, 129:146-147.
Rossi-Fanelli, A., Antonini, E., and Caputo, A., 1961, Studies on the relations between molecular and functional properties of hemoglobin, I. The effect of salts on the molecular weight of human hemoglobin, J. Biol. Chem., 236:391-396.
Schuster, K.-D., and Pflug, K.P., 1988, The overall fractionation effect of isotopic oxygen molecules during oxygen transport and utilization in humans, submitted for publication in: "Oxygen Transport to Tissue", XI, K. Rakusan, G.B. Biro, T.K. Goldstick, and Z. Turek, eds., Plenum Publishing Corporation, New York.

THE EFFECT OF HYPERCHYLOMICRONAEMIA ON OXYGEN AFFINITY IN HUMAN BLOOD

M. J. Poss and I. S. Longmuir

Department of Biochemistry
North Carolina State University
Raleigh, NC 27695-7622 U.S.A.

INTRODUCTION

There is a clear relationship between hyperchylomicronaemia (HC) and myocardial ischaemia. One mechanism for a causal relationship between the two is the occlusion of coronary arteries by atheromotous plaques. However, coronary artery disease progresses slowly and myocardial ischaemia is a rapid event which is not always preceded by atheroma.

Angina pectoris has been precipitated in patients following ingestion of a large fatty meal (Kuo and Joyner, 1955). Regan et al. (1961) suggested that postprandial lipemia impedes oxygen transport to tissues. Others (Fukuzaki et al., 1975; Ishikawa, 1979) have shown from clinical and experimental studies that HC might initiate myocardial ischaemia.

Ditzel and Dyerberg (1977) have demonstrated that red blood cells obtained from patients with HC exhibit a greatly increased affinity for O_2. This phenomenon would significantly impair O_2 delivery to the myocardium and might provide an additional mechanism for the incidence of myocardial ischaemia.

They observed that the P-50 of blood obtained from subjects with essential HC was markedly decreased. This change could be reversed by incubating the red cells in normal donor plasma. Normal red cells incubated in lactescent plasma also showed an increased affinity for O_2.

It was postulated that this event might be explained if HC affected the hydrogen ion concentration across the red cell membrane. We have explored this possibility and have demonstrated that a leftward shift in the oxygen dissociation curve corresponds with a rise in intracellular pH.

MATERIALS AND METHODS

The method of Longmuir and Chow (1970) was employed to verify the leftward shift in the oxygen dissociation curve observed by Ditzel and Dyerberg.

The apparatus consisted of a water-jacketed, glass cell, fitted with

413

a rubber stopper through which pO_2 and pH electrodes were inserted. Inlet and outlet needles allowed for the addition of solutions and a stirrer bar permitted mixing. The instrument was calibrated with air equilibrated, phosphate buffered, succinate solution, to which 300 µl of heart muscle prep (1963) were added so that O_2 was consumed linearly from a high value down to a zero value.

Blood was obtained from an antecubital vein of the same subject (project approved by NCSU's I.R.B.) and treated with K-EDTA (1.5 mg/ml) to prevent coagulation. Heavy whipping cream (MacFarlane et al., 1941; MacLagan and Billimoria, 1956) or emulsified safflower or coconut oil (Shafiroff and Frank, 1947) was added to approximate an acute, clinically recognized level of hyperchylomicronaemia (6000 mg/dl).

The pO_2 was continuously recorded. The ΔpO_2 was measured over a range of initial pO_2 values. This ΔpO_2 could not be explained by electrode drift or poisoning since the instrument was calibrated before and after each experiment. Any O_2 consumption which occurred over time could be corrected for by extrapolation. The difference in slopes was indicative of a decrease in pO_2 and a corresponding leftward shift in the oxygen dissociation curve. Therefore the O_2 which has disappeared must be bound to haemoglobin. Normal (nonhyperlipidaemic) blood produced a linear trace reflecting only a slow O_2 consumption over time.

The freeze/thaw method of Hilpert et al. (1963) was employed to measure intracellular pH. Intracellular pH was examined at the completion of the experiment for the HC sample and a parallel, normal sample of blood.

RESULTS

A slow, significant decrease in pO_2 of blood (37°C, at in vivo pH) was measured when lipid emulsions were added (Fig. 1). The ΔpO_2 at all values reflected a 6.0 mm Hg leftward shift of the oxygen dissociation curve at P-50 after 3.6 hrs. Polyunsaturated emulsions produced essentially the same shift as saturated emulsions (4.5 vs. 6.0 mm Hg). Normal blood revealed no ΔpO_2 other than that due to oxygen consumption (Fig. 2).

During this period the pH gradient across the red cell membrane was largely abolished in blood plus lipid emulsions. No such change was found in normal blood (Table 1).

DISCUSSION

Ditzel and Dyerberg first described a marked left shift in the oxygen dissociation curve in HC patients using a Duvelleroy dissociation curve analyzer (DCA-1). However, Robertson et al. (1978) was unable to reproduce this shift using a mixing technique to obtain P-50. It was suggested that the Duvelleroy dissociation curve analyzer contains a pO_2 electrode which is adversely affected by the accumulation of lipid. Furthermore, it is unable to compensate for O_2 consumption by whole blood over time.

Although the mixing technique of Robertson et al. did not show a shift, it is complicated by an ambiguous blood-gas correction factor which may have masked any changes that might have occurred. In addition, the incubation studies involving normal red cells and lipemic plasma were performed for too short a period (only 30 min.). See Figure 1.

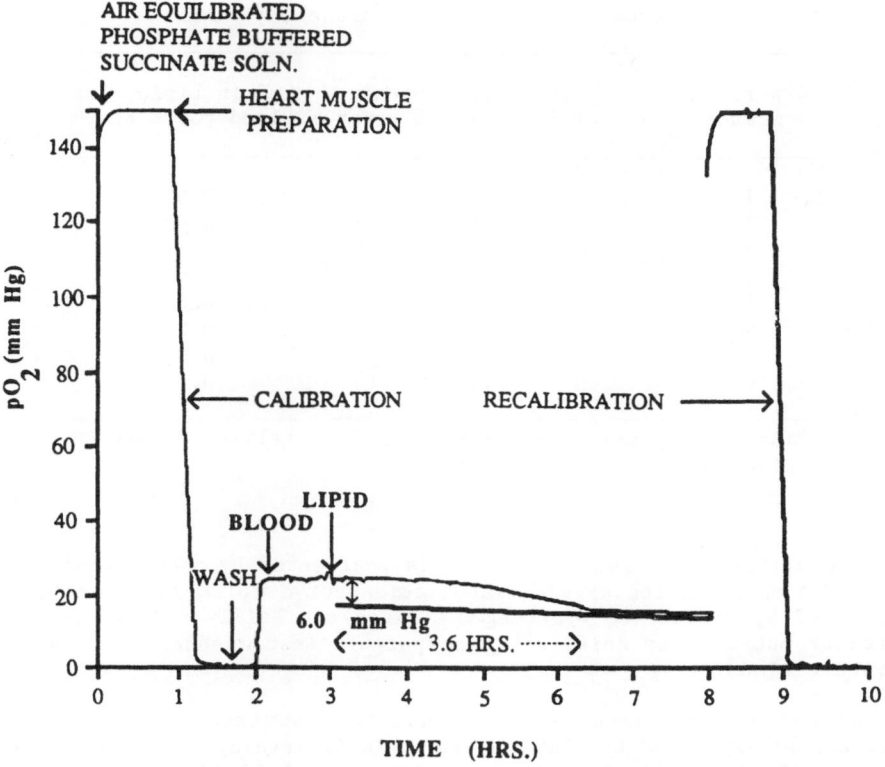

Fig. 1. Model experiment for a 6% hyperchylomicronaemic emulsion, $pO_2 \sim P-50$, 37°C, at _in vivo_ pH. Note $\Delta pO_2 \sim 6.0$ mm Hg is independent of O_2 consumption over 3.6 hrs.

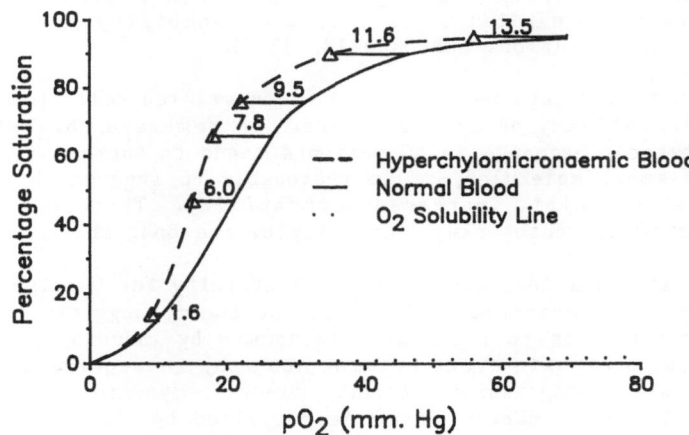

Fig. 2. Oxygen dissociation curves of hyperchylomicronaemic blood compared with normal blood. Lines representing differences between the curves are parallel with the O_2 solubility line.

Table 1. The effect of 6% hyperchylomicronaemia on ΔpH
across the red cell membrane

Experiment Number	ΔpH in the Absence of Lipid for 6 hrs.[a]	ΔpH in Lipid Plus Blood for 6 hrs.[a]
1	0.23	0.02
2	0.21	0.08
3	0.20	0.01
4	0.18	0.04
5	0.30	0.03
6	0.26	0.03
7	0.20	0.02
8	0.26	0.10

[a]The varying value of ΔpH results from varying % saturation.

The method of Longmuir and Chow is unaffected by these problems and revealed a shift in the oxygen dissociation curve which is comparable to that found by Ditzel and Dyerberg. Cooksey and Reilly (1978) have shown a similar but smaller shift using comparable instrumentation to examine HC rabbit blood.

Ditzel and Dyerberg reported that the increased affinity for O_2 could not be explained by elevated haemoglobin levels, decreased red cell 2,3-diphosphoglycerate or an increase in arterial blood pH.

They postulated that exposure to HC might physically modify the composition of the red cell membrane. It has been shown that membrane cholesterol can equilibrate with HC plasma within the same time period (~4 hrs.) that our experiments revealed an increase in O_2 affinity (Hagerman and Gould, 1951; Murphy, 1962). Bagdade and Ways (1970) have observed that this exchange produced cholesterol-deficient red cells in blood obtained from essential HC patients. Furthermore, cholesterol-deficient red cells exhibit an increased permeability to several non-electrolytes and Na^+ (Kroes and Ostwald, 1971).

Our results indicate that exposure of normal red cells to HC affects membrane permeability of hydrogen ions. If membrane cholesterol and plasma cholesterol exchange in HC patients leads to increased permeability of other small molecules, it is reasonable to suggest that hydrogen ions might also exhibit increased permeability. Therefore a rise in intracellular pH increases oxygen affinity by the Bohr shift.

This substantial increase in red cell affinity for O_2 might significantly starve the myocardium of O_2. Current theory suggests that hyperlipidaemia contributes to myocardial ischaemia by inducing atheroma and thereby causes a hypoxic event brought about by inadequate blood flow. This is a slow and progressive process. However, myocardial ischaemia is an acute event and frequently cannot be explained by the accumulation of atheromatous plaques that narrow the lumen of coronary vessels. Our studies indicate that HC increases the affinity of red blood cells for O_2. This is an acute and dramatic event which might provide an additional mechanism for myocardial ischaemia.

SUMMARY

Evidence for a leftward shift of the oxygen dissociation curve in HC patients was first described by Ditzel and Dyerberg. These results were challenged by Robertson et al., who saw no shift using a different technique to obtain P-50. Both methods are open to possible errors. In order to clarify the mechanism whereby HC blood shows an increased affinity for O_2, the method of Longmuir and Chow was used since it is unaffected by these problems.

Exposure of blood to various lipid emulsions was shown to exhibit a slow fall in pO_2 due to increased binding of oxygen to haemoglobin. During the same period the pH difference across the cell membrane is largely abolished. As a result the oxygen dissociation curve is displaced to the left by the Bohr shift.

This is an acute and dramatic event. We postulate that this increased affinity of red blood cells for O_2 may predispose patients in the HC state to myocardial ischaemia by this mechanism.

REFERENCES

Bagdade, J. D., and Ways, P. O., 1970, Erythrocyte membrane lipid composition in exogenous hypertriglyceridemia, J. Lab. Clin. Med., 75:53.

Coleman, C. H., and Longmuir, I. S., 1963, A new method for registration of oxyhemoglobin dissociation curves, J. Appl. Physiol., 18:420.

Cooksey, J. D., and Reilly, P., 1978, Effect of hypertriglyceridemia on the hemoglobin-oxygen dissociation curve, J. Surg. Res., 25:70.

Ditzel, J., and Dyerberg, J., 1977, The oxyhemoglobin dissociation curve in patients with familial hyperchylomicronemia, J. Lab. Clin. Med., 89:573.

Fukuzaki, H., Okamoto, R., and Matsuo, T., 1975, Studies on pathophysiological effects of postalimentary lipemia in patients with ischemic heart disease, Jpn. Circ. J., 39:31.

Hagerman, J. S., and Gould, R. G., 1951, The in vitro interchange of cholesterol between plasma and red cells, Proc. Soc. Exp. Biol. Med., 78:329.

Hilpert, P., Fleischmann, R. G., Kempe, D., and Bartels, H., 1963, The Bohr effect related to blood and erythrocyte pH, Am. J. Physiol., 205:331.

Ishikawa, Y., 1979, Pathophysiological studies on postalimentary lipemia with special reference to the effect of exogenous hypertriglyceridemia on intramyocardial tissue oxygen tension, Kobe J. Med. Sci., 25:19.

Kroes, J., and Ostwald, R., 1971, Erythrocyte membranes--effect of increased cholesterol content on permeability, Biochim. Biophys. Acta, 249:647.

Kuo, P. T., and Joyner, C. R., 1955, Angina pectoris induced by fat ingestion in patients with coronary artery disease, J.A.M.A., 158:1008.

Longmuir, I. S., and Chow, J., 1970, Rapid method for determining effects of agents on oxyhemoglobin dissociation curves, J. Appl. Physiol., 28:343.

MacFarlane, J. W., Trevan, J. W., and Attwood, A. M., 1941, Participation of fat soluble substance in coagulation of the blood, J. Physiol., 99:7P.

MacLagan, N. F., and Billimoria, J. D., 1956, Food lipids and blood coagulation, Lancet, 1:235.

Murphy, J. R., 1962, Erythrocyte metabolism. IV. Equilibrium of choles-terol-4-C^{14} between erythrocytes and various treated sera, J. Lab. Clin. Med., 60:571.

Regan, T. J., Bianak, K., Gordon, S., DeFazip, V., and Hellems, H. K., 1961, Myocardial blood flow and heparin induced lipolysis, Circula-tion, 23:55.

Robertson, H. T., Chait, A., Hlastala, M. P., and Brunzell, J. D., 1978, Red cell oxygen affinity in severe hypertriglyceridemia, Proc. Soc. Exp. Biol. Med., 159:437.

Shafiroff, G. P., and Frank, C., 1947, A homologous emulsion of fat, protein and glucose for intravenous administration, Science, 106:474.

ALTERATIONS IN OXYHEMOGLOBIN DISSOCIATION CURVE DURING

NORMOXIC ACUTE NORMOVOLEMIC HEMODILUTION

A. Trouwborst, W.G.M. van den Broek, R. Tenbrinck,
T.H.N. Groenland, M. Bucx, and N.S. Faithfull
Department of Anaesthesiology, Erasmus University
Rotterdam, The Netherlands

INTRODUCTION

Shifts in the oxygen dissociation curve (O.D.C.) have
been taken into account when studying oxygen delivery to
tissue as a dependent factor of oxygen availability. It can
be assumed that changes in the oxyhemoglobin dissociation
characteristics of the blood will have a certain effect on
the amount of oxygen available (1,2).

A shift to the left of the O.D.C. limits oxygen delivery
when acute and when blood flow is limited (3) and induces
increased blood flow to several organs, probably to compensate for decreased tissue O_2 pressure (4). Perfusion of the
brain with low P50 blood induced a drop in oxygen consumption
(Vo_2), a decrease in mixed venous oxygen tension (PvO_2), an
increase in the mixed venous oxygen content (CvO_2) and caused
the E.E.G. to detoriate earlier (5); while reducing the P50
after coronary artery occlusion significantly increased the
degree of myocardial necrosis for the same degree of ischemia
(6). Studying oxygen delivery to tissue, it can be assumed
that oxygen bound to Hb at a tension of less than 35 torr (7-
9) should be considered relatively unavailable and that this
amount of unavailable oxygen is dependent on the characteristics and changes of the O.D.C. For instance, a shift to the
left increases the saturation of the hemoglobin with oxygen
at a certian PO_2 level, therefore the amount of unavailable
oxygen, impairing tissue oxygenation. This amount of "unavailable" oxygen can be described on the O.D.C. by determining the saturation at a PO_2 of 35 torr ("S35").

Subtraction of this portion of oxygen from the total
arterial oxygen content should give a better indication of
the real oxygen availability (O_2av) because alterations in
the O.D.C. and its influence on the oxygen unloading capacity
are inclosed. Introduction of the real oxygen availability
into the formula of the O_2 extraction ratio (ERav) should

give more insight into the ratio between oxygen influx into the tissue and oxygen consumption.

Nomenclature and formulae used:

P50   :PO_2 at which Hb is 50% saturated
S_{35} :Saturation of Hb at a PO_2 of 35 torr
O_2 flux   :Arterial oxygen content x cardiac output
O_2 availability: $\dfrac{SaO_2 - S_{35}}{SaO_2}$ x O_2 flux
(O_2av)

ER $:\dfrac{VO_2}{O_2\ flux}$

E.R.av $:\dfrac{VO_2}{O_2av}$

This study examined whether ERav predicts the critical point of total oxygen influx (O_2 flux) in relation to oxygen consumption.

Furthermore, as it is known that a chronically reduced hemoglobin concentration is compensated for by improved oxygen unloading, afforded by the increased P50 (10,11), the authors also studied the effect on the O.D.C. of normoxic acute normovolemic hemodilution.

MATERIAL AND METHODS

Six Yorkshire pigs, weighing 10-12 kg were used in the experimental procedure. Under Midazolam a Swan Ganz thermo-dilution catheter, as well as a catheter in the left femoral artery and an intravenous line, were placed. All animals were ventilated with air. Midazolam and Pancuronium were administered continuously. Total cardiovascular monitoring (Mennen, Horizon 2000) was performed in all animals.

Samples for blood gas analysis and for measurement of hemoglobin concentration were taken from both systemic arterial and pulmonary arterial catheters. Blood gas analysis and hemoglobin concentration were determined with the aid of the spectrophotometer O.S.M.3 and the ABL 330 (Radiometer, Copenhagen). Oxygen consumption (VO_2) was calculated as the product of the cardiac output and the arteriovenous oxygen content difference and measured simultaneously via a closed circuit rebreathing system.

The P50 and S35 of the mixed venous blood were calculated according to Siggaard-Andersen (12). The P50 and the steep part of the O.D.C. (pH and PCO_2 corrected) were also established according to the concept guidelines for measure-

ment of blood hemoglobin oxygen affinity of the International
Federation of Clinical Chemistry (based on the estimation of
a Hill plot of the O.D.C. via stepwise incubation of a sample
of desaturated blood with air, with correction of the measured PO_2 values to pH = 7.4 and pCO_2 = 40 mmHg). Acute normovolemic hemodilution was achieved in steps of 10 or 5
percent exchange of total blood volume. Withdrawal of blood
was accompanied by simultaneous infusion of Dextran 40, 50
g/l (Isodex).
A full set of data was obtained after each step of exchange
and the time between each step of blood exchange was 15 min.

RESULTS

      Total body oxygen consumption (Vo_2) started to decrease
after the modified oxygen extraction ratio (ERav) reached the
value of 1.07 (SEM: $\pm$ 0.043). With a further decrease in
Vo_2, the ERav did not exceed the value of 1.2 (fig. 1).

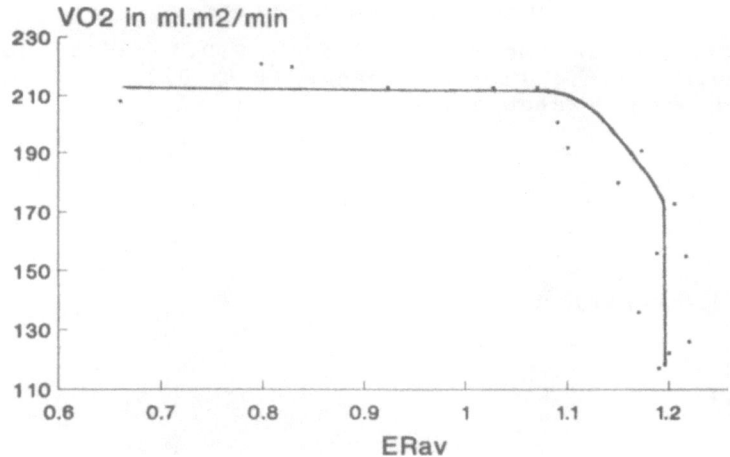

Fig.1: Total body oxygen consumption (VO_2) versus modified
        oxygen extraction ratio (ERav).

      Stepwise hemodilution was associated with a decrease in
the actual S35 (fig. 2) (P $<$ 0,05 at 20 percent of blood
exchange). The actual change in S35 correlated directly with
the change in the actual mixed venous Po_2 (fig. 3) (r-value:
0.96). No correlation was found with pH or PCO_2. During
hemodilution a shift and rotation in the steep part of the
slope of the O.D.C., after correction of the measured PO_2 for
pH and PCO_2, occurred (fig. 4). (P50: 32.9 $\pm$ 1.5 v.s. 37.1 $\pm$
1.4; n.$_{hill}$: 2.83 $\pm$ 0.13 v.s. 2.48 $\pm$ 0.12). Changes in the
classic oxygen extraction ratio (ER) were directly correlated
to the alterations of the actual S35 (r-value: 0.94)(fig. 5).

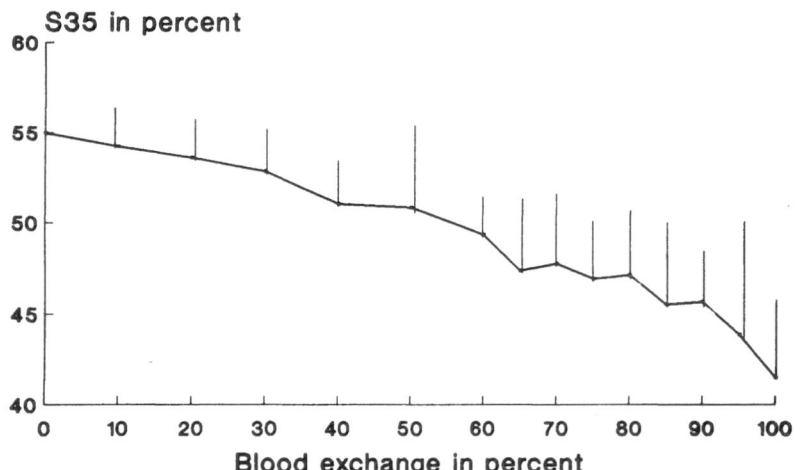

Fig. 2: Changes in the actual S35 (+ S.D.) during normoxic
normovolemic blood exchange (P < 0.05 at 20 percent
of blood exchange).

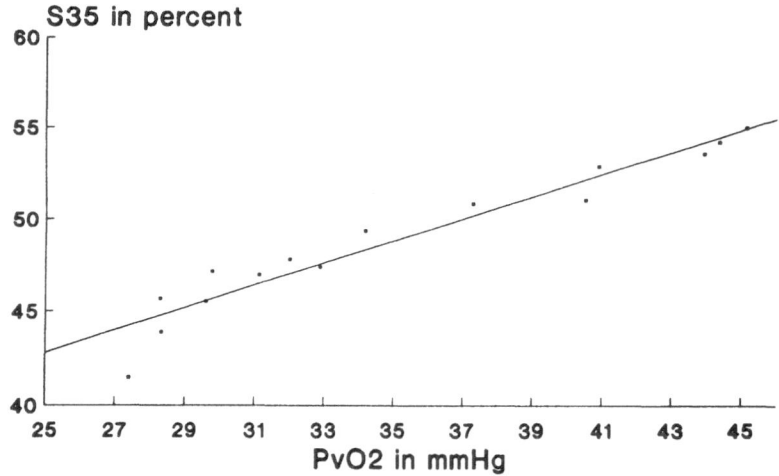

Fig. 3: Actual mixed venous oxygen tension (PvO$_2$) versus the
actual S35 during acute normovolemic hemodilution (r-
value: 0.96)

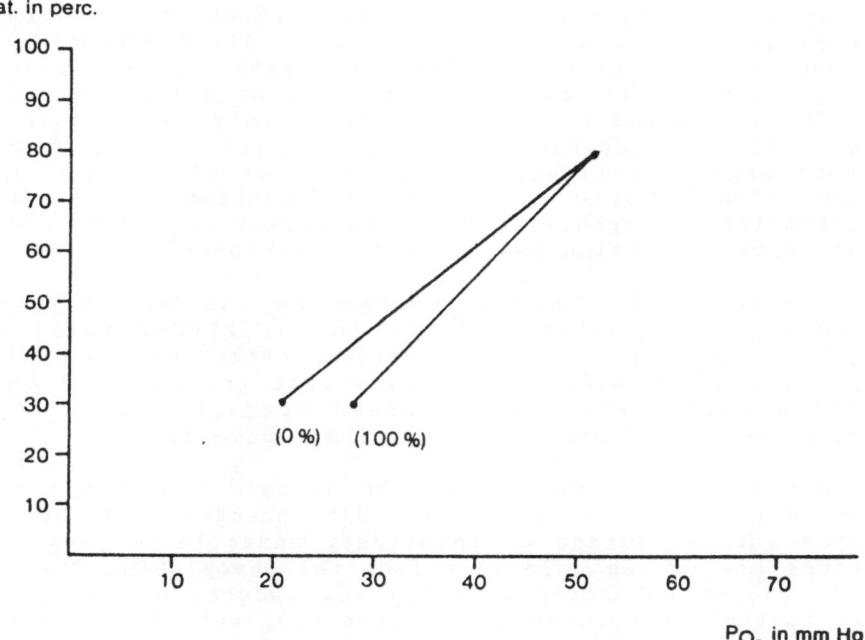

Fig. 4: Steep part of the oxyhemoglobin dissociation curve, corrected for pH and PCO_2, before (Hemoglobin = 10.09 gm%) and after 100 percent exchange of blood volume (Hemoglobin = 2.2 gm%)

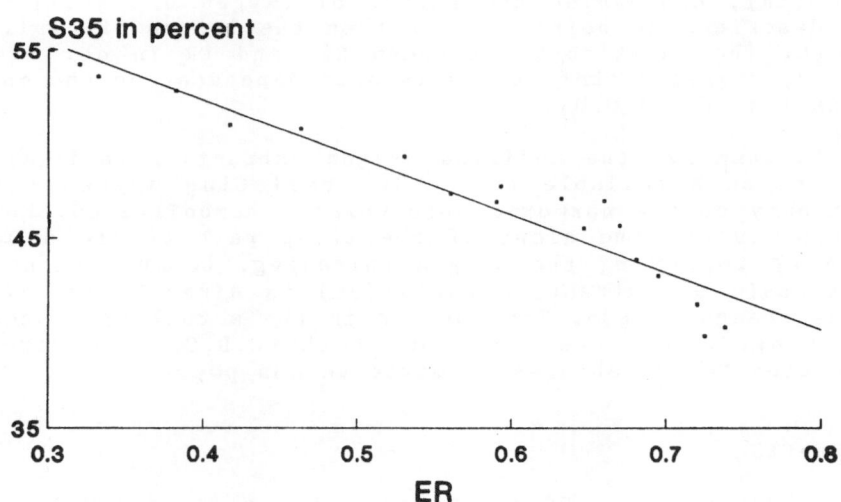

Fig. 5: Classix oxygen extraction ratio (ER) versus the actual S35 during acute normovolemic hemodilution (r-value: 0.94).

DISCUSSION

In the present study S35, the saturation at an oxygen tension of 35 torr at actual values of mixed venous blood, was used as a dynamic baseline for oxygen delivery calculations, assuming that oxygen bound to Hb at a tension of less than 35 torr should be considered relatively unavailable as a drop of the end-capillary PO_2 below 35 torr may be followed in some organs by tissue hypoxia because of limitation of oxygen diffusion into the tissue. Subtraction of the portion of unavailable oxygen from the total arterial oxygen content should give the real oxygen availability (O_2av).

Theoretically, when O_2av reaches the VO_2 (ERav = 1) oxygen supply dependence starts. In our study, total body oxygen consumption started to decrease after ERav reached the value of 1.07 ($\pm$ 0.03), suggesting that the modified O_2 extraction ratio (ERav) does indeed predict oxygen supply dependence during normoxic normovolemic hemodilution.

During acute hemodilution the dynamic baseline for the oxygen delivery, expressed by the S35, changed. This decreases the percentage of relatively unavailable oxygen and improves the oxygen unloading from the hemoglobin. The changes in oxygen unloading capacity was induced not only by a shift to the right, expressed by the P50, but also by a rotation of the oxyhemoglobin dissociation curve (O.D.C.) denoted by a change in n_{hill}. The interrelationship of the various physiologic factors affecting the oxyhemoglobin relationship is complex and multivariate. In our study no correlation was found between pH, PCO_2 or temperature.

A direct correlation was found between the changes in the actual mixed venous PO_2 (PvO_2) and the changes in the actual S35, suggesting that alterations in the O.D.C. compensate the decrease in end-capillary PO_2. Oxygen extraction ratio (ER), expressing the extent of oxygen unloading, has been described as being dependent on the length of capillary network. The relationship between S35 and ER in our study, however, suggests that the ER is also dependent on the characteristics of the O.D.C.

In summary, the modified oxygen extraction ratio (ERav) seems to be a reliable factor in predicting oxygen supply dependency during normoxic normovolemic hemodilution. Hemodilution alters the slope of the steep part of the O.D.C., probably improving the oxygen unloading, because change in the classic O_2 extraction ratio (ER) is directly correlated to the change in S35. The changes in the actual S35, used as an expression of the position of the O.D.C., are directly correlated to the changes in mixed venous pO_2.

REFERENCES

1. C.W. Bryan-Brown, SE-Min Baek, G. Makabali, and W.C. Shoemaker. Consumable oxygen: availability of oxygen in relation to oxyhemoglobin dissociation. Crit. Care Med., 1: 17 (1973).

2. M.J. Miller. Tissue oxygenation in clinical medicine: an historical review. Anaesth. Analg., 61: 527 (1982).

3. P.O. Malmberg, M.P. Hlastala, and R.D. Woodson. Effect of increased blood-oxygen affinity on oxygen transport in hemorrhagic shock. J. Appl. Physiol. 47: 889 (1979).

4. D.W Woodson, and S. Auerbach. Effect of increased oxygen affinity and anemia on cardiac output and its distribution. J. Appl. Physiol. 53: 1299 (1982).

5. D.W. Woodson, J.H. Fitzpatrick, D.J. Costello, and D.D. Gilbol. Increased blood oxygen affinity decreases canine brain oxygen consumption. J. Lab. Clin. Med. 100: 411 (1982).

6. G.A. Pantely, A.A. Oyama, J. Metcalfe, M.S. Lawson, and J.E. Welch. Improvement in the relationship between flow to ischemic myocardium and the extent of necrosis with glycolytic intermediates that decreases blood oxygen affinity in dogs. Circ. Res. 49: 395 (1981).

7. R. Weisel, R. Dennis, J. Manny, J. Mannietz, R. Valeri, and H. Hechtman. Adverse effects of transfusion therapy during abdominal aortic aneurysectomy. Surgery 83: 682 (1978).

8. M.B Nicotia, P.M. Stevens, J. Viroslav, and A.A. Alvarez. Physiologic evaluation of positive end expiratory pressure ventilation. Chest 64: 19 (1973).

9. R.R. Springer, and P.M. Stevens. The influence of PEEP on survival of patients in respiratiry failure. Am. J. of Med. 66: 196 (1979).

10. M.J. Edwards, and B. Canon. Oxygen transport during erythropoietic response to moderate blood loss. New Engl. J. of Med. 287: 115 (1972).

11. A. Blumberg, and H.R. Marti. Adaptation to anemia by decreased oxygen affinity of hemoglobin in patients on dialysis. Kidn. Int. I: 263 (1972).

12. O. Siggaard-Andersen. Determination and presentation of Acid-Base Data. Contr. Nephr. 21/20: 128 (1980).

CENTRAL NERVOUS SYSTEM

EFFECTS OF NIMODIPINE ON THE RESPONSES TO CEREBRAL ISCHEMIA IN THE MONGOLIAN GERBIL

Shlomo Cohen and Avraham Mayevsky

Department of Life Sciences, Bar Ilan University

Ramat Gan, 52100   Israel

Introduction

Brain ischemia has significant implications to the development of damage in patients undergoing stroke condition. In order to understand the mechanisms involved, a few animal models have been adopted. The Mongolian gerbil, having an incomplete circle of Willis, is widely used as a model for partial or complete ischemia (Levy and Brierley, 1974. Berry et al, 1975. Kobayashi et al, 1977. Mayevsky, 1978. Mies and Hossmann, 1983. Mayevsky et al, 1985). The disturbances of ion homeostasis in the brain under ischemia were recorded by a few groups using an extracellular K^+ electrode and more recently, Ca^{2+} electrode (Branston et al, 1977. Hansen and Zeuthen, 1981. Harris et al, 1981. Choki et al, 1984. Mayevsky et al, 1985. Mayevsky and Zarchin, 1987).

The involvement of calcium ions in the development of irreversible brain damage is a well-documented phenomenon (Siesjo, 1980. Farber, 1981. Nowicki et al, 1982). During the last few years the effects of a Dihydropyridine calcium antagonist-Nimodipine were studied in experimental as well as clinical conditions (Heffez and Passonneau, 1985. Gelmers, 1987. Milde et al, 1986. Vibulsresth et al, 1987). The evidence for the beneficial effect of Nimodipine on the responses to brain ischemia is not solid as yet.

The purpose of the present investigation was to develop an animal model in which the effects of drugs such as Nimodipine with a potential protection properties against cerebral ischemia could be tested.

Using the surface fiber optic fluorometry/reflectometry, the intramitochondrial NADH redox state was monitored simultaneously with the hemodynamic properties of the observed tissue. This technique provides two signals, namely the 450 nm fluorescence (NADH redox state) and the 366 nm reflectance trace(R) correlated mainly to the changes in tissue blood volume. We showed that under ischemia in the gerbil brain (Mayevsky and Zarchin, 1981) a large increase in the 366 nm reflectance signal (SRI= Secondary Reflectance Increase) was recorded about 2 minutes after decapitation. This change in the R trace was correlated to the large inos shift as recorded by surface electrodes (Mayevsky et al, 1985). We are reporting here the effects of Nimodipine on the responses to complete ischemia induced in the Mongolian gerbil by bilateral carotid artery occlusion.

Methods

Fluorometer/Reflectometer

In the present study, a two-channel DC fluorometer/reflectometer was used (Mayevsky, 1978. Mayevsky and Chance, 1982. Mayevsky, 1984). The source for the 366 nm excitation light was a 100 W air-cooled mercury lamp passing through the two excitation fiber optic bundles to the brain. The emitted light from the brain that was transmitted through another fiber optic bundle was split into a 90:10 ratio to measure the fluorescence light (F; 90%) and the reflcted light (R; 10%). The two fluorescence, the two reflectance and the two corrected fluorescence signals were recorded on a multichannel Grass instrument. The correction of the fluorescence signals for hemodynamic artifacts was done by substructing the R from the F signal in a 1:1 ratio (Mayevsky, 1984).

Animal Preparation

Male gerbils (50-70 gr) obtained from Tumblebrook Farms (Mass, USA) were anesthetized by IP injection of Equi-thesin (0.3 ml/100 gr) for the duration of the operation. Later on, a small volume of Equi-thesin was added (0.03-0.05 ml) every 30 minutes. The skull was exposed and two holes were drilled in the parietal bone to hold the light guide holder. Two pairs of stainless steel screws (used also as ECOG electrodes) were held in the parietal bone area and dental acrylic cement was used. The two common carotid arteries were isolated and ligatures of 3-0 silk thread were placed around them.

Nimodipine (0.5 mg/ml) was dissolved in the following solution composed of 969 gr Polyethylene-glycol 400, 60 gr Glycerine and 100 gr water.

Experimental Protocol

The gerbil was connected to the monitoring system 10 minutes after the operation. In order to test the redox state of the brain, a one-minute N_2 exposure was used. Only gerbils showing a typical increase response were used further on.

Each animal was exposed to 4-6 bilateral carotid arteries occlusions. The release of the anourysm clips was done after recording the typical SRI change in the reflectance trace and the range was 2-6 minutes. The interval between successive occlusions was about 60 minutes.

Injection of Nimodipine, placebo or saline solution was done 15-20 minutes before the third and the fifth carotid arteries occlusion. Each gerbil was injected with 0.1 ml of Nimodipine, placebo or saline solution. Also, a short N_2 exposure was done 5-10 minutes before the occlusions for maximum response recording. Body temperature was measured and adjusted to be in the physiological range. The ECOG measured bilaterally was used for the evaluation of the recovery after each ischemic episode.

Results

A total of 31 gerbils were used, 8 of which were controls, 11 treated with a placebo, and 12 treated with Nimodipine. Each animal was exposed to two ischemic control episodes so that baseline responses were obtained before injections were done.

Figure 1 shows two typical responses to bilateral carotid artery occlusion in a gerbil before and after placebo injection. Immediately after the occlusion, the redox state of the brain became reduced as seen on both hemispheres (CF_R, CF_L) and reached a plateau level within one minute. The reflectance trace shows a transient increase at the occlusion point which later on showed a large secondary increase (SRI). This large increase in the reflectance led to an artifact in the corrected fluorescence trace (increase or a decrease). Both the R and CF signals returned to the preischemic level after reperfusion was done (R + L open). As can be seen, the differences between the responses to ischmia before and after placebo injection were minimal as expected.

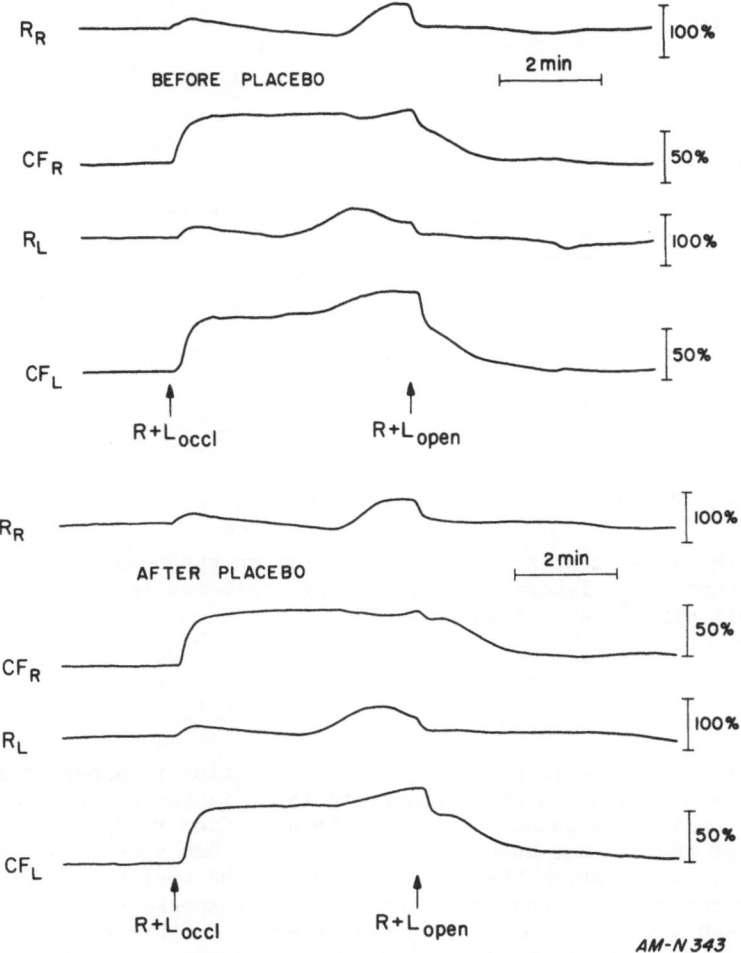

Figure 1: The effects of bilateral carotid artery occlusion on the metabolic and hemodynamic signals measured from the two hemispheres of the gerbil brain. Two ischemic episodes were induced before and after injection of placebo solution. Each point represents the mean and the standard error of it.

R_R, R_L -366 nm reflectance measured from the right and left hemispheres respectively.

CF_R, CF_L -450 nm corrected fluorescence measured from the right and left hemispheres respectively.

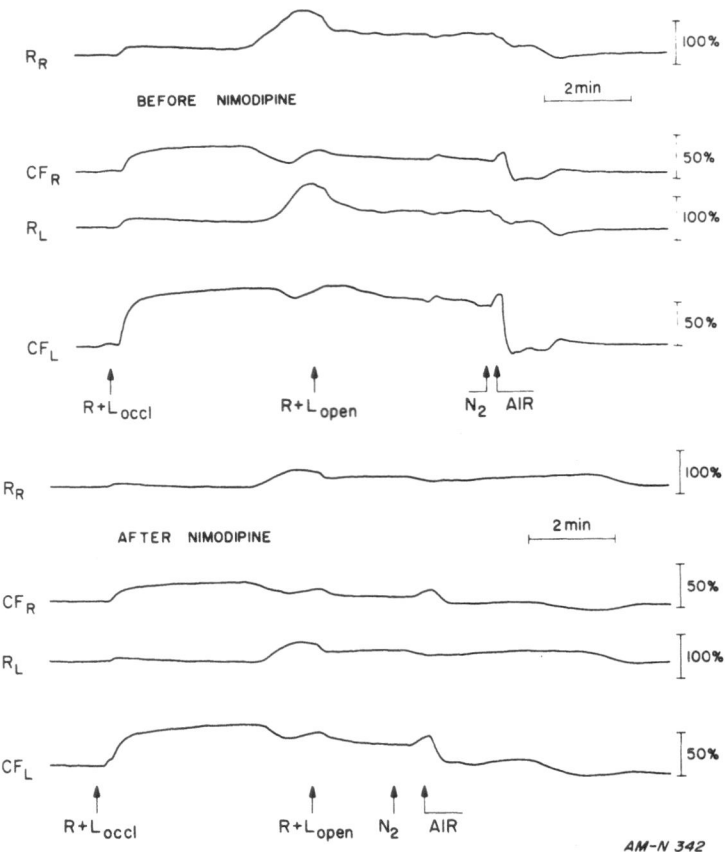

Figure 2: The effects of Nimodipine injection on the hemodynamic and redox state responses to bilateral carotid artery occlusion in the Mongolian gerbil. All symbols are as in Figure 1.

In Figure 2 the effects of Nimodipine injection is shown in another gerbil. The same pattern of responses to ischemia was recorded as shown in Figure 1. During the second ischemia episode (after Nimodipine) the amplitude of the SRI was much smaller in the two hemispheres due to the drug injection. Although the NADH response to the occlusion was different in the two hemispheres, the consistency of the response before and after the treatment was very good in each hemisphere. In this specific gerbil, the recovery from the ischemia was not spontaneous so we applied a short N_2 cycle to stimulate the recovery (empirical information).

In order to test the effects of Nimodipine in a statistical way, the analog signals were analyzed and quantitated for each ischemic episode. The following three parameters were calculated:

       1. Time to the appearance of the SRI
       2. Maximal amplitude of the SRI
       3. Maximal level of NADH during the ischemia (CF)

The results obtained are presented in Figures 3-5 and the general
explanation to the way of calculation is as follows: In each gerbil the
responses obtained in the second occlusion were used as a reference to the
other 4 occlusions and got a value of 100%. All other responses are
presented in percent change relative to the second occlusion. In all three
parameters and all occlusions the differences calculated between control
and placebo groups were very small and statistically insignificant. The
main effect was of Nimodipine as expected. As seen in Figure 3, the
Nimodipine delayed the appearance of the SRI by 10-40% as compared to the
second ischemic episode which served as a control. A slight decrease in
time was recorded in the control and placebo animals.

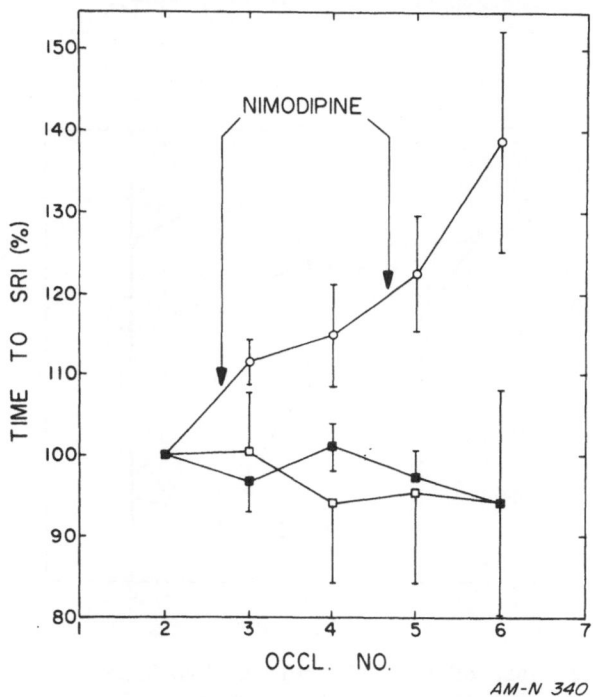

AM-N 340

Figure 3: The effects of calcium channel blocker-Nimodipine on the time in'
which SRI appeared under bilateral carotid artery occlusion in the gerbils.
SRI-Secondary Reflectance Increase; Open and filled squares represent
control and placebo injected animals respectively. Open circles represent
Nimodipine injected gerbils: the arrows are indicators of the two injection
points during the experiment.

    The amplitude of the change during the SRI was decreased dramatically
by the Nimodipine, as shown in Figure 4. Here also a small difference was
recorded between the control and placebo groups. While in the SRI response
the effect of Nimodipine was clear, the NADH response was not so clear. As
shown in Figure 5, the maximal increase in NADH (CF) in the ischemic
episodes decreased from the second to the last occlusion in the control and
placebo animals. In the Nimodipine treated group an initial increase in

the change was recorded which later on in the occlusion returned to the 100% level area. If one, consider the decrease in response in the control and placebo group than the Nimodipine group remain high as compared to the control and placebo groups through the 4 last occlusions. The differences between the Nimodipine group and the control or the placebo groups were significant through all the occlusions (p<0.05-p<0.01).

Discussion

The purpose of the present study was to develop a simple animal model in which the effects of calcium channel blocker could be analyzed on the level of microcirculation as well as the mitochondrial function. Using the surface fluorometry technique, two main parameters were measured, namely the redox state of the intramitochondrial space and the changes in blood volume of the same measurement site (Mayevsky, 1984).

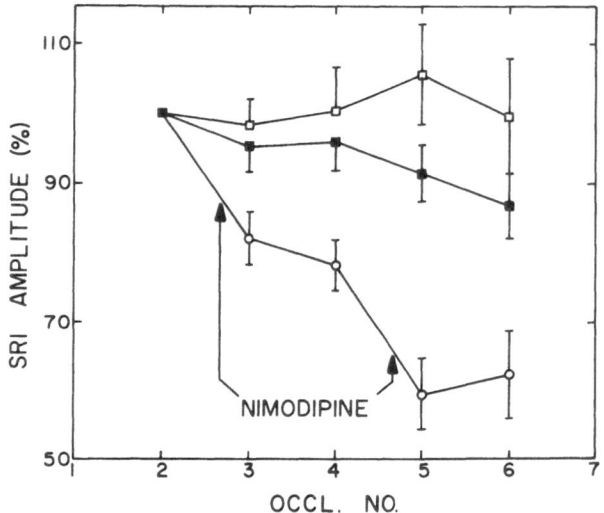

Figure 4: The responses of the SRI amplitude to Nimodipine under bilateral carotid artery occlusion in the Mongolian gerbil. Symbols are as in Figure 3.

In the gerbil model (Meriones Unguiculatus), occlusion of the two carotid arteries produced in most animals complete or close to a complete ischemia (Kobayashi et al, 1977. Mayevsky et al, 1985). The bilateral monitoring provides information on the two hemispheres increasing the validity of the results and also enables one to occlude one hemisphere and compare it to the unoccluded side in the same animal (Mayevsky and Zarchin, 1981). Two main phenomena were recorded under ischemic conditions, namely the decrease in oxygen availability (increase in NADH levels) and the large increase in the reflectance trace (SRI). As suggested and shown previously, the SRI appeared as a result of a massive vasoconstriction that occured when the large increase in extracellular K^+ recorded (Mayevsky et al, 1985.1986. Mayevsky and Zarchin, 1987). More recent studies showed that Ca^{2+} influx occured at the same time as the large increase in K_e^+ and the SRI (Mayevsky, Yoles and Zarchin, Unpublished results). Therefore, the SRI was chosen as an indirect parameter to easily evaluate the appearance of the general depolarization event that occured under ischemic conditions. Recently, two groups suggested the usage of ions shift and DC potential negativation as good parameters to evaluate the effects of calcium channel blockers (Dierking et al, 1987. Mori et al, 1987). Therefore, our approach is a complimentary one and could be done more easily in many animals. As shown in Figures 3 and 4, the two parameters calculated for the SRI event were affected significantly by the Nimodipine injection. The time for appearance of the SRI was delayed and the amplitude of the change was much smaller, suggesting the partial

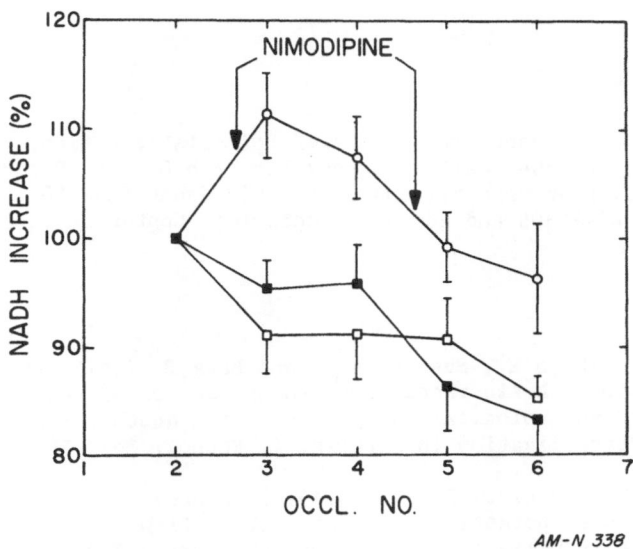

AM-N 338

Figure 5: The effects of Nimodipine on the NADH redox state responses under cerebral ischemia in the gerbil. Symbols are as in Figure 3.

inhibition of the Ca^{2+} influx to the vascular muscle cells thus decreasing the level of vasoconstriction under ischemia. Those changes in the SRI parameters continued to appear after the first and second injection of Nimodipine and also through all the occlusions.

The responses of the mitochondrial redox state to the injection of Nimodipine were more complicated. As seen in Figure 5, the increased change during the ischemia suggested that Nimodipine led to a more oxidized state of the mitochondria before the ischemia and therefore the capacity of reduction was larger later on. Another possibility is that the resting oxygen supply was not affected by the Nimodipine and the increased response during ischemia, was due to an uncoupling effect.

In summary, the results presented here suggest the beneficial effects of Nimodipine on the pathophysiology of ischemic insult. Nimodipine appears to affect cerebral blood flow and NADH levels in the cerebral cortex, as it was found from reflectance and fluorescence changes during and after bilateral occlusions of the carotides in the gerbil. These effects may be atributed to the Calcium antagonism effect of the Nimodipine. After 2-5 minutes of cerebral ischemia an increase in the amplitude of Reflctance signal (SRI - Secondary Reflectance Increase) is generally observed. This SRI reflects vasoconstriction of cortex blood vessels and it is atributed to membrane depolarization and K^+ leakage from cells. Maximum SRI during R+L ischemia, was significantly lower ($P<0.01$) in gerbils after 50 ug Nimodipine injection, relative to control or placebo treated animals. In Nimodipine treated animals the time needed to the beginning of SRI seems to be longer relatively to control or placebo treated animals. According to these preliminary results it seems that Nimodipine delayed the appearance of the massive depolarization and decreased its influence as reflected by the delay and the decrease of the SRI amplitude.

ACKNOWLEDGEMENTS

Supported by a grant from the Chief Scientist's Office, Ministry of Health, Israel; by the Health Sciences Research Center, Department of Life Sciences, Bar-Ilan University, and by the NIH Grant NS-22881, RR-02305; NIH SBIR Grant NS-22309 and Advanced Technology Center of South-Eastern Pennsylvania.

REFERENCES

Berry,K., Wisniewski,H.M., Svarzbein,L. and Baez,S. (1975) On the relationship of brain vasculature to production of neurological deficit and morphological changes following acute unilateral common carotid artery ligation in gerbils. J. Neurol. Sci. 25:75-92.

Branston,N.M., Strong,A.J., Symon,L. (1977) Extracellular potassium activity evoked potential and tissue blood flow. Relationship during progressive ischemia in baboon cerebral cortex. J.Neurol.Sci. 32:305-321.

Choki,J., Greenberg,J., Sclarsky,D. and Reivich,M. (1984) Correlation between brain surface potassium and glucose utilization after bilateral cerebral ischemia in the gerbil. Stroke. 15:851-857.

Dierking,H., Tegtmeier,F., Holler,M. and Peters,Th. (1987) Effects of Fluonarizine, Nimodipine and Diphenylhydantoine on K_e^+, Ca_e^+ and DC in the hypoxic cortex of the rat brain in vivo. J.CBF and Metabolism. 7: Supp 1, S156.

Farber,J.L. (1981) The role of calcium in cell death. Life Sciences. 29: 1289-1295.

Gelmers,H.J. (1987) Nimodipine in ischemic Stroke. Clinical Neuropharmacol. 10:412-422.

Hansen,A.J. and Zeuthen,T. (1981) Extracellular ion concentration during spreading depression and ischemia in the rat brain cortex. Acta.Physiol.Scand. 113:437-445.

Harris,R.J., Symon,L., Branston,N.M. and Bayhan,M. (1981) Changes in extracellular calcium activity in cerebral ischemia. J.CBF and Metabolism. 1:203-209.

Heffez,D.S. and Passonneau,J.V. (1985) Effet of Nimodipine on cerebral metabolism during ischemia and recirculation in the Mongolian gerbil. J.CBF and Metabolism. 5:523-528.

Kobayashi,M., Lust,W.D. and Passonneau,J.V. (1977). Concentrations of energy metabolites and cyclic nucleotides during after bilateral ischemia in the gerbil cerebral cortex. J.Neurochem. 29:53-59.

Levy,D.E. and Brierley,J.B. (1974) Communications between vertebrobasilar and carotid arterial circulations in the gerbil. Exp.Neurol. 45: 503-508.

Mayevsky,A. (1978) Pyridine nucleotide oxidation-reduction state of the cerebral cortex in the awake gerbil. J.Neurosci.Res. 3:369-374.

Mayevsky,A. (1978) Shedding light on the awake brain. In P.L. Dutton, J. Leigh and A. Scarpa. (eds) Frontiers in Bioenergetics from Electrons to Tissues. Vol. II. Academic Press, New York, pp.1467-1476.

Mayevsky,A. (1984) Brain NADH redox state monitored in vivo by fiber optic surface fluorometry. Brain.Res.Rev. 7:49-68.

Mayevsky,A. and Zarchin,N. (1981) The effects of unilateral carotid occlusion on the responses to decapitation in the gerbil brain. Brain Research. 206:155-160.

Mayevsky,A. and Chance,B. (1982) Intracellular oxidation reduction state measured in situ by multichannel fiber-optic-surface fluorometer. Science. 217:527-540.

Mayevsky,A. and Zarchin,N. (1987) Metabolic Ionic and electrical activities during and after incomplete or complete cerebral ischemia in the Mongolian gerbil. In: Oxygen Transport to Tissue IX, Silver,I.A. and Silver,A. (eds). Plenum Publishing Corp. pp.265-273.

Mayevsky,A., Friedli,C.M. and Reivich,M. (1985) Metabolic, ionic and electrical responses of the gerbil brain to ischemia. Am.J.Physiol. 248:R99-R107.

Mayevsky,A., Yoles,E. and Zarchin,N. (1986) Metabolic, ionic and electrical responses to oxygen deficiency in the newborn dog in vivo. In: Oxygen Transport to tissue VIII, I.S. Longmuir (ed). Plenum Press. pp. 261-269.

Mies,G. and Hossmann,K.A. (1983) A method for regional evaluation of potassium depletion during graded focal ischemia in gerbils. J.CBF and Metabolism.,3, Suppl. 1. S313-314.

Milde,L.N., Milde, J.H. and Michenfelder,J.D. (1986) Delayed treatment with Nimodipine improves cerebral blood flow after complete cerebral ischemia in the dog. J.CBF and Metabolism. 6:332-337.

Mori,K. Iwayama,K., Kawano,T$_2$ and Kaminogo,M. (1987) DC ptoential and extracellular K^+ and Ca^{2+} at critical levels of brain ischemia in cats. J.CBF and Metabolism. 7, Suppl. 1, S112.

Nowicki,J.P. Mackenzie,E.T. Young,A.R. (1982) Brain ischemia, calcium and calcium antagonists. Patholo.Biol. (Paris) 30:282-288.

Siesjo,B.K. (1981) Cell damage in the brain: a speculative synthesis. J.CBF and Metabolism. 1:155-185.

Vibulsresth,S. Dietrich,W.D., Busto,R. and Ginsberg,M.D. (1987) Failure of Nimodipine to prevent ischemic Neuronal damage in rats. Stroke. 18:210-216.

HEMATOCRIT CHANGES IN THE EXTRA- AND INTRAPARENCHYMAL CIRCULATION OF THE FELINE BRAIN CORTEX IN THE COURSE OF GLOBAL CEREBRAL ISCHEMIA

Andras Eke

2nd Department of Physiology, Semmelweis Medical University, Budapest, Hungary, and Department of Pathology, University of Alabama at Birmingham, Birmingham, AL, U.S.A.

INTRODUCTION

Collapse of the microcirculation is always among the factors contributing to the evolution of ischemic tissue damage in the brain. The approach to assess the progress of failure in the ischemic parenchymal circulation taken by the present study was to monitor the erythrocyte and plasma microflows along with local tissue hematocrit in a two dimensional plane at high topographical and temporal resolution by a television densitometric method[3,4] in the intact cerebrocortical tissue. Preliminary studies with this method have revealed significant heterogeneity of these parameters under control conditions. It was postulated that the decisively different microhemodynamic behavior of erythrocytes and plasma could focally be further accentuated under ischemic conditions (endothelial lesions, vasoaction, edema, obstruction and occlusion of the capillary network) and may lead to a dissociation of erythrocyte and plasma channels manifested by marked focal alterations in tissue hematocrit and an altered distribution pattern of microflows. To test this hypothesis, global cerebral ischemia was induced and maintained by adjusting the mean arterial blood pressure to 40 mmHg by controlled arterial hemorrhage.

METHODS

The experiments were carried out in alfa-Chloralose (50 mg/kg body weight, i.p.) anesthetized cats of either sexes. The weight of the animals ranged between 2.5 and 3.5 kgs. Global cerebral ischemia was induced and maintained by lowering the mean arterial blood pressure (MABP) by controlled arterial hemorrhage via a catheter introduced into the abdominal aorta and connected to a precision infusion/withdrawal pump, that was controlled by an error signal generated between the actual MABP measured at the origin of the right brachial artery and that of a requested pressure level. MABP was lowered in a stepwise manner form the control level to 100, 80, 60 mmHg where it was maintained within +/- 2 mmHg for a period of 5 minutes and finally to 40 mmHg where it was maintained until the ischemia was terminated. The criteria for terminating the ischemic stage were either the cessation of flow in the monitored area or shed blood completely having been reinfused for the MABP to be maintained at 40 mmHg. For a complete set of data analysis four experiments were selected with increasingly longer survival time, that is 15, 60, 90 and 120 minutes. Measurement of central arterial hematocrit by the capillary technique and blood gases in blood withdrawn via the aortic catheter were made at the time of imaging the above mentioned microcirculatory parameters, that is at the 5th minute of each hypotensive stage and at 15 minute interval on the ischemic level of 40 mmHg and following

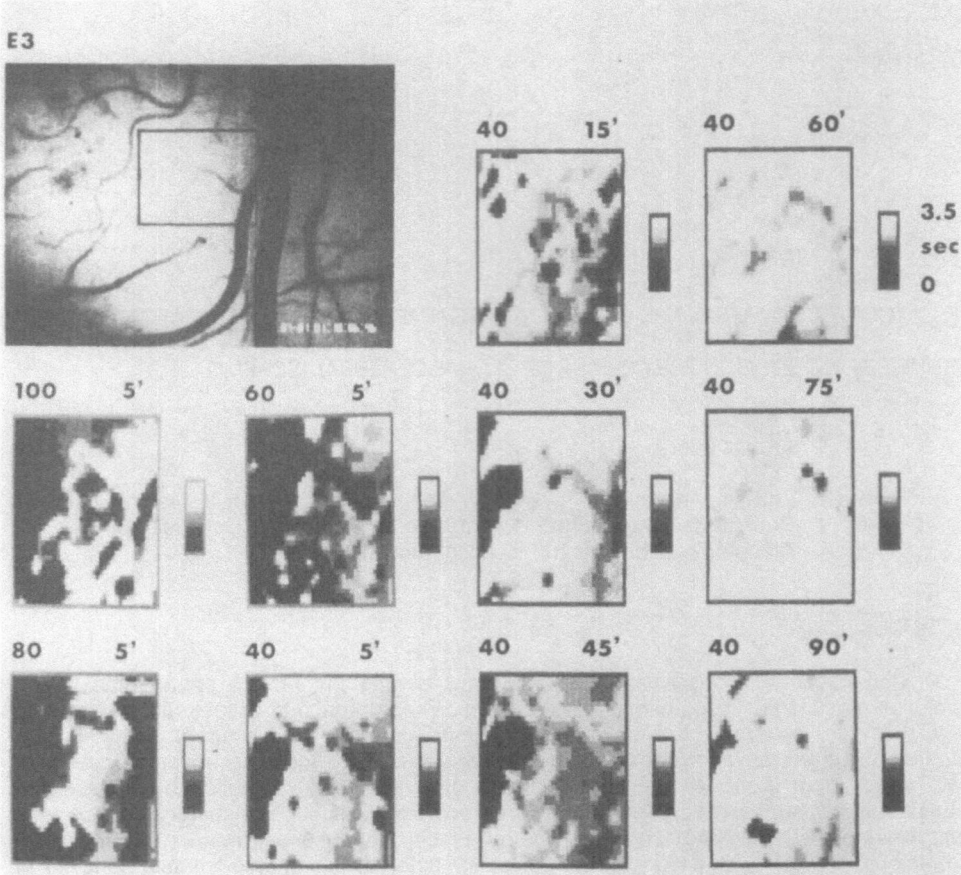

Fig. 1. Time sequence of changing local plasma mean transit time pattern in the feline brain cortex as the animal's mean arterial blood pressure is being lowered in a stepwise manner from 100, 80, 60 and finally 40 mmHg. A number of local microcirculatory parameters including plasma mean transit times has been mapped by the television reflectometric imaging method[3,4] within the enframed cerebrocortical area of 2.5 times 3 mm under white light epillumination and detection at one of the isosbestic point of hemoglobin (584 - 589 nm). The spatial resolution is 2500 location/7.6 square mm of cerebrocortical tissue. Above the upper right corner of each map, the actual systemic arterial pressure level (left, mmHg) and the time of the measurement within each pressure step (right, minutes) is indicated. The maps are intensity coded within the range of 0 sec (black) to 3.5 sec (white).

reperfusion. A nondestructive reflectometric method[3,4] that allows for repetitive imagining of cerebrocortical local hematocrit along with local erythrocyte and plasma flows has been utilized in these studies. The following microcirculatory parameters were measured: weighted hematocrit for the arterial system between the carotids and a pial artery feeding to the monitored cerebrocortical area (feed hematocrit), local erythrocyte and plasma volume, mean transit time, volume flow, local tissue hematocrit in a raster array of 2500 location/7.6 square mm of cerebrocortical tissue. A single tissue element of these microcirculatory parameter maps constitutes for a volume of 0.01 cubic mm. The time resoliton power of this imaging method is 1 image/min.

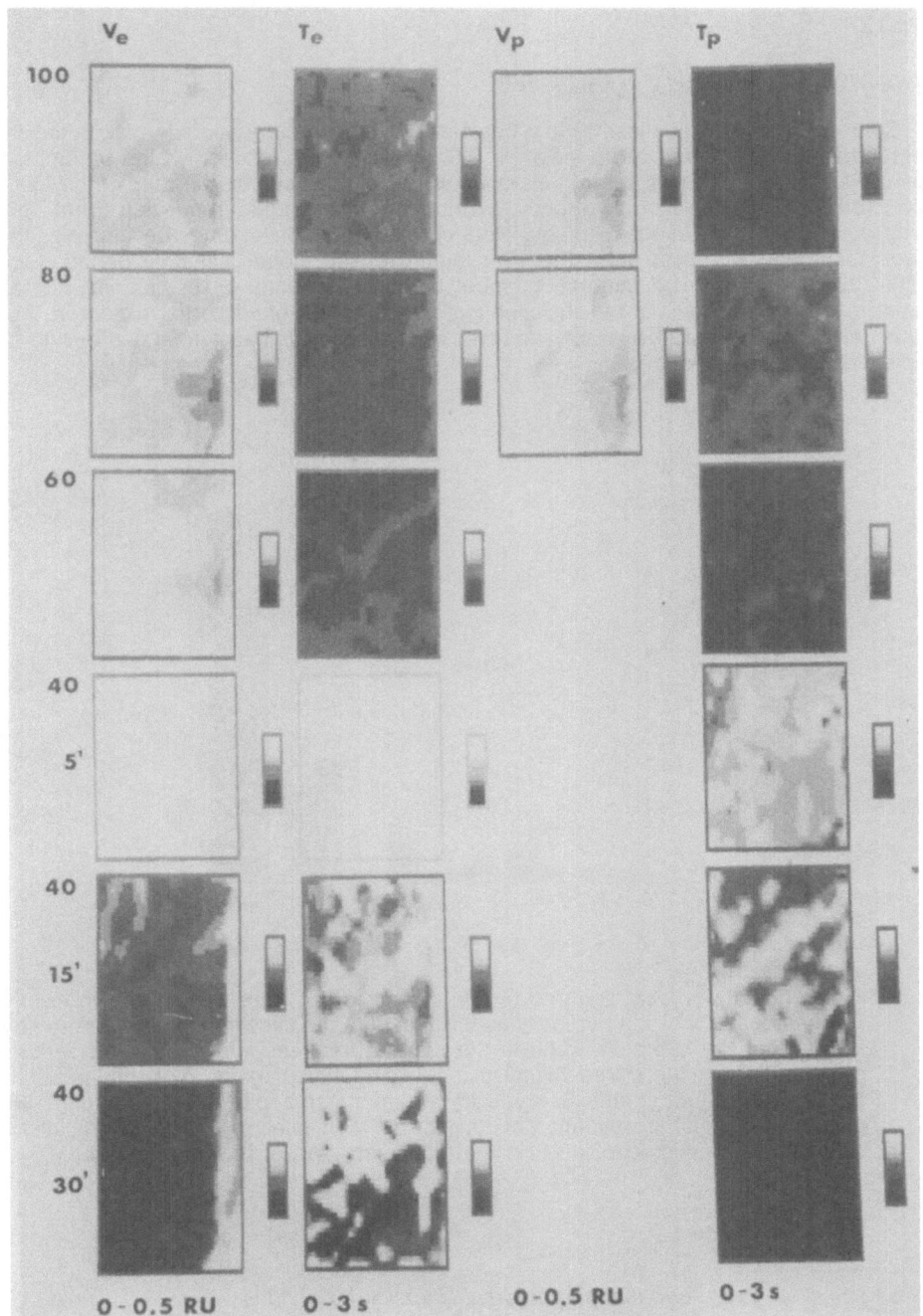

Fig. 2. Changes of local erythrocyte (Ve) and plasma volume (Vp), erythrocyte (Te) and plasma mean transit time (Tp) during the course of global cerebral ischemia induced and maintained by controlled arterial hemorrhage. Left to each map, the actual systemic arterial pressure level in mmHg's (top) and the time of the measurement in minutes within each pressure step is indicated (middle). The maps are intensity coded. Their range from black to white is indicated at the bottom.

RESULTS

Erythrocyte and Plasma Microflow Pattern

Both erythrocyte and plasma flow has been found heterogeneously distributed in cerebrocortical tissue areas of as small as 7.6 square mm as shown on the parameter maps in Fig. 1 and 2. The degree of heterogeneity were assessed by the ratio of the half-peak width to the peak of frequency distribution histograms generated from the parenchymal pixel data of these maps. Heterogeneity of both erythrocyte and plasma perfusion abruptly increased from the 15th minute of the global ischemic period (Fig. 3). By 30, 60, 90 or 120 minute of global cerebral ischemia it resulted in a total collapse of the microcirculation in the monitored area of the brain cortex except in the case of the animal with 120 min survival time, where ischemia was terminated due to a complete uptake of the shed blood.

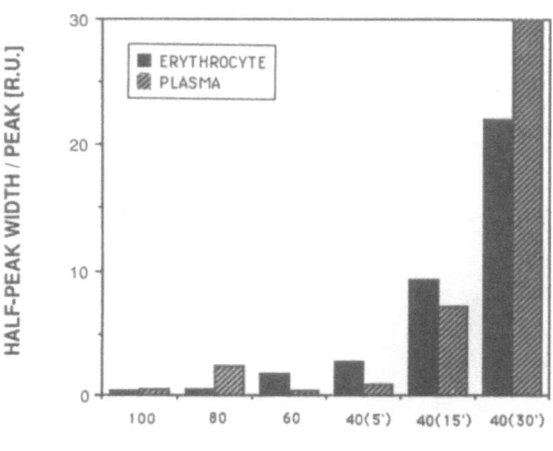

MEAN ARTERIAL BLOOD PRESSURE [mmHg]

Fig. 3.   Heterogeneity of local erythrocyte and plasma mean transit times shown in Fig. 2.. Heterogeneity of the perenchymal pixel data of the shown maps is assessed by the ratio of the half-peak width to the peak of frequency distribution histograms generated from the parenchymal pixel data of those maps. Note the abruptly increasing heterogeneity of erythrocyte and plasma perfusion from the 15th minute of the global ischemic period, that by 45 minites of this period results in collapse of the microcirculation in the monitored area of the brain cortex.

Time Sequence of Extra- and Intraparenchymal Hematocrit Changes

The time sequence of extra- and intraparenchymal hematocrit changes are shown in a normalized fashion in Figs. 4 and 5. In Fig. 4 feed hematocrit is shown normalized by the systemic arterial hematocrit as a funtion of time. For the analysis of local hematocrit changes in global cerebral ischemia, the purely parenchymal pixel data were filtered out from the images by masking it slightly below the level of optical density characteristic to visible pial arteries (of much higher optical density that typical to capillary, that is parenchymal areas).   As a result, some 500 - 600 individual parenchymal hematocrit value were then averaged per image and plotted against time (Fig. 5).   The control systemic arterial hematocrit (systemic hematocrit) was 39 (SEM=2).

FEED / SYSTEMIC HEMATOCRIT

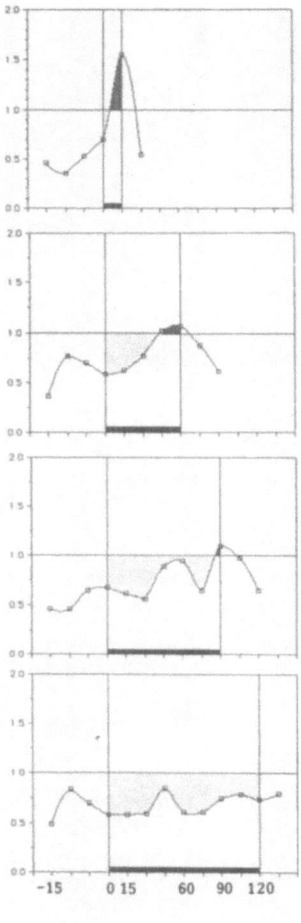

TIME [min]

Fig. 4. Ratio of the feed hematocrit to the monitored intraparenchymal circulatory system and that of the systemic arterial hematocrit in four representative experiments before and during global cerebral ischemia. From time -15 minutes to 0 minute global cerebral ischemia was induced by stepwise controled arterial hemorrhage from the control level of 100 mmHg down to 40 mmHg. Horizontal bars indicate the length of the time it took for the microcirculation to collapse (survival time). Lightly shaded area indicates the time span during global cerebral ischemia when the feed hematocrit was lower than that of the systemic arterial due to the Fahreus-effect. Narrowing of this area is a sign of increasing hemoconcentration in the extraparenchymal system feeding the cerebrocortical tissue. Darkly shaded area indicates when the feed hematocrit exceeded the systemic arterial hematocrit.

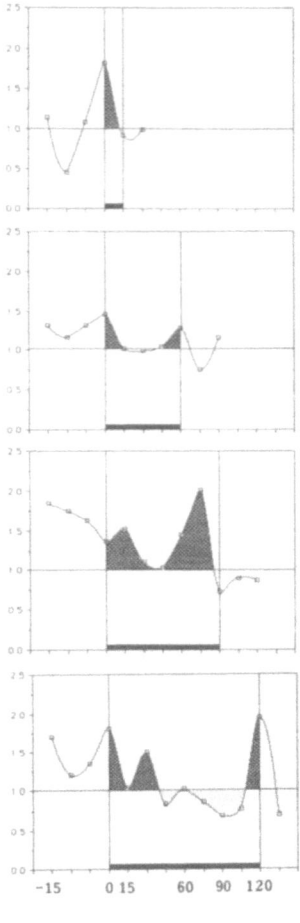

TIME [min]

Fig. 5. Ratio of the regionally integrated local hematocrit to the feed (pial) hematocrit
in four representative experiments before and during global cerebral ischemia.
From time -15 minutes to 0 minute global cerebral ischemia was induced by
stepwise controlled arterial hemorrhage from the control level of 100 mmHg
down to 40 mmHg. Horizontal bars indicate the length of the time it took for the
microcirculation to collapse (survival time).

The mean feed/systemic hematocrit ratio was 0.44 (SEM = .055) that can
certainly be attributed to the Fahreus effect along the increasingly smaller pial arteries
[1,5,6,7]. This effect was found to be gradually diminishing during the course of ischemia at
different rate and extent among the animals. Animals with shorter survival time showed
a rapid and marked increase in their feed hematocrit approaching to or exceeding the
central arterial hematocrit; Animals with longer survival time maintained a lower feed
hematocrit for increasingly longer periods of time.

The ratio of local tissue hematocrit to the systemic arterial hematocrit was found 0.67 (SEM=0.2). In other words, the blood supplying the tissue goes through extraparenchymal hemodilution and relative intraparenchymal hemoconcentration. Although this relative hemoconcentration varies during the induction phase of ischemia (100 mmHg - 40 mmHg), all of the animals entered with the condition of an initial hemoconcentration into the ischemic period (Fig. 5). As the pial hematocrit was rising with the progression of the global cerebral ischemia (Fig. 4), a compensatory hemodilution was be observed in the tissue (Fig. 5) that, however, got ultimately overridden by an abrupt rise in the local tissue hematocrit. The animal with the longest survival time (120 minutes) showed this compensatory hemodilution at the greates degree and for the longest duration. Under the second phase of tissue hemoconcentration the microcirculation collapsed rapidly (Fig. 6). Local cerebrocortical erythrocyte and plasma volume flows markedly decreased during global cerebral ischemia except in the case of the longest survival, where both actually increased. Following prompt reperfusion, besides the erythrocyte and plasma flow having been restituted, the hematocrit values got restored, too, to their value measured just preceding the abrupt rise in the mean local tissue hematocrit.

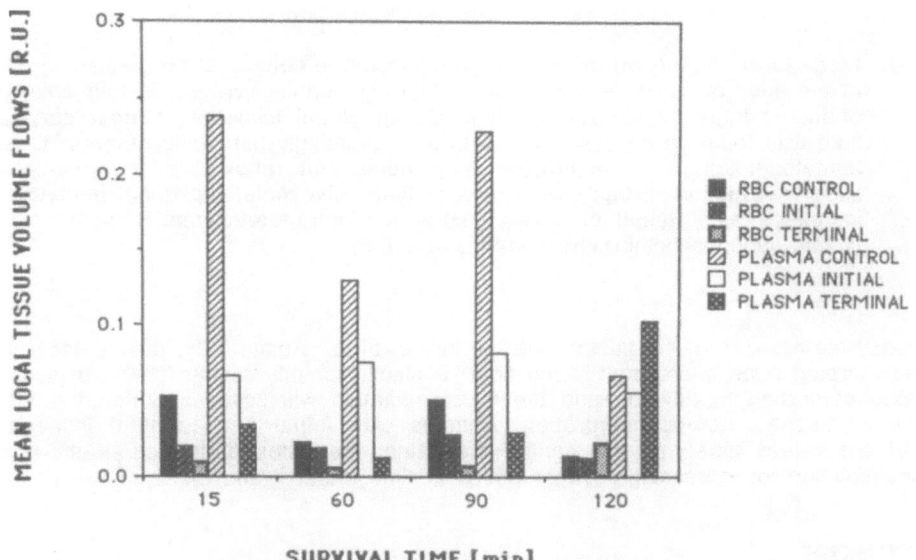

Fig. 6. Mean local red blood cell (RBC) and plasma volume flows measured in four animals from within parenchymal microareas in control, 5 minutes after reaching the 40 mmHg systemic blood pressure level (initial) and at the survival time of the microcirculation (terminal), that is the time on the 40 mmHg level when either the microcirculation collapsed (in animals with survival times of 15, 60, 90 min), or while microcirculation was still perfused but shed blood completely retaken by the animal (120 min).

For the mean tissue hematocrits does not significantly differed under the phases of initital and terminal hemoconcentration (35 % and 41.1 % with respective SEMs of 5.5 % and 5.1 %) (Fig. 7), one may well assume that either the hemorheological properties - including increased viscosity, rouleau formation and aggregation, white blood cell thrombosis etc. - of the flowing blood or geometry of the capillary network has been altered pathologically during this terminal hemoconcentration period that immediately

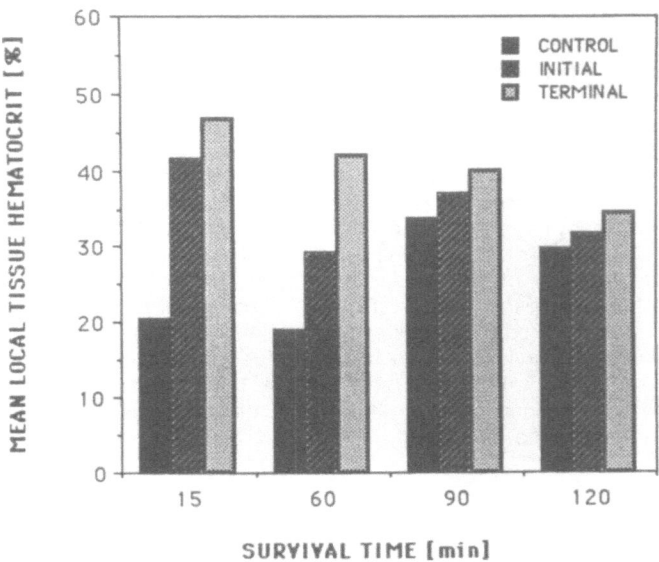

Fig. 7. Mean local cerebrocortical hematocrit measured in control, at the beginning and at the end of the ischemic period (initital and terminal values) in four animals of increasingly longer survival time during global ischemia. Longer survival time was found to be associated with an increasingly narrowing range of these hematocrit values. The terminal local hematocrit showed a strong inverse linear correlation (r=0.98) with survival time. Also note, that local hematocrit increases as a result of several extra- and intraparenchymal hemodynamic factors during ischemia (also see Figs. 8 and 9).

precedes the collapse of the microcirculation in the area. Again, if the events occur at an accelerated pace, there may be no time available for this second period of tissue hemoconcentration to develop and the microcirculation will collapse during the first phase of tissue hemoconcentration. Animals with initial and terminal ischemic hematocrit values closer to the control maintained perfusion in the cerebrocortical microcirculation for increasingly longer priods of time (Figs. 7 and 8).

CONCLUSIONS

The newly perfected method of repetitive reflectometric imaging of local tissue hematocrit and other local microcirculatory parameters in superficial layers of the feline brain cortex[3,4] proved suitable for revealing some potentially significant mechanisms pertinent to the pial and intraparenchymal circulation that can lead to the collapse of extra- and intraparenchymal circulation during global cerebral ischemia. Namely, under control conditions hematocrit of blood - flowing through the cerebral arterial vasculature of successively smaller diameter from the carotids to all the way down to the branches of the pial arterial network of around 100 - 200 micra in diameter - drops drastically most likely owing to the Fahreus-effect[1,5,6,7] to the level of around 44% of that in the central arterial circulation. Under these control conditions the tissue hematocrit is higher than that of the pial, that is its feed hematocrit. The control ratio of the cerebrocortical mean local tissue hematocrit and the central arterial hematocrit was found around 0.67 which is comparable to the data reported in the literature[2,8,9,10,11,12]. Under ischemic conditions, most likely due to diminishing Fahreus-effect, the complex and interrelated effects of a decreasing linear velocity of red and white blood cells, red and white blood cell aggregation, rouleau formation, there is an increasing hemoconcentration along the carotid-pial system. The rate of this extraparenchymal event is the highest in case of short survival and lowest when survival time is long. The change in this rate can be interpreted as extraparenchymal adaptation that aproaches its limit when the rate is very low keeping the feed hematocrit at the control level, namely 0.5 time the actual systemic hematocrit (Fig. 9).

The propagation of this effect to the microcirculation, however, can and seemed to be delayed by a temporarily hemodilution in the tissue. If time permits, the microcirculation shows adaptation in form of a temporary tissue hemodilution. It can collapse any time during the course of this adaptation depending on the extraparenchymal condition. Namely, in wake of rapidly rising feed hematocrit and insufficient tissue hemodilution, it collapses earlier. The intraparenchymal adaptation reaches its limits when the ratio of the local/feed hematocrit approaches 2.0 (Fig. 9).

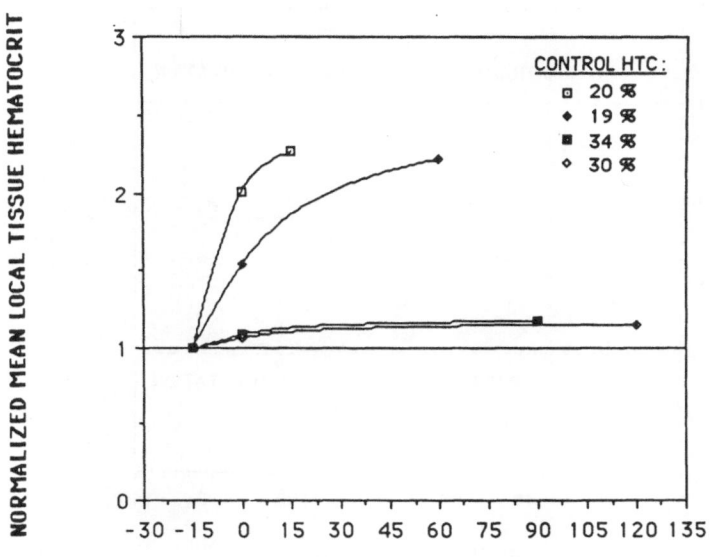

SURVIVAL TIME [min]

Fig. 8. Local tissue hematocrits shown in Fig. 7. normalized by their respective controls plotted as a funtion of survival time. Animals with initial and terminal ischemic hematocrit values closer to the control maintained perfusion in the cerebrocortical microcirculation for increasingly longer priods of time.

Since the mean tissue hematocrit was only slightly higher under these circumstances than during the initial phase of tissue hemoconcentration (Fig. 7), it may well demonstrate the importance of the hemodynamic and rheological factors - enumerated above in addition to those like tissue edema, thrombosis and embolization in the capillaries, role of endothelial surface properties, etc. that are specific to events occurring only in the microcirculation - in the evolution of the ischemic deterioration of the extra and intraparenchymal circulation leading to subsequent tissue damage and loss of neuronal functions.

SUMMARY

Based on distinctly different hemodynamic behavior of erythrocytes and plasma affecting properties of the circulating blood in vivo, such as apparent viscosity, flow resistance, axial streaming of erythrocytes, plasma skimming, etc., hematocrit (Htc) can have an apparent impact on tissue perfusion. Hematocrit also shows a diameter dependent decrease along the extraparenchymal arterial vascular routes that levels off being markedly lower in the microcirculation than in the central arterial blood. It was postulated that the impact of Hct may become a critical aspect of the macro- and microcirculatory compensatory mechanisms under ischemic conditions, when excessive fluid shifts between the extra- and intravascular compartments can in fact alter both systemic and local Htc, and when a decreased perfusion pressure sets the stage for sluggish flow velocities at which orientation of erythrocytes in the plasma stream can abruptly change and impair the macro- and microcirculation alike. To test this hypothesis, systemic (Htc$s$), feed (Htc$f$) and local hematocrit (Htc$l$) were simultaneously monitored in anesthetized and mechanically ventillated cats from the

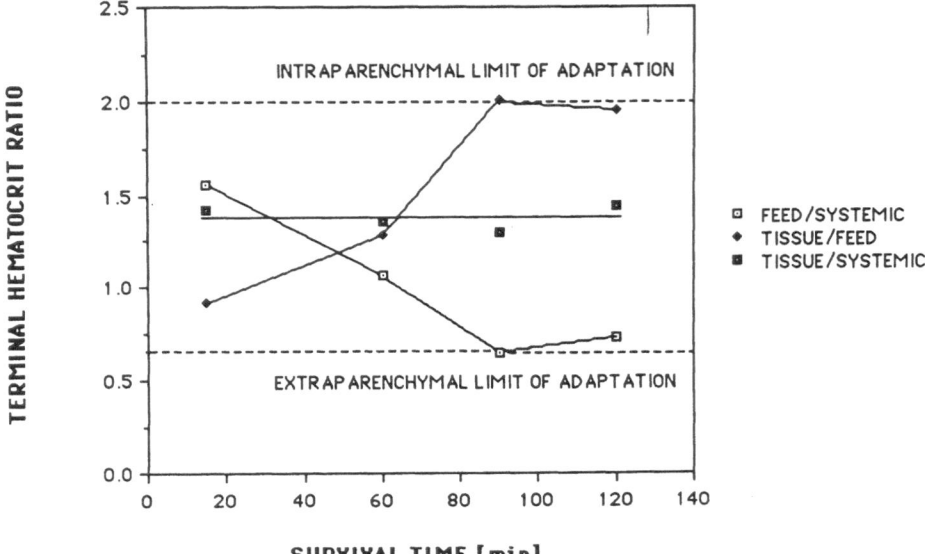

Fig. 9. Ratios of terminally determined hematocrit values as a function of survival time. Three hematocrit ratios were calculated. Feed/Systemic characterizing extraparenchymal mechanisms (primarily the Fahreus effect within the carotis-rete-pia system), Tissue/Feed reflecting hematocrit adjustments within the monitored intraparenchymal circulation and Tissue/Systemic indicating overall changes of hematocrit within the lumped extra- and intraparenchymal system. See Figs. 4 and 5 for the time sequence of these parameters leading up to the sessation of circulation in the area.

abdominal aorta, a pial artery of 100 micra in diameter and 500-600 cerebrocortical microareas of 0.01 cubic mm each respectively, by a television densitometric method[3,4] while global cerebral ischemia was induced and maintained by adjusting the systemic mean arterial blood pressure to 40-50 mmHg by controlled arterial hemorrhage. Global cerebral ischemia was terminated when cerebrocortical microcirculation collapsed or shed blood completely got reinfused to the animal. The data show that under control conditions Htc$f$ is 44 % of Htc$s$, while hemoconcentration in the tissue brings Htc$l$ up to 67% of Htc$s$. Under ischemic conditions, in cases of short survival time, the extraparenchymal arterial hemoconcentration can not be sufficiently compensated by intraparenchymal hemodilution and the microcirculation collapses under the conditions of lowering or moderately rising local tissue hematocrit. In case of longer survival, the rate of extraparenchymal hemoconcentration is increasingly lower and therefore the intraparenchymal hemodilution becomes more effective and prolonged. Due to factors most likely pertinent to the tissue proper, microcirculation collapses under abruptly developing secondary tissue hemoconcentration. Since terminal Htc*l* was only slightly higher than that at the beginning of the ischemic episode, attention to other hemodynamic and rheological factors in the microcirculation - not directly influenced by Htc - have been turned to. Such factors, like progressive capillary thrombosis, capillary embolization by white blood cells in cerebral ischemia are planned to be identified.

REFERENCES

1. J. H. Barbee, G. R. Cokelet, The Fahreus effect, *Microvasc. Res.* 3:6-16 (1971).
2. J. E. Cremer, M. P. Seville, Regional brain blood flow, blood volume, and hematocrit values in the adult rat, *J. CBF and Metabol.* 3:254-256 (1983).
3. A. Eke, Repetitive mapping of tissue hematocrit over microareas of the brain cortex, *Int. J. Microcirc. Clin. Exp.* 3(3/4):548 (1984).
4. A. Eke, Reflectometric imaging of local tissue hematocrit in the cat brain cortex, *in:* "Cerebral Hyperemia and Ischemia: From the Standpoint of Cerebral Blood Volume," M. Tomita, T Sawada, H. Naritomi and W.D. Heiss, ed., Elsevier Science Publishers BV, Amsterdam, New York, Oxford (1988).
5. R. Fahreus, The suspension stability of blood, *Phys. Rev.* 9:241-274 (1929).
6. P. Gaethgens , K. K. Albrecht, F. Kreutz, Fahreus effect and cell screening during tube flow of human blood. I. Effect of variation of flow, *Biorheology* 15:147-154 (1978).
7. P. C. Johnson , J. Blaschke, K. S. Burton et. al., Influence of flow variations on capillary hematocrit in mesentery, *Am. J. Physiol.* 221(1):105-112 (1971).
8. A. A. Lammertsma, D.J. Brooks, R.P. Beany et.al., In vivo measurement of regional cerebral hematocrit using positron emission tomography, *J. CBF and Metabol.* 4:317-322 (1984).
9. O. A. Larsen, N. Lassen, Cerebral hematocrit in normal man, *J. Appl. Physiol.* 19(4):571-574 (1964).
10. W. H. Oldendorf, M. Kitano, S. Shimizu, et.al., Hematocrit of the human cranial pool, *Circ. Res.* 17:532-539 (1965).
11. F. Sakai, K. Nakazawa, Y. Tazaki et.al., Regional cerebral blood volume and hematocrit measured in normal human volunteers by single-photon emission tomography, *J. CBF and Metabol.* 5:207-213 (1985).
12. F. H. Sklar, E. F. Burke Jr., T. W. Langfitt , Cerebral blood volume: values obtained with 51Cr-labeled red blood cells and RISA, *J. Appl. Physiol.* 24:79-82 (1968).

TISSUE OXYGEN TENSION IN THE CEREBRAL CORTEX OF THE RABBIT

M. Fennema, J.N. Wessel[*], N.S. Faithfull[**], and
W. Erdmann

Departments of Anaesthesia and [*]Neurology, Erasmus
University, Rotterdam, The Netherlands and
[**]University Department of Anaesthetics, Hope
Hospital, Manchester, U.K.

INTRODUCTION

For the normal function and survival of an organism, a
continuous supply of energy is necessary. Without energy nearly
all physiological processes, such as electrolyte pumping, could
not take place; in other words, Cannon's 'homeostasis' would
not exist. On the cellular level, oxydative phosphorylation
produces energy. This process is maintained by an almost
continuous supply of oxygen. Therefore oxygen is the decisive
vital parameter in (human) life. Disturbance in tissue
respiration causes a decrease in cellular function, reversible
functional breakdown and, finally, irreversible cell death. One
of the most common hazards during the perinatal period is
hypoxia. Prolonged hypoxia may cause permanent brain damage.
Oxygen transport to tissue occurs in three steps:

1. oxygen uptake in the lung
2. oxygen transport in blood
3. diffusion of oxygen from the capillaries to the tissue and
   through the tissue to the cells

When pulmonary gas exchange, cardiac output and oxygen
transport capacity of the blood are within the normal range,
oxygen supply to the tissue is furthermore dependent on the
following microphysiological parameters:

1. distribution/perfusion ratio of the capillary meshwork
2. Oxygen consumption of the cell
3. Oxygen diffusion parameters from capillaries to tissue,
   through the tissue and across the cell membrane into the
   cell

Normally, capillary perfusion and oxygen consumption are in balance and oxygen supply to the cell is continuously auto-regulated to its needs. Figure 1 shows that (micro-) collaterals between capillaries of different arterioles can help to maintain an adequate supply of oxygen to all areas. However, many pathophysiological conditions can severely interfere with this balance, for example: changes in blood PO_2, PCO_2, or pH; acute or chronic occlusion of blood vessels, edema, hypothermia and certain pharmacological drugs. From figure 1, it can be seen that if, for instance, one of the arterioles is completely blocked, certain parts of the capillary meshwork might not be sufficiently supplied with

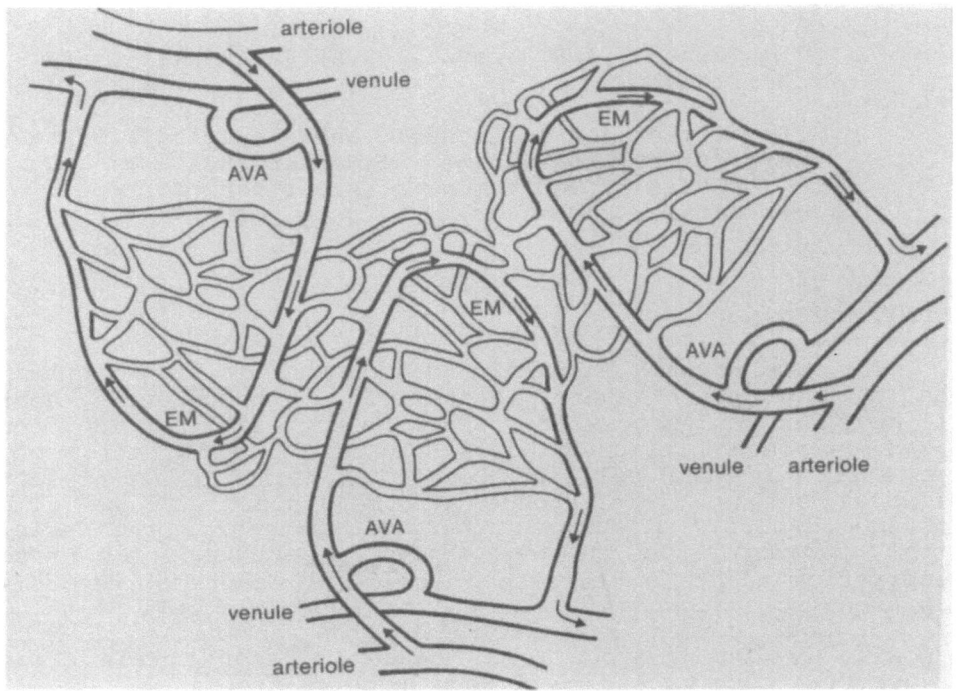

Figure 1. Two-dimensional schematic drawing of a micro-circulatory unit.
AVA = arteriovenous anastomosis
EM = end of metarteriole

oxygenated blood, even though the other arterioles might partially compensate. If the distance between this micro-area and the nearest well-oxygenated capillary is too great, the oxygen pressure could fall so low as to cause cell death (Figure 2). Chen et al. (1978) and Erdmann (1978) have hypothesised that an autoregulative mechanism exists in the cell membrane which enables it to maintain an intracellular PO_2 at a (low) value of 4-8 mm Hg. Later studies by Erdmann et al. (1988) showed that this intracellular PO_2 is only maintained between intercellular PO_2 values of about 9 and 50 mm Hg (in aplysia giant neurons). Their studies showed that the relation-ship between intra- and intercellular PO_2 is not static, thus making prediction of PO_2 profiles by computer models difficult.

This study was performed to test the reliability of the glass-gold microelectrodes in vivo, to assess the cerebral PO_2 profile under normal conditions of anaesthesia and to investigate the effects of hypoxia, hyperoxia and hypercapnia on cerebral tissue PO_2 and its autoregulation. The studies were performed in the rabbit because of parallel studies taking place in the same species on hypoxia and behavioural disorders.

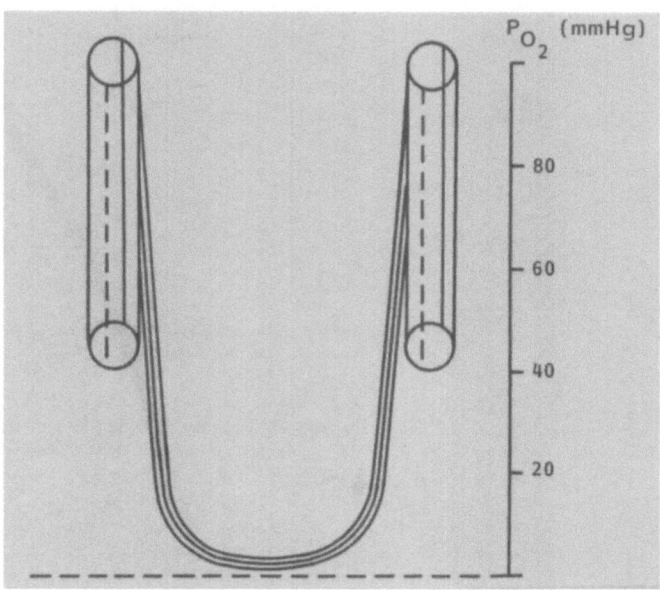

Figure 2. Oxygen distribution between two capillaries. The tissue PO_2 in those areas which are not in the immediate neighbourhood of a (well-oxygenated) capillary may be very low.

MATERIAL AND METHODS

A study was performed in adult albino New Zealand rabbits ($\pm$ 2.5 kg). Polarographic techniques were employed to measure oxygen partial pressure (PO_2). When a noble metal electrode is negatively polarised, oxygen is reduced, causing a current to flow through the measuring circuit. At -800 mV, this current is directly proportional to the oxygen concentration, or the partial pressure if measured in a medium with a constant O_2-solubility coefficient, as described by Erdmann et al (1970). Briefly, these electrodes are made as follows (Figure 3): a glass-covered gold wire of approximately 10 microns diameter is glued with silver lacquer to a copper wire and this combination is inserted into a tapered glass capillary (2 mm diameter) so that the gold protrudes from the tapered end by about 0.5 mm. The gold wire is glued into the capillary, while the copper wire is also glued to the capillary so that it cannot be pulled out. The tip of the gold electrode is covered with an oxygen permeable plastic membrane (Rhoflex) so that the electrode measures in a medium with a constant O_2-solubility coefficient. This ensures correct partial pressure measurements. This membrane also prevents protein poisoning and minimises convection disturbances. A detailed description of the fabrication of the electrodes and electronic circuitry involved is described elsewhere (Fennema and Erdmann 1985).

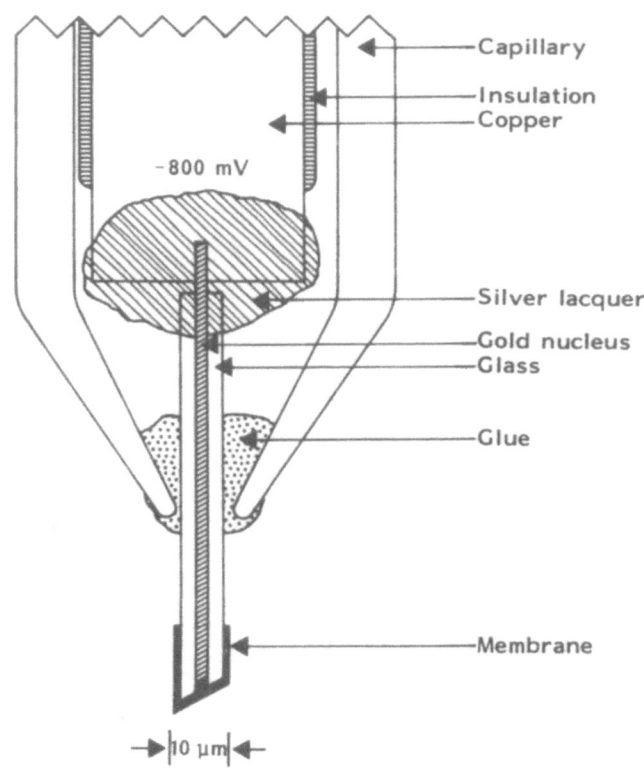

Figure 3. Oxygen microelectrode. For description see text.

An hour before each series of measurements, a microelectrode was connected to the polarisation voltage, using a Ag/AgCl wire as reference electrode. This system was connected to a Knick nano-amperemeter which amplified the current measured so that the readings could be recorded on a Ricka Denki recorder model KA 60. After stabilisation, the electrode was calibrated in a 37° C saline solution, through which gases containing 2 or 5% oxygen in nitrogen could be bubbled for equilibration. The electrodes were recalibrated after the experiment, and the results discarded if the drift was more than 7.5% per hour.

Because the electrodes break with any lateral mechanical movement, the animal was treated to preclude all possible movement. This was achieved in the following manner: A few days before the experiment the rabbit was subjected to induction of anaesthesia with halothane and operated on under strict sterile conditions. After exposing the frontal portion of the skull, a bolt was fixed to the skull by means of dental cement. The grip of the cement was strengthened by four small bolts, screwed into the skull around the large bolt head. The animal was then allowed to recuperate for three days. For the actual experiment (4th day), induction of anaesthesia was performed with thiopental (30 mg/kg i.v.) and the rabbit was intubated. Ventilation was maintained with 30% oxygen and 70% nitrous oxide using a Keuskamp infant ventilator, set to a frequency of 40 per minute. Muscle relaxation was achieved with administration of pancuronium (2 mg/hour) via an ear vein. The head of the rabbit was firmly attached to a modified stereotactic holder (own design) by means of the 'unicorn' bolt and the electrode attached to a micromanipulator as a fixed part of the stereotactic holder; thus movement of the rabbit's head, in relation to the electrode, was impossible.

Two small holes were drilled into the skull to facilitate EEG recordings. Arterial pressure was monitored by means of a catheter inserted in one of the femoral arteries. To check adequate ventilation, end expiratory CO_2 was measured by capnography at the entrance of the intubation tube. The tubing leading to the capnograph was $\pm$ 2 m long with a total volume of $\pm$ 2 ml, accounting for a time delay in registration of about 3 seconds when using a suction rate of 40 ml/min.

A burr hole was drilled above the right visual cortex. After removal of the dura mater, a thin layer of paraffine was applied to the cortex, supplemented by a layer of agar-agar to prevent O_2 leakage and minimise movement of the cortex due to ventilation. The electrode was calibrated and inserted into the cortex with a micromanipulator under visual guidance of a stereo-microscope. The reference electrode was inserted into a muscle exposed by the surgical wound. All parameters were registered and recorded on a mingograph recorder (Siemens).

Because the currents measured are so small (10^{-8} to 10^{-12} amperes), the experiments were performed in a Faraday cage, while the ventilatory and registration apparatus were outside the cage to eliminate feedback noise and electrical interference. This meant that the length of tubing from the ventilator to the animal was about 2 m. Consequently, there was an enlarged system volume and a rather long latency period following changes in inspiratory gas concentrations.

RESULTS AND DISCUSSION

On insertion of the oxygen electrode into the cortex, with the aid of the micromanipulator, a PO_2-profile can be produced (Figure 4). This profile can be reasonably well reproduced by withdrawing and re-inserting the electrode. Changes in the PO_2 profile can, in part, be attributed to damage caused by the needle electrode. Another cause of differences in the profile are due to the fact that PO_2 is in a state of constant change, though these oscillations are relatively small providing that there are no changes in other parameters. With an FiO_2 of 0.3, the average PO_2 was about 9 mm Hg (Figure 5). Nearly all measurements were below 25 mm Hg; measurements above 50 mm Hg were rare. These higher measurements are probably due to the proximity of the electrode to an arteriole (Figure 4).

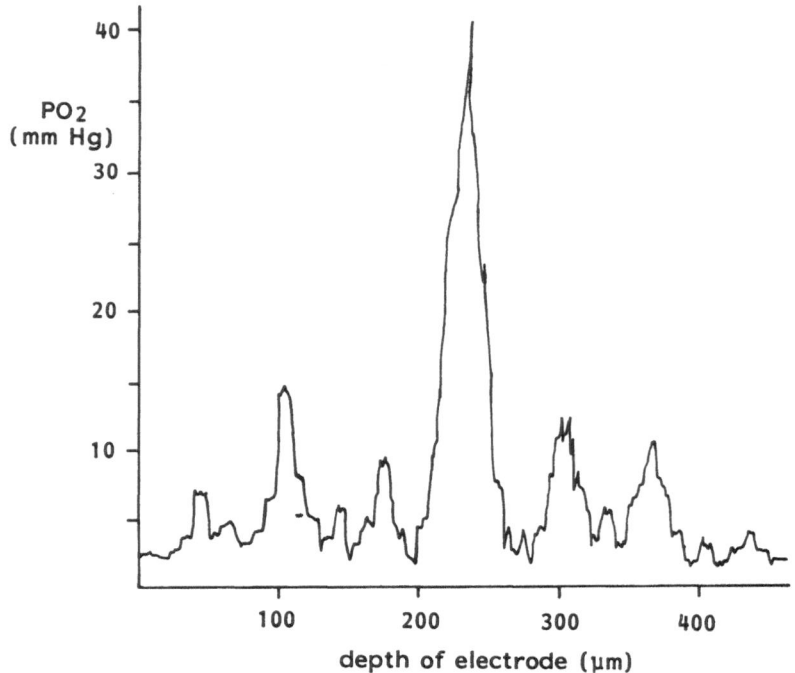

Figure 4. Typical example of a PO_2 profile in the cortex of a rabbit. Note the high sensitivity of the electrode.

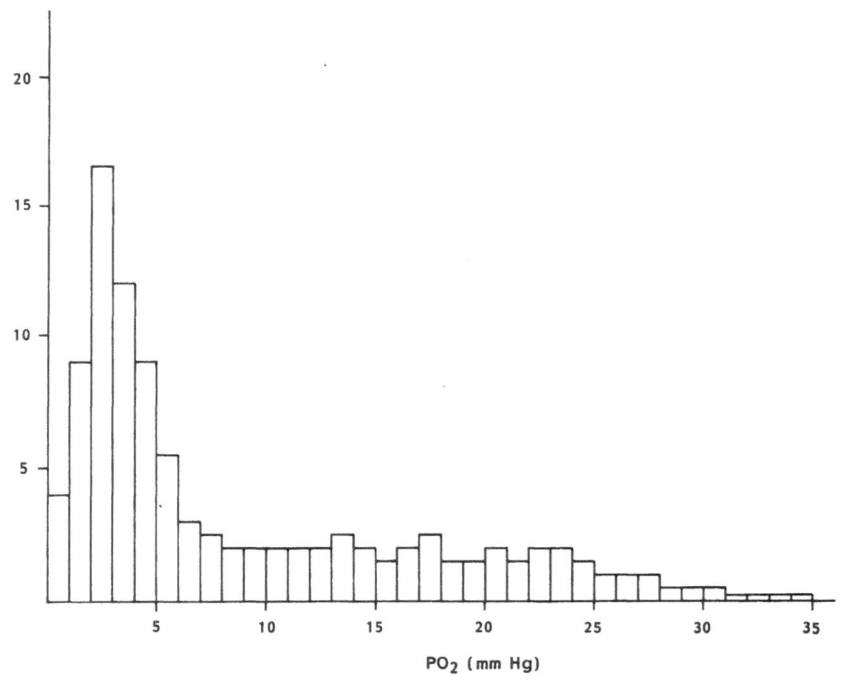

Figure 5. Frequency distribution of cortical PO_2's in a rabbit.

Although PO$_2$'s vary in the broad range, local cerebral PO$_2$'s in different regions and different animals showed the same characteristic changes when subjected to the same changes in some general physiological parameter. It should be noted that these electrodes can measure values less than 1 mm Hg. In these studies measurements of even lower values were not necessary as mitochondrial activity ceases at 0.4 mm Hg (Chance et al., 1957).

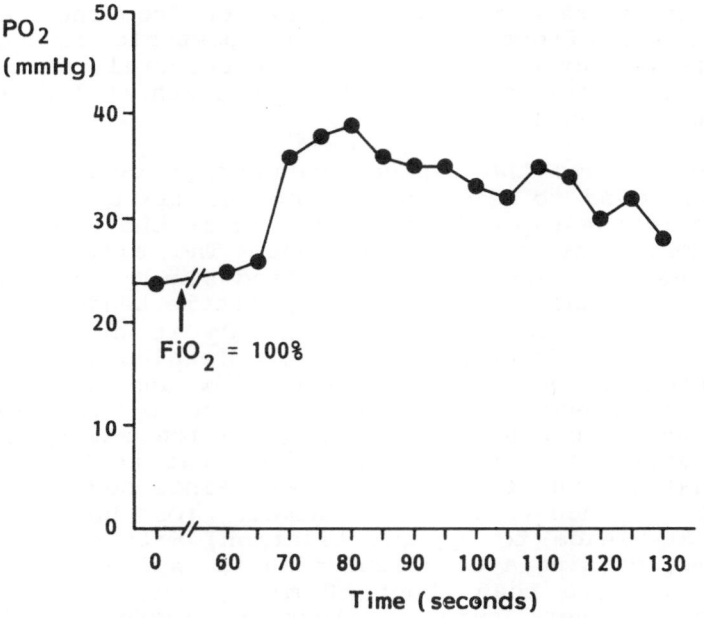

Figure 6. An example of change in cerebral PO$_2$ in response to a change in FiO$_2$ from 0.3 to 1.0, in a rabbit.

Figure 6 shows the typical pattern of cerebral tissue PO$_2$ following a change from an FiO$_2$ of 0.3 to 1.0. The arterial pressure and expiratory CO$_2$ remained constant. Whilst still ventilated with 30% O$_2$, the tissue PO$_2$ in this animal was $\pm$ 23 mm Hg. This high value is probably due to the electrode being close to a well-oxygenated arteriole. Measurements in a poorly-oxygenated area will produce less marked effects of hyperoxia (Faithfull et al, 1985). The period of more than one minute delay after the increase in inspiratory O$_2$ is partly due to the enlarged system volume of the tubing system. In addition, there is a time lag before the increased FiO$_2$ actually has an effect on the capillary and tissue PO$_2$. After a preliminary almost two-fold increase in PO$_2$, which remained constant for about 10 seconds, the PO$_2$ gradually decreased to approximately 120% its original value. This effect is probably induced by auto-regulatory vasoconstriction. Guyton (1986) suggests that this is possibly due to a direct relationship between tissue PO$_2$ and precapillary sphincter tension; an increase in PO$_2$ leads to an increase in the sphincter smooth muscle O$_2$, causing an increase in contraction and a decrease in blood flow. This in turn will lead to a decrease in tissue PO$_2$. The reverse effect occurs during hypoxia. Probably more important during hypoxia is the release of adenosine, causing vasodilation (Berne and Rubio, 1969; Stam and de Jong, 1977). Figure 6 shows an oscillatory drop in PO_2 as the autoregulatory mechanism attempts to maintain equilibrium. These oscillations were often observed in this study and other experiments (Faithfull et al., 1985). Obviously these autoregulative mechanisms have their limits. For example when the FiO_2 was changed from 0.3 to 0.1, the PO_2 of the electrode shown in fig. 6 dropped from 21 mm Hg to 13,5 mm Hg. Faithfull et al (1986) showed the autoregulative changes of PO_2 in the heart after ischaemic hypoxia. This ischaemic area could recuperate by receiving oxygen from the surrounding area (the steal effect). This is not possible during hypoxic hypoxia. Indeed our experiments in the cerebral cortex tend to show that autoregulation during acute hypoxia is less efficient than during hyperoxia.

Figure 7 shows the typical pattern of cerebral PO_2 when 12% CO_2 was added to the inspiratory gas mixture. This effect can be seen in the rapid increase to 16% in end expiratory CO_2, with a time delay of $\pm$ 10 seconds. The arterial pressure remained constant. PO_2 values increased slowly to nearly three times their original value and showed little tendency to return to normal values. This increase in PO_2 is due to a direct effect of CO_2 and H^+ on blood vessels, leading to vasodilation and a consequent increase in blood flow and tissue PO_2. In normal circumstances this increase in blood flow would eliminate CO_2, diminishing the vasodilation stimulus so that PO_2 would decrease. In this study CO_2 was kept artificially high, so that maximum vasodilation was maintained. It should be noted that in non-ventilated animals (or humans) a high arterial PCO_2 leads to hyperventilation, with an increase in tidal volume, causing a transient rise in expiratory CO_2. Acute CO_2 values of more than about 80 mm Hg causes depression of respiration, rather than a further increase. This can ultimately lead to death.

This study demonstrates that, as is generally accepted, CO_2 stimulates an increase in blood flow. This effect is stronger than the effect of hypoxia. A level of 12% inspiratory CO_2 will lead to an increase in cerebral tissue PO_2. During the relatively short period of observation ($\pm$ 5 min) PO_2 values remained high; no compensation for the vasodilatory effect of CO_2 was observed.

CONCLUSION

The above findings can be explained by relatively simple microphysiological mechanisms. These mechanisms provide an explanation for the changes in cerebral intercellular oxygenation. The changes described above were all acute changes. It is well known that an organism can adjust to a slow decrease in oxygen (Longmuir, 1987). Erdmann et al. (1979) and Clark et al. (1978) showed that intracellular PO_2 can be regulated by changes in the tissue O_2 diffusion coefficient. More investigations with, amongst others, the aid of oxygen microelectrodes are needed to help explain the mechanisms involved in the causes of cell death (and survival) in critical situations.

SUMMARY

   Polarographic techniques were employed to measure oxygen
partial pressure using 10 micron glass-protected gold
microelectrodes. When inserting the electrode into the cortex,
a PO_2-profile is produced. The average PO_2 was about 9 mm Hg.
Nearly all measurements were below 25 mm Hg and measurements
above 50 mm Hg were rare. When the FiO_2 was increased from 0.3
to 1.0, tissue PO_2 increased, then gradually decreased. This is
probably due to vasoconstriction of pre-capillary sphincters.
Acute hypoxia showed the opposite effect, but the
autoregulation does not seem to be so effective. When CO_2 was
added to the inspiratory gas mixture the PO_2 increased and
showed little tendency to return to normal values. This
increase in PO_2 is due to the direct effect of CO_2 and H^+ on
the blood vessels, causing vasodilation, and therefore an
increase in blood flow and tissue oxygenation.

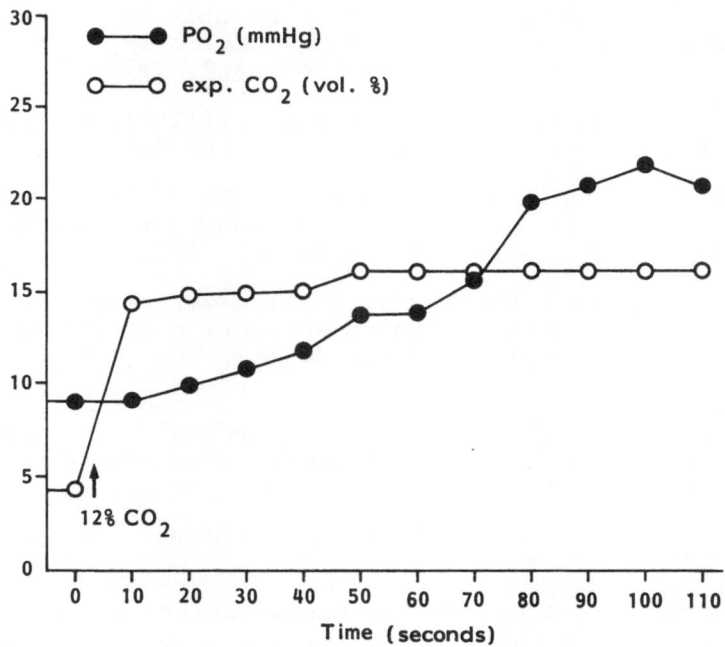

Figure 7. An example of cerebral tissue PO_2 in a rabbit in
         relation to end expiratory CO_2.

REFERENCES

Berne R.M. and Rubio R. (1969). Acute coronary occlusion: Early
changes that induce coronary dilation and the development of
collateral circulation. Am. J. Cardiol. 24, 766.

Chance B., Schoener B. and Schindler F. (1957). Cellular oxygen
requirements. Fed. Proceedings 16, 671.

Chen C., Erdmann W. and Halsey J. (1978). The sensitivity of aplysia giant neurons to changes in extracellular and intracellular PO_2. Adv. Exp. Med. Biol. <u>94</u>, 691.

Clark D.K., Erdmann W., Halsey J.H. and Strong E. (1978). Oxygen diffusion, conductivity and solubility coefficients in the microarea of the brain. Adv. Exp. Med. Biol. <u>94</u>, 697.

Erdmann W., Krell W., Metzger H. and Nixdorf I. (1970). Production of a standardised gold microelectrode for measurement of tissue PO_2. Pflugers Arch. Ges. Physiol. 319.

Erdmann W. (1978). Microelectrode studies in the brain of fetal rats and new born rats. Adv. Exp. Med. Biol. <u>94</u>, 455.

Erdmann W., Salt P., Agoston S. and Langrehr D. (1979). The importance of new microphysiological methods in determining cellular oxygen metabolism in anaesthesia. Acta Anaesth. Belg. <u>30</u>, 51.

Erdmann W., Chen C.F., Armbruster S. and Lachmann B. (1988). Autoregulation of intracellular PO_2 in aplysia giant neurons and response of neural activity following changes in extracellular PO_2. Adv. Exp. Med. Biol. <u>222</u>, 397.

Faithfull N.S., Fennema M. and Erdmann W. (1985). Oxygen supply to the myocardium. Adv. Exp. Med. Biol. <u>180</u>, 411.

Faithfull N.S., Fennema M., Erdmann W., Dhasmana M. and Eilers G. (1986). The effects of acute ischaemia on intramyocardial oxygen tensions. Adv. Exp. Med. Biol. <u>200</u>, 339.

Fennema M. and Erdmann W. (1985). On line determination of PO_2, oxygen diffusion, tissue perfusion and action potentials with a 10 micron electrode system. Proc. Sensor '85, Network Events. <u>5</u>, 8.2.1.

Guyton A.C. (1986). In: Textbook of medical physiology, 7th ed., Chapter 20: 233, Saunders Publ. Philadelphia, London, Toronto.

Longmuir I.S., (1987). Adaption to hypoxia, Adv. Exp. Med. Biol. <u>215</u>, 249.

Stam H. and de Jong J-W. (1977). Sephadex-induced reduction of coronary flow in the isolated rat heart: A model for ischaemic heart disease. J. Molec. Cell. Cardiol. <u>9</u>, 633.

THE EFFECT OF CALCIUM ENTRY BLOCKER S-EMOPAMIL ON CEREBROCORTICAL METABO-BOLISM AND BLOOD FLOW CHANGES EVOKED BY GRADED HYPOTENSION

A.G.B. Kovách, L.T. Nguyen, L. Pék, L. Dezsi,
and ZS. Lohinai

Experimental Research Department and 2nd Institute of Physiology, Semmelweis Medical University, Budapest, Ulloi ut
78/A, Hungary

INTRODUCTION

Calcium is an extremely important intracellular electrolyte which plays a fundamental role in the regulation of many aspects of cell metabolism and activity (Somlyo, 1984; Somlyo et al., 1985; Trump et al., 1984; Siesjo, 1981; Fleckenstein, 1988; Vanhoutte et al., 1988). It has long been known that calcification of tissues is associated with cell death. Regardless of the type of injury, the loss of calcium homeostasis, leads to calcium overload, either by impaired energy metabolism or plasma membrane alterations. This elevated intracellular calcium concentration is responsible for cytoskeletal modifications, the activation of phospholipases which cause perpetuation of membrane damage and finally, mitochondrial calcification.

During the past years there has been an increase in the use of calcium entry blockers in the prevention of tissue injury and the effects of cerebral ischemia. Experimental brain damage in these studies was usually produced by focal or global brain ischemia (Wauquier et al., 1988). Kovach et al (1984; 1988) studied the effect of different calcium entry blockers in standardized hemorrhagic shock and demonstrated that between the different verapamil derivatives and some other calcium antagonists, emopamil showed the most powerful protection. The majority of the emopamil-pretreated animals survived hemorrhagic shock which was irreversible in untreated animals. These results were confirmed in hemorrhagic and burn shock studies on rats, dogs and cats.

Defects in central nervous function following severe blood loss were clearly recognized early in the literature (see Kovach and Sandor, 1976 for review), but later it was assumed that owing to its well developed autoregulatory mechanisms, the brain's vital functions were protected during hypovolemic and other types of shock. Observations made in the last decades have established that the central nervous system is, in fact, vulnerable in prolonged hemorrhagic hypotension and shock (Kovach, 1961; 1977; 1988). It has been demonstrated that spontaneous electrocortical activity (ECoG) and somatosensory evoked potentials disappear during hypovolemia and do not return after retransfusion (Kovach, 1961; see Kovach and Sandor, 1976 for review). The demonstrated metabolic, microcirculatory-blood flow and functional alterations also confirm that the central nervous system is seriously affected in shock, and the development of the brain damage can contribute to irreversibility (Kovach, 1988).

(S)-emopamil [(2S)-isopropyl-5-(methylphenethylamino)-2-phenylvale-ronitrile- hydrochloride] pretreatment exerted a pronounced protective effect also against the development of ECoG changes in hemorrhagic shock. Seven of nine dogs did not develop isoelectric ECoG, which occurs in all untreated dogs undergoing the shock procedure. Emopamil protection was also demonstrated when the drug was administered 90 minutes after blood loss had been started (Kovach et al., 1968; Ivanics et al., 1988). (S)-emopmil significantly reduced total oxygen consumption in the dog (Nguyen et al., 1988). In emopamil-treated cats cerebral lactate concentration is significantly smaller and intracellular acidosis is less pronounced during brain ischemia than in the untreated animals (Bielenberg et al., 1987; Ligeti et al., 1987). (S)-emopamil increases cerebral blood flow by 100% in the rat according to autoradiographic measurements (Raschack et al., 1988; Szabo, 1988).

In contrast to other calcium entry blockers, this new verapamil derivative crosses the blood-brain barrier easily (Szabo, 1988), therefore it may significantly alter metabolism of cerebral parenchymal cells, in addition to affecting the cerebral circulation.

The purpose of our present experiments was to clarify the effect of (S)-emopamil on cerebrocortical oxygen delivery, metabolism and blood flow during stepwise decrements in arterial blood pressure. In this way, varying severities of brain ischemia, -from mild to severe-, were compared. Mean arterial blood pressure was decreased in steps of 20 mmHg, and each blood pressure level was maintained for 20 min. The lowest arterial blood pressure value was 30 mmHg. Following this step, the shed blood was reinfused, and recovery of untreated and emopamil-treated animals was compared. Cerebrocortical pO_2, blood flow with two methods, cerebral arterio-venous O_2 difference, NAD/NADH redox state, and blood volume were simultaneously monitored.

MATERIAL AND METHODS

The experiments were carried out on 20 cats of both sexes, weighing 2.4-4.0 kg, anesthetized intravenously by 50-60 mg/kg alpha-D-glucochloralose (Merck). If it was necessary, additional doses of chloralose were injected. After tracheostomy, the animals were placed on a Harvard cat respirator and immobilised with gallamine triethiodide (Flaxedil, 5 mg/kg, American Cyanamid). End-tidal CO_2 and mean arterial blood pressure (MABP) were continuously monitored. Arterial blood gases and pH were measured periodically and the ventillation was adjusted to keep arterial pO_2 and pCO_2 close to 100 and 35 mmHg, respectively. Temperature of the animals was continuously monitored and maintained at 37°C by a thermostatically controlled heating lamp. The cats were divided into 2 groups: I. Control (11 cats) and II. (S)-Emopamil-treated (intravenously 1 mg/kg, 9 cats).

Surgical procedures

The trachea and both femoral arteries were exposed and cannulated. In some animals the sagittal sinus was exposed and a needle inserted for cerebral venous blood sampling. The head of the animal was mounted in a DAVID-KOPF stereotaxic head-holder and the skin and muscles were removed from both sides of the skull. On both hemispheres a hole 12 mm in diameter was drilled in the parietal bone, above the suprasylvian gyrus, and the dura was carefully removed. A plastic ring-surrounded glass window was placed and cemented into the holes. Electrodes for measuring cortical surface pO_2 and blood flow (hydrogen clearance) were mounted into one of the windows.[2]

Routinely measured parameters

Rectal temperature was maintained at 38°C by an infra-red lamp and Yellow-Spring temperature regulator, Model 73/A. Systemic arterial blood pressure was measured in one of the exposed femoral arteries and monitored using a STATHAM P23D transducer. End-tidal CO_2 was measured using a KIPP-ZONNEN Capnograph, and recorded on a GRASS polygraph.

Surface-electrode measurements

The surface electrodes were mounted into one of the plastic window holders: 2 for locally generated hydrogen clearance, 2 recessed (covered) oxygen (gold or platinum) electrodes (0.5 mm). Most of the electrodes were prepared in Bristol by I. Silver. Polarizing units were built in the workshop of our Institute. In some of the studies another type of electrode probe was used developed by Silver and Chance in Philadelphia. The fluororeflectometric light guide (quartz fibre optic, Mainz) was in the center of the window, surrounded by platinum electrodes (6-8) for PO_2 and hydrogen measurements. From the PO_2 electrodes those which measured PO_2 in a range of 30-40 mmHg were selected and recorded during the experiment. The cerebral $A-VO_2$ was calculated from the arterial and sagittal sinus venous pO_2 values.

Optical measurements, NADH fluorescence and UV reflectance

Cerebrocortical NADH florescence and UV reflectance were measured through the implanted window by a modified fluororeflectometer originally developed by Chance et al., (1958). The brain cortex was excited by a 366 nm wavelength light beam carried by a light guide. The emitted NADH fluorescence (450 nm) and the reflected light (the sum of the scattered and reflected light) were measured by photomultiplier tubes. (NADH- but not NAD-molecules are fluorescent at 450 nm). The intensity of the reflected light decreases with the increase in the blood content of the brain cortex (e.g. vasodilatation). The degree of oxygenation of the hemoglobin molecules do not change the intensity of the reflected light. The blood present in the surface vessels of the cortex scatters the fluorescent (450 nm) light. This fact is causing the so called hemodynamic artifact. To avoid this artifact a correction method was used. An artificial hemodilution was produced by the bolus injection of 0.2 ml oxygenated dextran solution into the lingual artery. This minute amount of saline does not cause metabolic changes in the brain cells, so the change in the fluorescent signal is only due to the change in reflectance. So a correction factor can be calculated: this factor equals the ratio of the hemodilution-induced fluorescence and reflectance responses (Harbig et al. 1976):

$$C = \frac{\Delta F}{\Delta R}$$

When this factor is determined, it is possible to detect the NADH concentration dependent NADH fluorescence by the corrective unit of the fluororeflectometer. The fluororeflectometer was developed and built in the Dept. of Physiology, Budapest (Kovach et al, 1983).

(S)-emopamil (1 mg/kg) was introduced by a slow (3 min) intravenous infusion. It did not influence the systemic blood pressure, respiration and blood gases.

Mean arterial blood pressure and bleeding volume

Mean arterial blood pressure (MABP) was decreased to the same extent by bleeding the two groups. Each MABP step was kept for 20 minutes. The bleeding volume was somewhat higher in the (S)-emopamil group but it was significantly different only at 30 mmHg. Maximal bleeding volume at 30 mmHg in control group was 33.4 ± 3.6. In the (S)-emopamil treated group it was 42.3 ± 2.1 ml/kg. After retransfusion, MABP was significantly higher in the control group for 60 minutes compared to the original value. In the (S)-emopamil group MABP returned to normal after retransfusion. The difference between the two groups was significant (Figure 1).

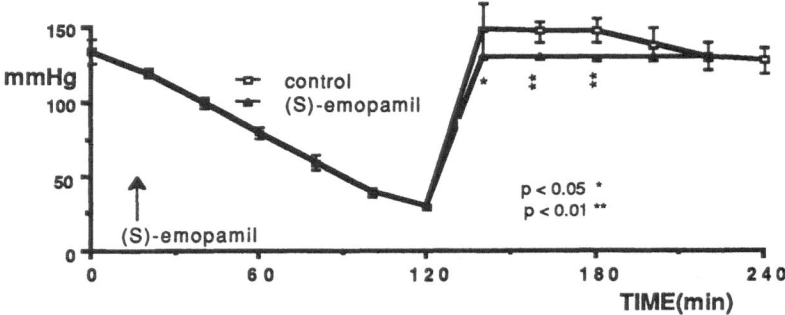

Figure 1. Mean arterial blood pressure changes during stepwise bleeding and after retransfusion in control and (S)-emopamil pretreated cats.

Cerebrocortical blood flow

Cerebrocortical blood flow was measured by two methods. On Figure 2 the results of microflow (locally generated hydrogen clearance) measurements are shown. It can be seen that after (S)-emopamil administration, cortical microflow increased significantly despite the reduction in MABP by bleeding. During stepwise lowering of MABP, blood flow autoregulation can be observed. Microflow is higher in the (S)-emopamil group during the bleeding phase. Immediately after retransfusion flow returns to initial values but in the control group flow continuously decreases. Flow is significantly higher in the (S)-emopamil treated group.

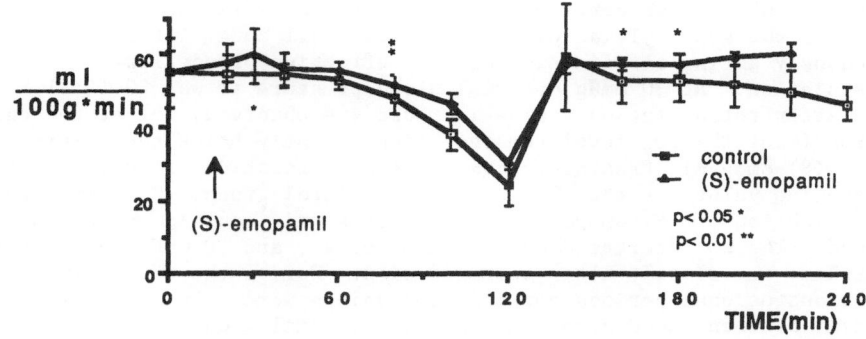

Figure 2. Cerebrocortical microflow, measured by locally generated
hydrogen clearance, in control and (S)-emopamil treated cats
during stepwise decreased MABP and after retransfusion.

Figure 3 shows the results of cerebrocortical blood flow, measured by
UV reflectance and dextran flush washout curves (Eke et al, 1979). The
flow is measured in a circular cortical volume of 2 mm diameter and
approximately 150 micron depth. 1 mg/kg (S)-emopamil increased the flow
by 30%, this increase was highly significant. In the control group the
lower limit of blood flow autoregulation is 60 mmHg. Below this pressure
level flow decreases in parallel with arterial pressure. At 30 mmHg blood
pressure, flow is 25 ml/100g min. After retransfusion, blood flow remains
low, - around 35 ml/min - in the untreated group, during the 80 minute
observation period. Cortical blood flow in the (S)-emopamil treated group
is significantly higher during the whole experiment. The lowest flow
value at 30 mmHg is 44 ml/100g min.; significantly higher then control.
After retransfusion flow returns back to initial values in the
(S)-emopamil group.

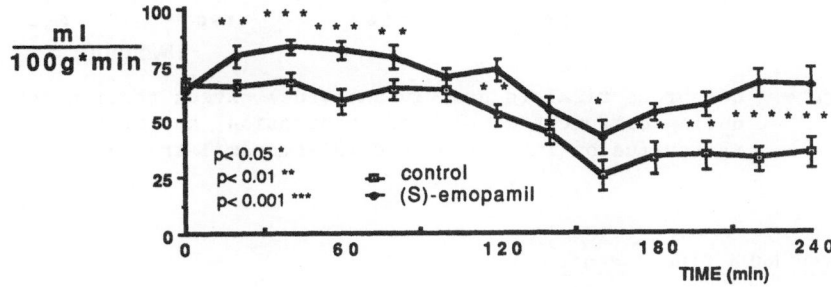

Figure 3. Cerebrocortical blood flow, measured with UV reflectance and
flush washout curve analysis in control and (S)-emopamil
treated cats during stepwise decreased blood pressure and
after retransfusion.

Cortical pO_2

Results of the cerebrocortical pO_2 measurements are presented in

Figure 4. The average cortical pO_2 in control cats, measured on the surface of the suprasylvian gyrus was 37.59 ± 5.1 mmHg. It decreased continuously during bleeding in spite of effective blood flow autoregulation. At 30 mmHg arterial blood pressure it was 20.1 ± 1.9 mmHg. After retransfusion a slow increase was observed, but at the end of the experiment the pO_2 level remained significantly below the initial value. (S)-emopamil treatment enhanced the restoration of cortical pO_2. The baseline value was the same as in the control group. Shortly after the application of (S)-emopamil a small but significant increase was observed. The pO_2 decreased only at MABP of 40, and 30 mmHg, and returned to baseline immediately after retransfusion. During the whole hypotensive and retransfusional periods cortical pO_2 values were significantly different between the control and the (S)-emopamil groups.

Cerebral arterio-venous oxygen difference

The cerebral A-V O_2 difference before bleeding was 6.4 ± 0.6 ml/dl in both groups. During stepwise bleeding, A-VO_2 in the control group started to increase already at 100 mmHg blood pressure level, it was 13.8 ± 1.8 at 40 mmHg and 15.1 at 30 mmHg MABP. After retransfusion A-V O_2 decreased but remained significantly higher than in the treated group until the end of the experiment. (S)-emopamil pretreatment prevented the early widening of A-VO_2. The A-VO_2 started to increase at 60 mmHg MABP. Shortly after restoration of MABP, A-V O_2 returned to the baseline value. During the whole postischemic period there was significant difference in A-VO_2 between the control and the (S)-emopamil treated groups.

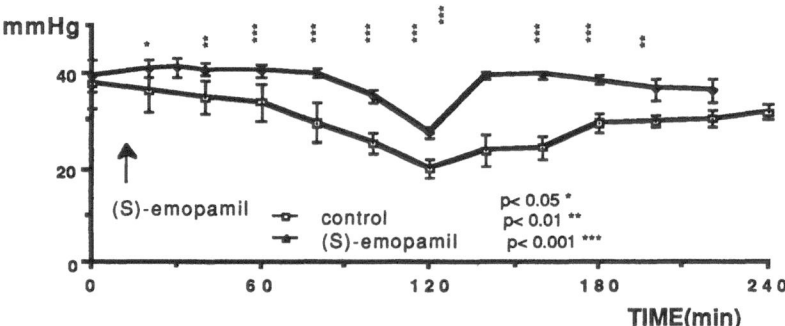

Figure 4. Cerebrocortical (suprasylvian gyrus) oxygen tension changes during different degrees of hypotension, and after retransfusion in control and (S)-emopamil treated cats.

Corrected NADH fluorescence

Results of cerebrocortical NADH fluorescence are presented in Figure 6. During bleeding the corrected fluorescence signal (NADH cc.) increased simultaneously with the decrease in MABP in the untreated group. The maximal reduction was 42 ± 2.2%. After retransfusion the NADH of the cerebral cortex remained reduced. (S)-emopamil pretreatment changed the picture totally. NADH reduction at 30 mmHg was only 50% of the control group (21.9 ± 2%) and after retransfusion NADH was immediately reoxidized to baseline.

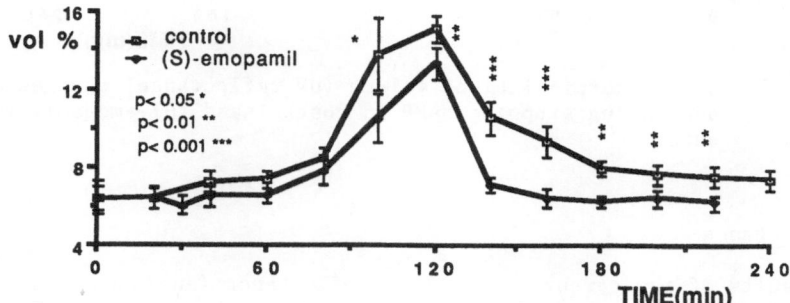

Figure 5.   Cerebral A-VO$_2$ difference during stepwise decreasing MABP
            and after retransfusion in control and (S)-emopamil treated
            cats.

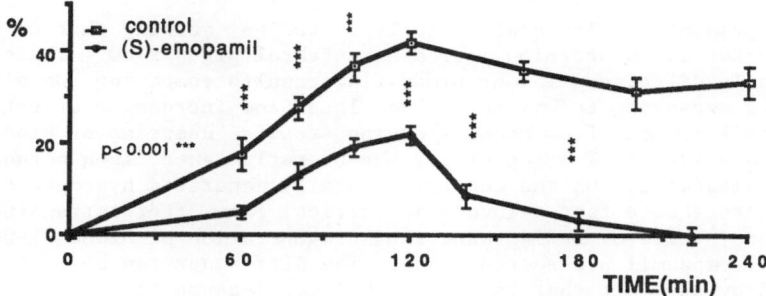

Figure 6.   Cerebrocortical NADH fluorescence changes during stepwise
            decreasing mean arterial blood pressure in control and (S)-
            emopamil treated cats.

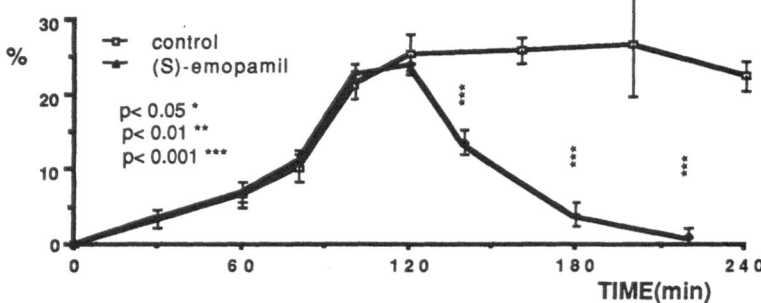

Figure 7. Cerebrocortical blood volume (UV reflectance) changes during decreasing stepwise MABP in control and (S)-emopamil treated cats.

UV reflectance

Results of the cerebrocortical UV reflectance (an index of cortical blood volume) are summarized in Figure 7. During hypovolemia, UV reflectance increased by 25% in both groups. A dramatic difference was observed after retransfusion. While in the control series the blood volume remained high, in the (S)-emopamil pretreated group it returned to baseline. The difference was statistically highly significant.

DISCUSSION

The protective effect of calcium entry blockers can be explained either by their vascular effects increasing cerebral blood flow, and by this means oxygen delivery or by metabolic interventions decreasing the calcium load of the neurons.

Our present results confirm earlier studies, showing that calcium entry blocker administration increases cerebral blood flow. There is a considerable difference in our blood flow results comparing two different blood flow measuring techniques. The blood flow increasing effect of (S)-emopamil and the flow changes during stepwise decrease of blood pressure are identical using either the UV reflectance flush method, or hydrogen clearance. On the contrary, locally generated hydrogen clearance measurements show a fast recovery in cortical flow after retransfusion in both groups, while UV reflectance shows regeneration of blood flow, only in the (S)-emopamil pretreated group. The difference can be most probably explained by the fact that the two techniques measure flow in different tissue volumes. UV reflectance measures the average flow of 10-20 times larger tissue volume than hydrogen washout method. A similar difference was found in the blood volume measurements, during the recovery phase; there was no recovery in cortical blood volume in the untreated group, but fast recovery was observed in the (S)-emopamil treated group.

The higher brain cortex surface pO_2 measurements suggest a more adequate tissue oxygenation in the (S)-emopamil treated group, both during stepwise decrease in blood pressure and during recovery period. The higher pO_2 values in the (S)-emopamil group can be explained also by a decreased tissue oxygen consumption. In fact it was demonstrated (Nguyen et al, 1988) that (S)-emopamil decreases the total oxygen consumption of the body in dog. Our present measurements of the cerebral arterio-venous O_2 difference confirm this possibility. The cerebral A-V O_2 difference widens less in the (S)-emopamil treated cats during hypovolemia, and return to baseline faster than in the untreated cats. The metabolic impact can be supported further by our finding that NAD reduction is ameliorated in the (S)-emopamil group, and is recovered after retransfusion.

SUMMARY

Our studies demonstrate that in the explanation of the protective effect of the calcium antagonist (S)-emopamil the possibility of beneficial metabolic causes (lower O_2 consumption) must also be considered beside of blood flow increasing effects.

It is suggested that (S)-emopamil may be a useful drug for the treatment of cerebro-ischemic disorders.

REFERENCES

Bielenberg, G.W., Beck T., Sauer, D., Burniol, M., and Kriegelstein, J., 1987, Effects of cerebroprotective agents on cerebral blood flow and on postischemic energy metabolism in the rat brain, J. Cereb. Blood Flow Metabol., 7:480.

Chance, B., and Baltscheffsky, H., 1958, Respiratory enzymes in oxidative phosphorylation. VII. Binding of intramitochondrial reduced pyridine nucleotide, J. Biol. Chem., 233:736.

Eke, A., Hutiray, Gy., and Kovach, A.G.B., 1979, Induced hemodilution detected by reflectometry for measuring microregional blood flow and blood volume in cat brain cortex, Am. J. Physiol., 236:H759.

Fleckenstein, A., 1988, Historical overview. The calcium channel of the heart, Ann. N.Y. Acad. Sci., 522:1.

Harbig, K., Chance, B., Kovach, A.G.B., and Reivich, M., 1976, In vivo measurement of pyridine nucleotide fluorescence from cat brain cortex, J. Appl. Physiol., 41:480.

Ivanics, T., Kovach, A.G.B., Tolgyessy, L., Nguyen, L.T., Pek, L., and Lohinai, Zs., 1988, personal communication.

Kovach, A.G.B., 1961, Comments to Dr Levenson's lecture in connection to nutritional and metabolic aspects of shock, Fed. Proc., 20(suppl. 9):122.

Kovach, A.G.B., and Sandor, P., 1976, Cerebral blood flow and brain function during hypotension and shock, Annu. Rev. Physiol., 38:571.

Kovach, A.G.B., 1977, Cerebral hemodynamic and metabolic alterations in hypovolemic shock, Adv. Exp. Med. Biol., 78:343.

Kovach, A.G.B., Nguyen, L.T., Tolgyessy, L., Pek, L., Ivanics, T., Ikrenyi, K., 1988, Effect of (S)-emopamil on the deterioration of cerebral bio-electrical activity evoked by standardised hemorrhagic shock, in: "Pharmacology of Cerebral Ischemia", J. Kriegelstein, ed., Elsevier, Amsterdam, in press.

Ligeti, L., Osbakken, M., Subramanian, H., Kovach, A.G.B., Leigh, J.S., Chance, B., 1987, ^{31}P and ^{1}H NMR spectroscopy to study the effects of gallopamil on brain ischemia, Magn. Reson. Med., 4:441.

Nguyen, L.T., Kovach, A.G.B., Tolgyessy, L., and Ikrenyi, K., 1988, personal communication.

Raschak, M., Unger, L., and Szabo, L., 1988, (S)-emopamil, a brain protecting calcium antagonist of the verapamil group for the treatment of cerebrovascular disorders, in: "Pharmacology of Cerebral Ischemia", J. Kriegelstein, ed., Elsevier, Amsterdam, in press.

Siesjo, B.K., 1981, Cell damage in the brain: a speculative synthesis, J. Cereb. Blood Flow Metabol., 1:155.

Somlyo, A.P., 1984, Cell physiology: cellular site of calcium regulation (news), Nature, 309:516.

Somlyo, A.P., Urbanics, R., Vadasz, G., Kovach, A.G.B., and Somlyo, A.V., 1985, Mitochondrial calcium and cellular electrolytes in brain cortex frozen in situ: electron probe analysis, Biochem. Biophys. Res. Commun., 132:1071.

Szabo, L., 1988, Effect of (S)-emopamil on post-ischemic cerebral blood flow and glucose utilisation in the rat, in: "Pharmacology of Cerebral Ischemia", J. Kriegelstein, ed., Elsevier, Amsterdam, in press.

Trump, B.F., Berezesky, I.K., Sato, T., Laiho, K.U., Phelps, P.C., and DeClaris, N., 1984, Cell calcium, cell injury and cell death, Environ. Health Perspect., 57:281.

Vanhoutte, P.M., Paoletti, R., and Govoni, S., 1988, Calcium Antagonists, Pharmacology and Clinical Research, Ann. N.Y. Acad. Sci., 522.

Wauquier, A., Ashton, D., and Clincke, G.H.C., 1988, Brain ischemia as a target for Ca^{2+} entry blockers, Ann. N.Y. Acad. Sci., 522:478.

CHANGES IN REGIONAL CEREBRAL BLOOD FLOW AND SUCROSE SPACE

AFTER 3-4 WEEKS OF HYPOBARIC HYPOXIA (0.5 ATM)

Joseph C. LaManna, Kimberly A. McCracken, and
Kingman P. Strohl[*]

Depts. of Neurology & Physiology/Biophysics, and
[*]Dept. of Medicine, Case Western Reserve University
School of Medicine, Cleveland, Ohio 44106, U.S.A.

INTRODUCTION

The brain is exquisitely sensitive to changes in its oxygen supply. Indeed, oxygen delivery to the brain is regulated on a moment to moment basis. While the brain depends on oxygen for oxidative energy metabolism, there is also the danger of excess exposure to oxygen and the attendant problems of oxygen toxicity. Consequently, there must be mechanisms which regulate regional cerebral blood flow that optimize the contradictory requirements of energy demand and oxygen toxicity. Thus, regional blood flow is usually kept low so that brain tissue oxygen tension is low (Sick et al., 1982) to avoid toxicity until functional requirements produce increased neuronal activity which are accompanied by local increases in blood flow to accommodate the resultant increased energy demand (Kreisman et al., 1981; LaManna et al., 1987). This hypothesis is a possible explanation for the paradoxical observations which indicate that the brain is always "on the brink" of hypoxia (Rosenthal et al., 1976) despite a demonstrable large reserve blood flow capacity. The mechanisms of local control of oxygen delivery seem to be set so that there is just sufficient oxygen for immediate energy demand but no more. Hypoxic and hypoxemic conditions will obviously force a response from the regulatory mechanisms in order that the balance between oxygen delivery and consumption be maintained (LaManna et al., 1984).

In addition to the immediate, short-term responses triggered by acute hypoxia, (e.g., LaManna et al., 1984; Kintner et al., 1984; Rapoport et al., 1986; Javaheri, 1986; Behar et al., 1983; Kintner et al., 1983), there are also long-term mechanisms which can be activated by persistent hypoxia. All of these responses can involve systemic as well as cerebral mechanisms. The effects of acute and chronic hypoxia on systemic variables have been reviewed (Dempsey, and Forster, 1982). These authors point out that there is a time course of acclimatization of these systemic variables to continued hypoxia, as well as, a time course upon return to normoxia. Two of the more readily observable changes are: 1) an immediate and continued hyperventilation; and 2) an increase in relative red blood cell content. This latter increase in hematocrit requires a week of constant exposure for maximum effect, and thus this response cannot be applied to acute hypoxic exposure.

In contrast to the systemic data, there is much less available information concerning the effect of exposure to chronic hypoxia on control of regional cerebral blood flow or other cerebrovascular variables. One effect that has been shown was an increased leak through the blood-brain barrier of larger proteins, possibly through stimulated increased pinocytotic activity (Dux et al., 1984). It has also been suggested that increased tissue lactate production with chronic hypoxic exposure occurs as a compensatory change which allows the tissue pHi to remain constant even though carbon dioxide tension has fallen due to increased tissue washout and hyperventilation (Musch et al., 1983). Thus, there is data in the literature to suggest significant tissue compensatory reactions to chronic hypoxia. Our own preliminary results, reported in this paper, suggest a significant increase in regional cerebral blood flow without increased blood volume in chronic hypoxia, in contrast to increased blood flow with increased blood volume as occurs in acute hypoxia (Shockley, and LaManna, 1988).

METHODS

Male Wistar rats (250 - 350 g) were exposed to chronic hypoxia by keeping them in a hypobaric chamber at 0.5 ATM for 3 - 4 weeks. Control rats were kept in the same room just outside the chamber. In addition to the control group of rats, there were 3 experimental groups. Group I: rats exposed to chronic hypoxia without return to normoxia. Group II: rats exposed to chronic hypoxia and then returned to normoxia for four hours. Group III: rats exposed to chronic hypoxia and then returned to normoxia for 24 hours. Regional cerebral blood flow and blood volume were then determined by the indicator-fraction method.

The specific details of the technique have been given (LaManna, and Harik, 1986). Rats, after having been exposed to chronic hypoxia together with their appropriate controls, were initially anesthetized with chloral hydrate (400 mg/kg, i.p.), so that cannulae could be inserted into a tail artery and into the right atrium of the heart via the external jugular vein. The skin was infiltrated with local anesthetic solution and then sutured. Rats were allowed to recover from anesthesia and restrained in plaster casts for at least three hours after surgery, at which time they were fully awake. Rectal temperature was monitored in all rats and kept near 37 °C by feedback controlled infrared lamps. Samples drawn from the tail cannula were used to determine arterial blood gases, pH, hematocrit and plasma glucose concentration. Blood pressure was monitored through this cannula also. The arterial cannula was connected to a syringe fitted to a pump calibrated to withdraw blood at a rate of 1.60 ml/min. Within seconds of starting this pump, a 150-ul bolus containing radiolabeled tracers (in Ringer's solution buffered with 10 mM HEPES to pH 7.4) was injected into the right atrium. Ten seconds after the atrial injection, rats were decapitated, the withdrawal pump stopped simultaneously and the arterial cannula removed. The withdrawn blood was then transferred to a tared vial, weighed and aliquots measured for radioisotope content. The brain was rapidly removed from the skull and bilateral samples from the frontal cortex, parietal cortex, hippocampus, cerebellum and striatum were weighed and radioisotope content determined by β-scintillation spectroscopy. A sample of blood oozing from the foramen magnum was collected in heparinized tubes, and an aliquot of the plasma was counted to estimate the radioactive content of the cerebral intravascular compartment at decapitation (Sage et al., 1981).

Blood Flow. To measure regional cerebral blood flow, 10 uCi of butanol ([^{14}C]n-butanol, 1 Ci/mol) was included in the bolus. Regional brain blood flow (ml/100g/min) was calculated by the indicator-fractionation method (Sage et al., 1981) without correction for incomplete butanol extraction according to the equation:

$$BF = \frac{F_S \times {}^{14}C_{brain}}{{}^{14}C_S \times brain\ wt} \times 100$$

where F_S is the calibrated withdrawal rate of the syringe (ml/min), ${}^{14}C_{brain}$ is the radioactivity of the brain sample (dpm), and ${}^{14}C_S$ is the radioactivity of the withdrawn blood (dpm); Brain weight is expressed in grams. The effect of incomplete extraction of butanol means that the reported blood flows are about 10% low for flows up to 250 ml/100g/min and 25% low for flows up to about 350 ml/100g/min (Sage et al., 1981; Shockley, and LaManna, 1988).

Residual Vascular Volume. Regional vascular space was estimated by adding 30 uCi of ${}^{3}H$-sucrose ([fructose-1-${}^{3}H$], 14 Ci/mmol, evaporated to dryness under reduced pressure just before use to remove any volatile ${}^{3}H$ contaminants) to the bolus. The volume of distribution (ul/g) was determined for each region by dividing the radioactive content of ${}^{3}H$-sucrose in brain samples (dpm/g) by the concentration (dpm/ul) of ${}^{3}H$-sucrose in the plasma of venous blood oozing from the severed head.

RESULTS AND DISCUSSION

Table 1 compares the systemic physiological variables among the different control and experimental groups. The primary systemic effects of chronic hypoxic exposure can be seen in experimental Group I, the rats which were not re-exposed to normal oxygen. The most characteristic findings in this "no recovery" group are the decreased arterial oxygen, increased hematocrit, and decreased arterial CO_2.

From the table, it can be seen that the change in hematocrit persists even after 4 and 24 hours of return to normobaric normoxia. Arterial CO_2 remains low at 4 hours, but has returned to control levels at 24 hours of recovery. All of the experimental animals exhibit some degree of acidosis compared to control rats.

Table 1. Systemic Physiological Variables

Systemic Variable	Normoxic Control	No Recovery	4 Hrs Recovery	24 Hrs Recovery
N	(10)	(3)	(5)	(5)
Body Wt (g)	337 ± 4	403 ± 4[b]	355 ± 13	344 ± 3
Body Temp (°C)	37 ± 1	37 ± 1	37 ± 1	37 ± 1
MAP (mmHg)	120 ± 2	----	127 ± 4	122 ± 4
Glucose (mg%)	191 ± 18	184 ± 10	179 ± 19	180 ± 11
pCO_2 (torr)	36 ± 1	23 ± 2[b]	29 ± 2[a]	37 ± 1
pO_2 (torr)	92 ± 2	55 ± 11[b]	96 ± 4	89 ± 2
pH	7.41 ± .01	7.33 ± .15	7.29 ± .01[b]	7.34 ± .02[b]
HCT (%)	47 ± 2	65 ± 3[b]	76 ± 1[b]	65 ± 1[b]

Significance with respect to the normoxic control group was determined by the t-test, two-tailed, corrected for multiple comparisons: [a]p < 0.05; [b]p < 0.01. Values are mean ± sem; N = number of animals in a group.

Table 2. Regional Brain Blood Flow (ml/100g/min)

Region (N)	Normoxic Control (10)	No Recovery (3)	4 Hrs Recovery (5)	24 Hrs Recovery (5)
Cerebral Cortex				
Frontal	118 ± 13	205 ± 35[a]	376 ± 42[b]	129 ± 11
Parietal	128 ± 14	219 ± 34[a]	391 ± 44[b]	128 ± 10
Hippocampus	81 ± 6	181 ± 31[b]	297 ± 35[b]	99 ± 12
Cerebellum	79 ± 8	114 ± 18	201 ± 23[b]	74 ± 10
Striatum	111 ± 10	205 ± 34[b]	359 ± 37[b]	112 ± 11

Significance with respect to the normoxic control group was determined by the t-test, two-tailed, corrected for multiple comparisons: [a]$p < 0.05$; [b]$p < 0.01$. Values are mean ± sem.

Greatly increased blood flow was observed in all brain regions in the chronic hypoxic group of rats compared to control (Table 2). The increased blood flow was present to an even larger extent after 4 hours of normoxic recovery; but had returned to control levels by 24 hours.

Table 3. Regional Brain Sucrose Space (ul/g)

Region (N)	Normoxic Control (10)	No Recovery (2)	4 Hrs Recovery (5)	24 Hrs Recovery (5)
Cerebral Cortex				
Frontal	18 ± 3	(14)	13 ± 2	21 ± 4
Parietal	20 ± 3	(14)	14 ± 2	22 ± 4
Hippocampus	21 ± 3	(12)	15 ± 2	22 ± 4
Cerebellum	23 ± 2	(16)	19 ± 3	26 ± 4
Striatum	16 ± 2	(12)	12 ± 3	16 ± 2

Significance with respect to the normoxic control group was determined by the t-test, two-tailed, corrected for multiple comparisons: [a]$p < 0.05$; [b]$p < 0.01$. Values are mean ± sem. No significant differences were detected. Data values in parentheses are the mean of two values only.

Thus, cerebral blood flow was significantly altered by chronic hypoxia in rats even after they had been returned to normoxic conditions for 4 hours. The higher cerebral blood flow, in the presence of lower arterial CO_2 concentration (Table 1), suggests that there must be a change in the normal relationship between CO_2 and blood flow. The lower CO_2 is due to hyperventilation stimulated by hypoxia. In a normoxic rat never exposed to chronic hypoxia, hyperventilation to this level would result in significantly decreased blood flow. Alternatively, the increased blood hydrogen ion concentration may be responsible for the increased blood flow, although whether protons are acting directly on the vasculature or through neuronally mediated mechanisms is not apparent.

There was no indication in these animals of a large vasodilation. The sucrose space data in Table 3 show, if anything, the opposite effect. The regional sucrose space in the chronic hypoxic/no recovery and 4 hour recovery groups were, on the whole, lower than the normoxic controls, although not reaching statistical significance in either group. We might expect to find a lower sucrose space because most of the space indicated by this marker is vascular plasma volume, and a lower plasma volume could be entirely accounted for if the systemic increase in large vessel hematocrit was also reflected as a similar change in small vessel and capillary hematocrit. In any case, the finding is not compatible with increased capillary blood volume.

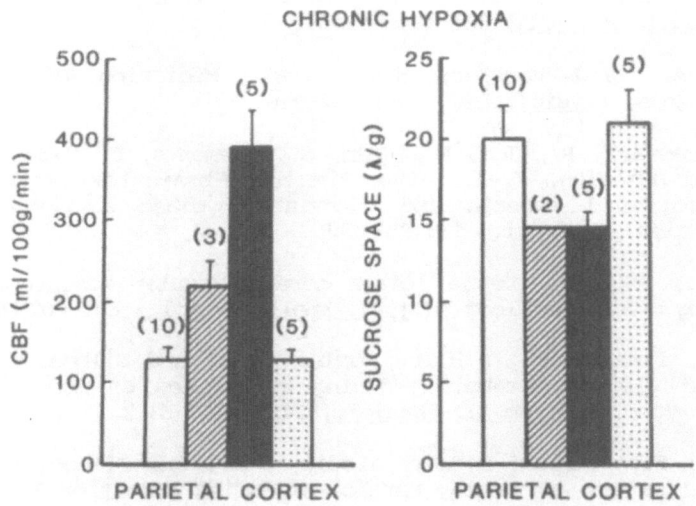

CHRONIC HYPOXIA

Figure 1. This figure summarizes the effect of chronic hypoxia and re-exposure to normoxia using the parietal cerebral cortex as an example. Blood flow is shown at the left; Sucrose space on the right. Open bars are normoxic controls; hatched bars are the no recovery group I; group II, 4 hours of re-oxygenation are shown as solid bars; and, the stippled bars are group III rats, 24 hours of recovery after 3 - 4 weeks of chronic hypobaric hypoxia.

SUMMARY

In summary, we can come to a number of meaningful conclusions regarding chronic exposure to hypobaric hypoxia in rats (refer to Figure 1). First, despite an increased hematocrit, and thus increased oxygen carrying capacity, regional cerebral blood flow is elevated after 4 weeks of chronic hypobaric hypoxia. This elevation in blood flow occurs even though the rat hyperventilates to lower than normal arterial CO_2 content which would ordinarily decrease cerebral blood flow. Second, although blood flow is increased in both chronic and acute hypoxia, the increases can not be through similar mechanisms since in the acute hypoxic condition there is also an increase in local blood volume that is absent in the chronic response. Third, the effect of chronic hypoxic exposure on cerebral blood flow persists for at least 4 hours after the animal is returned to normobaric normoxia. Fourth, sometime between 4 and 24 hours of recovery is necessary to reverse the effect of chronic hypoxia on cerebral blood flow. One day after having been returned to normobaric normoxia cerebral blood flow had returned to control. On the other hand, hematocrit was still elevated in these rats. Thus, the change in hematocrit does not seem to be associated in any mechanistic manner with the blood flow response.

REFERENCES

Behar, K.L., den Hollander, J.A., Stromski, M.E., Ogino, T., Shulman, R.G., Petroff, O.A.C., and Prichard, J.W., 1983, High-resolution ^{1}H nuclear magnetic resonance study of cerebral hypoxia in vivo, Proc. Natl. Acad. Sci. USA, 80: 4945-4948.

Dempsey, J.A., and Forster, H.V., 1982, Mediation of ventilatory adaptations, Physiol. Rev., 62: 262-346.

Dux, E., Temesvári, P., Joó, F., Adám, G., Clements, F., Dux, L., Hideg, J., and Hossmann, K.-A., 1984, The blood-brain barrier in hypoxia: ultrastructural aspects and adenylate cyclase activity of brain capillaries, Neurosci., 12: 951-958.

Javaheri, S., 1986, Hypoxemia lowers cerebrovascular resistance without changing brain and blood [H^+], J. Appl. Physiol., 60: 802-808.

Kintner, D., Fitzpatrick, J.H.Jr., Louie, J.A., and Gilboe, D.D., 1983, Cerebral glucose metabolism during 30 minutes of moderate hypoxia and reoxygenation, Am. J. Physiol., 245: E365-E372.

Kintner, D., Fitzpatrick, J.H.Jr., Louie, J.A., and Gilboe, D.D., 1984, Cerebral oxygen and energy metabolism during and after 30 minutes of moderate hypoxia, Am. J. Physiol., 247: E475-E482.

Kreisman, N.R., Sick, T.J., LaManna, J.C., and Rosenthal, M., 1981, Local tissue oxygen tension - cytochrome a,a_3 redox relationships in rat cerebral cortex in vivo, Br. Res., 218: 161-174.

LaManna, J.C., and Harik, S.I., 1986, Regional studies of blood-brain barrier transport of glucose and leucine in awake and anesthetized rats, J. Cereb. Blood Flow Metab., 6: 717-723.

LaManna, J.C., Light, A.I., Peretsman, S.J., and Rosenthal, M., 1984, Oxygen insufficiency during hypoxic hypoxia in rat brain cortex, Br. Res., 293: 313-318.

LaManna, J.C., Sick, T.J., Pikarsky, S.M., and Rosenthal, M., 1987, Detection of an oxidizable fraction of cytochrome oxidase in intact rat brain, Am. J. Physiol., 253: C477-C483.

Musch, T.I, Dempsey, J.A., Smith, C.A., Mitchell, G.S., and Bateman, N.T., 1983, Metabolic acids and [H^+] regulation in brain tissue during acclimatization to chronic hypoxia, J. Appl. Physiol., 55: 1486-1495.

Rapoport, S.I., Lust, W.D., and Fredericks, W.R., 1986, Effects of hypoxia on rat brain metabolism: unilateral *in vivo* carotid infusion, Exptl. Neurol., 91: 319-330.

Rosenthal, M., LaManna, J.C., Jöbsis, F.F., Levasseur, J.E., Kontos, H.A., and Patterson, J.L.Jr., 1976, Effects of respiratory gases on cytochrome a in intact cerebral cortex: is there a critical PO_2, Br. Res., 108: 143-154.

Sage, J.I., Van Uitert, R.L., and Duffy, T.E., 1981, Simultaneous measurement of cerebral blood flow and unidirectional movement of substances across the blood-brain barrier: theory, method, and application to leucine, J. Neurochem., 36: 1731-1738.

Shockley, R.P., and LaManna, J.C., 1988, Determination of rat cerebral cortical blood volume changes by capillary mean transit time analysis during hypoxia, hypercapnia, and hyperventilation, Br. Res., 454: 170-178.

Sick, T.J., Lutz, P.L., LaManna, J.C., and Rosenthal, M., 1982, Comparative brain oxygenation and mitochondrial redox activity in turtles and rats, J. Appl. Physiol., 53: 1354-1359.

EFFECT OF EMOPAMIL ON CEREBROCORTICAL MICROCIRCULATION DURING HYPOXIA AND REACTIVE HYPEREMIA AND ON $[K^+]_e$, pH, pO_2 CHANGES DURING AND AFTER N_2 ANOXIA

R. Urbanics and A.G.B. Kovach

Experimental Research Dept. and II. Institute
of Physiology, Semmelweis Medical School
1082 Budapest, Hungary, Pf. 448

INTRODUCTION

Ischemic and hypoxic changes in the central nervous system are commonly accompanied with different clinical disorders, as stroke, head injury, intracranial pressure increase, status epilepticus, etc. In the functional impairment these symptoms are very important components, often the leading symptoms and have serious impact on the final outcome of disorders.

The integrity of cell membranes is essential for the maintenance of normal neuronal function. Hypoxia and/or ischemia disturbs the metabolism, disintegrates the neural membranes and this leads to perturbation of ion homeostasis. The intracellular Ca^{++} accumulation seems to be the final reason of irreversible cell damage (Siesjö, 1981, 1988). Metabolic and vascular changes are involved in the calcium loading of the cells. In the Ca^{++} loaded cells the endoplasmic reticulum and the mitochondria take up the excess Ca^{++} and by this it participates in the regulation of cytosolic free Ca^{++} concentration. The calcium sequestering process in mitochondria in the damaged cells with overwhelming Ca^{++} load disrupts the metabolic process of these organelles (Somlyo, 1984; Somlyo et al., 1985). This process further aggravates the disrupted metabolic function, inhibits the respiration and the final outcome is the cell death. The increased ionized calcium plays a central role in these reactions, triggering catabolic as well as vascular events.

Calcium channel blocking agents are possible natural choices for cerebral protection. The exact mechanism of action of these drugs is not yet known but several proposed effect is mentioned. On of them is decrease of lactate level following ischemia (Ligeti et al., 1987; Bielenberg et al., 1986), which is known one of the most detrimental factor (Renchrona et al., 1981). The importance of maintained perfusion during and after hypoxic, ischemic insults is well documented (Hossmann, 1982). Emopamil has a vasodilating property as it is demonstrated by Bielenberg et al. (1986). Calcium entry blockers diminish the

NAD reducing effect of anoxia and epilepsy (Kovach et al., 1983), can influence the endothelium, decrease the vasoconstriction, diminishing platelet aggregation and so improve postischemic hypoperfusion. Kazda et al., (1983) postulated that these drugs can prevent the "rigor lock" of the precapillary sphincters.

The questions to be addressed were the following: do the calcium antagonists show their effects only during severe ischemia and/or anoxia, to prevent the imminent cell death or display their potentially beneficial effects during moderate hypoxia too? Is the therapeutic effect of calcium blocker Emopamil primarily vascular or metabolic?

The purpose of our experiments was twofold. First, to investigate the effects of calcium channel blocking agent (S)-emopamil - which is able to cross the blood brain barrier (Szabo et al., 1988) - on cerebral blood flow under moderate hypoxia and study the effect on reactive hyperemia evoked by direct cortical stimulation under normoxic and hypoxic conditions. Second, testing of the drug during severe, transient hypoxia on vascular and metabolic indicators as blood pressure, cerebrocortical pO_2, extracellular pH and potassium activity $[K^+]_e$.

METHODS

The studies were done on two separate group of cats, weighting 2.5 -3.5 kg. The moderate hypoxic study (seven animals) was performed under pentobarbital sodium (Nembutal, 40 mg/kg ip.) anesthesia. The N_2 anoxic group (twelve animals) were anesthetized by chloralose (45 mg/kg alpha-chloralose iv.). Both femoral arteries, a femoral vein and the trachea were cannulated. After tracheostomy the animals were immobilized with gallamine triethiodide (Flaxedil, 5 mg/kg, American Cyanamid) and respirated with a Harvard ventilator. Endtidal CO_2 and blood pressure were monitored continuously. Blood gases and pH were checked periodically. Hypoxic episodes were achieved by a 10% O_2 and 90% N_2 gas mixture inhalation. Anoxia was produced by 95% N_2, 5% CO_2 gas mixture inhalation for 3 minutes. Temperature of the animals was continuously monitored and maintained at 37 °C by a thermostatically controlled heating lamp. The head of the animal was mounted in a stereotaxic holder. A 12 mm diameter disc was removed from the left side of the skull by hand trephine over the suprasylvian gyrus. The dura was carefully dissected and a multiwire surface sensor for continuous measuring of microflow by local hydrogen clearance (Lübbers and Stosseck, 1970) was positioned on a cortical area free of visible blood vessels. The outer diameter of the sensor was 1.6 mm and included four independent measuring sites (25μm Pt wires) (Leniger-Follert and Lübbers, 1977). Direct electrical stimulation was achieved by two silver wires placed on the surface of the brain at a distance of about 5 mm. The stimulating parameters were: 5, 10 and 15 V amplitude, 1 ms pulse duration, 20 Hz frequency and 5 sec.

In the N_2 anoxic studies potassium sensitive surface electrodes were constructed from 20 μm silicon rubber membrane incorporating 3 % valinomycine glued to the end of glass capillary tubing (O.D. 0.9 mm; I.D. 0.5 mm) with silicon rubber adhesive and filled with 0.5 M KCl. The glass pH miniprobes (CORNING 0150) were the same size as the potassium sensitive ones. A separate, Ag/AgCl element served as reference (O.D. 1 mm) for both ionsensitive electrodes. Oxygen tension was measured with multiwire 25 µm platinum-iridium surface electrodes with glass insulation.

To protect the brain from loss of heat and drying up the surface of the brain after positioning the electrodes was covered by warmed paraffin oil.

EXPERIMENTAL PROTOCOL

Moderate hypoxic group

After completion of surgery the microflow measuring electrodes were polarized at least for one hour to obtain stable readings and minimizing the electrode shifts before the H_2 generation (generating current 0.08 µA) started. Direct cortical stimulation was made with the above mentioned parameters, and repeated under normoxic and hypoxic condition (10% O_2 inhalation) when the microflow readings reached a new steady state. Arterial blood samples were taken prior to stimulation in normoxia and before and at the end of the hypoxic periods. 30 min after reaching the control level 1 mg/kg Emopamil (KNOLL, Ludwigshafen am Rhein, FRG) was injected slowly, intravenously. One and two hours after the treatment the stimulation was repeated in normoxic and hypoxic conditions.

N_2 anoxic group

After mounting and stabilizing the ion sensitive electrodes anoxia was produced for 3 minutes (6 cats). In the treated group after the control period the drug, (S)-emopamil was infused at a dose of 1 mg/ kg body weight, 30 minutes prior to the anoxia (6 animals). The Emopamil studies were performed on cats which did not experience, N_2 anoxia before the drug administration.

DATA ANALYSIS

The obtained microflow readings showed different values from experiment to experiment and sensor to sensor, depending on the measuring site, and the distance of the measuring elements from the invisible capillaries. To better match the microflow data, after sacrificing the animals with a bolus of saturated KCl we maintained the continuous H_2 generation and the obtained stable values represented the zero flow values. The values measured in the control period represented the 100%. The mean of four sensors in the same experiment was used as one value for the analysis of the data. The transformed data were statistically analyzed and the simple paired t test was used to compare the data obtained in the normoxic and hypoxic as well as during the treatment period.

Since in the anoxic studies the electrode measurements during the second insult (in part of the experiments) differed
from the first, in the data analysis we used only the first set
of data. For the statistical analysis the t test for unpaired
data was used.

RESULTS

Emopamil, introduced with a slow intravenous infusion
(dose: 1 mg/kg) did not influence the systemic blood pressure.
Blood gas values were in the normal range. These values after
Emopamil treatment were not significantly different.

Microflow data of seven cats are summarized in the first
figure. Emopamil dilated the cerebral vessels and significantly increased local cerebral blood flow in the cortex (Fig.
1/a). The arterial pO_2 value in the control period was 97.3 ±
5.6 (mean ± SD) and the treatment did not changed it significantly. The arterial pO_2 values during 10 % O_2 inhalation
were: 26.5 ± 2.6 during hypoxia in the control period, 27.9 ±
3.7 and 24.4 ± 3.4 mmHg one and two hours after the Emopamil
treatment during the hypoxic insult, respectively. The values
were not significantly different. The flow increase during
normoxia following 5 V and 10 V stimulation and after 5 V activation during hypoxia was significantly higher in the treated
animals.

The duration of reactive hyperemia (Fig. 1/b) was significantly longer one and two hours after the Emopamil treatment
during hypoxic insult following 15 V stimulation and two hour
after the treatment during normoxia after 15 V stimulation.
Except the 5 V stimulation evoked reactions all others showed a
tendency of treatment effect.

The effect of acute severe N_2 inhalation hypoxia on Mean
Arterial Blood Pressure (MABP) is shown in the Fig. 2. The control animals, starting with a MABP 122.2±12.8 mmHg increased to
136.3 ± 2.5 mmHg one minute and to a maximal value of 144.3 ±
2.64, 2 minutes after starting N_2 respiration. This change is
statistically significant (p < 0.05). One minute after re-oxygenation MABP dropped to 109.8 ± 3.3 and later it returned to
the initial values. The infusion of 1 mg/kg (S)-emopamil did
not change MABP. The arterial blood pressure was 120.2 ± 3.8
mmHg 30 minutes after drug administration. The hypoxia in (S)-
emopamil-pretreated cats had a much smaller and not significant
BP raising effect. As it can be seen in Fig. 2, there is a
significant difference in blood pressure response to hypoxia
between the control and pretreated groups.

Results of the cerebrocortical pO_2 measurements are presented in Fig. 3. The Average cortical pO_2 in untreated cats,
measured on the surface of the suprasylvian gyrus was 29.8 ± 5
mmHg. 30 seconds after hypoxia started it was already significantly reduced to 8.5 ± 2.3 mmHg . The rapid fall of cortical
oxygen tension progressed at the same speed, and at the end of
the first minute it was only 1.1 ± 0.3 mmHg. During

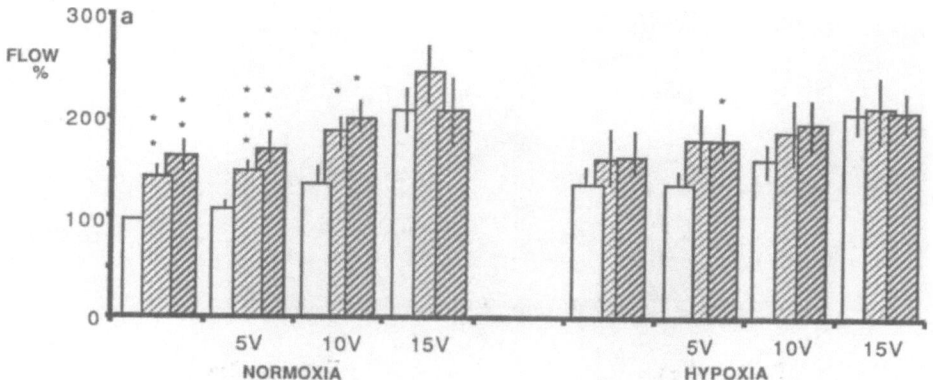

Effect of Emopamil on microflow changes during normoxic and hypoxic conditions and during direct cortical activation of the cat brain cortex. Values are given as mean ± SD. Different from corresponding control values (paired t test) :** = p<0.01, * = p<0.05

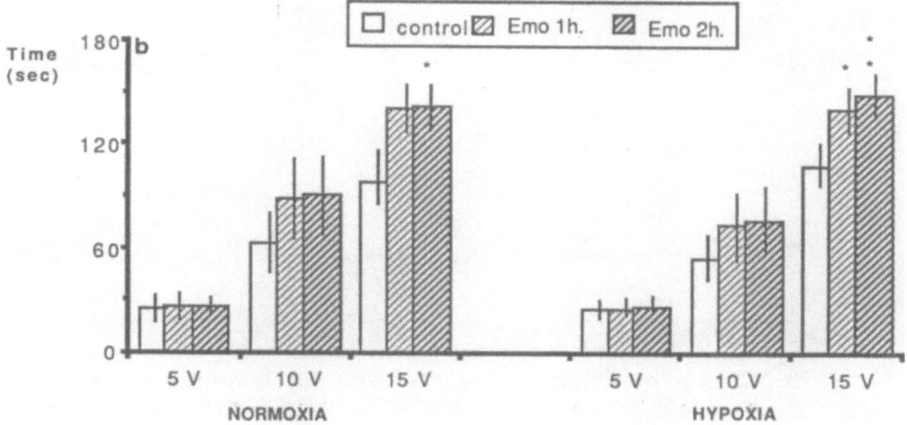

Effect of Emopamil on the duration of reactive hyperemia. The reactions were evoked by direct cortical activation on the cat brain cortex.

Fig. 1/a: upper panel.   Fig. 1/b: lower panel

the next two minutes pO_2 reached the zero level. The rate of the cortical pO_2 decrease was slower in the (S)-emopamil pre-treated cats. From the control value of 29 ± 2.4, it was 15.4 ± 2.4, 5.1 ± 1.5 and 1.7 ± 0.5, 30 seconds, one and two minutes after starting N_2 anoxia, respectively. There was a significant difference both in the rate of decrease and in the restoration (p < 0.001) of cortical pO_2.

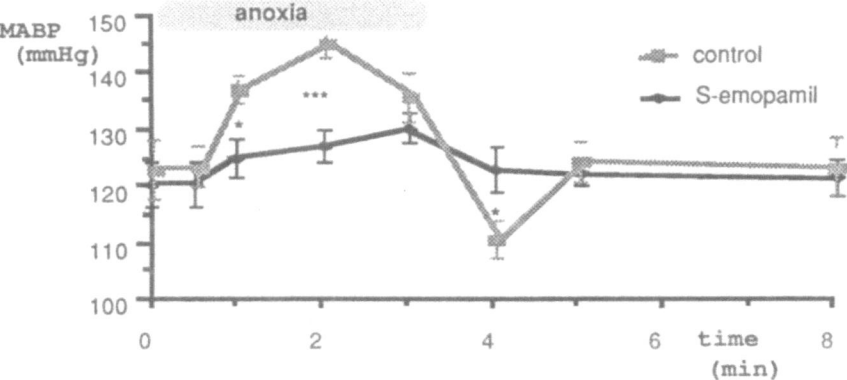

Fig. 2. Effect of S-Emopamil on MABP during and after
anoxia. Solid line: Average of Emopamil
treated animals. Dashed line: control group.
p< 0.05: *, p< 0.01: **, p< 0.001: ***

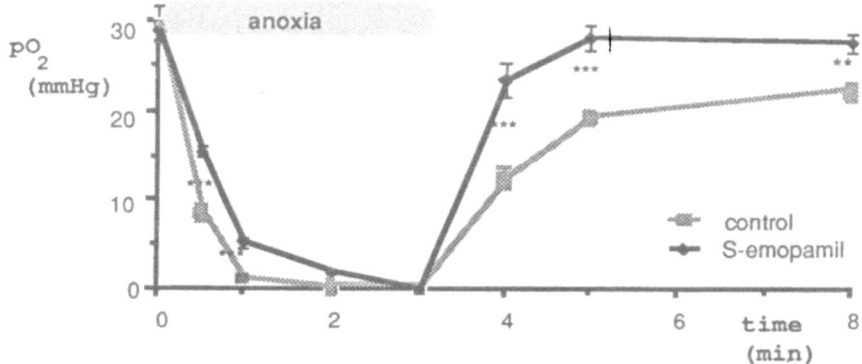

Fig. 3. Cerebrocortical oxygen tension changes during
and after nitrogen anoxia. The rate of decrease
and return is significantly different in the two
groups.

The cerebrocortical surface <u>extracellular potassium activ-</u>
<u>ity</u>, $[K^+]_e$ was 3 ± 0.1 mM in the control group before N_2 res-
piration (Fig.4). In the first minute of hypoxia a small progressive rise of 0.88 mM was shown. A new phase of rapid increase of $[K^+]_e$ appeared within 2 minutes. The value in the
second minute was 15.4 ± 6.9, in the third 19.23 ± 2.8 mM.
The $[K^+]_e$ decreased slowly after air respiration. Re-oxygenation values of potassium concentrations were : 16.9 ± 1.1, 9.8
± 1.1 and 5.5 ± 0.5 mM at 1, 2 and 5 minutes resp. (S)-
emopamil pretreatment changed the reaction and diminished the
increase in $[K^+]_e$. In this group the control value was $3.03 \pm$
0.08, maximal value 11.9 ± 2.8 mM. A rapid recovery

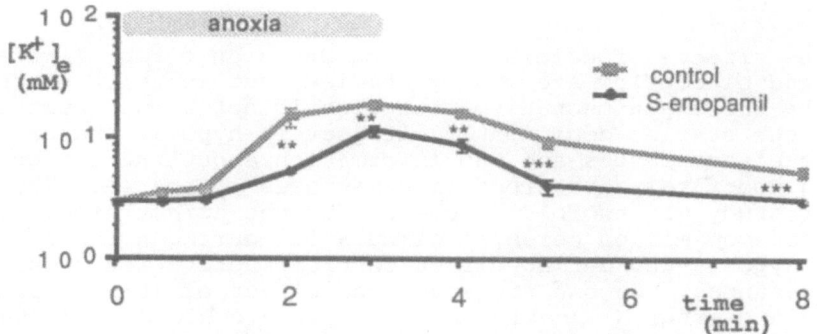

Fig. 4. Extracellular potassium activity changes
measured by a surface ionselective sensor on
the suprasylvian gyrus.

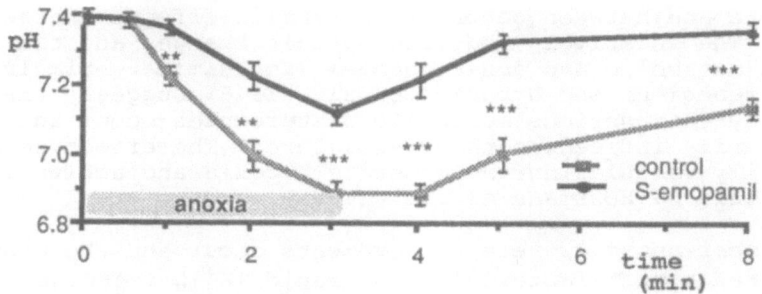

Fig. 5.Extracellular pH changes during and after
nitrogen anoxia. S-Emopamil treatment significantly diminished the acidosis and increased
the recovery rate during reoxygenation.
p< 0.05: *, p< 0.01: **, p< 0.001: ***,

occurred after air respiration, reaching the control level after 5 minutes . The difference between the control and (S)-emo-
pamil pretreated group is statistically significant. (S)-emopa-
mil pretreatment protected against the hypoxic $[K^+]_e$ rise and
accelerated recovery.

The cerebrocortical surface pH value in control animals was
7.39 ± 0.03 (see Fig. 5). A significant decrease was seen one
minute after N_2 anoxia started. It reached 7.23 ± 0.07, 7.00 ±
0.09, and 6.89 ± 09 at 1, 2 and 3 minutes of anoxia, respectively. The cerebrocortical pH after air respiration showed
a very slow recovery. The cortical pH values were 6.89 ± 0.07,
7.01 ± 0.13 and 7.14 ± 0.08 at 1, 2, and 5 minutes after
oxygenation. In the (S)-emopamil pretreated group during the
first minute of hypoxia, cortical surface pH remained unchanged. Two and three minute pH values were 7.23 ± 01 and 7.12
± 0.08. Recovery after oxygen inhalation was very rapid. The
respective pH values were 7.22 ± 0.11, 7.33 ± 0.07 and 7.36 ±
0.07 at 1, 2, and 5 minutes of reoxygenation. The pH values
were significantly different from the untreated group during
and after hypoxia.

DISCUSSION

The effects of calcium blocking drugs on cerebral metabolism and blood flow are widely studied, but in spite of the efforts the exact mechanism of action is not known. Furthermore, the studies deal mostly with severe hypoxic and/or ischemic circumstances and the information about the action during moderately disturbed circumstances is sparse. The aim of our study was twofold: investigate the effect of calcium blocker emopamil on cerebral circulation during normoxia, moderate hypoxia and during direct cortical activation under these circumstances. Second, to study the effect of (S)-emopamil during and after N_2 anoxia on the systemic blood pressure, on the cerebrocortical $[K^+]_e$, pH and pO_2 changes.

From our experimental arrangements and data we cannot distinguish between direct effects of the drug on cerebral vasculature and between secondary metabolic effects on the blood flow. The relatively moderate hypoxic changes and the stimulus evoked metabolic and ionic changes (for similar stimulus evoked ionic reactions see Urbanics et al., 1978) suggests that it is unlikely that serious metabolic disturbances occur and secondarily influenced the vasculature. The effect seems to be primarily vascular during moderate hypoxia and activation, as also found by Hossmann et al., (1985).

Under anoxia severe derangements occur and the disturbance undoubtedly multifactorial. The rapid $[K^+]_e$ increase reflects the failure of the Na^+-K^+-ATPase and the pH decrease is a good indicator of severe metabolic disturbances. These changes rise the cytosolic and intramitochondrial Ca^{++} and it causes metabolic and membrane aberrations. The significant improvement of both indicators after (S)-emopamil pretreatment suggests that the effect is primarily metabolic in the neuronal elements, and this is in agreement with the conclusion of Bielenberg et al., (1986). At the same time the effect on vasculature is very likely, and the significant MABP and pO_2 changes support this hypothesis.

REFERENCES

Bielenberg, G. W., Beck, T., Haubruck, H., Krieglstein, J., 1986, Effects of calcium entry blocker emopamil on postischemic energy metabolism of the isolated perfused rat brain and on local cerebral blood flow in the conscious rat. in: "Pharmacology of Cerebral Ischemia," Krieglstein, J. ed., Amsterdam, Elsevier Science Publishers, 309-315,

Hossmann, K. A., 1982, Treatment of experimental cerebral ischemia, J. Cereb. Blood Flow Metabol. Vol. 2, 275-297.

Hossmann, K. A., Grosse Ophoff B, Schmidt-Kastner R, Oschlies U., 1985, Mitochondrial calcium sequestration in cortical and hypocampal neurons after prolonged ischemia of the cat brain. Acta Neuropath. (Berlin) Vol. 68, 230-238.

Kazda, S., Knorr, A., Towart, R., 1983, in: "Pharmacology Vol. 5/2". (P. A. Van Zwieten and E. Schonbaum, Eds.), p. 83, Gustav Fischer Verlag, New York.

Kovach, A. G. B. , Dora, E., Sedlacsek, S., Koller, A., 1983, Effect of the organic calcium antagonist D-600 on cerebro-cortical redox responses evoked by adenosine, anoxia and epilepsy, J. Cereb. Blood Flow Metabol. Vol. 3, 51-61.

Leniger-Follert, E., Lübbers, D.W., 1977, Arzneimittelforsch. (Drug Res.) 27, 1517.

Ligeti, L., Osbakken, M. D., Subramanian, H. V., Kovach, A. G. B., Leigh, J. S., Chance, B., 1987, Effect of emopamil on cerebral lactate and pH changes during ischemia. Magn. Reson. Med. Vol. 4, 441-561.

Lübbers, D. W., Stosseck, K., 1970, Quantitative Bestimmung der lokalen Durchblutung durch elekrtrochemisch im Gewebe erzeugten Wasserstoff, Naturwissenschaften 57, 311-312.

Renchrona, S., Rosen, I., Siesjö, B. K., 1981, Brain lactic acidosis and ischemic cell damage: I. biochemistry and neu-rophysiology, J. Cereb. Blood Flow Metabol. Vol. 1, 297-311.

Siesjö, B. K., 1981, Cell damage in the brain: a speculative synthesis. J. Cereb. Blood Flow Metabol. Vol. 1, 155-185.

Somlyo, A. P., 1984, Cellular site of calcium regulation. Nature, Vol. 309, 516-517.

Somlyo, A. P., Urbanics, R., Vadasz, G., Kovach, A. G. B., Somlyo, A. V., 1985, Mitochondrial calcium and cellular elec-trolytes in the brain cortex frozen *in situ* : electron probe analysis. Biochem. Biophys. Res. Commun., Vol. 132(3): 1071-1078.

Szabo, L., Hoffmann, H. P., Raschack, M., Unger, L., 1988, (S)-emopamil: a new Ca^{++} antagonist of the Verapamil group with high cerebral availability and antiischemic properties. Soc. Neurosci. Abstr. 14:

Urbanics, R., Leniger-Follert, E., Lübbers, D. W.,1978, Time course of changes of extracellular H+ and K+ activities during and after direct electrical stimulation, Pflügers Arch. Vol. 378, 47-53.

CARDIOVASCULAR SYSTEM

LIMITATION OF LONG CHAIN FATTY ACID OXIDATION IN VOLUME OVER-LOADED RAT HEARTS

Z. El Alaoui-Talibi and J. Moravec

Laboratoire d'Energétique et de Cardiologie

Cellulaire, INSERM, UER de Médecine et de

Pharmacie, 21000 Dijon, France

The oxidation of long chain fatty acids provides up to 70 per cent of metabolic energy to well oxygenated myocardium (CRASS et al 1969, NEELY and MORGAN 1974). This contribution to the energy production may further increase during acute exercise (CRASS et al 1969, ORAM et al 1973). However, for the rate of long chain fatty acid oxidation to be easily adjusted to the instantaneous energy requirements of the heart, the cytosolic activation and transport of activated long chain fatty acids from the cytosol to the mitochondria should not be limited. In other words, functional and unlimited carnitine acylCoA transferase and carnitine-acylcarnitine translocase (PANDE 1973, PANDE 1975, BREMER 1983) are required in order to promote the high rate of cytosolic long chain acyls translocation into the mitochondria and that of free carnitine recycling back to the cytosol (Fig.1). High myocardial content of L-carnitine is therefore necessary for the predominance of lipid metabolism in the heart to occur (NEELY and MORGAN 1974, VARY and NEELY 1982). In mechanically overloaded hearts, it has been suggested that the low tissue levels of L-carnitine, which regularly occur during the development of cardiac hypertrophy (WITTELS and SPANN 1968, REVIS and CAMERON 1979, REIBEL et al 1983, BOWE et al 1984), may impair the long chain fatty acid utilisation (BISHOP and ALTSCHULD 1971, WITTELS and SPANN 1968). This in turn could limit both the overall energy turnover and mechanical performance of chronically overloaded hearts.

In this work, we tried to assess this possibility using the hearts of rats exposed to 3 month-old aorto-caval fistula (MORAVEC 1980, BOWE et al 1984). The hearts were perfused in vitro under conditions of moderate and high work loads and their ability to oxidize the exogenous long chain fatty acids (U-14C palmitate) was evaluated from the 14CO2 production data. In a parallel series of experiments, the ability of control and mechanically overloaded hearts to oxidize 1-14C octanoate was also tested in order to discriminate between two prospective causes of the impaired long chain fatty acid oxidation: (1) the limitation of carnitine-acyl-carnitine

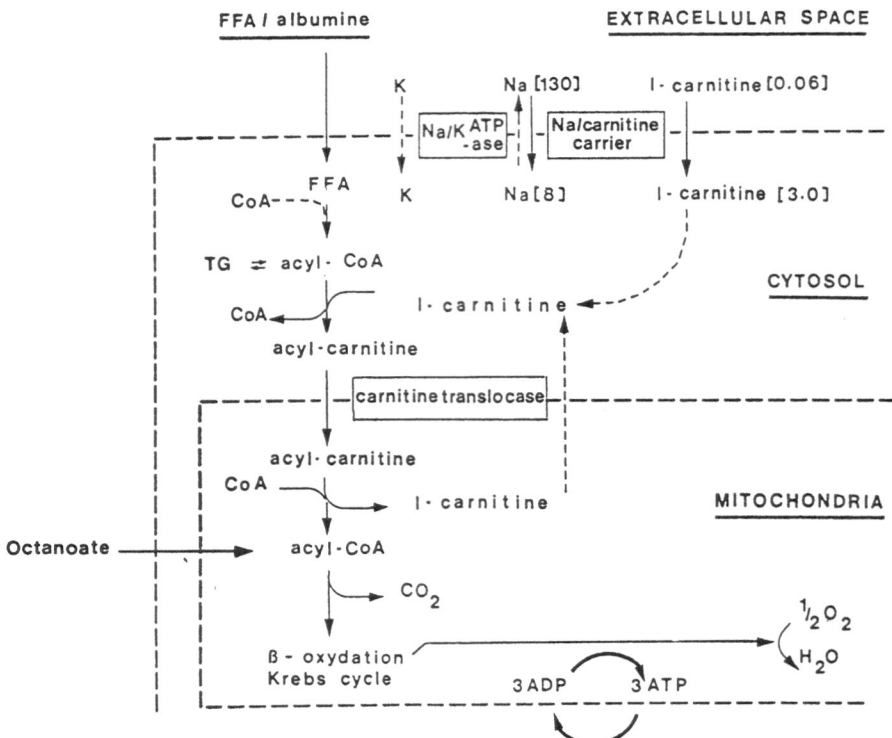

Fig 1. Mechanisms involved in the control of FFA utilisation. The activation of cytosolic FFA requires the presence of free CoA. The penetration of long chain fatty acylCoA into the mitochondria depends on cytosolic carnitine acylCoA transferase and mitochondrial carnitine-acylcarnitine translocase. Relatively high tissue levels of L-carnitine are necessary in order to avoid the substrate limitation of oxidation. A decrease of tissue L-carnitine results in a lowered carnitine-to-CoA ratio which favors FFA incorporation into tissue triacylglycerols and impairs the cytosolic carnitine-acylCoA transferase. Under normal conditions, the high tissue content of L-carnitine is maintained by its active transport from the blood. The latter has been shown defective in mechanically overloaded hearts whose tissue content of L-carnitine is depressed by about 30 to 40 per cent. translocase and/or (2) a defect of oxidation per se.

Methods:

A chronic volume overload was surgically induced in two month-old female rats of Wistar strain by opening of the aorto-caval fistula (HATT et al 1980). The sham-operated animals were used as controls. Three months after the surgery, the animals were sacrificed and their hearts used for the in vitro experiments.

After excision, the hearts were rapidly attached to an aortic cannula and perfused for 10 min with a bicarbonate buffer containing 11 mM glucose. They were then recirculated via the left atrium with 200 ml of the same solution containing 11 mM glucose and 1.2 mM palmitate (2.4 mM octanoate) bound to 3 per cent bovine albumine. The hearts were then allowed to perform mechanical work under conditions of moderate work load (10 Torr preload, 70 Torr afterload for 20 min). After this period, the work load was increased by clamping of the aortic outflow tract and the mechanical performance (pressure work) and 14CO2 production were measured during a further 20 min (high work load).

The 14CO2 production was evaluated from the cumulative curves of 14CO2 radioactivity detected in both the perfusion medium (14CO2 present as bicarbonate) and the hyamine hydroxide trap (gaseous 14CO2) placed at the outlet from the oxygenator. Aliquots of the hyamine hydroxide were inrtroduced directly into the Safefluor scintillation liquid. Samples of

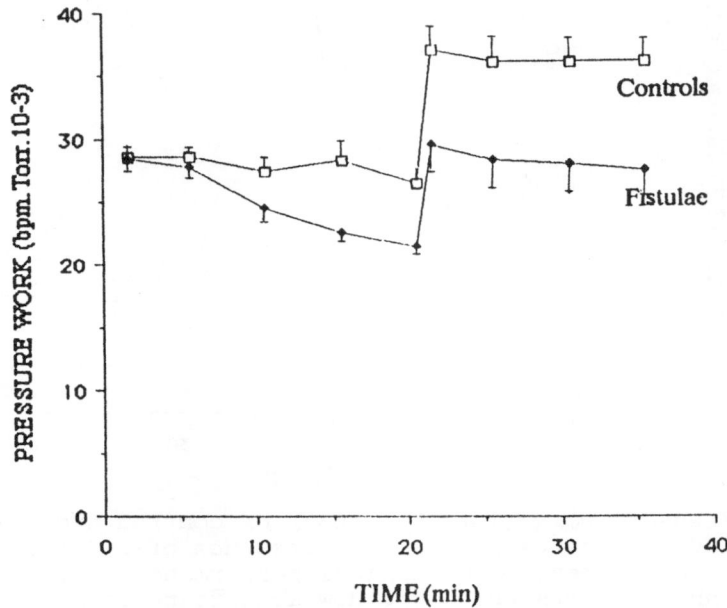

Fig 2. Pressure work (Torr.bpm.10-3) observed in control and volume overloaded rat hearts perfused in presence of 11mM glucose and 1.2mM palmitate under conditions of a moderate (0-20min) and high (20-40min) work loads. Note impaired mechanical performance of the hypertrophied hearts unable to sustain the acute overload. Mean + SEM for n=5.

the perfusion medium were first placed, without contact with air, in 25ml stoppered flasks containing 1 ml of 1N sodium hydroxide and, in the central wells, 1ml of 1 M hyamine hydroxide. They were then acidified by addition of 1 ml of 9N H2SO4 and gently stirred for 2 hrs in order to determine 14CO2 present as bicarbonate (LOPASCHUK et al 1986). At the end of the experiments, the hearts were removed from the perfusion canulae, the atria were dissected and the right and left ventricles were freeze clamped. The frozen tissue was lyophilized in order to determine the ventricular dry weight.

The mechanical activity of the atrially perfused hearts was evaluated in terms of pressure work (LOPASCHUK et al 1986). The work related to the ventricular ejections is not measurable after the clamp of the aortic outflow.

Results:

The dry weight of control and mechanically overloaded hearts equaled respectively 157+14 mg and 296+38 mg dry wt. The mechanical performance of the volume overloaded hearts perfused in presence of 1.2 mM palmitate (Fig 2) is

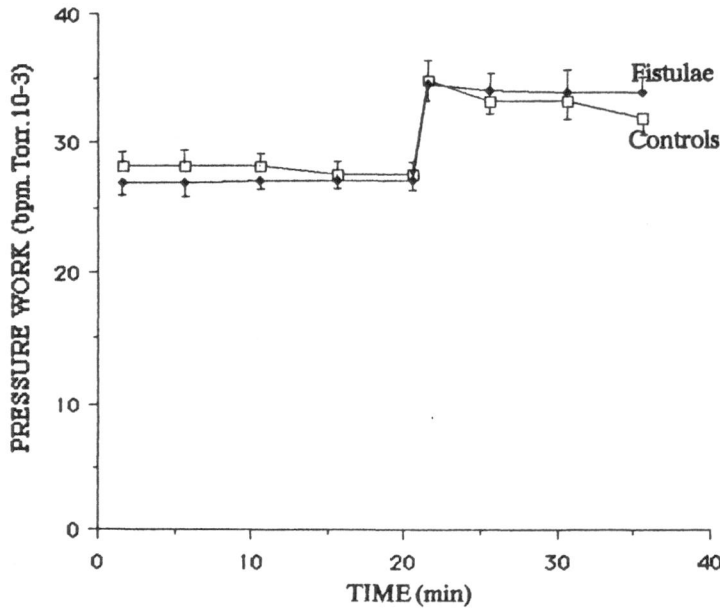

Fig 3. Pressure work (Torr.bpm.10-3) of control and volume overloaded rat hearts perfused in presence of 2.4mM octanoate. Note the improvement of mechanical performance of the hypertrophied hearts which are now able to perform without failure even at high work loads. Mean + SEM for n=5.

significantly depressed compared to that of controls. While the control hearts remain stable under conditions of both low and high work loads, a progressive decline of the pressure work can be noted in hearts of rats with aorto-caval fistula. In contrast, the mechanical performance of the volume overloaded hearts remains comparable to that of controls if 2.4 mM octanoate is used as the major exogenous substrate (Fig 3). This improvement of mechanical activity, as observed in presence of a short chain fatty acid, fits well with the oxygen consumption data. While under conditions of the moderate work load the oxygen consumption of control and overloaded hearts perfused in presence of 2.4 mM octanoate is comparable (respectively 59.5 + 6.5 and 59.0 + 6.6 umol.g-1 dry wt.min-1), a significant decrease of the QO2 occurs in the overloaded hearts when 1.2 mM palmitate is used (27.0 + 4.3 intead of 46.6 + 5.0 umol.g-1dry wt. min-1).

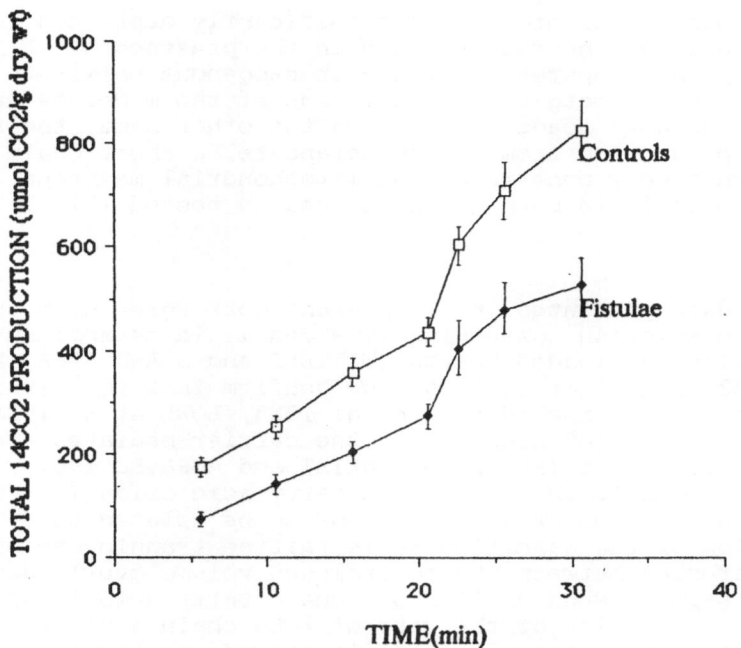

Fig 4. 14CO2 production in presence of 1.2mM   U-14C palmitate
Note decreased palmitate oxidation in volume overloaded hearts
during both moderate and high work loads. Mean + SEM for n=5.

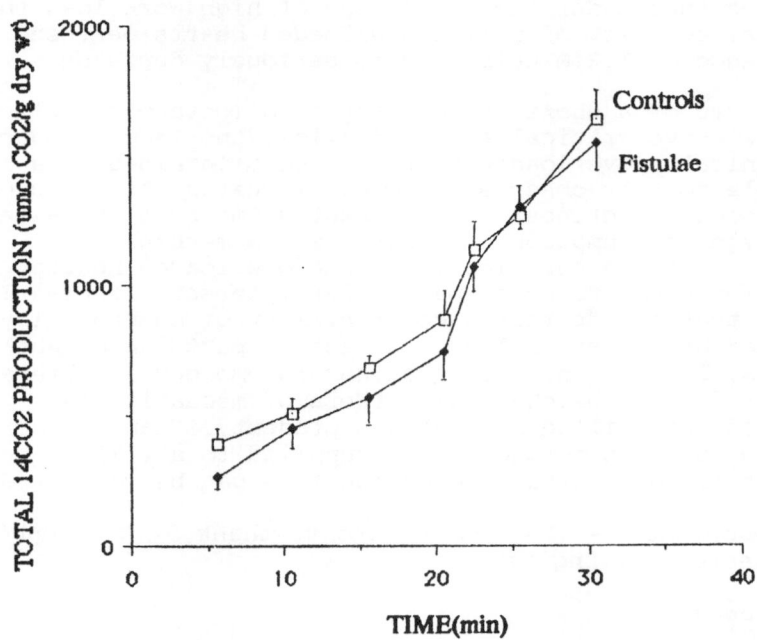

Fig 5. 14CO2 production in presence of 2.4mM 1-14C octanoate.
In these conditions the respective rates of 14CO2 production
are comparable in both control and volume overloaded hearts.
Mean + SEM for n=5.

We found 14CO2 production significantly depressed in volume overloaded hearts perfused in the presence of U-14C palmitate. This apparent decrease in exogenous palmitate oxidation occurs both under conditions of the moderate and, mainly, high work loads (Fig 4). On the other hand, the rate of 14CO2 production from 1-14C octanoate, a short chain fatty acid which freely penetrates the mitochondrial membrane, is quite comparable in both groups of hearts tested (Fig 5).

Discussion:

The data presented in the present work agree with the previous prediction concerning impaired lipid metabolism in mechanically overloaded hearts (WITTELS and SPANN 1968, REIBEL et al 1983, REIBEL et al 1986). We confirm that low tissue levels of L-carnitine (REIBEL et al 1983, BOWE et al 1984), secondary to altered kinetics of the carrier mediated carnitine transport (EL ALAOUI-TALIBI and MORAVEC 1987), effectively impair the long chain fatty acid oxidation. This limitation of 14CO2 production seems to be related to a dysfunction of the carnitine-acylcarnitine translocase since the differences between the control and volume overloaded hearts disappear when a 14C short chain fatty acid is used. The above limitation of the rate of long chain acylCoA translocation into the mitochondria becomes particularly serious under the conditions of high work loads when long chain fatty acids constitute the major source of metabolic energy. It has been suggested (NEELY and MORGAN 1974) that under these conditions, even in control hearts, the overall energy turnover is in fact determined by the rate of carnitine- acylcarnitine exchange which lags behind the rate of mitochondrial oxidations per se. Therefore it is not surprising that under the conditions of high work load the mechanical activity of volume overloaded hearts perfused in the presence of 1.2mM palmitate is seriously depressed.

The use of a short chain fatty acid (octanoate) circumvents the above critical step and allows the 14CO2 production of mechanically overloaded hearts to be maintained at a level comparable to that observed in control hearts. At the same time, exogenous octanoate significantly improves the steady state oxygen consumption and restores the mechanical performance of the chronically volume overloaded hearts. This observation seems to us of particular interest. It clearly suggests that the decreased contractility of mechanically overloaded hearts may relie, at least in part, on metabolic disorders. Despite the fact that several molecular alterations (considered to be biochemical markers of mechanically overloaded and failing hearts) are present, if an appropriate substrate (such as octanoate) is supplied to a failing heart, then most of the contractile dysfunction can be corrected.

Acknowledgements — We would like to thank Dr.S. Campbell who corrected our English.

Bibliography

1 Crass, M.F. Mac Caskill, E.S. and Shipp, J.C. (1969) Amer. J. Physiol. 216, 1569-1576.
2 Neely, J.R. and Morgan, H.E. (1974) Annual. Rev. Physiol. 36, 413-459.

3 Oram, J.F. Bennetch, S.L. and Neely, J.R. (1973) J. Biol. Chem. 248, 5259-5309.

4 Pande, S.V. (1973) Biochim. Biophys. Acta 306, 15-20

5 Pande, S.V. (1975) Proc. Natl. Acad. Sci. 72, 883-887.

6 Bremer, J. (1983) Physiol.Rev. 63, 1420-1480.

7 Vary, T.C. and Neely, J.R. (1982) Amer. J. Physiol. 242, H585-H592.

8 Wittels, B. and Span, J.F.J. (1968) J. Clin. Invest. 47, 1787-1793.

9 Revis, N.W. and Cameron, A.J.V. (1979) Metabol. Clin. Exp. 28, 801-813.

10 Reibel, D.K. Uboh, C.E. and Kent, R.L. (1983) Amer. J. Physiol. 244, H839-H843.

11 Bowe, C. Nzonzi, J. Corsin, A. Moravec, J. and Feuvray, D. (1984) Pflügers Arch. 102, 317-320.

12 Moravec, J. (1980) FEBS Lett. 113, 134-137.

13 Lopaschuk, G.D. Hansen, C.A. and Neely, J.R. (1986) Amer. J. Physiol. 250, H351-H356.

14 Hatt, P.Y. Rakusan, K. Gastineau, P. Laplace, M. and Cluzeaud, F. (1980) 75, 105-108.

15 Moravec, J. and El Alaoui-Talibi, Z. (1986) Fed. Proc. 46, 1402a.

HYPOXIA IN SALINE-PERFUSED HEART MAY BE DUE TO

MISMATCH OF REGIONAL METABOLISM AND PERFUSION

Johannes H.G.M. van Beek

Laboratory for Physiology, Free University, van der Boechorststraat 7
1081 BT Amsterdam, the Netherlands

INTRODUCTION

A number of observations reported in the literature indicate that the average myocardial tissue oxygen tension is substantially lower than the mean coronary venous oxygen tension. Measurements with small oxygen electrodes done by Schubert et al. (1978) in the saline-perfused cat heart show that large regions have oxygen tensions far below the average venous oxygen tension. Gayeski et al. (1988) found that the saturation of myoglobin with O_2 in heart and skeletal muscle indicated a P_{O_2} of the order of 5 mmHg. Rose et al. (1988) state that there are "two and only two explanations for these anomalies": diffusional shunting of O_2 and a high diffusional resistance for O_2 between blood vessel lumen and tissue.

In this paper we show that oxygen supply limits oxygen consumption in the saline-perfused rabbit heart in spite of high arterial and venous oxygen tensions and that this might be explained by perfusion heterogeneity entailing mismatch between flow and metabolism. Such mismatch constitutes a third explanation for the existence of low tissue P_{O_2} that cannot yet be discarded.

OXYGEN SUPPLY LIMITS OXYGEN UPTAKE IN SALINE-PERFUSED HEART

In order to investigate whether oxygen supply limits oxygen consumption in the saline-perfused heart, we perfused eight rabbit hearts according to Langendorff at 37°C with Tyrode solution. The perfusate contained 10 μM adenosine, so that the hearts were maximally vasodilated (Van Beek et al., submitted). The left ventricle of the heart was vented to minimize pressure development and keep metabolic demand constant. The pulmonary artery was cannulated and part of the coronary venous outflow was drawn through a cuvette which contained an O_2 electrode. Oxygen uptake was calculated from the arterial-to-venous O_2 concentration gradient and coronary flow. The perfusate P_{O_2} was lowered by about 10% from around 680 mmHg at constant perfusion flow. This resulted in each case in a decrease in oxygen uptake (see Figure 1). The oxygen uptake returned to its original level when the P_{O_2} was brought back to its control value. On average there was a 5.8% decrease in oxygen uptake in the heart per 10% decrease in P_{O_2}. In our experiments there was no change in perfusion pressure when the perfusate P_{O_2} was lowered by 10% and the resulting decrease in oxygen consumption means that oxygen supply limits oxygen consumption in the saline-perfused heart.

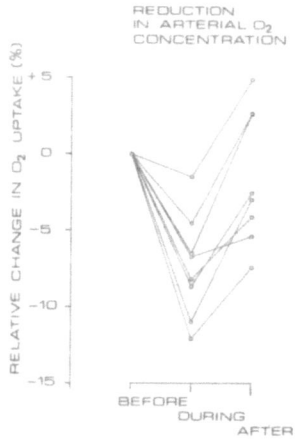

Figure 1. Relative change in oxygen uptake resulting from a 10% reduction in perfusate P_{O_2} in Tyrode-perfused rabbit hearts. Both perfusion flow and perfusion pressure were constant.

Paradise et al. (1984) had already reported that oxygen uptake in saline-perfused heart decreases after small reductions in arterial O_2 concentration. Although they ascribed this result to inadequate oxygenation, the outcome of their experiment might have been partially due to the decreases in perfusion pressure following hypoxic vasodilation since their preparation was perfused at constant flow. This effect of perfusion pressure had been found by Arnold et al. (1968), and was called the garden hose effect. In order to investigate the effect of changes in perfusion pressure further, we lowered the perfusion flow in the same eight rabbit hearts mentioned above by 10%. This resulted in a 4.4% reduction in oxygen uptake. In this case the reduction in perfusion pressure by about 10% came on top of the 10% reduction in oxygen supply due to diminished flow. However, the reduction in oxygen uptake was not greater than in the case of the 10% perfusate P_{O_2} reduction where there was no perfusion pressure change. We conclude that the effect of the reduction of perfusion pressure was not big compared with the effect of the reduction in oxygen supply.

EXPLANATIONS FOR OXYGEN-SUPPLY LIMITATION OF OXYGEN UPTAKE

The finding that oxygen supply limits oxygen consumption is amazing in view of the fact that arterial P_{O_2} was around 680 mmHg and the venous P_{O_2} around 350 mmHg. The mitochondrial cytochrome c oxidase reaction has generally been found to be saturated with oxygen for P_{O_2} values below 10 mmHg in studies on isolated myocytes (Wittenberg and Robinson, 1981; Kennedy and Jones, 1986). An explanation must be found for the large difference between the large vessel P_{O_2} on one hand and the low critical P_{O_2} for the cell on the other.

Diffusional shunting of oxygen from the arterial to the venous side of the coronary circulation would mask low capillary P_{O_2}'s. This means that a major part of the oxygen entering the heart bypasses the myocytes while keeping the venous P_{O_2} high. The high diffusional resistance between vessel lumen and mitochondrion, that exists according to Rose et al. (1988) and Gayeski et al. (1988), should also be investigated. The last hypothesis we will consider states that there is a mismatch between local flow and oxygen consumption leading to the simultaneous existence of hypoxic and well-oxygenated regions.

DIFFUSIONAL SHUNTING

Since the arterioles and venules in the heart show a countercurrent arrangement, the difference in P_{O_2} between the arterial and the venous vessels might drive a diffusional flux of oxygen that enters the venous perfusate and is transported out of the heart with the venous outflow. We call this diffusional shunting. This means that part of the oxygen entering the heart bypasses the myocytes. The end-capillary P_{O_2} is lower than the venous P_{O_2}. The shunt pathway is short and thereby fast, so that it can be expected that an indicator that shunts appears in the venous perfusate before a simultaneously injected indicator that has to follow the intravascular route. We did experiments based on this idea in Tyrode-perfused isolated rabbit hearts. The perfusate concentration of oxygen and indocyanine green-tagged albumin were changed simultaneously using a perfusion system that allowed fast changes of perfusate composition by switching between two solutions that could be mixed with the main part of the perfusion flow (Van Beek and Elzinga, 1987). The coronary venous concentration changes were measured by drawing part of the pulmonary artery outflow through a cuvette containing a fast O_2 electrode and a fiber-optic catheter connected to a densitometer for measuring indocyanine green concentration. The response time of the O_2 and dye measurements were matched.

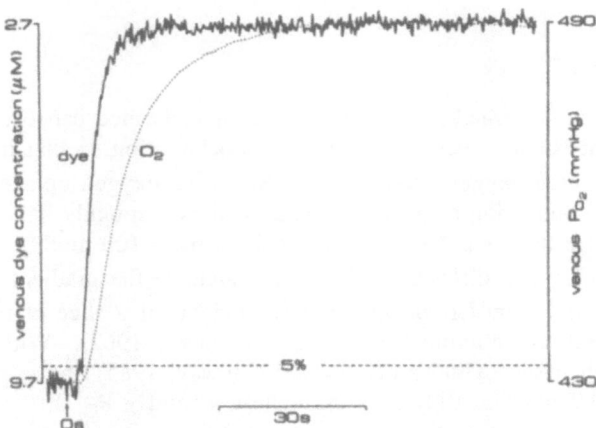

Figure 2. Venous time course of oxygen and indocyanine green-tagged albumin measured in an isolated perfused rabbit heart. At 0 sec the switch in the perfusion system was activated and perfusate P_{O_2} and dye concentration changed simultaneously. Note that P_{O_2} starts to change after the dye has emerged in the venous effluent.

Differences in time course found at the exit of the heart (see Figure 2) must therefore result from the passage of the indicators through the heart. In 7 hearts we found that the P_{O_2} did not change before the arrival of the dye, both for upward and downward steps in O_2 (Van Beek and Elzinga, 1987).

This experiment is a good test for the existence of shunt diffusion pathways only if it can be shown that diffusion from arterial to venous side takes less time than the travel of intravascular indicator. We have calculated that the diffusion flux from arteriole to venule, which are less than 100 μm apart in the myocardium, starts to change less than 0.33 sec after a change in the arteriole, and the change in flux is halfway in less than 1 sec. At places where the arteriolar and venular ends of certain capillaries are less than 20 μm apart these times would be less than 0.013 and 0.04 sec respectively. The diffusion times for shunt pathways are short with respect to the mean transit time of the capillary bed, estimated to be 1.4 sec in this preparation (Van Beek and Elzinga, 1987). Since we did not find any O_2 changes before arrival of the intravascular indicator, we conclude that shunt diffusion of oxygen is negligible in this preparation.

Rose and Goresky (1985) found no emergence of tracer oxygen before simultaneously injected labeled red blood cells in blood-perfused dog heart, and concluded that diffusional shunting of oxygen played no role. Katz and Feigl (1987) found that diffusional shunting of CO_2 was very small in the blood-perfused dog heart. Weiss and Sinha (1978) found spectrophotometrically that the oxygen saturation of hemoglobin did not decrease significantly when they followed intramyocardial arterioles, making it unlikely that diffusional shunting plays an important role for arterioles larger than 40 μm. Duijst et al. (1988) found that about 60% of injected heat entering the dog heart via the coronary circulation is shunted. Taking into account that the diffusivity of O_2 is two orders of magnitude lower than that of heat, it was estimated that less than 1% of O_2 entering the heart will shunt.

We conclude that shunting of oxygen is very small in both the blood-perfused and saline-perfused heart and cannot explain our finding that oxygen supply limits oxygen consumption in the presence of a high venous P_{O_2}.

DIFFUSIONAL RESISTANCE

Here we examine whether a high diffusional resistance can cause a sufficiently large P_{O_2} gradient between vessel lumen and mitochondrion so that it can explain the oxygen supply-limited oxygen consumption. Since the oxygen uptake in the hearts in our experiments, contracting against a minimal load, was roughly half that found in the heart *in situ*, where the available P_{O_2} gradient is at most 100 mmHg, we could expect at most a 50 mmHg P_{O_2} difference. Next we calculate the gradient from the classic Krogh (1919) model. The Krogh diffusion coefficient for a slice of rat heart tissue at 37°C is $1.9 \cdot 10^{-5}$ ml/(cm·min·mmHg) (Grote and Thews, 1962). With a radius of 17 μm for the Krogh tissue cylinder (Weiss and Conway, 1985) and a measured oxygen consumption of 4.7 ml/(min·100g), we calculate a 4 mmHg P_{O_2} drop towards the periphery of the Krogh cylinder. This is far too small to explain the finding that oxygen supply limits oxygen consumption. However, Rose and Goresky (1985) derived from indicator dilution curves of tracer oxygen that the diffusional resistance between the red cell and the myocyte interior was higher than predicted by the Krogh model. They estimated the capillary permeability for oxygen to be $4.6 \cdot 10^{-3}$ cm·sec^{-1}. Since the capillary surface area is 500 cm^2/g), the permeability surface product is 140 cm^3·g^{-1}·min^{-1}. In order to accommodate a flux of oxygen of 4.7 ml·min^{-1}·100g^{-1}, a concentration difference of $3.36 \cdot 10^4$ ml·ml^{-1} is necessary, which corresponds to a P_{O_2} difference of about 11 mmHg. This gradient is still very small compared with the venous P_{O_2} of 350 mmHg. The difference between the prediction of the Krogh model and Rose and Goresky's finding might lie in the fact that the Krogh diffusion coefficient was measured across a slice of tissue (Grote and Thews, 1962) where diffusion can take place through the interstitial space, and where diffusion resistance across the vascular wall and from interstitium into the cell is not included in the measurement. Such a gradient is still small compared with a venous P_{O_2} of 350 mmHg. The conclusion would be that the diffusional gradients over the capillary wall cannot explain the finding that oxygen supply limits oxygen consumption, unless tissue edema and capillary obstruction as found in the saline perfused heart (Poche et al., 1971) make diffusional resistance appreciably higher.

MISMATCH BETWEEN LOCAL FLOW AND OXYGEN DEMAND

Here we will investigate whether heterogeneity of local flow and metabolism can give rise to sufficiently low tissue P_{O_2}'s and can thereby cause the limitation of oxygen uptake by oxygen supply. When the local oxygen consumption to flow ratio varies as a result of flow heterogeneity, the end-capillary P_{O_2} could also be heterogeneous. When the variation in local oxygen consumption parallels the variation of flow, so that regions with a low oxygen demand have a low flow and vice versa, the flow heterogeneity would not cause heterogeneity of end-capillary P_{O_2}.

Here we will discuss a hypothetical example of mismatch between flow and metabolism in saline-perfused heart and calculate the venous oxygen tension and mean tissue oxygen tension. We consider a heart that shows a homogeneous O_2 consumption and extracts one third of the oxygen that enters. To calculate the venous and end-capillary P_{O_2} we use Fick's formula, oxygen uptake = flow·$(C_{a_{O_2}} - C_{v_{O_2}})$, where $C_{a_{O_2}}$ and $C_{v_{O_2}}$ denote the arterial and venous oxygen concentrations respectively. The Fick equation entails that O_2 extraction, $(Ca_{O_2} - Cv_{O_2})/Ca_{O_2}$, is inversely proportional to flow at constant oxygen consumption. To calculate the average tissue P_{O_2} we use the classic Krogh model, where oxygen uptake is homogeneous along the capillary wall and where there is no longitudinal diffusion. Since we are concerned here with the effects of perfusion heterogeneity, we will assume that the diffusivity of O_2 is infinitely high, so that diffusion gradients are zero. We will consider the situation in which perfusate flow is very heterogeneous: 80% of the mass of the heart receives specific blood flow (i.e. blood flow per unit mass) at one third of the overall average rate, and 20% receives blood flow at 3.6667 times the average rate. The local oxygen consumption per unit weight of tissue is the same for both regions, as is tissue density. The assumption is such that the capillary P_{O_2} has just reached 0 mmHg at the end of the capillary in the lower flow region for an arterial P_{O_2} of 680 mmHg. In the higher flow region the extraction is 0.3333/3.6667 times that in the higher flow region so that end-capillary P_{O_2} is 680 - 61.8 = 618.2 mmHg. The mixed coronary venous P_{O_2} should be calculated by weighting by flow:

$$P_{v_{O_2}} = 0.8 \cdot (0.3333) \cdot 0 + 0.2 \cdot (3.6667 \cdot) \cdot 618.2 = 453.3 \, mmHg.$$

Although this is a high value it is occasionally measured in saline-perfused rabbit hearts (see Figure 2). The average tissue P_{O_2} should be calculated by weighting by volume:

$$P_{t,o2} = 0.80 \cdot (0.5 \cdot 680 + 0.5 \cdot 0) + 0.20 \cdot (0.5 \cdot 680 + 0.5 \cdot 618.2) = 401.8 \, mmHg.$$

The mean tissue P_{O_2} for each region is here calculated by averaging the arterial and venous P_{O_2} according to the Krogh model, since the radial diffusional gradient is assumed to be zero. The result shows that the average tissue P_{O_2} can be lower than the mixed coronary venous P_{O_2}. This is due to the fact that the end-capillary P_{O_2}'s are weighted by flow to obtain the mixed venous P_{O_2}, but the average tissue P_{O_2} is obtained by volume weighting the average P_{O_2} of the regions. This example shows

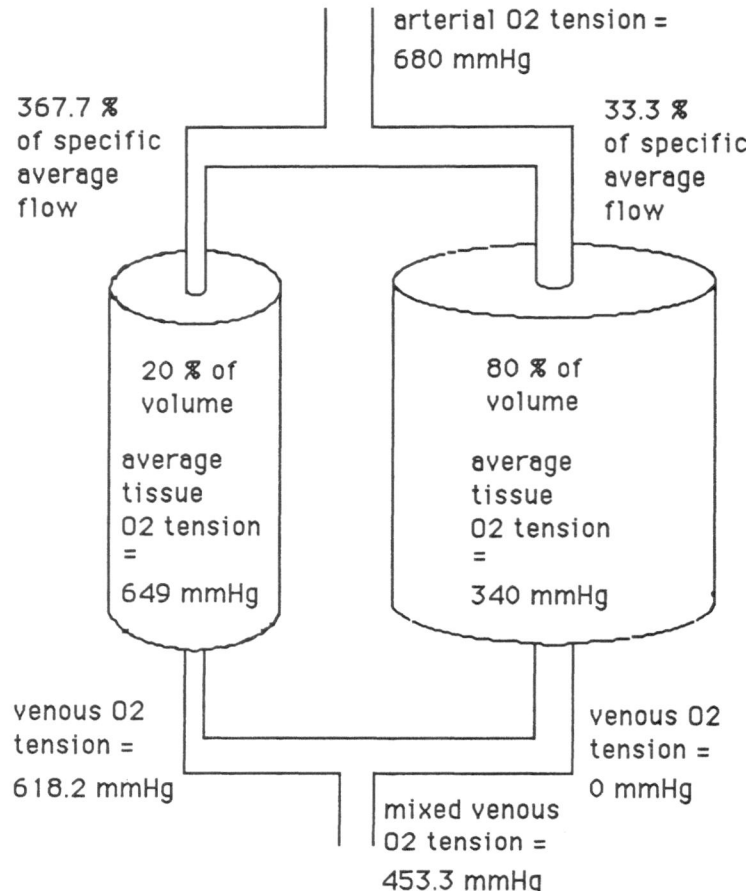

arterial O2 tension =
680 mmHg

367.7 %
of specific
average
flow

33.3 %
of specific
average
flow

20 % of
volume

average
tissue
O2 tension
=
649 mmHg

80 % of
volume

average
tissue
O2 tension
=
340 mmHg

venous O2
tension =
618.2 mmHg

venous O2
tension =
0 mmHg

mixed venous
O2 tension =
453.3 mmHg

Figure 3. Hypothetical example of mismatch between perfusion and metabolism. Two regions have the same specific oxygen consumption (i.e. oxygen consumption per unit mass). One region has a low specific perfusate flow; the other has a high flow. The regions are represented by Krogh cylinders. The regional venous P_{O_2} differs tremendously between the two regions. The mixed venous P_{O_2} is higher than the average tissue P_{O_2} which is 401.8 mmHg. In actual myocardium the two regions might be fragmented and intermingled. Most probably there would be a broad distribution of flow.

that mismatch of flow and metabolism could lead to "convective shunting": high flow regions keep the venous P_{O_2} high, while low flow regions are hypoxic at the same time.

The preceding example had a fairly strong perfusion heterogeneity. Another example was constructed along similar lines, but this time it fits the conditions in the rabbit hearts of our experimental series on oxygen supply limitation of oxygen consumption. The venous P_{O_2} is 350 mmHg. Here 58% of the saline-perfused heart has a flow of 0.485 times the average, so that the capillary P_{O_2} has just become zero at the end of the capillary. The remaining 42% has a flow of 1.711 times average, which yields an end-capillary P_{O_2} of 487.1 mmHg. The average tissue P_{O_2} is calculated to be 442.3 mmHg. When the arterial P_{O_2} is lowered by 10% with perfusion flow constant, the oxygen consumption in the low flow region is forced to go down by 10%. Since this applies to 58% of the total consumed oxygen this means that oxygen consumption goes down by 5.8%. The example shows that mismatch of perfusion and metabolism could explain the outcome of our experiments on oxygen supply limitation of oxygen uptake. The real distribution of flow is presumably more bell-shaped. However, metabolism-perfusion mismatch might lead to hypoxic regions according to the same principles. Although these examples identify mismatch of metabolism and perfusion as a possible explanation, this hypothesis still has to be proven. It will be difficult to measure end-capillary or local venular oxygen tension in the saline-perfused heart. It might be possible to measure local uptake of substrates for energy metabolism in relation with local flow, and this might yield an indirect indication of mismatch between local flow and metabolism. In any case, Kuikka et al. (1986) measured the local flow in saline-perfused rabbit hearts by assessing the deposition of microspheres in small pieces of the heart. They found a relative dispersion (RD = standard deviation/mean) of flow which was 30-40% when the heart was divided in 0.03 to 0.19 gram pieces. However, the RD for flow has been found to increase with higher spatial resolution of the measurement (cf. Van Beek et al., this volume), so that the relevant RD for calculations on oxygen transport might be higher. An appreciable perfusion heterogeneity seems to be present. Indeed, NADH fluorescence shows discrete hypoxic regions in saline-perfused rat heart (Steenbergen et al., 1977).

Oxygen supply-limited oxygen consumption has been found in saline-perfused heart, and it is possible that it also exists in blood-perfused heart. Bergman et al. (1979) did experiments on isolated rabbit hearts. When they added red blood cells to the saline perfusate oxygen uptake increased. When they increased the hematocrit from 25 to 40% oxygen uptake increased by 60%, while flow stayed nearly constant. This finding indicates that oxygen supply might limit oxygen consumption even in the blood-perfused heart. In blood-perfused dog heart it was found that small reductions in coronary arterial oxygen content led to decreases in cardiac oxygen consumption (Scharf et al., 1975). However, this finding differs from that in another study (Powers and Powell, 1973).

A marked heterogeneity of flow has been found in the blood-perfused heart (King et al., 1985). Weiss and Sinha (1978) have found that the oxygen saturation of hemoglobin in small venules in dog heart varies from below 10% to above 60%, indicating that there is a marked spread in the ratio of local oxygen uptake to flow. The spread in end-capillary P_{O_2}'s might even be bigger than the spread found at the venular level since variations would tend to average out. Two hypotheses for the variation of local venular oxygen saturation come to mind. The variation in oxygen saturation might result from the relation between the limited local permeability and flow, as might be the case when the hypothesis of Rose and Goresky (1985) applies. In this case local oxygen consumption follows the limitations set by the local permeability and the tissue P_{O_2} will be very low. The second hypothesis is that local metabolism is actually set by the functional needs of the cell, but that the relation between local metabolic demand and local flow shows a certain degree of mismatch. The studies in the saline-perfused heart point to the second hypothesis since the diffusional limitation would be of less consequence in face of the high oxygen tensions in the big vessels. Of course, when metabolism-perfusion mismatch exists, this leads to low intracapillary P_{O_2}'s, and high diffusion resistance becomes important again. It should, however, be remarked that the saline-perfused heart may not be an adequate model for the blood-perfused heart, and capillary closure and edema may invalidate the argument.

CONCLUSION

In summary, we have found that oxygen uptake in the saline-perfused heart is limited by oxygen supply, although the arterial and venous oxygen tensions in these hearts were very high. A number of studies make it unlikely that diffusional shunting can explain these findings. Diffusional resistance between vascular lumen and myocyte interior might be of insufficient magnitude to explain the oxygen supply limitation, unless diffusional resistance is markedly increased during saline perfusion. Convective shunting due to mismatch between local flow and metabolism may lead to a high venous oxygen tension, masking low end-capillary oxygen tensions in certain regions in the heart. The underperfused regions may thus be hypoxic. It is possible that a mismatch between local blood flow and metabolic demand also exists to some extent in the normal blood-perfused heart.

(This work has been partially supported by ZWO, the Netherlands Organization for Scientific Research.)

REFERENCES

Arnold, G., F. Kosche, E. Miessner, A. Neitzert, W. Lochner. The importance of the perfusion pressure in the coronary arteries for the contractility and the oxygen consumption of the heart. Pflügers Arch 299: 339-356, 1968.

van Beek, J.H.G.M., and G. Elzinga. Diffusional shunting of oxygen in saline-perfused isolated rabbit heart is negligible. Pflügers Arch. 410: 263-271, 1987.

van Beek, J.H.G.M., P. Bouma, and N. Westerhof. Oxygen uptake in saline-perfused rabbit heart is decreased to a similar extent during reductions in flow and in oxygen concentration. (submitted)

van Beek, J.H.G.M., S.A. Roger and J.B. Bassingthwaighte. Fractal networks explain regional myocardial flow heterogeneity. (this volume)

Bergmann, S.R, R.E. Clark, and B.E. Sobel. An improved isolated heart preparation for external assessment of myocardial metabolism. Am. J. Physiol. 236: H644-H651, 1979.

Duijst, P., J.H.G.M. van Beek, G.H.M. ten Velden, G. Elzinga and N. Westerhof. Shunting of heat in the canine myocardium. In: Cardiac Metabolism and Flow. Thesis of P. Duijst, Free University Press, Amsterdam, pp. 42-87, 1988.

Gayeski, T.E.J., W.J. Federspiel, and C.R. Honig. A graphical analysis of the influence of red cell transit time, carrier-free layer thickness, and intracellular P_{O_2} on blood-tissue O_2 transport. Adv. Exp. Med. Biol. 222: 25-35, 1988

Grote, J., and G. Thews. Die Bedingungen für die Sauerstoffversorgung des Herzmuskelgewebes. Pflügers Arch. 276: 142-165, 1962.

Katz, S.A., and E.O. Feigl. Little carbon dioxide diffusional shunting in coronary circulation. Am. J. Physiol. 253: H614-H625, 1987.

Kennedy, F.G., and D.P. Jones. Oxygen dependence of mitochondrial function in isolated rat cardiac myocytes. Am. J. Physiol. 250: C374-C383, 1986.

King, R.B., J.B. Bassingthwaighte, J.R.S. Hales, and L.B. Rowell. Stability of heterogeneity of myocardial blood flow in normal awake baboons. Circ. Res. 57: 285-295, 1985.

Krogh, A. The number and distribution of capillaries in muscles with calculations of the oxygen pressure head necessary for supplying the tissue. J. Physiol. 52: 409-415, 1919.

Kuikka, J.M. Levin, and J.B. Bassingthwaighte. Multiple tracer dilution estimates of D- and 2-deoxy-D-glucose uptake by the heart. Am. J. Physiol. 250: H29-H42, 1986.

Paradise, N.F., J.M. Surmitis, and C.L. Mackall. O_2 reserve of left ventricle of isolated, saline-perfused rabbit heart. Am. J. Physiol. 247: H861-H868, 1984.

Poche, R., G. Arnold, and D. Gahlen. Uber den Einfluß des Perfusionsdruckes im Coronarsystem des stillgestellten, aerob perfundierten, isolierten Meerschweinchenherzens auf Stoffwechsel und Feinstruktur des Herzmuskels. Virchows Arch. (Zellpath.) 8: 252-266, 1971.

Powers, E.P., and W.P. Powell, Jr. Effect of arterial hypoxia on myocardial oxygen consumption. Circ. Res. 33: 749-756, 1973.

Rose, C.P., and C.A. Goresky. Limitations of tracer oxygen uptake in the canine coronary circulation. Circ. Res. 56: 57-71, 1985.

Rose C.P., C.A. Goresky, G.C. Bach, J.B. Bassingthwaighte, and S. Little. In vivo comparison of non-gaseous metabolite and oxygen transport in the heart. Adv. Exp. Med. Biol. 222: 45-54, 1988.

Scharf, S.M., S. Permutt, and B. Bromberger-Barnea. Effects of hypoxic and CO hypoxia on isolated hearts. J. Appl. Physiol. 39: 752-758, 1975.

Schubert, R.W., W.J. Whalen, and P. Nair. Myocardial P_{O_2} distribution: relationship to coronary autoregulation. Am. J. Physiol. 234: H361-H370, 1978.

Steenbergen, C., G. Deleeuw, C. Barlow, B. Chance, and J.R. Williamson. Heterogeneity of the hypoxic state in perfused rat heart. Circ. Res. 41: 606-615, 1977.

Weiss, H.R., and R.S. Conway. Morphometric study of the total and perfused arteriolar and capillary network of the rabbit left ventricle. Cardiovasc. Res. 19: 343-354, 1985.

Weiss, H.R., and A.K. Sinha. Regional oxygen saturation of small arteries and veins in the canine myocardium. Circ. Res. 42: 119-126, 1978.

Wittenberg, B.A., and T.F. Robinson. Oxygen requirements, morphology, cell coat and membrane permeability of calcium-tolerant myocytes from hearts of adult rats. Cell Tissue Res. 216: 231-251, 1981.

OXYGEN DELIVERY AND PERFORMANCE IN THE ISOLATED, PERFUSED RAT HEART:

COMPARISON OF PERFUSION WITH AQUEOUS AND PERFLUOROCARBON-CONTAINING MEDIA

G. P. Biro, M. Masika and B. Korecky

Department of Physiology, University of Ottawa
Ottawa, Ontario, Canada. K1H 8M5

INTRODUCTION
     The isolated, Langendorff-perfused rat heart is used extensively in biomedical investigation. This preparation is generally perfused with aqueous buffers oxygenated with high-oxygen gas mixtures. In spite of high pO_2, because of the poor solubility of oxygen in water, the oxygen content of such perfusates is quite low: generally not exceeding 1.5 ml/dl. Adequate oxygen delivery to the the myocytes can only be maintained under these conditions by nearly maximal coronary vasodilation.

     Perfluorocarbons (P.F.C.'s) are chemically inert and their oxygen solubility is roughly 20-times that of water (Biro and Blais, 1987). This renders them potentially useful as "blood substitutes" and various products have been used as perfusate in isolated heart preparations (Segal and Rendig, 1982; Tomera and Geyer, 1982; Chemnitius et al., 1985; Segal and Ensunsa, 1988). These studies show that PFC-based perfusates support isolated heart function better and/or longer than the routinely used Krebs-Ringer medium.

     We used a newer PFC preparation based on perfluoro-octylbromide which is reasonably stable at room temperature in substantially higher concentration than previous products (FC-43, Fluosol-DA, etc.). Since isolated hearts perfused with blood exhibit contractile performance which is substantially better than that shown by buffer-perfused hearts, the latter may be considered to be oxygen-limited because of the buffer's low oxygen content. We tested this hypothesis by comparing oxygen supply and contractile performance in Langendorff-perfused hearts, using conventional Krebs-Ringer-bicarbonate and P.F.O.B-containing perfusates.

METHODS

     Langendorff preparations were made using hearts from rats weighing 220-280 g. Krebs-Ringer-bicarbonate buffer (KRB) was oxygenated in a temperature-controlled reservoir ($37^{o}C$). In another reservoir a 50:50% mixture of perfluoro-octylbromide (PFOB) and KRB was oxygenated. The PFOB preparations contained 50% PFOB by weight (or approximately 26% by volume). Its emulsion stabilizer was lecithin. The 50:50% mixture was made with K.R.B., to assure that this perfusate's electrolyte composition approximated that of the K.R.B.

The hearts were suspended by the aortic root on a cannula attached to a bubble trap. A roller-pump drew perfusate from the reservoir and pumped it into the temperature-controlled bubble trap and from there, into the aortic root. Through a sidearm, perfusion pressure was monitored with a mercury manometer and recorded through a Statham P23 Db transducer, on a Grass four-channel recorder. A small balloon was inserted in the left ventricular cavity and inflated with saline to produce end-diastolic pressure in the range of 1-5 cm H_2O. Systolic pressure and its derivative were recorded from the balloon, through a Statham P23Db transducer, on the recorder.

The removal of the heart from the rat and its mounting on the cannula was accomplished in less than two minutes. Perfusion was re-established and spontaneous contractions rapidly resumed. The heart was stimulated using silver electrodes attached to the atria, at 274 beats/minute.

Initial perfusion with KRB was established at a pumping rate to maintain perfusion pressure at 50 ± 2 mm Hg. This rate of perfusion could be maintained in control hearts for 75-90 minutes, during which time perfusion pressure rose slowly, the total rise not exceeding 12 mm Hg. At the end of this period the hearts gained an average weight of 15%. All PFOB experiments were completed in less than seventy minutes.

The experimental protocol consisted of the following (summarized in Figure 1):

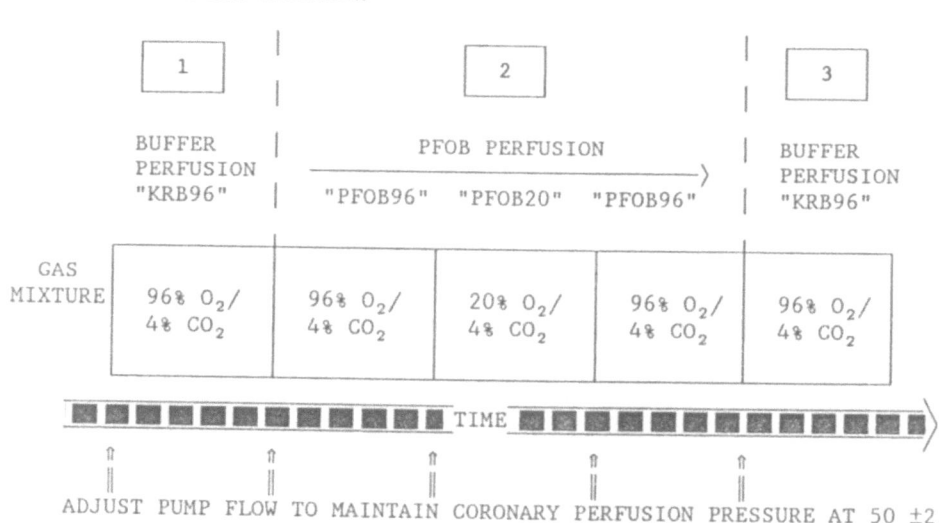

Figure 1

Schematic summary of the experimental protocol, along the time axis indicated near the bottom. At times indicated by the vertical arrows, pump perfusion flow was adjusted after changing perfusate, to maintain perfusion pressure at about 50 mm Hg. During period 1, the hearts were perfused with Krebs-Ringer-bicarbonate buffer equilibrated with $96\%O_2/4\%CO_2$ (abbreviated KRB96). During period 2, the hearts were perfused with perfluoro-octylbromide (abbreviated as PFOB) equilibrated gas mixtures as indicated. During period 3, the hearts were perfused as in period 1.

During the initial control period the heart was perfused with K.R.B. bubbled with 96% O_2/4%CO_2. After the measurements were completed, perfusion was switched to PFOB which had been bubbled with 96%O_2/4%CO_2. As soon as this perfusate reached the heart, perfusion pressure rose, indicating coronary vasoconstriction. The pump speed was reduced so that perfusion pressure returned to the previous level. Pump flow (and therefore coronary flow) was determined by timed collections of effluent. After a steady state was reached, measurements were obtained. After this, PFOB bubbled with 20%O_2/4%CO_2 was used as perfusate. This necessitated adjustment of pump flow because perfusion pressure fell somewhat. After the appropriate adjustments, a new steady state was reached and the measurements were repeated. Thereafter, perfusate was switched again to PFOB bubbled with 96%O_2/4%CO_2. Pump flow was readjusted as necessitated by the change in perfusion pressure and the measurements were repeated.

Finally, the perfusate was switched to KRB bubbled with 96%O_2/4%CO_2 and measurements were completed.

The pO_2, pCO_2 and pH of the perfusates were measured with an IL 813 blood gas analyser (Instrumentation Laboratories). pH was adjusted as appropriate by the addition of acid or base. From the pO_2, using concentrations and appropriate solubility factors, oxygen content was calculated in the perfusates.

RESULTS

The characteristics of the perfusates are summarized in Table I. The PFOB/KRB mixture bubbled with 96% O_2 contains about five times as much oxygen as KRB alone. The PFOB/KRB mixture bubbled with 20% O_2 contains 67% more oxygen than KRB alone, when the latter is bubbled with 96% O_2.

The hemodynamic variables and left ventricular performance parameters, are summarized in Tables II and III respectively. Perfusion could be maintained within 10% of the control setting by appropriate adjustment of the pump. Switching perfusate from KRB96 to PFOB96 necessitated an average 68% reduction in pump flow. Estimated coronary vascular resistance rose by an average 218%. In spite of the marked reduction of pump flow, myocardial oxygen delivery was estimated to have increased by 74%. This was associated with a marked improvement of the mechanical performance of the heart: left ventricular systolic pressure increased by 50% and the rate of pressure development (+dP/dt) increased by 38% (Table III); both of these increments were statistically significant ($p < 0.05$).

When perfusate was switched, from PFOB96 to PFOB20, there was a small fall in perfusion pressure requiring a small adjustment in pump flow; for the group, this averaged 21%. Since the oxygen content of this perfusate was only 30% of that of PFOB96, the small increase in coronary flow failed to compensate for the large decrease in perfusate oxygen concentration: oxygen delivery was reduced by 66%. This level was 59% of the oxygen delivery when KRB96 was used as perfusate. In spite of this reduction in oxygen delivery, left ventricular mechanical performance was not significantly lower than that during perfusion with KRB96, but this was reduced when compared to that seen during perfusion with PFOB96 (Table III).

Reversal of the perfusate, from PFOB20 to PFOB96 resulted in full reversal of the changes observed in the previous switch.

Table I
PERFUSATE CHARACTERISTICS
(mean±S.D.)

PERFUSATE.

	KRB96	PFOB96	PFOB20	PFOB96	KRB96
pO_2 (mm Hg)	516 ±61	481 ±186	172 ±29	562 ±14	505 ±53
pCO_2 (mm Hg)	28 ±2	27 ±3	29 ±1	29 ±1	28 ±1
pH	7.453 ±.079	7.411 ±.076	7.370 ±.018	7.362 ±.019	7.459 ±.062
estimated O_2 content (ml/dl)	1.55 ±0.17	8.26 ±0.88	2.59 ±0.44	8.26 ±0.23	1.52 ±0.18

Table I:
    Statistical summary of the characteristics (mean ± SD) of the different perfusates used in the experiments. KRB96: Krebs-Ringer-bicarbonate buffer equilibrated with $96\%O_2/4\%CO_2$ gas mixture. PFOB96 and PFOB20: perfluoro-octylbromide equilibrated with $96\%O_2/4\%CO_2$ and $20\%O_2/4\%CO_2/76\%N_2$, respectively. For further details see text.

Table II
HEMODYNAMICS

PERFUSATE

	KRB96	PFOB96	PFOB20	PFOB96	KRB96
coronary perfusion pressure (mm Hg)	50.0 ±1.2	56.3 ±1.7	56.8 ±3.0	58.2 ±1.2	52.8 ±1.5
pump flow (ml/min)	14.6 ±0.9	4.7 ±0.8	5.7 ±0.5	4.7 ±0.7	10.8 ±0.8
coronary vascular resistance (relative)	1.00	3.18 ±0.37	2.95 ±0.26	3.79 ±0.38	1.40 ±0.14
estimated O_2 delivery (ml/min)	22.3 ±2.1	38.8 ±3.6	13.2 ±0.7	39.1 ±2.8	16.9 ±1.11
increase in weight					15.3 ±2.2%

Table II:
    Statistical summary of the hemodynamic variables measured (mean ± SEM) during perfusion with different perfusates (for abbreviations, see legend to Table I). Coronary vascular resistance was calculated as the ratio of pump flow pressure and expressed as the ratio of that during the control period (:KRB96). Oxygen delivery was calculated as the product of the estimated oxygen content of the perfusate and of coronary flow.

Table III

	PERFUSATE				
	KRB96	PFOB96	PFOB20	PROB96	KRB96
L. V. Systolic pressure (mm Hg)	98.0 ±2.6	148.2 ±12.7	97.0 ±10.8	147.0 ±20.6	103.9 ±12.7
P	——> <.05	——> =.07		——> >.10	
L. V. (+dP/dt) (mm Hg/sec.)	3260 ±62	4564 ±456	2836 ±391	4433 ±880	2608 ± 815
P	——> <.05	——> =.06		——> >.15	

Table III:
  Statistical summary of the variables (mean ± SEM) characterising left ventricular performance during perfusion with the perfusates (for abbreviations, see legend to Table I.). The significance of the comparisons, by means of t-Test, are indicated by the arrows.

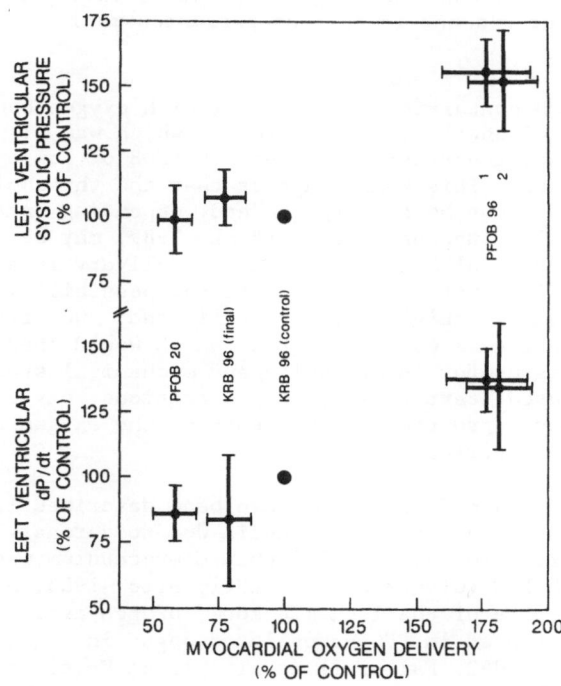

Figure 2

  Statistical summary of the change (mean ± SEM) in the variables characterizing left ventricular performance, as a function of myocardial oxygen delivery. All variables are expressed as percentage of the control value during the first perfusion with KRB96 (for abbreviations, see Table I.).

Lastly, switching perfusate, from PFOB96, to KRB96 resulted in nearly complete reversal. The pump flow required to maintain perfusion pressure was 25% less than that during the initial control period; mechanical performance was not significantly impaired.

The changes in oxygen delivery and performance are summarized in Figure 2. These suggest that dP/dt is more sensitive to myocardial oxygen delivery than left ventricular systolic pressure.

DISCUSSION

The second-generation fluorocarbon preparations exemplified by Fluosol-DA have the disadvantage that their PFC-concentration is relatively low. As a result, the amount of oxygen present in the PFC-phase is only three times that present in the aqueous phase (Biro and Blais, 1987). In contrast, the newer fluorocarbon preparations made from PFOB can be manufactured to higher PFC-concentrations. This increases the relative proportion of oxygen supplied by the PFC-phase and reduces the extent of dilution when the preparation is administered in vivo. The purpose of this study was to determine the effects of this new preparation upon the organ of largest oxygen extraction, the heart, and to assess whether such a preparation can be used as a perfusion medium for isolated hearts. The exact composition of the PFOB preparation is different from that of KRB. For this reason, the perfusate used was a 50:50% mixture of PFOB and KRB. This mixture contains about 23% PFOB by volume and still compares favourably with the PFC concentration of 11-14% (v/v) present in Fluosol-DA. As a result, the oxygen content of this perfusion medium is several times that present in aqueous perfusion media.

Using a PFOB-containing perfusate of high oxygen content, we observed a level of contractile performance which was significantly higher than that seen during perfusion with either PFOB of low oxygen content or conventional buffer. This would suggest that the threshold level of oxygen delivery may not be a sharp one and, depending on the nature of the perfusate, perfusion inhomogeneities (Beek, 1989) may be evened out. Our finding that substantially augmented oxygen delivery is accompanied by improved contractile performance suggests the possibility that buffer perfused hearts may be marginally oxygen-limited. Our findings also suggest that a perfusate containing PFC, which would thereby have greater oxygen content, is useful in metabolic and mechanical studies on isolated Langendorff-perfused hearts. In such preparations, oxygen supply could be varied at will, by appropriate adjustments of the oxygen content of the gas oxygenating the perfusate.

Several unexplained phenomena have been described in PFC-perfused hearts. In hearts with regional ischemia due to coronary occlusion, considerable short term salvage of ischemic myocardium was observed after hemodilution with PFC (Glogar et al., 1981; Biro, 1983; Nunn et al., 1983). In the centre of the ischemic zone, oxygen availability was substantially increased in PFC-hemodiluted dogs, in spite of poor blood flow (Rude et al., 1982; Faithfull et al., 1986; Vogel et al., 1989). Some of the unexplained improvements in oxygen supply observed with PFC's have been attributed to "facilitated" diffusion (Faithfull and Cain, 1988) or the ability of small PFC particles to penetrate where red cells cannot (Rude et al., 1982). While there is little direct evidence for either of these effects on a macroscopic scale, the evidence is strong that PFC's may influence oxygen supply to tissue in ways which require further investigation.

SUMMARY

Perfluoro-octylbromide-containing perfusate is capable of supporting the contractile activity of isolated hearts. Because of the high oxygen content of such perfusate, "normal" mechanical function is maintained by such hearts with substantially reduced coronary flow.

ACKNOWLEDGEMENTS

PFOB was generously donated by Dr. D. Long, Fluoromed Pharmaceutical Inc, La Mesa, Ca. Financial support for some of these investigations was provided by the Medical Research Council of Canada to B. Korecky.

REFERENCES

Beek, J. H. G. M. and Westerhof, N., 1989. Mismatch of regional myocardial metabolism and perfusion is a possible cause of tissue hypoxia. This issue.

Biro, G. P. and Blais, P. 1987. Perfluorocarbon blood substitutes. C. R. C. Crit. Rev. Oncol. Hematol. 6:311.

Biro, G. P., 1983. Fluorocarbon and dextran hemodilution in myocardial ischemia. Can. J. Surg., 26:163.

Deutschmann, W., Linder, E., and Deutschlander, N., 1984. Perfluorochemical perfusion of the isolated guinea pig heart. Pharmacology, 28:336.

Glogar, D. H., Kloner, R. A., Muller, J., DeBoer, L. W. V., and Braunwald, E., 1981. Fluorocarbons reduce ischemic damage after coronary occlusion. Science, 211:1439.

Faithfull, N. S., Erdman, W., Fennema, M., and Kok, A. 1986. The effects of hemodilution with fluorocarbons or dextran on oxygen tensions in the acutely ischeamic myocardium. Brit. J. Anaesth. 58:1031

Faithfull, N. S. and Cain S. M., 1988. Cardiorespiratory consequences of fluorocarbon reactions in dogs. Biomat. Art. Cells. and Art. Organs 16:643.

Nunn, G. R., Dance, G., Peters, J., and Cohn, L. H., 1983. Effect of fluorocarbon exchange transfusion on myocardial infarct size in dogs. Am. J. Cardiol., 52:203.

Rude, R. E., Glogar, D., Khuri, S. F., Kloner, R. F., Karaffa, S., Muller, J. E., Clark, L. C. jr., and Braunwald, E. 1982., Effects of intravenous fluorocarbons during and without oxygen enhancement on acute myocardial ischemic injury assessed by intramyocardial gas tensions. Am. Heart J., 103:986.

Segal, L. D. and Ensunsa, J. L., 1988. Albumin improves the stability and longevity of perfluorochemical-perfused hearts. Am. J. Physiol., 254:H1105.

515

Segal, L. D. and Rendig, S. V., 1982. Isolated working rat heart perfusion with perfluorochemical emulsion. Fluosol-43. <u>Am. J. Physiol.</u>, 242 (<u>Heart Circ. Physiol.</u> 11):H485.

Tomera, J. F. and Geyer, r. P., 1982. Perfluorochemical perfusion of the rat heart: in vivo and in vitro. <u>J. Molec. Cell. Cardiol.</u>, 14:573.

Vogel, H., Gunther, H., Harrison, D. K., Anderer, W., Kessler, M., and Peter, K., 1989. Hemodilution and myocardial oxygen supply. The influence of Fluosol-DA. <u>This issue</u>.

THE EFFECT OF IABP VENTRICULAR CONTRACTILITY OF THE NORMAL AND ISCHEMIC CANINE HEART ASSESSED IN SITU BY T-Emax

Chen Jie, Jiang Yisheng*, Fang Fuzen, Zeng Jun,
Zhao Min, Zhang Weilian* and Zhang Lifan

Laboratory of Applied and Systems Physiology and
*Department of Cardiovasvular Surgery, The Fourth
Military Medical University, Xi'an, China

Intra-aortic balloon pumping (IABP) has been shown to be effective in supporting the failing circulation through two alternating hemodynamic effects, i.e., a diastolic augmentation, and a reduction of the impedance to left ventricular ejection. These effects improve the myocardial oxygen supply/demand ratio and thereby increase cardiac performance (Norman and Igo, 1985). Of the three determinants, i.e., preload, after-load and contractile state, only the last reflects the intrinsic performance of the myocardium. For a better understanding of the effectiveness and the physiological mechanism of IABP, the effects of IABP on the contractile state of the in situ heart must be fully elucidated. It is not surprising that there have been some confusing and contradictory reports from previous studies (Mullins et al., 1971; Rose et al. 1979; Grover et al., 1979; Norman an Igo, 1985), since most of the indices used to assess the contractile state during IABP are described as load- and heart rate-dependent (Munch and Downey, 1981; Norman and Igo, 1985). However, the slope (Emax) of the End-Systolic Pressure Volume Relation (ESPVR) was proposed as an index of ventricular contractility, being independent of preload and after load (Suga et al., 1973; Suga and Sagawa, 1974; Sagawa, 1978). In addition, the transient slope of the end-systolic pressure-volume line (T-Emax) was reported to be a useful index for assessment of left ventricular contractility of the in situ canine heart (Igarashi and Suga, 1986). To our knowledge, T-Emax has not been used so far to evaluate the effect of IABP on either normal or ischemic heart. The present study was undertaken to elucidate the effect of IABP on T-Emax of both normal and ischemic canine heart in situ. For comparison, we also investigated simultaneously the changes of the maximum rate of rise of left ventricular pressure (dP/dtmax), an isovolumic phase index of contractility commonly used by previous authors (Mullins et al., 1971,; Grover et al., 1979; Okada and Nakamura, 1986) to assess contractility changes caused by IABP.

METHODS

Surgical Preparation and Physiological Measurements   A total of 16 mongrel dogs, ranging in weight from 11 to 20 kg (mean 14.5 $\pm$ 2.9 (SD) kg) were anesthetized with intravenously administered sodium pentobarbital (30 mg/kg) and ventilated through an endotracheal tube by a positive pressure respirator with room air. A left thoracotomy was performed and the heart was suspended in a pericardial cradle. Left ventricular pressure was measured through a polyethylene cannula inserted via a stab wound in the ventricular apex and connected to a Statham pressure transducer(P 23 ID). The first derivative of left ventricular pressure, dP/dt, was obtained with an analog pressure pulse wave differentiating unit (ED 601 G, Nihon Kohden, Japan) connected to the output of the pressure channel. Aortic pressure was measured by means of a polyethylene cannula inserted via the left subclavian artery into the aortic arch and connected to another Statham pressure transducer (P 23 ID). Aortic flow was measured by an electromagnetic flow probe placed around the root of the aorta and connected to a flowmeter (MFV-1200, Nihon Kohden). The left anterior descending coronary artery was dissected free, just beyond the 1st diagonal branch, for later ligation to induce regional myocardial ischemia. The intra-aortic balloon for counterpulsation was inserted via the left external iliac artery and advanced to a position just distal to the origin of the left subclavian artery. A polyethylene catheter was inserted into the femoral vein for administration of fluid and drugs as needed. The analog outputs were appropriately calibrated, recorded on a multichannel thermal pen recorder (Polygraph RM-6000, Nihon Kohden) and stored on magnetic tapes.

IABP Device   Two types of balloon were used: (1) single balloons, 13 ml in volume and 1 cm in diameter, and (2) dual-chambered balloons, 17 to 21 ml in volume and 1 cm in diameter for the working balloon and 1.3 cm for the occlusion balloon. The intra-aortic balloons were connected to a cardiac assist console (Datascope System 83, Datascope, Paramus, NJ, USA), which monitored aortic pressure and a standard lead II electrocardiogram, either of which could be used to trigger inflation of the intra-aortic balloon just at the beginning of diastole, and deflation of the balloon just before systole.

Aortic Occlusion   For each T-Emax determination, the ascending aorta between the flow probe and the brachiocephalic artery was abruptly occluded in diastole with an aortic snare, after steady-state ejecting contractions, to produce an isovolumic contraction. The reason for using an aortic snare, instead of vascular forceps was to avoid any interference with the stability of the electromagnetic probe from metal forceps. The criteria we adopted for a successful isovolumic contraction were: (1) no ejection in the aortic flow and its time-integral tracings, (2) smooth sinusoidal left ventricular pressure and (3) a monotonic fall of aortic pressure and therefore, in the case of IABP, an interruption in balloon pumping if the aortic pressure signal was being used to trigger the counterpulsation (Fig. 1). Each occlusion was performed when the respirator had been stopped in an expiratory phase for about 10 sec. Hemodynamic changes before, during and after each successful isovolumic contraction were recorded and stored for data analysis. For further details, see Igarashi and Suga, 1986.

Experimental Protocol   Two groups of animals were studied.
    Group 1.   Six dogs underwent regional myocardial ischemia induced by ligation of the left anterior descending coronary artery without IABP treatment. Briefly, after completion of the surgical preparation, 15 min were allowed for hemodynamic stabilization. Control measurements were then obtained, and the coronary artery branch was ligated. Repeated measurements were made at 30 and 60 min after ligation. To prevent ventricular arrhythmias, a bolus of lidocaine (4 mg/kg) was given before the ligation and a lidocaine infusion (1 mg/min) was continued during the period of coronary artery ligation.
    Group 2.   In ten dogs, measurements of hemodynamic changes resulting from IABP during the control state and in the presence of myocardial is-

chemia were carried out. First the animals underwent IABP for 15 min and the measurements were made before and after 15 min of IABP. Then the IABP was gradually reduced from balloon pumping with every beat to every other beat,

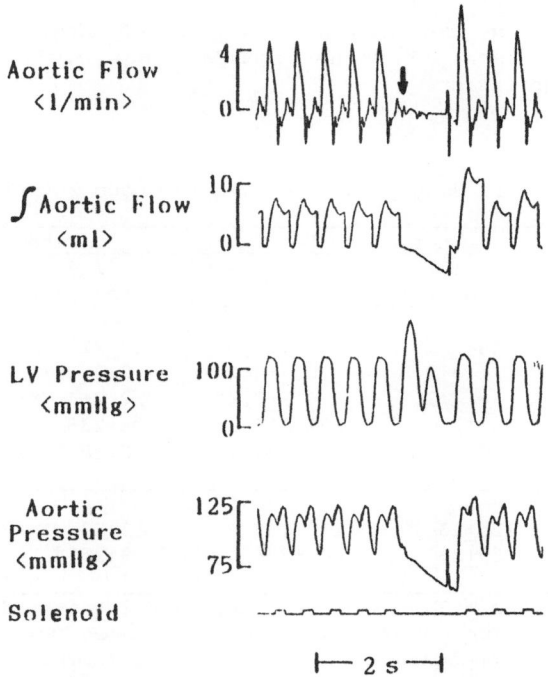

Aortic Flow
<l/min>

∫Aortic Flow
<ml>

LV Pressure
<mmHg>

Aortic
Pressure
<mmHg>

Solenoid

⊢— 2 s —⊣

Fig. 1.   An example of tracings of steady-state ejecting contractions and an interposed isovolumic contraction during IABP. The arrow indicates the onset of aortic occlusion. Note the smooth sinusoidal left ventricular pressure wave, the monotonic fall of aortic pressure, the stopped aortic flow (the slightly negative aortic flow and the time-integrated aortic flow as shown in the tracings was caused by the baseline shift which was corrected during computation), and an absence of IABP during isovolumic contraction.

then to every third beat, and finally it was stopped. After evaluation of the effects of counterpulsation in the control state, 15 min was allowed for the recovery of hemodynamics. Thereafter, regional ischemia was induced by ligation of the coronary artery and the IABP was reactivated 30 min later. Measurements were repeated immediately before and at 30 min of counterpulsation in the presence of acute ischemia. Countermeasures taken to prevent ventricular arrhythmias were the same as in Group 1.

After each experiment, the left ventricle including the interventricular septum was weighed. The left ventricle was 73 ± 14 (SD) g, ranging from 45 to 98 g.

Data Analysis   The stored signals were analyzed with a microcomputer
(IBM PC/XT). We digitalized the reproduced signals at a sampling

frequency of 250 Hz, located the characteristic points with an on-screen graphic expanding and cursor moving technique and the required data were calculated (Table 1). For T-Emax computation, the reproduced signals were digitized at 500 Hz and filtered by a 5-point moving averaging method. We programmed the computer to determine the transient end-systolic pressure-volume line and calculate its slope (T-Emax) (Fig. 2). For further details, see Igarashi and Suga 1986.

<u>Statistics</u>   Data are presented as mean $\pm$ SD. Student's t-test was used for paired comparison. P values smaller than 0.05 were considered statistically significant.

Table 1   Hemodynamic changes in Group 2 (n=10)

	normal state		ischemic state	
	1st control	1st IABP	2nd control	2nd IABP
HR (beat/min)	164 $\pm$ 19	162$\pm$19	147$\pm$22**	144$\pm$24
$\int$AoF (ml/beat)	11.3$\pm$2.8	10.7$\pm$4.2	9.3$\pm$3.1*	9.9 $\pm$3.1
LVEDP (mmHg)	6.4$\pm$1.3	5.8$\pm$1.4*	12.9$\pm$4.7**	10.2$\pm$4.7''
SP (mmHg)	135$\pm$14	122$\pm$16*	122$\pm$23*	109$\pm$24''
EVR	0.90$\pm$0.10	1.07$\pm$0.13**	0.88$\pm$0.12	1.05$\pm$0.12''
ESPe (mmHg)	128$\pm$13	119$\pm$14*	115$\pm$27	98$\pm$26''
ESPi (mmHg)	259$\pm$29	242$\pm$27*	179$\pm$26**	166$\pm$26''

(1) Values are means $\pm$ SD. HR=heart rate; $\int$AoF=integrated aortic flow; LVEDP=left ventricular end diastolic pressure;   SP=systolic pressure;   EVR=endocardial viability ratio (Norman and Igo, 1985); ESP=left ventricular end-systolic pressure for determining T-Emax; e=ejecting contraction; i=isovolumic contraction.
(2) Different from 1st control: *, P<0.05;   **, P<0.01. different from 2nd control: '. P<0.05; '', P<0.01.

RESULTS
     Table 1 summarizes the hemodynamic changes in group 2 animals. Differences between the parameters of the first control run and those of the second control run illustrate the changes resulting from acute myocardial ischemia. With the onset of ischemia, heart rate, integrated aortic flow, and systolic pressure decreased significantly; while left ventricular end diastolic pressure increased remarkably. These changes were shown by the paired t-test to be either statistically highly significant (P<0.01), or significant (P<0.05).

As illustrated in Table 1, the hemodynamic effects of IABP in both the normal state and ischemia show a similar trend. With the onset of IABP, LVEDP and SP decreased (normal state, P<0.05; ischemic state, P<0.01), endocardial viability ratio (EVR) increased significantly (in both states, P<0.01), indicating that IABP is effective in unloading the ventricle and improving the myocardial oxygen supply and demand relation in both normal and ischemic conditions. There were no significant changes in heart rate or integrated aortic flow during IABP in either state.

Fig. 2, presents four sets of pressure-volume trajectories for the last ejecting contraction and the first isovolumic contraction obtained during four runs, namely: first control, IABP, second control and ischemia with IABP. It shows clearly the changes of T-Emax (i.e., transient slope of the end-systolic pressure-volume line) resulting from regional ischemia and IABP under different conditions. The relevant data needed for calculating the T-Emax value for group 2, i.e. end-systolic pressure of the last ejecting contraction (ESPe) and ESP of the isovolumic contraction (ESPi) are included in Table 1.

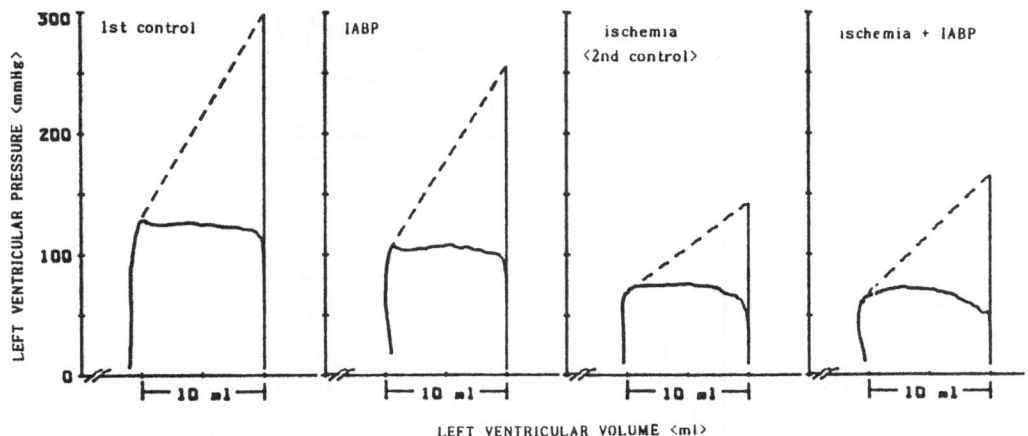

Fig. 2. An example of pressure-volume trajectories for the last ejecting contraction and the first isovolumic contraction showing the effects of regional ischemia and IABP. The broken line is the end-systolic pressure-volume line. Its slope is the transient Emax, T-Emax.

Fig. 3 illustrates how, in group 1 animals, T-Emax and dP/dtmax changed with time after ligation of the left anterior descending coronary

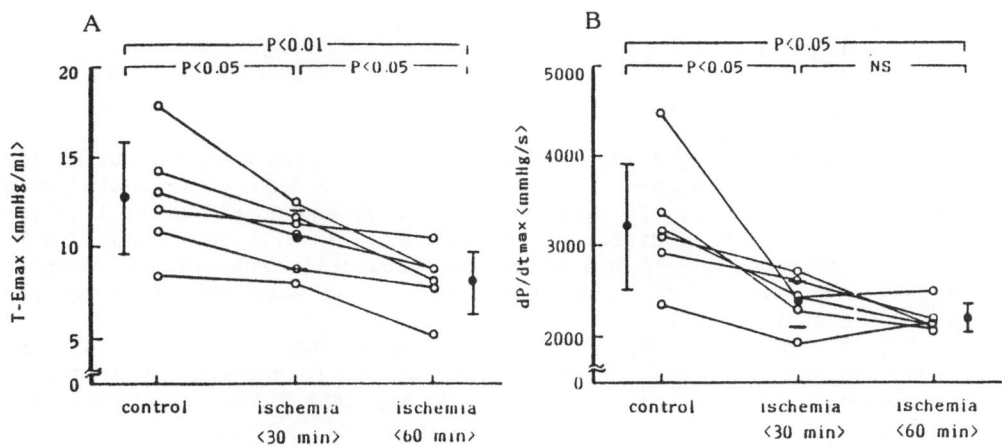

Fig. 3. Plots of grouped data of T-Emax (A) and dP/dtmax (B) for group 1.
Each point represents group mean. Vertical bars represent SD.
See text for details.

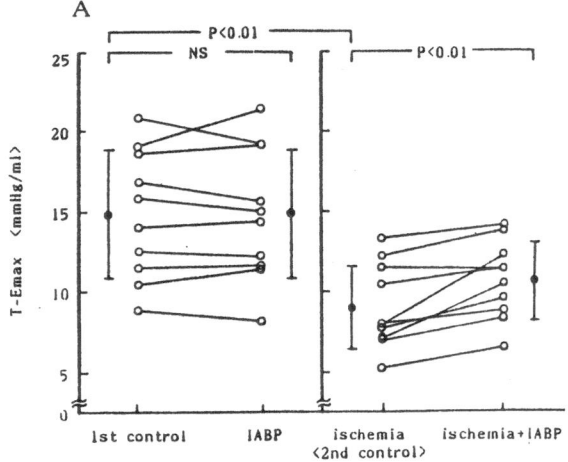

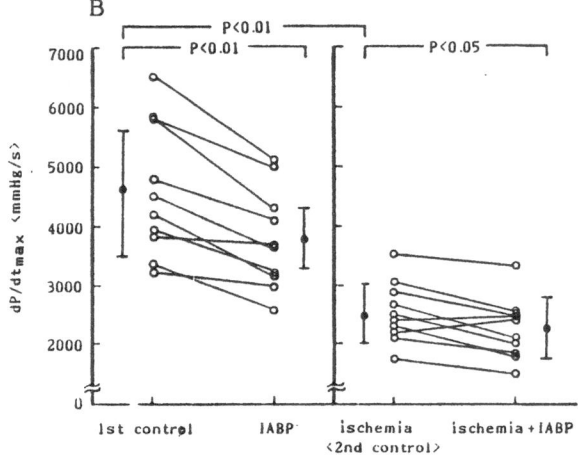

Fig.4. Plots of grouped data of T-Emax (A) and dP/dtmax (B)
for group 2.
Note: see Fig. 3.

artery. As shown in Fig. 3, A, T-Emax decreased progressively during the
60 min period of ischemia. However, while dP/dtmax, decreased in a
similar way to T-Emax, the date obtained at 30 and 60 min after ligation
were not significantly different (P>0.05, Fig. 3, B). The changes in T-
Emax and dP/dtmax at 30 min of acute myocardial ischemia for group 2
(Fig.4) were similar to those for group 1 and were confirmed by the
paired t-test (P<0.01).

Fig. 4 depicts the changes of T-Emax and dP/dtmax induced by IABP intervention and shows how the changes in the two parameters deviated from
each other. As shown in Fig. 4, A, during IABP T-Emax showed no significant changes (P>0.05) in the normal state and increased significantly
(P<0.01) in the ischemic state. In contrast, as Fig.4, B shows, dP/dtmax
decreased significantly during IABP in both normal (P<0.01) and ischemic
(P<0.05) states.

DISCUSSION

Two important aspects of this work should be emphasized. First, taking T-Emax as a sensitive index for assessment of the ventricular contractile state of the in situ heart, we have examined the changes of T-Emax induced by IABP in both normal and ischemic canine heart. Second, we have shown that· T-Emax changes reflect more properly the changes of contractility during IABP in both conditions as predicted by previous works (Urschel et al., 1970; Norman and Igo, 1985), suggesting the positive inotropic effect of IABP acts only on the ischemic heart. On the contrary, owing to its load-dependence (Mason et al., 1969), dP/dtmax will usually underestimate the myocardial contractility during IABP.

Previous authors have proposed the slope (Emax) of the end-systolic pressure-volume relation (ESPVR) as an index of myocardial contractility, in view of its almost complete independence of preload, afterload and heart rate (Suga et al, 1973; Suga and Sagawa, 1974; Sagawa, 1978). However, it is difficult to maintain constant contractility while producing variably loaded steady-state contractions and to measure the absolute volume changes accurately in an in situ ventricle. To circumvent these difficulties and extend the applicability of Emax to the in situ heart, a new method for quick determination of the transient Emax (T-Emax) by an abrupt aortic occlusion in the in situ canine heart has been established and examined with both positive and negative inotropic agents (Igarashi and suga, 1986). Although this new method has some advantages over previous methods, a few problems still remain. For instance, it has been shown in isolated canine left ventricle preparations that T-Emax is influenced by the ejection fraction and contractile state when conventional Emax is assessed by T-Emax (Igarashi et al., 1987). It is likely that the situation is more complicated in the in situ heart. Nevertheless, our results are of special interest, since there is no direct information concerning the applicability of T-Emax (or Emax) for the assessment of the changes of contractile state resulting from any mechanical circulatory assist system.

We have shown that in the in situ canine heart, during a 60 min period of regional ischemia resulting from coronary artery occlusion, T-Emax decreased by 32 % at 30 min and by 36 % at 60 min. However, our results with T-Emax are not compatible with the findings reported by Sunagawa et al. (1983) and Little et al. (1987). They have shown that the major effect of acute regional ischemia on the ESPVR in the physiological pressure range is a parallel shift to the right, i.e., a shift in the volume axis intercept without a change in the slope, Emax; this is in striking contrast to the effect of global ischemia under which only Emax is changed without a substantial change in the intercept. As has been pointed out by Sunagawa et al. (1983) in in situ conditions, the left ventricle is ejecting under the complex influence of the neurohumoral control of contractility, variable loads, direct mechanical effects of adjacent organs and perfusing conditions of peripheral organs. Their results cannot be directly extended to and compared with those from our experiments in which the period of ischemia was longer (up to 60 min vs. 3-5 min in the works of Sunagawa et al., 1983 and Little et al., 1987); the lidocaine infusion (Côté et al., 1973) may also have contributed to the difference mentioned above. Furthermore, when T-Emax is used to represent the slope of the ESPVR, T-Emax does not reflect any changes in the intercept if they do occur. Although it is impossible at the moment to elucidate thoroughly the physiological meaning of the decrease in T-Emax during regional ischemia observed in the present work, we still consider that the decrease in T-Emax basically reflects a decrease in overall ventricular elastance in which the reduced contractility of the non-ischemic region may play an important role. This is supported by the fact that T-Emax decreased progressively with time during the 60 min ischemic period and reversed as IABP was activated.

The intra-aortic balloon pumping device is an intravascular volume-displacement apparatus in series with the heart and has been widely used in patients with heart failure. Activation of the IABP results in a decrease in myocardial oxygen demand (reduced preload and afterload), and increase in myocardial oxygen supply and increase in external cardiac performance (Norman and Igo 1985).However, the concomitant effects on left ventricular performance are still not well understood (Rose et al., 1979). We attribute the conflicting results to the fact that isovolumic and ejection phase indices (such as dP/dtmax), are sensitive to loading conditions and thus unable to reflect the contractile state properly. In the present study, adopting the sensitive index of contractility, T-Emax, to assess conventional Emax, we have demonstrated in the in situ canine heart that, during the activation of IABP, there was no significant change of T-Emax in the normal state and a significant increase by 18 % in T-Emax in the ischemic state. Our results suggest that the concomitant effects of IABP on the myocardial contractile state are dependent on the basal state of the in situ heart, and cause an increase in contractility only in the ischemic heart. A similar suggestion was made previously by Urschel et al. (1970) and Norman and Igo (1985). In contrast, the loading-dependent index of contractility, dP/dtmax, fell by 18 % and 10 % during IABP in the normal and ischemic states, respectively. This may indicate an underestimate of contractility during IABP. Because of the limitations of the index, T-Emax, this preliminary result showing the positive inotropic influence of IABP on the in situ ischemic canine heart should be further investigated with more sophisticated methods.

SUMMARY

The effects of IABP on the contractile state of the in situ canine heart was evaluated under control conditions and conditions of regional ischemia by determining the transient slope of the end-systolic pressure-volume line, T-Emax. For comparison, the conventional index reflecting contractility, dP/dtmax and other related hemodynamic parameters were also assessed. The results showed that, T-Emax, rather than dP/dtmax, reflected more appropriately the changes in contractile state during activation of the IABP under both conditions . It suggests that IABP has a positive inotropic influence only in the ischemic state and this is a result of the improvement in the myocardial oxygen supply and demand relation. In addition, the limitations of the new indicator, T-Emax, is discussed.   .

REFERENCES

Côté P, Basile J and Schroeder J S (1973) hemodynamic interaction of procainamide and lidocaine after experimental myocardial infarction. Am J Cardiol 32:937.

Grover F L, Fewel J G, Joseph V, Ghidoni J J, Arom K V, Norton J B and Trinkle J K (1979) Effects of aortic balloon pumping during cardiopulmonary bypass on myocardial perfusion, metabolism and contractility. Chest 75:37.

Igarashi Y and Suga H (1986) Assessment of slope of end-systolic pressure- volume line of in situ dog heart. Am J Physiol 250:H685

Igarashi Y, Goto Y, Yamada O, Ishii T and Suga H (1987) Transient vs. steady end-systolic pressure-volume relation in dog left ventricle. Am J Physiol 252:H998.

Little W C, Park R C and Freeman G L (1987) Effects of regional ischemia and ventricular pacing on LV dP/dtmax-end-diastolic volume relation. Am J Physiol 252:H933.

Mason D T (1969) Usefulness and limitations of rate of rise of intraventricular pressure (dP/dt) in the evaluation of myocardial contractility in man. Am J Cardiol 23:516

Mullins C B, Sugg W L, Kennelly B M, Jones D C and Mitchell J H (1971) Effect of arterial counterpulsation on left ventricular volume and pressure. Am J Physiol 220:694.

Munch D F and Downey J M (1981) Regulation of Myocardial contractility. in: Cardiac Pharmacology, pp 3, Academic Press.

Norman J C and Igo S R (1985) Mechanical circulatory assistance: established (IABP) and evolving (LVAD). A narrative summary. Thorac Cardiovasc Surgeon. 33:133.

Okada M and Nakamura K (1986) Intraaortic balloon pumping - current status of IABP and artificial heart in Japan. Jpn J Thorac Surg 39:172.

Rose E A, Marrin C A S, Bregman D and Spotnize H M (1979) Left ventricular mechanics of counterpulsation and left heart bypass, individually and in combination. J Thorac Cardiovasc Surg 77:127.

Sagawa K (1978) The ventricular pressure-volume diagram revisited. Cir Res 43:677.

Suga H, Sagawa K and Shoukas A A (1973) Load independence of the instantaneous pressure-volume ratio of canine left ventricle and effects of epinephrine and heart rate on the ratio. Cir Res 32:314.

Suga H and Sagawa K (1974) Instantaneous pressure-volume relationships and their ratio in the excised, supported canine left ventricle. Cir. Res. 35: 117.

Sunagawa K, Maughan W L and Sagawa K (1983) Effect of regional ischemia on the left ventricular end-systolic pressure-volume relationship of isolated canine hearts. Cir Res 52:170.

Urschel C W, Eber L, Forrester J, Matloff J, Carpenter R and Sonnenblick E (1970) Alteration of mechanical performance of the ventricle by intraaortic balloon counterpulsation. Am J Cardiol 25:546.

THE INFLUENCE OF DIFFERENT ANESTHETICS ON THE OXYGEN DELIVERY TO AND CONSUMPTION OF THE HEART

G.J.van Daal, B.Lachmann, W.Schairer, R.Tenbrinck,
L.J.van Woerkens#, P.Verdouw#, and W.Erdmann

Depts. of Anesthesiology and Experimental Cardiology
Thorax centre #, Erasmus University, Postbus 1738
3000 DR Rotterdam, The Netherlands

Introduction

It is well known that some routinely used anesthetics, e.g. halothane, have a severely depressive effect on the cardiovascular system (1,2,3). These depressive effects include: decreased myocardial contractility, decreased systolic as well as diastolic blood pressure as well as a decreased cardiac output. Studies in cell cultures demonstrate, that halothane reduces oxygen consumption and contraction rate of the heart myocyt (4) and affects the uptake and availability of Ca2+ (5,6).

It is not yet known whether the effects of halothane in vivo are caused by a decreased oxygen consumption or oxygen uptake from the blood by the heart myocyt or by a decreased delivery of oxygen to the myocard. Therefore the present study was carried out to evaluate the effect of halothane-anesthesia in comparison to 70% nitrous oxide and the combination of pentothal with fentanyl on hemodynamics, oxygen delivery to the myocard and oxygen consumption of the myocard. Furthermore we were interested in getting information on local perfusion of different regions of the myocard during different methods of anesthesia.

Materials and methods

The experiments were performed in 7 young Yorkshire pigs (24-26 kg). After an overnight fast the pigs were sedated with 120 mg azaperidone i.m. and anesthesised with 150 mg metomidate i.v.. After intubation the animals were ventilated (Servo ventilator 900C, Siemens Elema, Sweden) with the following mixtures: $O_2:N_2O$ (1:2), $O_2:N_2O$ (1:2) and 1% halothane. Volume and frequency were set, that an arterial PCO_2 between 35 and 45 mm Hg resulted (ABL-3, Radiometer, Copenhagen). Catheters were placed in the superior caval vein and in the descending aorta. Then the animals were thoracotomised and Millar microtipped catheters (Millar Instruments Houston, Houston, Texas, USA) were placed in the left ventricle and aorta for measuring blood pressures. Ascending aortic blood flow was measured with an electromagnetic flowprobe (Skalar, Delft, The Netherlands) around the vessel. The left atrial appendage was catheterized for the injection of microspheres.Cardiac output was derived by adding myocardial blood flow (measured with radioactive microspheres; see later) to ascending aortic blood flow. Oxygen saturation and haemoglobin concentration were determined in blood samples witdrawn from the abdominal aorta and the great cardiac vein (OSM2, Radiometer, Copenhagen, Denmark).

Oxygen delivery and consumption

Left ventricular (LV) oxygen consumption was calculated by multiplying the difference between the aortic oxygen content and that of the great cardiac vein by LV-myocardial blood flow. Oxygen delivery to the left ventricle was calculated by multiplying the aortic oxygen content per milliliter and LV-myocardial blood flow per 100g myocard per minute.

Regional blood flow

Regional perfusion of the various parts of the myocard was determined by the radioactive microsphere method (7). Microspheres were injected in random order via a cannula inserted into the left atrial appendage. To calibrate flow measurements, an arterial reference blood sample was withdrawn (10 ml/min) starting 10 s before and continuing until 1 min. after completion of each microsphere injection. At the end of each experiment the animal was killed and the various tissues of the heart were dissected out, weighed and placed in plastic vials for counting radioactivity. Data were processed by use of a set of computer programmes described elsewhere (8).

Data presentation and statistical analysis

All data in the text and illustrations are presented as means + standard error of mean. For statistical analysis the Student-Newman Keuls test was applied once an analysis of variance had revealed that the samples represented different populations (9). Statistical significance was accepted at $P < 0.05$ (two-tailed).

RESULTS

Systemic haemodynamics

Application of 1% halothane in the inhalation mixture reduced systolic blood pressure by 32%, diastolic blood pressure by 42%, heartrate by 18% (Fig. 1), cardiac output by 27% (Fig. 2) and contractility (LVdP/dt) by 62% (Fig. 3) in comparison to 70% nitrous oxide. Halothane reduced mean arterial blood pressure was by 37% (Fig. 4), while in the same time systemic vascular resistance was reduced by 14% (Fig. 5) in comparison to 70% nitrous oxide.

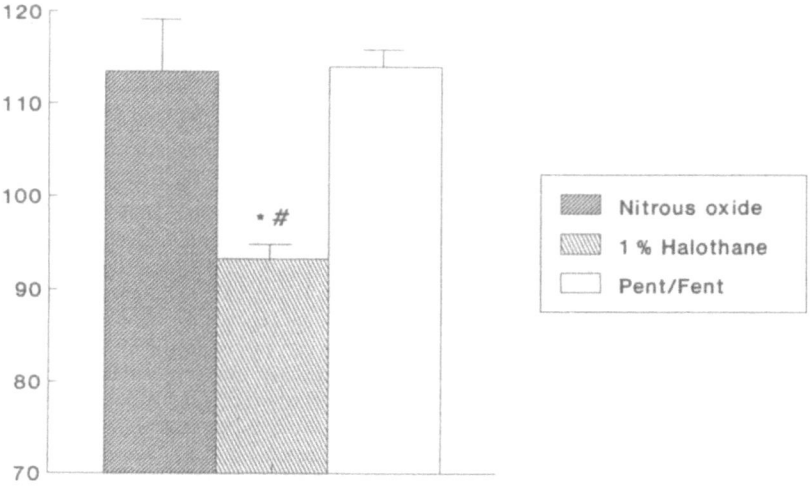

Fig. 1. Heart rate in beats.min^{-1}.

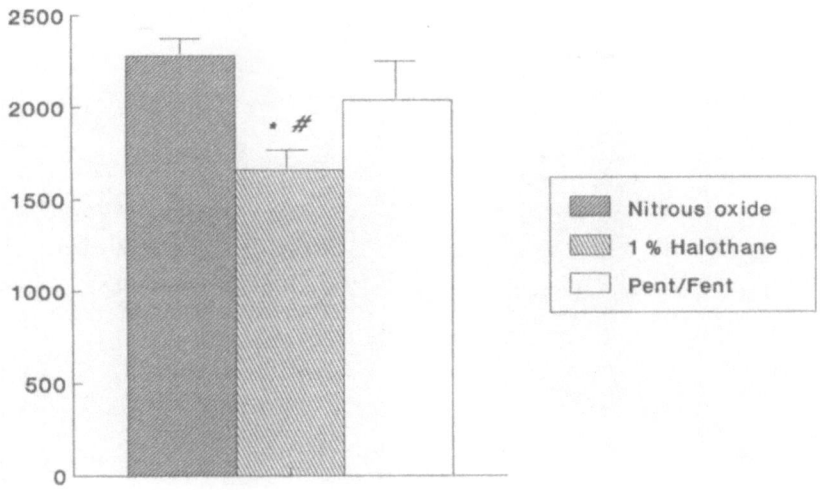

Fig. 2. Cardiac output in ml.min^{-1}

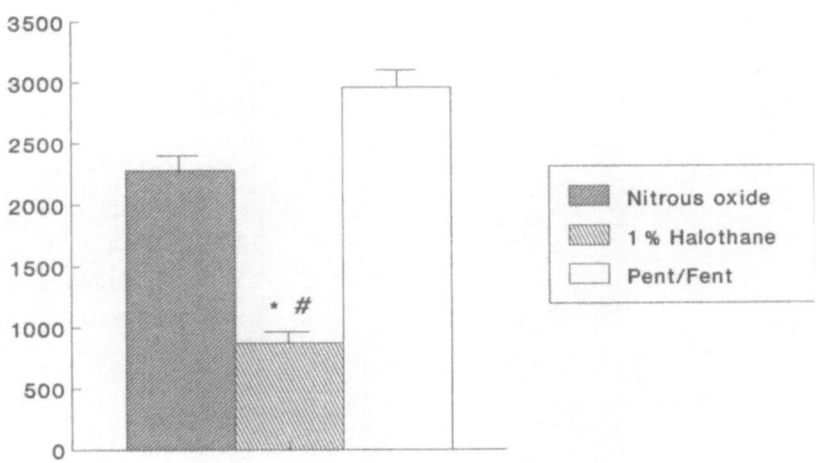

Fig. 3. Left ventricular contractility (dP/dt) in mm Hg.s^{-1}

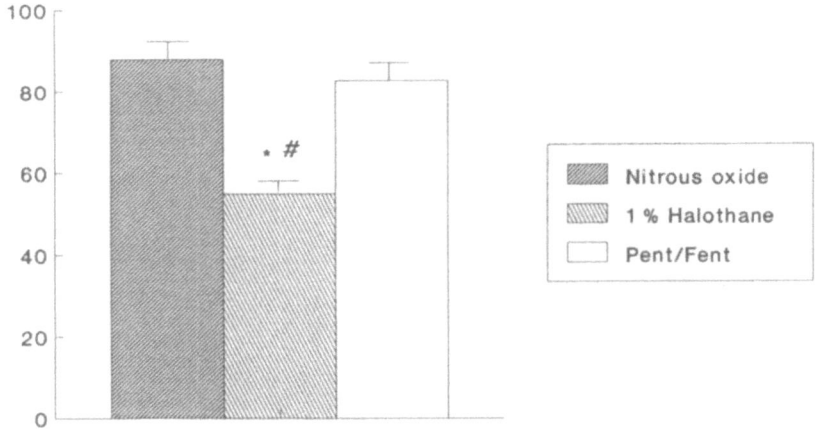

Fig. 4. Mean arterial blood pressure in mm Hg.

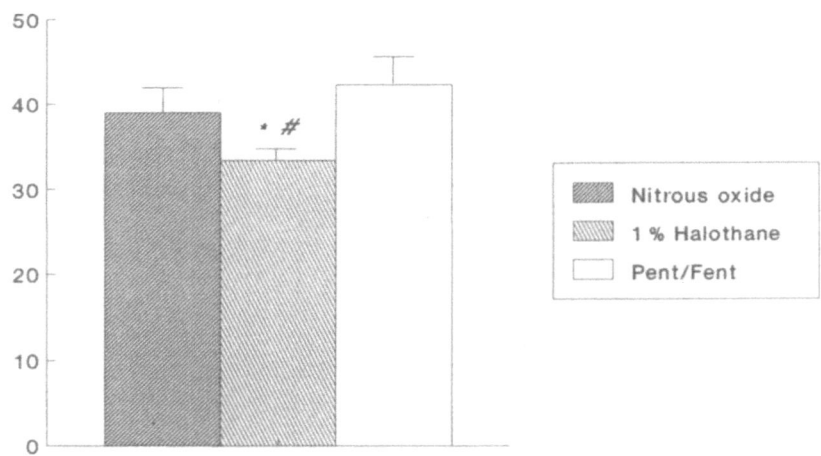

Fig. 5. Total peripheral resistance in mm Hg.l^{-1}.min.

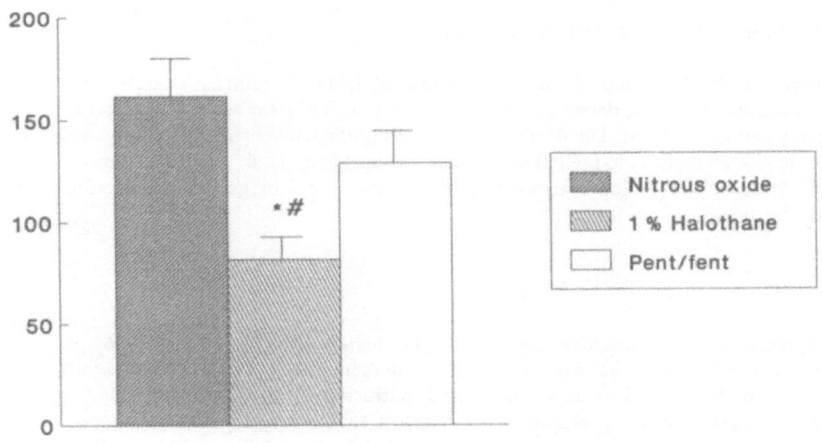

Fig. 6. Left ventricular perfusion in ml.min^{-1}.100g^{-1}.

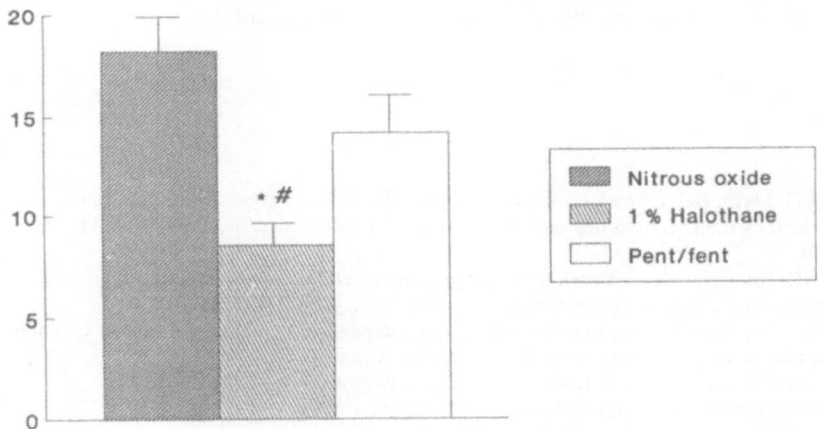

Fig. 7. Oxygen delivery to the left ventricle in ml.min^{-1}.100g^{-1}.

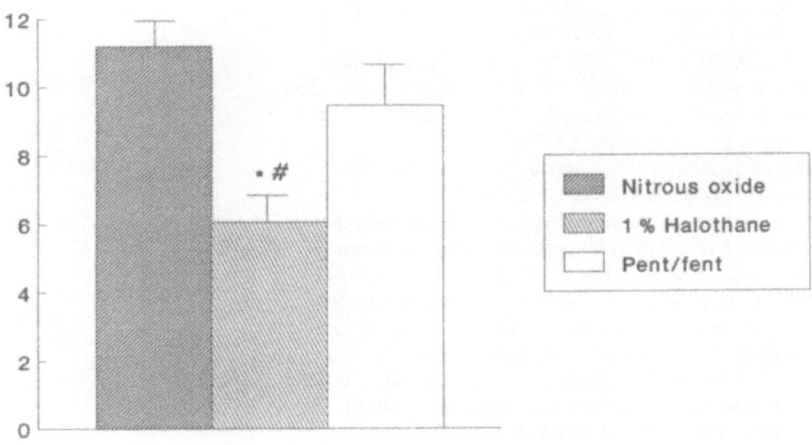

Fig. 8. Oxygen consumption by the left ventricle in ml.min-1.100g-1.

Regional perfusion of the heart and oxygen consumption

When compared with the results obtained with penthotal/fentanyl or nitrous oxide alone, halothane causes a significant decrease in the perfusion (50%) of total heart. This effect is seen in every part of the heart. The decreased perfusion of the left ventricle (Fig. 6) caused a decrease in oxygen delivery to the left ventricular myocard (Fig. 7). In parallel to the decreased oxygen delivery to the heart we also found a decrease in oxygen consumption of the left ventricular myocard (Fig. 8).

DISCUSSION

Several explanations for the negative inotropic effect of halothane, were reported in the literature. E.g. it is discussed that halothane dissolves in membranes of heart cells and thus effects Ca^{2+} availability which leads to a decreased contractility (4,10). Our results demonstrate further a reduced perfusion of the myocard, followed by a decreased oxygen delivery to the heart, in comparison to N_2O and pentothal/fentanyl. This might imply that the oxygen delivery is a limiting factor in cardiac metabolism during halothane anesthesia in swine. That means that an important factor for the cardiodepressive effect of halothane is the decreased perfusion of the left ventricle.

From these results we conclude, that in patients, who are at risk for developing myocardial ischeamia, the usage of halothane should be avoided to prevent myocardial infarction.

References

1. Sontagg H, Merin RG, Donath H, Radke J, Schenk HD. (1976) Myocardial function and metabolism in the conscious dog and during halothane anesthesia. Anesthesiology.: 51: 204-210.
2. Merin RG, Kumazawa T, Luka HL. (1979) Myocardial metabolism and oxygenation in man awake and during halothane anesthesia. Anesthesiology.: 44: 402-415.
3. Merin RG, Verdouw PD, De Jong JW. (1977) Dose-dependent depression of cardiac function and metabolism by halothane in swine (Sus scrofa). Anesthesiology 46: 417-423.
4. Albrecht RF, Miletich DJ, Dinsmore P. (1985). Comparative metabolic effects of halothane and enflurane in rat heart cell culture. J. Cardiovasc. Pharmacol.: 7: 799-804.
5. Komai H, Rusy BF. (1987). Negative inotropic effects of isoflurane and halothane in rabbit papillary muscles. Anesth. Analg.: 66: 29-33.
6. Malinconico SM, Hartzell CR, McCarl RL. (1983). Effect of calcium on halothane-depressed beating in heart cells in culture. Mol. Pharmacol.: 23: 417-423.
7. Saxena PR, Verdouw PD (1985). 5-Carboxamidetryptamine, a compound with high affinity for 5-hydroxytryptamine binding sites, dilates arterioles and constricts arteriovenous anastomoses. Br. J. Pharmac.: 84: 533-544.
8. Saxena PR, Schamhardt HC, Forsyth RP, Loeve J. (1980). Computer programs for the radioactive microsphere technique. Determination of regional blood flows and other haemodynamic variables in different experimental circumstances. Comp. Prog. Biomed.: 12: 63-84.
9. Saxena PR. (1985). An interactive computer programme for data management and parametric and non-parametric statistical analysis. Proc. Br. Pharmacol. Soc.: Edinburg: p. D3.
10. Malincino SM, Hartzell CR, McCarl RL. (1983). Effect of calcium on halothane-depressed beating in heart cells in culture. Mol. Pharmacol.: 23: 417-423.

MECHANISM OF HEMOGLOBIN-INDUCED SPASM IN THE ISOLATED MIDDLE CEREBRAL ARTERY OF THE CAT

Eors Dora*, Erzsebet Feher**, Maria Farago, Ildiko Horvath, and Csaba Szabo

*NIAAA, Rockville and Exp. Res. Dept. and 2nd Dept. of Physiol. and **2nd Dept. of Anatomy, Semmelweis Univ. Med. Sch., Budapest

INTRODUCTION

Cerebral vasospasm is frequently associated with subarachnoid hemorrhage. Despite extensive efforts to identify measures to dilate the narrowed arteries, no reliable method exists for preventing or reversing vasospasm (Tanishima, 1980; Wellum et al., 1980: Awad et al., 1987; Nakagomi et al., 1987; Schuier and Ulrich, 1987). One of the major reasons for this inadequacy is the lack of understanding the pathomechanism of vasospasm. Although the cause of vasospasm following subarachnoid hemorrhage is likely to be multifactorial, the present study was focused on the possible involvement of calcium mobilizing vascular smooth muscle receptors and neural mechanisms in the spastic effect of hemoglobin in the middle cerebral artery.

METHODS

Cats anesthetized by nembutal (30 mg/kg, b.w.) and exanguinated were used for this study. The brain was quickly removed from the skull and pieces of mesenterial tissue containing small branches of the mesenteric artery (outer diameter approx. 300-500 μm) were excised. The brain and the mesenterial tissue pieces were placed into ice-cold Na^+-Krebs solution composed of (in mM): NaCl 119, KCl 4.6, $CaCl_2$ 1.5, $MgCl_2$ 1.2, $NaHCO_3$ 15, NaH_2PO_4 1.2, glucose 6 (Hogestatt et al., 1983). Microsurgical methods and a Zeiss operational microscope were used to remove and clean the vessels. Then, 1-3 mm long segments were cut from the vessels and placed into a double walled and thermostated (37 $^{\circ}$C) tissue chamber. The tissue chamber was filled with Na^+-Krebs solution and bubbled with a gas mixture containing 5% CO_2 and 95% O_2. The pH of the Na^+-Krebs solution was approx. 7.4. The 1-3 mm long vessel segments were mounted on two L-shaped specimen holders and their isometric tension was measured with Grass FT-03 force transducers (Hogestatt et al., 1983). Using a micromanipulator the vessel was stretched to get a tension between 400 and 600 mg, and incubated at this tension for 60 to 90 min. Six measurements were made simultaneously. The viability of the vessels has been tested with repeated applications of 10^{-6} mol/l norepinephrine and acetylcholine.

When the vessels showed steady response to norepinephrine and acetylcholine the actual experiment was started. All measurements were carried out in the presence of 5×10^{-6} mol/l indomethacine and 5×10^{-7} mol/l propranolol. Appropriate concentrations of the drugs were injected into the tissue chamber with Hamilton microsyringes. All drugs (norepinephrine, serotonine, acetylcholine, atropine, PGF_{2alpha}, indomethacine, propranolol) were obtained from Sigma. Hemoglobin was prepared from the carefully oxygenated arterial blood of the cat. The arterial blood was centrifuged at 2000 g and washed several times with Na^+-Krebs solution. Then, red blood cells were hemolysed with distilled water and diluted in Na^+-Krebs solution. Hemoglobin concentration of the stock solutions was measured in every experiment.

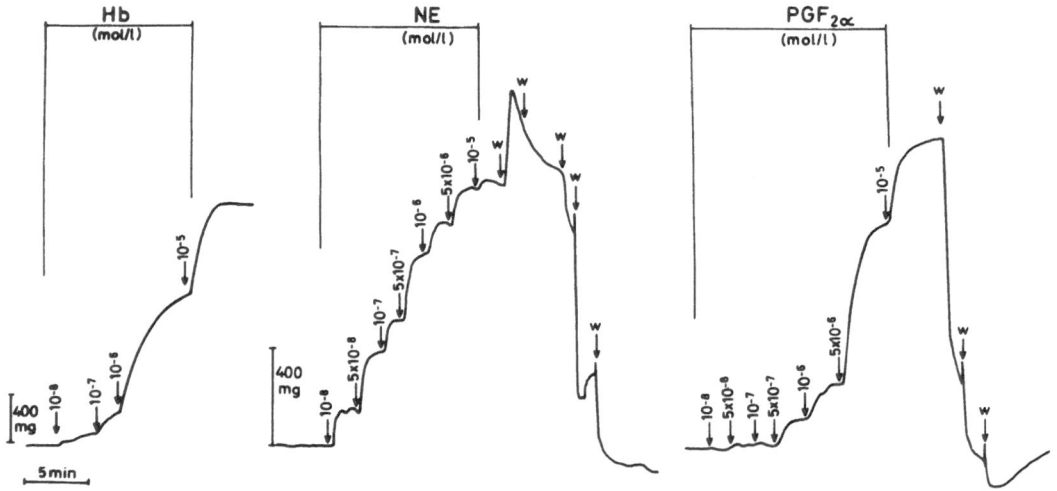

Fig. 1 Dose-response curves of hemoglobin (Hb), norepinephrine (NE), and PGF_{2alpha} in the middle cerebral artery in a single experiment. Calibrations of tension and time are shown in the left bottom corner of the Fig. Note the change in gain following hemoglobin application. Small w letters and arrows indicate the washout of drug from the tissue chamber, i.e. replacement of the fluid with fresh Na^+-Krebs solution.

In some of the experiments we varied potassium ion concentrations of the Na^+-Krebs solution. This was done so that appropriate concentrations of NaCl were replaced with KCl in the standard Na^+-Krebs solution. The isotonic K^+-Krebs solution contained 127 mM KCl and no NaCl.

In the electron microscopic studies the vessels were in situ fixed in the tissue chamber with Karnowsky fixative (120 min), and for further processing stored in buffered glutaraldehyde solution at 4 °C. A JEOL 100B

electron microscope was used to analyse changes in number and quality of presynaptic vesicles.
For statistical analysis of the data the Student t-test was applied.

RESULTS

Fig. 1 shows the effects of cumlative concentrations of hemoglobin, norepinephrine, and PGF_{2alpha} in the middle cerebral artery in a single experiment. As one can see hemoglobin produced much greater contraction than norepinephrine or PGF_{2alpha}. It is also worth to note that even 10^{-8} mol/l hemoglobin contracted the middle cerebral artery.

In agreement with available data from the literature (Tanishima, 1980) hemoglobin was far more potent in contracting the cerebral artery than the mesenteric artery. Beside that, we found that we found that the middle cerebral artery was more sensitive to hemoglobin than the mesenteric artery. 10^{-7} mol/l hemoglobin in the middle cerebral artery produced approximately 40% of the maximum contraction, but its contractile effect in the mesenteric artery was negligible.

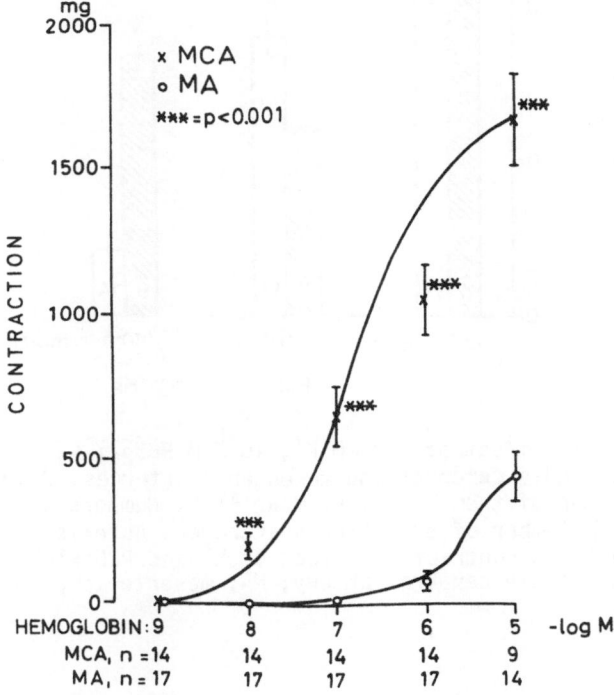

Fig. 2 Dose-response curves of hemoglobin in the middle cerebral and mesenteric arteries. Each point represents mean $\pm$ SE. n shows the number of experiments averaged. Asterisks show the significant differences in the contractile effect of hemoglobin in the middle cerebral and mesenteric arteries. MCA: middle cerebral artery; MA: mesenteric artery.

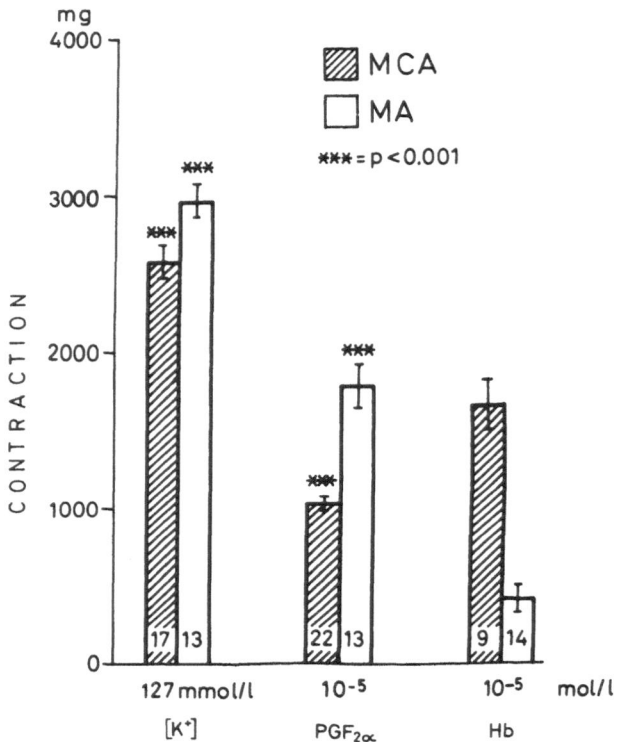

Fig. 3 Contractile effect of 127 mM K^+, 10^{-5} M PGF_{2alpha}, and 10^{-5} M hemoglobin in the middle cerebral and mesenteric arteries. Columnes and vertical bars on the top of them represent mean ± SE. Numbers written inside the columnes show the number of experiments averaged. Asterisks mark the significant differences in contractile effect of K^+ and PGF_2alpha as compared to that of Hb. MCA: middle cerebral artery; MA: mesenteric artery.

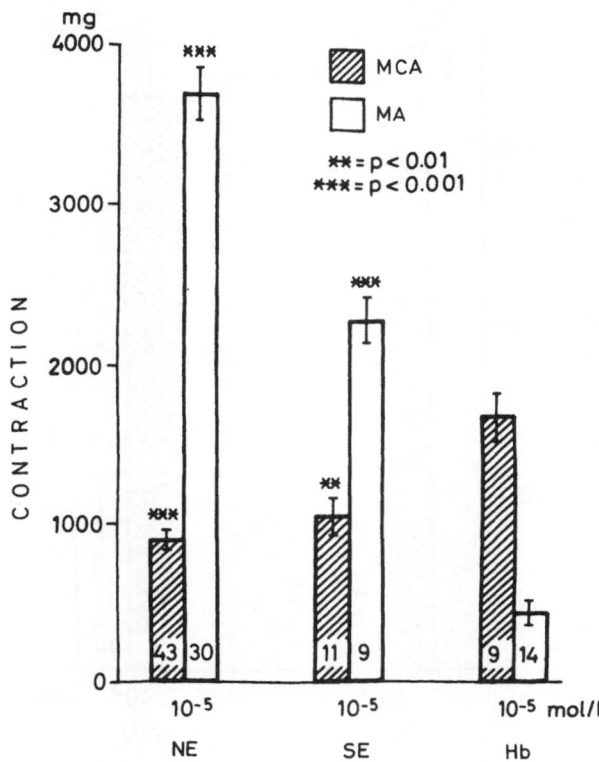

Fig. 4 Contractile effect of 10^{-5} M norepinephrine (NE), 10^{-5} M serotonine (SE), and 10^{-5} M hemoglobin (Hb) in the middle cerebral and mesenteric arteries. Columnes and vertical bars on the top of them represent mean $\pm$ SE. Numbers written inside the columnes show the number of experiments averaged. Asterisks mark the significant differences in contractile effect of NE and SE as compared to that of Hb. MCA: middle cerebral artery; MA: mesenteric artery.

In order to elucidate the underlying cause of the greatly differing contractile effect of hemoglobin in the middle cerebral and mesenteric arteries,we tested the effects of K^+-Krebs solution and various contractile receptor agonists. As Fig. 3 shows,however, K^+-Krebs solution produced similar contraction in the middle cerebral and mesenteric arteries. Furthermore, contrary to hemoglobin, PGF_{2alpha} induced greater contraction in the mesenteric than in the middle cerebral artery.

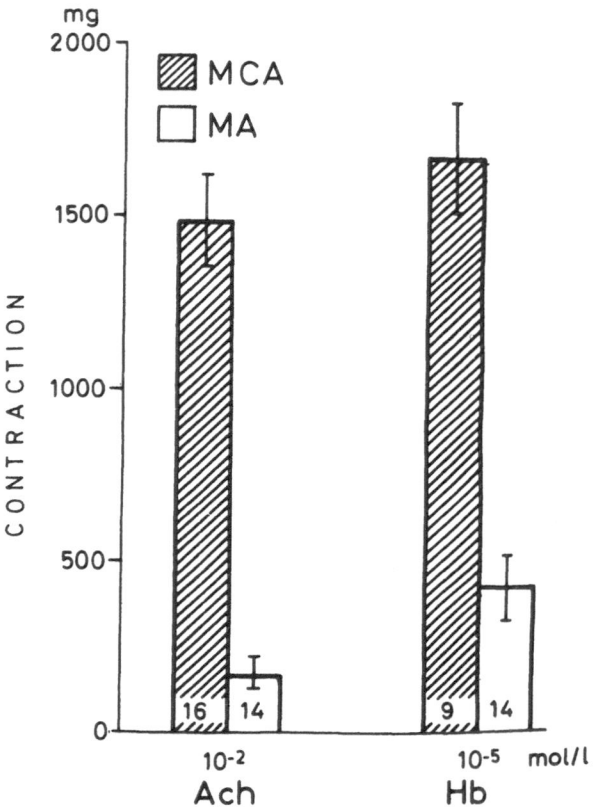

Fig. 5 Contractile effect of 10^{-2} M acetylcholine (Ach) and 10^{-5} M hemoglobin (Hb) in the middle cerebral and mesenteric arteries. Columnes and vertical bars on the top of them represent mean ± SE. Numbers written inside the columnes show the number of experiments averaged. Asterisks mark significant differences between Ach- and Hb-induced contractions. MCA: middle cerebral artery; MA: mesenteric artery.

Norepinephrine and serotonine, even at pharmacologically high concentration produced significantly smaller contraction in the middle cerebral artery than did the same concentration of hemoglobin (Fig. 4). At the same time, the contractile effect of these receptor agonists was much greater in the mesenteric artery than that of hemoglobin. The findings shown in Figs. 3 and 4 indirectly indicate that increase of extracellular potassium level and/or activation of PGF_{2alpha},alpha-adrenergic and serotoninergic receptors cannot account for the disparate contractile potency of hemoglobin in the middle cerebral and mesenteric arteries.

Interestingly, we revealed no difference between Ach- and Hb-induced contractions (Fig. 5). 10^{-2} M acetylcholine produced similar magnitude of contraction than 10^{-5} M hemoglobin in both vessels, but while the contraction
was considerable in the middle cerebral artery it was negligible in the mesenteric artery. Since these findings indicated the involvement of a musca-

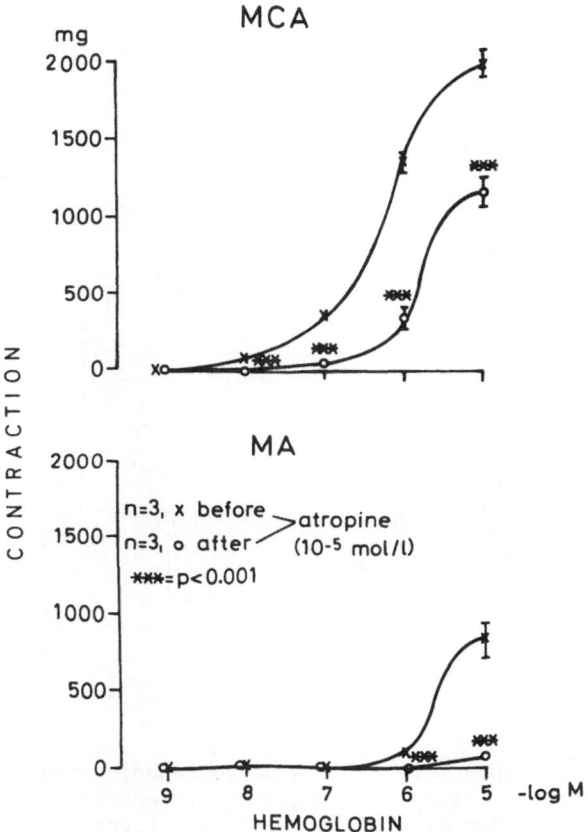

Fig. 6 Contractile effect of hemoglobin before and after 10^{-5} M atropine
in the middle cerebral and mesenteric arteries. Results are expressed as
mean $\pm$ SE; n shows the number of experiments averaged. Asterisks mark the
significant differences between values obtained before and after atropine
treatment. MCA: middle cerebral artery; MA: mesenteric artery.

rinic cholinergic mechanism in the vasospastic effect of hemoglobin we tested the effect of 10^{-5} M atropine pretreatment on hemoglobin-induced spasm.
As Fig. 6 shows, 10^{-5} atropin almost fully prevented and markedly attenuated
the spastic effect of hemoglobin in the mesenteric artery and middle cerebral artery, respectively.

The inhibitory effect of 10^{-5} M atropine on hemoglobin-induced spasm was specific since the same concentration of atropine did not alter discernibly the contractile effect of norepinephrine in the middle cerebral artery (Fig. 7).

Our findings with atropine strongly suggest, therefore, that hemoglobin may induce vasospasm mainly via a neural mechanism which encounters the release of acetylcholine from presynaptic terminals and its binding to muscarinic receptors. In order to see how hemoglobin affects presynaptic nerve terminals in the adventitial layer of the middle cerebral artery, segments of this vessel had been treated in vitro with 10^{-5} M hemoglobin

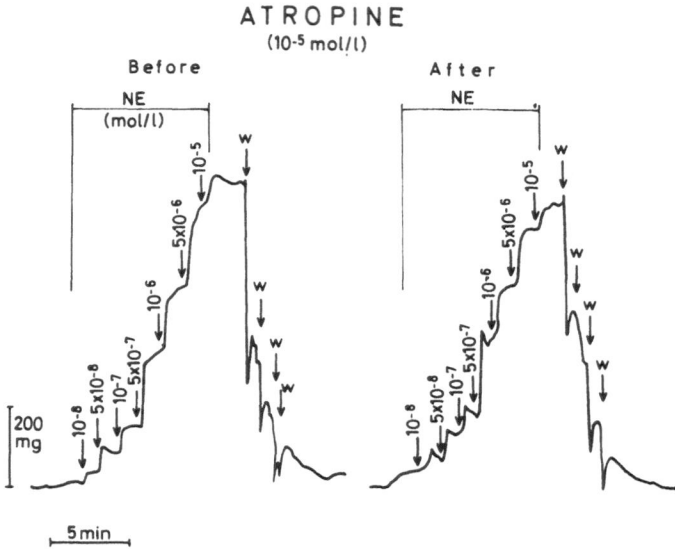

Fig. 7 Effect of atropine (10^{-5} M) on norepinephrine-induced contraction in the middle cerebral artery. Calibrations of tension and time are shown at the left bottom corner of the Fig. Small w letters and arrows indicate the replacement of drug containing Na$^+$-Krebs solution with fresh one.

and fixed with Karnowsky fixative. Electron microscopic analysis of these pretreated vessels revealed emptying of large dark vesicles in the presynaptic terminals. These vesicles are known to contain peptidergic neurotransmitters and neuromodulators. No remarkable changes were observed in the number and outlook of the so-called small vesicles that contain norepinephrine or acetylcholine (not shown). 10^{-2} M acetylcholine had similar, but somewhat less pronounced effect than 10^{-5} M hemoglobin on the synaptic vesicles of the middle cerebral artery (not shown).

DISCUSSION

In the present study the underlying mechanism of hemoglobin-induced cerebral vasospasm has been investigated. It was found that hemoglobin contracts the middle cerebral and mesenteric arteries in a concentration dependent manner, but the contractile potency of hemoglobin was much more pronounced in the cerebral than the mesenteric vessel. In addition dose-response curve of hemoglobin in the mesenteric artery was shifted to the right as compared to that in the middle cerebral artery.

Available data (Tanishima, 1980) give no explanation for the disparate potency of hemoglobin in inducing vasospasm in the brain and other organs. Since a better understanding of this differing potency may help to clear the pathomechanism of cerebral vasospasm associated with subarachnoid hemorrhage, we compared the effects of various contractile agents on the middle cerebral and mesenteric arteries. Our comparative investigations indicated that an increase in extracellular potassium ion concentration, an enhanced production of cyclooxygenase-derived vasoconstricting agents, and/or the activation of constrictory adrenergic and serotoninergic receptors play no or negligible role in cerebral vasospasm induced by hemoglobin. These results agree with available data from the literature, where investigators used appropriate enzyme inhibitors and receptor antagonists (Tanishima, 1980; Boullin et al., 1983).

We revealed a strict relationship between acetylcholine- and hemoglobin-induced contraction in both vessels, indicating that a muscarinic cholinergic neural mechanism may be involved in the vasospastic effect of hemoglobin. 10^{-2} M acetylcholine and 10^{-5} M hemoglobin produced similar magnitude of contractions in the middle cerebral and mesenteric arteries, respectively. However, the contractions in the mesenteric artery were negligible as compared to that in the middle cerebral artery. The differing reactivity of the two vascular beds studied can be attributed either to a more pronounced endothelium-mediated relaxation in the mesenteric artery, or to a smaller number of calcium mobilizing muscarinic receptors in the mesenteric vascular smooth muscle. The usage of pharmacologically high concentrations of acetylcholine was necessitated to overcome the indirect, endothelium-mediated vasodilatory effect of acetylcholine (Furchgott, 1983). Supporting our hypothesis concerning the contribution of a muscarinic cholinergic mechanism in the vasospastic effect of hemoglobin, we were able to show first in the literature that atropine significantly attenuates the constrictory effect of hemoglobin.
In addition, we discovered that hemoglobin and acetylcholine released peptidergic neurotransmitters and modulators from nerve terminals that are located in the adventitia of the middle cerebral artery.

On the basis of these findings, it is suggested that the activation of a cholinergic-peptidergic neural mechanism in the wall of cerebral arteries by hemoglobin greatly contributes to the development of cerebral vasospasm associated with subarachnoid hemorrhage.

SUMMARY

a./ Great regional differences were observed in the vasospastic effect of hemoglobin. Namely, hemoglobin was far more potent in the middle cerebral artery than in the mesenteric artery.

b./ The role of potassium ions as well as activation of calcium mobilizing adrenergic, serotoninergic and PGF_{2alpha} smooth muscle receptors were excluded in the mechanism of cerebral vasospasm induced by hemoglobin.

c./ First in the literature we showed that atropine significantly attenuates the vasospastic effect of hemoglobin, and hemoglobin releases peptidergic neurotransmitters and modulators from nerve terminals that are located in the adventitia of the middle cerebral artery.

REFERENCES

Awad, I. A., Carter, L. P., Spetzler, R. F., Medina, M., and Williams, F.
W., 1987, Clinical vasospasm after subarachnoid hemorrhage: res-
ponse to hypervolemic hemodilution and arterial hypertension,
Stroke, 18:365.
Boullin, D., Tagari, P., DuBoulay, G., Aitken, V., and Hughes, J. T.,
1983, The role of hemoglobin in the etiology of cerebral vasospasm.
An in vivo study of baboons, J. Neurosurg., 59:231.
Furchgott, R. F., 1983, Role of endothelium in responses of vascular smooth
muscle, Circ. Res., 53:357.
Hogestatt, E. D., Andersson, K- E., and Edvinsson, L., 1983, Mechanical
properties of rat cerebral arteries as studied by a sensitive device for recording of mechanical activity in isolated small blood
vessels, Acta Physiol. Scand., 117:49.
Nakagomi, T., Kassel, N. F., Sasaki, T., Fujiwara, S., Lehman, R. M., and
Torner, J. C., 1987, Impairment of endothelium-dependent vasodilation induced by acetylcholine and adenosine triphosphate following
experimental subarachnoid hemorrhage, Stroke, 18:482.
Schuier, F. J. and Ulrich, F., 1987, Global cerebral blood flow and energy
metabolism in experimental subarachnoid hemorrhage before and after treatment with calcium antagonist, in: "Stroke and Microcirculation", J. Cervos-Navarro and R. Ferszt, eds., Raven Press, New
York.
Tanishima, T., 1980, Cerebral vasospasm: contractile activity of hemoglobin
in isolated canine basilar arteries, J. Neurosurg., 53:787.
Wellum, G. R., Irvine, T. W., and Zervas, N. T., 1980, Dose responses of
cerebral arteries of the dog, rabbit, and man to human hemoglobin
in vitro, J. Neurosurg., 53:486.

VARIATION IN AXIAL VELOCITY PROFILE OF RED CELLS

PASSING THROUGH A SINGLE CAPILLARY

Christopher G. Ellis, Karel Tyml and Barbara K. Strang

Department of Medical Biophysics
University of Western Ontario, and
The John P. Robarts Research Institute
London, Ontario, Canada

INTRODUCTION

Recently, we described an interactive computer graphics system for quantifying the velocity of individual red cells in capillaries in vivo (Ellis et al., 1987). This technique was developed in order to investigate our observation that, at specific sites in some capillaries, the velocity of individual red cells appeared to be consistantly faster or slower than that of other cells in the same capillary. We have proposed that these variations in velocity were due to irregularities in the capillary lumenal cross-sectional area along the length of the capillary segment. The red cell velocity at any point in the capillary should be inversely proportional to the cross-sectional area available for flow, assuming that the red cell velocity is proportional to the blood flowrate.

Secomb (1987) has investigated the flow-dependent rheological properties of blood in capillaries to determine whether this might cause local variations in capillary resistance. Various studies have reported a graded response in the fraction of flowing capillaries at low perfusion pressures which implies that the flow resistance between capillaries must be varying. The results of Secomb's theoretical analysis indicates that local constrictions in capillaries have the potential, at low driving pressures, for causing flow cessation due to the energies required for the elastic deformation of the cells. The purpose of this study was to determine whether there are irregularities in the capillary lumen of sufficient magnitude and frequency to support Secomb's hypothesis.

METHODS

Using an intravital microscopic set-up described previously (Tyml, 1986), video recordings were made of red cell flow in capillaries at the surface of frog sartorious and rat gracilis muscles at rest. The muscles were covered with either a glass cover slip or saran wrap to isolate the tissue from the surrounding environment. Capillaries were selected which contained at least 4 red cells in clear focus over the entire analysis region. Capillary segments from 200 to 600 μm in length in the frog (n=6) and from 80 to 110 μm in length in the rat (n=3) were analyzed for time intervals of 5 to 10 seconds.

The interactive graphics system was used to acquire velocity and coordinate information on every cell passing through each capillary segment as previously described (Ellis et al., 1987). The video tape was replayed on a broadcast quality slowmotion video system (Sony BVU820 or Panasonic AU650 MII). The x-y coordinate location of every cell in the designated segment was recorded every 6th video field using the interactive graphics system. From this coordinate information, the velocity of every cell could be determined, as a function of time and location in the capillary, $v(t,z)$. To extract from this raw velocity data the axial velocity profile, i.e. the variation of red cell velocity with position, $V(z)$, the temporal variability had to be removed. This was accomplished by dividing the velocity of each red cell in the segment at one point in time by the mean velocity of all red cells in that segment at that point in time, i.e. $v(t,z)/\overline{V}(t)$. Any remaining variation in this normalized velocity between cells was assumed to be due to the location of the cell in the capillary. The location of a cell was determined as the distance along the capillary axis relative to a starting location positioned as close as possible to the beginning of the segment. The normalized or relative velocity data was then plotted as a function of the axial location of the cell in the capillary (see Fig. 1a).

To test for regions of significant local variations in velocity, the capillary segment was divided into 15 μm sections. The mean relative velocity value for each section was then tested to determine whether the mean was significantly different from the overall mean for all values, i.e. overall mean = 1, using the Dunnett Test.

RESULTS

All capillaries analyzed to date, in both frog and rat, have shown regions which have relative velocity values significantly different from the mean. Figure 1a shows an axial velocity profile for a single capillary in the frog based on the passage of 44 red cells over a 10 second time period. Although there is considerable scatter in the data, with individual measurements ranging between 0.2 and 1.98, there is a consistent pattern which indicates that the red cell velocity did vary with position along the capillary. This observation is confirmed in Figure 1b which shows the mean relative velocity with corresponding 95% confidence intervals for each 15 μm long section of the capillary. There are from 24-51 velocity values in each 15 μm section. There are two regions with relative velocity values significantly greater than the mean, both 30 μm long, and two regions with values significantly smaller than the mean, one 15 μm long and the other 60 μm long. The maximum relative velocity for a 15 μm section was 1.30 and the minimum was 0.857.

Figure 2 shows histograms of this data compiled for the 6 capillaries in the frog. Panel "a" and "b" show the length of the regions which were significantly below and above the mean respectively. Panels "c" and "d" show the distribution of velocity values

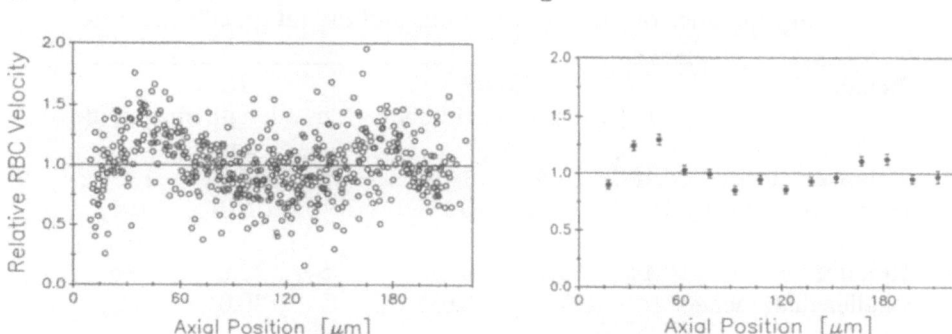

Fig. 1 Axial velocity profile for a capillary at surface of frog sartorious muscle. *Panel a* shows the normalized velocity values for 44 cells which passes through the capillary in 10 seconds. *Panel b* shows the mean velocity for each 15 μm section of capillary. The error bars represent the 95% confidence interval about the mean. If the 95% confidence interval does not overlap the solid line at a relative velocity of 1.0, then the section shows a significant deviation from the overall mean for the capillary.

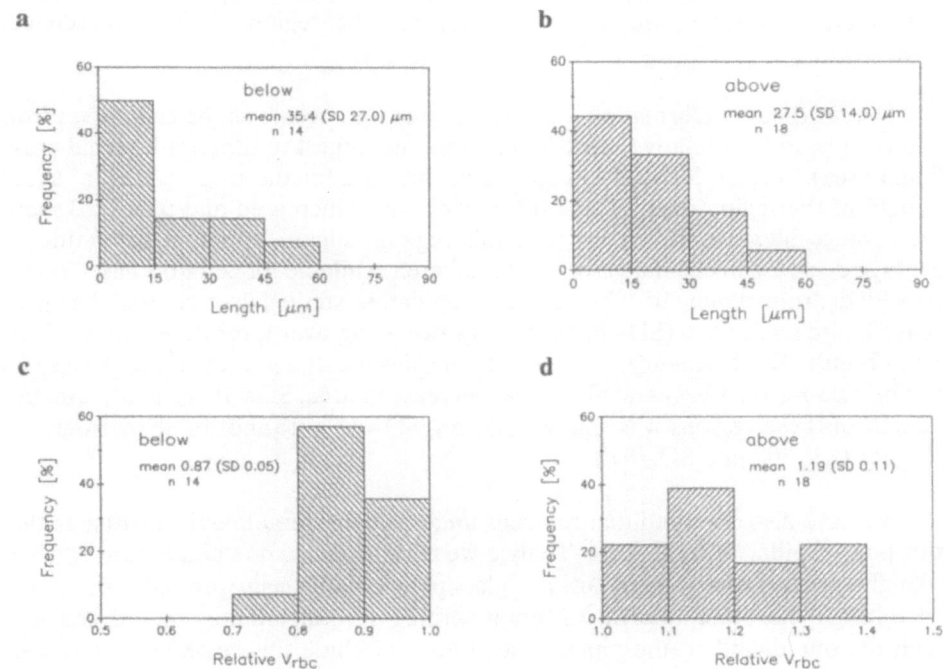

Fig. 2 Histograms combiled from 6 capillaries in the frog sartorious muscle. *Panels a and b* show the length of the regions with relative velocity significantly below and above the mean, respectively. *Panels c and d* show the magnitude of the relative velocity values in these regions

TABLE 1: Summary of statistical analysis of relative RBC velocity versus position in capillaries of the frog sartorious and the rat gracilis muscles.

Region		FROG			RAT	
	n	length μm	magnitude	n	length μm	magnitude
Relative velocity greater than mean	18	27.5 (14.0)	1.19 (0.11)	6	17.5 (6.0)	1.15 (0.06)
Relative velocity smaller than mean	14	35.4 (27.0)	0.87 (0.05)	5	27.0 (20.0)	0.86 (0.02)
Transition from minimum to maximum relative velocity	24	28.1 (14.8)	19.9% (10.3%)	5	19.0 (15.0)	33.5% (12.0%)

for these regions. A summary of the statistics for these histograms and for the rat data are given in Table 1. The regions with increased cell velocity appear to be narrower with a greater deviation from the mean than are the regions with decreased cell velocity.

The maximum deformation a cell would undergo occurs as the cell passes from a region of low to high relative velocity (i.e. from the largest to smallest lumenal cross-sectional area). Figure 3 shows histograms of this data for the frog; panel "a" shows the length of the region over which the velocity values increased and panel "b" shows the percentage increase in velocity over this region. Figure 3c shows the estimated percentage decrease in capillary cross-sectional area available for cell flow based on the velocity data from panel "b". For the six capillaries studied in the frog the mean decrease in area was 16% (SD = 6.8%, n = 24), occurring over a mean length of 28 μm (SD = 14.2 μm). The frequency of these irregularities per 100 μm was 1.2 (SD = 0.4) for frog. The data for the rat was similar (30% decrease in area, SD = 10%, n = 5), although the length of these regions was smaller (19 μm, SD = 15 μm) and the their frequency was higher (1.9/100 μm, SD = 0.6).

We have also observed that red cells undergo considerable deformation as they pass through capillary bifurcations. To date we have analyzed one bifurcation in which one daughter vessel was occluded using a glass pipette while cells were allowed to flow freely through the other branch. After recording several minutes of red cell flow through the one daugther, the pipette was used to occlude this vessel and cells were allowed to flow through the previously occluded vessel. Using the same analysis strategy as described above we were able to determine the axial velocity profile for the parent and two daughter vessels. Figure 4 shows the axial velocity profile for the parent capillary (15 to approximately 165 μm), the bifurcation region (approximately 165 to 195 μm) and one daughter capillary. There was an 87% increase in cell velocity as the

a

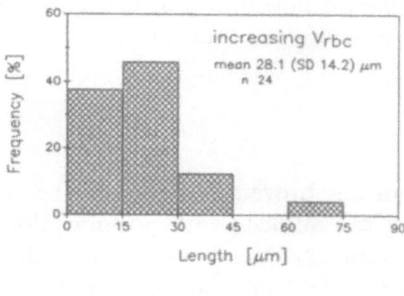

b

c

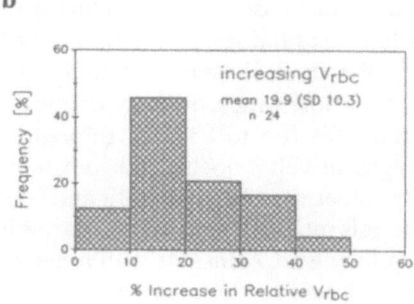
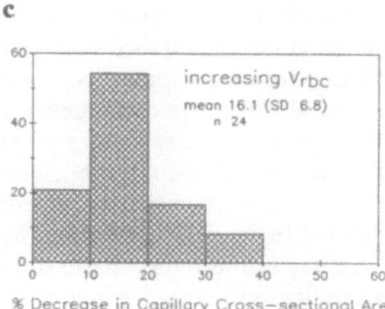

Fig. 3 *Panels a and b* show histograms of the length of the regions over which there was a significant increase in velocity and the percentage increase in velocity of these regions, respectively. *Panel c* shows a histogram of the estimated percentage decrease in lumenal area available for flow computed using the relative velocity data.

a

b

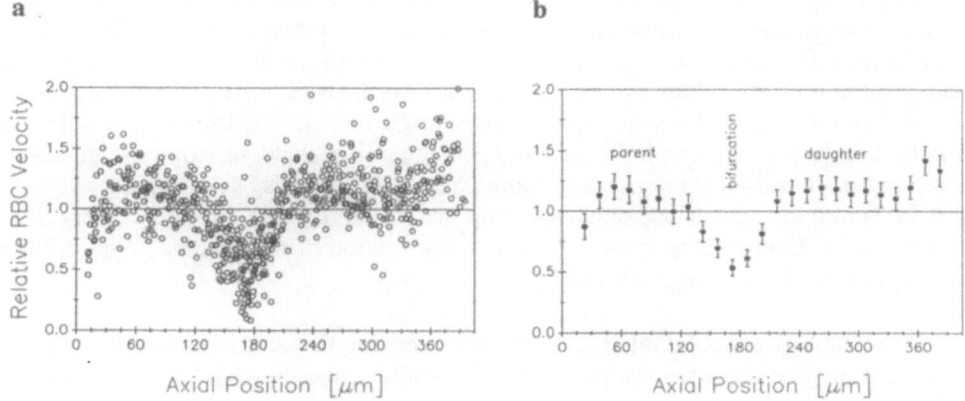

Fig. 4 Axial velocity profile for a capillary bifurcation including approximately 150 μm of the parent capillary and 195 μm of one daughter capillary. *Panel a* shows the relative velocity values for the individual red cells passing through these vessels. *Panel b* shows the mean value for each 15 μm section along these vessels. The error bars represent the 95% confidence interval about the mean value.

red cells left the bifurcation region and entered the daughter vessel. This corresponded to a 54% decrease in cross-sectional area available for flow over a distance of 45 μm. The other branch showed a smaller decrease in area of 45% over a 60 μm length.

DISCUSSION

The interactive computer graphics system has proved to be a very sensitive method for detecting persistent changes in red cell velocity with position along a capillary. The error in a single velocity measurement depends upon our ability to detect the location of the red cell in the capillary and the total displacement of the cell from one analysis frame to the next. The error can be decreased by using a larger time increment between analysis frames but this also decreases the spatial resolution of the method due to the increased cell displacement. We have chosen a "sampling rate" which corresponds to approximately a 10% error in a velocity measurement on average. By following the velocity history of numerous red cells (up to 83 cells through one capillary) we are able to detect significant changes in cell velocity with position as small as 5%. The scatter in the individual velocity values seen in Figures 1a and 5a can be attributed to a number of sources in addition to this measurement error. Variability in cell size, collisions between the cell and the capillary wall, collisions with other cells and variability in the relative velocity between the cell and surrounding plasma will all contribute to this scatter.

Do these velocity variations accurately reflect changes in the capillary lumenal cross-sectional area? If the relationship between red cell and plasma velocity were constant, independent of changes in vessel diameter and cell spacing, then one could compute the relative changes in cross-sectional area. However, it is likely that the cell velocity relative to that of the plasma will not be constant (as the capillary cross-sectional area becomes smaller the ratio of the plasma velocity to cell velocity will increase towards 1.0). Thus the axial velocity profile for red cells can only be used to obtain an approximate estimate for the relative cross-sectional area along the capillary. In addition the absolute size of the capillary lumen must be considered. Our interpretation of the relative velocity values has assumed that the red cell velocity will increase in the region of a constriction. However, if the constriction is too small or the driving pressure is too low, individual red cells may slow down or stop momentarily in the constriction rather than speedup (Note: all of the red cells in the capillary segment should simultaneously slow down or stop as a cell passes through a very small constriction). Under these conditions our interpretation of the axial velocity profile in this region would no longer be valid.

Secomb proposed that the energy required for the elastic deformation of red cells as they pass though a constriction in a capillary may be sufficient to cause flow cessation at low driving pressure. His analysis was based on a mean red cell velocity of 10 μm/s and four constriction in 100 μm reducing the capillary diameter from 5 to 4 μm, a 36% reduction in area available for flow. Our analysis to date has shown that there are approximately two constrictions per 100 μm in the rat gracilis muscle of approximately the magnitude assumed by Secomb. In the frog sartorious muscle there are fewer irregularities of smaller magnitude (note however that the nucleated frog red cells are less deformable than mammalian cells). Our analysis under conditions of normal driving pressure has provided evidence in support of Secombs hypothesis.

Further experiments are required at low driving pressure to determine whether these irregularities are of sufficient magnitude and frequency to cause flow cessation in vivo. In addition, our results from one bifurcation provides evidence that the red cell must undergo even greater deformation as it passes through a capillary bifurcation than it will in most other sections of the capillary network. Thus the bifurcation in conjunction with irregularities in the capillary lumen may play a significant role in the flow-dependent rheological properties of blood, and may be responsible, in part, for the observed spatial heterogeniety of red cell velocity between capillaries.

SUMMARY

We have used an analysis of the velocity of individual red cells as the cells pass through a capillary in order to estimate the variability in cross-sectional area of the capillary lumen available for flow along the length of the vessel. The purpose of the study was to determine if there were irregularities of sufficient magnitude and frequency to support Secomb's hypothesis that local constrictions in the capillary lumen could hinder blood flow at low driving pressure, due to the energy required to deform red cells as they pass through the constriction. All capillary segments analyzed to date, in both rat and frog, have shown regions where the velocity of individual cells is consistantly faster or slower than that of the mean velocity of all other cells in the same segment. There are approximately two constrictions per 100 μm in the rat and one per 100 μm in the frog. On average these constrictions appear to reduce the cross-sectional area by 30% in the rat and 16% in the frog. These results provide evidence in support of Secomb's hypothesis. In addition, our results from one bifurcation indicate that the capillary lumen increases in cross-sectional area as one moves from the parent vessel to the region of the bifurcation. Downstream of the bifurcation the lumen rapidly decreases in area by 45 to 54%. Thus a red cell must undergo even greater deformation as it passes through a capillary bifurcation than it will in most other sections of the capillary network. Any model of the flow dependent rheological properties of blood should take into account the unique geometry of the capillary bifurcation. Our results point toward further documentation of the axial velocity profiles of capillary segments and bifurcations, as well as new studies at low perfusion rates to test Secomb's hypothesis directly.

ACKNOWLEDGEMENTS

We wish to thank Wayne Burgess who developed the interactive computer graphics system, and Keith Ellis and Mark Noss who worked on initial stages of the analysis strategy and who spent long hours acquiring the coordinate data. This research was funded by a HSFO grant to Dr. Ellis and an MRC grant to Dr. Tyml.

REFERENCES:

Ellis, C.G., Tyml, K. and Burgess, W. (1987). Quantification of red cell movement in microvessels: a new application of interactive computer graphics. Microvasc. Res. 33: 428-432.

Secomb, T. (1987). Flow-dependent rheological properties of blood in capillaries. Microvasc. Res. 34: 46-58.

Tyml, K. (1986). Capillary recruitment and heterogeneity of microvascular flow in skeletal muscle before and after contraction. Microvasc. Res. 32: 84-89.

^{31}P NMR STUDIES OF THE METABOLIC STATUS OF PIG HEARTS

PRESERVED FOR TRANSPLANTATION

G. V. Forester*, J. K. Saunders, G. W. Mainwood**, K. W. Butler,
J. R. Scott, Hilje Paradis, O.Z. Roy and Roxanne Deslauriers

*Division of Electrical Engineering, Division of Biological
Sciences, National Research Council of Canada, Ottawa, Canada, K1A
OR6, **Dept. of Physiology, School of Medicine, Univ. of Ottawa

INTRODUCTION

Preservation of the donor heart during transport is of paramount importance for a successful transplantation procedure [1,2]. An increase in the preservation window from the current 4 - 5 hours to over 24 hours [3,4] would lead to more compatible cross-matches and wider geographical harvest of transplantable hearts. In spite of the large amount of research on the development of solutions and modalities, there is still much to be learned to achieve optimal organ preservation. Preservation solutions have evolved from simple high-potassium buffers to complex recipes [3,4].

Nuclear magnetic resonance (NMR) techniques are becoming increasingly useful in studies of cardiac metabolism [5,6]. We have used ^{31}P NMR to evaluate the effect of temperature [7] and buffer concentration [8] on the preservation of high-energy phosphates in human atrial appendages. The work is now extended to an animal model using the isolated cardioplegically-arrested pig heart to evaluate the effect of temperature and slow perfusion [4] on the energy status of the myocardium. These data are correlated with functional recovery of the heart upon reperfusion in vitro at 37°C.

MATERIALS AND METHODS

Experimental Preparation.

Pigs (6 - 15 kg body weight) were obtained from Agriculture Canada. The animals were presedated with Valium (1 - 2 mL) and anaesthetized with Ketamine (2 - 6 mL, 200 - 600 mg/Kg). Halothane was given by mask (3%) with 50/50% O_2/N_2 at 2 L/min. total flow. A tracheotomy was performed and the animals placed on a ventilator. The tidal volume was approximately 300 - 600 mL/min. and the ventilation rate was 10 - 15/min. Halothane was maintained at a level of 1.5%. A midline sternotomy was performed and the pericardium retracted. The great vessels were cleared of tissue, the brachiocephalic artery was isolated with a loose ligature and subclavian artery was ligated. Two mL of heparin (1000 I.U./mL) were injected into the inferior vena cava. The heart was rapidly removed and placed in ice cold saline. The brachiocephalic artery was cannulated and perfused retrograde in the Langendorff mode with commercial cardioplegic solution (Plegisol, Abbott) at 4°C. The heart was cleared of tissue. A latex fluid-filled balloon was placed into the left ventricle via the left atrium in order to follow the left ventricular developed pressure. The heart was transported (1.7 km) to the NMR facility on ice.

NMR Spectroscopy

NMR spectra were obtained using a Bruker Medspec 4.7/30 spectrometer operating at 4.7 Tesla with a 30 cm bore. The heart was inserted into a jacketed container which was then placed onto a custom-built 2.5 cm surface coil. The magnet was shimmed using the proton free induction decay. For ^{31}P NMR spectroscopy, a Cyclops pulse sequence was used. The pulse length was adjusted to give the maximum signal, usually 100 μs at 180 W., 120 scans were acquired for each spectrum with a 5 s. recycle delay. Sensitivity was enhanced using a 20 Hz line broadening. Peak heights were measured after removal of the broad baseline component. The intracellular pH was estimated from the chemical shift of P_i. Calibration curves were obtained at 4°, 6°, 12° and 37°C. using a standard solution containing 3 mM lactate, 3 mM P_i, 3 mM phosphocreatine (PCr), 1.5 mM ATP, 120 mM KCl and 2mM $MgCl_2$ at pH 7.0. The pH values were measured at room temperature and pH varied from 5.5 to 8.0. The heart temperature was controlled by circulating water in the probe jacket.

Biochemical Materials and Assays

The Krebs-Henseleit solution for heart reperfusion contained NaCl 118 mM, KCl 4.7 mM, $MgSO_4$ 0.6 mM, $CaCl_2$ 1.8 mM, $NaHCO_3$ 20 mM, glucose 11 mM and KH_2PO_4 1.2 mM. The solution was equilibrated with 95% O_2, 5% CO_2 the pH was 7.4. The cardioplegic solution Plegisol (Abbott) contained NaCl 110 mM, $CaCl_2$ 1.2 mM, $MgCl_2$ 16 mM and KCl 16 mM. Prior to use, KCl (10 mM) and $NaHCO_3$ (10 mM) were added to the solution, the pH was 7.4. The Wicomb solution [4] for slow flow perfusion (constant pressure of 25 cm H_2O or constant flow of ca. 45 mL/min.) contained NaCl 136.4 mM, $MgSO_4$ 14.1 mM, $CaCl_2$ 1.1 mM, KH_2PO_4 1.8 mM, K_2HPO_4 6.4 mM, glucose 11.1 mM, sucrose 7.3 mM, glycerol 136.8 mM, taurine 4.0 mM, procaine hydrochloride 1 g/L, chlorpromazine 0.0031 g/L, phenoxybenzamine 0.0025 g/L and was bubbled with oxygen.

Samples taken from the pig heart were quick-frozen using Wollenberger clamps, placed in liquid nitrogen and stored at -70°C prior to extraction. Samples were assayed as described previously [7].

Functional Testing

At the end of the preservation period, hearts were reperfused with Krebs-Henseleit buffer, with or without Dextran (0.5-2%) using constant flow (150 mL/min/100 g heart) or constant pressure (60 cm - 1 m of H_2O) of perfusate. For both protocols, the jacket of the chamber containing the heart was warmed to 37°C and the cardioplegic solution was washed out with Krebs-Henseleit buffer containing 11 mM glucose at pH 7.4. When using a constant flow pump, the reperfusion sequence was 6 minutes at 40 mL/min., 6 minutes at 80 mL/min. and 6 minutes at 120 mL/min. for a 100 gram heart. The hearts were perfused for 1.5 hours.Control hearts were reperfused without having been subjected to hypothermic preservation.

After rewarming to 37° C. , the heart was examined for natural atrial activity and ventricular fibrillation. If the heart displayed only atrial activity it was defibrillated in the magnet with indwelling electrodes (30-60 joules). The heart rate and developed pressure were monitored with a pressure transducer (Gould). Using the above described system, freshly perfused control hearts showed a rate-pressure product (RPP = rate x developed pressure) of ca. 6000 mmHg/min.

At the end of the experiment, four samples (left atrium, left ventricle, right ventricle, right atrium) were taken for biochemical analyses from hearts which displayed contractile activity (RPP >1000 mmHg/min).

RESULTS

Typical NMR spectra obtained for pig hearts preserved at 5°C. in Plegisol are plotted in Figure 1. Figure 2 shows the relative peak heights of the PCr, ß ATP and P_i resonances as well as the intracellular pH from two pig hearts preserved in cardioplegic solution (Plegisol) at 5° C. for 5 and 12 hours. In all spectra, the height of the ß resonance of ATP was set at 100. In the absence of an internal concentration standard, ATP was used since this peak does not decrease until there has been a significant decrease in PCr [9]. Figure 3 shows results for hearts preserved at 12° C. for 5 and 12 hours. PCr decreases exponentially with preservation time, ATP remains relatively stable until the disappearance of PCr, and then decreases linearly with time. At 12° C, PCr remains visible for 3 - 4 hours after removal of the heart. At 5° C, the PCr resonance remains for 5 - 6 hours. Whether kept at 5° or 12°C, over 70% of the ATP observable at the beginning of the experiment remains in hearts preserved for 5 hours (5°C: 79% ± 7, n=3; 12°C: 72% ±8, n=3). Preservation for 12 hours accentuates the trends observed above. After 12 hours at 5°C, over 40% (44% ± 3, n=3) of the initial ATP remains visible. However, after the same time at 12° C. , little ATP is detectable in the spectra (14% ± 3, n=3). During preservation, the pH drops to ca. 6.5 in hearts preserved for 5 hours. After 12 hours of preservation, the pH is 6.2. The pH decrease reaches a plateau after approximately 8 hours (Table I).

When the hearts were tested functionally after hypothermic storage with Plegisol, those with NMR-visible ATP contents greater than 60% of the initial level could be successfully restored to function. On the other hand, hearts which lost more than 60% of their ATP contents did not function upon reperfusion. Contracting hearts generally had an average intracellular pH ≥ 7.0, hearts which could not be restored to function had a pH < 6.9. In all cases, ATP levels dropped upon reperfusion. The increase in PCr levels upon reperfusion is attributed to the prese..ce of oxygen in the perfusate. Table II shows the ATP levels determined enzymatically following 1.5 hours of reperfusion.

Figure 4 a) demonstrates results of a preservation experiment with continuously flowing preservation solution [4]. Both PCr (77% ± 8, n=5) and ATP (67% ± 13, n=5) are well preserved after 24 hours. Figure 4 b) shows the pH measured during this experiment. The preservation solution contains phosphate and the increase in the P_i peak seen in Figure 4 a) is partly due to the accumulation of perfusion fluid in the probe around the heart or in the cavities within the heart. Although the phosphate resonance from the perfusate can mask that of intracellular P_i used to monitor pH, the system is sensitive to alterations in metabolic parameters. We noted that in one experiment, a heart was not perfused for the first 5 hours, following which perfusion was established; during the ischemic period there was a decrease in the PCr resonance with a corresponding increase in that from P_i. When perfusion was established, PCr increased, P_i decreased and ATP levels remained fairly constant. The pH decreased during the ischemic period and increased with perfusion.

Functional testing of hearts preserved with Wicomb solution under continuous perfusion showed greater variability in contractile function than did hearts preserved under no-flow cardioplegic conditions. In all cases, reperfusion led to a loss of NMR-observable ATP and PCr. These observations are currently under investigation.

TABLE I

pH VALUES[+] OBSERVED in PRESERVED PIG HEARTS PRIOR to REPERFUSION

Time	Temp.	pH		Δ pH
Hours	°C	Initial *	Final	
5	5	7.3 ± 0.3 (n=4)	6.5 ± 0.2 (n=4)	0.8
5	12	7.3 ± 0.2 (n=5)	6.4 ± 0.3 (n=5)	0.9
12	5	7.3 ± 0.3 (n=4)	6.2 ± 0.3 (n=4)	1.1
12	12	7.3 ± 0.1 (n=3)	6.2 ± 0.0 (n=3)	1.1

[+] Intracellular pH values are determined from chemical shift of P_i in the ^{31}P NMR spectra.
* Initial pH represents the value measured 1.5 hours after severing the aorta.

TABLE II

BIOCHEMICAL ASSAY OF REPERFUSED PRESERVED HEART

TEMPERATURE °C	PRESERVATION TIME Hours	ATP μmol/g dry weight	PCr	%DryWeight
In Situ* 37 (3)	0	21 ± 1.6	18 ± 1.6	19
5 (3)	5	17 ± 1.0	43 ± 3.3	12
5 (3)	12	10 ± 1.5	21 ± 4.1	10
12 (3)	5	13 ± 1.7	34 ± 4.1	12

Values are Means ± S.E.M.   ( ) represents number of heart samples
* Blood was present in these tissues

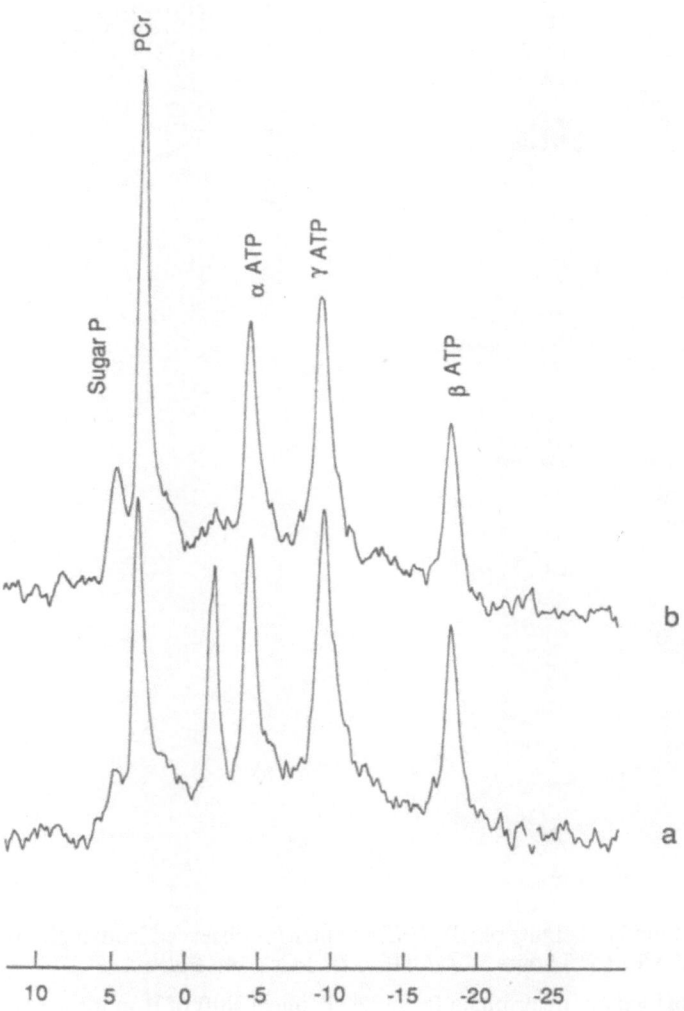

Figure 1. ^{31}P NMR spectra of a pig heart preserved at 5°C with Plegisol for a) 1 hour and b) for 5 hours. Other conditions are described in Materials and Methods section.

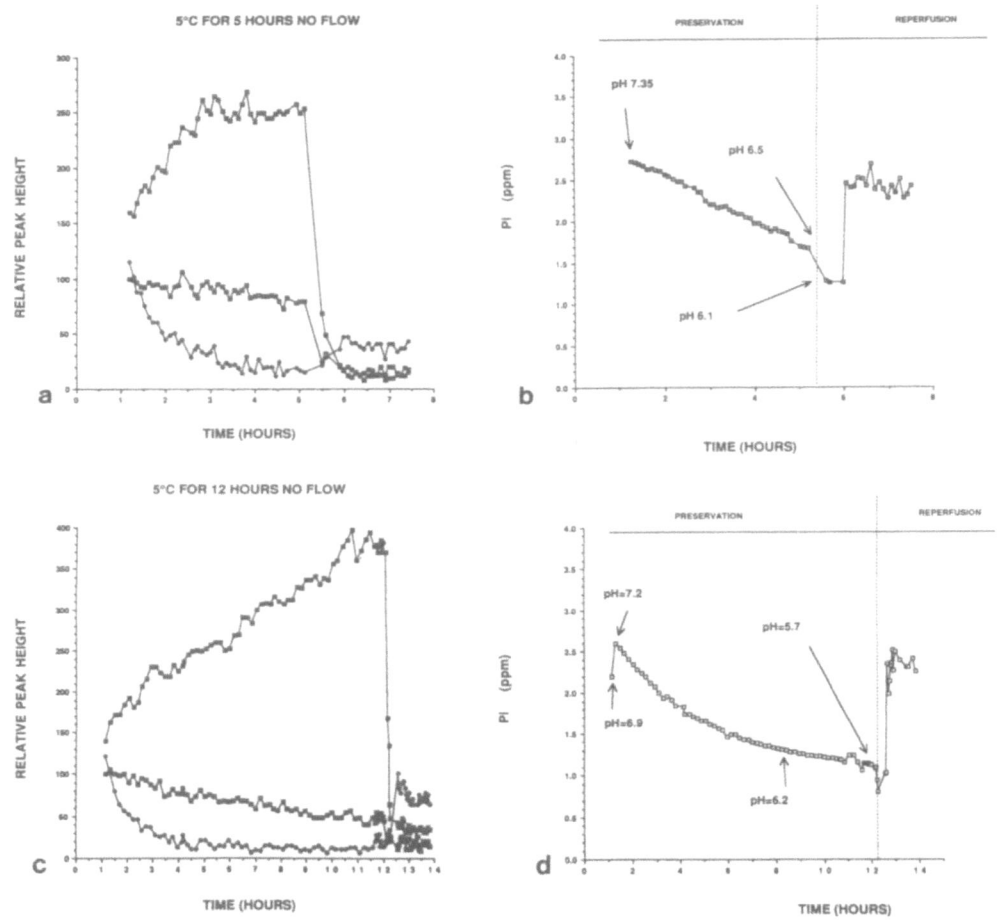

Figure 2a. Relative heights of ³¹P NMR resonances observed from a pig heart preserved with Plegisol at 5°C. for 5 hours. ❑ ATP ; ■ PCr ; ◆ P$_i$.

Figure 2b. pH values determined from the chemical shift of P$_i$ in the spectra.

Figure 2c. Relative heights of ³¹P NMR resonances observed from a pig heart preserved with Plegisol at 5°C. for 12 hours. ❑ ATP ; ■ PCr ; ◆ P$_i$.

Figure 2d. pH values determined from the chemical shift of P$_i$ in the spectra.

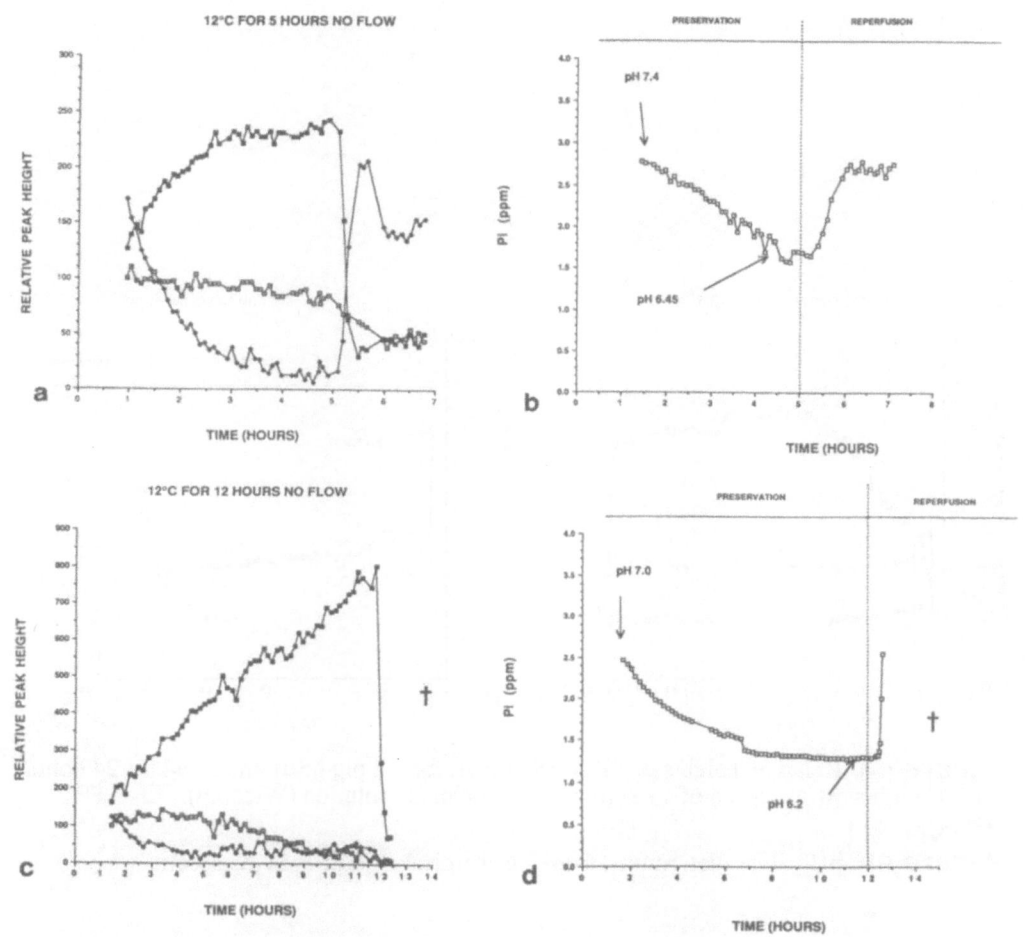

Figure 3a. Relative heights of ^{31}P NMR resonances observed from a pig heart preserved with Plegisol at 12°C. for 5 hours. ❑ ATP ; ■ PCr ; ◆ P_i .

Figure 3b. pH values determined from the chemical shift of P_i in the spectra.

Figure 3c. Relative heights of ^{31}P NMR resonances observed from a pig heart preserved with Plegisol at 12°C. for 12 hours. ❑ ATP ; ■ PCr ; ◆ P_i .

Figure 3d. pH values determined from the chemical shift of P_i in the spectra.

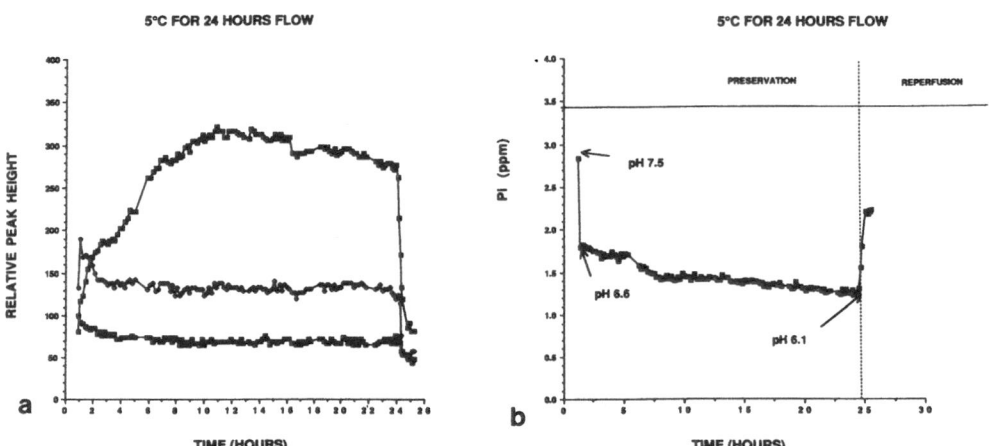

Figure 4a. Relative heights of ^{31}P NMR signals from a pig heart preserved for 24 hours at 5°C with slow perfusion of a modified cardioplegic solution (Wicomb). ❏ ATP ; ■ PCr ; ◆ P_i .

Figure 4b. pH values determined from the chemical shift of P_i in the spectra.

DISCUSSION

The pig heart model was chosen because its morphology and coronary dynamics are similar to those of the human. In addition, it can be similar in size to that of the human. The results in the Figures cannot be directly quantitated due to the short recycle time used for the NMR measurements (5 s.) which lead to partial saturation of the peaks. There is however little change in the relaxation times of P_i, PCr and ß ATP at the different temperatures examined (R. Deslauriers et al, unpublished). The results of the different experiments can therefore be compared with each other. The data for ATP are the most quantitative, since the recycle time for the experiment was approximately 3 times the spin-lattice relaxation time.

These studies suggest that current clinical conditions used for heart transplants (hypothermic preservation for up to 4 hours) allow almost complete preservation of ATP levels. Preservation of pig hearts for 5 hours at either 5° or 12°C is consistent with good preservation of ATP levels. However, preservation for 12 hours under ischemic conditions shows marked differences between the results obtained at 5° and 12° C. Under the latter conditions, little ATP (14%) is NMR-visible at the end of the experimental period.

Slow perfusion of a modified cardioplegic solution (Wicomb) [4] maintains PCr and ATP levels more adequately than do ischemic preservation conditions. The maintenance of high levels of PCr may contribute to preservation of ATP by preventing loss of phosphate from the cell and ultimately loss of nucleotide precursors. A further beneficial effect may be removal of glycolytic products [10].

^{31}P NMR studies [11,12] of heart preservation have shown that oxygen delivery and buffer capacity are two important parameters in preservation of high-energy phosphates. In ischemic human atrial appendages, ATP preservation is greatly improved by high external buffer concentration [8]. This may in part be simulated by the slow perfusion with Wicomb solution which limits the pH decrease in the tissue. Preservation with Plegisol, even with the addition of 10 mM bicarbonate, results in a continuous decrease in pH (to a value of 6.1) under ischemic preservation conditions at low temperature.

One limitation of this type of study is that NMR may selectively monitor cytosolic ATP, with mitochondrial ATP remaining NMR-invisible [13,14]. We have preliminary data indicating that preservation of high levels of NMR-detectable ATP (> 50% of starting values) correlates with functional recovery of the heart upon reperfusion at 37° C. This data must be treated with some caution because while preservation of high energy phosphates appears to be a necessary condition for functional recovery, it may not be a sufficient condition for recovery.

SUMMARY

^{31}P NMR spectroscopy has been used to evaluate the metabolic status of cardioplegically arrested pig hearts. Hearts were stored with Plegisol for up to 12 hours at either 5°C or 12°C. Results indicated that the ATP content of hearts could be maintained (>70% of initial values) for up to 5 hours in the ischemic storage state. The ATP loss was greater at 12°C. PCr was lost exponentially under the same conditions. Functional testing by reperfusing the stored hearts in vitro indicated a good correlation between the ATP content and survivability of the preparations. Twenty - four hour preservation of pig hearts using slow perfusion with a modified cardioplegic solution (Wicomb) allowed for preservation of both PCr and ATP. in all cases, reperfusion of hearts revealed a loss of NMR- visible ATP and PCr.

REFERENCES

[1]   R. Hetzer, H. Warnecke, and S. Schuler, (1986) The Donor Heart: Procurement, Selection, and Preservation. Transpl. Proc. 28, 27-30.

[2]   J.M. Levett, and R.B. Karp, (1985) Heart Transplantation, Surgical Clinics of North America, 65, 613-635.

[3]   J.M. Burt and J.G. Copeland, (1986) Myocardial function after preservation for 24 hours, J. Thorac. Cardiovasc. Surg., 92, 238-246.

[4]   W.N. Wicomb, D. Novitzky, D.K.C. Cooper, and A.G. Rose, (1986) Forty-eight Hours Hypothermic Perfusion Storage of Pig and Baboon Hearts, J. Surg. Res., 40, 276-284.

[5]   C.J.A. Van Echteld, and T.J.C. Ruigrok, (1987) Nuclear Magnetic Resonance Spectroscopy in Experimental Cardiology in "Non-invasive Imaging of Cardiac Metabolism" Dordrecht Martinus Nijhoff Publ., 265-277.

[6]   G.A. Elgavish, (1987) NMR Spectroscopy of the Intact Heart in "Biological Magnetic Resonance", L.J. Berliner and J. Reuben, eds., Plenum Press, New York, 81-127.

[7]   R. Deslauriers, W.J. Keon, S. Lareau, D. Moir, J.K. Saunders, I.C.P. Smith, K. Whitehead, and G.W. Mainwood, (1989) Preservation of High Energy Phosphates in Human Myocardium - A ^{31}P NMR Study of the Effect of Temperature on Atrial Appendages. J. Thorac. Cardiovasc. Surg., (in press)

[8]   G.W. Mainwood, S. Lareau, K. Whitehead, W.J. Keon, and R. Deslauriers, (1989) The Effects of Temperature and Buffer Concentration on the Metabolism of Human Atrial Appendages Measured by ^{31}P NMR, Adv. in Exptl. Med. & Biol. - Oxygen Transport to Tissues , 12, K. Rakusan, G. Biro & T. Goldstick, eds., Plenum Press, N.Y. (in press).

[9]   R.J. Connett, (1988) Analysis of Metabolic Control: New Insights Using Scaled Creatine Kinase Model, Am. J. Physiol., 254, R949-R959.

[10]   J.R. Neely, and L.W. Grotyohann, (1984) Role of Glycolytic Products in Damage to Ischemic Myocardium, Circ. Res., 55, 816-824.

[11]   T.A. English, J. Foreman, D.G. Gadian, D.E. Pegg, D. Wheeldon, and S.R. Williams, (1988) Three Solutions for Preservation of the Rabbit Heart at 0° C., A Comparison with Phosphorus-31 Nuclear Magnetic Resonance Spectroscopy, J. Thorac. Cardiovasc. Surg., 96, 54-61.

[12]   L.D. Shorr, R.T. Thompson, G.D. Marsh, F.M. Keith, A.A. Driedger, and D.J. Magilligan, Improved Preservation of Isolated Rabbit Hearts Using Oxygenated Cardioplegic Solutions. A ^{31}P NMRS Study. (private communication).

[13]   H. Takami, H. Matsuda, S. Oumi, H. Watari, E. Furuya, K. Tagawa, and Y. Kawashima, (1987) Assessment of Possible Compartmentalization of Myocardial ATP and Characteristics of Depletion Under Ischemia Utilizing ^{31}P NMR, Circulation, 76, 247 .

[14]   H. Takami, E. Furuya, K. Tagawa, Y. Seo, M. Murakami, H. Watari, H. Matsuda, H. Hirose, and Y. Kawashima, (1988) NMR-Invible ATP in Rat Heart and Its Changes in Ischemia, J. Biochem., 104, 35-39.

OXYGEN SUPPLY OF THE MYOCARDIUM

Frank K.H., Kessler M., Appelbaum K., Dümmler W., Zündorf J., Höper J., Klövekorn W.P.*, and Sebening F.*

Institut für Physiologie und Kardiologie, Universität Erlangen-Nürnberg, Erlangen, West-Germany

*Deutsches Herzzentrum München, München, West-Germany

INTRODUCTION

Hypokinetic zones in the myocardium develop when the oxygen supply of the myocytes is diminished. For quite a long time it was supposed that a local depletion of energy rich phosphates might be the cause of disturbances of wall motion in the critically perfused myocardium.

Our aim was to answer the question whether local hypokinesia is induced by other factors such as tissue hypoxia and whether hypokinesia could thus be a protective mechanism which might be triggered by local tissue hypoxia.

In order to clarify this question the oxygen supply of the myocardium of anaesthetized dogs was determined in our laboratory by measuring the local pO_2 with the multiwire surface pO_2 electrode (Kessler et al. 1986, Kessler et al. 1988) and the intracapillary hemoglobin oxygenation was recorded with the Erlangen micro-lightguide spectrophotometer EMPHO I (Frank et al. 1988). In these experiments the ultrasonic transit time technique was used to assess the regional wall motion (Hagel et al. 1975). In the light of results from these studies, investigations of the intracapillary hemoglobin oxygenation were performed in patients undergoing open heart bypass surgery.

METHODS

The EMPHO I consists of four modules: the light source, the micro-lightguide cable, the detection device and the computing system. A schematical drawing of the whole apparatus is shown in Fig. 1.

The Xenon high pressure arc lamp (XBO 75 W/2 Osram) (1) is mounted in a lamp housing which is cooled by water. The convectional cooling system does not need pumps or ventilators. The light from the Xenon arc lamp is collimated and focused onto the incident plane of the illuminating micro-lightguide fibre (3) by two lenses (2). The luminous cone of the lens assembly is compatible with the acceptance angle of the micro-lightguides.

The illuminating micro-lightguide transfers the light to the tissue
surface and illuminates the tissue. The remitted light is collected by the
detecting micro-lightguides (4) and transmitted to the detection unit of
the EMPHO I. The detection device consists of a motor-driven bandpass
interference filterdisk (5) (Anders, Naaburg) which serves as a
monochromating unit. The filter disk is mounted, together with a segmented
wheel (8) for the decoding of the rotation angle, on the axle of a
micromotor. The monochromated light is transmitted to the photomultiplier
tube (7) by use of a fluid light guide (6). After amplification and AD-
conversion the spectra are stored, processed and displayed by an AT-
compatible computer. The AD-conversion is initiated by a reset signal and
triggered by TTL impulses which indicate wavelength equidistant revolution
angles. Both impulses are generated by the decoding electronics (9).

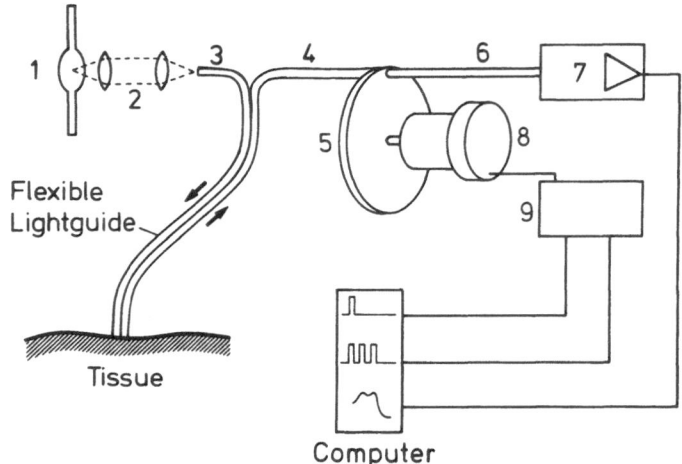

Fig. 1. The Erlangen micro-lightguide spectrophotometer consists of a
        Xenon arc lamp (1), a focusing lens system (2), the micro-
        lightguide device (3 and 4), an interference filter disk (5),
        a fluid light guide (6), a photomultiplier tube (7), a decoder
        wheel (8) coupled to a motor, a decoding electronic device (9)
        and a computer.

RESULTS

   In 15 patients intracapillary hemoglobin oxygenation was measured
within 12-15 different areas (1,5 x 1,5 cm²) of the anterior wall of the
left ventricle. At each location a histogram of the intracapillary HbO_2
distribution was measured (n = 210). Due to the high sampling velocity of
70 spectra per second, the measurement of one histogram took only 3
seconds.

   Three HbO_2 histograms measured in a control area of the human
myocardium are shown in Fig. 2. The Figure demonstrates a typical local
heterogeneity of distribution patterns produced by different oxygenation
levels of intracapillary Hb. HbO_2 histograms measured in the anterior wall
of left ventricle of a patient who suffered from a three vessel disease
are shown in Figure 3. The HbO_2 values of the two lower histograms reveal
critically low values between 0-25%.

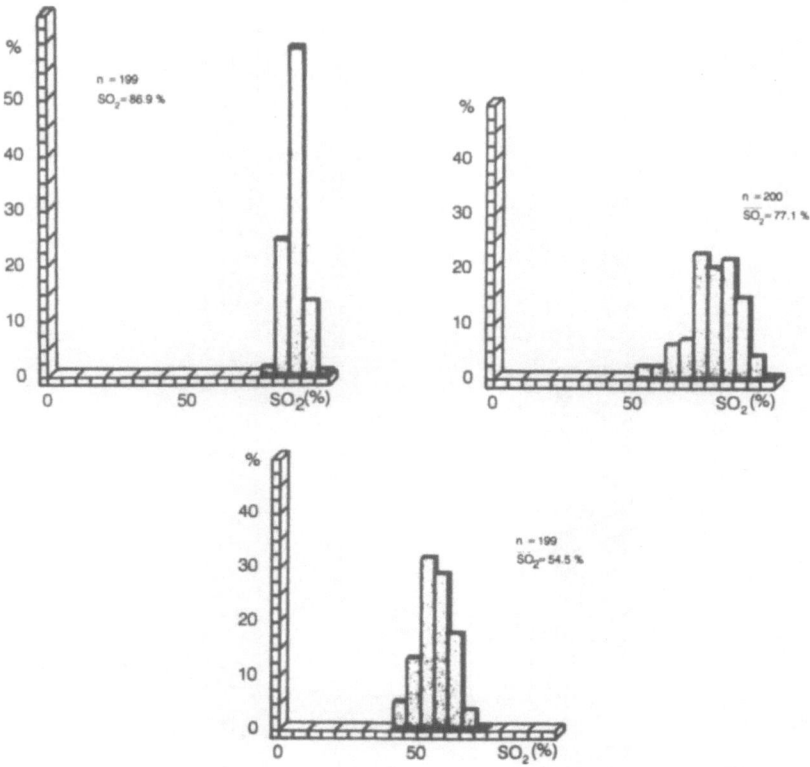

Fig. 2. Histograms of intracapillary hemoglobin oxygenation measured
in a control area of human myocardium. The Figure demonstrates
a typical local heterogeneity of the different oxygenation
levels.

The findings shown in Fig. 4 were discovered in the myocardium of the
left anterior wall of the heart of a 71 years old female patient. She
suffered from angina induced by the slightest exertion. The angiogram
revealed a long subtotal occlusion of the ramus intraventricularis
anterior. The right coronary artery was stenosed by 50%. The anterior wall
showed slight hypokinesia. The lowest values of the HbO_2 histograms,
indicated by the black columns, display critical values of less then 10%
HbO_2. The 10% values are indicated by the small bars at the left of the
respective areas. After the surgical intervention the low HbO_2 values
increased to 50% HbO_2 as shown in Fig. 5.

DISCUSSION

Our results indicate that hypokinetic zones in the myocardium develop
when the hemoglobin oxygenation at the venous end of the capillaries falls
below a critical threshold between 20 and 10%. Calculations using the
Fick's laws and a mean oxygen uptake rate typical for the beating heart
indicate that these oxygenation values of intracapillary hemoglobin
produce hypoxic pO_2 values in between two capillaries. Even in patients
suffering from severe angina, very few values were found in this range.
These findings are in agreement with the lowest pO_2 values found by
Kessler (1986) using the multiwire surface electrode during critical
stenosis of the left coronary artery. The catchment volume of this
electrode is so small (Harrison, personal communication) that the lowest

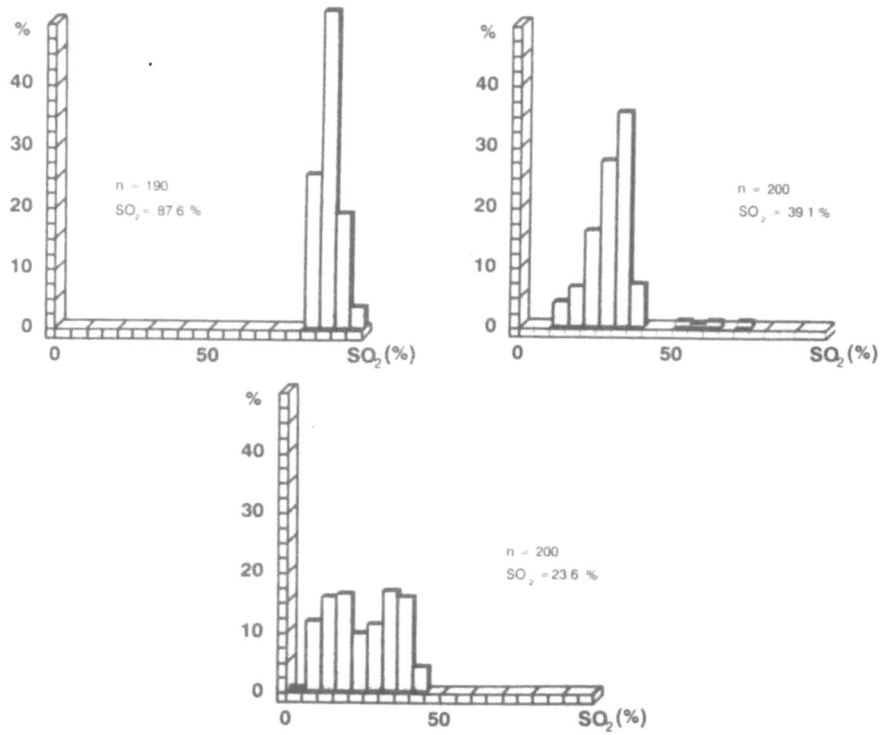

Fig. 3. Typical HbO₂ histograms measured in the anterior wall of the left ventricle of a patient who suffered from a three vessel disease. The HbO₂ values of the two lower histograms reveal critically low values between 0-25%.

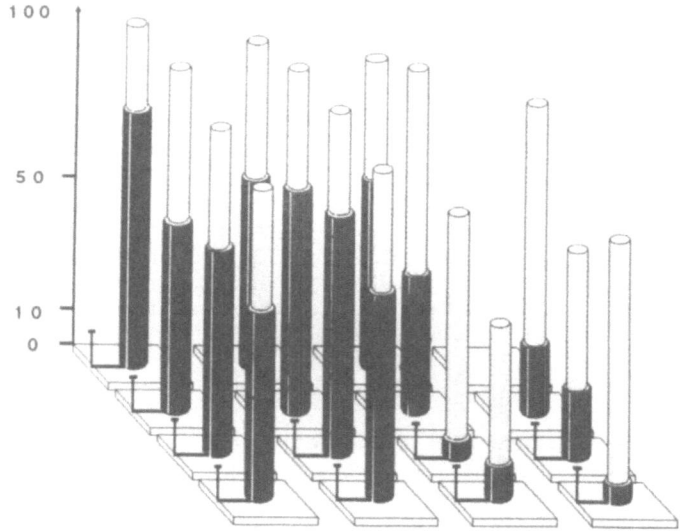

Fig. 4. Oxygen supply of the myocardium before bypass surgery. The black columns indicate the lowest and the white columns show the highest values of the intracapillary HbO₂ histograms as measured at 15 different locations at the surface of the left ventricle of a patient who suffered from a three vessel disease. The bars at the left of the columns indicate to 10% HbO₂.

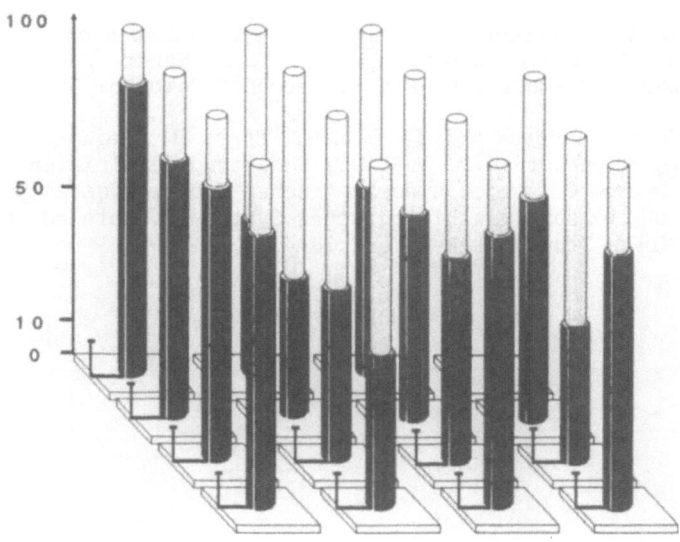

Fig. 5. Oxygen supply of the myocardium after the bypass surgery. The black columns indicate the lowest and the white columns show the highest values of the intracapillary HbO₂ histograms, as measured at 15 different locations at the surface of the left ventricle of a patient which suffered from a three vessel disease. The bars at the left of the columns indicate the values of 10% HbO₂.

values from histograms represent those at the "lethal corner", i.e. of intracellular and interstitial spaces at the venous end of the capillaries.

Under control conditions the lowest pO₂ values measured at the surface of the beating heart lie in the range from 12 to 15 mm Hg. Lower pO₂ values such as those reported by Honig and Gayeski (1988) were found in pronounced pathophysiological situations in our experiments.

The results from the measurements of the intracapillary hemoglobin oxygenation in the hearts of coronary patients indicate that quantitativ tissue spectrophotometry detects the oxygen supply of the myocardium in patients. For the first time the heart surgeon has a possibility of performing the preoperative diagnosis and to verify the result of the surgical interventions.

REFERENCES

Frank K.H., Kessler M., Appelbaum K., Dümmler W. The Erlangen micro-lightguide spectrophotometer EMPHO I submitted to Phys. Med. Biol.

Gayeski T.E.J., Federspiel W.J., Honig C.R. A graphical analysis of the influence of red cell transit time, carrier free layer thickness, and intracellular pO₂ on blood-tissue O₂ transport Adv. Exp. Med. Biol. 222: 25

Hagl S., Heimisch W., Meisner H., Erben R., Franklin D., Sebening F. Ultraschallverfahren zur direkten Erfassung der regionalen Myokardfunktion Thoraxchirurgie 1975 23 291-297

Kessler M., Harrison D.K., Höper J. 1986 Tissues oxygen measurements in: Microcirculatory Techniques C.H. Baker, W.L. Natruk eds. Akademic Press Inc. New York London Toronto: 391

Kessler M., Anderer W., Frank K.H., Höper J., Harrison D.K., Richter H., Klövekorn W.P. 1988 Die Bedeutung der lokalen O_2-Versorgung für das Entstehen hypokinetischer Myokardareale in: Funktionsanalyse biologischer Systeme J.Grote ed. Gustav Fischer Verlag Stuttgart New York 18: 215

THE OXYGEN DEPENDENCE OF THE ENERGY STATE OF CARDIAC TISSUE:

^{31}P-NMR AND OPTICAL MEASUREMENT OF MYOGLOBIN IN PERFUSED RAT HEART

H. Fukuda, H.Yasuda, S. Shimokawa[*], and M. Tamura[**]

Department of Cardiovascular Medicine, School of Medicine
[*]NMR Laboratory, Hokkaido University, [**]Research Institute
of Applied Electricity, Sapporo, Japan.

INTRODUCTION

The intregrity of cardiac function depends on the energy of the tissue where ATP is produced mainly by oxidative phosphorylation in the mitochondria. Thus, a critical oxygen concentration has been postulated below which function cannot be maintained. Araki et al. (1983) found oxygen consumption fell when cardiac tissue oxygen concentration fell below 10 ωmM. However, energy state can be governed by factors other than oxygen concentration (Owen and Wilson, 1974). In the present study, we performed simultaneously measuring the energy state of isolated hearts, by ^{31}P-NMR, and estimating the tissue oxygen concentration by optical spectroscopy. Similar _in situ_ measurements in brain were reported previously (Tamura et al., 1988).

MATERIALS AND METHODS

Heart perfusion:
    Hearts were removed from heparinized male Albino rats (250-300 g) which had been anesthetized with an intraperitoneal injection of pentobarbital (Nembutal, 10 mg/100 g). The heart was cannulated through the aorta and perfused by the Langendorff method with Krebs-Hensselbit-bicarbonate buffer, containing 10 mM glucose, at 30°C. The perfusate was gassed in separate containers with 95%O_2/5%CO_2 and 95%N_2/5%CO_2. A variable-speed pump was used to mix the perfusates in varying proportions, to produce hypoxia. Perfusion pressure was kept at 80 cm H_2O; the perfusion system is illustrated in Figure 1.

^{31}P-NMR spectroscopy:
    Our sample tube was a double-walled glass cylinder (external diameter, 25 mm; internal diameter, 20 mm), schematically illustrated in Figure 1. Deuterium oxide (D_2O) was to the external chamber to achieve frequency lock. Observations were made on a Bruker 200 (4.7 Telsa) magnetic resonance spectrometer. ^{31}P spectra were obtained at 81.075 MHz without photon decoupling. Radiofrequency pulses of 90° were used with one second intervals. One spectrum consisted of 300 F.I.D.s acquired over five minutes. An exponential apodization was applied to the F.I.D. to improve the signal-to-noise ratio.

Intracellular pH was calculated from the chemical shift of inorganic phosphate (P_i).

Optical measurements:

Dual wavelength reflectance spectrophotometry (Araki et al., 1983) was used. > 500nm wavelength light was guided onto the left ventricular surface by flexible plastic optical fibers of 5 m length. Reflected light at 620 and 580 nm was conducted back to the detection system.

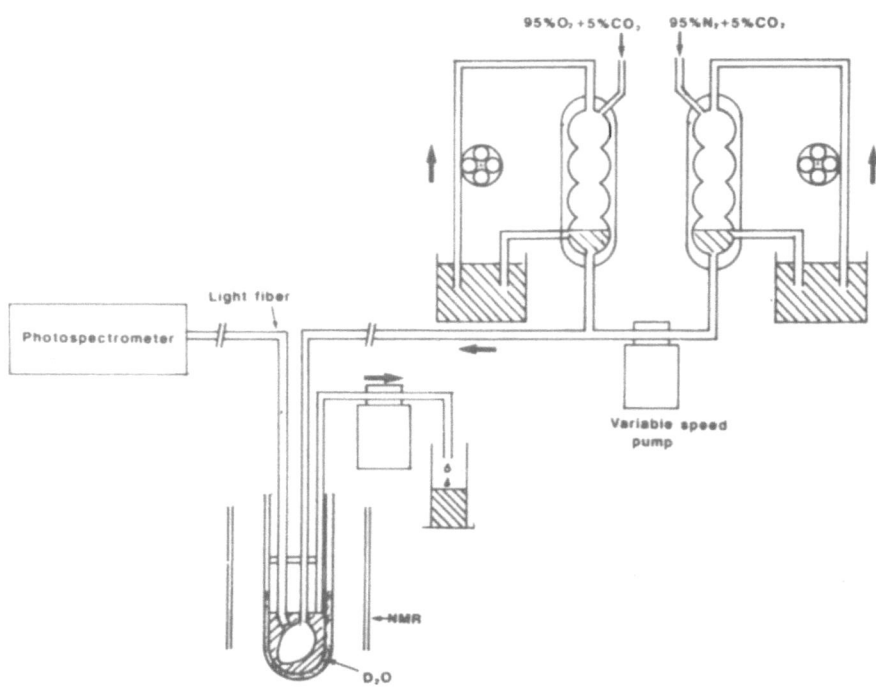

Figure 1.
Block diagram for simultaneous measurements of ^{31}P-NMR and myoglobin absorbance of perfused rat heart.

RESULTS

Figure 2 shows a typical example of the simultaneously recorded ^{31}P-NMR spectrum and the absorbance change of myoglobin when the oxygen concentration of the perfusate was changed by changing perfusate proportion. During NMR measurements, a steady state could be achieved in our experimental conditions. The spectrum during normoxia, indicated by I in Figure 2, is that of the typical wall-energized one where the ratio of creatine phosphate (CrP) to inorganic phophate (P_i) was >10. During hypoxia, indicated by II in Figure 2, when 70% of the myoglobin was deoxygenated the NMR measurement indicated a large increase in P_i and fall in CrP. During anoxia, indicated by III in Figure 2, the

concentration of CrP was less than half that if P_i. The P_i-peak
increased more than five-fold and shifted to a lower frequency indicating
a decrease in pH. ATP was also slightly decreased. Restoration of
normoxia, indicated by IV in Figure 2, restored the spectrum seen in
Phase I. Myoglobin absorption also recovered to the original level.

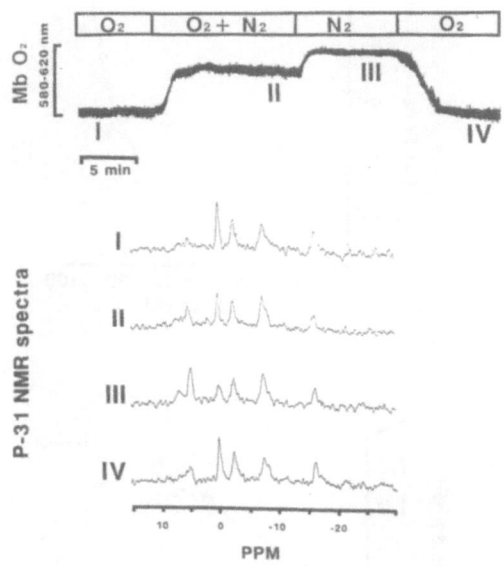

Figure 2.
Simultaneous recording of myoglobin oxygenation state and [31]P-
NMR spectra. The I-IV in ^{31}P-NMR spectra were taken at the stages of
top trace.

The data obtained with all the hearts are summarized in Figure 3.
The fall in CrP/P_i ratio appeared to occur at about 10% deoxygenation of
myoglobin whereas the pH shift seemed to occur at about 30%
deoxymyoglobin.

Similar experiments were performed with different heart
preparations, and the data were summarized in Figure 3. The CrP/P_i
started to fail at 10% deoxygenation of myoglobin (Figure 3, top) whereas
the pH shift started to occur at 30% deoxygenation (Figure 3, bottom).

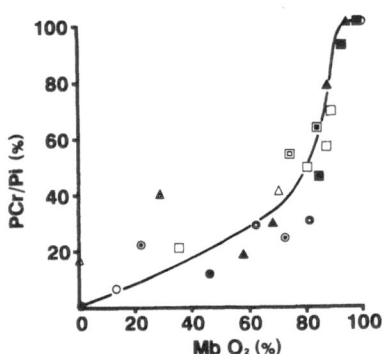

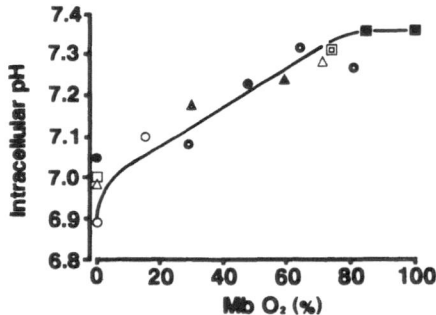

Figure 3.

The relationships between CrP/P$_i$ (top), pH-shift (bottom) and degree of myoglobin oxygenation state. Different symbols denote different heart preparations.

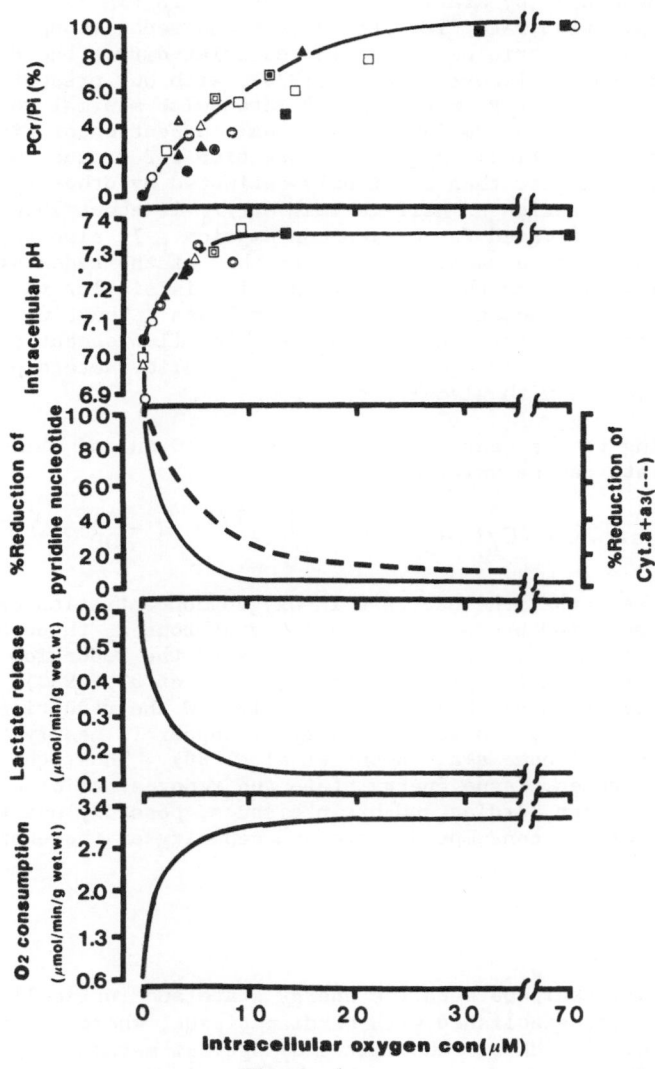

Figure 4.
The oxygen dependence of various biochemical parameters in cardiac tissue. The curves for oxygen consumption, lactate release and redox changes of pyridine nucleotide and of cytochrome a+a₃ were replotted from Araki et al., 1983 and Tamura et al., 1978, respectively.

DISCUSSION

It is well known that intracellular oxygen concentration is the critical determinant for tissue energy metabolism. In the perfused heart, using myoglobin oxygenation estimated by spectroscopic means, our results suggest that the critical oxygen concentration is of the order of 10 mM. This value is about ten times higher than that estimated for isolated mitochondria by Araki et al., (1983). In Figure 4 we attempt to correlate respiration rate, lactate-release and redox-changes of cytochrome a+a$_3$ and pyridine nucleotide calculated from the data of Araki et al., (1983) and of Tamura et al., (1978), with our present findings. It appears that the CrP/P$_i$ ratio was affected with minimal reduction of the tissue oxygen concentration. Thus visual inspection of Figure 4 suggests that the critical oxygen concentration for tissue energy state is substantially higher than previously estimated by other variables. The oxygen concentration at half-maximal CrP/P$_i$ is about 10 μM, about five times that estimated for oxygen consumption. It also appears that the CrP/P$_i$ behaved in a manner similar to that of the redox state of cytochrome a+a$_3$, whereas the curve of pH shift is similar to those of pyridine nucleotide reduction and lactate release. Thus, we concluded that mitochondrial pyridine nucleotide, a major fluorescent chromophore in cardiac tissue, was in equilibrium with cystolic lactate-pyruvate ratio and therefore with glycolysis.

According to the near-equilibrium theory (Owen and Wilson, 1974), respiration rate can be written as:

$$- \frac{dO_2}{dt} = \text{constant} \cdot (\text{Cyt.a}_3^{3+}) \cdot \left(\frac{\text{NADH}}{\text{NAD}} \right)^{1/2} \cdot \left(\frac{\text{ATP}}{\text{ADP P}_i} \right)^{-3/2} \cdot (O_2)$$

Figure 4 suggests that the decrease in oxygen concentration can be primarily compensated by the increase in cytochrome a$_3$ through the change in phosphate potential, where CrP/P$_i$ paralleled the reduction of cytochrome a+a$_3$. In the previous work (Tamura et al. 1983), the decrease in CrP of intact brain tissue parlleled the reduction of copper in cytochrome a+a$_3$. The oxygen affinity of copper is nearly 5-fold higher than that of heme a+a$_3$ (Hoshi et al. 1988). Thus, the relationship between tissue energy state and redox state of cytochrome a+a$_3$ differs between cardiac and brain tissues, possibly due to the difference of oxygen consumption rate and capacity of the energy reservoir.

SUMMARY

The relationship between the energy state and intracellular oxygen concentration was established with cardiac tissue, where the former could be monitored by ^{31}P-NMR and the latter by optical method for myoglobin absorption. The ratio of creatinephosphate to inorganic phosphate, PCr/Pi, started to fall at higher intracellular oxygen concentration than oxygen consumption (10ÎM). The pH shift measured by frequency shift of Pi paralleled the increase of lactate-release, of which half maximum was ~3ÎM. The critical oxygen concentrations differed nearly 5-fold when determined by energy state and oxygen consumption.

REFERENCES

Araki, R., Tamura, M., and Yamazaki, I. 1983. The effect of intracellular oxygen concentration on lactate release, pyridine nucleotide reduction, and respiration rate in the cardiac tissue. Circ. Res., 53:448.

Hoshi, Y., Hazeki, O., and Tamura, M. 1989. Oxygen characteristics of heme and copper in cytochrome oxidase. <u>This issue.</u>

Owen, C.S., and Wilson, D.F. 1974. Control of respiration by the mitochondrial phosphorylation state. <u>Arch. Biochem. Biophys.</u>, 161:581.

Tamura, M., Hazeki, O., Nioka, S., Chance, B., and Smith, D.S. 1988. The simultaneous measurements of tissue oxygen concentration and energy state by near-infrared and nuclear magnetic resonance spectroscopy. <u>Adv. Exp. Med. Biol.</u>, 222:359.

Tamura, M., Oshino, N., Chance, B., and Silver, I. 1978. Optical measurements of intracellular oxygen concentration of rat heart in vitro. <u>Arch. Biochem. Biophys.</u>, 191:8.

THE INTERACTION BETWEEN OXYGEN AND VASCULAR WALL

J. Grote[1], G. Siegel[2], K. Zimmer[1], and A. Adler[2]

1 Institute of Physiology I, University of Bonn,
D-5300 Bonn 1, Germany
2 Institute of Physiology, The Free University of Berlin
D-1000 Berlin 33, Germany

INTRODUCTION

During the last decade the important role of endothelium in the local control of vascular smooth muscle function has become more and more evident. In response to various chemical and physical stimuli, vascular endothelial cells synthesize and release substances which can induce changes in tone of the underlying smooth muscle cells (1,5,17,20,24,25,27). In addition, due to the uptake and the enzymatic conversion or breakdown of several circulating vasoactive substances, endothelium influences their activity in vascular smooth muscle (s. 25). Oxygen metabolites are able to influence or even disrupt these functions. Various reactive intermediates of oxygen metabolism cause characteristic changes in the metabolism of the vascular endothelial cells and induce e.g. an increased production of certain arachidonic acid metabolites (4,11,19,25). In vitro experiments suggest that oxidizing free radicals facilitate the release of the endothelium-derived relaxing factor(s) (EDRF) and a smooth muscle relaxation while superoxide anions depress the EDRF mediated decrease in smooth muscle tone (s. 25). Following exposure to a variety of free radicals, endothelial cell lesions, most frequently in the area of intercellular junctions, and also cytolysis were observed (11-13).

The present experiments were designed to investigate the oxygen-dependent reactions of membrane potential and tone of vascular smooth muscle and to elucidate the role of the endothelial cells in the responses of the smooth musculature to hyperoxia and hypoxia.

METHODS

The investigations were performed on isolated strips of canine carotid arteries. The vessels were excised from dogs immediately after death. After removing the adventitial connective tissue under a dissection microscope, vessel segments 4 to 5 mm in length were cut across to obtain circumferential strips. In some preparations the endothelium was gently denuded. Complete removal of the endothelium was evaluated after staining by light-microscopy at the conclusion of the experiments. The vessel strips were placed in a tissue bath perfused at constant flow (10 ml/min) with a modified Krebs-Ringer solution (21,22) maintained at 37 $^\circ$C. The physiological salt solution was equilibrated with gas mixtures containing 5% CO_2 and varried O_2 and N_2 fractions. The oxygen tension of the superfusion medium was continuously monitored close to the vessel strip with PO_2 microelectrodes (8). The experiments started under hyperoxic conditions with oxygen tensions of approx. 550 mmHg. After completion of the initial membrane potential and force generation measurements, the oxygen tension of the superfusion medium was stepwise decreased and, after reaching minimal values between approx. 20 and 30 mmHg, again increased to the control level by changing between superfusion solutions with varying oxygen tension. The CO_2 tension remained constant between 35 and 40 mmHg. The application period of each test solution was 15 minutes. Intermittent measurements of tissue PO_2 in the endothelial cells as well as of the underlying smooth muscle cells were performed using multiwire surface PO_2 microelectrodes (6). The membrane potential of the smooth muscle cells was intracellularly recorded with glass microelectrodes (21,22). At each PO_2 level at least 10 impalements were performed from the luminal side of the blood vessel preparations, and the derived values averaged. For determination of the force generation, the vessel strips were attached to an isometric force transducer. The resting tension was adjusted to 2 g. Since the initial tension values differed slightly in the various preparations, a normalization to a uniform level was performed (23).

RESULTS AND DISCUSSION

Under hyperoxia and normocapnia conditions a mean membrane potential of -65 mV was determined in the smooth muscle cells of intact preparations with endothelium. The corresponding smooth muscle tension was 2 g. As summarized in Fig. 1, the stepwise lowering of the oxygen tension in the superfusion medium caused a hyperpolarization of the cell membrane and a decrease in smooth muscle tone as soon as values below approx. 150 mmHg were attained. A first reaction of the vessel preparations to changes in PO_2 occured after 1-2 minutes, the total effect was obtained after 10 minutes at the latest. The further decrease in PO_2 induced a dose-dependent hyperpolarization and relaxation. Maximal hyperpolarization of the smooth muscle cell membrane and a minimal smooth muscle tension were observed with mean values of -74 mV and 1.3 g, respectively, at oxygen tensions of approx. 35 mmHg. Tissue PO_2 measurements, performed simultaneously in the superficial cell layers, resulted in values 10 to 15 mmHg below the ambient oxygen tension. Pronounced hypoxia with PO_2 values of the superfusion medium between 20 and 30 mmHg induced reduction of the hyperpolarization or even a depolarization of the cell membrane and a subsequent active force generation. The smooth muscle tension increased up to the level observed during hyperoxia. On stepwise return of oxygen tension to the initial levels, the changes in membrane potential and smooth muscle tension were reversible. As in preceeding experiments (8), membrane potential and force generation were strictly coupled.

Several authors described a comparable oxygen dependent relaxation or delatation of vascular smooth muscle preparations during mild hypoxia (1,10,25,26). Since biosynthesis of eicosanoids in various tissues is stimulated at low intracellular oxygen tensions (14,16) and an increase in the release of prostacyclin and prostaglandin E_2 were observed during hypoxia experiments with perfused vessel segments or vascular strips, it was hypothesized that release of prostaglandins from endothelial cells mediates vascular oxygen reactivity (1,8,18). Several in vivo and in vitro studies, however, provided evidence against this hypothesis (9,10,26). When stimulated by various neurohumoral mediators as well as by mechanical forces, the endothelial cells release a short-lived humoral agent which mediates vasorelaxation (5,17,25). One of these endothelium-derived relaxing factors (EDRF) has been described as nitric oxide (17). Possibly the EDRF mediates the observed reactions of the vascular smooth muscle cells under the conditions of mild hypoxia. During pronounced intracellular hypoxia with oxygen tensions below 10-15 mmHg, the cyclooxygenase activity is suppressed (14). Simultaneously, the endothelial cells release one or more diffusible vasoconstrictor substances (20,25,27). An endothelium-derived constricting factor (EDCF), endothelin, has recently been identified as a 21-residue peptide (27).

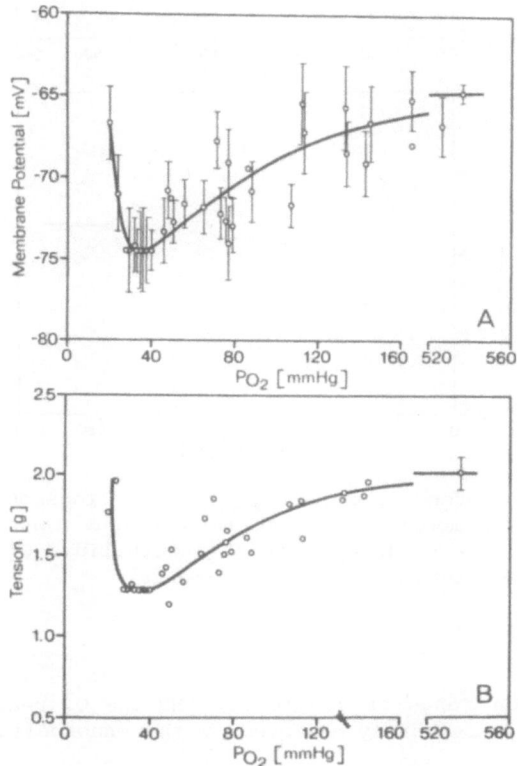

Figure 1. Effect of stepwise decrease in oxygen tension of the superfusion medium on membrane potential (A) and tension (B) of canine carotid artery strips with endothelium.
Given are mean values +SEM determined during 8 experiments.

Since the hypoxia-induced as well as the anoxia-induced changes in vascular smooth muscle tone have been described to be endothelium-dependent (1,20,25), the experiments were repeated on vessel preparations without endothelium. Mechanical removal of the endothelium caused a depolarization of the cell membrane and an active force generation of the vascular smooth muscle. Under hyperoxia conditions a depolarization of about 4 mV and an increase in tension of about 1.2 g was found. As can be seen in Fig. 2, the stepwise lowering of oxygen tension in the superfusion medium up to 35 mmHg induced in preparations without endothelium only a slight hyperpolarization of maximal 5 mV. The respective mean values of the membrane potential were -60 and -64 mV. The simultaneously determined smooth muscle tensions amounted to 3.2 and 2.9 g, respectively. Both results provide evidence that in

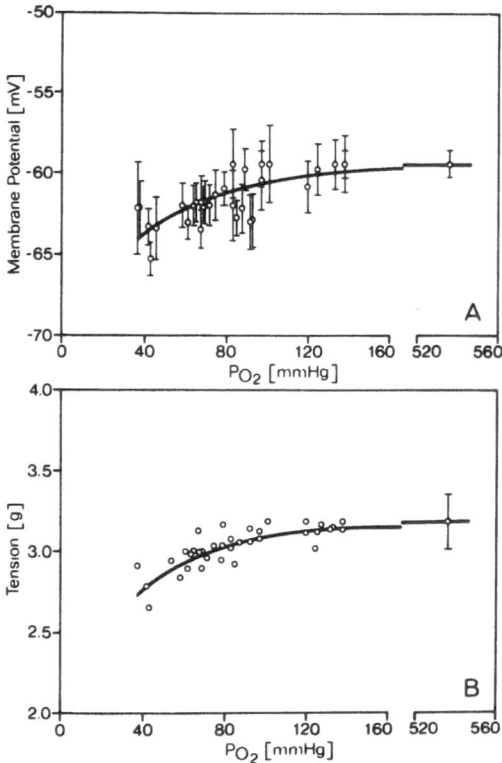

Figure 2. Effect of stepwise decrease in oxygen tension of the superfusion medium on membrane potential (A) and tension (B) of canine carotid artery strips without endothelium. Given are mean values ±SEM determined during 7 experiments.

the oxygen tension range of 550 to 35 mmHg the O_2 reactivity of vascular smooth muscle cells is mainly mediated by the endothelium.

To test the assumption whether the changes in membrane potential and smooth muscle tension observed during hypoxia are induced by changes in prostanoid production and release of the endothelial cells, comparable vascular strip experiments were performed after application of the cyclo-oxygenase inhibitor indomethacin (10^{-5} M). Under hyperoxia conditions, indomethacin induced a slight depolarization of the membrane by about 2 mV as well as an increase in smooth muscle tension of 0.3 g. Comparable augmentation of smooth muscle tone following indomethacin treatment has been reported by Busse et al. (1) and Rubanyi and Vanhoutte (20). The decrease in the oxygen tension of the superfusion medium induced the typical PO_2-dependent reactions of the vascular smooth muscle cells as shown in Fig. 3. However, when comparing the membrane potential and tension values measured in the indomethacin treated preparations with the respective values of the untreated ones a small but significant reduction in the hypoxia-induced hyperpolarization and relaxation of the smooth muscle cells becomes obvious. In addition, the ambient oxygen tension with which maximal hyperpolarization and relaxation occur, was shifted to higher values between 55 and 65 mmHg. At oxygen tensions below approx. 50-60 mmHg, a pronounced depolarization and a subsequent increase in smooth muscle tone were observed (Fig. 3).

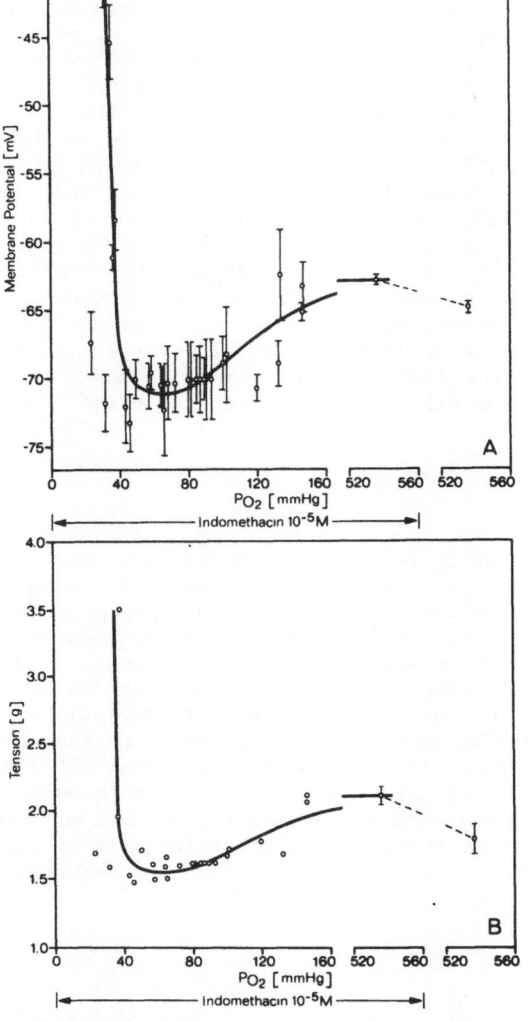

Figure 3. Effect of stepwise decrease in oxygen tension of the superfusion medium on membrane potential (A) and tension (B) of canine carotid artery strips with endothelium in the presence of indomethacin (10^{-5} M). Given are mean values $\pm$ SEM determined during 7 experiments.

The results of the three groups of experiments provide convincing evidence that the PO_2-dependent changes in membrane potential and tone of vascular smooth muscle cells as observed during mild and pronounced hypoxia are mediated mainly by the endothelial cells. The role of prostanoids in the vascular oxygen reactivity seems to be of minor importance.

The present results permit no explanation of the causal mechanisms underlying the oxygen-dependent reactions of membrane potential and tone of vascular smooth muscle cells. According to our findings as well as to results of various authors the release in EDRF and possibly prostanoids seem to be causally related to the hyperpolarization of the membrane and the decrease in force generation during mild hypoxia (s. 1,7,8,17,18,20,24,

25). An increase in the K^+ permeability, a decrease in the Na^+ permeability or a stimulation of the electrogenic Na^+ outward pump can be taken in consideration as reasons for the change in membrane potential (2,21). Prostacyclin e.g. induces hyperpolarisation due to an increase in the K^+ permebilility of the cell membrane and a stimulation of the electrogenic Na^+ outward pump (23). The smooth muscle relaxation simultanously observed may be attributed to a decrease in the intracellular Ca^{2+} activity caused by an increased absorption of calcium ions (21) or by the inactivation of recently described Ca^{2+} channels (15). In addition, the hypoxia induced vasodilatation may be caused by the stimulation of the guanylate cyclase. The production of cyclic GMP and an activation of cyclic GMP-dependent protein kinase are discussed to be responsible for the relaxation mediated by endothelium-dependent factors (24,25). The membrane depolarisation and contraction of the smooth muscle cells observed during pronounced hypoxia seem to be caused by the release of an endothelium-derived contracting factor (20,25,27). According to the results of Yanagisawa et al. (27) the action of the EDCF, endothelin, is closely associated with the Ca^{2+} influx through the dihydropyridine-sensitive Ca^{2+} channels.

SUMMARY

In vascular strips of canine carotid arteries stepwise lowering of oxygen tension from hyperoxic levels of 550 mmHg to 20 mmHg caused in preparations with endothelium a dose-dependent hyperpolarisation and relaxation of smooth muscle cells when oxygen tensions between approximately 150 mmHg and 35 mmHg were attained. Pronounced hypoxia with oxygen tensions below 30 mmHg induced a depolarisation and an increase in force generation. During comparable investigations on vessel preparations without endothelium only a slight hyperpolarisation and relaxation of the smooth muscle were observed when decreasing the oxygen tension from 550 mmHg to approx. 35 mmHg. In the presence of indomethacine (10^{-5} M) a small but significant reduction in the hypoxia-induced hyperpolarisation and decrease in smooth muscle tone was found in intact vascular strips with endothelium. Depolarisation and contraction occured at oxygen tensions below approx. 50-60 mmHg.

References

1. R. Busse, U. Förstermann, H. Matsuda, and U. Pohl, The role of prostaglandins in the endothelium-mediated vasodilatory response to hypoxia, Pflügers Arch. 401:77-83 (1984).
2. R. Detar, Mechanism of physiological hypoxia-induced depression of vascular smooth muscle contraction, Am. J. Physiol. 238:H761-H769 (1980).
3. J.E. Faber, P.D. Harris, and J.G. Joshua, Microvascular response to blockade of prostaglandin synthesis in rat skeletal muscle, Am. J. Physiol. 243: H51-H60 (1982).
4. B.A. Freeman, S.L. Young, and J.D. Crapo, Liposome-mediated augmentation of superoxide dismutase in endothelial cells prevents oxygen toxicity, J. Biol. Chem. 258:12534-12542 (1983).
5. R.F. Furchgott and J.V. Zawadzki, The obligatory role of endothelial cells in the relaxation of arterial smooth muscle by acetylcholine, Nature 288:373-376 (1980).
6. J. Grote, K. Zimmer, and R. Schubert, Effects of severe arterial hypocapnia on regional blood flow regulation, tissue PO_2 and metabolism in the brain cortex of cats, Pflügers Arch. 391:195-199 (1981).

7. J. Grote and R. Schubert, Regulation of cerebral perfusion and PO_2 in normal and edematous brain tissue, in: J.A. Loeppky, M.L. Riedesel, (eds.), Oxygen Transport to Human Tissue, Elsevier North Holland, Amsterdam, New York, Oxford, pp.169-178 (1982).

8. J. Grote, G. Siegel, K. Zimmer, and A. Adler, The influence of oxygen tension on membrane potential and tone of canine carotid artery smooth muscle, Adv. Exp. Med. Biol. 222:481-487 (1988).

9. T.H. Hintze and G. Kaley, Prostaglandins and the control of blood flow in the canine myocardium, Circ. Res. 40:313-320 (1977).

10. W.F. Jackson, Prostaglandins do not mediate arteriolar oxygen reactivity, Am. J. Physiol. 250:H1102-H1108 (1986).

11. R.M. Jackson, D.B. Chandler, and J.D. Fulmer, Production of arachidonic acid metabolites by endothelial cells in hyperoxia, J. Appl. Physiol. 61:584-591 (1986).

12. H.A. Kontos and M.L. Hess, Oxygen radicals and vascular damage, Adv. Exp. Med. Biol. 161:365-375 (1983).

13. F.S. Lamb, C.M. King, K. Harrell, W. Burkel, and R.C. Webb, Free radical-mediated endothelial damage in blood vessels after electrical stimulation, Am. J. Physiol. 252: H1041-H1046 (1987).

14. W.E.M. Lands, J. Sauter, and G.W. Stone, Oxygen requirements for prostaglandin biosynthesis, Prostaglandins Med. 1:117-120 (1978).

15. G. Loirand, P. Pacaud, C. Mironneau, and J. Mironneau, Evidence for two distinct calcium channels in rat vascular smooth muscle cells in short-term primary culture, Pflügers Arch. 407:566-568 (1986).

16. G. Markelonis and J. Garbus, Alterations of intracellular oxidative metabolism as stimuli evoking prostaglandin biosynthesis, Prostaglandins 10:1087-1106 (1975).

17. R.M.J. Palmer, A.G. Ferrige, and S. Moncada, Nitric oxide release accounts for the biological activity of endothelium-derived relaxing factor, Nature 327:524-526 (1987).

18. J.D. Pickard, Role of prostaglandins and arachidonic acid derivatives in the coupling of cerebral blood flow to cerebral metabolism, J. Cereb. Blood Flow Metab. 1:361-384(1981).

19. G.M. Rosen and B.A. Freeman, Detection of superoxide generated by endothelial cells, Proc. Natl. Acad. Sci. USA, 81:7269-7273 (1984).

20. G.M. Rubanyi and P.M. Vanhoutte, Hypoxia releases a vasoconstrictor substance from the canine vascular endothelium, J. Physiol. 364: 45-56 (1985).

21. G. Siegel, Membranphysiologische Grundlagen der peripheren Gefäßregulation, Physiologie aktuell 1:31-52 (1986).

22. G. Siegel, R. Ehehalt, and H.P. Koepchen, Membrane potential and relaxation in vascular smooth muscle, in: P.M. Vanhoutte, I. Leusen, (eds.), Mechanisms of Vasodilatation, Karger, Basel, pp. 56-72 (1978).

23. G. Siegel, G. Stock, F. Schnalke, and B. Litza, Electrical and mechanical effects of prostacyclin in the canine carotid artery, in: R.J. Gryglewski, G. Stock, (eds.), Prostacyclin and its Stable Analogue Iloprost, Springer, Berlin,pp. 143-149 (1987).

24. P.M. Vanhoutte, The end of the quest?, Nature 327:459-460 (1987).

25. P.M. Vanhoutte, G.M. Rubanyi, V.M. Miller, and D.S. Houston, Modulation of vascular smooth muscle contraction by the endothelium, Ann. Rev. Physiol. 48:307-320 (1987).

26. E.P. Wei, E.F Ellis and H.A. Kontos, Role of prostaglandins in pial arteriolar response to CO_2 and hypoxia, Am. J. Physiol. 238:H226-H230 (1980).

27. M. Yanagisawa, H. Kurihara, S. Kimura, Y. Tomobe, M. Kobayashi, Y. Mitsui, Y. Yazaki, K. Goto, and T. Masaki, A novel potent vasoconstrictor peptide produced by vascular endothelial cells, Nature 332:411-415 (1988).

REGULATION OF CAPILLARY BLOOD FLOW: A NEW CONCEPT

D.K. Harrison, S. Birkenhake, N. Hagen, S. Knauf and M. Kessler

Institut für Physiologie und Kardiologie der Universität Erlangen-Nürnberg, Waldstrasse 6, 8520 Erlangen/FRG

INTRODUCTION

Last year we presented preliminary data which indicated a considerable inhomogeneity of capillary blood flow in skeletal muscle as measured both in the sartorius muscle of dogs and in the vastus medialis muscle in rabbits (Harrison et al., 1988). Measurements of capillary blood flow with particular emphasis on its distribution at rest, using hydrogen clearance techniques, during hypoxaemia and in contracting muscle led us to propose a two compartment model for the distribution of capillary blood flow in skeletal muscle.

This paper presents further data concerning the relationship between local oxygen supply and capillary blood flow, and a model for their local regulation is proposed.

METHODS

Experiments were carried out in the sartorius muscle in dogs and the vastus medialis muscle in rabbits. The animals were anaesthetised (without the use of barbiturates), relaxed and artificially ventilated. Blood pressure, heart rate blood gas and acid base status were monitored throughout the course of the experiments.

Tissue pO_2 was measured using the multiwire surface electrode of Kessler and Lübbers. The same type of electrode was used for recording local hydrogen clearance for the measurement of capillary blood flow. Needle electrodes were used to measure intravenous and intraarterial hydrogen clearance curves. These curves were analysed using standard curve peeling techniques.

HYPOXAEMIA AND STIMULATION EXPERIMENTS

Hypoxaemia was induced in the first series of experiments in dogs in a single step down to a mean arterial pO_2 of 32 mmHg in order to investigate changes in regional and capillary blood flow in relation to the local tissue oxygen supply. In a third series, the muscle was stimulated electrically at increasing frequencies via the femoral nerve.

Figure 1 shows the cumulative pO_2 histograms measured under control conditions and during hypoxaemia. Note that how the histogram becomes much steeper during hypoxaemia, but less than 5% of values lay below 5 mmHg despite the low arterial pO_2. Figure 2 shows cumulative pO_2 histograms from the stimulation experiments. Again changes in slope are observed, but even at 8 Hz stimulation only 0.5% of values are to be found within the 0-5 mmHg class.

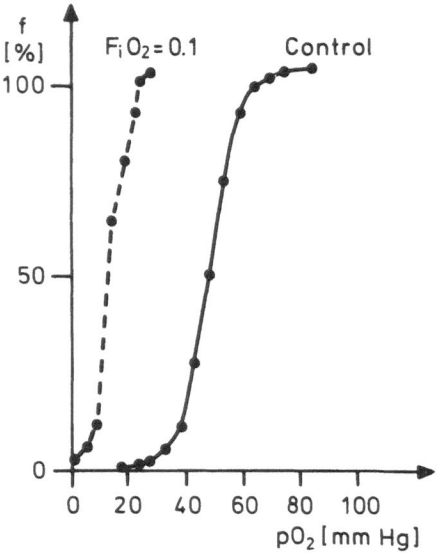

Fig. 1. Cumulative pO_2 histograms measured under control conditions and at a mean arterial pO_2 of 32 $\pm$ 4 (S.D.) mmHg. Each histogram represents at least 624 measurements.

Figure 3 summarises the results of femoral artery flow and capillary blood flow measurements from all experiments. It can be seen that a significant increase in arterial flow was accompanied by a slight, but not statistically significant decrease in capillary flow during hypoxaemia. On the other hand, stimulation induced a highly significant increase in capillary blood flow together with a slight, but not significant, increase in arterial flow.

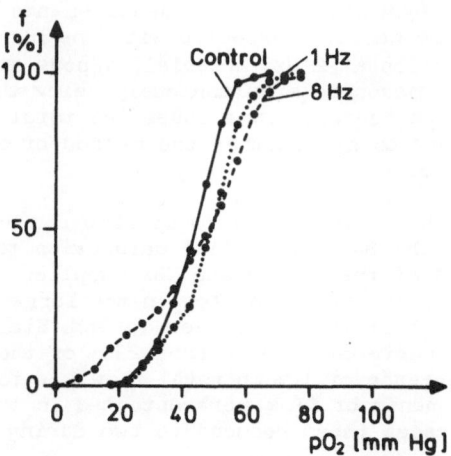

Fig. 2. Cumulative pO_2 histograms measured at rest and during stimulation of the muscle at 1 and 8 Hz. Each histogram contains at least 520 measurements.

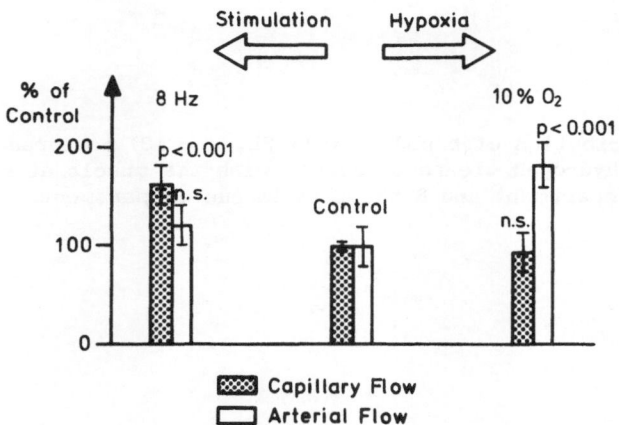

Fig. 3. Changes in blood flow in femoral artery and capillary blood flow as a result of stimulating the muscle at 8 Hz (left) and during severe hypoxaemia.

INTRAVASCULAR HYDROGEN CLEARANCE MEASUREMENTS

In order to investigate the discrepancy between the capillary blood flow measured at the surface of the muscle and the arterial flow (see Fig. 3) intravascular hydrogen clearance measurements were carried out in the vastus medialis muscle of rabbits with the muscle at rest and during direct stimulation at 2 Hz. Arterial, venous and tissue hydrogen clearance curves were measured simultaneously using the 30 second partial saturation technique. In some cases the total saturation technique was used in order to see whether the method of application influenced the results.

Figure 4 shows that two components of flow can be detected in the resting muscle using the 30 sec. partial saturation technique. 77% of the flow supplies 35% of the tissue and 23% supplies 65%. When the muscle was stimulated, the relative flow in the larger tissue compartment increased from 35% of the total flow to 96%. Simultaneously, the flow in the other compartment reduced from 220% of the total flow to 99%. A small mean decrease of 10% in total flow was found during stimulation. Three components of flow were detected in the three saturation experiments, in all cases these reduced to two during stimulation.

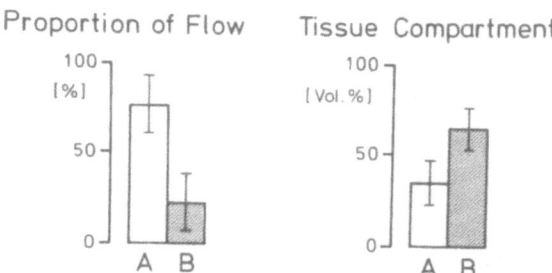

Fig. 4: Distribution of blood flow ($\pm$ SD, n = 13) measured from the intravenous hydrogen clearance curves with the muscle at rest. A is the high flow compartment and B the main tissue compartment.

DISCUSSION

Our results indicate very clearly the existence of local mechanisms for redistributing capillary blood flow in skeletal muscle in response to an increase in the demand for oxygen or a reduction in its supply.

pO_2 topograms recorded in our institute by Beier (1987) by moving the multiwire electrode in 50 μm steps across the surface of the gracilis muscle in rats have recently been discussed by Kessler et al., 1988. These topograms indicate functional structures and provided interesting information concerning principles behind the regulation of capillary blood flow and local oxygen supply. Areas of quite uniform pO_2 distribution are to be found and their extent is determined by the distribution of the lengths and numbers of capillaries within the microvascular unit supplying it. However, there are considerable differences in the pO_2 levels between these capillary supply units. This is reflected in the relatively broad frequency histograms measured under control conditions (see for example Figs. 1 and 2).

These topogram structures are included in the block diagram for capillary flow illustrated in Fig. 5 based on our measurements. The heterogeneity between average supply units is small relative to the high flow regions and thus only two compartments are detected in the intravenous clearances. Heterogeneities exist not only in terms of high flow capillaries – reflected in the topographical structure as areas of

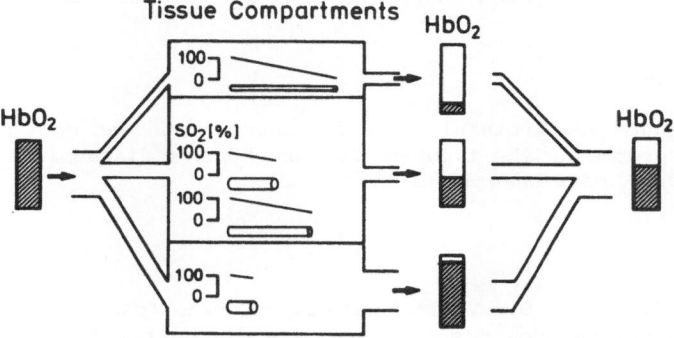

Fig. 5. A model based on our experimental results illustrating that capillaries have various diameters and lengths which give rise to heterogeneities of capillary blood flow measured in our experiments, and hence a topogram structure. The associated levels of haemoglobin saturation (HbO_2) emerging from the compartments and in mixed venous blood are shown.

high pO_2 – but also in terms of longer, presumably narrower capillaries (Potter and Groom, 1983; Skalak and Schmid-Schönbein, 1986). The cells at the venous end of such capillaries may function as a form of early warning system should the oxygen supply fall or consumption increase to a level that would cause anoxia within the tissue (Frank et al., 1989). A convergence of signals from these cells induces redistributions of flow between the supply units. This mechanism is the first, local reaction which mobilises the oxygen reserves which are available in the immediate vicinity in the high flow regions should the demand increase as a result of contraction. Should the demand further increase, for example as the result of the contraction of a group of muscles, then the cardiac output increases to ensure that adequate reserves of oxygen are maintained.

The way in which local signals may be generated and propagated are discussed by Kessler et al., (1984, 1986) based on evidence that local anoxia induces an increase in flow and modulates oxygen consumption (Höper and Kessler, 1981). Briefly, changes in the activity of monoamine oxidase B (MAO B) located on the outer wall of the mitochondrial membrane produces a local signal which propagates along the capillary to the nearest smooth muscle cell. One mechanism involved in this signal chain is probably a change in intercellular calcium permeability due to a reduced pO_2 (Höper, 1984).

CONCLUSIONS

The results of many experiments have shown that under the key feature of the regulation of blood flow in skeletal muscle is a functional heterogeneity characterised by: 1) Topogram structures comprising 2) high flow compartments which are a local oxygen reserve and 3) low flow regions which have a protective, early warning function. These are regulated by a highly efficient system of local signals which ensure that the local oxygen supply is always adequate to meet the demands of all cells, even those at the venous end of capillaries.

ACKNOWLEDGEMENTS

The authors are grateful to Gundi Schuster who was most patient and skillful in preparing the figures. Dr Jens Höper's discussions, comments and criticisms have been invaluable.

REFERENCES

Beier, I., 1987, Die Verteilung des Sauerstoffpartialdruckes an der Oberfläche des Musculus gracilis der Ratte. Dissertation, Friedrich-Alexander Universtität Erlangen/Nürnberg.

Frank, K.H., Kessler, M., Appelbaum, W., Dümmler, J., Zündorf, J., Anderer, W., Höper, J., Klövekorn, W.P. and Sebening, F., 1989, Oxygen supply of the myocardium. This volume.

Harrison, D.K., Birkenhake, S., Knauf,S., Hagen, N., Beier, I. and Kessler, M., 1988, The role of high flow capillary channels in the local oxygen supply to skeletal muscle. Adv. Exp. Med. Biol., 222:623

Höper, J. and Kessler, M., 1981, pO_2 and sodium dependant mechanism regulating liver blood flow. In "Oxygen Transport to Tissue", A.G.B. Kovach, E. Dora, M. Kessler and I.A. Silver, eds., Adv. Physiol. Sci., 25, Budapest/London: Akademiai Kiado/Pergamon.

Höper, J., 1984, Einfluss der Sauerstoffversorgung auf das Membranpotential und die intercelluläre elektrische Koppelung von Hepatocyten. Dissertation, Friedrich-Alexander Universität Erlangen/Nürnberg.

Kessler, M. and Lübbers, D.W., 1966, Aufbau und Anwendungsmöglichkeit versschiedener PO_2-Elektroden. Pflüg. Arch. ges. Physiol., 291:R82.

Kessler, M., Höper, J., Harrison, D.K., Skolasinska, K., Klövekorn, W.P., Sebening, F., Volkholz, H.J., Beier, I., Kernbach, C., Rettig, V. and Richter, H., 1984, Tissue O_2 supply under normal and pathological conditions. Adv. Exp. Med. Biol., 169:69.

Kessler, M., Harrison, D.K. and Höper, J., 1986, Tissue oxygen measurement techniques. In: "Microcirculatory Technology", C.H. Baker and W.L. Nastuk, eds., New York: Academic Press.

Kessler, M., Anderer, W., Frank, K.H., Höper, J., Harrison, D.K., Richter, H. and Klövekorn, W.P., 1988, Die Bedeutung der lokalen O_2-Versorgung für das Entstehen hypokinetischer Myokardareale. Funktionsanalyse biologischer Systeme, 18:215 Stuttgart: Fischer,

Skalak, T.C. and Schmid-Schönbein, G.W. (1986), The microvasculature in skeletal muscle IV. A model of the capillary network. Microvasc. Res., 32:333.

Potter, R.F. and Groom, A.C., 1983, Capillary diameter and geometry in cardiac and skeletal muscle studied by means of corrosion casts. Microvasc. Res., 25:68.

PRECAPILLARY O_2 LOSS AND ARTERIOVENOUS O_2 DIFFUSION SHUNT ARE BELOW LIMIT OF DETECTION IN MYOCARDIUM

C.R. Honig and T.E.J. Gayeski

The University of Rochester, School of Medicine and Dentistry
601 Elmwood Avenue
Rochester, NY 14642

Is venous PO_2 an index of tissue PO_2, of end capillary PO_2, or neither of these because of diffusive shunting from arterioles to venules? To evaluate diffusive shunting and venous PO_2 in myocardium we determined hemoglobin (Hb) saturation in arterioles and venules and myoglobin (Mb) saturation in individual subepicardial myocytes by use of a cryomicrospectrophotometric method (Gayeski, 1981, Degner and Gayeski, 1987).

METHODS

Animals were anesthetized with pentobarbital, and ventilated to hold blood gases and arterial pH within normal limits. Arterial pressure and heart rate were monitored, and blood losses during thoracotomy were replaced with dextran or donor blood. Hearts were frozen in situ by applying a copper heat sink of appropriate size (Honig and Gayeski, 1987, Gayeski and Honig in press). The hearts were removed while in firm contact with the heat sink and immersed in liquid N_2. Approximately 10 ms were required to reach 0°C 100 μm from the surface, a rate of freezing sufficient to trap steep O_2 gradients (Gayeski and Honig 1986, 1988). Samples were trimmed under liquid N_2 and transferred to the cold stage of a Leitz MPVI microspectrophotometer regulated at -110°. Mean saturation of myoglobin (Mb) was invariant with depth up to 800 μm from the pericardial surface (Gayeski and Honig, in press). All measurements to be described were made within 500 μm of the pericardium. The calculated error in Mb and Hb saturation attributable to freezing is <0.1%.

Determination of Saturation and PO_2

Mb and Hb saturations were determined with a four-wavelength method that takes account of the effect of scattered light (Gayeski, 1981). Cytochromes had no detectable effect on absorption at the wavelengths used even in hearts deliberately rendered anoxic (Honig and Gayeski, 1987; Gayeski and Honig, in press). The PO_2 in equilibrium with Mb was calculated from saturation assuming a Mb P_{50} of 5.3 torr at 37°C (Gayeski, 1981). This PO_2 can be interpreted as the PO_2 to which cytochrome a,a3 is exposed since the ΔPO_2 between cytosol and mitochondria is <0.05 torr (Clark et al., 1987). Hb saturation determined with the 4-wavelength method is the same between 10% and 45% hematocrit (Degner and Gayeski, 1987). Saturations can therefore be determined in microvessels despite variability of dynamic hematocrit in vivo. Venule PO_2 was estimated from Hb saturation and the oxydissociation curves of dog or rat blood.

Spatial Resolution

Light is collected from a very small volume of quick-frozen tissue chiefly because of reflection and refraction in the ice crystals of the sample. (Gayeski, 1981; Gayeski and Honig, 1986 and in press). When approaching a vessel from a cardiac myocyte, presence of Hb could not be detected until the leading edge of a 2 x 2 μm measuring diaphragm was < 2μm from the vessel wall. The change in signal was equally sharp when the wall was approached from the interior of the vessel. An optical model predicts that the optical catchment volume extends 1.5 μm beyond the perimeter of the measuring diaphragm of the photometer and 1.5 μm deep to the sample surface. Mb saturation was determined with a 2 x 2 μm measuring diaphragm positioned at the center of the cell profile. Since cardiac myocytes were 10 - 18 μm in diameter the Hb in contiguous vessels was well outside the catchment volume for Mb measurements. A 6 x 6 μm measuring diaphragm was used in microvessels, except when searching for saturation gradients within a vessel. For that purpose a 2 x 6 μm measuring diaphragm was positioned within the image of the vessel with its long axis parallel to the wall. With the leading edge as close as to the wall as possible measurements correspond to saturation 3 μm from the wall.

Sampling

Probability distributions for $PmbO_2$ were based on 50 cells, each at least 10 cell diameters from any other cell chosen. Hb saturation measurements were limited to vessels ≥16 μm in diameter. Arterioles and venules could be readily distinguished when paired; the former were smaller in diameter with thicker walls and circular rather than elliptical profile in cross-section. Since subepicardial arterioles are sparse (Bassingthwaighte et al., 1974; Grayson et al., 1974), all those encountered were included in the sample. All venules larger than 100 μm were sampled, as well as any grossly visible ones in the pericardial fat and connective tissue. Intramural venules with diameters ≥16 μm but <100 μm were selected at random.

RESULTS AND DISCUSSION

Relation Between Venous PO_2 and Cell PO_2

Figure 1 shows the probability distribution for 100 myocytes, 50 from a dog and 50 from a rat. The distribution of PO_2 in 83 subepicardial venules from the same two hearts is shown for comparison. Virtually identical distributions of intracellular PO_2 were obtained from 18 additional hearts from 5 species (Gayeski and Honig, in press). Thus the PO_2 distributions in Figure 1 are representative. There is little overlap of the two distributions; the maximum cell PO_2 encountered was lower than PO_2 in 93% of venules sampled, and median venule PO_2 was more than 5 times median cell PO_2.

There are two possible explanations for the difference (ΔPO_2) between blood and tissue. 1. A large drop in PO_2 exists over the short distance between the red cell and sarcolemma. Such steep O_2 gradients are predicted by recent clearance experiments (Rose and Goresky, 1985; Rose et al., 1988), by mathematical models, (Federspiel and Popel, 1986; Groebe and Thews, 1986), and by spectrophotometric measurements in red skeletal muscle (Gayeski and Honig, 1986, 1988). If alternative #1 is correct *a red cell does not equilibrate with tissue* in passage through a capillary. 2. An alternative explanation for Figure 1 is that venous PO_2 is not representative of end-capillary PO_2 because of diffusive shunting of O_2 from arterioles to venules (Pittman and Duling, 1977; Schubert et al., 1978; Roth and Feigl, 1981; Harris, 1986). Diffusive shunting is

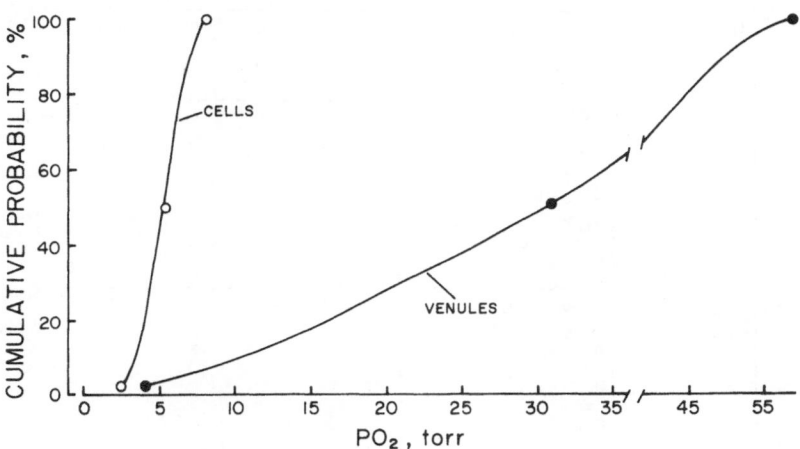

Figure 1. Comparison of PO_2 in 100 subepicardial myocytes with PO_2 in 83 subepicardial venules. Note change in scale of abscissa.

consistent with the anatomy of the coronary microcirculation. Arterioles 10 μm in diameter and larger are accompanied by venules to form a countercurrent pair or triplet. The larger vessels run together for several mm. (Bassingthwaighte et al., 1974; Grayson et al., 1984). Moreover, substantial diffusive shunting of heat and inert gases has been observed in myocardium (Roth and Feigl, 1981; Bassingthwaighte et al., 1984; Piiper et al., 1985), and precapillary O_2 losses have been clearly demonstrated in skeletal muscle (Duling and Berne, 1970; Pittman and Duling, 1977; Kuo and Pittman, 1988; Honig and Gayeski, submitted). We sought to determine whether alternative 1 or alternative 2 is correct by quantifying precapillary O_2 losses and diffusive shunting of O_2 in subepicardium.

Precapillary O_2 Losses

Our principal results are summarized in Figure 2; see legend for conventions.

Slopes of linear least squares regressions of arteriolar saturation on vessel diameter were not different from zero, and with one exception saturation in every arteriole was within 5% of saturation in the aorta.

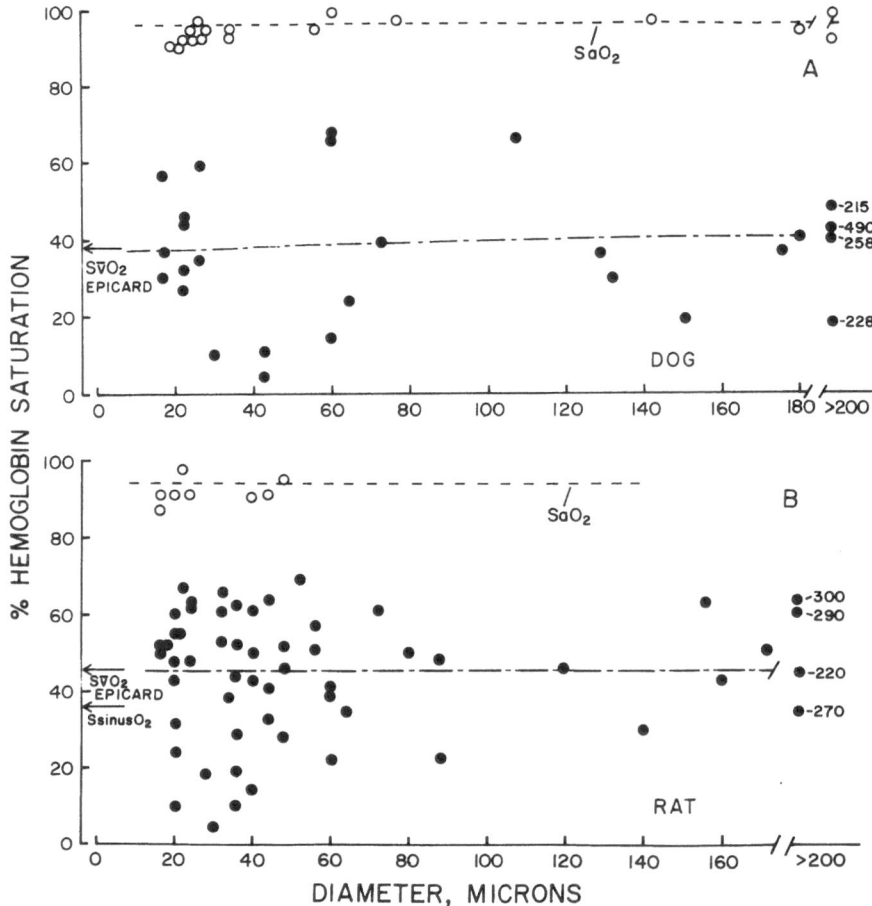

Figure 2. Measurements were obtained in one dog heart and one rat heart. O = arterioles, ● = intramural venules; numbers indicate diameters of venules beyond the range of the graph. ---- = saturation in aorta determined at 37°C with a Co-Oximeter.—— - —— - —— = least squares regression of saturation on venule diameter.

Thus *precapillary O_2 losses were within the error of the cryospectrophotometric method.* Even if all the precapillary O_2 flux were shunted to venules the rise in venous saturation would be orders of magnitude too small to account for the data in Figure 1.

In skeletal muscle O_2 diffusion from arterioles accounts for up to 25% of $\dot{V}O_2$ at very low flow at rest, and 10-15% of $\dot{V}O_2$ at flows and rates of cytochrome turnover comparable to those in hearts of anesthetized animals (Honig and Gayeski, submitted). Virtually all this O_2 is from vessels <40 μm in diameter. The fact that the same methods could not detect precapillary O_2 loss in subepicardium could be accounted for by the geometry of the arteriolar network. Since arterioles are short and sparse in subepicardium (Bassingthwaighte et al, 1974), the aggregate arteriolar surface area is small. Consequently, the resistance of the arteriolar network to O_2 diffusion should be large in subepicardium. Low arteriolar density in subepicardium limits the aggregate cross-sectional area, resulting in high red cell velocities at normal or high volume flow. The combination of high red cell velocity and short red cell path length generates short red cell precapillary transit times; there is simply *not enough time for a significant amount of O_2 to diffuse across the limited surface area of the network* (Gayeski et al., 1988).

Diffusive Shunting

Relation to Venule Diameter. Saturation in venules was markedly heterogeneous, as reported by Weiss et al. (1978). If a physiologically significant O_2 shunt were present saturation should increase with venule diameter. However, slopes of linear least squares regressions of venule saturation on diameter were not statistically different from zero, and the mean saturation in the population sampled was the same as saturation in the macroscopic epicardial veins. We again conclude that O_2 diffusion from coronary arterioles to venules is far too small to account for the difference between cell and venule PO_2 shown in Figure 1.

Distribution of Saturation Within a Venule. Pittman and Duling (1977) determined saturation in arterioles and venules in hamster cheek pouch in vivo. Saturation in a venule was found to be higher where it crossed over an adjacent arteriole. We confirmed their finding in frozen dog and rat gracilis muscles. However, a detectable influence of an arteriole on saturation in the paired venule was observed only if the two vessels were less than 15 μm apart. The observed effect of separation distance is predicted by a mathematical model of diffusive O_2 shunting (Clark et al., personal communication). All vessel pairs encountered in the hearts used for Figures 1 and 2 were separated by more than 15 μm. In coronary casts prepared by Bassingthwaighte et al. (1974) separation distances were 30-80 μm, an arrangement that precludes significant shunting of O_2 despite countercurrent blood flow. Even at measuring sites as close as possible to the venule wall opposite an arteriole saturations did not exceed saturations elsewhere in the venule. In the example shown in Figure 3 saturations closest to the arteriole were among the lowest found. The saturation gradient at upper right in Figure 3 illustrates the limited range of diffusive interactions in blood. Since small coronary venules are extremely short (Grayson et al., 1974) heterogeneity of saturation within a large venule is probably caused by incomplete mixing of confluent streams. Similar saturation gradients were found in macroscopic epicardial veins and in the coronary sinus, suggesting that catheter sampling of effluent blood could introduce significant error in estimates of O_2 extraction and VO_2.

Interpretation of Venous PO_2

The foregoing results demonstrate that venous PO_2 should closely approximate mean end-capillary PO_2. The large ΔPO_2 between blood and tissue shown in Figure 1 therefore identifies the capillary as a major site of resistance to blood-tissue O_2 transport, as predicted by Gayeski (1981). This resistance is considered from the experimental standpoint by Gayeski and Honig (1986) and by Rose and associates (1985, 1988). Theoretical analyses of O_2 mass transport consistent with experimental data have been reported by Federspiel and Popel (1986) and by Groebe and Thews (1986). A practical consequence of the steep blood-tissue O_2 gradient is that *coronary venous PO_2 cannot be used to judge the adequacy of O_2 supply.*

Significance for O_2 Gradients from Epi to Endocardium

In the rat the coronary sinus was included in the tissue beneath the heat sink. Saturation in the sinus was significantly lower than in subepicardial venules, as might be expected from the reported transmural gradient in O_2 extraction (Weiss et al., 1978). Since subendocardial flow is equal to or slightly greater than subepicardial flow (Feigl, 1983), lower saturation in subendocardial venules has been interpreted to

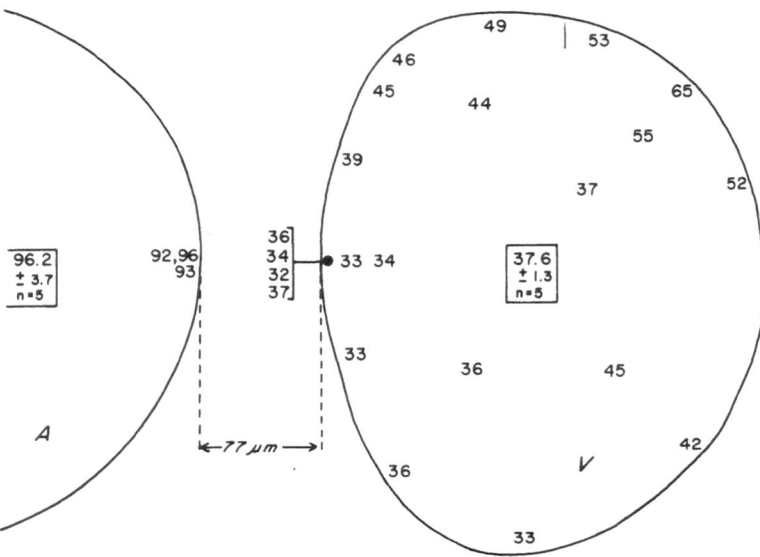

Figure 3. Distribution of saturations in cross-sectional profile of a 170 μm venule situated 77 μm from its paired 160 μm arteriole.

indicate a transmural gradient in $\dot{V}O_2$ (Weiss et al., 1978). This interpretation depends on absence of arteriovenous O_2 shunting between the long, countercurrent distributing vessels that penetrate from epi to endocardium. Since precapillary O_2 losses and arteriovenous diffusive shunting could not be detected even in small coronary vessels, our data strongly support the view that $\dot{V}O_2$ is indeed higher in the deeper layers of the wall.

Comparison with Results of Others

Results with O_2 electrodes depend qualitatively and quantitatively on the ratio of tip diameter to cell diameter (Honig and Gayeski, 1987). The smaller the tip the better. Whalen et al. (1973) used microelectrodes with tip diameter 1-3 μm to determine PO_2 in subepicardium. Only sites at which intracellular action potentials could be recorded were included in the analysis. Median PO_2 was >5 torr and

<10 torr, only slightly higher than the median PO_2 in our study. Coburn et al. (1973) calculated intracellular PO_2 for dog heart from the blood-tissue partition of CO. The mean PO_2 so estimated was 4-6 torr. Rose and Goresky (1985) estimated tissue PO_2 in dog heart at 1/3 vascular PO_2 from clearance of O_2^{18} in multiple indicator dilution experiments. Thus four independent methods yield cell PO_2 well below coronary venous PO_2. Since diffusive shunting can be ruled out, it is difficult to escape the conclusion that the O_2 gradient between blood and tissue is indeed large in myocardium.

We are by no means the first to state that the ΔPO_2 between blood and tissue is not caused by diffusive shunting. Huhmann et al. (1967) observed that PO_2 in the outflow from a saline perfused heart approaches zero asymptotically as PaO_2 falls, a result that precludes a significant arteriovenous diffusion shunt. Rose et al. (1985, 1988) could not detect O_2^{18} in the venous outflow from blood perfused hearts prior to the arrival of labelled red cells. This is strong evidence against an arteriovenous diffusion shunt. It is not conclusive, however, because similar experiments with inert gases that are indeed diffusively shunted indicate that precession of O_2 could be very small and difficult to detect (Feigl and Roth, 1981). The principal difference between the behaviors of O_2 and inert gases relates to the enormous binding capacity of Hb. The influences of Hb binding, red cell transit time, and arteriovenous separation distance have been evaluated for dog gracilis muscle (Clark et al., personal communication). Their analytical model indicates that arteriovenous diffusive shunting is negligible in both heart and red skeletal muscles. Comparably low estimates of the extent of diffusive shunting of O_2 have been obtained with compartment models (Roth and Wade, 1986, Rose et al. 1988).

Sites of Resistance to O_2 Mass Transport

The foregoing indicates that a large drop in PO_2 exists over the short distance between the surface of a red cell and the surface of a myocyte. This steep gradient denotes the principal site of resistance to O_2 mass transport. The PO_2 gradients within red myocytes are small by comparison (Gayeski and Honig, 1986, 1988 in press). The barrier-like behavior of the microcirculation is largely due to O_2 binding to Hb (Rose et al., 1985, 1988; Groebe and Thews, 1986). However, according to Rose et al. (1988) low O_2 diffusivity through endothelium is also a factor. We believe that a low O_2 diffusion coefficient need not be assumed. Instead the steep O_2 gradient between blood and tissue could be caused by high flux density through the carrier-free region between the red cell and sarcolemma (Gayeski, 1981; Federspiel and Popel, 1986; Groebe and Thews, 1986; Gayeski et al., 1988). Experiments and models designed to discriminate between these alternative hypotheses are essential for further progress in O_2 transport.

SUMMARY

1. Mean intracellular PO_2 is much lower than mean venous PO_2 in subepicardium.
2. The drop in Hb saturation between aorta and terminal arterioles is within the 5% error of our method.
3. Arteriolar O_2 has no effect on saturation in paired countercurrent venules in myocardium.
4. Saturation in coronary venules is independent of venule diameter and indistinguishable from saturation in macroscopic epicardial veins.

5. Since diffusive O_2 shunting is negligible and PO_2 is approximately linearly related to saturations over the observed range, mean coronary venous PO_2 should closely approximate mean-end capillary PO_2.
6. O_2 mass transport from blood to tissue requires a steep PO_2 gradient between the capillary and the surface of a tissue cell.

ACKNOWLEDGEMENTS

We thank our colleagues, A. Clark, P. Clark and K. Groebe, for many helpful discussions, and for sharing results of their mathematical model of precapillary O_2 losses. Our research is supported by Grant HLB03290 from the U.S. Public Health Service.

REFERENCES

Bassingthwaighte, J. B., Yipintsoi, T., and Harvey, R. B., 1974, Microvasculature of the dog left ventricle myocardium, Microvasc. Res., 7:229.

Bassingthwaighte, J. B., Yipintsoi, T., and Knoop, T. J., 1984, Diffusional arteriovenous shunting in the heart, Microvasc. Res., 28:233.

Clark, A., Clark, P. A. A., Connett, R. J., Gayeski, T. E. J., and Honig, C. R., How large is the drop in PO_2 between cytosol and mitochondrial? Am. J. Physiol., 252:C583, 1987.

Coburn, R. F., Ploegmakers, F., and Gondrii, P., and R. Abboud., 1973, Myocardial myoglobin O_2 tension, Am. J. Physiol., 224:870.

Degner, F., and Gayeski, T. E. J., 1987, A comparison of a four wavelength analysis and multicomponent wavelength analysis applied to determination of hemoglobin saturation, Adv. Exper. Med. Biol., 215:153.

Duling, B. R., and Berne, R. M., 1970, Longitudinal gradients in periarteriolar oxygen tension, Circ. Res., 27:669.

Federspiel, W. J., and Popel, A. S., 1986, A theoretical analysis of the effect of the particulate nature of blood on oxygen release in capillaries, Microvasc. Res., 32:164.

Feigl, E. O., 1983, Coronary physiology, Physiol. Rev., 63:1.

Gayeski, T. E. J., 1981, A Cryogenic Microspectrophotometric Method for Measuring Myoglobin Saturation in Subcellular Volumes; Application to Resting Dog Muscle, (Ph.D. Thesis), Rochester, NY: University of Rochester, Univ. Microfilms, No. DA9224720, Ann Arbor, MI.

Gayeski, T. E. J., Federspiel, W. J., and Honig, C. R., 1988, A graphical analysis of the influence of red cell transit time, carrier-free layer thickness, and intracellular PO_2 on blood-tissue O_2 transport, Adv. Exper. Med. Biol., 222:25.

Gayeski, T. E. J., and Honig, C.R., 1986, O_2 gradients from sarcolemma to cell interior in a red muscle at maximal VO_2, Am. J. Physiol., 251:789.

Gayeski, T. E. J., and Honig, C. R., 1988, Intracellular PO_2 in individual cardiac myocytes in dog, cat, rabbit, ferret and rat, Am. J. Physiol., in press.

Gayeski, T.E.J., and Honig, C.R., 1988, Intracellular PO_2 in long axis of individual fibers in working dog gracilis muscle, Am. J. Physiol., 254:H1179.

Grayson, J., Davidson, J. W., Fitzgerald-Finch, A., and Scott, C., 1974, The functional morphology of the coronary microcirculation in the dog, Microvasc. Res., 8:20.

Groebe, K., and Thews, G., 1986, Theoretical analysis of oxygen supply to contracted skeletal muscle, <u>Adv. Exper. Med. Biol.</u>, 200:495.

Harris, P. D., 1986, Movement of oxygen in skeletal muscle., <u>News Physiol. Sci.</u>, 1:147.

Honig, C. R., and Gayeski, T. E. J., 1987, Comparison of intracellular PO_2 and conditions for blood-tissue O_2 transport in heart and working red skeletal muscle, <u>Adv. Exper. Med. Biol.</u>, 215:309.

Huhmann, W., and Niesel, W., 1967, Untersuchungen über die Bedingungen für die Sauerstoffversorgung des Myocards an perfundierten rattenherzen. I. Zür frage nach dem vorliegen einer Gegenstromversorgung. <u>Pflüger's Arch.</u>, 294:250.

Kuo, L. and, Pittman, R. N., 1988, Effect of hemodilution on oxygen transport in arteriolar networks of hamster striated muscle, <u>Am. J. Physiol.</u>, 254:H331.

Piiper, J., Meyer, M., and Scheid, P., 1984, Dual role of diffusion in tissue gas exchange: blood-tissue equilibration and diffusion shunt, <u>Resp. Physiol.</u>, 56:131.

Pittman, R. N., and Duling, B. R., 1977, The determination of oxygen availability in the microcirculation, <u>in</u>: "<u>Oxygen and Physiological Function</u>," F.F. Jöbsis, ed., Prof. Infor. Library, Dallas, Texas.

Rose, C. P., and Goresky, C. A., 1985, Limitations of tracer oxygen uptake in the canine circulation, <u>Circ. Res.</u>, 56:57.

Rose, C. P., Goresky, C. A., Bach, G. G. Bassingthwaighte, J. B., and Little, S., 1988, In vivo comparison of non-gaseous metabolite and oxygen transport in the heart, <u>Adv. Exp. Med. Biol.</u>, 222:45.

Roth, A. C., and Feigl, E. O., 1981, Diffusional shunting in the canine myocardium, <u>Circ. Res.</u>, 48:470.

Roth, A. C., and Wade, K., 1986, The effects of transmural transport in the microcirculation: A two gas species model, <u>Microvasc. Res.</u>, 32:64.

Schubert, R. W., Whalen, W. J., and Nair, P., 1978, Myocardial PO_2 distribution: Relationship to coronary autoregulation, <u>Am. J. Physiol.</u>, 234:H361.

Weiss, H. R., Neubauer, J. A., Lipp, J. A., and Sinha, A. K., 1978, Quantitative determination of regional oxygen consumption in the dog heart, <u>Circ. Res.</u>, 42:394.

Weiss, H. R., and Sinha, A. K., 1978, Regional oxygen saturation of small arteries and veins in the canine myocardium, <u>Circ. Res.</u>, 42:119.

Whalen, W. J., Nair, P., and Buerk, D., 1973, O_2 tension in the beating cat heart in situ, <u>in</u>: "<u>O_2 Supply</u>," M. Kessler, D. F. Bruley, L. C. Clark, Jr., D. W. Lübbers, I. K. Silver, and J. Strauss, eds., University Park Press, Baltimore, Maryland.

REGIONAL DIFFERENCES IN THE REGULATION OF CONTRACTION-RELAXATION MACHINERY OF VASCULAR SMOOTH MUSCLE

Arisztid G. B. Kovach, Eors Dora, Maria Farago, Ildiko Hor-Ildiko Horvath, and Csaba Szabo

Exp. Res. Dept. and 2nd Dept. of Physiol., Semmelweis Univ. Med. Sch., Budapest

INTRODUCTION

It is well-established that there are marked differences in the basal tone and reactivity to various vasoactive agents of the cerebral and mesenteric arterial vascular beds. While cerebral arteries have considerable, mesenteric arteries have negligible basal tone. It seems likely that the metabolic factors are the most important controllers of cerebral circulation, but they seem far less important in regulating intestinal blood flow. Furthermore, the neural control is probably much more pronounced in the intestinal than in the cerebral circulation. These differences urged us to compare the effects of various contractile and dilatory agents on similar sizes of middle cerebral and mesenteric arteries of the cat. To induce contraction we used potassium ions as well as various agonists of vascular smooth muscle calcium mobilizing receptors. Relaxation of the vascular smooth muscle was achieved by drugs acting via the endothelium (acetylcholine, adenosine triphosphate) and directly on the smooth muscle (adenosine) (Furchgott, 1983).

METHODS

Cats anesthetized by nembutal (30 mg/kg, b.w.) and exanguinated were used for this study. The brain was quickly removed from the skull and pieces of mesenterial tissue containing small branches of the mesenteric artery (diameter approx. 300-500 μm) were excised. The brain and the mesenterial tissue pieces were placed into ice-cold Na^+-Krebs solution and stored at 4 ^oC in a refrigerator. The standard Na^+-Krebs solution composed of (in mM): NaCl 119, KCl 4.6, $CaCl_2$ 1.5, $MgCl_2$ 1.2, $NaHCO_3$ 15, NaH_2PO_4 1.2, glucose 6 (Hogestatt et al., 1983). Microsurgical methods and a Zeiss operational microscope were used to remove and clean the middle cerebral and mesenteric arteries. Then, 1-3 mm long segments were cut from the vessels and placed into a double walled and thermostated (37 ^oC) tissue chamber. The tissue chamber was filled with Na^+-Krebs solution and bubbled with a gas mixture containing 5% CO_2 and 95% O_2. The pH of the Na^+-Krebs solution was approximately 7.4. The 1-3 mm long vessel segments were mounted on two L-shaped specimen holders and their isometric tension was measured by Grass FT-03 transducers (Hogestatt et al, 1983). Six parallel measurements were made simultaneously. Using a micromanipulator the vessel was stretched to get a tension between 400-600 mg, and incubated at this tension for 60 to 90 min. The viability of the vessels has been tested with repeated application of 10^{-6} mol/l norepinephrine and acetylcholine. When the vessels showed a steady response to norepinephrine and acetylcholine the actual experiment was started. All measurements were carried out in the presence of 5×10^{-6} mol/l indomethacine and 5×10^{-7} mol/l propranolol. Appropriate concentrations of the drugs were injected into the tissue chamber with Hamilton mic-

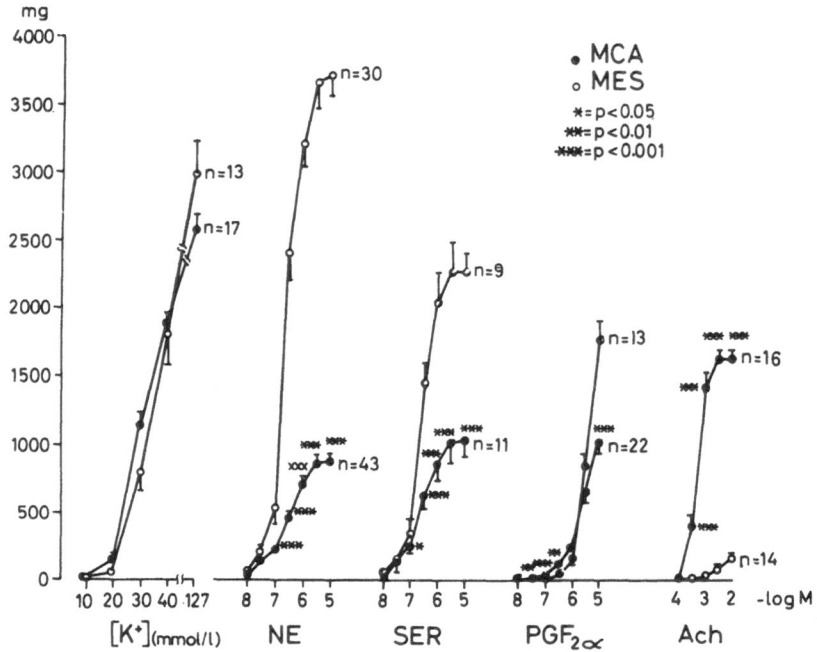

Fig. 1 Contractile effects of potassium ions (K$^+$), norepinephrine (NE), serotonine (SE), PGF$_{2alpha}$, and acetylcholine (Ach) in the middle cerebral (MCA) and mesenteric arteries (MA). Each point represents mean $\pm$ SEM. The number of experiments averaged is shown by n. Asterisks mark the significant differences between the two vascular beds.

rosyringes. All drugs (norepinephrine, serotonine, acetylcholine, PGF$_{2alpha}$, adenosine, adenosine triphosphate, indomethacine, propranolol) were obtained from Sigma. In the experiments where potassium ion concentration of the Na$^+$-Krebs solution was varied we simply replaced appropriate concentrations of NaCl with KCl. The isotonic K$^+$-Krebs solution contained 127 mM KCl and no NaCl
For statistical analysis of the data the Student t-test was applied.

RESULTS

    Fig. 1 shows that although elevation of potassium ion concentration resulted in similar contractile responses in the two vascular beds, other contractile responses were greatly different. Norepinephrine, serotonine, and PGF_{2alpha} in the concentration range of 10^{-6}-10^{-5} mol/l produced more pronounced contractions in the mesenteric artery than in the middle cerebral artery. Contrary, acetylcholine was a more potent contractile agent in

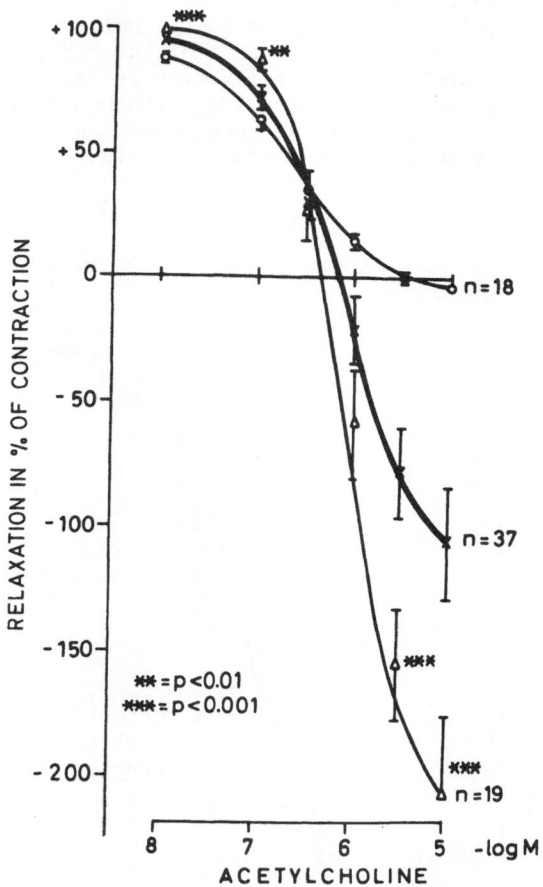

Fig. 2 Acetylcholine-induced relaxation in the middle cerebral artery. Each point represents mean ± SEM; n shows the number of experiments averaged. The relaxation is expressed as % of contraction elicited by $5x10^{-7}$-10^{-6} mol/l norepinephrine.

the middle cerebral artery than in the mesenteric artery. To induce contractions with acetylcholine we had to apply it at extremely high concentrations, because at low concentrations it relaxes vascular smooth muscle via an endothelium-dependent pathway (Furchgott, 1983).

In the concentration range of 10^{-8}-10^{-5} acetylcholine relaxed the precontracted middle cerebral artery (Fig. 2). However, the magnitude of relaxation depended very much on the tone that the vessels gained before the contractile agent and consecutively,acetylcholine have been applied.

Accordingly, if a considerable tone developed the dose-response curve of acetylcholine showed a marked undershot. If there was no or negligible spontaneous tone, acetylcholine induced no or minimal undershot. In Fig. 2 the thick line shows the composite dose-response curve of acetylcholine. The two other curves were derived from the composite one according to the magnitude of undershot. The asterisks show that the undershooting vessels were somewhat less sensitive to 10^{-8}-10^{7} mol/l acetylcholine than the ones had no undershot.

In Fig. 3 the dilatory potencies of acetylcholine, adenosine, and adenosine triphosphate are compared in case of the middle cerebral artery. To avoid base line problems, the relaxation is expressed as % of the maximum relaxation response. As one can see, although these dilatory agents produce relaxation via different mechanisms their dose response curves did not differ from each other significantly.

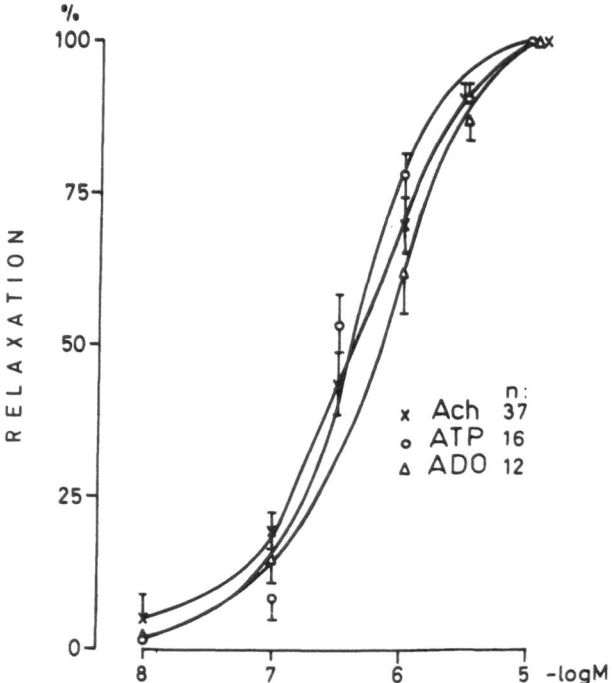

Fig. 3 Dilatory potencies of acetylcholine (Ach), adenosine (ADO), and adenosine triphosphate (ATP) in the middle cerebral artery. Each point represents mean $\pm$ SEM; n shows the number of experiments averaged. Ordinate shows relaxation in % of the maximum response.

In Figs. 4 and 5 relaxing potencies of acetylcholine and adenosine, acetylcholine and adenosine triphosphate are compared. As one can see, their dilatory effect is very similar and this similarity is not altered by the magnitude of undershot.

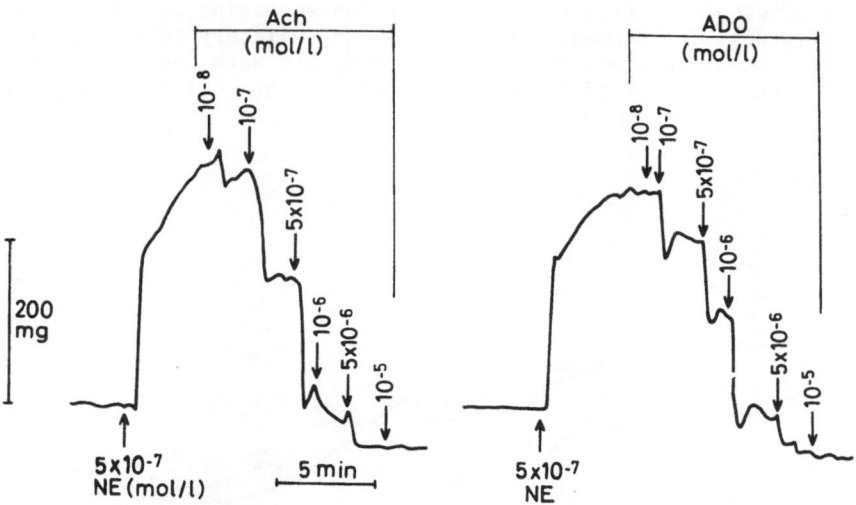

Fig. 4 Dose-response curves of acetylcholine (Ach) and adenosine (ADO) in a typical experiment in the middle cerebral artery. 5×10^{-7} mol/l norepinephrine (NE) was used to contract the vessel before Ach or ADO has been applied. Calibrations of tension and time are shown at the Ach responses.

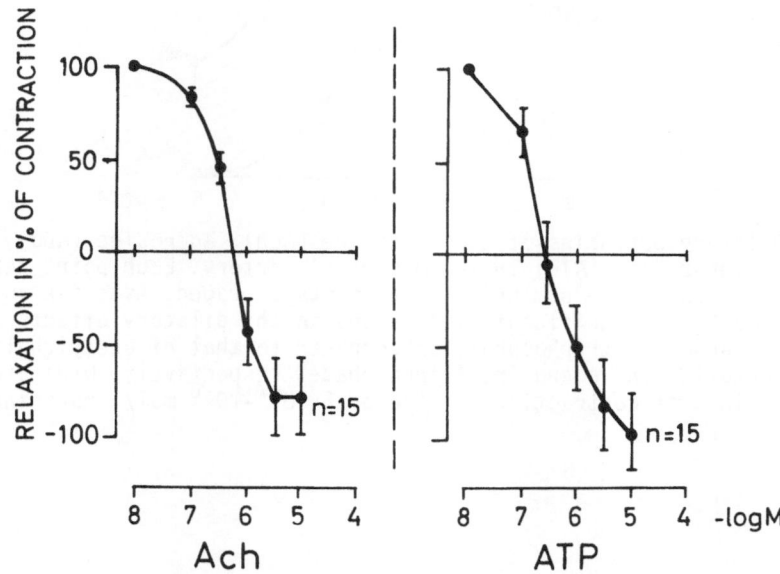

Fig. 5 Dose-response curves of acetylcholine (Ach) and adenosine triphosphate (ATP) obtained in the same middle cerebral arteries. Each point represents mean ± SEM; n number of experiments averaged.

Contrary to the middle cerebral artery, spontaneous tone never developed in the mesenteric artery. Accordingly, acetylcholine and the other dilatory substances tested did not elicit undershooting dose-response curves (Fig. 6). In the mesenteric artery, the dilatory potencies of acetylcholine, adenosine, and adenosine triphosphate were greatly different, acetylcholine being the most, adenosine triphosphate being the least potent dilator.

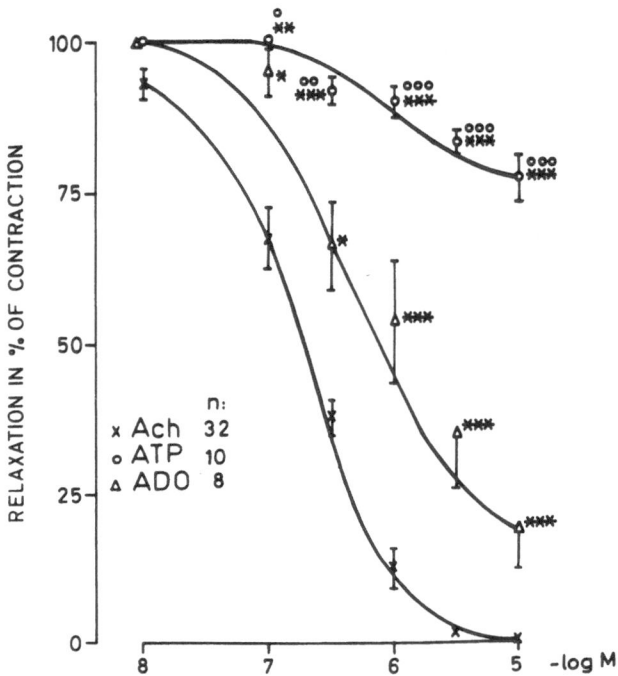

Fig. 6 Dilatory potencies of acetylcholine (Ach), adenosine (ADO), and adenosine triphosphate (ATP) in the mesenteric artery. Each point represents mean ± SEM; n shows the number of experiments averaged. Asterisk and small o letters mark the significant differences in the dilatory effects of adenosine and adenosine triphosphate as compared to that of acetylcholine and between adenosine and adenosine triphosphate, respectively. Ordinate shows relaxation in % of contraction induced by 5×10^{-7}-10^{-6} mol/l norepinephrine.

Fig. 7 summarizes our findings concerning the dilatory potencies of acetylcholine, adenosine triphosphate, and adenosine in the middle cerebral and mesenteric arteries. Acetylcholine was an equally potent dilator in these two vascular beds, but due to the marked spontaneous tone, it induced undershooting dose response curve ,as an average ,in the middle cerebral artery. The dilatory potency of adenosine triphosphate in the mesenteric artery was negligible as compared to that in the middle cerebral artery. Finally, adenosine was slightly more potent dilator in the middle cerebral artery than in the mesenteric artery.

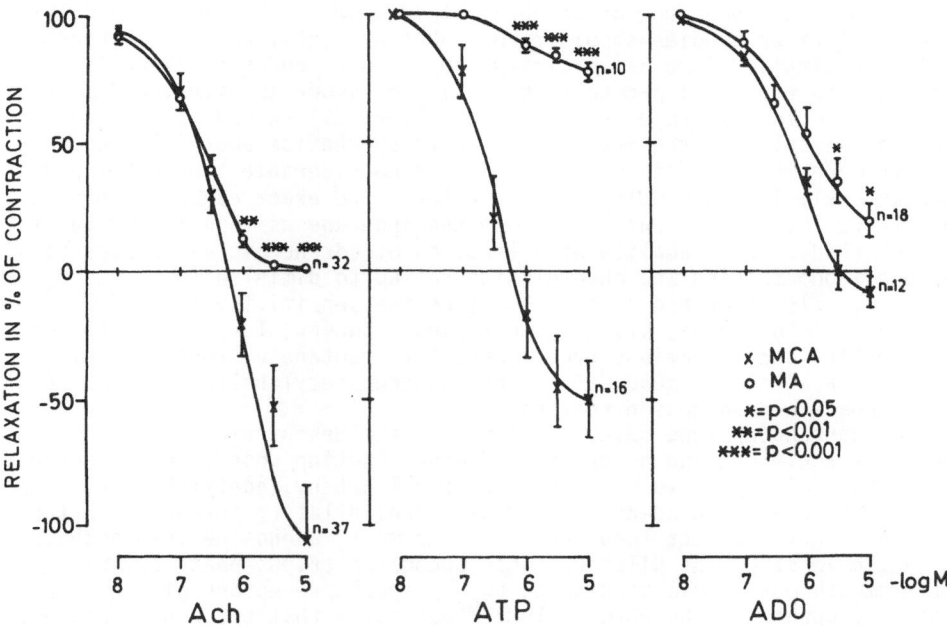

Fig. 7 Comparison of the dilatory potencies of acetylcholine (Ach), adenosine triphosphate (ATP), and adenosine (ADO) in the middle cerebral and mesenteric arteries. Each point represents mean ± SEM; n shows the number of experiments averaged. Asterisks mark the significant differences between the two vascular beds. MCA: middle cerebral artery; MA: mesenteric artery.

DISCUSSION

In the present study we compared the effects of various contractile and dilatory agents on the middle cerebral and mesenteric arteries of the cat. It was found that although K^+-Krebs solution elicits similar contractile responses in these vascular beds, norepinephrine, serotonine and PGF_{2alpha} induce much larger contractions in the mesenteric artery than in the middle cerebral artery. These findings correspond to available data in the literature (Skarby et al., 1983; Toda and Fujita, 1973; Toda, 1978), and the greater sensitivity of the mesenteric vascular smooth muscle to these contractile agents can be attributed probably to more numerous receptors and greater increase in free calcium in the cytosol.

Acetylcholine, applied at pharmacologically high concentrations, contracted both vascular beds but the contraction was much greater in the middle cerebral than in the mesenteric artery. The disparate contractile potencies of acetylcholine can be attributed to scarcity of smooth muscle muscarinic receptors or a very efficient endothelium-mediated relaxation in the mesenteric artery. Acetylcholine induces contraction and relaxation at the same time via binding to muscarinic receptors of the smooth muscle and endothelial cells, respectively (Furchgott, 1983).

Acetylcholine, adenosine triphisphate, and adenosine led to statistically not significantly different concentration-response curves in the middle cerebral artery. In theory, this means that acetylcholine could be as important regulator of cerebral blood flow as the other two dilators. However, further studies should reveal whether acetylcholine can reach that concentration in close vicinity to vascular endothelial cells in the brain in vivo that proved to be efficient under in vitro conditions. This well could be, since Parnavelas et al. (1985) showed that vascular endothelial cells in the brain are able to synthetise acetylcholine.

In more than 50% of our experiments a considerable tone developed spontaneously in the middle cerebral artery. The exact cause of this is not cleared by the present study, but the spontaneous tone is not due to some arachidonate metabolite or activation of adrenergic alpha receptors, because indomethacine and phentolamine failed to diminish it.
In the vessels which had spontaneous tone the sensitivity to 10^{-8}-10^{-7} mol/l acetylcholine was slightly decreased. However, larger than this acetylcholine concentrations even ceased the spontaneous tone. In this context, we also not found difference between acetylcholine, adenosine triphosphate and adenosine actions.

A spontaneous tone never developed in the mesenteric artery and consequently acetylcholine never induced undershooting dose response curve in this vessel. Contrary to the middle cerebral artery, acetylcholine, adenosine triphosphate and adenosine had different dilatory potencies in the mesenteric artery, acetylcholine being the most, adenosine triphosphate being the least potent dilator. Since adenosine triphosphate relaxes vascular smooth muscle via binding to P_{2gamma}-purinoreceptors of the endothelium (Hopwood and Burnstock, 1987), we assume that the negligible dilatory effect of adenosine triphosphate in the mesenteric artery can be attributed to scarcity of P_{2gamma} receptors.

SUMMARY

In the present study we revealed substantial differences in the regulation of the contraction-relaxation machinery of the middle cerebral and mesenteric arteries.
a./ Although K^+-Krebs solution resulted in similar contractile responses in both vascular beds, norepinephrine, serotonine and PGF_{2alpha} were more potent in inducing contraction in the mesenteric artery than in the middle cerebral artery. Contrary, extremely high concentrations of acetylcholine produced negligible contractions in the mesenteric artery as compared to the middle cerebral artery.
b./ In more than 50% of the cases a considerable spontaneous tone developed in the middle cerebral artery, which was not due to the activation of adrenergic alpha receptors or some arachidonic acid metabolite.
A similar phenomenon never occurred in the mesenteric artery.
c./ Acetylcholine, adenosine triphosphate and adenosine brought about similar dilatory concentration-response curves in the middle cerebral artery, but greatly different ones in the mesenteric artery where acetylcholine was the most, adenosine triphosphate the least potent dilator.

REFERENCES

Furchgott, R. F., 1983, Role of endothelium in responses of vascular smooth muscle, Circ. Res., 53:557.

Hopwood, A. M.,and Burnstock, G., 1987, ATP mediates coronary vasoconstriction via P_{2x}-purinoceptors and coronary vasodilatation via P_{2gamma}-purinoceptors in the isolated perfused heart, Eur. J. Pharmacol., 136:49.

Hogestatt, E. D., Andersson, K. E., and Edvinsson, L., 1983, Mechanical properties of rat cerebral arteries as studied by a sensitive device for recording of mechanical activity in isolated small blood vessels, Acta Physiol. Scand., 117:49.

Parnavelas, J. G., Kelly, W., and Burnstock, G., 1985, Ultrastructural localization of choline acetyltransferase in vascular endothelial cells in rat brain, Nature, 315:424.

Skarby, T. V. C., Andersson, K. E., and Edvinsson, L., 1983, Pharmacological characterization of postjunctional alpha-adrenoceptors in isolated feline cerebral and peripheral arteries, Acta Physiol. Scand. 117:63.

Toda, N., and Fujita, Y., 1973, Responsiveness of isolated cerebral and peripheral arteries to serotonine, norepinephrine, and transmural electrical stimulation. Circ. Res., 33:98.

Toda, N., 1978, Responses of isolated cerebral and peripheral arteries to vasoconstricting agents, in: "Neurogenic Control of the Brain Circulation", Ch. Owman and L. Edvinsson, eds., Pergamon Press, Oxford.

THE EFFECTS OF TEMPERATURE AND BUFFER CONCENTRATION ON THE METABOLISM OF HUMAN ATRIAL APPENDAGES MEASURED BY ^{31}P AND ^{1}H NMR

Graham W. Mainwood, Sylvain Lareau[+], Kristine Whitehead[+], Wilbert J. Keon[*] and Roxanne Deslauriers[+]

Department of Physiology, University of Ottawa, Ottawa, Ontario K1H 8M5,
[+]Division of Biological Sciences, National Research Council of Canada, Ottawa, Ontario K1A OR6
[*]Department of Cardiothoracic Surgery, University of Ottawa Heart Institute, Ottawa, Ontario K1Y 4E9

INTRODUCTION

Animal hearts are commonly used as models for human tissue in studies defining conditions for the preservation of heart tissue destined for transplantion. Human atrial appendages, removed during open heart surgery, have been used on occasion to determine the optimum temperature for the preservation of grafts prior to transplant [1,2]. We use ^{31}P and ^{1}H nuclear magnetic resonance (NMR) to follow the time course of metabolite levels in human atrial appendages to gain a better understanding of tissue energetics under various preservation conditions.

Previous NMR studies of human atrial appendages [1] in 0.9% NaCl solution at 1, 4, 12 and 20°C have shown that over an 8 hour time period ATP levels are better preserved at 1° and 4° than at 12° or 20°C. The work of Keon et al. [2] has shown that contractility of atrial trabeculae is better preserved at 12°C than at 4°C. NMR studies have shown a large drop in internal pH which may reduce the rate of ATP generation by glycolysis. In order to investigate this further, we measured ATP preservation in the presence of different concentrations of external buffer. The rationale was to provide a large extracellular proton sink for lactic acid efflux. We chose to do these experiments at what appears to be the optimal temperature for function recovery, i.e. 12°C.

MATERIALS AND METHODS

Human atrial appendages were obtained from consenting adult patients undergoing cardiac bypass surgery. Upon the removal from the patient, pieces of atrial appendage weighing 0.4 - 0.6 g were immediately placed in cardioplegic solution (Plegisol, Abbott) at 12°C for a period of 5 minutes. The tissue was transported to the NMR facility in 5 ml of one of the following solutions: **1.** 0.9 % NaCl **2.** modified Krebs-Heinseleit solution containing 20 mM PIPES buffer (pH 7.4) **3.** modified Krebs-Heinseleit solution containing 100 mM PIPES buffer (pH 7.4). The modified Krebs-Henseleit solution was composed of 4.74 mM KCl, 2.54 mM $CaCl_2$, 1.20 mM $MgSO_4$, PIPES buffer, 3.6% (W/V) dextran (60,000 - 90,000 dalton), and 5% D_2O (pH 7.4). The concentration of NaCl was 114 mM in solution 2 and 60 mM in solution 3. All the solutions contained 2.5 mM ethylphosphonic acid as an external pH indicator. PIPES buffer was chosen since it has an appropriate pK_a value (6.8), it does not chelate divalent ions to a significant extent and it does not have any measurable biological effects.

Samples were then placed in 10 mm NMR tubes with 1.2 ml of the appropriate fresh solution for subsequent NMR studies. A capillary containing 25 mM aminomethylphosphonic acid and 100 mM formic acid (pH 7.0) was added as an internal reference.

^{31}P and ^{1}H NMR spectra were acquired alternately for 20 - 22 hours using a Bruker AM-360 spectrometer. The temperature was controlled electronically to ± 1°C. The ^{31}P spectra were obtained using a 60° pulse angle (9 µs) and a 1 second recycling time. The number of scans was 1200. A line broadening of 25 Hz was used for the Fourier transform of the spectra. The ^{1}H spectra were acquired using a spin-echo sequence based on the water suppressing 1331 [3] pulse sequence. The pulse delay of the 1331 sequence was 400 µs and the τ value for the spin-echo was 400 ms. The recycling time was 3 seconds and the number of scans was 64. A 10 Hz line broadening was used for the Fourier transform of the ^{1}H spectra.

Intracellular and extracellular pH were calculated from the internal P_i and the external ethylphosphonate chemical shifts respectively, using curves calibrated at 4° and 12°C, from pH 5.0 to 8.0 using a model solution. The composition of the model solution was 1 mM PCr, 0.5 mM ATP, 1 mM NaH_2PO_4, 2 mM $MgCl_2$, 120 mM KCl, 10 mM PIPES buffer, 1 mM lactate and 5% D_2O. ^{31}P and ^{1}H NMR spectra of this reference solution (pH 7.0) were run before each human heart experiment to obtain an external standard for calibration purposes.

We have previously determined [1] that the lactate signal observed by ^{1}H NMR originates mainly from the bathing solution, even when the amount of lactate in the tissue is significantly higher. The amount of lactate released by the tissue was therefore calculated by correlating the NMR peak height of the last NMR spectrum with the amount of lactate measured biochemically in the bathing solution. The conversion factor obtained was then used to calculate the amount of external lactate in previous spectra.

Phosphorus metabolites were estimated by correlating their peak heights with those of the reference capillary. By comparison with the standard solution, we estimated that the reference capillary gave a signal equivalent to 1.5 µmoles of P_i. Using peak height, we could then calculate the amount of internal P_i in µmoles per gram of wet tissue. The PCr content was measured in the same way. Values were then corrected for the difference in peak height due to the difference in linewidth. ATP was calculated from the height of the β-ATP peak. A calibration curve using the model solution and the reference capillary was used to correct the peak height values for the line broadening observed with the decreasing pH of the solution over time.

Biochemical and chemical assays for ATP, phosphocreatine (PCr), creatine, glucose-6-phosphate, lactate, inorganic phosphate (P_i) and total phosphate were performed as previously described [1].

RESULTS

The results of the first series of experiments using 0.9% NaCl at 4 different temperatures (1, 4, 12 and 20°C) are summarised in Table 1.

TABLE 1. Energy demand of atrial myocardium and the equivalent O_2 consumption rate[1].

Temp (°C)	ATP loss[2]	Lactate[3] production	ATP[4] generated	ATP utilisation	Equivalent O_2 consumption (µl/g.min)
1°	7.34	43.3	65.1	72.4	0.27
4°	7.78	52.3	78.4	86.2	0.32
12°	12.22	105.8	158.7	170.9	0.64
20°	19.76	211.9	317.8	337.6	1.26

1 Based on a myocyte mass of 46% of total wet tissue mass. All values are given in $nmole.g^{-1}$ (myocyte mass).min^{-1}.
2 Calculated on the basis of a quantitative conversion of ADP to ATP by adenylate kinase.
3 Based on the rate of increase of NMR visible lactate, assuming that 46% of the total lactate is NMR visible.
4 Assuming 1.5 mole of ATP produced per mole of lactate from glycolysis.

The estimates of energy demand at these temperatures are based on the rate of fall of ATP and the rate of ATP generation by glycolysis. The values are corrected for the connective tissue component estimated by morphometric analysis [1]. The results are thus expressed in terms of ATP turnover per gram of atrial muscle tissue. The estimated ATP turnover falls to a very low level (70 nmole g^{-1} min^{-1}) at 1°C. The O_2 uptake required to meet this demand is only about 2% of the O_2 uptake of typical arrested mammalian myocardium at 37°C. It is interesting to note that the critical depth for O_2 diffusion under these conditions is about 3.2 mm. A 6 mm thick myocardial layer could then be adequatly supplied with O_2 by diffusion in an O_2 atmosphere.

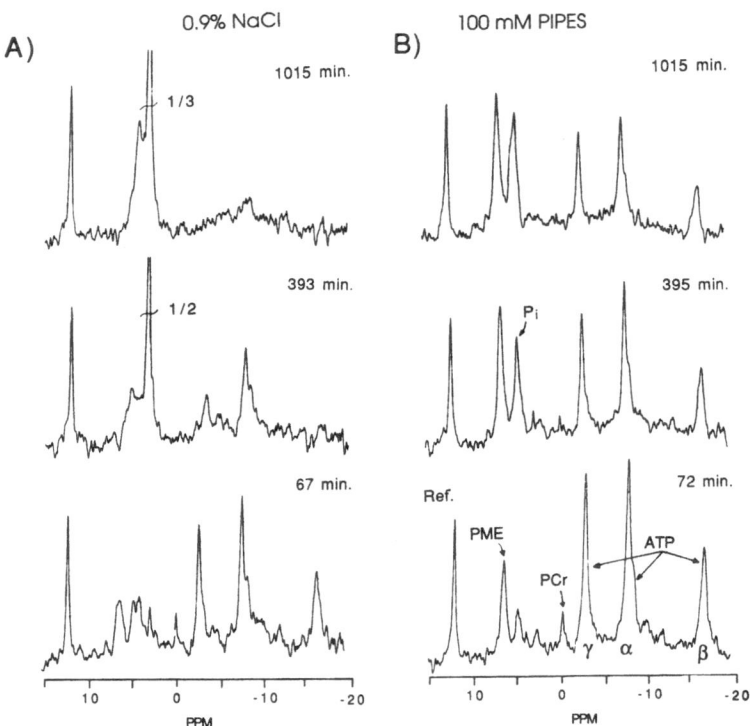

Figure 1. ^{31}P NMR spectra of two human atrial appendages, one (0.48 g) in 0.9% NaCl (Column A), the other (0.59 g) in 100 mM PIPES-K-H solution (B). Peaks are identified in the lower right hand spectrum (PME = phosphate monoester). Inorganic phosphate (Pi) shows a large increase in 0.9% NaCl and is off scale at 393 and 1015 min. The fraction of the total peak height is indicated on these peaks. ATP has virtually disappeared at 393 min. in A, the remaining peaks probably being those of ADP and NAD.

In these experiments, ATP fell to half of its initial value in about 300 min. at 12°C and 500 min. at 4°C. The rate of lactate production was much greater at 12°C than at 4°C and this was associated with a much steeper fall in pH. After 300 min. at 12°C, the mean pH_i fell to 6.2 while at 4°C, it was 6.6. It was reasonable to suppose that this very low pH would suppress glycolysis and that if the fall could be reduced, glycolysis would be better maintained and ATP would fall less rapidly. This was the objective of the second series of experiments using different buffer concentrations at a constant temperature.

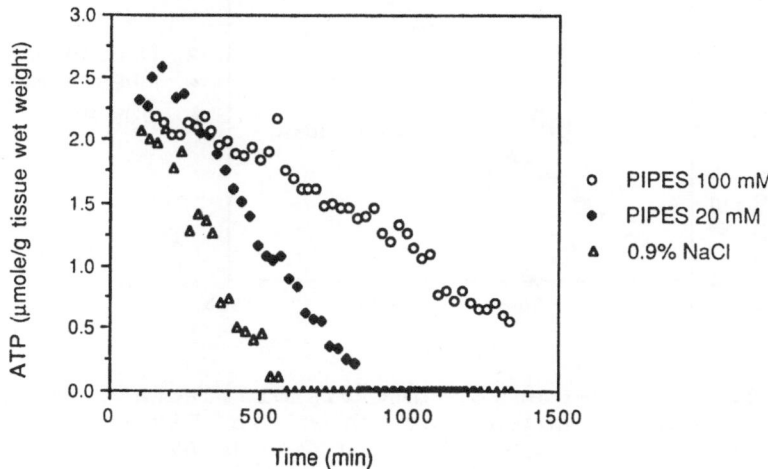

Figure 2. Mean ATP values from 12 atrial appendages. Three in 0.9% NaCl, six in 20 mM PIPES and three in 100 mM PIPES. In 100 mM PIPES, there is no detectable change in ATP in the first 10 hours, while in 20 mM PIPES it is about half decayed and in 0.9% NaCl, it is undetectable.

Figure 1 shows the ^{31}P spectra of 2 atrial appendages, one of them in 0.9% NaCl (a) and the other in 100 mM PIPES buffer (b). The first spectrum acquired at the earliest time shows similar levels of metabolites when expressed in μmole g^{-1} tissue wet weight. PCr is often, but not always present in the first spectrum obtained. The values observed are always less than 1 μmole g^{-1} tissue wet weight. After about 400 min., ATP has almost completely disappeared in A and a large increase of P_i is observed. In 100 mM PIPES, the ATP level is much better preserved, with only a small increase in phosphate. After 1000 min., ATP is only reduced to half the initial value in the appendage incubated in 100 mM PIPES, whereas in 0.9% NaCl, ATP has completely disappeared.

Figure 2 shows mean values of ATP preservation in 6 appendages in 0.9% NaCl compared with 3 in 20 mM PIPES and 3 in 100 mM PIPES buffer. The first NMR spectra obtained from the samples incubated in these solutions (60 - 100 min after removal from the patient) do not differ significantly. Average ATP values are in the range of 2.3 ± 0.7 µmole/g tissue wet weight and P_i values vary from 1 to 2 µmole/g tissue wet weight.

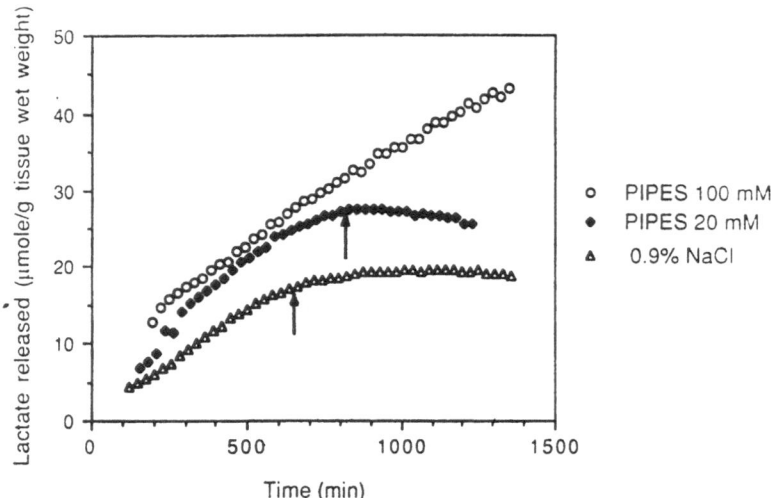

Figure 3. Mean values of NMR visible lactate in the 12 atrial appendages shown in Figure 2. Arrows indicate the times at which ATP is no longer detectable.

The time required for ATP to fall to half of its initial level increases from 300 min. in 0.9% NaCl to 500 min. in 20 mM PIPES and 900 min in 100 mM PIPES. Furthermore, the ATP levels in 100 mM PIPES show little decline in the first 10 hours.

This increased preservation of ATP is related to the increase of NMR visible lactate at the higher buffer concentration shown in Figure 3. This Figure shows that in 0.9% NaCl and 20 mM PIPES, lactate levels reach a plateau at the same time that ATP is no longer visible by NMR. In 100 mM PIPES, lactate continues to increase for the entire observation period ($\approx$ 22 hours).

The idea that glycolysis is suppressed by intracellular acidosis is supported by the following results. In saline and in 20 mM PIPES buffer, the internal pH of the tissue drops relatively fast to reach values below 6.0 after 300 min. of incubation. It is difficult to assess intracellular pH accurately in a 100 mM PIPES because P_i is produced slowly and at the same time leaks out of the cells. The pH indicated by the chemical shift is therefore partly due to an extracellular component. However, the measurements suggest that the internal pH remains fairly constant during the first 8 - 10 hours of the observation period and does not fall below 6.8 throughout the whole experimental period.

DISCUSSION

The ^{31}P spectra of human myocardium are not typical of spectra obtained from the ventricular muscle of other species [4]. The main difference is the low value of PCr as compared to ATP. The fact that the P_i peak is also small indicates that the PCr is not just lost by hydrolysis. The values observed here are consistent with previous studies [1] in which we found a mean value of 0.8 μmoles g^{-1} of PCr in specimens quickly frozen with Wollenberger clamps immediately after removal in the operating theatre. The samples contained twice as much ATP (1.8 μmole g^{-1}) as PCr and the total creatine content was 1.9 μmoles g^{-1}. The creatine to ATP ratio is therefore much lower than its usual value of 3 to 1 [5].

We previously showed by morphometric analysis that myocytes comprise only 46% of the total mass of the atrial appendage, the rest being composed mainly of collagen. Correcting for the tissue composition gives values close to 4 μmoles g^{-1} for ATP and about 1.6 μmoles g^{-1} for PCr. These ATP values are similar to those found in rabbit, rat and pig heart [5]. The initial uncorrected values of ATP (2.3 μmoles g^{-1}) estimated by NMR using the calibration procedure described in methods are about 22% higher than the mean values obtained in frozen tissue by HPLC or enzymatic assay. This could either be due to random variation in tissue composition or myocardial ATP content or it could be due to a difference inherent in the calibration method. Since in these experiment we are concerned with changes from the initial values, this difference does not affect our conclusion.

The main finding of these studies is that ATP preservation in isolated ischemic human atrial tissue is greatly improved by a high external buffer concentration. This improved preservation is associated with increased glycolysis and a much smaller fall in intracellular pH. During the first 500 min., the rate of appearance of NMR visible lactate is about 40% greater in 100 mM PIPES than in 0.9% NaCl. If the ratio between NMR visible lactate and total lactate remains the same with different buffer concentrations (i.e. about 45% of the total lactate is NMR visible) then ATP generated by glycolysis is about 220 nmoles g^{-1} min^{-1} at 12°C. Since the ATP concentration in the tissue remains fairly constant during this period the rate of ATP utilization must be equal to the rate of production. The estimated utilization rate of ATP in 0.9% NaCl is about 170 nmoles g^{-1} min^{-1} (Table 1). The utilisation rate in 100 mM PIPES is therefore about 30% greater than in NaCl. Its seems likely that this is due to the higher intracellular pH.

Lowering the temperature from 12°C to 4°C improves the preservation of ATP as does an increase in buffer concentration, but these studies show that the mechanism is quite different. The improved preservation at 4°C is associated with a reduction of about 50% in ATP utilisation and a lower rate of glycolysis. With an increase in buffer concentration, ATP is maintained in the face of increased ATP utilisation. Futhermore, glycolysis proceeds at a higher rate and is maintained for a longer time with what appears to be a reduced feedback drive (ADP + P_i).

A practical implication of this study is that cardioplegic solutions may give improved preservation with increased buffer capacity. One of the currently used cardioplegic solutions, St. Thomas' solution 2 (Plegisol), has a very low buffer capacity. This is because the buffer used (bicarbonate) is not effective in a closed system because of its low pK value.

It may be argued that the human tissue used in these experiments is abnormal since the patients have heart disease, they have been subjected to different drug regimes and they are taken from an older (generally 45 - 65) age group. While it is true that this represents a very selected group, there are three lines of evidence to suggest that this tissue functions normally and is not pathological. The first piece of evidence is that the contractile behaviour of trabeculae isolated from this tissue gives a developed force equivalent to that of mammalian preparations [2]. Secondly, we have found no evidence of morphological damage with either light or electron microscopy in the atrial appendages examined. Thirdly, the patients who undergo bypass surgery show good recovery with no signs of chronic myocardial damage. The drugs and anaesthetics used are of course chosen so as not to suppress cardiac function and are probably diluted out to insignificant levels in the initial wash.

In spite of these arguments it must be admitted that the tissue samples do represent a selected population. It is possible that atrial appendages taken from healthy young adults would show quantitative biochemical differences. It is recognised that atrial and ventricular myocardium differ in several respects. Studies in other species show that contraction velocity and myosin isoenzymes differ in the two tissues. Our observations based on a very limited number of human ventricular samples indicate relatively higher creatine phosphate levels but similar values of ATP. We have not yet been able to carry out studies on ATP preservation in different buffer concentrations in ventricular tissue because of the limited supply.

SUMMARY

In this study we measured changes in intracellular ATP and pH together with lactate production in isolated ischemic human atrial tissue. The measurements were made using ^{31}P and ^1H NMR. ATP preservation is improved as temperature is reduced from 20°C to 1°C because of a progressive decrease in energy demand. At a constant temperature (12°C), ATP preservation is improved by increasing the extracellular buffer capacity with PIPES buffer at concentrations up to 100 mM. Under these conditions, energy demand appears to increase but the ATP level is kept relatively constant for periods of 10 hours or longer. This appears to be due to a tighter regulation between supply and demand in which glycolysis is driven faster at relatively lower ADP and P_i levels. This tight regulation may be attributed to the better maintenance of intracellular pH.

ACKNOWLEDGEMENTS

This work was supported by a grant from the Ontario Heart and Stroke Foundation. Our thanks to M. St.Jean and P. Douglas for capable technical help.

REFERENCES

(1)   R. Deslauriers, W. J. Keon, S. Lareau, D. Moir, J. K. Saunders, I. C. P. Smith, K. Whitehead & G. W. Mainwood, Cardiac hypothermia: ^{31}P & ^{1}H NMR studies of human myocardial tissue, <u>J. Thorac. & Cardiovasc. Surg.</u>, In Press.

(2)   W. J. Keon, P. J. Hendry, G. C. Taichman & G. W. Mainwood, Cardiac Transplantation: The ideal temperature for graft transplant. <u>Ann. Thorac. Surg.</u> 46: 337-341 (1988).

(3)   P. J. Hore, Solvent supression in Fourier transform nuclear magnetic resonance, <u>J. Magn. Reson.</u> 55: 283-300 (1983).

(4)   I. A. Bailey, G. K. Radda, A. M. L. Seymour,& S. R. Williams, The effects of insulin on myocardial metabolism and acidosis in normoxia and ischemia, <u>Biochim. Biophys. Acta</u> 720: 17-27 (1982).

(5)   R. J. Connett, Analysis of metabolic control: New insights using scaled creatine kinase model, <u>Am. J. Physiol.</u> 254: R949-R959 (1988).

EFFECTS OF LEUKOCYTE-DERIVED OXIDANTS ON SARCOLEMMAL

NA,K,ATP-ASE AND CALCIUM TRANSPORT

T. Matsuoka, T. Yanagishita and K. J. Kako*

Department of Physiology, University of Ottawa
Ottawa, Ontario, K1H 8M5, Canada

INTRODUCTION

The mitochondrial electron transport chain is capable of reducing oxygen directly to water. However, 5 % of the oxygen consumption of tissues proceed by a univalent pathway in which superoxide anion, hydrogen peroxide and hydroxyl radicals are produced (for review, Thompson & Hess, 1986). Although hydroxyl radicals are very reactive, and therefore harmful, the cell is equipped with enzymes to metabolize superoxide anion and hydrogen peroxide to water, thereby bypassing the formation of hydroxyl radicals. These are superoxide dismutase, catalase and glutathione peroxidase, serving as part of physiological defense mechanisms (Thompson & Hess, 1986). Therefore, these intermediates of oxygen reduction play a pathogenic role only when their production is increased and/or when the cellular defense is reduced. Evidence has accumulated implicating oxy radical generation as an important factor in tissue injury caused by ischemia-reperfusion (Bolli, 1988; Burton, 1988; Kako et al., 1988, for reviews). Although the exact source of free radicals has not been settled, recent studies with spin resonance spectroscopy suggested it to be the endothelial cell (Zweier et al., 1988).

In response to activation, polymorphonuclear leukocytes undergo a respiratory burst; oxygen consumption increases, and superoxide anions and hydrogen peroxide are generated. Myeloperoxidase contained in granules of neutrophils catalyzes reactions to form hypochlorous acid (HOCl), which is a powerful oxidant (Weiss & Lobuglio, 1982). Several reports describe that neutrophils migrate to the damaged area of the myocardium and exacerbate ischemic-reperfusion injury (Werns & Lucchesi, 1988, for review).

Therefore, available evidence is in favour of the view that exogenous oxy radicals can act on cardiac cells and affect their function. Consequently, we investigated in our previous studies, the effects of representative oxidants, HOCl and hydrogen peroxide, on isolated cardiomyocytes. HOCl was chosen because, as stated above, it is produced by neutrophils and thus its oxidizing action must participate in tissue reactions. Hydrogen peroxide was selected because it can diffuse easily, even across the cell membrane, and therefore, it can simulate actions of both exogenous and endogenous oxidants.

---

* address for correspondence

RESULTS AND DISCUSSION

Experiment with Myocytes

Our results indicated that adult rat heart myocytes were readily labelled with radioactive calcium; 2.5 mM hydrogen peroxide raised the content of rapidly exchangeable intracellular calcium twofold, whereas addition of 1 – 30 mM HOCl decreased the calcium content. Therefore, it is unlikely that these oxidants caused non-specific changes in membrane permeability. Hydrogen peroxide-induced increase in calcium content was dependent on the medium Na^+, pH and temperature, but was not significantly changed by addition of calcium antagonists or amiloride. Binding of

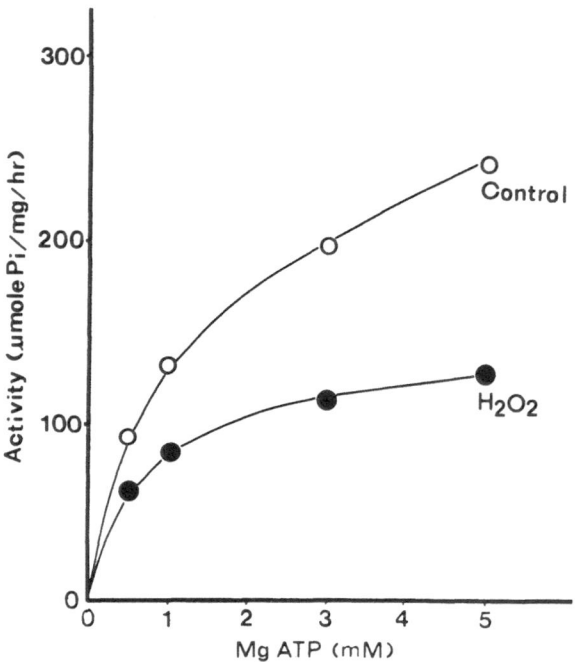

Fig.1. Effect of hydrogen peroxide on the kinetics of hydrolysis of ATP by Na,K,ATPase purified from porcine kidney medulla.
     The incubation was carried out in the presence of 0.5 mM hydrogen peroxide for 30 min at $37^\circ C$. The reaction was stopped by an addition of catalase and ATPase was assayed with various concentrations (abscissa) of ATP.

ouabain to the sarcolemma of myocytes was suppressed by hydrogen peroxide. The results indicated that hydrogen peroxide inhibited Na, K, ATPase, resulting in an increase in [Na]i and enhancement of sarcolemmal Na^+ dependent calcium influx (Kaminishi et al., 1988). In a separate study we confirmed that hydrogen peroxide plus Fe ions stimulated sarcolemmal sodium/calcium exchange activity (Kato & Kako, 1988).

By contrast, HOCl decreased the calcium content of rapidly exchangeable pool below control levels and accelerated calcium efflux from myocytes. Calcium uptake and calcium ATPase of isolated sarcoplasmic reticulum (SR) fraction were highly susceptible to the action of HOCl. Calcium uptake by intracellular sites, studied with myocytes permeabilized with saponin, was inhibited by HOCl. Therefore, these results indicated that HOCl inhibited the SR calcium pump and mobilized intracellular calcium, resulting in an increased calcium efflux from and decline in calcium uptake by myocytes (Kaminishi et al., 1988).

Experiment with Na,K,ATPase Preparation

We continued our investigation into effects of hydrogen peroxide on Na,K,ATPase, purified from porcine kidney outer medulla by using sodium dodecylsulfate. Fig.1 shows the results of experiments, in which Na,K,ATPase preparation was incubated for 30 min at 37°C in

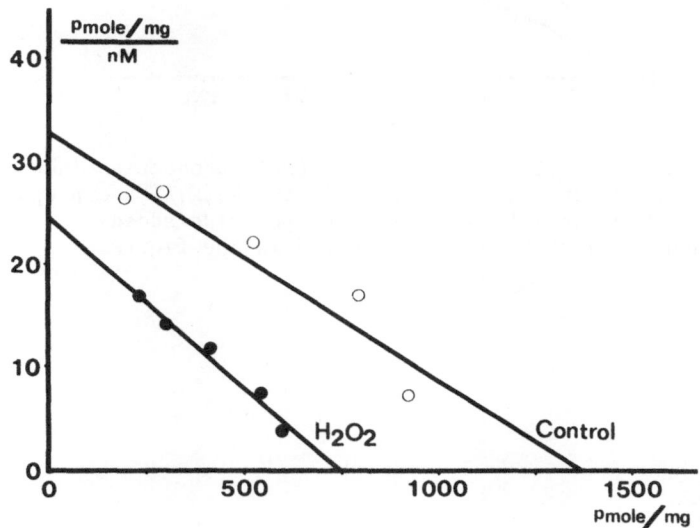

Fig.2.   Scatchard plot of [^{3}H]ouabain binding to Na,K,ATPase preparation. The ordinate indicates the bound/free ouabain ratio and the abscissa indicates bound ouabain per mg protein. The incubation was carried out in the presence or absence of 5 mM hydrogen peroxide plus Fe chelate for 1 h.

KCl–MgCl$_2$–imidazole buffer containing 0.5 mM hydrogen peroxide. It is evident that hydrogen peroxide inhibited ATPase activity, irrespective of the ATP concentration in the assay. Similar modifications of Na,K,ATPase by other oxidants have been reported previously (Skou & Norby, 1979). Fig.2 illustrates that the number of ouabain binding sites of Na,K,ATPase was decreased by 5 mM hydrogen peroxide plus Fe chelate. These results indicated that hydrogen peroxide–derived free radicals indeed inhibited the molecular activity of Na,K,ATPase and reduced the pump sites. The Na,K,ATPase preparation contains not only enzyme protein but also membrane lipid components. Therefore, the action of oxidant was not limited to the protein, but involved lipid peroxidation (Fig.3). Greater amounts of malondialdehyde, an end product of lipid peroxidation, were detected as the hydrogen peroxide concentration was raised (Fig.3).

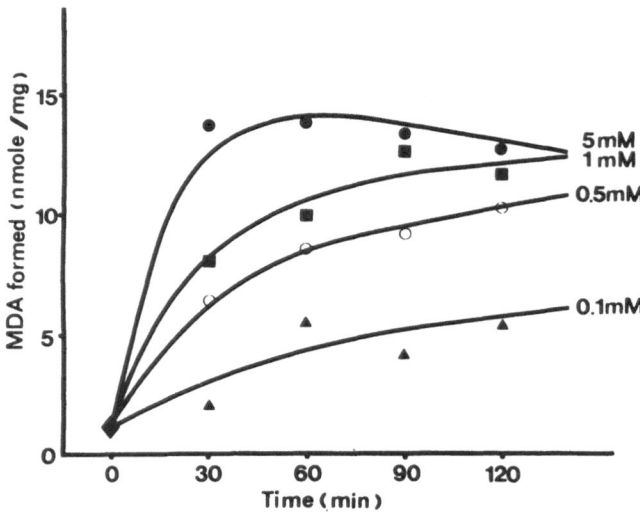

Fig.3. Time course of malondialdehyde (MDA) production induced by
hydrogen peroxide-Fe chelate in the Na,K,ATPase preparation.
Concentrations of hydrogen peroxide added to the
incubation mixture are indicated in the figure.

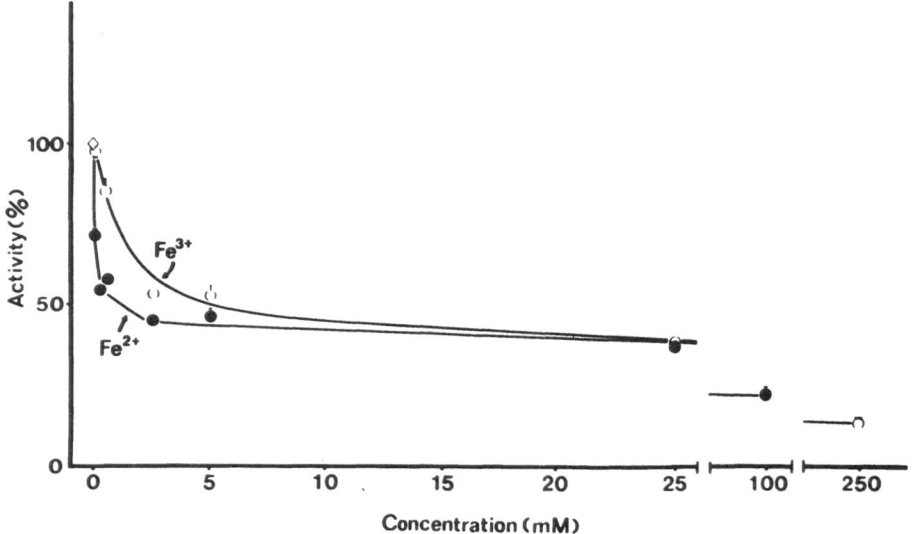

Fig.4. Effect of hydrogen peroxide on Na,K,ATPase prepared from
porcine kidney medulla.
Na,K,ATPase preparation was incubated with hydrogen
peroxide (abscissa) and Fe ions for 30 min at 22°C.

The dose–response relationship with hydrogen peroxide plus Fe ions is illustrated in Fig.4, which shows the results obtained by using partially purified Na,K,ATPase preparation. The relationship seems to be biphasic, in that low concentrations of hydrogen peroxide strongly inhibited the enzyme activity, whereas high concentrations showed only modest inhibitory effects. It is likely that the action of low concentrations of hydrogen peroxide is mediated by site–specifically generated, Fe–catalyzed free radical species, since transitional metal ions that are bound to macromolecules catalyze reactive radical formation from hydrogen peroxide. The radicals initiate chain reactions in the immediate vicinity (Shinar et al., 1983). By contrast, high concentrations of hydrogen peroxide either cause random free radical formation in the bulk phase, or suppress radical

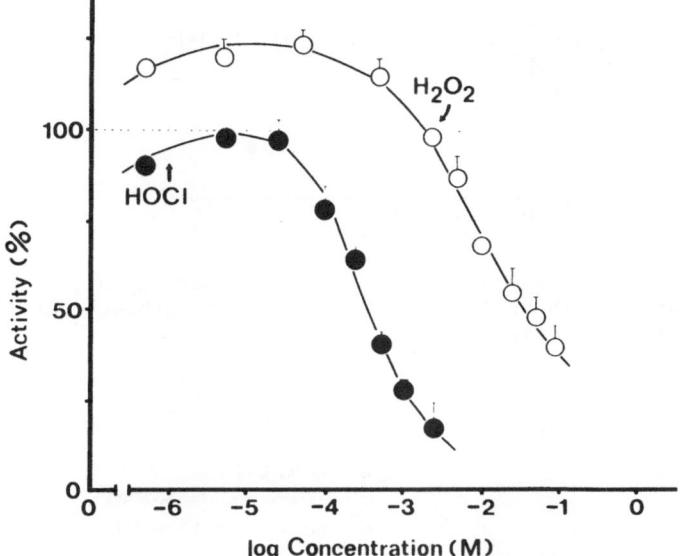

Fig.5. Effects of hydrogen peroxide and HOCl on myocyte Na,K,ATPase.
        Rat heart myocytes were permeabilized with saponin, and were incubated with oxidant (abscissa) for 30 min at 22°C. The control ATPase activity was $2.9 \pm 0.1 \times 10^{-13}$ mol Pi/cell.min; n = 5.

formation, hence the effectiveness of its action is curtailed (Imlay & Linn, 1988). However, the exact mechanism is as yet not defined.

Experiment with Chemically Skinned Cardiomyocytes

We studied further oxidant–induced modification of Na,K,ATPase by using isolated myocytes. Na,K,ATPase activity was determined with myocytes following permeabilization with saponin. The latter procedure made catalytic sites (existing on the intracellular surface of the sarcolemmal membrane) accessible for the assay, while maintaining ouabain binding sites (existing on the extracellular surface) exposed. The results revealed that the Na,K,ATPase of myocytes was relatively resistant to the action of hydrogen peroxide, i.e., half maximal inhibition required approximately 40 mM hydrogen peroxide (Fig.5). Part of the reason for this

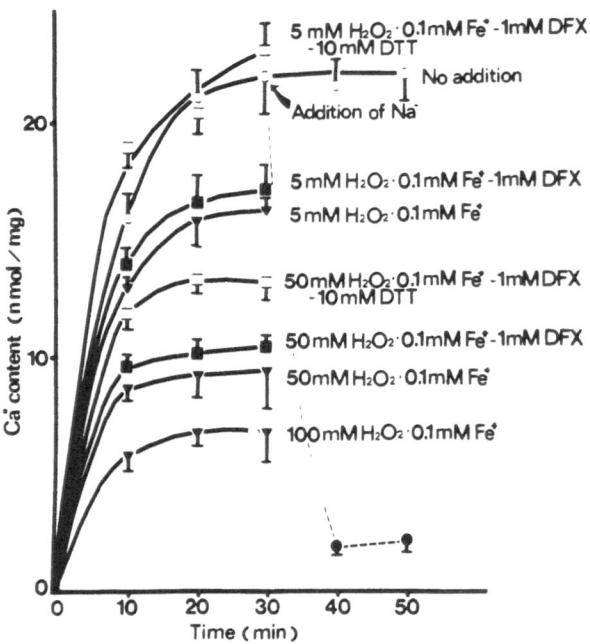

Fig.6. Inhibition of sarcolemmal calcium transport by hydrogen peroxide.
        Antagonism by deferoxamine (DFX) and dithiothreitol (DTT) is
        also illustrated.

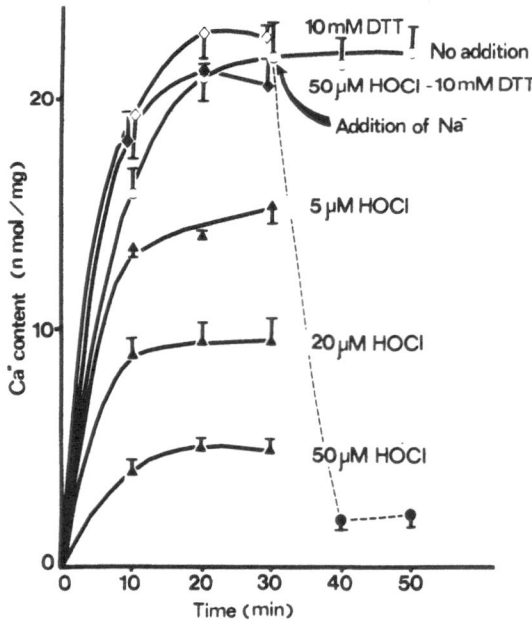

Fig.7. Inhibition of sarcolemmal calcium transport by HOCl.
        Note that concentrations are 3 orders of magnitude lower than
        those in Fig.6.

low sensitivity is that amino acids, buffer and albumin, which were present in the cell suspension medium could scavenge added oxidants. Whether or not the effect of hydrogen peroxide was mediated by a Fe-catalyzed reaction is not known. By contrast, half maximal inhibition by HOCl was approximately 0.4 mM. Interestingly, very low concentrations of hydrogen peroxide potentiated enzyme activity (Fig.5). Enhancement of Rb uptake, a measure of Na pump activity, by cultured lens epithelial cells under similar experimental conditions was reported recently by Spector et al.(1988). These results suggest that the Na pump function is modulated by oxidation-reduction states of amino acid residues of the molecule and that the ATPase in situ is in a reduced state.

Experiment with Sarcolemmal Preparation

We examined next the calcium pump function of an isolated canine sarcolemmal preparation. Our previous results, quoted above, indicated that HOCl raised the cytosolic calcium concentration as a result of calcium mobilization from intracellular stores and not as a result of increased net calcium influx into myocytes. Additionally, our measurements showed that calcium efflux from myocytes was accelerated by HOCl. For these reasons, effects of oxidants on sarcolemmal calcium transport were directly determined. Canine sarcolemmal fractions were prepared by the method of Slaughter et al.(1983). The fraction exhibited high Na,K,ATPase activities (160.7 $\pm$ 18.9 μmol Pi/mg protein.h; n = 5) and low percentages of leaky vesicles (13.1 $\pm$ 2.1 %; n = 4). Calcium uptake by sarcolemmal vesicles was found to be suppressed by HOCl as well as by hydrogen peroxide (Figs.6 & 7). Again the sensitivity to hydrogen peroxide was 3 orders of magnitude below that to HOCl. Figs. 6 & 7 demonstrate in addition that the calcium that was taken up by sarcolemmal vesicles was nearly completely released by the addition of Na^+ (i.e., via sodium/calcium exchange). These results do not seem to agree with our earlier results showing an accelerated calcium efflux by HOCl. However, it is possible that although the sarcolemmal calcium pump was inhibited by HOCl, sodium/calcium exchange was sufficiently functional to transport calcium outwards. An alternative possibility is that the sarcolemmal permeability was raised by oxidants. These assumptions are being verified by the experiment currently undertaken in our laboratory. Finally, Fig.6 indicates that deferoxamine plus dithiothreitol prevented hydrogen peroxide-induced impairment in sarcolemmal calcium transport only when the oxidative insult was not severe, whereas dithiothreitol effectively antagonized HOCl effects (Fig.7).

SUMMARY

Our study demonstrated that the Na,K,ATPase activity and ouabain binding sites were reduced by oxidants. Sarcolemmal calcium transport was also inhibited by hydrogen peroxide and HOCl. The action of HOCl on the sarcolemmal functions was 2 - 3 orders of magnitude more powerful than that of hydrogen peroxide. Effects of hydrogen peroxide consisted of two components, i.e., the first, highly sensitive one, most probably mediated by Fe-catalyzed, site-specific free radical formation, and the second, less potent action by (high concentrations of) hydrogen peroxide. Finally, very low concentrations of hydrogen peroxide potentiated Na,K,ATPase activities when assayed using myocytes.

REFERENCES

Bolli,R., 1988, Oxygen-derived free radicals and postischemic myocardial dysfunction ("stunned myocardium"). J.Am.Coll.Cardiol. 12:239.

Burton,K.P., 1988, Evidence of direct toxic effects of free radicals on the myocardium. Free Radicals Med.Biol. 4:14.

Imlay,J.A., and Linn,S., 1988, DNA damage and oxygen radical toxicity. Science 240:1302.

Kako K.J., Kato,M., Matsuoka,T., and Mustapha,A., 1988, The depression of membrane-bound Na^+K^+ATPase activity induced by free radicals and by ischemia of the kidney. Am.J.Physiol. 254:C330.

Kaminishi,T., Matsuoka,T., Yanagishita,T., and Kako,K.J., 1989, Increase versus decrease of calcium uptake by isolated heart cells by H_2O_2 versus HOCl. Am.J.Physiol. (in press)

Kato,M., and Kako,K.J., 1988, Na^+/Ca^{2+} exchange of isolated sarcolemmal membrane: Effects of free radicals, insulin and insulin deficiency. Mol.Cell.Biochem. 83:15.

Shinar,E., Navok,T., and Chevion,M., 1983, The analogous mechanisms of enzymatic inactivation induced by ascorbate and superoxide in the presence of copper. J.Biol.Chem. 258:14778.

Skou,J.C., and Norby,J.G., eds., 1979, "Na,K,ATPase. Structure and Kinetics," Academic Press, New York.

Slaughter,R.S., Sutko,J.L., and Reeves,T.R., 1983, Equilibrium calcium-calcium exchange in cardiac sarcolemmal vesicles. J.Biol.Chem. 258:3183.

Spector,A., Yan,G., Huang,R.C., McDermott,M.J., Gascoyne,P.R.C., and Pigiet,V., 1988, The effect of H_2O_2 upon thioredoxin-enriched lens epithelial cells. J.Biol.Chem. 263:4984.

Thompson,J.A., and Hess,M.L., 1986, The oxygen free radical system: A fundamental mechanism in the production of myocardial necrosis. Progr.Cardiovasc.Dis. 6:449.

Weiss,S.J., and Lobuglio,A.F., 1982, Phagocyte-generated oxygen metabolites and cellular injury. Lab.Invest. 47:5.

Werns,S.W., and Lucchesi,B.R., 1988, Leukocytes, oxygen radicals and myocardial injury due to ischemia and reperfusion. Free Radicals Biol.Med. 4:31.

Zweier,J.L., Kuppsamy,P., and Lutty,G.A., 1988, Measurement of endothelial cell free radical generation: Evidence for a central mechanism of free radical injury in postischemic tissues. Proc.Natl.Acad.Sci.USA. 85:4046.

HEMODYNAMIC RESPONSES TO VASOACTIVE COMPOUNDS IN

CHRONICALLY ALCOHOL TREATED RATS

H. M. Rhee, J. L. Valentine, D. Hendrix, M. Schweisthal[*] and M. Soria[**]

Departments of Pharmacology, Anatomy[*], and Psychiatry[**]
Oral Roberts University
School of Medicine
Tulsa, OK 74137

INTRODUCTION

Recent evidence suggests that substantial usage of ethanol (hereafter termed alcohol) for a long period of time is positively correlated with hypertension in animal models[7] as well as in epidemiologic human investigations[5,6,9,12,15,26]. Although variables such as smoking, nutritional status, stress and genetic predisposition make human data difficult to interpret, the evidence for the link between substantial amount of alcohol use and high blood pressure is substantial. For example, more than 50% of the alcoholics admitted for detoxification have high blood pressure, and their elevated pressure was returned toward normal when they withdrew the alcohol consumption[26]. However, biochemical or pharmacological mechanisms underlying the positive correlation are poorly understood.[2]

If alcohol increases blood pressure, one might anticipate an increased general sympathetic nerve discharge in alcoholics. Therefore, in our previous study[23] effect of acute alcohol on renal sympathetic nerve activity was examined in anesthetized rabbits. In the study, intravenous alcohol reduced the renal nerve activity and increased the duration of the nerve suppression. This might be related to the well-known effect of alcohol as a local anesthetic after a bolus injection. The objective of this research, however, was to investigate the chronic dietary alcohol effect on arterial blood pressure and sympathetic renal nerve activity. In addition, in these chronic alcoholic animals we also examined the effect of alcohol on arterial baroreceptor function because of published negative effect of alcohol on baroreflex system[1].

MATERIALS AND METHODS

Ethanol Administration and Grouping of Animals

Young male Sprague-Dawley rats (Charles Rivers, Mass.) initially weighing 180-220 g were randomly divided into three groups: rats fed on solid regular chow, control liquid diet and alcohol liquid diet group.

Control solid diet was rodent blox (8604-00) from Wayne Pet Food,
Chicago, Ill. Alcohol diet consisted of 36% calorie from alcohol, which
was supplied by Ziegler Food, Inc., in a liquid form, according to Lieber
and DeCarli[17,18]. Control rats were fed a comparable liquid diet in which
sucrose was substituted for alcohol. All animals were housed in standard
laboratory conditions (12 h light cycle) and had access to food and water
ad libitum.

Animal Preparation and Physiological Recording

Rats were anesthetized with intraperitoneal pentobarbital
(30 mg/kg). The left carotid artery and jugular vein were then cannulated for the measurement of arterial pressure and subsequent delivery of
drugs. The rats were allowed to respire spontaneously through a tracheostomy, and were maintained at 37°C with a temperature regulator (Gorman
Rupp, Bellville, OH). Arterial pressure was monitored by a Statham P23
pressure transducer coupled to a pressure processor amplifier (Gould
Instruments, Cleveland, OH, model 13-4615-52) which computes the systolic, mean, diastolic and pulse pressures. Heart rate was monitored by
a Gould Biotec ECG amplifier (model 13-4615-65).

After the basic surgery, the animals were placed on their right side,
and the kidney was exposed retroperitoneally. A branch of the left renal
nerve was carefully dissected from the surrounding connective tissue,
cleaned and suspended on a bipolar electrode in a pool of mineral oil.
Multiunit renal nerve activity (RNA) was monitored by a Gould universal
amplifier (model 13-4615-56) and was integrated with respect to time by a
Gould Integrator (model 13-4615-70) in 2-sec intervals. All of the above
parameters were recorded on a Gould 16-channel electrostatic recorder
(model ES 1000) as reported previously[24].

In order to test the validity of the nerve preparation, a reflex
withdrawal of sympathetic discharge was produced by an i.v. injection of
norepinephrine. Elimination of recorded activity after the rise in blood
pressure verified the measurement of multiunit nerve activity and that
ongoing activity was predominately sympathetic[22,24].

Drug Treatment and Determination of Baroreceptor Sensitivity

Phenylephrine was used to monitor the reflexogenic bradycardia by
giving the drug in bolus doses of 0.5, 1, 2 or 4 µg/kg in ascending order
with a 5-minute interval between doses for the three groups of rats.
Nitroprusside (2, 4, 8 and 16 µg/kg) was also used to establish the relationship between drug-induced hypotension and baroreceptor mediated
tachycardia in the three groups. Graded increases or decreases in arterial blood pressure evoked by injecting serial doses of phenylephrine or
nitroprusside were recorded continuously. The peak changes in mean blood
pressure (ΔBP) together with the reflexogenic change in heart rate (ΔHR)
were used to compute the ratio $\Delta HR/\Delta BP$, expressed as beat per min per
mmHg. The ratio was used as an index of baroreceptor sensitivity in each
group of rats as described by Rothbaum et al.[25].

Statistical Analysis

Each rat served as its own control to calculate percent changes after
each treatment with drugs. The mean values for arterial pressure, heart
rate and RNA were expressed as a function of the doses of phenylephrine,
or nitroprusside. The RNA was expressed as a percentage change in integration per min from the relevant control experiments or the duration (in

630

sec) of RNA suppression. Each datum point was analyzed by unpaired Student "t" test. The dose-response curves were analyzed by the analysis of variance, and significant difference was considered as P is less than .05.

RESULTS

Hemodynamic Effects of Alcohol

Effects of solid rat chow (Wayne Pet Food), liquid diet and liquid diet containing alcohol on hemodynamic parameters are summarized in Table 1. Liquid diet did not have a significant different effect on both blood pressure and heart rate. Liquid diet containing alcohol significantly reduced mean and diastolic blood pressure compared to those of rats on solid diet. However, the difference was not significant from the values of rats on liquid diet. Pulse pressure and heart rate were not different in the rats on the three types of diet.

Table 1. Effects of Diet and Alcohol on Cardiovascular Parameters in Intact Rats[a]

Parameters Treatment	N[b]	SP (mmHg)	MP (mmHg)	DP (mmHg)	PP (mmHg)	HR (BPm)
Solid diet[c]	12	96.8±3.9	84.9±3.8	78.2±3.7	20.1±3.0	265±4.7
Liquid diet[c]	5	90.0±2.4	75.0±2.9	65.2±3.3	22.8±2.2	252±4.3
Alcohol	4	81.5±3.6	63.8±3.1[d]	55.0±2.9[d]	24.8±2.7	274±8.1

[a]All values are expressed in mean ± S.E.
[b]Indicates number of animals used.
[c]See the "Method" for specific diet regimens used.
[d]Indicates p < 0.05, compared to solid diet group.

Responses to Pressor (Phenylephrine) Treatment

In order to test arterial baroreceptor function in these rats, phenylephrine, a pure adrenergic alpha receptor agonist, was injected and corresponding changes in blood pressures, heart rate and renal sympathetic nerve activity were recorded (fig. 1). Phenylephrine increased mean pressure in a dose dependent manner in control experiments. There was a slight, but gradual reduction in heart rate, due to a presumable reflex originating from baroreceptors. There was a dose-dependent reduction in sympathetic renal nerve activity in response to phenylephrine-induced hypertension. Diet containing alcohol did not produce remarkably different effects on either blood pressure, heart rate or renal nerve activity.

However, careful evaluation of the relationship between blood pressure and doses of phenylephrine in the two groups of rats revealed that alcohol-treated rats did respond differently to the vasoconstrictor from the control rats (Tab. 2). There was no significant difference between the rats on solid and liquid diet in blood pressure regressional lines;

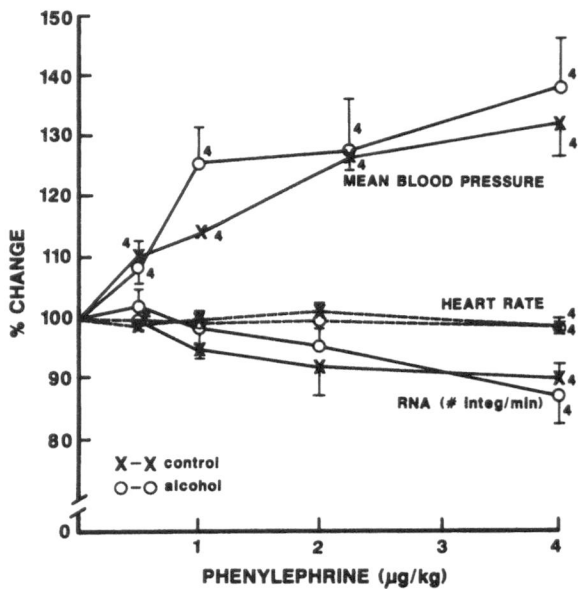

Fig. 1. Effects of chronic alcohol on blood pressure, heart rate, and renal nerve activity (RNA): phenylephrine effects. Young Sprague-Dawley rats were instrumented as described under "Methods." Phenylephrine was injected intravenously to the rats fed on control liquid diet or alcohol diet after 30 min stabilization period as described under Methods. Each parameter is expressed as percent change from the value at the control period. Renal nerve activity is expressed as number of integration per minute before the calculation of percent change. Numbers indicate numbers of animals used. Vertical bars indicate standard error of the mean.

that is, the slope, correlation coefficient or intercept. Rats on liquid alcohol diet had a distinctly greater response to phenylephrine challenge, although correlation coefficient or intercept was not different from the two control groups. In response to vasodepressor challenge induced by nitroprusside, the alcohol-treated group had a significantly reduced reaction to nitroprusside (Tab. 2).

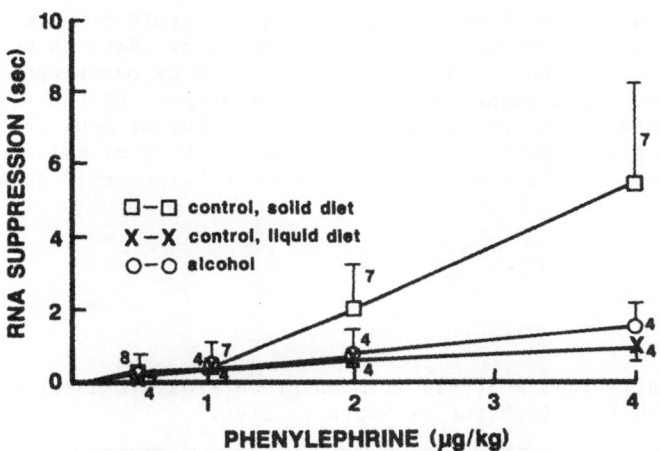

Fig. 2. Chronic alcohol effect on the duration of suppression of renal
nerve activity (RNA) in intact rats. Indicated doses of phenylephrine were administered as in Figure 1. The duration of RNA
suppression as a result of baroreflex response to phenylephrine
was determined in seconds. Numbers indicate the number of animals used and vertical bars indicated standard error of the
mean.

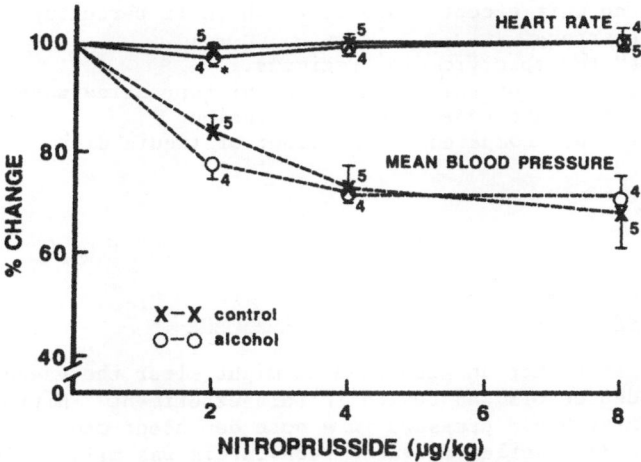

Fig. 3. Lack of chronic alcohol treatment on the response to nitroprusside in intact rats. Responses to nitroprusside on heart rate
and mean blood pressure were monitored as in Figure 1 in the
rats fed on either liquid diet or chronic alcohol. Numbers
indicate numbers of animals used and vertical bars indicate
standard error of the mean.

A sudden increase in blood pressure induced by phenylephrine not only reduced renal nerve activity, but also effectively blocked the discharge. The duration of RNA suppression as a result of hypertensive action of phenylephrine was compared in figure 2. Rats on control solid diet were sensitive to the hypertension induced by phenylephrine as indicated by a prolonged suppression of RNA discharge. At the 4 µg/kg dose of phenylephrine, RNA was effectively blocked for as long as 5 sec. There was no significant difference in the duration of RNA suppression between control liquid diet group and the alcohol-treated group (fig. 2).

TABLE 2. Statistical Analysis of Mean Blood Pressure after the Treatment with Phenylephrine or Nitroprusside.[a]

Treatment	Group	Solid Diet[b]	Liquid Diet[b]	Alcohol
Phenylephrine[c]	Slope	5.02	5.24	7.14[d]
	Correlation Coefficient	0.86	0.95	0.88
	Intercept	111.0	107.5	111.5
Nitroprusside[c]	Slope	2.05	2.29	0.76[d]
	Correlation Coefficient	0.94	0.91	0.76
	Intercept	77.6	85.9	77.5

[a]Values have no unit except "intercept" which is percentage change in mean blood pressure.
[b]See "Methods" for specific diet regimens.
[c]Three or four doses of phenylephrine or nitroprusside were used as in method to produce hypotension or hypertension.
[d]Indicate $p < 0.05$, compared to the group of liquid diet.

Responses to Depressor (Nitroprusside) Treatment

The possibility that an alcohol diet might alter the vascular response to vasodepressors was tested in this experiment. Nitroprusside decreased arterial blood pressure in a dose dependent manner in the control rats (fig. 3). Reflex-mediated tachycardia was mild since the reduction of blood pressure after nitroprusside treatment was moderate. There was essentially no difference in blood pressure or heart rate in the control and alcohol-treated groups (fig. 3). During the hypotension induced by the treatment of nitroprusside, potential elevation of sympathetic renal nerve activity in the control and alcohol groups were compared (fig. 4). At the 4 µg/kg dose of nitroprusside, RNA was significantly reduced in alcohol-treated rats. Overall slopes of the two groups after nitroprusside treatment was significantly lower than the slope with the liquid diet as indicated above (Tab. 2).

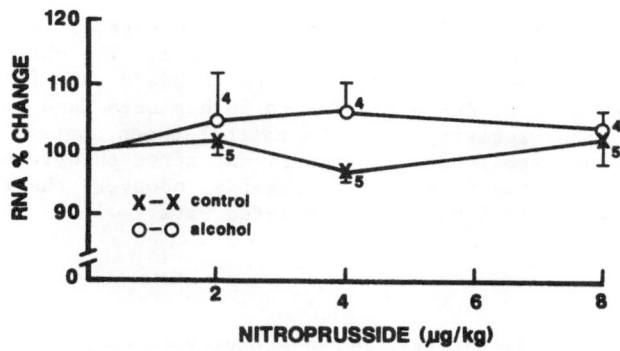

Fig. 4. Effects of chronic alcohol on renal nerve activity after the administration of nitroprusside in rats. Nitroprusside was administered as in Figure 3 to the control or chronically alcohol fed rats. In these rats, renal nerve activity (RNA) was compared by the number of integration per minute as described under "Methods." Numbers indicate numbers of animals used and vertical bars indicate standard error of the mean.

TABLE 3. Baroreceptor Responses After Phenylephrine or Nitroprusside in Rat Fed on Control Liquid Diet and Alcohol Treated Rats.[a]

Treatment	Dose µg/kg	Control Liquid Diet			Alcohol Treated Group[b]		
		ΔBP[c] mmHg	ΔHR[d] Bpm	ΔHR/ΔBP[c] Bpm/mmHg	ΔBP mmHg	ΔHR Bpm	ΔHR/ΔBP Bpm/mmHg
Phenyl-ephrine	0.5	8.5	-0.8	-0.09	5.2	-1.2	-0.23
	1.0	10.8	-3.0	-0.03	18.5	-1.5	-0.08
	2.0	20.8	-3.5	-0.17	24.0	-1.8	-0.08
	4.0	27.4	-6.8	-0.25	37.2	-4.5	-0.12
Mean±S.E.				-0.14±0.05			-0.13±0.04
Nitro-prusside	2.0	-11.4	1.2	-0.11	- 7.8	2.0	-0.26
	4.0	-17.4	2.0	-0.11	-14.8	4.0	-0.27
	8.0	-20.6	4.4	-0.21	-15.0	6.5	-0.43
	16.0	-29.4	7.6	-0.26	-22.0	7.2	-0.33
Mean±S.E.				-0.17±0.04			-0.32±0.39*

[a]All values are expressed in Mean ± S.E.
[b]See "Methods" for specific diet regimens.
[c]Indicates changes in mmHg after the treatment with either phenylephrine or nitroprusside.
[d]Indicates changes in beats per min after the treatment with either phenylephrine or nitroprusside.
[e]Indicates ratio of ΔHR/ΔBP as an index of baroreceptors sensitivity.

Effect of Chronic Alcohol on Baroreceptor Sensitivity

Baroreceptor function was tested in the rats, utilizing phenylephrine and nitroprusside as summarized in Table 3. The changes in mean pressure (ΔBP) and corresponding alterations in heart rate (ΔHR) were used to calculate the ratio of ΔHR/ΔBP. The ratio has been used as a sensitive index of baroreceptor activity[1,25]. The ratio between control rats and alcohol-fed rats was not statistically significant after phenylephrine challenge. After the treatment with nitroprusside, however, the ratio was increased significantly ($p < 0.05$) in alcohol-fed rats.

DISCUSSION

Effects of Alcohol on Sympathetic Nerve Activity

Considerable experimental and clinical evidence supports the positive correlation between chronic alcohol consumption and hypertension (see Introduction). However, the experimental evidence that defines underlying mechanism(s) by which chronic alcohol consumption causes a rise in arterial blood pressure is scanty. One of the most important facts in all types of human hypertension is sympathetic hyperactivity. Thus, it is understandable that many investigators have focused their attention on the level of catecholamines in hypertensive alcoholics. Alcohol is known to increase the release of catecholamines from the adrenal medulla[16,20]. Alcohol also increases urinary excretion of catecholamines in alcoholics[10,27] and in experimental animals[21]. These findings were confirmed recently with more refined techniques[11,19]. However, many studies do not confirm these findings. For example, hypertensive alcoholics had normal plasma concentrations of catecholamines, angiotensin, and aldosterone.[3]

In the present study we were interested in determining renal sympathetic nerve activity in alcoholic rats instead of determining plasma catecholamines for several reasons. Anatomically renal efferent nerves innervate renal tissues which in turn control renal sodium excretion and renin release[4]. Stimulation of intrarenal mechanoreceptors and chemoreceptors induces an increase in afferent nerve activity, which is closely associated with changes in blood pressure. Renal denervation lowers blood pressure and prevents the development of hypertension, which also is known to decrease sodium and water reabsorption[8]. Chronic renal artery stimulation or continuous intrarenal norepinephrine infusion in conscious dogs also produced sustained hypertension[8,13] which suggests a definite role of the renal artery in the pathogenesis of hypertension. Furthermore, in both SHR of the Okamoto strain and the DOCA-salt hypertensive models, increased renal efferent nerve activity lead to the development of hypertension by causing enhanced urinary sodium retention[14,28]. Additionally we could correlate blood pressure and renal nerve activity by the routine assay developed in our laboratory[24].

Basal discharge rate of the left renal nerve after pentobarbital anesthesia was not statistically different between the control group and alcohol-fed rats. It is well known that anesthetic agents do affect the number of physiological parameters, particularly nerve discharge rate. Determination of nerve activity from a small autonomic C fiber is not impossible in conscious animal, but was not technically easy in our hands. It will be necessary to compare critically the basal rate of renal sympathetic discharge rate in the control and alcohol-fed rats without anesthetic agents to prove the involvement of sympathetic nerve in alcohol-induced hypertension. However, rats on control solid diet had significantly greater sensitivity in renal nerve activity after phenylephrine compared to the rats that were on liquid control diet. This suggests that the liquid control diet may not be the perfect control diet, although it may be superior from the nutritional standpoint.

Effect of Alcohol on Baroreceptor Function and Vascular Reactivity

There was no statistically significant difference in blood pressure of the control and alcohol-fed rats (data not shown). Thus, we tested vascular reactivity to vasoactive compounds in the two groups of rats. Phenylephrine, a pure adrenergic agonist, increased blood pressure in the two groups of rats with a corresponding decrease in sympathetic renal nerve activity and heart rate (fig. 1). Response to phenylephrine was remarkably greater in the alcohol-fed rats without a corresponding decrease in either heart rate or renal nerve activity (Table 2). This indicates there may be an impairment or at least reduction in baroreflex sensitivity in alcohol-fed rats. It is not immediately clear why the alcohol-fed rats had an elevated blood pressure unless we assume there might be an alteration in resistance blood vessel reactivity to catecholamines.

There was no clear distinction between the two groups of rats after the treatment of vasodepressor challenge with nitroprusside. In response to nitroprusside, however, alcohol-fed rats did not increase renal nerve activity as we expected as a result of baroreflex responses (fig. 4). It is noted in this experiment both the two types of rats had very low response to the two agents because an elevation of blood pressure by 20 to 30 mmHg should reduce heart rate at least 20 to 25% of initial heart rate[23]. Although the reason is not readily available, liquid diet that we used may have to do with the discrepancy between our study and others[1].

This dampened response to both phenylephrine and nitroprusside is illustrated clearly in Table 3. Phenylephrine at the dose of 4 μg/kg increased blood pressure with a little reduction in heart rate. The responses after nitroprusside were also similarly reduced in the increase of heart rate in both the control and alcohol-fed rats. This reduced response may be in part responsible for the failure of this study to differentiate alcohol-fed rats from the control rats after phenylephrine treatment. However, although the response to nitroprusside was low, alcohol increased baroreceptor sensitivity ($p < 0.05$). This suggests that alcohol may not greatly affect baroreceptor sensitivity.

In summary, this study does not support the contention that alcohol increases blood pressure, even after chronic (60 days) use in experimental animals. Many experimental liquid diets that have been used as a universal control may not necessarily be the best control diets since the liquid diet certainly produced different vascular effects in this present study. Alcohol-fed rats demonstrated a greater sensitivity to phenylephrine. This observation is difficult to explain. Alcohol appears not to impair baroreceptor function, at least in experimental animals.

REFERENCES

1. Abdel-Rahman A-R A, Dar, MS, Wooles WR. Effect of chronic ethanol administration on arterial baroreceptor function and pressor and depressor, responsiveness in rats. J Pharmacol Exp Ther 1985; 232:194-201

2. Altura BM, Carella A, Altura BT. Acetaldehyde on vascular smooth muscle: possible role in vasodilator action of ethanol. Europ J of Pharmacol 1978; 52:73-83.

3. Arkwright PD, Beilin LJ, Rouse L, Vandongen R. The pressor effect of moderate alcohol consumption in man: a search for mechanism. Circulation 1982; 66:515-519.

4. Barajas L, Muller J. The innervation of the juxtaglomerular apparatus and surrounding tubules: a quantitative analysis by serial section electron microscopy. J Ultrasound Res 1973; 43:107.

5. Barboriak JJ, Rimm AA, Anderson AJ, Schmidhoffer M, Tristani FE. Coronary artery occlusion and alcohol intake. Brit Heart J 1977; 39:289-293.

6. Celentano DD, Martinez RM, McQueen DV. The association of alcohol consumption and hypertension. Prev Med 1981; 10:590-602.

7. Chan TCK, Sutter MC. Chronic ethanol consumption and blood pressure. Life Sci 1983; 33:1965-1973.

8. DiBona GF, Zambaski EJ, Aguilera AJ, Kaloyanides GJ. Neurogenic control of renal tubular sodium reabsorption in the dog. Circ Res 1977; 40 (Suppl. 11):1-127.

9. Dyer AR, Stamler J, Paul O, Beckson DM. Alcohol consumption, cardiovascular risk factors, and mortality in two Chicago epidemiological studies. Circ 1977; 56:1067-1074.

10. Giacobini E, Izikowitz S, Wegmann A. Urinary norepinephrine and epinephrine excretion in delirium tremens. A.M.A. Arch Gen Psychiat 1960b; 3:289-296.

11. Greenspon AJ, Stang JM, Lewis RP, Schaal SF. Provocation of ventricular tachycardia after consumption of alcohol. N Eng J Med 1979; 301:1049-1050.

12. Kannel WB, Sorlie P. Hypertension in Framingham. In: Epidemiology and Control of Hypertension, Paul O ed. Stratton Intercontinental Medical Book Corp., New York, 1975, pp. 553-590.

13. Katholi RE, Carey RM, Ayers CR, Vaughan ED, Yancey MR, Morton CL. Production of sustained hypertension by chronic intrarenal norepinephrine infusion in conscious dogs. Circ Res 1977; 40 (Suppl. I):1-118.

14. Katholi RE, Naftilan AJ, Oparil S. Importance of renal sympathetic tone in the development of DOCA-salt hypertension in the rat. Hypertension 1980; 2:266.

15. Klatsky AL, Friedman GD, Siegelaub AB, Gerard MJ. Alcohol consumption and blood pressure: Kaiser-permanente multiphasic health examination data. N Engl J Med 1977; 296:1194-1200.

16. Klingman GI, Goodall, McC. Urinary epinephrine and leverterenol excretion during acute sumethal alcohol intoxication in dogs. J Pharmacol Exp Ther 1957; 121:313-318.

17. Lieber CS, DeCarli, LM: The feeding of alcohol in liquid diets: Two decades of application and 1982 update. Alcohol Clin Exp Res 6:523-531, 1982.

18. Lieber CS. Metabolic effects of ethanol on the liver and other digestive organs. In: Alcohol and the GI Tract. Clinics in Gastroenterology, Leevy CM ed. Volume 10, Number 2, London, W. B. Saunders, 1981, pp. 315-342.

19. Ogota, M, Mendelson JH, Mello NK, Majchrowicz E. Adrenal function and alcoholism II. Catecholamines. Psychosom Med 1971; 33:159-180.

20. Perman ES. The effect of ethyl alcohol on the secretion from the adrenal medulla of the cat. Acta Physiol Scand 1960; 48:323-328.

21. Perman ES. Effect of ethanol and hydration on the urinary excretion of the adrenaline and noradrenaline and on the blood sugar of rats. Acta Physiol Scand 1961; 51:68-74.

22. Peterson DF, Coote JH, Gilbey MP, Futuro-Neto HA. Differential pattern of sympathetic outflow during upper airway stimulation with smoke. Am J Physiol 1983; 245:R433-R437.

23. Rhee, HM. Effects of ethanol on sympathetic renal nerve activity and baroreceptor sensitivity in anesthetized rabbits. Neurotoxicol 1986; 7(2):279-286.

24. Rhee HM, Eulie PJ, Peterson DF. Suppression of renal nerve activity by methionine enkephalin in anesthetized rabbits. J Pharmacol Exp Ther 1985; 234:534-537.

25. Rothbaum DA, Shaw DJ, Angell CS, Shock NW. Age difference in baro-receptor response of rats. J. Gerontol. 1974; 29:488-492.

26. Sounders JB, Beevers DG, Paton A. Alcohol-induced hypertension. Lancet 1981; 2:653-656

27. Schenker VJ, Kissin B, Maynard LS, Schenker AC. Adrenal hormones and amine metabolism in alcoholism. Psychosom Med 1966; 28:564-569.

28. Winternitz SR, Katholi RE, Oparil S. Role of the renal symapthetic nerves in the development and maintenance of hypertension in the spontaneously hypertensive rat. J Clin Invest 1980; 66:971.

DOES THE AREA AT RISK OF INFARCTION CHANGE OVER TIME?

Reena Sandhu and G.P. Biro

Department of Physiology, Faculty of Health Sciences
University of Ottawa, 451 Smyth Rd., Ottawa, Ontario
Canada, K1H 8M5

INTRODUCTION

The size of the hypoperfused zone (area at risk) produced by occluding any given coronary artery shows considerable variation among untreated animals (1). Since the amount of myocardium that will infarct is, to a large extent, dependant upon the size of the area at risk (AAR) (1-3), infarct sizes also show considerable variation. Therefore, it has become conventional to express infarct size as a percentage of the AAR in order to standardize this variable.

In the past, post-mortem isolated heart and *in vivo* techniques have been used to identify the AAR. The post-mortem techniques identify only the normal perfusion territory of the occluded artery and do not take into consideration collateral flow to the risk zone (4,5). The *in vivo* techniques are said to be measuring the "physiological area at risk" because they take into account hemodynamic conditions and collateral flow to the risk zone, thus identifying an area that is truly ischemic and at risk of undergoing infarction (5).

It has not yet been established whether the different *in vivo* techniques used to measure the AAR delineate the same area of myocardium. Equally unclear, is the manner in which collateral flow to the risk zone changes over the course of infarct development in the untreated animal. It would stand to reason that if collateral flow does change spontaneously over this initial period, it may be reflected in a change in the size of the physiological AAR.

The purpose of this investigation is two fold. Firstly, to determine whether the physiological AAR as measured by two commonly used *in vivo* techniques is the same and secondly, to determine if the size of the AAR changes from 10 minutes to three hours post-occlusion in open chested dogs. The two "tracers" used to measure AAR size were ^{99m}Tc macroaggregates and Monastral blue dye.

^{99m}Tc macroaggregates have an average diameter of 20 μm, quite similar to that of the plastic microspheres used to determine regional myocardial blood flow (15μm). Consequently, these macroaggregates distribute themselves in the circulation in a manner similar to that of the plastic microspheres. The use of ^{99m}Tc macroaggregates to identify the physiological AAR has been widespread in studies aimed at finding out the effect of pharmacological agents on infarct size limitation. This is principally due to the fact that macroaggregates, unlike the pre-terminal dyes, can be injected at any point during the experiment and thus reflect the status of collateral flow to the ischemic region prior to the administration of drugs (4). The half lives of both the ^{99m}Tc and the albumin macroaggregates is very short. Consequently, this method of assessing the size of the AAR can only be used in experiments of less than 24 hours duration.

The Monastral blue staining involves the injection of a blue dye into the systemic circulation. Dye injection stains all tissue not supplied by the occluded artery deep blue. Within the occluded bed dye penetration is only permitted where collateral flow is sufficient to permit its entry. Since dye injection is eventually fatal, this can only be done just prior to the conclusion of the experiment. Subsequent autoradiographic and photographic visualization revealed the AAR determined by ^{99m}Tc macroaggregate injection and Monastral Blue injection respectively.

METHODS AND MATERIALS

Experimental Preparation and Measurements

Mongrel dogs of either sex weighing 14-31 kg were used for this study. Eighteen hours prior to the planned start of surgery, food but not water was withheld. Animals were initially pre-medicated with Pre-mix (Atropine [0.03 mg/kg]/Demerol [2.5 mg/kg]/Acepromazine [0.1 mg/kg] i.m.) and subsequently anesthetized with Somnotol (26 mg/kg) intravenously. Supplemental anesthesia was administered during the experiment when needed. The dogs were intubated with a cuffed endotracheal tube and ventilated by a Penlon Volume Ventilator (Model No. AV-500) on a Boyle Anesthetic machine delivering a mixture of 20% nitrous oxide and pure oxygen at a rate of twelve breaths/minute with a tidal volume between 350 - 600 mls. Arterial pH, pO_2 and pCO_2 were periodically monitored throughout the experimental period and maintained between 7.2 - 7.3, 160 - 225 mm Hg and 33 - 39 mm Hg, respectively by adjusting tidal volume.

A Swan Ganz catheter was placed in the left jugular vein for the purpose of monitoring right atrial (RA), pulmonary arterial (PA) and capillary wedge (CW) pressures. A No. 7 Cardix catheter was placed in the right carotid artery for recording left ventricular (LV) pressure. A cannula was placed in the right jugular vein for intravenous infusion of saline (12-15 ml/kg/hour). A catheter was also placed in the femoral artery for recording arterial pressure and for obtaining blood samples. A catheter in the femoral vein was also used for obtaining blood samples. Lead II of the ECG was continually recorded throughout the experiment.

A left thoracotomy through the fourth or fifth intercostal space was performed, the lungs were retracted and the pericardium was opened. A Levine Feeding Tube was inserted into a branch of the pulmonary vein for the purpose of injecting microspheres, macroaggregates, Monastral blue dye and for measuring left atrial (LA) pressure. A No. 2 vascular silk was placed around the Left Anterior Descending coronary artery (LAD) at a point immediately above the first diagonal branch. After allowing approximately twenty minutes to establish a stable hemodynamic state, baseline measurements of RA, PA, CW, LA, LV, arterial pressures, heart rate and blood gases were determined. All pressures were referenced to mid-chest level and were measured using a Gould Statham transducer (Model No. P23Db) on a four-channel Grass recorder.

Experimental Groups

The protocol followed for the two experimental groups is described below and summarized in Figure 1.

GROUP 1- Microspheres were injected five minutes prior to and ten minutes following LAD occlusion in order to determine myocardial blood flow at these times. Immediately after the second microsphere injection, ^{99m}Tc macroaggregates were injected followed by Monastral blue dye in order to delineate the extent of the AAR at ten minutes.

GROUP 2- The protocol was similar to GROUP 1 except that the LAD artery was tied for three hours and the Monastral blue dye was injected at the end of the occlusion period. In addition, a third microsphere was injected three hours post-occlusion to ascertain the level of flow to normal and ischemic tissue.

Hemodynamic measurements were made just prior to the second microsphere injection in both groups. In GROUP 2, these measurements were subsequently made every half hour throughout the experiment. If an animal fibrillated during the course of an experiment, a defibrillator was used (Hewlett Packard, Model No. 42110-A).

Experimental Groups

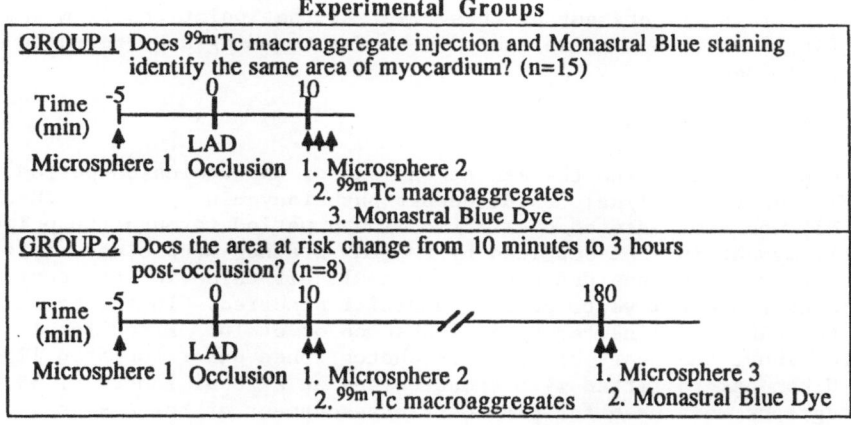

Figure 1

Following Monastral blue injection, the heart was excised, rinsed in saline and sliced into six mm thick transverse sections from apex to base using an electric meat slicer. The heart slices were then used to determine the AAR as defined by ^{99m}Tc macroaggregate and Monastral blue injection and also for determination of regional myocardial blood flow.

Measurements and Tissue Preparation - AAR Determination

^{99m}Tc Macroaggregates

Details of this methodology are described by Deboer et. al. (4). Briefly, one million albumin macroaggregates (3M Co., St Paul, MN.) labelled with ^{99m}Tc [0.35 mCi/kg) in a volume of nine mls of saline were injected into the pulmonary vein in order to delineate the extent of the hypoperfused zone ten minutes post-occlusion. Macroaggregate injection was done over a period of twenty seconds and was followed by a ten ml flush of saline. The macroaggregates were shaken vigorously by hand prior to injection to prevent clumping and this was confirmed periodically by microscopic inspection. Labeling efficiency, as assessed by Nucleopore size filtration, was 85%.

Autoradiography was performed by placing these slices in KODAK X-OMATIC x-ray cassettes with sensitive medium intensifying screens. High speed CHROMEX x-ray film (Dupont) was placed in the cassette and exposed for eighteen hours at $4^{\circ}C$ (to prevent tissue distortion or shrinkage). The film was developed by hand using KODAK GBX developer and fixer.

Once developed, the x-ray film revealed the extent of trapping of ^{99m}Tc labelled macroaggregates and thus delineated the physiological risk zone at the time of macroaggregate injection, i.e. ten minutes following LAD occlusion.

Only ^{99m}Tc labelled macroaggregates and not radiolabelled microspheres trapped in the circulation fogged the x-ray film to any appreciable extent. This was due to the fact that the macroaggregates contained approximately 2000X the amount of radioactivity contained in the microspheres. The extent of x-ray film fogging produced by radiolabelled microspheres was assessed by allowing some heart slices to sit for 72 hours. At this time only 0.024% of ^{99m}Tc originally injected remains in the sample. Since the isotopes with which the microspheres are labelled have much longer half-lives (65-127 days), almost all of the radioactivity contained in the microspheres is still present at this time. At that point the film was exposed for 18 hours but these second set of autoradiographs revealed virtually no fogging.

Monastral Blue Dye

In order to determine the AAR at the end of the occlusion period, Monastral blue ([0.5 ml/kg], 25.8% copper phthalocyanine pigment, (Heubach, Newark, NJ) was administered over a two minute period through the pulmonary vein until myocardium not supplied by the LAD stained deep blue. This sometimes resulted in the death of the animal. If this did not occur, KCl was administered until ventricular standstill occurred. In the heart slices, the AAR is delineated by it's absence of stain. Prior to autoradiography, the heart slices were photographed under tungsten light using a Nikon camera fitted with a macro lens at a focal length of 38 cm. The photographs were used for planimetric assessment of the AAR as identified by the Monastral Blue staining technique.

Both autoradiographs and photographs were analyzed by planimetry using an IBAS-1 (Interactive Image Based Analysis System) with IBAS-1 standard software (Version 5.41) to determine the outlines of the area occupied by the left ventricle and the AAR. Area at risk size was determined by dividing AAR of a slice by the area occupied by the left ventricle in the same slice. The values for the sequential slices were then added to provide total risk zone areas for each heart. All risk zone values are expressed as percentages.

Regional myocardial blood flow (RMBF)

Regional myocardial blood flow was measured prior to LAD occlusion, ten minutes and three hours after occlusion with $15\mu m$ microspheres labelled with either ^{113}Sn, ^{85}Sr, ^{46}Sc, or ^{57}Co (DuPont-NEN Canada Inc., Montreal, Canada). Microsphere suspensions were vortexed for ten minutes prior to injection and lack of aggregation was confirmed by periodic microscopic inspection.

One million microspheres diluted in 17 ml saline and 0.01% Tween 80 (Fisher Scientific Co) were injected into the pulmonary vein over twenty seconds followed by a ten ml flush of warm saline. Five seconds before microsphere injection, arterial blood withdrawal was begun from the femoral artery at a constant rate of nine mls/min over two minutes using a dual infusion/withdrawal pump (Harvard Apparatus Co.). The reference withdrawal catheter was filled with one ml heparin saline [500 U/ml].

After photography and autoradiography, transmural sections from the non-ischemic (normal), lateral, and central ischemic left ventricular myocardium were divided into endocardial and epicardial layers as diagrammed below:

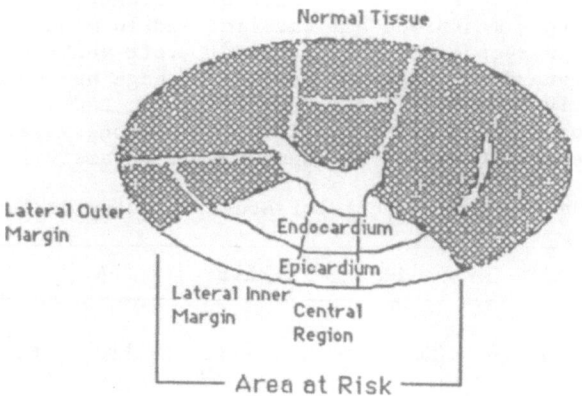

Figure 2

The samples were weighed (the sample weights ranged from 0.50-0.80 grams) and microsphere content was determined by counting radioactivity in a five channel LKB 1282 Compu-Gamma spectrometer. The window settings of the spectrometer were set for optimal counting at the isotopes' main photo-peaks. Blood flow was then calculated by comparing microsphere content of the individual samples to that of the reference blood (6).

Regional myocardial blood flow was calculated according to the formula:

myocardial tissue flow=
$$\frac{\text{tissue radioactivity X reference blood flow}}{\text{radioactivity of reference blood}}$$

Blood flow in all sections was expressed as a percentage of control (microsphere 1) flow in the same section.

Statistical Analysis

All data are expressed as mean ± SEM. A Student t-test for unpaired data was used to analyze hemodynamic data. A regression analysis was performed on the area at risk as measured by the two different techniques in both group and the slopes of the regression lines were compared to one by using the t-distribution to test for significant differences. A two-way analysis of variance was used to examine the blood flow data from the two groups followed by the Scheffe method for multiple comparisons when analysis of variance yielded significant differences. Statistical significance was defined as a p value of less than 0.05.

RESULTS

Two of the 25 dogs used in this study died from ventricular
fibrillation within 15 minutes of coronary occlusion. Both of these dogs
belonged to the three hour occlusion group. The following results were
analysed with data from the remaining 23 dogs. None of these animals
fibrillated during the study.

Hemodynamic Changes

The hemodynamic effects of ligation are summarized in Table 1. As
group 1 and group 2 animals did not differ significantly in any of the
hemodynamic parameters measured, all baseline and 10 minutes post-occlusion
hemodynamic values were pooled. Mean arterial, left ventricular, left and
right atrial, pulmonary arterial and capillary wedge pressures did not
change significantly from baseline values following LAD occlusion.
Injection of radioactive microspheres into the pulmonary vein resulted in
no significant changes in any of the hemodynamic parameters measured.

Table 1-Hemodynamic Effects of Acute Coronary Occlusion

	MAP	HR	MLVP	PA	RA	CWP
Pre-occlusion:	114.5±3.3	124.0±3.6	54.1±3.1	13.1±0.6	6.1±0.9	2.8±0.5
Post-occlusion:						
10 minutes	107.0±3.7	123.6±4.1	52.4±3.1	13.7±0.5	6.8±0.8	3.4±0.5
90 minutes	108.0±3.7	122.2±4.1	54.0±3.2	14.8±0.7	6.7±0.8	3.0±0.6
180 minutes	111.6±4.0	199.8±4.8	54.8±3.5	15.0±0.9	7.6±0.8	3.5±0.5

Abbreviations-MAP=mean arterial pressure, HR=heart rate, MVLP=mean left
ventricular pressure, PA=pulmonary arterial pressure, RA=right atrial
pressure, CWP=capillary wedge pressure. All pressures were measured in
mm Hg

None of the hemodynamic parameters measured changed significantly from
baseline.

Gross features of the AAR

In groups 1 and 2 (i.e. animals subjected to 10 minutes and 3 hours of
LAD occlusion, respectively) both techniques clearly demarcated the lateral
boundary that existed between well perfused, normal myocardium and the
ischemic myocardium within the AAR. Although this boundary was sharp, it
was often irregular.

Relationship between the Area at Risk size as determined by the two
different techniques

In each of the two groups, one animal exibited no AAR as identified by
either technique. In both groups, the relationship between the AAR size as
determined by the two techniques (Figure 3) was linear (p < 0.01). The
slope of the regression lines for both groups was 1.1 which is not
significantly different from 1.0 indicating that neither technique
consistently overestimates the size of the AAR. Figure 4 shows that in
group 1, the mean AAR size as determined by ^{99m}Tc macroaggregate injection

(25.48 ± 5.76% of the size of the left ventricle) was not significantly different from that determined using dye injection (24.34 ± 5.13%). Similarly, group 2 animals showed no significant differences between the mean AAR size as determined by ^{99m}Tc macroaggregate injection (22.35 ± 6.51%) and dye injection (24.67 ± 5.94%). Since the mean area at risk size as well as the varinces in the determinations were similar by the two methods, this suggests that ^{99m}Tc macroaggregate autoradiography and the dye method measure the same area of myocardium.

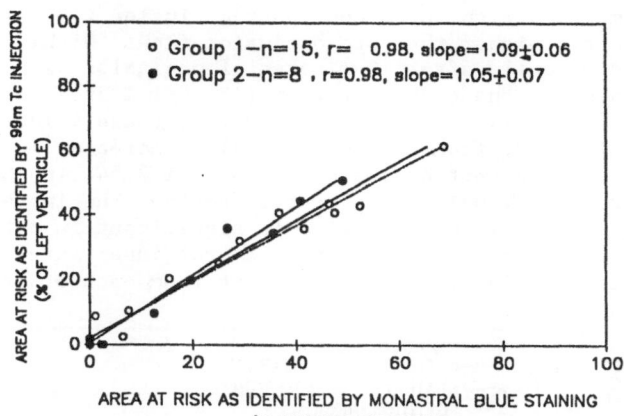

Figure 3-Correlation of area at risk size as determined by ^{99m}Tc macroaggregate injection and Monastral blue dye injection. For both groups, the relationship was linear (p<0.01). The slopes of the regression lines for both groups was 1.1, which is not significantly different from one, indicating that neither technique consistently overestimates the size of the area at risk.

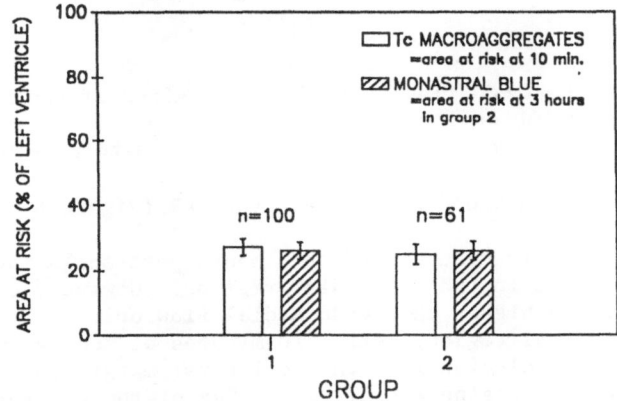

*note: "n" refers to the number of heart slices analysed

Figure 4-Mean area at risk size as measured by two different techniques in both experimental groups. There were no significant differences in the mean area at risk sizes as determined by the two techniques in either group 1 or group 2.

All blood flow values are expressed as a percentage of pre-occlusion flow within that same section. Pre-occlusion and 10 minute post-occlusion flows for groups 1 and 2 were pooled as these were not significantly different. Figure 5 shows that blood flow to the normal tissue increases significantly to 120.70 ± 8.42% 10 minutes after occlusion and, in group 2 animals, proceeds to increase to 142.86 ± 9.57% at 3 hours post-occlusion. This increase was also significant. A similar trend was seen in the lateral outer margin, with three hour post-occlusion flow (127.73 ± 4.90%) being significantly higher than pre-occlusion flow. Within the AAR, blood flow falls precipitously after 10 minutes of occlusion with the decline in flow being greater to the central region (15.30 ± 1.54%) than to the lateral inner margin (19.11 ± 1.71%). However, these values are not significantly different from each other. Flow increases in both of these regions at three hours post occlusion to 36.47 ± 2.54% in the lateral inner margin and to 27.71 ± 2.04% in the central region. The increase in flow from the corresponding 10 minute value is significant in both of these regions. At three hours post-occlusion, lateral inner margin and central region flows are not significantly different from each other.

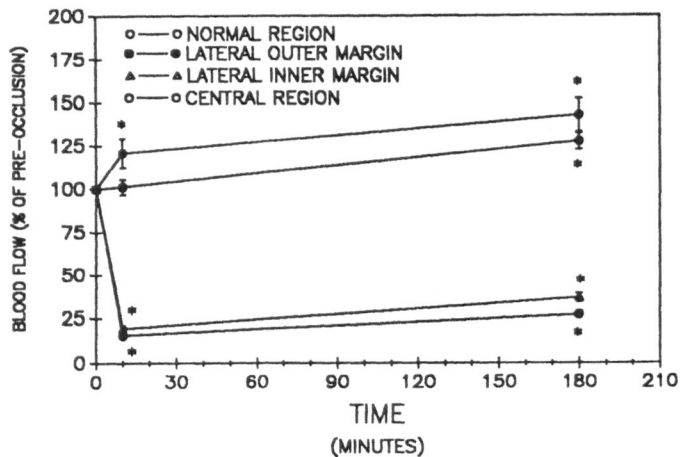

Figure 5-Myocardial blood flow to different regions of the heart following LAD occlusion.
* indicates a p value <0.05 when compared to pre-occlusion flow.

Transmural variations in blood flow within the AAR (Figure 6)

Within the AAR at 10 minutes and at 3 hours post-occlusion, blood flow was consistently higher in the epicardial regions. However, epicardial flow was significantly higher than endocardial flow only at 3 hours post-occlusion in the central region. After 10 minutes of occlusion, flow was 22.47 ± 2.57% of pre-occlusion flow in the lateral margin epicardium. At 3 hours, the flow in this region was 41.79 ± 3.84% of the control value; the latter being significantly greater than the 10 minute post-occlusion value, suggesting that over the 3 hour occlusion period, flow to this region increased. In the central epicardium, flow increased from 17.98 ± 2.21% at 10 minutes post-occlusion to 35.26 ± 2.67% at three hours post-occlusion. This increase in flow from 10 minutes to 3 hours was also significant. Lateral inner margin endocardial flow also showed a significant increase from 10 minutes to three hours post-occlusion.

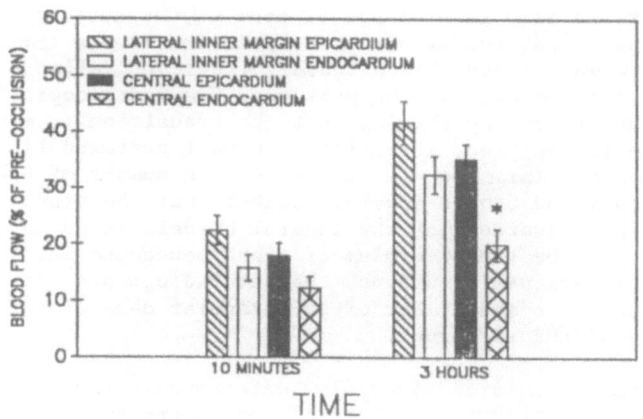

Figure 6-Collateral Flow within the Area at Risk following LAD occlusion.
* indicates endocardial flow significantly lower than epicardial
flow at the same time.
In all regions except the central endocardium, flow increased
significantly at three hours from the corresponding 10 minute value.

DISCUSSION

From this study we can conclude that in the open chested dog model 1)
99mTc macroaggregate injection delineates the same area of myocardium as
Monastral blue injection when 99mTc macroaggregates are injected
immediately before the Monastral blue dye. 2) The area at risk of infarction,
as measured by these two techniques, does not change spontaneously over a
three hour occlusion period.

In the past, the use of the 99mTc macroaggregate injection technique
to delineate the *in vivo* AAR has allowed researchers to circumvent the
problem that the size of the AAR measured at the end of the occlusion period
could be different from that determined just after an occlusion because of a
change in collateral flow to these border zones. As a result, the much
simpler and cheaper dye injection techniques, which also have the advantage
of minimizing the experimenter's exposure to radioactivity, lost their
popularity as these techniques could only be used at the end of an
experiment. Since our study shows that both techniques are identifying the
same area of myocardium, and that the size of the AAR does not change
spontaneously over a three hour occlusion period, it may be preferable to
utilize the Monastral blue staining technique to identlfy the AAR in short-
term occlusion studies in which collateral flow to the risk zone is not
altered by therapeutic intervention.

Inherent in the idea that it is possible to change the size of the
area at risk spontaneously is the notion that collateral flow to the risk
zone is not homogeneous or constant. Some have suggested that the central
ischemic area is surrounded by a region of tissue which is characterized
by intermediate reductions in blood flow and ischemia (7,8). These 'border
zones' would result if there was preferential perfusion from the non-
occluded vasculature that borders the ischemic region in the lateral and
transmural planes (9,10). If flow to these regions is adequate at the time
of occlusion, or if flow to these regions increases over the course of
infarction to a level compatible with cell survival, spontaneous "salvage"
in these regions may result.

The existence of a lateral border zone is the subject of much controversy. A lateral border could arise if there were intramural anastamoses between adjacent blood vessels (10-12). Yellon et. al. (7,8) has presented evidence against the presence of a macroscopically identifiable border zone by showing that the transition from normal to ischemic tissue is sharp but irregular with well perfused tissue lying next to ischemic tissue. This view is supported by a number of others (13-16). Studies by Murdock and others (14,17) suggest that the transition zone of intermediate flow reductions at the lateral borders is no more than 3 mm in width in either the subepicardium or the subendocardium. In our studies, gross macroscopic evaluation of both the autoradiographs and the photographs revealed a sharp, but often irregular demarcation of the lateral borders of the risk zone.

Our regional myocardial blood flow data reveals that no significant differences in blood flow to the lateral inner margin and the central region exist at both post-occlusion times. Therefore, it appears that our data supports the concept that flow to the risk zone is relatively homogeneous. However, lateral inner margin flow was consistently higher than central region flows at 10 minutes and at 3 hours post-occlusion. This may perhaps be a result of the fact that the microsphere method for determining blood flow has a limited resolving power. Limitations in defining the border zone may also be imposed by it's irregular nature. The combinatio of these two introduces a significant sampling error. Therefore, when obtaining a sample from the lateral inner margin, it becomes virtually impossible to exclude some adjacent normal tissue in the sample. As a result, a sample from the lateral inner margin may contain a mixture of normal and ischemic cells which would subsequently result in an apparently higher flow to this region.

Although our observation of a relatively homogeneously perfused risk zone is consistent with that reported by others (13,14), our data does not support the concept that collateral flow to the risk zone is constant from 10 minutes to 3 hours post-occlusion. Our data show that flow the lateral inner margin and the central region increases significantly from 10 minutes to 3 hours. Since blood flow to the normal tissue increases significantly from 10 minutes to three hours, contamination error in the lateral inner margin may also possibly serve to explain why blood flow increases in these regions over the same period. In the central region, the mean flow increase was about 10%. It is known that the accuracy of the microsphere method for determining blood flow is ± 7-10%. Therefore, it remains unclear whether an increase in flow to this region is real or apparent. Theoretically speaking, it is unlikely that collateral flow to the risk zone increases over this period given the weight of evidence that suggests that, in the absence of other hemodynamic changes, all collateral vessels are maximally dilated within minutes of a complete occlusion and that further growth of collaterals does not occur until at least six hours of occlusion (19).

In the transmural plane, it is recognized that flow to the subepicardium is higher than flow to the subendocardium. This is due to the fact that inter-arterial connections are more prevalent in the subepicardium than in the subendocardium (9,10,17). In our studies, collateral flow is consistently higher in the subepicardium as compared to the subendocardium in both the lateral inner margin and in the central region at both 10 minutes and three hours post-occlusion. It is thus evident that our blood flow data support the existence of a transmural border zone.

It therefore appears that if a real increase in flow to the area at risk occurs from 10 minutes to 3 hours post-occlusion in the untreated

animal, this increase is insufficient to produce a shift in the size of
the AAR. It may however be possible to change the size of the AAR
by the application of an agent capable of increasing or decreasing
collateral flow to the risk zone. Studies presently underway in our
laboratory are aimed at finding out whether nicardipine, a potent
vasodilator is capable of changing the size of the area at risk .

REFERENCES

1. J. E. Lowe, K. A. Reimer and R. B. Jennings, Infarct Size as a
   Function of the Amount of Myocardium at Risk. Am. J. Pathol.,
   90:363-380 (1978).

2. S. Koyanagi, C. L. Eastham, D. G.Harrison, and M. L. Marcus,
   Transmural Variation in the Relationship between Myocardial Infarct
   Size and Risk Area. Am. J. Physiol., 242 (Heart Circ. Physiol. II):
   H867-H874 (1982).

3. B. I. Jugdutt, G. M. Hutchins, B. H. Bulkley and L. C. Becker,
   Myocardial Infarction in the Conscious Dog. Three-Dimensional
   Mapping of Infarct Size, Collateral Flow and Region at Risk.
   Circulation. 60:1141-1150 (1979).

4. L. W. Deboer, W. H. Strauss, R. A. Kloner, R. E. Rude, R. F. Davis, P.
   R. Maroko and E. Braunwald, Autoradiographic Method for Measuring
   the Ischemic Myocardium at Risk: Effects of Verapamil on Infarct
   Size after Experimental Coronary Artery Occlusion. Proc. Nat. Acad.
   Sci., 77:6119-6123 (1980).

5. L. G. T. Ribeiro, Influence of the Extent of the Zone at Risk on the
   Effectiveness of Drugs in Reducing Infarct Size. Circulation.,
   66(2):181-186 (1982).

6. R. J. Bartrum Jr, D. M. Berkowitz, N. K. Hollenberg, A Simple
   Radioactive Microsphere Method for Measuring Regional Flow and
   Cardiac Output. Investigative Radiol., 9:126-132 (1974)

7. D. M. Yellon, D. J. Hearse, R. Chrome, J. Grannell and R. K. H. Wyse,
   Characterization of the Lateral Interface Between Normal and
   Ischemic Tissue in the Canine Heart During Evolving Myocardial
   Infarction. Am. J. Cardiol., 47:1233-1239 (1981).

8. D. M. Yellon, D. J. Hearse, Temporal and Spatial Characteristics of
   Evolving Cell Injury in Regional Myocardial Ischemia in the Dog.
   J. Am. Coll. Cardiol., 2:661-670 (1983).

9. M. J. Janse, Where is the Salvable Zone? in: "Therapeutic Approaches
   to Myocardial Infarct Size Limitation", D. J. Hearse et. al. eds,
   Raven Press, New York, pp 61-79 (1984).

10. D. J. Hearse, Why are we still in doubt about Infarct Size Limitation?
    The Experimentalist's Viewpoint in: Therapeutic Approaches to
    Myocardial Infarct Size Limitation, D. J. Hearse et. al. eds, Raven
    Press, New York, pp 17-41 (1984).

11. K. Przyklenk, M. T. Vivaldi, O. A. Malcolm, F. J. Schoen and R.A.
    Kloner, Capillary Anastamoses between the Left Anterior Descending
    and Circumflex Circulations in the Canine Heart: Possible
    Importance during Coronary Artery Occlusion. Microvasc. Res.,
    31:54-65 (1984).

12. E. M. Okun, S. M. Factor and E. S. Kirk, End Capillary Loops in the Heart: An Explanation for Discrete Myocardial Infarction Without Border Zones. Science, 79:565-566 (1979).

13. M. Fukunami, D. M. Yellon, Y. Kudoh, M. P. Maxwell, R. K. H. Wyse and D. J. Hearse, Spatial and Temporal Characteristics of the Transmural Distribution of Collateral Flow and Energy Metabolism during Regional Myocardial Ischemia in the Dog. Can. J. Cardiol., 3(2):94-103 (1987).

14. R. H. Murdock Jr, D. M. Harlan, J. J. Morris III, W. W. Pryor Jr. and F. R. Cobb, Transitional Blood Flow Zones between Ischemic and Non-Ischemic Myocardium in the Awake Dog. Analysis based on the Distribution of the Intramural Vasculature. Circ. Res., 52:451-459 (1983).

15. P. O. Sjöquist, G. Duker and O. Almgren, Distribution of the Collateral Blood Flow at the Lateral Border of the Ischemic Myocardium after Acute Coronary Occlusion in the Pig and the Dog. Basic Res. Cardiol., 79:164-175 (1984).

16. S. M. Factor, E. H. Sonnenblick and E. S. Kirk, The Histological Border Zone of Acute Myocardial Infarction-Islands or Penninsulas? Am. J. Pathol., 92:111-124 (1978).

17. F. R. Cobb, A. Chu, : Myocardial Infarction and Risk Region Relationships: Evaluation by Direct and Noninvasive Methods Prog. Cardiovasc. Dis., 30(5)323-348 (1988).

18. J. M. Downey, : Why the Endocardium? in: Therapeutic Approaches to Myocardial Infarct Size Limitation, D. J. Hearse et. al. eds, Raven Press, New York, pp 17-41 (1984).

19. M. L. Cohen, : Coronary Collaterals and Luminal Communications in Experimental Animals after Recent and Chronic Coronary Occlusions : Changes in Histology and Flow in "Coronary Collaterals: Clinical and Experimental Observations", Futura Publishing Co., New York, pp 289-305, 307-308, 317-322 (1985)

ACKNOWLEDGEMENTS

These investigations were supported by a grant in aid from the Heart and Stroke Foundation of Ontario (AN 1009).

Much thanks goes out to Ms. Miza Davie and Mr. and Mrs. D. Mauldin for assistance in conducting the experiments and preparing this manuscript.

652

HEMODILUTION AND MYOCARDIAL OXYGEN SUPPLY. THE INFLUENCE OF FLUOSOL-DA

H.Vogel[1], H. Günther[2], D.K. Harrison[2], W. Anderer[2], M. Kessler[2] and K. Peter[1]

Institut für Anaesthesiologie der Universität München/FRG[1]
Institut für Physiologie und Kardiologie der Universität
Erlangen – Nürnberg/FRG[2]

INTRODUCTION

Attempts to minimize the risk of transmission of infectious diseases have again focussed clinical interest on isovolemic hemodilution (Messmer 1978). Being the only organ which has to increase its specific work in order to compensate for acute anemia, the heart is in end effect the limiting organ for an extreme hemodilution (Sunder-Plassmann et al. 1976). Coronary vasodilation leading to a maximal increase in coronary flow compromises the coronary reserve. Increases of oxygen demand or further hemodilution will lead to myocardial hypoxia and consequently to myocardial failure (v.Restorff et al. 1975).

The breakpoint, where subendocardial ischemia starts to develop, is a hematocrit of less than 15 Vol % . This empirical limit of isovolemic hemodilution was found using colloidal plasma substitutes without specific oxygen carrying properties (Brazier et al. 1974, Kettler et al. 1976).

Perfluorochemicals (PFC) are compounds with a high physical solubility for gases including oxygen (Clark, Gollan 1966). In Fluosol-DA (FDA), two PFCs are emulsified in such a way that the intravenous administration as an oxygen carrying blood substitute is possible (Naito, Yokoyama 1978).

After acute experimental coronary occlusion, PFC – emulsions have proven to increase oxygen supply to the ischemic area (Biro 1982, Rude et al. 1982, Faithfull et al. 1986). However, no information is available about myocardial oxygen supply under an extreme isovolemic hemodilution with this oxygen carrying blood substitute.

Today, with special techniques, we are able to measure myocardial pO_2 and myocardial microflow directly on the beating heart. This allows us to answer the following questions:
1. How does an extreme isovolemic hemodilution with a conventional colloidal blood substitute influence oxygen supply and microflow to the myocardium?
2. Is it possible to exceed the limits of isovolemic hemodilution with Fluosol-DA without endangering myocardial oxygen supply?

MATERIALS AND METHODS

In an open chest preparation of 16 mongrel dogs (anesthesia: piritramid / flunitrazepam , ventilation: IPPV, air/oxygen, F_iO_2 = 0.3) the following parameters were measured:

p_mO_2 (mm Hg) = Myocardial pO_2 (polarographically with the Multiwire Surface Electrode).

MF (units) = Myocardial microflow (local hydrogen clearance also with a modified Multiwire Surface Electrode (Harrison 1985)).

The multiwire sensors were kept in contact with the myocardial surface by means of a highly flexible silicone-rubber-disc, which was attached to the epicardium with atraumatic sutures (Kessler et al. 1983).

CF (ml/min) = Coronary flow in the LAD (Electromagnetic flow probe)

AP, LP (mmHg) = Aortic and left ventricular pressures (tip manometers)

BGA = Blood gas analysis and oxygen contents (arterial and coronary venous)

After measurement of these values under control conditions (Hct 38) the animals were isovolemically hemodiluted with Hydroxyethylstarch 40 000 (HES), until a Hct of 15 Vol % was reached. After a period of stabilisation, all parameters were measured again.

In the second part of the study, the hemodiluted animals were ventilated with pure oxygen. Again all measurements were performed under ventilation with F_iO_2 = 1.0, and an isovolemic hemodilution with Fluosol-DA 20 % (FDA) was carried out until a hematocrit of 7 - 10 Vol % had been reached. After administration of FDA, control measurements of all parameters were once again carried out.

Student's t test for paired values was used for statistical analysis of the data.

RESULTS AND DISCUSSION

1. Hemodilution with Hydroxyethylstarch (HES)

Under isovolemic hemodilution with HES to a Hct of 15 %, the myocardial pO_2 decreased considerably, but no hypoxic values could be observed in the pO_2 - histograms (Fig. 1). Arterial pO_2 remained constant, whereas coronary venous pO_2 showed a slight but still significant (p 0.05) decrease from a mean value of 29 to 26 mm Hg. No changes were observed in heart rate, mean aortic pressure, and myocardial oxygen uptake, while left ventricular contractilty and filling pressures showed a modest increase.

Under hemodilution with HES, coronary flow increased by 270 % as compared to the control value (p 0.005). In contrast, myocardial microflow as measured with the local hydrogen clearance technique showed an increase of only 18 % of control (see Table 1).

Table 1. The most important parameters under isovolemic hemodilution with HES. MAP = mean aortic blood pressure, p_aO_2 = arterial, and $p_{cv}O_2$ = coronary venous oxygen tensions (mean $\pm$ SD).

Parameter	Control	Hemodilution HES	p
Hct (%)	37.8 $\pm$ 7.5	14.7 $\pm$ 2.3	0.001
MAP (mm Hg)	101.5 $\pm$ 11	98.5 $\pm$ 8	n.s.
CF (ml/min)	45 $\pm$ 8.5	165 $\pm$ 42	0.005
MF (units)	1.98 $\pm$ 0.3	2.33 $\pm$ 0.37	0.01
p_aO_2 (mm Hg)	113 $\pm$ 11	115.2 $\pm$ 18.4	n.s.
$p_{cv}O_2$ (mm Hg)	28.7 $\pm$ 4	25.8 $\pm$ 3.2	0.05

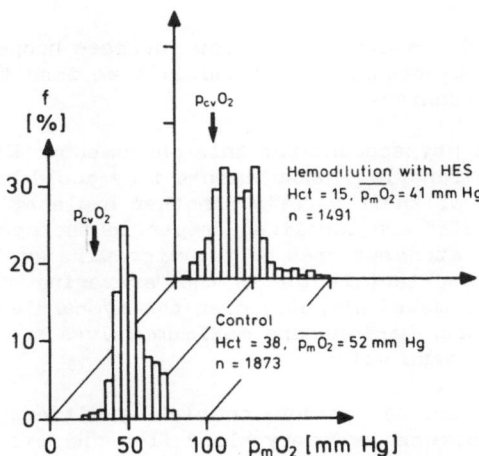

Fig. 1. pO_2-Histograms of the surface of the myocardium under control conditions and after isovolemic hemodilution with HES (Hct 15). Mean myocardial pO_2 decreases under hemodilution, but no hypoxic values are found. (n = number of measurements, $p_{cv}O_2$ = cor. ven. pO_2)

The huge discrepancy between coronary flow and tissue perfusion which is illustrated in Fig. 2. seems to have several reasons: First, there may be a preferential perfusion of so-called "high flow channels". These are relatively short and/or wide capillaries, which in spite of their small frequency can transport a considerable proportion of total organ blood flow.The existence and frequency distribution of such capillaries has been demonstrated by morphometric studies (Potter 1983, Skalak 1986), their functional signification could be proved by Harrison (Harrison 1988).

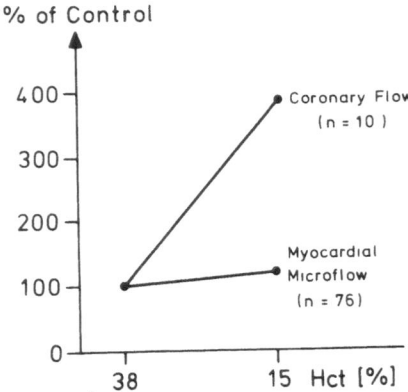

Fig. 2. Electromagnetically measured coronary flow (CF) and myocardial microflow (MF) under extreme isovolemic hemodilution with HES.

Nevertheless, if the dilutional flow increase happened mostly through the "high flow channels", why couldn't we find them with the hydrogen clearance technique?

Several reasons may account for this phenomenon: With the total number of 76 single microflow-measurements they could have been "overlooked" because of their rarity. Another explanation would be that the "high-flow-channels" are localized deeper in the myocardium and are thus outside of the catchment area of the microwire surface H2-electrode. A third important factor is the "averaging effect" of a microelectrode, being moved slightly over the myocardial surface with every beat of the heart. Minimum and maximum values are thus levelled off in favour of the mean values.

There may, however, be another, completely different explanation for the discrepancy between coronary blood flow and microflow:

A redistribution of flow in favour of deeper layers of the myocardium.

This apparently is in contradiction with the results of authors working with radioactive labelled microspheres, who observed severe decreases of subendocardial perfusion under extreme isovolemic hemodilution (Brazier 1974, Buckberg 1975). Electrocardiographic and histologic signs of subendocardial hypoxia seem to confirm these observations (Brazier 1974).

On the other hand it is known that shunting of microspheres, which leads to an underestimation of flow, occurs mainly in the subendocardial layers and is also related to the amount of coronary flow (Marshall 1976). The observed decrease of subendocardial flow under hemodilution might thus be merely a consequence of these two factors.

But what is then the reason for the subendocardial hypoxia under extreme hemodilution? New results of tissue photometry indicate another direction (Fig. 3).

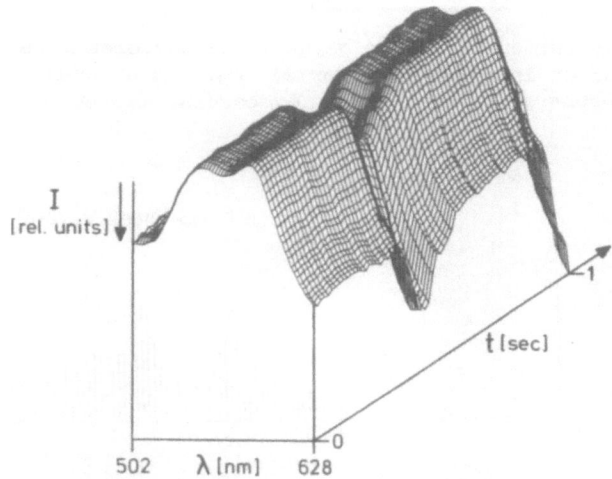

Fig. 3. Sequence of hemoglobin spectra recorded at the surface of the myocardium during a cardiac cycle. Hemoglobin concentration in the myocardium shows a significant decrease during the systole. Recordings were made with the Erlangen Microlightguide Spectrophotometer (Frank 1985).

Subendocardial Anemia

The systolic decrease of myocardial hemoglobin concentration (Fig.3) is caused by the increase of myocardial wall tension during systole, which pumps the blood out of the capillaries into the venous system.

However, systolic wall tension is highest in the subendocardial layer (Brandi 1969) and therefore the systolic decline of hemoglobin concentration should be most severe in the subendocardium. This would decrease the "mean hemoglobin concentration" in the subendocardial layer and would lead to a relative "subendocardial anemia".

Under normal conditions, the "subendocardial anemia" may not be of great importance as it can easily be compensated by an elevated tissue perfusion in the subendocardium (contradictory findings with microspheres are no real argument against this - we discussed this point above). However, under borderline conditions, such as extreme anemia (hemodilution) or possibly a coronary artery stenosis, where there is no coronary reserve left, a "subendocardial anemia" may be crucial for the tissue concerned. This would also quite well explain the well-known pathological-anatomical findings of subendocardial myocardial necrosis.

During the diastole, where the majority of coronary flow takes place, the endo-/epicardial ratio of myocardial tension is simply reversed compared with during the systole: the intramural pressure is highest in the subepicardial and lowest in the subendocardial layers (Stein 1980). This can be taken as another index for a preferential perfusion of the inner layers of the heart compensating for a physiological, relative "subendocardial anemia".

2. Hemodilution with Fluosol-DA 20%

Under ventilation with pure oxygen, the isovolemic hemodilution with FDA caused an increase in arterial pO_2, and an even more pronounced increase in coronary venous pO_2 and myocardial tissue pO_2 (Fig. 4).

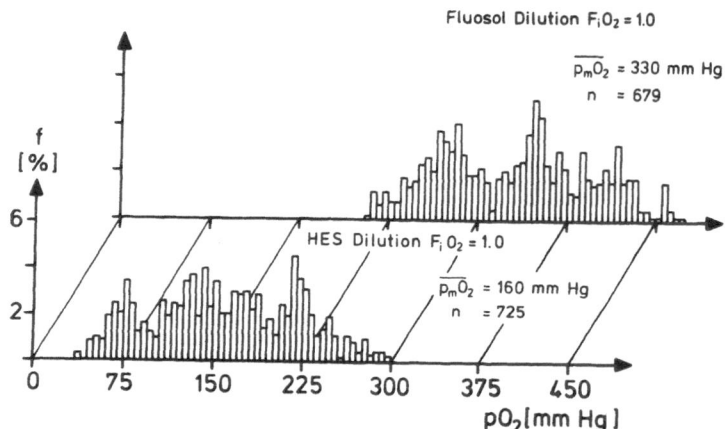

Fig. 4. pO_2 histograms of the myocardium under hemodilution with Hydroxyethylstarch and Fluosol-DA 20 %. Ventilation in both cases with pure oxygen. n = number of measurements, pmO2 = mean myocardial pO_2. Hemodilution with FDA results in an increase of myocardial oxygen tension.

The modest increase in arterial oxgen tension can be explained by a decrease of intrapulmonary right-to-left shunts, or an elevated mixed venous pO_2, or as a consequence of both.

It is, however, not sufficient, to explain the dramatic increases in coronary venous and in myocardial tissue oxygen tension.

One possible explanation is the linear oxygen dissociation curve of FDA which, in contrast to the sigmoid O_2-dissociation curve of blood, allows a relevant oxygen discharge already at higher tissue pO_2-levels (Fig. 5).

A second reason may be the considerable increases of both coronary flow and myocardial microflow (see Table 2).

Table 2. The most important parameters under hemodilution with HES and FDA: ventilation with pure oxygen. For abbreviations see Table 1: Fct = Fluocrit.

Parameter	Hemodilution HES	Hemodilution FDA	p
Hct (%)	14.9 ± 2.5	7.2 ± 2.1	0.001
Fct (%)	0	8.0 ± 1.7	–
MAP (mmHg)	98.3 ± 10.0	96.7 ± 8.4	n.s.
CF (ml/min)	128 ± 43	168 ± 53	0.05
MF (units)	2.40 ± 0.43	2.89 ± 0.57	0.01
P_aO_2 (mmHg)	443 ± 81	516 ± 57	0.01
$P_{cv}O_2$ (mmHg)	29.5 ± 2.9	49.0 ± 14.3	0.01

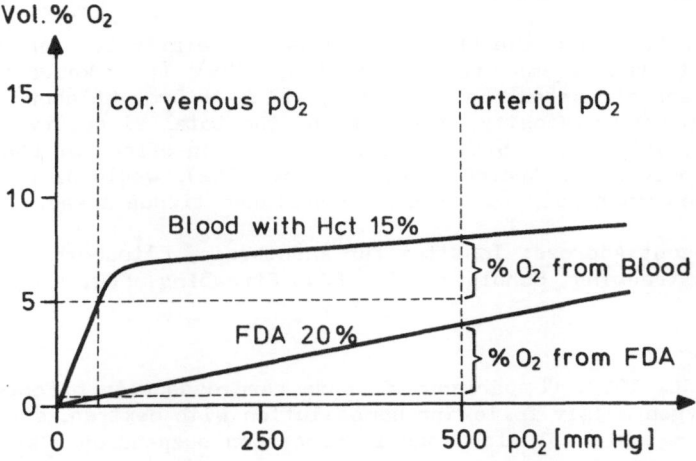

Fig. 5. O_2 dissociation curves of diluted blood and FDA. Arterial and coronary venous pO_2 values represent the measured data.

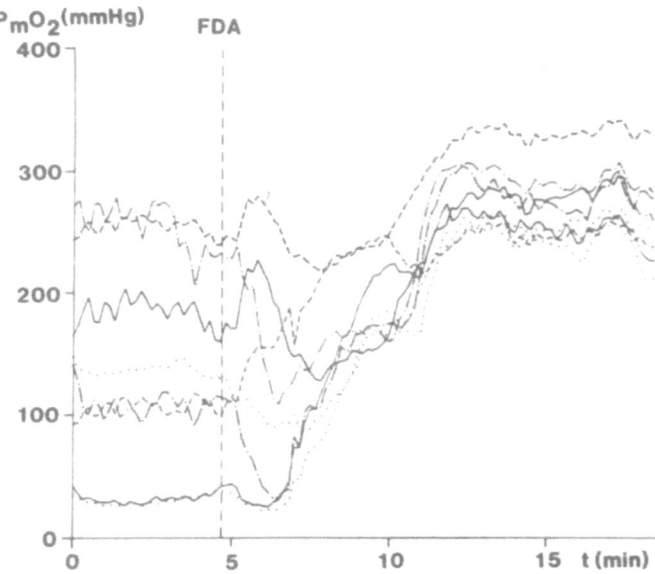

Fig. 6. Continuous recording of myocardial oxygen tension on the beating
dog heart under administration of FDA. The animal had been pre-
hemodiluted with Hydroxyethylstarch. Ventilation with $F_iO_2 = 1.0$

The third possible reason for the increase in myocardial pO_2 is a
microrheological one. Figure 6 shows the behaviour of the myocardial pO_2
during hemodilution with FDA in a single experiment. There is not only a
general increase of myocardial oxygen tension, but also a redistribution
of oxygen supply in favour of the poorly oxygenated measuring sites.

The redistribution could be explained by plasma skimming where the
0.12 micron droplets of the PFC-emulsion can transport the oxygen to
erythrocyte-free perfused capillaries.

On the other hand, the PFC-emulsion has a relatively high viscosity
as compared to the plasma viscosity (Naito, 1978). It is known that the
increased shear stress on the red cells, which is brought about by an
increase of plasma viscosity, can decrease the total viscosity of whole
blood (Chien, 1972). This blood-viscosity lowering-effect of FDA, which
has already been demonstrated in-vitro (Biro 1982), would also be able
to aid the oxygen supply to the poorly perfused tissue areas.

Author's Present Address: Institut für Anästhesie, Elisabeth
Krankenhaus Straubing, Schulgasse 20, 8440 Straubing, FRG.

REFERENCES

    Biro, G.P., 1982, Comparison of acute cardiovascular effects and
        oxygen supply following hemodilution with dextran, strom-free
        hemoglobin solution, and fluorocarbon suspension. Cardiovasc.
        Res. 16:194.
    Brandi, G., McGregor, M., 1969, Intramural pressure in the left
        ventricle of the dog. Cardiovasc. Res. 3:472.
    Brazier, J., Cooper, N., Maloney, J.V., Buckberg, G., 1974, The
        adequacy of myocardial oxygen delivery in acute normovolemic
        anemia. Surgery 75:508.
    Chien, S., 1972, Present state of blood rheology. In:
        "Hemodilution. Theoretical basis and clinical application", K.
        Messmer and H. Schmid-Schönbein, eds., Karger, Basel CH.

Clark, L.C., Gollan, F., 1966, Survival of mammals breathing organic liquids equilibrated with oxygen at atmospheric pressure. Science 152:1755

Faithfull, N.S., Erdmann, W., Fennema, M., Kok, A., 1986, Effect of hemodilution with fluorocarbons or dextran on oxygen tensions in the acutely ischemic myocardium. Brit. J. Anaesth. 58: 1031.

Frank, K.H., 1985, Optische Streuung an biologischen Partikeln und Zellen. Thesis, Friedrich-Alexander Universität Erlangen-Nürnberg, FRG.

Harrison, D.K., Birkenhake, S., Knauf, S., Hagen, N., Beier, I., and Kessler, M., 1988, The role of high flow capillary channels in the local oxygen supply to skeletal muscle. Adv. Exp. Med. Biol. 222:623.

Harrison, D.K., Birkenhake, S., Hagen, N., Knauf, S., and Kessler, M., 1989, Regulation of capillary blood flow: A new concept. This volume.

Kessler, M., Höper, J., Krumme, B.A., 1976, Monitoring of tissue perfusion and cellular function. Anesthesiology. 44:184.

Kessler, M., Klövekorn W.P., Höper, J., Sebening, F., Brunner, M., Frank, K.H., Harrison, D.K., Kernbach, C., Anderer, W., Richter, H. and Ellermann, R., 1983, Local oxygen supply and regional wall motion of the dog's heart during critical stenosis of the LAD. Adv. Exp. Med. Biol. 169:331.

Kettler, D., Hellberg, K., Klaess, G., Kontokolias, J.S., Loos, W., and de Vivie, R., 1976, Hämodynamik, Sauerstoffbedarf und Sauerstoffversorgung des Herzens unter isovolämischer Hämodilution. Anaesthesist 25:131.

Marshall, W.G., Boatman, G.B., Dickerson, G., Perlin, A., Todd, E.P., and Utley, J.R., 1976, Shunting, release, and distribution of nine and fifteen micron spheres in myocardium. Surgery., 79,6:631.

Messmer, K.F.W., 1987, Acceptable hematocrits in surgical patients. World J. Surg., 11:41.

Naito, R., and Yokoyama, K., 1978, In: "Perflurochemical Blood Substitutes." Techn. Information, Series No 5. The Green Cross Corporation, Osaka, Japan, 1978.

Potter, R.F., Groom, A.C., 1983, Capillary diameter and geometry in cardiac and skeletal muscle studied by means of corrosion casts. Microvasc. Res. 25:68.

v. Restorff, W., Höfling, B., Holtz, J., Bassenge, E., 1975, Effect of increased blood fluidity through hemodilution on coronary circulation at rest and during exercise in dogs. Pflügers Arch. ges. Phys., 357:15.

Rude , R.E., Glogar, D., Khuri, S.F. Kloner, R.A., Karaffa, S., Muller, J.E., Clark, L.C., Braunwald, E., 1982, Effects of intravenous fluorocarbons during and without oxygen enhancement on acute myocardial ischemic injury assessed by measurement of intramyocardial gas tensions. Amer. Heart J., 103:986.

Skalak, T.C., and Schmid-Schönbein, G.W., 1986, The microvasculature in skeletal muscle IV. A model of the capillary network. Microvasc. Res., 32:333.

Stein, P.D., Marzilli, M., Sabbah, H.N., and Lee, T., 1980, Systolic and diastolic pressure gradients within the left ventricular wall. Am. J. Physiol. 238 (Heart Circ. Physiol. 7):H 625.

Sunder-Plassmann, L., Kloevekorn, W.P., Messmer, K., 1976, Pre-operative hemodilution. Basic adaption mechanisms and limitations in clinical application. Anaesthesist, 25:124.

THE EFFECT OF ACETYLCHOLINE ON REGIONAL MYOCARDIAL O_2 CONSUMPTION AND

CORONARY BLOOD FLOW IN THE RABBIT HEART

Harvey R. Weiss and Bat-Ami Acad

University of Medicine and Dentistry of New Jersey
Robert Wood Johnson Medical School
Heart and Brain Circulation Laboratory
Department of Physiology and Biophysics
675 Hoes Lane
Piscataway, New Jersey 08854-5635
U.S.A.

ABSTRACT

    The effect of acetylcholine on the regional coronary perfusion and O_2 consumption was determined in anesthetized open chest rabbits in order to compare its direct vasodilatory effects with the metabolic vasoconstriction it induces. Experiments were conducted in seven untreated controls and seven rabbits which were infused with acetylcholine (1 μg/kg/min). Myocardial blood flow was determined before and during acetylcholine infusion using radioactive microspheres. Regional arterial and venous O_2 saturation was analyzed microspectrophotometrically. Acetylcholine reduced heart rate by 30% and significantly depressed the arterial systolic and diastolic blood pressure. The mean O_2 consumption was significantly reduced with acetylcholine from 9.6±2.0 to 6.1±3.6 ml O_2/min/100 g. Coronary blood flow decreased uniformly across the left ventricular wall by about 50% and resistance to flow increased by 42% despite potential direct cholinergic vasodilation. O_2 extraction was not affected during acetylcholine infusion. It is concluded that the acetylcholine infusion directly decreased myocardial O_2 consumption which in turn lowered the coronary blood flow and increased the resistance. The decreased flow was related to a reduced metabolic demand rather than a direct result of lowered aortic blood pressure. Unaffected myocardial O_2 extraction also suggests that blood flow and metabolism were matched. This indicates that direct cholinergic vasodilation of the coronary vasculature does not allow a greater reduction in metabolism than flow in the anesthetized open chest rabbit heart during acetylcholine infusion.

INTRODUCTION

    Administration of acetylcholine reduces myocardial metabolism. Most studies have shown that acetylcholine or vagal stimulation cause a negative chronotropic effect and a possible decrease in myocardial contractility[1,2,3]. This would reduce coronary blood flow by metabolic mechanisms[4,5]. Since the subendocardial layers are predominantly perfused during diastole[6], acetylcholine induced bradycardia might alter the inner/outer flow ratio. Intracoronary infusion of acetylcholine could induce transmural changes in coronary flow distribution even when the heart rate was held constant[3]. The effects of acetylcholine on regional myocardial metabolism have not been determined.

Acetylcholine can also directly lower coronary blood flow. This effect on coronary blood flow can be independent of hemodynamic changes[3]. The direct coronary vasodilating action of acetylcholine has been demonstrated in many in vivo studies[3,7,8,9]. This effect could reduce the metabolically induced vasoconstriction caused by acetylcholine.

The purpose of the present study was to compare the contribution of these factors, i.e. the reduced O_2 demand and the direct vasodilating effect of acetylcholine, to the changes in regional coronary perfusion, and O_2 consumption. Coronary blood flow was determined using labeled microspheres. Microspectrophotometric observations of small myocardial arteries and veins were performed in order to determine regional O_2 extraction and O_2 consumption as well as study acetylcholine's effect on the microvascular distribution of oxygenation.

METHODS

New Zealand white rabbits were anesthetized with sodium pentobarbital (30 mg/kg, i.v.). Catheters were inserted into a femoral artery and vein. The trachea was cannulated and artificial respiration was instituted such that eucapnia was maintained. A left thoracotomy was performed at the fifth intercostal space. A catheter was inserted into the left atrium for injection of radioactive microspheres. After allowing time for stabilization, blood pressure and heart rate were recorded via the arterial catheter. Blood samples obtained from the femoral artery were analyzed for blood gases and pH electrometrically and hemoglobin was determined spectrophotometrically using a cyanmethemoglobin method. Regional blood flow was then determined by using radioactive microspheres. For the coronary blood flow determinations, approximately 7.5×10^5 microspheres, 15 ± 3 μm in diameter, labeled with ^{85}Sr or ^{141}Ce were shaken for 5 min and injected as a 0.2 ml bolus into the left atrial catheter. A reference sample method was used for the flow determinations[10]. The blood sample was withdrawn from the femoral artery with a peristaltic pump set at 1 ml/min. Withdrawal was begun 30 sec before injection and was continued for 3 min.

After the initial flow measurement, the animals were divided into two groups: 1) untreated controls and 2) acetylcholine treated. In the treated group, acetylcholine (1 μg/kg/min) was infused through the femoral vein with a Harvard infusion pump. Ten min after the infusion was begun, the final blood gas, pressure and myocardial blood flow determinations were made. After this measurement, the hearts were rapidly removed by cutting them at the atrioventricular ring with shears. They were then quick-frozen and stored in liquid N_2.

A three-wavelength microspectrophotometric method was used in order to measure O_2 saturation in frozen arterial and venous blood vessels[11,12]. Hearts were cut on a band saw at -20°C and adjacent plugs were obtained from the left ventricular free wall for flow measurements. Twenty-five micra sections were cut on a microtome at -20°C in a nitrogen atmosphere. They were transferred to glass slides and covered with degassed silicone oil and a coverglass. These slides were placed on a Zeiss microspectrophotometer fitted with an N_2-flushed cold stage to obtain readings of optical density at 560, 524, and 507 nm. The size of the measuring spot was 8 μm. Readings were obtained to determine O_2 saturation in the first five arteries and five veins, 20-100 μm in diameter, found in the left ventricular subepicardium and subendocardium.

For coronary blood flow determinations, the activity of the microspheres in these regions and in the reference blood samples was determined. Blood flows were calculated from the formula $Fu = Fk * Cu/Ck$, where Fu is the flow to any organ, Fk is the flow to a reference organ, Cu is the radioactivity in any organ, and Ck is the radioactivity in the reference organ[10]. Blood flow was expressed in ml/min/100 g of tissue. Coronary vascular resistance was calculated by dividing mean arterial pressure by coronary blood flow.

The O_2 content of the blood was obtained by multiplying the percentage O_2 saturation by the hemoglobin concentration times 1.36 ml O_2/g hemoglobin. The difference between the average arterial and venous O_2 contents was then obtained. By use of the Fick principle, the paired product of O_2 extraction and blood flow was obtained to determine O_2 consumption on a regional basis within the heart. The regional ratio of O_2 supply to O_2 consumption was determined by dividing the local O_2 supply by local O_2 consumption. Because of the nature of the measurement, O_2 saturations were measured only at the end of the experiment.

A factorial analysis of variance was used to determine whether differences existed in hemodynamics or blood gas parameters before and after acetylcholine infusion and between these parameters in the control and the treated group. This analysis was also used to determine differences between the regions for O_2 supply and consumption parameters. Duncan's procedure was used for multiple comparisons. A value of $P < .05$ was accepted as significant. Values are presented as mean±S.D.

RESULTS

The hemodynamic and blood gas parameters for both groups are shown in Table 1. The initial values for these parameters were similar in the two groups. Infusion of acetylcholine significantly reduced heart rate from 254±44 to 176±40 beats/min. Systolic and diastolic blood pressure were also significantly depressed by acetylcholine. The mean arterial blood pressure dropped from 85±7 mm Hg before to 56±10 mm Hg during acetylcholine infusion. Arterial blood gases and pH were maintained within normal limits for anesthetized rabbits in both groups.

Regional coronary blood flow and coronary vascular resistance in the control and in the treated group before and during acetylcholine infusion are shown in figure 1. The coronary blood flow and vascular resistance values were similar before infusion in the two groups. Acetylcholine reduced the flow in the examined regions of the treated group by about 50%. The decrease in blood flow was similar in both the subepicardium and subendocardium. The concomitant elevation in vascular resistance of the coronary bed was approximately 42%.

Figure 2 shows the distribution on a regional basis of O_2 saturation measurements in small arteries and veins of the control and the acetylcholine treated group. Arterial O_2 saturation values were similar for both groups in both the subepicardial and subendocardial regions. In the control group, O_2 saturation was found to be significantly lower for the subendocardial small veins than for the subepicardial veins. Acetylcholine did not significantly change the average O_2 saturations observed in the small veins, although the subepicardial vs. subendocardial difference was not significant in the treated group.

TABLE 1
The Effect of Acetylcholine Infusion on the Hemodynamic
and Blood Gas Parameters.

| | Control | Acetylcholine Treated | |
		Before Infusion	During
Infusion			
Systolic	107	109	72 +
blood pressure (mm Hg)	±23	±10	±10
Diastolic	76	74	48 +
blood pressure (mm Hg)	±18	± 8	±10
Mean Blood Pressure	87	85	56 +
(mm Hg)	±19	± 7	±10
Heart Rate	254	254	176 +
(beats/min)	±27	±44	±40
PO_2	68	76	70
(mm Hg)	± 9	± 4	± 5
PCO_2	36	35	37
(mm Hg)	± 1	± 2	± 3
pH	7.54	7.54	7.41
	±.06	±.06	±.18

+ Significantly different from the preinfusion value.

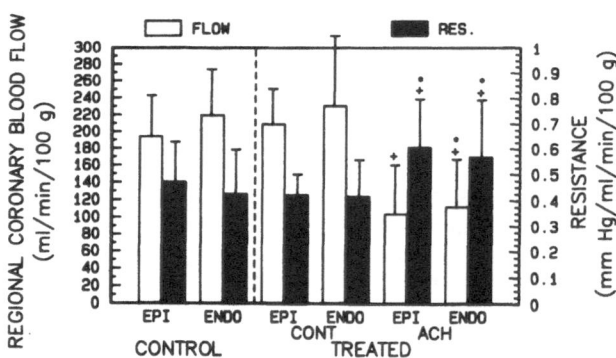

Figure 1. The effect of acetylcholine infusion on regional coronary blood
flow (ml/min/100 g) and resistance (mm Hg/ml/min/100 g).
+ Significantly different from the preinfusion value.
* Significantly different from the control value.

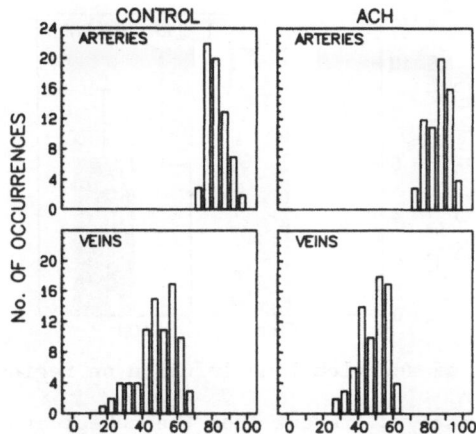

Figure 2. Histograms of O_2 saturations determined in small myocardial
arteries and veins in control and acetylcholine treated heart.

In both groups, O_2 extraction tended to be higher in the
subendocardium than in the subepicardium, figure 3. This difference was
significant for the control group only. There were no significant
differences in O_2 extraction between the control and acetylcholine treated
groups.

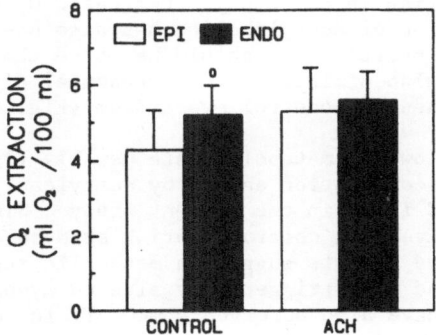

Figure 3. The effect of acetylcholine infusion on regional myocardial O_2
extraction.
° significantly different from the subepicardial value.

Oxygen consumption for the subendocardium in the control group was
significantly higher as compared to the subepicardial layer, figure 4. In
comparison to the control rabbits, the overall O_2 consumption was
significantly depressed in the acetylcholine treated group. This
reduction was mostly pronounced in the subendocardium minimizing the
transmural difference during acetylcholine infusion.

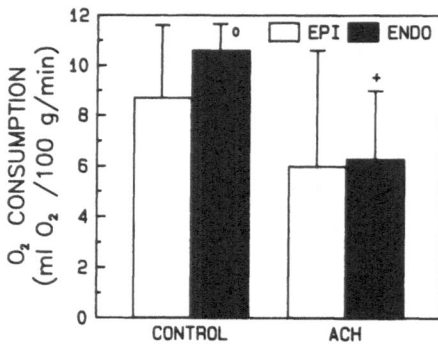

Figure 4. The effect of acetylcholine infusion on regional myocardial O_2 consumption.
    o significantly different from the subepicardial value.
    + significantly different from control.

    In spite of the differences in O_2 consumption between the groups, no differences in the O_2 supply to consumption ratio existed when the two groups were compared. The control O_2 supply to consumption ratio was lower in the subendocardial layer (2.4±0.5) than in the subepicardial region (2.9±0.6). With acetylcholine, the subendocardial and subepicardial ratios were not significantly different from the control group, 2.3±0.3 and 2.5±0.4, respectively.

DISCUSSION

    Acetylcholine administration caused bradycardia and arterial hypotension. These reductions in cardiac work led to a decreased myocardial O_2 consumption in our study. Decreased O_2 consumption following administration of acetylcholine has also been observed in an isolated perfused rat heart[13]. It should be noted that on a regional basis, acetylcholine also diminished the transmural differences in O_2 consumption that are seen in control myocardium (Fig. 4).

    Coronary blood flow and metabolism are usually tightly linked[1]. The reduced myocardial O_2 consumption caused by acetylcholine also led to a reduced coronary blood flow, in the current study. Similar effects on coronary blood flow have been observed during bradycardia in experimental animals[14,15,16]. These reports suggest a metabolic regulatory mechanism which affects the blood flow triggered by altered myocardial O_2 utilization. Others have also observed some fall in coronary blood flow with acetylcholine[3,17].

    No regional differences in flow distribution were observed with acetylcholine, although others have observed some regional redistribution[3]. In the current report, a similar decrease in blood flow was observed in both layers of the myocardium. This result is in contrast to some studies in dogs that reported preferential vasodilation in the subendocardium with acetylcholine[18]. This phenomenon was found with acetylcholine even when the heart rate was held constant[3]. Altered transmural distribution of flow in favor of the subendocardium was also observed during bradycardia which was not induced by acetylcholine[6]. It is possible that the arterial pressure drop in our study diminished the difference in O_2 consumption and thus coronary flow distribution across left ventricular wall that is usually seen in bradycardia and or acetylcholine treatment. Also there may be a species difference as the other studies were performed in dog.

It is widely accepted that parasympathetic activation can affect the coronary vasculature independent of hemodynamic changes. In vivo studies performed on dog hearts have demonstrated vasodilation with acetylcholine, i.e. acetylcholine dilates large and coronary resistance vessels when administered intravenously[19]. Increases in coronary blood flow have been observed during intracoronary infusion of acetylcholine[1,3]. This muscarinic receptor mediated effect was independent of metabolic and adrenergic mechanisms[19]. Vasodilating effects of acetylcholine have been demonstrated in different species[7,8,9] including rabbits[20]. Using low concentration of the agonist, vasodilation has been also observed in the beating rat heart[21], although in rat heart acetylcholine can also produce vasoconstriction[17].

Despite the potential direct vasodilatory effect of acetylcholine infusion, we observed a decrease in overall coronary blood flow. There was an approximately 42% increase in coronary resistance in our study. This suggested that the decreased flow was a consequence of reduced metabolic demand. A drop in arterial blood pressure occurred simultaneously with the decrease in flow. It is possible that the metabolic vasoconstriction competed with the neuronal and autoregulatory vasodilation. Coronary blood flow in rabbits was found to be related in almost linear fashion with pressure decrease during hemorrhage[20]. However, some autoregulatory capacity has also been demonstrated in the rabbit myocardium[22].

Myocardial O_2 saturation measurements in small arteries and veins showed that the O_2 extraction was not changed during acetylcholine treatment in the current study. This result may indicate that in spite of the reduced blood supply, the tissue was not flow limited. The unaffected O_2 supply to consumption ratio further suggests that the lowered coronary blood flow was a result of metabolic alterations rather than pressure dependency. It should be pointed out that the direct influence of acetylcholine on the coronary vasculature might also have an effect on the O_2 supply to consumption ratio. Further, the loss of regional differences in O_2 extraction and consumption could indicate a regional difference in the number of cholinergic receptors or their effect in the left ventricle.

In summary, we studied the effects of acetylcholine on the regional coronary blood flow and myocardial O_2 consumption in order to compare its direct vasodilatory effects with the metabolic vasoconstriction it induces. Acetylcholine reduced heart rate and depressed the arterial blood pressure. The mean O_2 consumption was significantly reduced with acetylcholine. Coronary blood flow decreased uniformly across the left ventricular wall by about 50%. O_2 extraction was not affected by acetylcholine infusion. It can be concluded that the acetylcholine infusion directly decreased myocardial O_2 consumption, which lowered the coronary blood flow. The decreased flow was related to a reduced metabolic demands rather than a direct result of lowered blood pressure. Unaffected myocardial O_2 extraction also suggested that blood flow and metabolism were matched in the anesthetized open chest rabbit heart. This indicated that direct cholinergic vasodilation of the coronary vasculature did not allow a greater reduction in metabolism than flow during acetylcholine infusion.

ACKNOWLEDGEMENTS

This work was supported in part by U.S.P.H.S. grant HL26919 and a grant-in-aid from the American Heart Association, New Jersey Affiliate.

REFERENCES

1. E. O. Feigl, Coronary physiology, Physiol Rev. 63:1-205 (1983).

2. M. N. Levy, H. Zieske, Comparison of the cardiac effects of vagus nerve stimulation and acetylcholine infusions, Am. J. Physiol. 216:890-897 (1969).

3. J. V. O. Reid, et al., Parasympathetic control of transmural coronary blood flow in dogs, Am. J. Physiol. 249:H337-H343 (1985).

4. D. Laurent, et al., Effects of heart rate on coronary flow and cardiac oxygen consumption, Am. J. Physiol. 185:355-364 (1956).

5. H. R. Weiss, Regional oxygen consumption and supply in the dog heart: effect of atrial pacing, Am. J. Physiol. 236:H231-H237 (1979).

6. G. D. Buckberg, et al., Variable effects of heart rate on phasic and regional left ventricular muscle blood flow in anesthetized dogs, Cardiovasc. Res. 9:1-11 (1975).

7. M. R. Blumenthal, et al., Effects of acetylcholine on the heart, Am J. Physiol. 214:1280-1287 (1968).

8. A. M. Brown, Motor innervation of the coronary arteries of the cat, J. Physiol (London) 198:311-328 (1968).

9. S. F. Vatner, et al., Coronary dynamics in unrestrained conscious baboons, Am. J. Physiol. 221:1396-1401 (1971).

10. G. D. Buckberg, et al., Some sources of error in measuring regional blood flow with radioactive microspheres, J. Appl. Physiol. 31:598-604 (1971).

11. A. K. Sinha, et al., Blood oxygen saturation determination in frozen tissue, Microvasc Res 13:133-144 (1977).

12. H. R. Weiss, A. K. Sinha, Regional O_2 saturation of small arteries and veins in the canine myocardium, Circ Res 42:119-126 (1978).

13. E. M. Nuutinen, et al., The effect of cholinergic agonists on coronary flow rate and oxygen consumption in isolated perfused rat heart. J. Moll. Cell Cardiol. 17, 31-42 (1985).

14. J. Kedem, et al., An experimental approach for evaluation of the O_2 balance in local myocardial regions in vivo, Quart. J. Exp. Physiol. 66:501-514 (1981).

15. B. Pitt, D. E. Gregg, Coronary hemodynamic effects of increasing ventricular rate in the unanesthetized dog, Circ Res 22:753-761 (1968).

16. S. White, et al., Effects of altering ventricular rate on blood flow distribution in conscious dogs, Am. J. Physiol. 221:1402-1407 (1971).

17. J. G. Dobson, Jr., The effect of acetylcholine, ischemia and anoxia on rat heart purine cyclic nucleotides and contractility, <u>Circ Res</u> 49:912-922 (1981).

18. G. J. Gross, et al., Transmural distribution of blood flow during activation of coronary muscarinic receptors, <u>Am. J. Physiol.</u> 240:H941-H946 (1981).

19. D. A. Cox, et al., Effects of acetylcholine on large and small coronary arteries in conscious dogs, <u>J. Pharmacol. Exp. Ther.</u> 225:764-769 (1983).

20. J. Grayson, D. Mendel, Myocardial blood flow in the rabbit, <u>Am. J. Physiol</u> 200:968-974 (1961).

21. K. Sakai, Coronary vasoconstriction by locally administered acetylcholine, carbachol and bethanechol in isolated, donor perfused, rat hearts, <u>Br. J. Pharmacol.</u> 68:625-632 (1980).

22. G. J. Grover, H. R. Weiss, Coronary adjustments to graded hypotension in rabbits, <u>Circulatory Shock</u> 23:71-80, (1987).

INFLUENCES OF DIFFERENT ROUTINELY USED MUSCLE RELAXANTS ON OXYGEN DELIVERY TO AND OXYGEN CONSUMPTION BY THE HEART DURING XENON-ANESTHESIA

L.J.van Woerkens#, B.Lachmann, G.J.van Daal, W.Schairer,
R.Tenbrinck, P.D.Verdouw#, and W.Erdmann

Depts. of Anesthesiology and Experimental Cardiology
Thorax Centre #, Erasmus University, Postbus 1738
3000 DR Rotterdam, The Netherlands

Introduction

To date there are no known studies in which the influence of routinely used muscle relaxants on the heart metabolism were studied, avoiding the cardiodepressive effects of halothane in combination with nitrous oxide or fentanyl (1,2,3). In earlier investigations we have shown Xenon to be an effective anesthetic gas with potent analgesic properties. Further on we could demonstrate that Xenon anesthesia has virtually no side-effects on cardiocirculatory parameters (4,5). Therefore it is possible to study the influence of different muscle relaxants on cardiac metabolism, without interaction with other anesthetics.

Materials and methods

After an overnight fast 9 Yorkshire pigs (16-24kg) were premedicated with 0.5 mg atropine-sulphate i.m., 15 mg Midazolam i.m. and 10 mg/kg ketamine-HCl. After administration of 15 mg/kg thiopental i.v. they were intubated and connected to a closed circuit anesthesia system, which consists of Siemens Servo Ventilator 900C and a modified Servo Anesthesia Circle 985 (Siemens-Elema, Solna, Sweden[1]). During thoracotomy and preparation the ventilation was done with 100% oxygen.
After thoracotomy an electromagnetic flowprobe (Skalar, Delft, The Netherlands) was placed around the ascending aorta to measure ascending aortic blood flow. A 7F catheter was placed in the superior caval vein for administration of muscle relaxants, while another 7F catheter was placed in the inferior caval vein for the infusion of Haemaccel (Behringwerke A.G., Marburg, FRG) to replace bloodloss. Left ventricular and aortic blood pressure were measured with 8F micro-tipped catheters (Millar Instr., Houston, Texas, USA). An 8F catheter was placed in the descending aorta to withdraw arterial blood samples and measure arterial blood pressure. The great cardiac vein was cannulated to withdraw venous bloodsamples and the left atrial appendage was catheterized for the administration of microspheres.

---

1. We like to thank Siemens-Elema, Solna, Sweden for providing us with the anesthesia
   system and Xenon.

Then the ventilatorsetting was switched to closed circuit anesthesia for ventilation with a mixture of oxygen (30%) and Xenon (70%). Respiratory rate and tidal volume were adjusted to keep arterial PCO_2 within normal limits (35-45 mmHg, ABL-3, Radiometer, Copenhagen, Denmark). Baseline values were collected after a stabilisation period of 15 minutes.

Tissue perfusion was measured using carbonized radioactive microspheres (15 ± 1 [sd] um in diameter) labelled with ^{141}Ce, ^{113}Sn, ^{103}Ru, ^{95}Nb or ^{46}Sc (NEN Chemicals Gmbh, Dreieich, FRG). Details of the radioactive microspere method and of the calculation of flow data have been reported elsewhere (6,7).

Myocardial oxygen consumption was calculated by multiplying the difference in the oxygen content of the arterial and coronary venous blood by coronary blood flow.

Protocol

After preparation of the experimental setup muscle relaxants were administered in a random fashion with a stabilisation period of at least one hour before data collection. Relaxation was measured every 15 seconds and the degree of neuromuscular block was estimated using supramaximal nerve stimulation and quantitation of the evoked elektromyographic response. The neuromuscular block was continously held between 95 and 100%, for each musclerelaxant, during the full time of the experiment in each pig.
After killing the animals at the end of the experiment with an i.v. overdose of pentobarbital. The various tissues of the heart were dissected out after a 24 hours fixation period in 10% formaline and placed in plastic vials. The radioactivity in these vials was counted for 10 min. in a gamma-scintillation counter (Packard, model 5986) equipped with a multichannel analyser (Contrac) using suitable windows for discriminating the different isotopes (7).

Calculations and statistical analysis

All data are presented as mean + SEM. The significance of the differences between the variables and the different groups was evaluated using the Student-Newman Keul test after an analysis of variance (randomized block design) had revealed that the samples represented different populations. A P < 0.05 (two-tailed) was considered to be statistically significant.

Results

All muscle relaxants that we used in our experiment caused a significant increase in oxygen delivery to the left ventricle when compared with baseline values (pancuronium 60%, suxamethonium 41%, atracurium 29% and vecuronium 49%) (Fig. 1).
Only pancuronium and vecuronium showed a significantly increased oxygen consumption (52% and 40%), while the increase caused by suxamethonium (32%) and atracurium (10%) did not show statistical significance (Fig. 1).

All muscle relaxants caused a significant increase in perfusion of the right ventricle (pancuronium 60%, suxamethonium 56%, atracurium 43% and vecuronium 48%) (Fig. 2). In the left ventricle the increased perfusion caused by pancuronium (70%), suxamethonium (48%) and vecuronium (64%) showed to be statistically significant, atracurium caused a non-significant increase of 37% (Fig. 2).

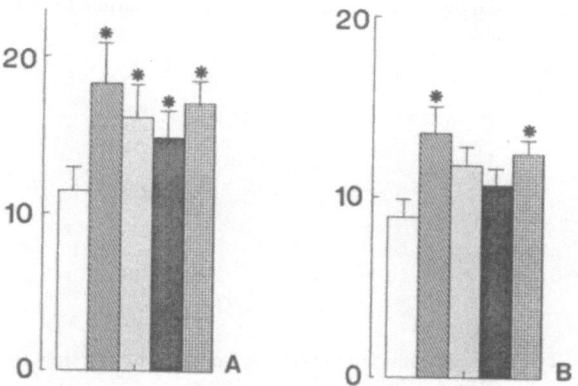

Fig. 1. Oxygen delivery to the left ventricle (A) and oxygen consumption of the left ventricle (B) in ml.min^{-1}.100g^{-1}. * P < 0.05 versus baseline.

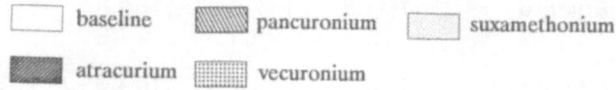

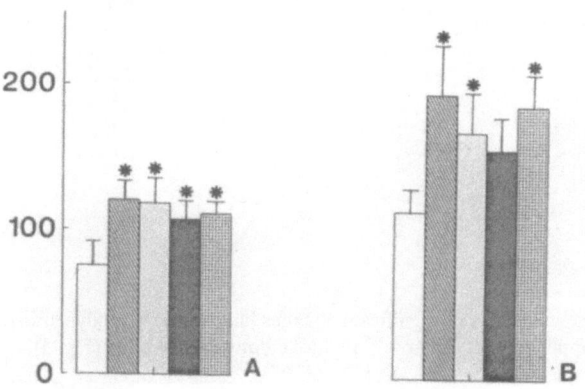

Fig. 2. Perfusion of the right (A) and left (B) in ml.min^{-1}.100g^{-1}.
* P < 0.05 versus baseline.

Left ventricle specified for different tissue layers

All muscle relaxants caused a significantly increased perfusion of the endocard (pancuronium 69%, suxamethonium 50%, atracurium 45% and vecuronium 68%) (Table 1). Pancuronium and vecuronium caused a significantly increased perfusion of the epicard (76% and 63%), the increased perfusions caused by suxamethonium (49%) and atracurium (41%) did not show statistical significance (Table 1).

Table 1. Left ventricular perfusion, specified for different tissue layers in ml.min^{-1}.100g^{-1}, values are mean and SEM.

	baseline	pancur.	suxameth.	atracur.	vecur
Epicard	101.8	117.8	150.0	142.4	165.3
	12.7	28.1	21.8	20.0	17.7
Mesocard	125.5	211.6	184.3	163.6	204.9
	16.0	40.4	30.0	23.6	26.5
Endocard	128.9	217.7	193.2	186.8	217.4
	18.6	40.8	27.1	26.1	27.9

Hemodynamics

Compared to baseline values no significant changes in systemic hemodynamic parameters were measured after application of the above mentioned muscle relaxants (Fig. 3).

Discussion

Muscle relaxants can influence several cardiovascular parameters. Pancuronium has vagolytic and sympaticomimetic properties which tend to antagonise the vagally mediated bradycardia resulting from fentanyl administration (8). Pancuronium may exaggerate circulatory responses to noxious stimuli in patients, which may be extra harmful in patients with limited coronary reserve (9). Vecuronium and atracurium are relatively new muscle relaxants without vagolytic and sympaticomimetic effects. The effect of succinylcholine on the autonomic nervous system even complicates the interpretation of measured parameters from the cardiovascularory system. But up to now there are no data available that demonstrate that the described effects are

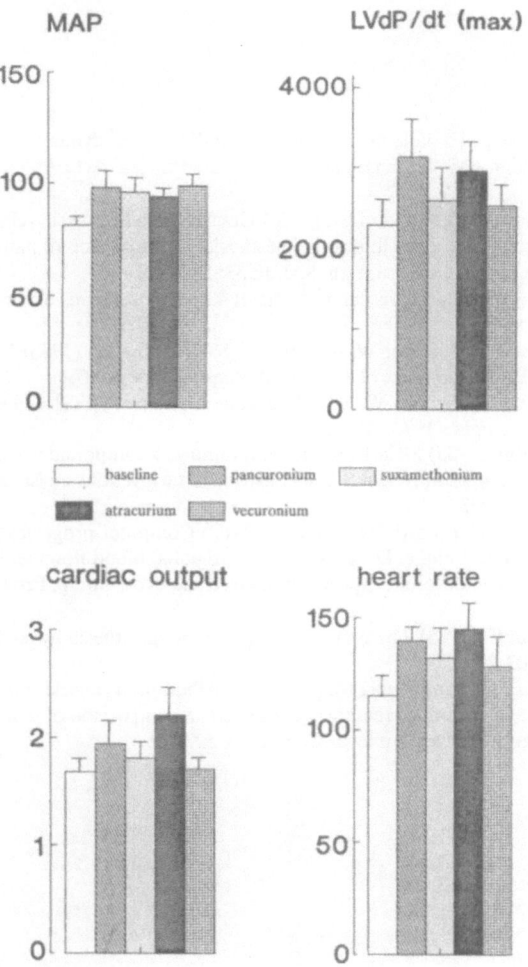

Fig. 3. Systemic hemodynamics; cardiac output in l.min^{-1}, heart rate in beats.min^{-1}, mean arterial pressure (MAP) in mm Hg, contractility (LVdP/dt [max]) in mm Hg.sec^{-1}.

realy caused by the farmacological effect of the muscle relaxants themselves and not caused by the combined effect of muscle relaxants and anesthetics.

A prerequisite to study the effects of muscle relaxants on different cardiovascular parameters is therefore an anesthesia technique, which has no influence on these parameters. Such a technique became recently available. Using Xenon in a concentration of 70% we could demonstrate that Xenon does not influence any cardiovascular parameters (4,5).

Our results clearly demonstrate that the tested muscle relaxants have no negative effects on oxygen delivery to the heart or on left ventricular oxygen consumption. We conclude that the combination of Xenon and the above mentioned muscle relaxants can safely be used as a routine in anesthesia and that it might be usable in patients with ischeamic heart disease.

References

1. Meretoja OA, Takkunen O, Heikilla H, Wegelius U. (1985) Haemodynamic response to nitrous oxide during high-dose fentanyl pancuronium anaesthesia. Acta Anaesthesiol. Scand.: 29: 137-141.
2. Heinonen J, Salmenpera M, Suomivuori M. (1986) Contribution of muscle relaxant to the haemodynamic course of high-dose fentanyl anaesthesia: a comparison of pancuronium, vecuronium and atracurium. Can. Anaesth. Soc. J.: 33: 597-605.
3. Eisele JH, Smith NT. (1972) Cardiovascular effects of 40 percent nitrous oxide in man. Anesth. Analg.: 51: 956-962.
4. Lachmann B, Trouwborst A, Schairer W, Armbruster S, Erdmann W. (1988) Xenon and its analgesic effect.: 9th World congress of Anaesthesiologists 1988: A0079.
5. Lachmann B, Verdouw PD, Schairer W, Van Woerkens LJ, Van Daal GJ. (1988): 9th World congress of Anaesthesiologists 1988: A0242.
6. Saxena PR, Verdouw PD. (1985) 5-Carboxamide tryptamine, a compound with high affinity for 5-HT1 binding sites, dilates arterioles and constricts arteriovenous anastomoses. Br. J. Pharmacology: 84: 533.
7. Saxena PR, Schamhardt HC, Forsyth RP, Loeve J. (1980) Computer programs for the radioactive microsphere technique. Determination of regional blood flows and other haemodynamic variables in different experimental circumstances. Comp. Prog. Biomed.: 12: 63-84.
8. Savarese JJ, Lowenstein E. (1985) The name of the game: no anesthesia by cookbook. Anesthesiology: 62: 703-705.
9. Heinonen J, Salmenpera M, Suomivuori M. (1986) Contribution of muscle relaxant to the haemodynamic course of high-dose fentanyl anaesthesia: a comparison of pancuronium, vecuronium and atracurium. Can. Anaesth. Soc. J.: 33: 597-605.

SKELETAL MUSCLE

CENTRAL REFLEX EFFECTS OF HYPOXIA ON MUSCLE OXYGENATION

D.L. Bredle, C.K Chapler*, and S.M. Cain

Dept. of Physiology and Biophysics, Univ. of Alabama
at Birmingham, Birmingham, AL, and *Dept. of
Physiology, Queen's Univ., Kingston, Ontario, Canada

INTRODUCTION

We postulated that whole body hypoxia causes O_2 demand to increase in skeletal muscle because hypoxia stimulates sympathetic release of catecholamines which are calorigenic. Such an effect would be masked in a typical hypoxia model in which regional O_2 uptake would also be limited by O_2 availability. Our first goal in these experiments was to circumvent this difficulty and demonstrate that our postulate was correct. The second goal for these experiments was to determine the strength and duration of centrally mediated vasoconstriction in muscle in the absence of local hypoxia. Previous studies (Cain and Chapler, 1979; 1980) have shown that muscle vascular resistance increased less than 20% in severe whole body hypoxia and that after 30 min of hypoxia, local dilatory factors had overcome any vasoconstriction. To meet both these goals, we maintained normoxic regional perfusion to hindlimb skeletal muscles of anesthetized dogs by use of a pump and membrane oxygenator while ventilating the animal with an hypoxic gas mixture. To show the role of innervation in regional hypoxic reactions, we compared metabolic and hemodynamic events in innervated and denervated limbs.

METHODS

Fourteen mongrel dogs were anesthetized with pentobarbital sodium (30 mg/kg), paralyzed with succinylcholine chloride, and pump-ventilated to keep end-tidal PCO_2 near 40 Torr. The left hindlimb, mostly skeletal muscle tissue, was vascularly isolated by two tourniquets tied tightly near the groin. The paw circulation was excluded by a separate tourniquet. The limb was pump-perfused at constant flow with a circuit originating in the contralateral femoral artery. A membrane oxygenator kept the limb arterial PO_2 ~100 Torr and the PCO_2 ~35 Torr. To assure arterial isolation to the leg, we ligated the left internal and external iliac and deep circumflex arteries. This isolation, as well as reactive hyperemia, was verified by brief femoral artery clamping before and after each experiment.

Simultaneous arterial and venous blood samples were drawn for both the whole body and limb and analyzed with an ABL-30 blood gas analyzer and IL 282 co-oximeter. Whole body gas exchange was calculated from expired volume and O_2 and CO_2 concentrations. Limb O_2 uptake ($\dot{V}O_2$) was calculated as the product of venous blood flow and $(a-v)O_2$ difference.

Resistance was determined as leg perfusion pressure/blood flow. O_2 extraction was $\dot{V}O_2/O_2$ delivery.

Eight dogs comprised an intact, innervated group. The limb muscles of a second group (n=6) were denervated by severing the sciatic and femoral nerves. Data were compared between and within groups by ANOVA and Duncans post-hoc test, and significance accepted at p <.05.

RESULTS

Global hypoxia reduced systemic arterial PO_2 to ~25 Torr and whole body $\dot{V}O_2$ fell ~30%. During this period, however, $\dot{V}O_2$ in the normoxic limbs of the intact group rose about 25% (Fig 1). This effect, while quite variable early in hypoxia, was consistent by the 25th minute and continued unabated throughout the 40 minutes of recovery. Denervation prevented any rise in limb O_2 uptake during the hypoxic period, but the first measurement in hypoxia was significantly greater than in the prehypoxic control period.

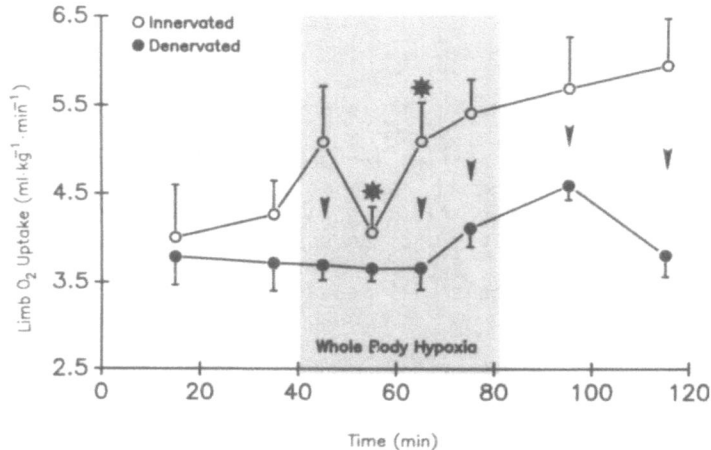

Fig.1. Hindlimb muscle O_2 uptake. Values are means ± SE. *, sig. diff. from preceding value in same group. ▼, sig. diff. between groups.

Limb resistance (Fig.2) approximately doubled in the innervated group during hypoxia and also increased significantly in the denervated group but not as much. Although resistance tended to slacken towards the end of hypoxia, it was still higher in the innervated group than it was before hypoxia. Limb perfusion pressure (Fig.3) similarly reached a higher peak in hypoxia in the innervated group than in the denervated. Limb blood flow and O_2 delivery (Figs. 4 & 5) both decreased somewhat in hypoxia despite the constant flow from the perfusion pump; however, limb O_2 delivery was always above the critical point of supply dependence. Limb O_2 extraction ratio never exceed 30%, on the average, in either group during hypoxia. Limb venous PO_2 decreased slightly in the hypoxic period, from 50 to 43 Torr in the innervated group and from 52 to 48 Torr in the denervated group.

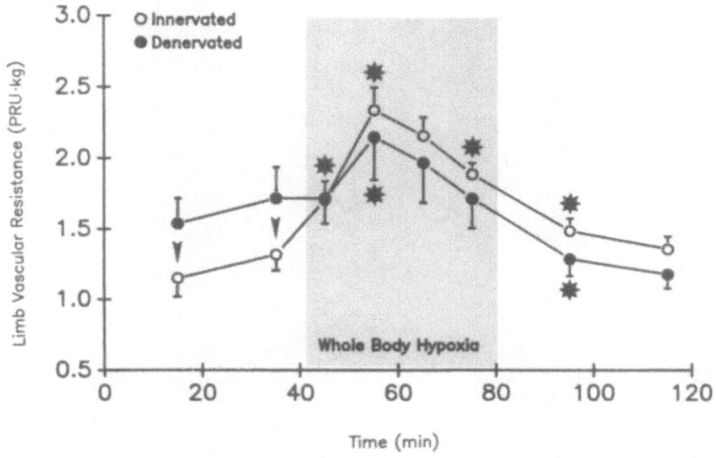

Fig.2. Hindlimb vascular resistance. Symbols same as Fig.1.

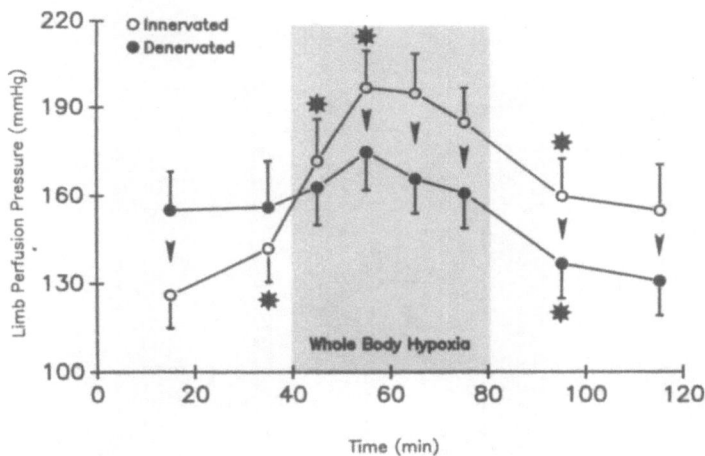

Fig.3. Hindlimb perfusion pressure. Symbols same as Fig.1.

DISCUSSION

In spite of some early variability in its response, muscle O_2 uptake clearly increased even as the whole body had to decrease its O_2 uptake because of supply limitations in hypoxia. Furthermore, sympathetic innervation to the limb was necessary for this to occur. Our first postulate was proven quite conclusively. We surmise that the delayed and temporary increase in limb O_2 uptake seen in the denervated group was due to increased circulating levels of catecholamines as a result of global hypoxia (Sylvester et al., 1979). In addition to denervation, we have

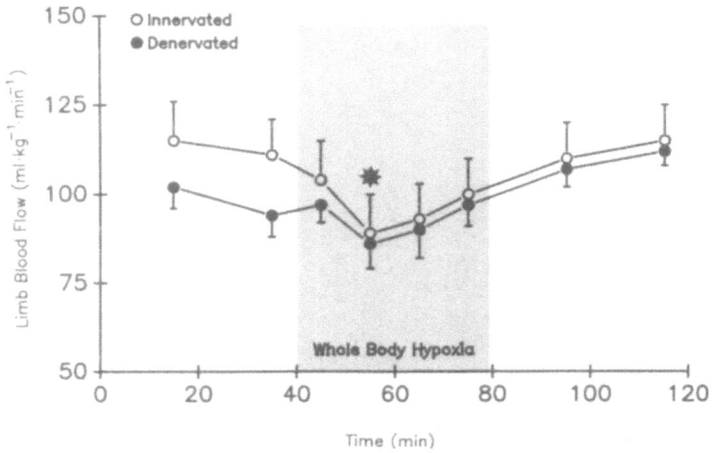

Fig.4. Hindlimb blood flow. Symbols same as Fig.1.

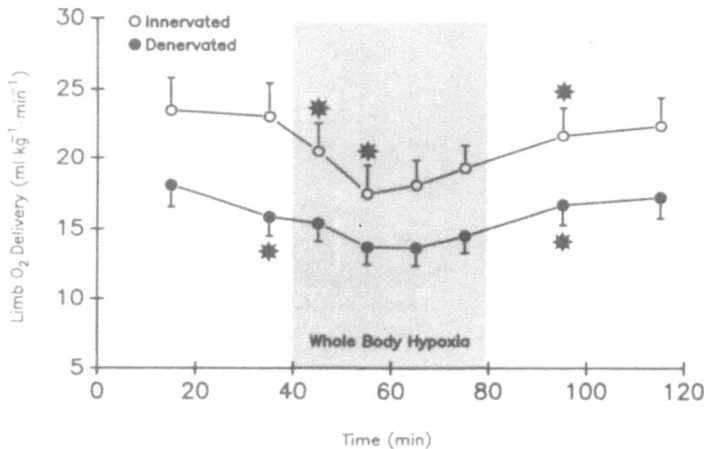

Fig.5. Hindlimb O₂ delivery. Symbols same as Fig.1.

observed that either specific β_2-adrenergic receptor blockade (Bredle et al., 1988a) or a non specific α-adrenergic blocker (Bredle et al., 1988b) would also prevent the calorigenic response in limb skeletal muscles. These results taken together suggest that both α- and β-receptor function, as well as intact innervation, are necessary for the response.

With intact innervation, the vasoconstrictor response of normoxic skeletal muscle to whole body hypoxia was several times greater than had been observed in dog limb muscles that were also allowed to become hypoxic (Cain and Chapler, 1979; 1980). Although some decrease in resistance was seen, vasoconstriction was sustained above the level in normoxia during the entire hypoxic period. No "autoregulatory escape" occurred and the slight decrease that was observed might have been due to decreased neurotransmitter liberation at sympathetic endings. Denervation did not prevent vasoconstriction as it did the increase in limb O_2 uptake. Furthermore, a-adrenergic blockade did not prevent vasoconstriction with global hypoxia so circulating catecholamines presumably were not responsible (Bredle et al., 1988b). This suggests that circulating levels of other vasoconstrictor agents such as vasopressin or angiotensin were increased during whole body hypoxia. The possibility must also be raised that the denervation procedure was only partially successful. However, Clonninger and Green (1955) have shown that the sciatic and femoral nerves carry >90% of the sympathetic innervation to the hindlimb muscles of the dog. We favor the first possibility as the more likely explanation. There is probably a hierarchy of vasoconstrictor effects, but all are overridden when the limb becomes hypoxic as the rest of the body with consequent low PO_2 and production of local metabolites to act as vasodilators. The far greater vasoconstriction seen in the intact group was very good evidence that this was the major factor for the difference between these results and those obtained from intact but hypoxic limb muscles (Cain and Chapler, 1979, 1980).

In summary, global hypoxia increased O_2 demand in muscle so that its O_2 uptake increased when its supply was not limiting. Intact innervation was required for most of this calorigenic effect of hypoxia. Vasoconstriction occurred despite denervation, though not quite as strongly as in innervated tissue, suggesting the role of other circulating vasoconstrictors when a-constriction was blocked. Full autoregulatory escape from hypoxic vasoconstriction did not occur in either innervated or denervated muscle as long it was maintained normoxic.

The work was supported by NHLBI Grant Nos. HL 14693, HL 26927, 5T32 HL 0755305 as well as funds from MRC of Canada.

REFERENCES

Bredle, D.L., Chapler, C.K., and S.M. Cain, 1988a, Metabolic and circulatory responses of normoxic skeletal muscle to whole body hypoxia, J. Appl. Physiol. (in press).
Bredle, D.L., Chapler, C.K., and S.M. Cain, 1988b, Metabolic and circulatory responses of normoxic skeletal muscle to whole body hypoxia and a-blockade. Physiologist (abstract, in press).
Cain, S.M. and Chapler, C.K., 1979, Oxygen extraction by canine hindlimb during hypoxic hypoxia, J. Appl. Physiol. 46:1023-1028.
Cain, S.M. and Chapler, C.K., 1980, O_2 extraction by canine hindlimb during a-adrenergic blockade and hypoxic hypoxia, J. Appl. Physiol. 48:630-635.
Clonninger, G.L. and Green, H.D., 1955, Pathways taken by the sympathetic chain to the vasculature of the hindleg muscles of the dog, Am. J. Physiol. 181:258-262.
Sylvester, J.T., Scharf, S.M., Gilbert, R.D., Fitzgerald, R.S., and Traystman, R.J., 1979, Hypoxic and CO hypoxia in dogs: hemodynamics, carotid reflexes, and catecholamines, Am. J. Physiol. 236:H22-H28.

HOW PHOSPHOCREATINE BUFFERS CYCLIC CHANGES IN ATP DEMAND IN WORKING MUSCLE.

C. Funk[1], A. Clark[1], Jr., and R. J. Connett[2]

[1]Department of Mechanical Engineering
[2]Department of Physiology
University of Rochester
Rochester, NY 14627, USA

INTRODUCTION

What is the role of phosphocreatine in energy metabolism of muscle tissue? The various roles that have been proposed over the last 50 years all fall into two categories: spatial transport and storage of high-energy phosphate. We analyse the function of phosphocreatine as a temporal buffer within a muscle cell on the basis of a simple model. Our model is based on the assumption of equilibrium of the creatine kinase reaction

$$Mg.ADP + PCr + H^+ \rightleftharpoons Mg.ATP + Cr.$$

Also, all species are allowed to diffuse freely between mitochondria and myofibrils. In the absence of any anatomical evidence and following the reasoning of MEYER et al.(1984), we do not incorporate compartmentation of ATP and ADP in mitochondrial or myofibrillar pools.

Previous studies have been mainly concerned with the need to keep ATP concentration constant. We show that, during a twitch stimulation protocol, variations in ATP, without the phosphocreatine buffer, would be about 3%. These variations in ATP would imply large variations in ADP and AMP concentrations. These compounds play a major role in matching ATP utilization and ATP production by controlling key enzyme cascades. It is therefore necessary to determine the magnitude of temporal fluctuations in the concentrations of controlling reactants to understand cell metabolism as a biologic system.

In this paper we present a model where the concentrations within the cell depend on time only: Based on an estimate of gradients in steady work, we neglect spatial variations. Our model allows us to relate the time variations of concentrations to ATP production and consumption. We then focus on how phosphocreatine buffers high frequency transients associated with variations in ATP consumption during a single muscle twitch.

PRELIMINARY CALCULATIONS

Notation. The state variables we are concerned with are the concentrations of

ATP, ADP, PCr, and Cr, which we denote by N_{ATP}, N_{ADP}, N_{PCr}, and N_{Cr}. The relevant diffusion coefficients are D_{ATP} $(=D_{ADP})$ and D_{PCr} $(=D_{Cr})$. Contraction occurs at the myofibrils when ATP is split into ADP and Pi, a reaction catalysed by $ATPase$, and the ATP consumption rate is denoted by Γ_c. For a given set of cytosolic concentrations, the mitochondria set up a certain ATP production rate, denoted by Γ_p, intimately linked to the oxygen consumption rate. The apparent equilibrium constant at given Mg^{2+} and pH for the creatine kinase reaction is denoted by K:

$$K = \frac{N_{ATP}N_{Cr}}{N_{ADP}N_{PCr}}. \tag{1}$$

Spatial variations. The geometry to be considered is fairly complex. Muscle fibers are made of roughly cylindrical long myofibrils separated by cytosolic tissue, with scattered mitochondria. To get an order of magnitude for the gradients in N_{ATP} and N_{PCr} between mitochondria and the center of myofibrils, we consider a simple one-dimensional slab model. A mitochondrion at one end delivers ATP, which then diffuses through a distance r towards the center of a myofibril which consumes ATP at a constant rate Γ_c. Exact integration of the general transport equations for this case gives the following variation across the cell:

$$\Delta(D_{PCr}N_{PCr} + D_{ATP}N_{ATP}) = \frac{\Gamma_c r^2}{2}. \tag{2}$$

Without the creatine kinase reaction, the variation in ATP would be

$$\Delta(N_{ATP}) = \frac{\Gamma_c\, r^2}{2D_{ATP}}. \tag{3}$$

Consider some typical numbers. We use $D_{ATP} = 1.5 \cdot 10^{-6} cm^2/s$ and $D_{PCr} = 2 \cdot 10^{-6}\ cm^2/s$ (MAINWOOD and RAKUSAN, 1982). Typical myofibrils have a 1 to 5 μm radius. We use an equilibrium constant K of 140 at 1 mM Mg^{2+} and pH 7 (MEYER. 1988). Then, for a myofibrillar $ATPase$ rate Γ_c of 0.5 mM/s in a cell containing 34 mM total creatine and 7 mM total adenine we find spatial variations in ATP concentration less than 0.01% in the presence of the creatine kinase reaction and no greater than 0.6% in its absence. These calculations agree on the order of magnitude of the variations with others we have done for more sophisticated geometries.

Therefore, phosphocreatine is hardly needed to buffer spatial variations of ATP concentration. We next examine temporal variations.

Temporal variations during a single twitch event. During a single twitch event, there is a phase of tension development, followed by a recovery phase. It is then reasonable to consider that the actual ATP consumption occurs only during the period of rising tension development in a twitch stimulation protocol.

Based on measurements of the time course of tension development in dog gracilis muscle (unpublished results), we evaluated this rising period to be $\approx 50ms$ for several twitch frequencies.

We modeled ATP consumption rate by constant non-zero consumption during a fraction f of each twitch followed by zero consumption for the rest of the twitch, so that the average consumption Γ_c fell in the physiological range of working muscle.

For simplicity, the ATP production rate was kept constant (equal to the mean ATP consumption rate). For a stimulation frequency F, the variation in N_{ATP} during a twitch without the creatine kinase reaction is

$$\Delta(N_{ATP}) = \frac{(1-f)\Gamma_c}{F}. \tag{4}$$

For a frequency of $2Hz$ and Γ_c of 0.5 mM/s, we get variations in ATP concentrations of the order of 3%, much larger in general than the spatial variations we found. Therefore, we focus on the buffering of temporal variations by phosphocreatine.

DERIVATION OF THE MODEL

Model equations. Model equations describing temporal variations for the concentrations are obtained from mass balances. We let Γ_p be the production rate of ATP at mitochondria, Γ_c be the consumption rate of ATP at myofibrils, and Γ_{CK} the production rate of ATP by the creatine kinase reaction. Then the mass balances are

$$\frac{dN_{ATP}}{dt} = \Gamma_p + \Gamma_{CK} - \Gamma_c = -\frac{dN_{ADP}}{dt},$$
$$\text{and} \quad \frac{dN_{PCr}}{dt} = -\Gamma_{CK} = -\frac{dN_{Cr}}{dt}. \tag{5}$$

The total adenine concentration $N_{AdT} = N_{ADP} + N_{ATP}$ and the total creatine concentration $N_{CrT} = N_{PCr} + N_{Cr}$ are both conserved. Equation (5) is to be supplemented with the equilibrium relation (1). The combined system has *one degree of freedom* and we choose to express all concentrations as a function of the *creatine charge*:

$$S_C = \frac{N_{PCr}}{N_{CrT}}.$$

In order to describe how phosphocreatine buffers changes in ATP concentrations, we define a *buffering factor* f_B as follows:

$$f_B = \frac{dN_{PCr}}{dN_{ATP}}. \tag{6}$$

We can express f_B in terms of the creatine charge, the equilibrium constant K and the ratio N_{AdT}/N_{CrT} :

$$f_B = \left(\frac{N_{CrT}}{N_{AdT}}\right) \frac{((K-1)S_C + 1)^2}{K}. \tag{7}$$

As the creatine charge gets very low, buffering becomes ineffective.
Example: N_{AdT}/N_{CrT} ranges from 0.15 to 0.25 depending on the species (CON-NETT,1988), and K is very large (about 140). Then, for $S_C = 0.03$, $f_B \approx 1$ (variations in phosphocreatine are equal to variations in ATP).
With this new notation the time course of S_C is governed by the equation:

$$N_{CrT}(1 + \frac{1}{f_B})\frac{dS_C}{dt} = \Gamma_p - \Gamma_c. \tag{8}$$

The other state variables expressed in terms of S_C are

$$
\begin{aligned}
N_{PCr} &= S_C N_{CrT}. \\
N_{Cr} &= (1 - S_C) N_{CrT}. \\
N_{ATP} &= \frac{K S_C}{1 + (K - 1) S_C} N_{AdT}. \\
N_{ADP} &= \frac{1 - S_C}{1 + (K - 1) S_C} N_{AdT}.
\end{aligned}
\tag{9}
$$

RESULTS

We used our model to simulate a series of twitches and to observe the time course of the creatine charge S_C (defined as N_{PCr}/N_{CrT}) and the variable S_A (defined as N_{ATP}/N_{AdT}). The results are shown in Figure 1, and discussed below.

The period of rising tension development, or ATP consumption was fixed to 10% of the twitch for a stimulation frequency of $2\,Hz$, and an average consumption rate of $0.5\,mM/s$ (moderate work).

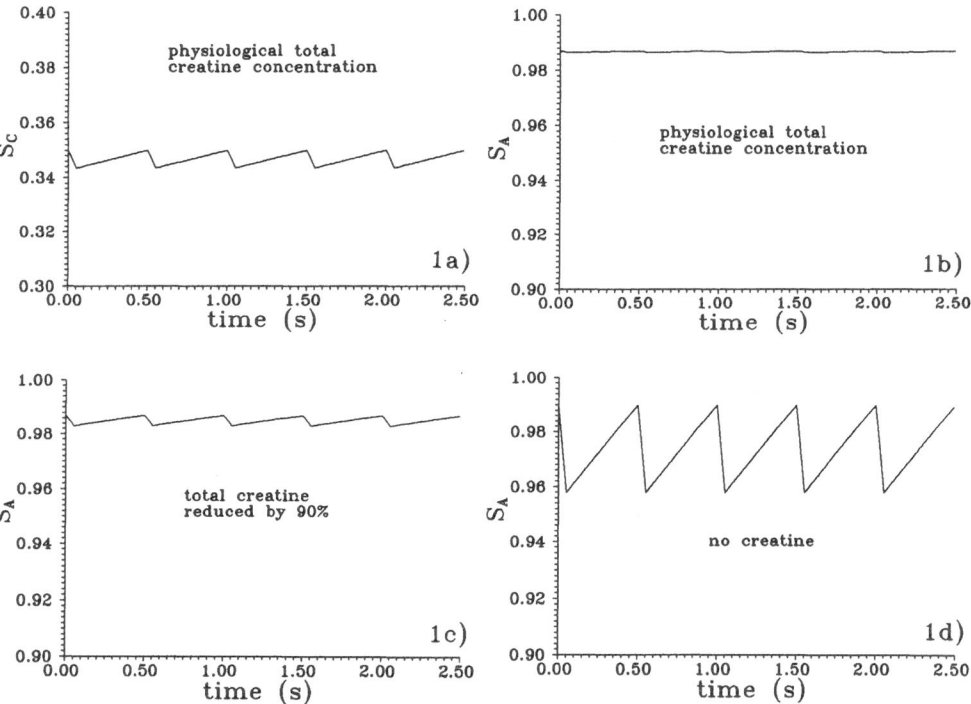

Figure 1. Time course during a twitch stimulation protocol of the adenine saturation $S_A = N_{ATP}/N_{AdT}$ and of the creatine charge $S_C = N_{PCr}/N_{CrT}$. The given conditions are a total adenine concentration N_{AdT} of $7mM$, an initial creatine concentration of 0.35, an average ATP consumption rate of $0.5mM/s$, a $2Hz$ twitch frequency. The fraction of the twitch during which ATP is actually consumed is 10% of twitch duration. The creatine kinase equilibrium constant K is 140. a) $S_C(t)$, total creatine concentration $N_{CrT} = 34mM$. b) $S_A(t)$, $N_{CrT} = 34mM$. c) $S_A(t)$, $N_{CrT} = 3.4mM$. d) $S_A(t)$, $N_{CrT} = 0$.

Figure 1a shows the time course of S_C under physiological total creatine concentration ($N_{CrT} = 34mM$, initial creatine charge of 0.35). S_C oscillates during a twitch event by 2%. Figure 1b shows the parallel time course of S_A under the same conditions. Variations in S_A are negligible (less than 0.1%).

Figure 1c shows what the time course of S_A would be if the total creatine concentration was reduced by 90%. We see that buffering is efficient even for this very small total creatine concentration. (0.4% variation in S_A during a twitch.)

Figure 1d shows what the time course of S_A would be in the absence of the creatine kinase reaction (total creatine content equals zero). We observe 3.3% variation in S_A during a twitch.

DISCUSSION

Phosphocreatine very efficiently buffers potential high frequency transients in ATP levels. In our example, without phosphocreatine, ATP concentration would vary by 3%. This variation would be amplified, via the adenylate kinase system, into very large variations in ADP and AMP concentrations: During a twitch, ADP concentration would change by a factor of 4, and AMP concentration would change by a factor of 16. ADP and AMP are important regulators of glycolytic flux at phosphofructokinase and pyruvate kinase. Thus phosphocreatine may play an important role in metabolic control by buffering even small variations in ATP, thereby preventing glycolysis from being inappropriately recruited. Temporal buffering may be essential for integrated function of cytosol and mitochondria as a biologic system.

SUMMARY

1. Spatial buffering is hardly necessary for maintaining ATP concentration.

2. Temporal buffering may be essential to defend concentration of *products of ATP hydrolysis*. ADP would vary 4-fold and AMP would vary 16-fold during a single twitch without buffering.

3. The principal function of the transphosphorylating reaction may be to buffer temporally the concentrations of controlling reactants.

Acknowledgements.
We thank Drs P.A.A. Clark, T.E.J. Gayeski, and C.R. Honig for many useful discussions of this work. We thank K. Groebe for many discussions and for his extensive help for the manuscript. We thank M.J. Kushmerick and R.S. Ballaban for their fruitful suggestions. We are grateful for financial support from the National Institutes of Health under grant numbers $HL37205$ and $AR36154$.

REFERENCES

ATKINSON, D. E., 1977, "Cellular Energy Metabolism and Its Regulation," Academic Press, New York.

CONNETT, R. J., 1988, Analysis of metabolic control: new insights using scaled creatine kinase model, Am.J.Physiol. 254:R949.

MAINWOOD, G. W. and RAKUSAN, K., 1982, A model for intracellular energy transport, *Can.J. Physiol.Pharmacol.* 60:98.

MEYER, R. A., 1988, A linear model of muscle respiration explains monoexponential phosphocreatine changes, *Am.J.Physiol.* 254:C548.

MEYER, R. A., SWEENEY, H. L. and KUSHMERICK, M., 1984, A simple analysis of the "phosphocreatine shuttle", *Am.J.Physiol.* 246:C365.

CAPILLARY BLOOD FLOW AND LOCAL OXYGEN SUPPLY IN SKELETAL MUSCLE DURING
ACUTE HAEMODILUTION WITH HYDROXYETHYL STARCH

D.K. Harrison, S. Birkenhake, N. Hagen and M. Kessler

Institut für Physiologie und Kardiologie der Universität
Erlangen-Nürnberg, Waldstrasse 6, 8520 Erlangen/FRG

INTRODUCTION

The use of colloidal infusions to maintain plasma volume in
patients has increased in recent years due to the risk of infections
from donated blood. The need for infusions which are effective over
relatively long periods led to the introduction of high molecular weight
solutions. However, by increasing the retention time in this way, the
viscosity of the blood is also raised and this may have a deleterious
effect on capillary blood flow.

Experiments on the beating dog heart during extreme haemodilution
with hydroxyethyl starch of molecular weight 40 000 (HES 40, Vogel at
al., this volume) have shown that changes in the pattern of capillary
blood flow distribution are induced in the myocardium when the haematocrit (Hct) is reduced to about 14 vol % (Hct 14). Under normal conditions the oxygen supply to the myocardium appeared to be maintained at
an adequate level, as indicated by pO_2 histograms measured using the
Kessler and Lübbers multiwire surface electrode (Kessler et al., 1986).

In view of the results obtained from capillary blood flow measurements in skeletal muscle at rest, during hypoxaemia and during electrical stimulation at increasing frequencies (Harrison et al., 1988; this
volume), further investigations were carried out to study the influence
of extreme haemodilution upon the tissue oxygen supply and distribution
of blood flow.

These preliminary experiments were carried out during haemodilution
with a 70 000 molecular weight hydroxyethyl starch (HES 70) in the
vastus medialis muscle of rabbits.

METHODS

Experiments were carried out in 7 anaesthetised, relaxed, artificially ventilated rabbits. Experience in our laboratory has shown that
the way in which the initial anaesthesia is performed is crucial to the
success of the subsequent experiment if physiological conditions are to
be maintained in the muscle of these highly sensitive animals. The
initial ketamine/xylazin injection was administered in a darkened, quiet

environment. During tracheotomy the head was placed in a box so that N_2O could be administered. After tracheotomy anaesthesia was maintained with Enflurane and N_2O. During the preparation fluid balance was maintained by infusion of Ringer solution until the haemodilution was performed.

Catheters were inserted to measure pressures in the left femoral artery and the jugular vein. The former catheter also served for obtaining samples for blood gas / acid-base analysis and measurement of Hct. The right vastus medialis muscle was exposed and freed of fascia. In addition, in some experiments the right femoral artery and vein were exposed and needle electrodes inserted for recording intravascular hydrogen clearance curves. Due to the frequency of arterial branches and the desire to disturb the regional blood supply as little as possible, it was not possible to place an electromagnetic flowprobe around the femoral artery. Regional blood flow was thus calculated from the intra-venous hydrogen clearance curves.

In addition to the global parameters listed above, tissue pO_2 and capillary blood flow (hydrogen clearance technique) were measured in the muscle. The haemodynamic parameters, ECG, hydrogen clearances and local pO_2 values were all recorded with the aid of Digital Equipment LSI/03 microcomputers.

After control measurements of all parameters, isovolemic haemo-dilution was carried out in two stages to mean haematocrits of 21% and 18%. Measurement of all parameters were recorded at both stages.

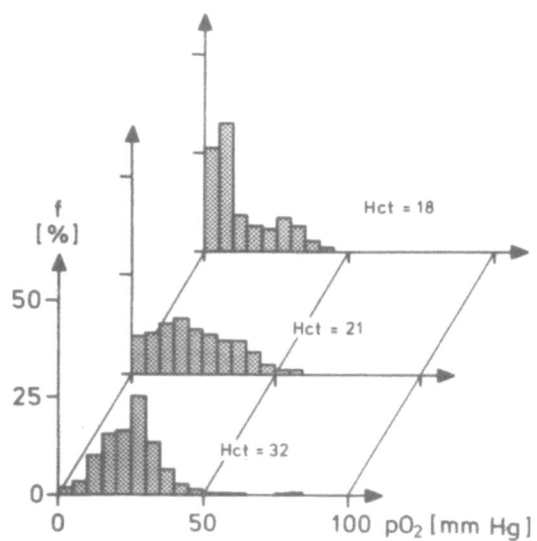

Fig. 1. Summated histograms from all experiments recorded under control conditions and after haemodilution to Hct 21% and Hct 18% (n greater than 448 in all histograms).

RESULTS

Figure 1 presents the frequency histograms from all experiments. A clear left shift is to be observed at each stage of dilution. At Hct

21%, 10% of pO$_2$ values are below 5 mmHg and at Hct 18% there are 28% of values in the lowest class. This implies a considerable degree of tissue anoxia.

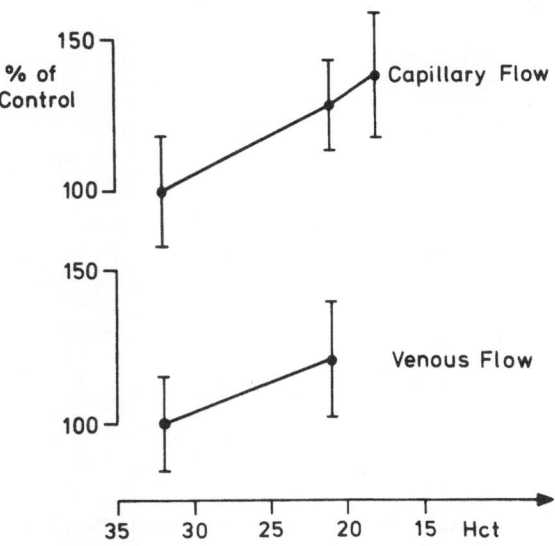

Fig. 2. Percentage changes (+ SEM) in capillary blood flow (above) and regional flow (below), calculated from the intravenous hydrogen clearance curves, as a result of haemodilution.

The results of capillary and regional flow measurements shown in Fig. 2 demonstrate small increases in both values.

The relative change in local and regional oxygen transport capacity was calculated by multiplying the Hb concentration by Hüfner's number and the change in regional or capillary blood flow. The results shown in Fig. 3 indicate that the transport capacity falls as a result of haemodilution to Hct 21% and falls even more steeply locally after the second haemodilution.

DISCUSSION

A surprising feature of these results is that haemodilution to a haematocrit of 21 % induces a pronounced left shift in the pO$_2$ histograms in the vastus medialis muscle. The decrease represents a 66 % reduction of red cells, equivalent to a haemodilution to Hct 30% in dogs or humans. This degree of haemodilution does not produce such an effect in dogs diluted with Dextran 70 (Kessler and Messmer, 1975). Figure 4 shows that near normal histograms can be found even after haemodilution to Hct 14% in the sartorius muscle of dogs with HES 40.

The reason for this apparent paradox becomes clear in the light of the results of local and regional flow measurements. Despite increases

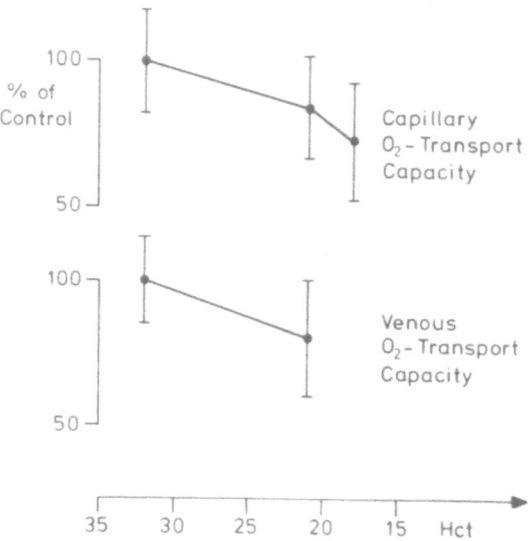

Fig. 3. Above: percentage change in local (capillary) oxygen transport
capacity. Below: percentage change in regional oxygen transport
capacity as a result of haemodilution.

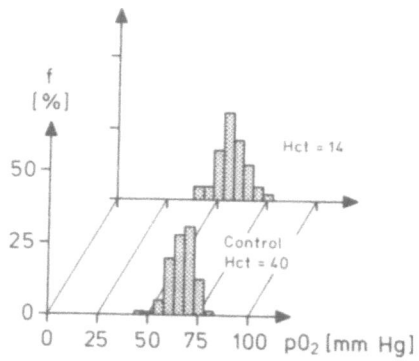

Fig. 4. Tissue pO_2 histograms recorded in the sartorius muscle of the
dog under control conditions and after haemodilution to Hct 14%
with HES 40 (n = 560).

of about 25% in both parameters (Fig. 2), this is insufficient to maintain the reduced oxygen capacity caused by a lower haematocrit (Fig. 3). In contrast, Fig. 5 shows that even at a haematocrit of 14% in the dog, an increase in cardiac output occurs which maintains a normal oxygen transport capacity.

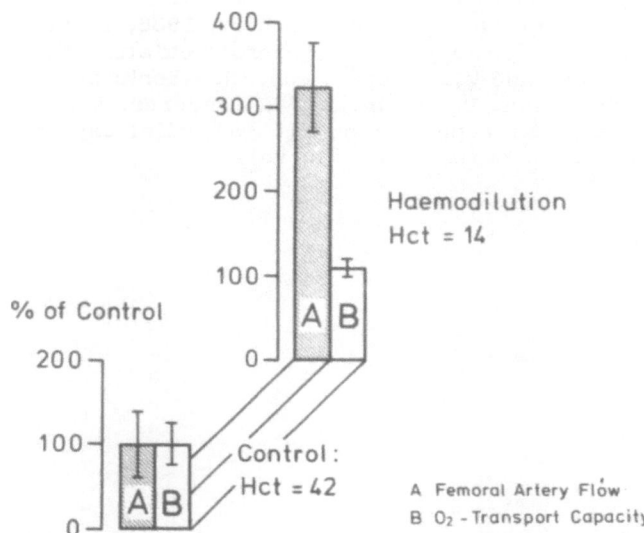

Fig. 5. A: percentage change in femoral artery blood flow in dogs as a result of isovolemic haemodilution to Hct 14% with HES 40. The change in oxygen transport capacity is shown in B.

Our results indicate a fundamental difference between the exercised dog and rabbit models for haemodilution. Caged, sedentary rabbits are unable to increase their cardiac output to compensate for the reduced oxygen capacity. They would therefore appear to be an ideal model for studying the effects of haemodilution in patients who indulge in no physical exercise or those with cardiac insufficiency.

Haemodilution in such patients, far from producing an increased oxygen transport capacity may indeed induce a critical oxygen supply situation in the peripheral circulation. Great care should be taken in such cases and it is essential that monitoring of local supply parameters be carried out during the haemodilution.

ACKNOWLEDGEMENTS

The authors are grateful to Gundi Schuster who was most patient and skillful in preparing the figures. The authors are also grateful to Dr. H. Beisbarth and Pfrimmer & Co., Erlangen for their support of this project.

REFERENCES

Harrison, D.K., Birkenhake, S., Knauf,S., Hagen, N., Beier, I. and
    Kessler, M., 1988, The role of high flow capillary channels in
    the local oxygen supply to skeletal muscle. Adv. Exp. Med.
    Biol., 222:623.

Harrison, D.K., Birkenhake, S., Hagen, N., Knauf,S. and Kessler,
    M., 1989, Regulation of capillary blood flow: A new concept.
    This volume.

Kessler, M. and Messmer, K., 1975, Tissue oxygenation during
    haemodilution. Biblthca Haemat., 41:16.

Kessler, M., Harrison, D.K. and Höper, J., 1986, Tissue oxygen
    measurement techniques. In: "Microcirculatory Technology",
    C.H. Baker and W.L. Nastuk, eds., New York: Academic Press.

Vogel, H., Harrison, D.K., Günther, H., Anderer, W., Kessler, M.,
    and Peter, K., Hemodilution and myocardial oxygen supply. The
    influence of Flousol DA. This volume.

INFLUENCE OF PROSTAGLANDIN E_1 ON MUSCULAR TISSUE-PO_2 IN PATIENTS WITH DIABETIC GANGRENE

Robert Heinrich[1], Hans-Werner Krawzak[2], and Helmut Strosche[2]

[1]Dep. of Medical-Geriatrics and [2]Dep. of Surgery
Ruhr-Universität-Bochum
Marienhospital D-4690 Herne 1
W-Germany

INTRODUCTION

A humid gangrene of the foot is a common complication in patients suffering from diabetes for many years. It is a multicausal disease which combines problems of neuropathy, macro- and especially micro-angiopathy. The surgical activities include vascular reconstruction, sympathectomy (with limited efficacy) and a whole spectrum of local measures. In spite of all therapeutic efforts a major amputation is often unavoidable.

In patients who are not indicated for vascular reconstructive surgery the aim of a conservative treatment is the reduction of local infection and -above all- the improvement of microcirculation. This for instance can be done by hemodilution or application of vasoactive drugs.

Prostaglandin E_1 is a potent short acting vasodilative and antiaggregatory substance. It was first used by CARLSON and ERIKSSON in 1973 in the treatment of patients with severe arterial occlusive disease to make rest pain ease off and to produce a complete cure of the ulcers. Because of its well-known rapid pulmonary inactivation the intra-arterial infusion of PGE_1 is recommended as the most effective way of administration.

Investigations of muscle-pO_2 using polarographic needle probes have only been reported on intravenous PGE_1 application (Ehrly et al., 1987, Krawzak et al., 1987).

It was the aim of our preliminary study to verify the influence of intra-arterial PGE_1 on the tissue pO_2 of the tibialis anterior muscle in patients suffering from diabetic gangrene.

PATIENTS AND METHOD

We measured 9 patients with diabetic gangrene of the foot. 6 of them were men and 3 women, aged from 63 up to 84 years (mean:71.3 yrs). All were suffering from insulin-dependent diabetes for about 19 years on an average. In none of the cases reconstructive vascular surgery was indicated. 4 patients suffered from hypertonus. Furthermore we found 2 cases of myocardial insufficiency, 2 of renal failure and 1 of hepatic cirrhosis. All patients were submitted to angiography. There have been 3 cases with no vascular occlusion and one case with an obstruction of the superficial femoral arteria. 5 other angiograms were showing occlusions of one or two of the lower leg arteries. The mean ankle/brachial pressure index was 0.87 (S.D. 0.32).

For the polarographic measurements of muscle pO_2 we used the "SIGMA-pO_2-Histograph/KIMOC" (Eppendorf Instruments, Hamburg, FRG) with a 300 μm needle probe according to the method developped by FLECKENSTEIN et al. (1984). The probes were inserted through a short flexible vein catheter approximately 10 cm below the medial knee-joint cavity and about 3 cm beside the tibial margin. We measured in a cranial direction at an angle of about 30° to the longitudinal axis of the muscle fibres. The microprocessor controlled probe advance was effected according to the so-called "Pilgerschrittverfahren". Every pO_2-histogram was calculated out of 200 single values.

The investigation of the patients took place shortly after their admission to hospital and before any specific conservative treatment had started. For the prostaglandin administration we used a totally implanted arterial access system (so-called "port-system") (Krawzak and Strosche, 1988). The catheter had been operatively inserted into the superficial femoral arteria two days before.

After a rest-period of 15 minutes blood gas analysis and basic hemodynamic parameters were recorded and the first pO_2-histogram was taken up. Thereafter the patients received Alprostadil (PGE₁-alpha-cyclodextrin) (Fa.Sanol Schwarz,

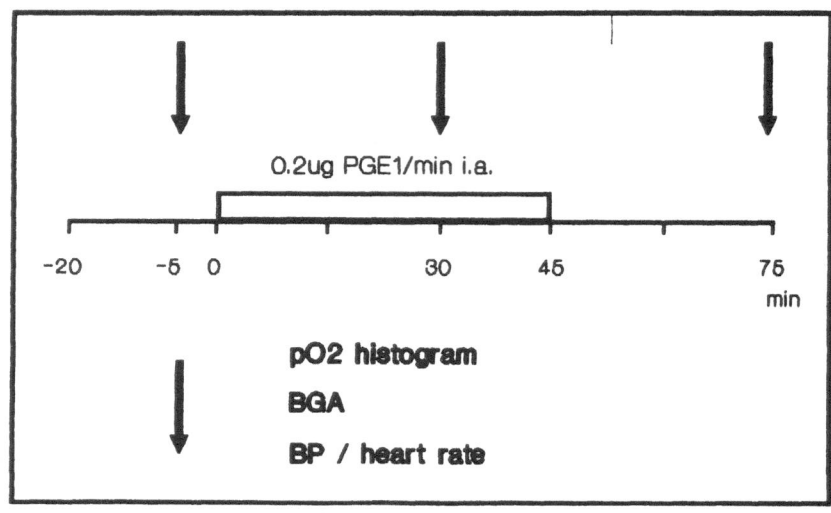

Fig. 1. Schematic presentation of the test procedure

Table 1. Mean pO2 of tibialis anterior muscle during the investigation

TISSUE-pO$_2$ (mmHg)	BEFORE	DURING	AFTER i.a. PGE$_1$
mean	20.3	18.2	19.4
S.D.	± 6.3	± 6.7	± 6.2

Table 2. Arterial blood gas analysis (BGA), blood pressure (BP) and heart rate (HR) before, during and after intra-arterial PGE$_1$-infusion

	BEFORE	DURING	AFTER i.a. PGE$_1$
BGA			
pH	7.448	7.460	7.477
pO$_2$ (mmHg)	69.8	66.7	71.5
pCO$_2$ (mmHg)	37.5	37.2	36.0
SBC (mmol/l)	25.2	25.7	25.9
BP			
syst (mmHg)	138.3	137.2	136.6
diast (mmHg)	70.0	68.3	70.5
HR (min^{-1})	81.3	82.2	80.8

Monheim, FRG) intra-arterially during a period of 45 min. The flow rate was 0.2 µg/min (corresponding to a dosage of approximately 3 ng/kg/min). The second pO$_2$-histogram was taken up 30 minutes after the first one. 30 minutes after the end of the infusion-period the final histogram was recorded (Fig.1). For the purpose of statistic analysis we used the WILCOXON-test.

RESULTS

Regarding the individual courses under PGE$_1$-application an increase of the muscle-pO$_2$ could only be found in 3 out of the 9. In 6 cases the tissue-pO$_2$ of the tibialis anterior muscle decreased. The mean pO$_2$-figure for all patients slowed down from 20.3 mmHg to 18.2 mmHg. 30 minutes after the end of the infusion-period tissue-pO$_2$ again minimally increased to a mean of 19.4 mmHg (Tab.1). None of these differences was significant below the 5% level.

In order to demonstrate changes in the configuration of the histograms single ones were pooled.

Compared to healthy individuals the results show a shift of the histograms to the left side at the beginning of the investigation. During and after PGE₁-infusion there is no change in the configuration of histograms worth mentioning (Fig.3).

Arterial blood gas analyses do not show any significant alteration during and after PGE₁-application. Blood pressure and heart rate too remain almost unaffected (Tab.2).

During the infusion-period 7 of the 9 patients complained of strong tension respectively pain in the treated leg. All of them showed a flush of the infused limb. Pain and flush disappeared shortly after the end of infusion.

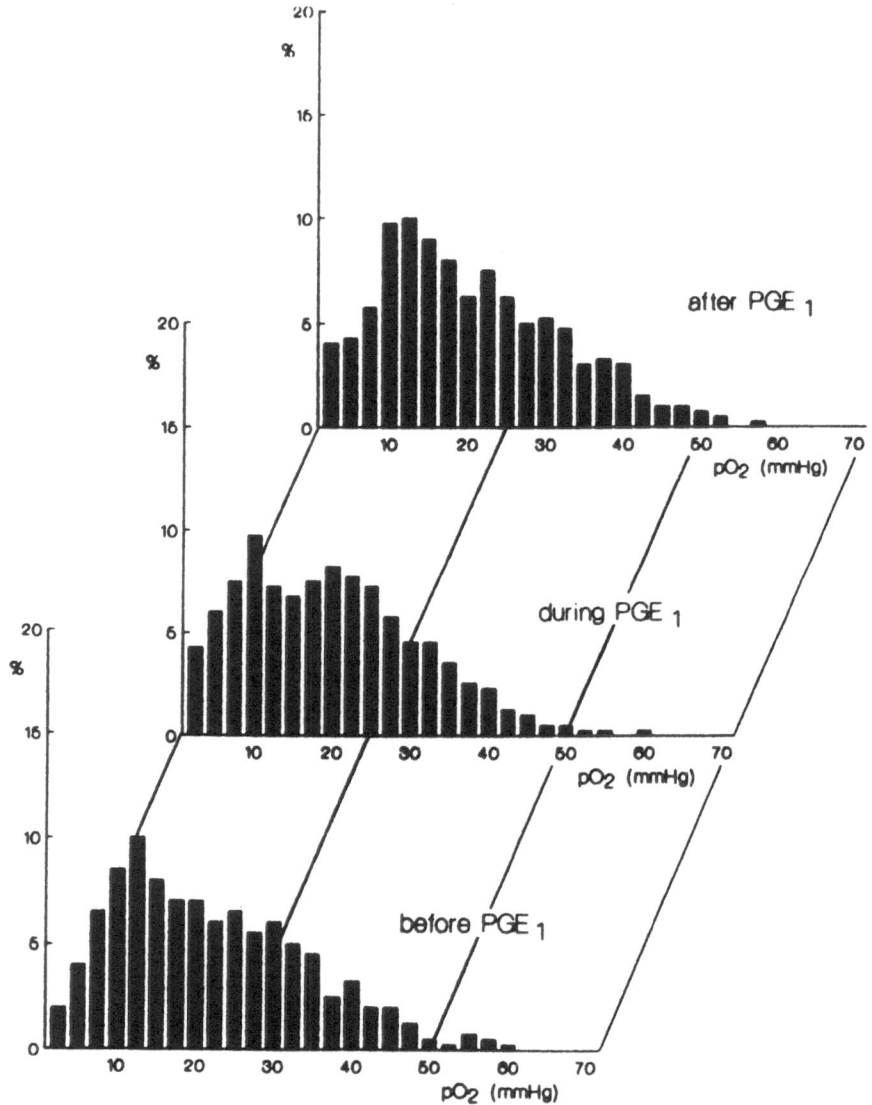

Fig. 2. Pooled histograms of tissue pO₂ of the tibialis anterior muscle of healthy subjects and of diabetic patients before, during and after intra-arterial application of PGE₁

DISCUSSION

Our investigations show that there is -in comparison with healthy subjects- a decrease of the tissue-pO_2 of the tibialis anterior muscle in patients suffering from a diabetic gangrene due to micro- and macroangiopathy.

Several studies show an increase of transcutaneous pO_2 in normal subjects and in patients with POD state II during intra-arterial PGE_1-administration (Creutzig et al., 1984, 1987). In our diabetics such positive effects can not be proved true by means of muscular pO2-measurements. Intra-arterial infusion of 0.2 µg prostaglandin E_1/min does not lead to any significant alteration of the muscle-pO_2 of the lower leg.

As an augmentation of total blood flow to the limb under intra-arterial infusion is known for PGE_1 (Brecht and Ayaz, 1985, Rexroth et al., 1985), the results may indicate a changed distribution of blood in favour of skin perfusion. A macroscopic sign for this might be the flush of the infused leg, which we observed during application.

SUMMARY

Intraarterial infusions of short acting vasoactive drugs have been established in the treatment of peripheral occlusive disease. We studied the effect of prostaglandine E_1 on the tissue pO_2 of the tibialis ant. muscle.

Using a fast responding polarographic pO_2-probe of the hypodermic needle type, tip diameter 300 µm , we made tissue pO_2 investigations in 9 patients with diabetic gangrene. pO_2-measurements were done before, during and after administration of 0.2 µg PGE_1/min i.a..

PGE_1 led to a slight decrease of the mean muscular pO_2 during application followed by a reincrease after the end of infusion.

It is known that intraarterial PGE_1 infusion leads to an increase of transcutaneous pO_2 of the forefoot. Our results show, that there is no increase of tissue pO_2 of the tibialis ant. muscle. The reason might be a changed distribution of blood flow in favour of an increased oxygenation of the skin due to PGE_1 administration.

REFERENCES

Brecht, T., Ayaz, M., 1985, Circulation parameters during intravenous and intra-arterial administration of increasing doses of prostaglandin E_1 in healthy subjects. Klin.Wochenschr., 63:1201

Carlson, L.A., Eriksson, I., 1973, Femoral-artery infusion of prostaglandin E_1 in severe peripheral vascular disease. Lancet, 1:155

Creutzig, A., Lux, M., Alexander, K., 1984, Muscle tissue oxygen pressure fields and transcutaneous oxygen pressure in healthy men during intra-arterial prostaglandin E_1 infusion. Inter.Angio., 3 (Suppl.):105

Creutzig, A., Caspary, L., Ranke, C., Kiessling, D., Wilkens, J., Fröhlich, J., Alexander, K., 1987, Transcutaner pO_2 and Laser Doppler Flux bei steigender Dosierung von intraarteriell und intravenös appliziertem Prostaglandin E_1. VASA, 16:114

Ehrly, A.M., Schenk, J., Saeger-Lorenz, K., 1987, Einfluß
    einer intravenösen Gabe von Prostaglandin E$_1$ auf den
    Muskelgewebesauerstoffdruck, die transkutanen Gasdruck-
    werte und die Fließeigenschaften des Blutes von
    Patienten im Stadium III und IV der chronischen
    arteriellen Verschlußkrankheit. <u>VASA</u>, Suppl. 20:196
Fleckenstein, W., Heinrich, R., Kersting, T., Schomerus, H.,
    Weiss, C., 1984, A new method for the bedside recording
    of tissue pO$_2$-histograms. <u>Verh. Dtsch. Ges. Inn. Med.</u>,
    90:439
Krawzak, H.-W,. Strosche, H., Heinrich, R., 1987, Zum
    Einfluß systemischer PGE$_1$-Infusion auf den Muskel-pO$_2$
    bei Patienten mit arterieller Verschlußkrankheit.
    <u>Klin.Wochenschr.</u>, 65:1004
Krawzak, H.-W., Strosche, H., 1988, Vollständig implan-
    tierbare Kathetersysteme zur arteriellen Extremi-
    täteninfusion. <u>Chirurg</u>, in print
Rexroth, W., Amendt, K., Römmele, U., Stein, U., Wagner, E.,
    Hild, R., 1985, Effekte von Prostaglandin E$_1$ auf
    Hämodynamik und Extremitätenstoffwechsel bei Gesunden
    und Patienten mit arterieller Verschlußkrankheit
    Stadium III-IV. <u>VASA</u>, 14:220

MAXIMAL O_2 UPTAKE LIMITATION IN CONTRACTING SKELETAL MUSCLE DURING CARBON MONOXIDE HYPOXIA.

C.E. King

School of Rehabilitation Therapy, Queen's University

Kingston, Ontario, Canada K7L 3N6

Previous studies have shown that at levels of approximately 20% carboxyhemoglobin (COHb), O_2 uptake (VO_2) during maximal exercise in humans, was preserved at normoxic levels (Ekblom and Huot, 1972; Pirnay et al., 1971; Vogel and Gleser, 1972). However, when severe carbon monoxide hypoxia (COH) (COHb=60%) was superimposed on submaximal contractions of the canine gastrocnemius muscle preparation, VO_2 fell below normoxic levels (King et al., 1987). Maximal O_2 uptake (MVO_2) is limited even at levels as low as 7% COHb (Horvath et al., 1975). At levels of 20% COHb, MVO_2 has been shown to be reduced by approximately 20% in both exercising humans (Ekblom and Huot, 1972; Pirnay et al., 1971; Vogel and Gleser, 1972) and dogs (Horstman et al., 1974). It would appear that the limitation of VO_2 during COH is dependent upon both the magnitude of the O_2 demnd ($\%MVO_2$) and severity of COH. The purpose of the present experiments, therefore, was to define the mechanisms of the VO_2 responses of the canine gastrocnemius muscle at rest, submaximal, and maximal contractions during normoxia and COH.

Mongrel dogs were anesthetized with pentobarbital sodium (30 mg/kg); additional doses were given as necessary. The animals were ventilated to maintain an end-tidal PCO_2 of approximately 35 mmHg. The left gastrocnemius muscle was surgically exposed and the popliteal artery and vein were isolated. The tendon was freed from its insertion, placed in a metal clamp and attached to a forced transducer. A two-channel cannula was placed in the popliteal vein. One channel contained an electromagnetic flow probe for monitoring muscle blood flow while the other channel was used to set an in situ zero and withdraw venous blood samples. A venous reservoir cannula was inserted into the right femoral vein for venous drainage of the muscle preparation. One end of a two channel catheter was placed in the right femoral artery and the other inserted into the popliteal artery supplying the muscle preparation. One channel allowed for autoperfusion of the muscle preparation while the other channel was in-line with a peristaltic pump. In this manner, blood flow to the muscle preparation could be mechanically altered. A pressure transducer was placed in line for measurement of muscle perfusion pressure. The muscle was then set at optimal length. A catheter was placed in the brachial artery for measurement of arterial blood pressure and for withdrawal of arterial blood samples.

Two groups of animals were studied; a normoxic group (n=6) and a COH group (n=6-8). COH was induced by ventilation with 1% CO in air until 60% COHb was achieved. The arterial O_2 concentration was reduced from a normoxic

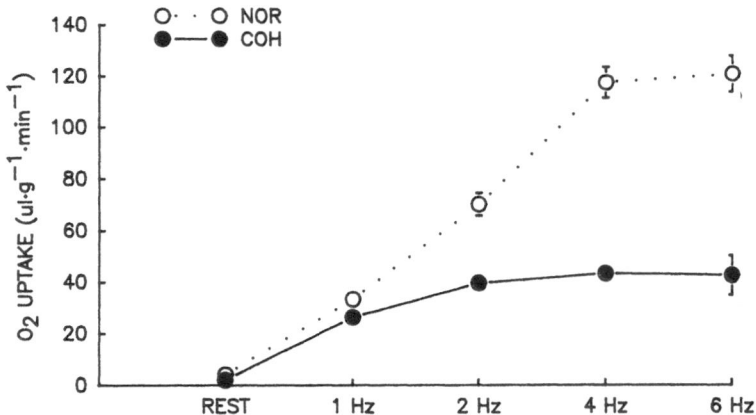

Figure 1.   Muscle O_2 uptake in $ul \cdot g^{-1} \cdot min^{-1}$ (means ± SE) o=normoxia (NOR), o=carbon monoxide hypoxia (COH), Hz=Hertz

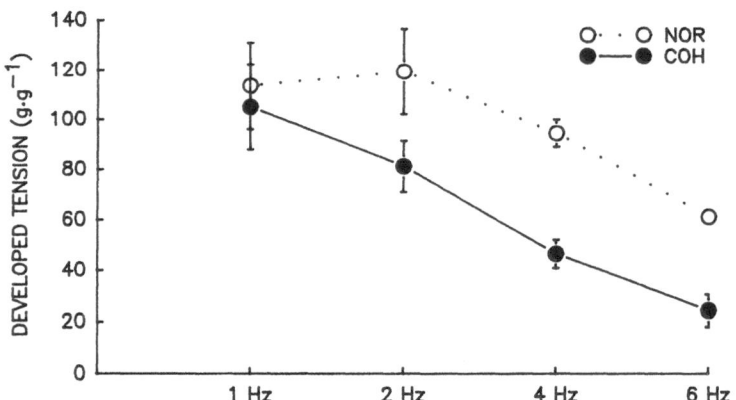

Figure 2.   Developed tension in grams of tension developed per gram of wet muscle weight (means ± SE). See Figure 1 for legend.

value of 19.5 to 8.3 volumes per cent during COH. The animals were then ventilated with 0.1% CO in air to maintain the desired COHb level. The muscle was studied at rest, and then stimulated to contract isometrically (0.5msec duration, 10 volts) for 5 min at rates of 1, 2, 4, and 6 Hertz(Hz). These rates represent approximately 30, 50, 90 and 100% MVO_2 in this muscle preparation. Arterial and muscle venous blood samples were obtained at rest and during the last minute of each contraction period. Mean arterial pressure, muscle perfusion pressure and blood flow were continuously monitored. Each stimulation period was followed by a recovery period of 25 min. All blood samples were analysed for PO_2, PCO_2, and pH using a BMS Mk11 Radiometer blood gas analyser and for O_2 concentration using an Instrumentation Laboratories 282 Co-oximeter. Muscle VO_2 was calculated using the Fick principle.

The values for muscle VO_2 are shown in Figure 1. During normoxia, VO_2 increased from a resting value of 4.2 ± 1.3 (SE) $ul \cdot g^{-1} \cdot min^{-1}$ to a maximal value of 120.5 ± 7.2 (SE) $ul \cdot g^{-1} \cdot min^{-1}$ at 6 Hz. In the COH group, VO_2 achieved a maximal value of only 39.7 ± 3.8 (SE) $ul \cdot g^{-1} \cdot min^{-1}$ at 2 Hz. Muscle VO_2 was significantly lower ($p < 0.05$) in the COH group than in the normoxic group at 2, 4, and 6 Hz.

The values for developed tension are shown in Figure 2. Developed tension averaged 113.4 ± 17.5 (SE) and 104.9 ± 17.2 (SE) $g \cdot g^{-1}$ in the normoxic and COH groups respectively at 1 Hz; these values were not significantly different. Developed tension decreased significantly in both groups at 4 and 6 Hz, but more so during COH so that the values for developed tension at 4 and 6 Hz during COH were significantly less than in the normoxic group.

VO_2 was analysed for the amount of O_2 consumed per gram of developed tension per twitch (Figure 3). At 1 and 2 Hz less O_2 was consumed per gram of developed tension in the COH group. This suggested that some of the ATP for tension development may have been supplied by anaerobic energy sources. At 4 and 6 Hz, however, there was no significant difference in the amount of O_2 per gram of tension developed between the two groups.

The values for blood flow and O_2 delivery are shown in Figure 4 and 5 respectively. Muscle blood flow did not differ between the two groups at rest. Muscle blood flow increased significantly during contractions in both groups but was significantly greater ($p < 0.05$) in the COH group at 1 and 2 Hz. Accordingly O_2 delivery was maintained at normoxic levels at rest and 1 and 2 Hz contractions during COH. Because the further increases in O_2 demand at 4 and 6 Hz were not accompanied by additional increases in blood flow, O_2 delivery was significantly lower ($p < 0.05$) in the COH group than in the normoxic group at 4 and 6 Hz.

The values for O_2 extraction are shown in Figure 6. O_2 extraction ratio is defined as the $(a-v)O_2$ difference divided by the arterial O_2 concentration. O_2 extraction increased from similar values at rest in both groups to 50 and 60% at 1 Hz in the normoxic and COH groups respectively. At 2 Hz, O_2 extraction increased further in the normoxic group to 0.74, but did not change in the COH group. As the contraction frequency was increased, no further increases in O_2 extraction were observed in either group. O_2 extraction in the COH group was significantly less ($p < 0.05$) than in the normoxic group at 2 and 4 Hz.

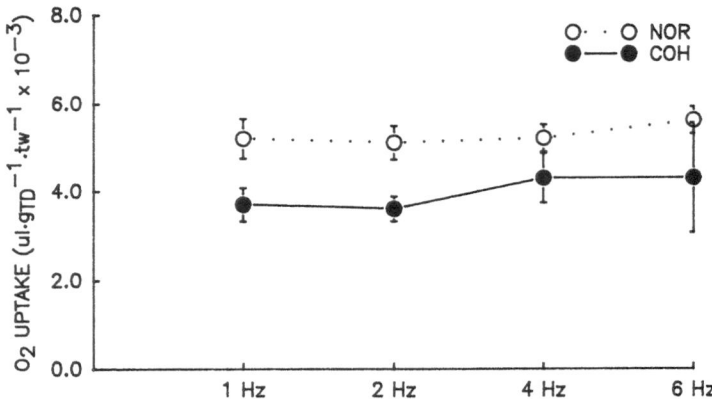

Figure 3. Muscle O_2 uptake in ul of O_2 consumed per gram of developed tension per twitch (tw). (means ± SE). See Figure 1 for legend.

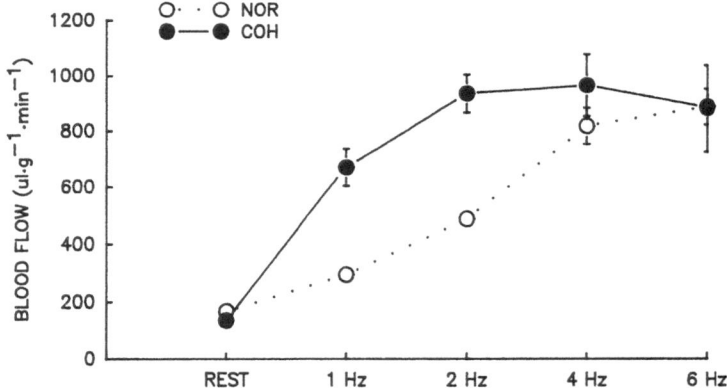

Figure 4. Muscle blood flow in $ul \cdot g^{-1} \cdot min^{-1}$ (means ± SE). See Figure 1 for legend.

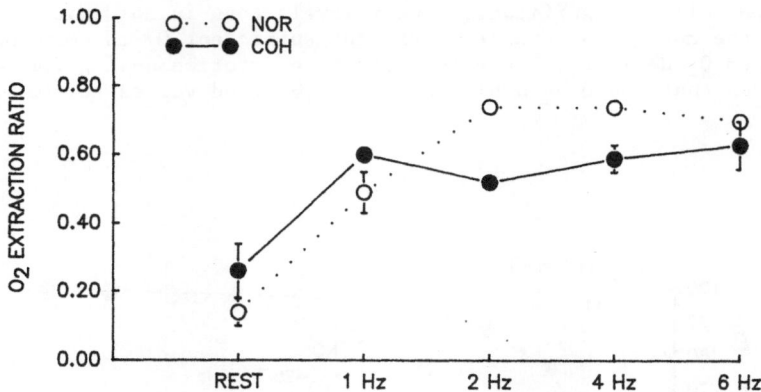

Figure 5. O_2 delivery in $ul \cdot g^{-1} \cdot min^{-1}$ (means ± SE)
See Figure 1 for legend.

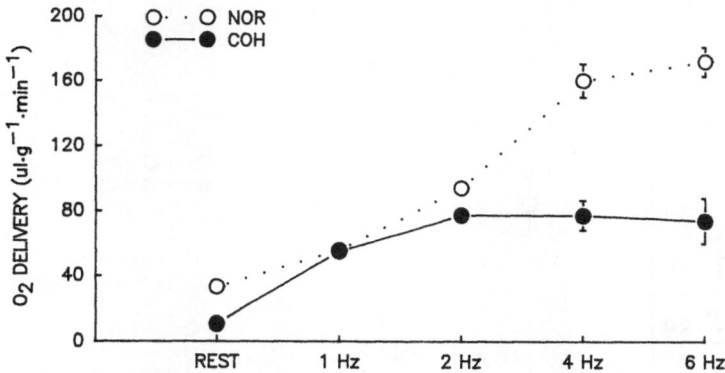

Figure 6. O_2 extraction ratio (means ± SE).
See Figure 1 for legend.

The oxyhemoglobin dissociation curves for the normoxic and COH groups are shown in Figure 7. During COH the P_{50} was reduced from a normoxic value of 28.7 mmHg to 11.5 mmHg. The leftward shift in the oxyhemoglobin dissociation curve may have been responsible for the lower values for O_2 extraction during COH at 2 and 4 Hz because less O_2 can be released from hemoglobin at any given PO_2 when the P_{50} is reduced. The values for muscle venous PO_2 during COH fell to significantly lower levels than in normoxia (Figure 8), however, the muscle was unable to extract sufficient O_2 to compensate for the reduced O_2 delivery. The values for O_2 extraction during COH were maximum values that could be achieved at the observed values for muscle venous PO_2.

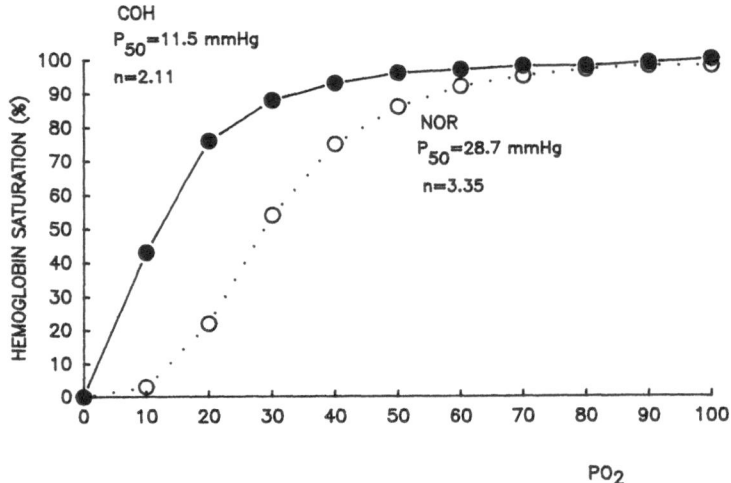

Figure 7.   Oxyhemoglobin dissociation curves for normoxia
and COH. See Figure 1 for legend.

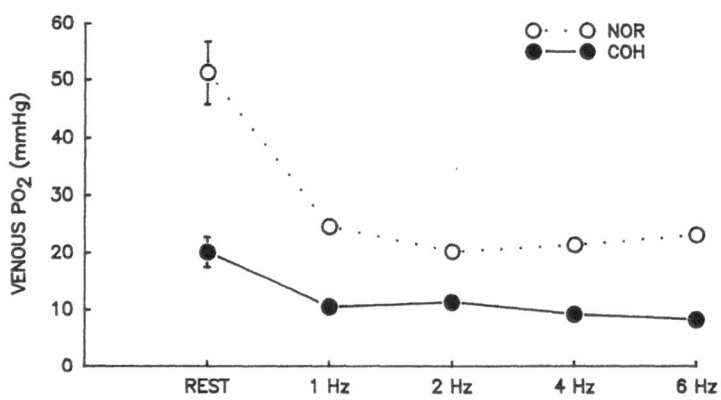

Figure 8.   Muscle venous PO_2 in mmHg. (means ± SE)
See Figure 1 for legend.

Three major findings have resulted from the present study. First, VO_2 in contracting skeletal muscle during severe COH (COHb=60%) is limited at 35% of normoxic VO_2. Second, during submaximal contractions of 30% MVO_2, blood flow and O_2 extraction increased sufficiently to maintain VO_2 at normoxic levels during COH. As O_2 demand was increased further (50% MVO_2), muscle VO_2 was compromised due to an O_2 extraction limitation. Third, at higher levels of O_2 demand (90% MVO_2) both O_2 delivery and O_2 extraction contributed to the O_2 limitation in contracting muscle during severe COH.

Carbon monoxide can interfere with cellular oxygenation in three ways. First, CO can combine with hemoglobin at any reduced sites, thereby decreasing the amount of O_2 carried in the arterial blood. Second, the affinity of hemoglobin for O_2 is increased which means that less O_2 is released from hemoglobin at any given O_2 tension. A third effect is the partial combination of CO with cellular heme proteins, namely myoglobin and cytochrome oxidase. The present study can address the first two effects of CO.

During severe hypoxic hypoxia, in which arterial O_2 concentration was reduced a similar amount as in the present study, values for maximal blood flow in the gastrocnemius muscle at 2 and 4 Hz were 1250 and 1500 $ul \cdot g^{-1} \cdot min^{-1}$ respectively (King et al., 1987). The maximal value for blood flow in this study was only 960 $ul \cdot g^{-1} \cdot min^{-1}$. Because mean arterial pressure is reduced during severe COH (King et al., 1984), muscle perfusion pressure was maintained at normoxic levels in the COH group. One question that arises is whether the blood flow could have increased further during COH if the perfusion pressure had not been maintained at a constant level. In an additional series of COH experiments (unpublished data), adenosine was given intrarterially (30 $ul \cdot kg^{-1} \cdot min^{-1}$) to cause maximal vasodilation in the muscle bed. Blood flow increased approximately 18% above that observed with constant perfusion pressure and no adenosine. The additional increase in blood flow was offset by decreases in O_2 extraction so that no improvement in VO_2 was observed at 2, 4, or 6 Hz. From these data, it would appear that the blood flow responses observed in the present study were the maximum that could occur to best compensate for the reduction in arterial O_2 concentration.

The findings of the present study and the adenosine experiments demonstrate the role of O_2 extraction in the VO_2 limitation in contracting muscle during severe COH. At 2 Hz, O_2 delivery was not different between the normoxic and COH groups but VO_2 was reduced during COH. This O_2 limitation was associated with a significantly lower O_2 extraction in the COH group. The oxyhemoglobin dissociation curve for the COH group is shown in Figure 7. At 2 Hz, muscle venous PO_2 was 11.3 mmHg and the O_2 extraction ratio was 0.53. If one reads the oxyhemoglobin saturation that would remain at this PVO_2 on the COH curve, it is clear that this was the maximal amount of O_2 that could have been released from hemoglobin at that given PO_2. At 4 and 6 Hz, muscle venous PO_2 decreased slightly to 9 and 8 mmHg respectively. These small reductions in PO_2 were sufficient to increase the O_2 extraction ratio to 0.59 and 0.63 respectively.

An VO_2 limitation due to a reduction in P_{50} is consistent with findings in other studies (King et al., 1987; Warley and Gutierrez, 1988). When COH was superimposed on prolonged 1 Hz contractions in the canine gastrocnemius muscle, the resultant VO_2 limitation was associated with the reduced P_{50} during COH (King et al., 1987). When the P_{50} was reduced to approximately 20 mmHg using sodium cyanate, the critical value for O_2 delivery was significantly increased. This indicated that the range of compensation for a reduction in O_2 delivery is severely curtailed when oxyhemoglobin affinity is increased. Further evidence of the O_2 extraction limitation with a reduced

P_{50}, was the inability of the tissue to extract more O_2 when blood flow was increased with the infusion of adenosine. The reciprocal reduction in O_2 extraction with an elevated blood flow could be explained by a decreased transit time coupled with the tighter binding of O_2 to hemoglobin (Gutierrez, 1986).

One last point to be discussed is the relationship of VO_2 to developed tension (TD). Both VO_2 and TD were reduced to a greater extent during COH as compared to normoxia. Figure 3 illustrates the amount of O_2 consumed per gram of developed tension. At 1 and 2 Hz, less O_2 was consumed per gram of developed tension during COH, even though O_2 delivery was maintained at normoxic levels. The most likely explanation for this finding is that some of the ATP for TD was supplied from anaerobic energy sources. At the higher contraction frequencies, the amount of O_2 consumed per gram of TD did not differ between the two groups indicating VO_2 and TD fell proportionately during COH. O_2 delivery was significantly reduced at 4 and 6 Hz which may indicate that at higher levels of O_2 demand, the fall in tension development in contracting muscle is solely a function of O_2 delivery and not the internal milieu of the muscle cell.

In summary, VO_2 in contracting skeletal muscle during severe COH (COHb=60%) is limited at 35% of normoxic MVO_2. At lower levels of O_2 demand (<50% MVO_2), the reduction in VO_2 is primarily associated with an O_2 extraction limitation. At higher levels of O_2 demand, both O_2 extraction and O_2 delivery contribute to the O_2 limitation during severe COH. Finally, at higher levels of O_2 demand, the proportionate fall in VO_2 and tension development suggests that the decline in tension was a function of the reduced O_2 delivery during severe COH.

REFERENCES

Ekblom, B., and Huot, R., 1972, Response of submaximal and maximal exercise at different levels of carboxyhemoglobin. Acta Physiol. Scand. 86: 474-482.

Gutierrez, G., 1986, The rate of O_2 release and its effect on capillary O_2 tension: a mathematical analysis. Resp. Physiol. 63:79-86.

Horstman, D.H., Gleser, M., Wolfe, D., Tryon, T., and Delehunt, J., 1974, Effects of hemoglobin reduction on VO_2max and related hemodynamics in exercising dogs. J. Appl. Physiol. 37:97-102.

Horvath, S.M., Raven, P.B., Dahms, T.E., and Gray, D.J., 1975, Maximal aerobic capacity at different levels of carboxyhemoglobin. J. Appl. Physiol. 38:300-303.

King, C.E., Cain, S.M., and Chapler, C.K., 1984, Whole body and hindlimb cardiovascular responses of the anesthetized dog during CO hypoxia. Can. J. Physiol. Pharm. 62:769-774.

King, C.E., Dodd, S.L., and Cain, S.M., 1987, Muscle O_2 deficit during hypoxia and two levels of O_2 demand. J. Appl. Physiol. 62:1382-1391.

King, C.E., Dodd, S.L., and Cain, S.M., 1987, O_2 delivery to contracting muscle during hypoxic or CO hypoxia. J. Appl. Physiol. 63:726-732.

Pirnay, R., Dujardin, J., Deroanne, R., and Petit, J.M., 1971, Muscular exercise during intoxication of carbon monoxide. J. Appl. Physiol. 31:573-575.

Vogel, J.A., and Gleser, M.A., 1972, Effect of carbon monoxide on O_2 transport during exercise. J. Appl. Physiol. 32:234-239.

Warley, A.R., and Gutierrez, G., 1988, Chronic administration of sodium cyanate decreases O_2 extraction ratio in dogs. J. Appl. Physiol. 63:1584-1590.

DEVELOPMENT OF MUSCULAR TISSUE PO$_2$ AFTER VASCULAR

RECONSTRUCTIVE SURGERY

Hans-Werner Krawzak[1], Robert Heinrich[2], and Helmut Strosche[1]

[1]Dep. of Surgery
[2]Dep. of Medical-Geriatrics
Ruhr-Universität-Bochum
Marienhospital   D-4690 Herne 1
W-Germany

INTRODUCTION

A measure for the success of vascular reconstructive surgery in patients with severe peripheral occlusive disease is the healing of the ulcers respectively the gangrene. In the state II -III according to FONTAINE a good therapeutical result first of all is estimated by an increase of the pain free walking distance. This improvement can take place unspecifically. There is no indispensable relation to the therapeutical procedures (Schoop, 1988). An increase of the ankle/brachial pressure index (doppler-index) (Marshall, 1984) is visible in most of the cases after surgery. But the doppler-index does not tell us anything about the improvement of microcirculation in sceletal muscles. Nevertheless this is the central point of all conservative and surgical efforts.

The measurement of tissue pO$_2$ has shown itself as sensitive to disturbances of microcirculation. It is based on the polarographic analysis of oxygen tissue pressure (Lübbers, 1981). In contrary to the surface electrodes macro- and micro- needle-probes are able to measure inside the tissue without any greater injury of the patient (Ehrly, 1978, Fleckenstein, 1984).

The aim of our preliminary study was to investigate the development of muscular tissue pO$_2$ after successful surgical intervention and furthermore to verify if there is a correlation between the increase of walking distance, doppler-index and muscular oxygenation.

PATIENTS AND METHOD

We measured 11 patients (10 men, 1 woman) -mean age 58.8 years- with peripheral occlusive disease state II - III

Table 1. Clinical data of investigated
         patients

n	11
age	58.8 yrs (41-75)
angiographic occlusion	
pelvis	7
thigh	4
lower leg	0
combined	0
surgery	
bypass	9
aorto-bifemoral	7
iliaco-femoral	1
femoro-popliteal	1
thrombo-endarteriectomy	2

according to FONTAINE who underwent vascular reconstructive
surgery. Among them there were 9 smokers and 4 diabetics.
For the type of angiographic occlusion and kind of surgery
see Tab.1.

The investigations of muscle pO_2, pain free walking
distance (at 120 steps/min), ankle/brachial pressure index
(doppler-index) and arterial blood gas analysis were done
under standardized conditions preoperatively, 2 weeks after
and 6 weeks after surgery.

For the polarographic measurements of muscle pO_2 we
used the "SIGMA-pO_2-Histograph/KIMOC" (Eppendorf Instruments, Hamburg, FRG) with a 300 μm needle probe according
to the method developped by FLECKENSTEIN et al.(1984). The
probes were inserted through a short flexible vein catheter
approximately 10 cm below the medial knee-joint cavity and
about 3 cm beside the tibial margin. We measured in a
cranial direction at an angle of about 30° to the longitudinal axis of the muscle fibres. The microprocessor
controlled probe advance was effected according to the so-
called "Pilgerschrittverfahren". Every pO_2-histogram was
calculated out of 200 single values.

RESULTS

In all cases ultrasonic examinations showed the good
function of the bypass grafts. According to that there was
an increase of the doppler-index shortly after surgery
reaching almost normal values (Fig.2, Tab.2).

Regarding the pooled pO_2 histogram before surgery, the
configuration was leftshifted with a mean muscle pO_2 of

18.0 mmHg. After surgery there was a continuous time dependent increase of the mean pO₂ up to 27.7 mmHg (Fig.2, Tab.2). Finally the pooled histogram showed a bell shaped pO₂ distribution (Fig.1).

During the observation period we found a continuous increase of the pain free walking distance in all patients (Fig.2, Tab.2). Arterial blood gas analyses did not show any remarkable alteration before and after surgery.

Table 2. Mean values and S.D. (n=11) of tissue pO₂, pain free walking distance and doppler-index before and after vascular reconstruction

	before	2 weeks after	6 weeks after surgery
tissue-pO2 (mmHg)	18.0 ±3.9	22.1 ±5.9	27.7 ±5.9
walking distance (m)	127.8 ±90.5	256.7 ±147.7	619.4 ±217.4
doppler-index	0.58 ±0.19	0.94 ±0.33	0.86 ±0.25

DISCUSSION

Our preliminary results show, that vascular reconstructive surgery leads to a prompt improvement of the doppler-index, which is independent from the pain free walking distance. From Pt-micro-needle-probe measurements of tissue pO₂ we know, that the muscle pO₂ in patients with a peripheral occlusive disease is already reduced at rest (Ehrly, 1975). By means of platinum multiwire surface electrodes Sunder-Plassmann (1981) could establish a normalization of pathologic oxygen distribution immediately after vascular surgery. In opposition to that there is a time dependent increase of the mean tissue pO₂ of the tibialis anterior muscle in our investigations. This development of tissue pO₂ coincides with an increasing pain free walking distance.

Repeated muscle pO₂ measurements after vascular reconstructive surgery might be a method to objectify the benefit of our surgical intervention.

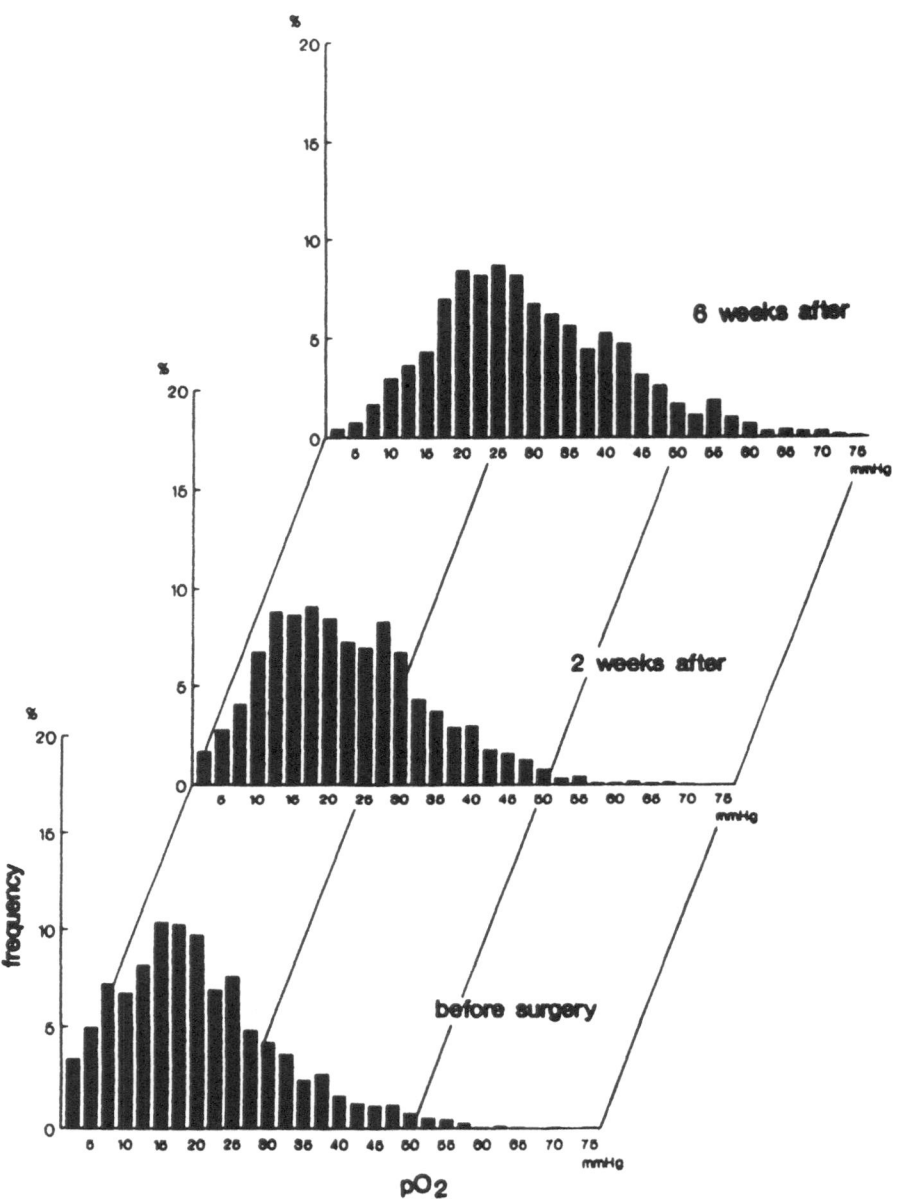

Fig. 1. Pooled tissue pO₂ histograms before and
after vascular reconstructive surgery.

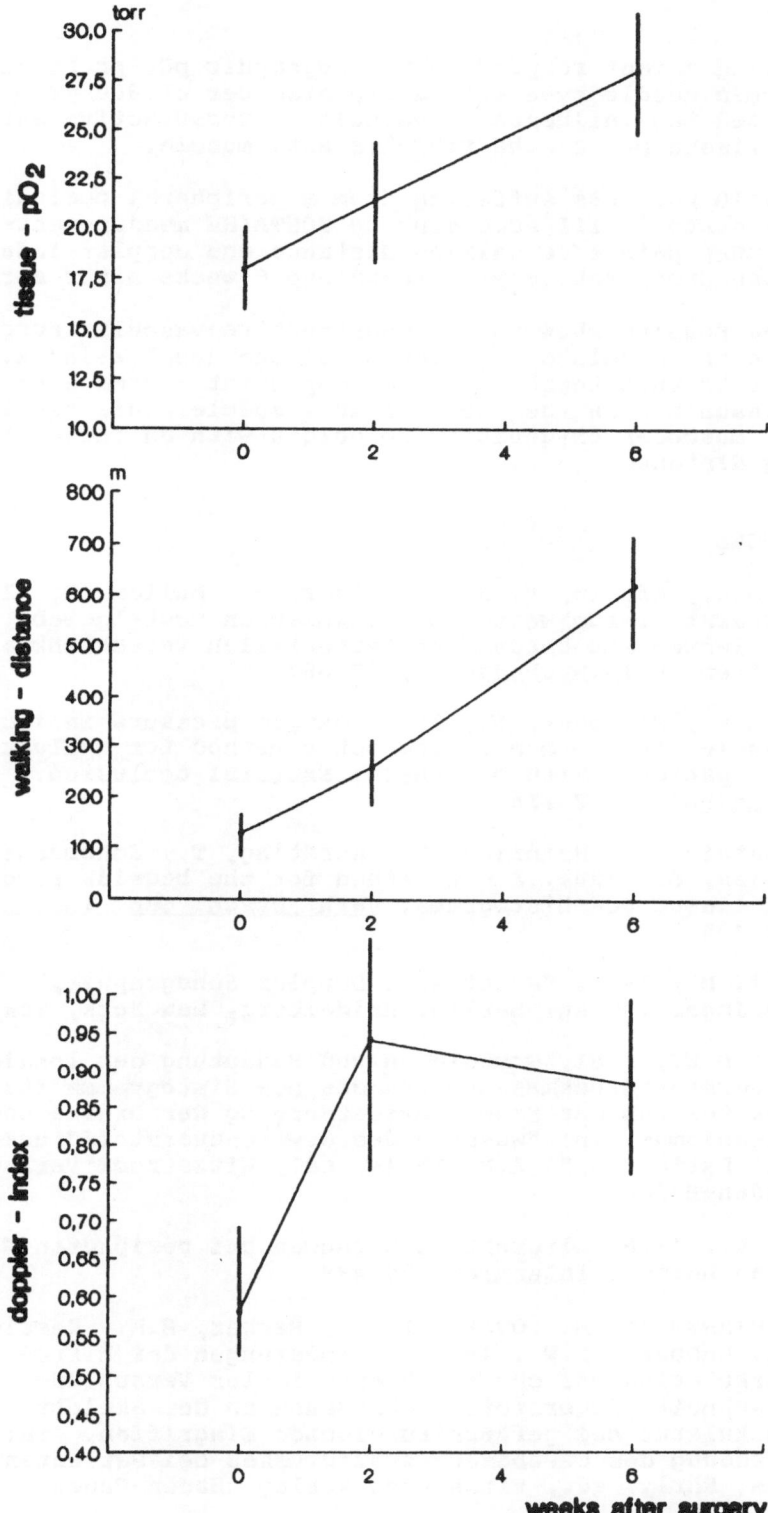

Fig. 2. Tissue pO₂, pain free walking distance
and doppler-index before, 2 weeks and
6 weeks after vascular surgery

SUMMARY

Using a fast responding polarographic pO₂-probe of the
hypodermic needle type with a tip diameter of 300 µm we investigated the influence of vascular reconstructive surgery
on the tissue pO₂ of the tibialis ant. muscle.

In 10 patients suffering from a peripheral occlusive
disease state II-III according to FONTAINE measurements of
tissue pO₂, pain free walking distance and doppler-index
were done preoperatively, 2 weeks and 6 weeks after surgery.

The results show that reconstructive vascular surgery
leads to an immediate improvement of the doppler-index. In
contrast to that there is a time dependent increase of the
mean tissue pO₂ in the tibialis ant. muscle. This development of muscular oxygenation coincides with an increasing
walking distance.

REFERENCES

Ehrly, A.M., Köhler, H.-J., Schröder, W., Müller, R., 1975,
    Sauerstoffdruckwerte im ischämischen Muskelgewebe von
    Patienten mit chronischen arteriellen Verschlußkrank-
    heiten. Klin.Wochenschr., 53:687

Ehrly, A.M., Schröder, W., 1978, Oxygen pressure in ischemic
    muscle tissue: a new diagnostic method for evaluating
    the patients with peripheral arterial occlusion.
    Vasc.Surg., 12:215

Fleckenstein, W., Heinrich, R., Kersting, T., Schomerus, H.,
    Weiss, C., 1984, A new method for the bedside recording
    of tissue pO₂-histograms. Verh. Dtsch. Ges. Inn. Med.,
    90:439

Marshall, M., 1984, "Praktische Doppler-Sonographie,"
    Springer Verlag, Berlin, Heidelberg, New York, Tokio

Lübbers, D.W., 1981, Grundlagen und Bedeutung der lokalen
    Sauerstoffdruckmessung und des pO₂-Histogramms für die
    Beurteilung der Sauerstoffversorgung der Organe und des
    Organismus, in: "Messung des Gewebesauerstoffdruckes
    bei Patienten," A.M. Ehrly, ed., Witzstrock Verlag,
    Baden-Baden

Schoop, W., 1988, Alternative Methoden bei peripheren Gefäß-
    krankheiten. Internist, 29:499

Sunder-Plassmann, L., Overkott, A., Becker, H.M., Kessler,
    M., Lübbers, D.W., 1981, Veränderungen der Mikro-
    zirkulation bei chronisch-arterieller Verschluß-
    krankheit. Sauerstoffdruckmessung an der Skelett-
    muskulatur bei gefäßchirurgischen Eingriffen, in:
    "Messung des Gewebesauerstoffdruckes bei Patienten,"
    A.M. Ehrly, ed., Witzstrock Verlag, Baden-Baden

SPATIAL AND TEMPORAL VARIABILITY OF BLOOD FLOW

IN STIMULATED DOG GASTROCNEMIUS MUSCLE

J. Piiper, C. Marconi, N. Heisler, M. Meyer, H. Weitz,
D.R. Pendergast and P. Cerretelli

Abteilung Physiologie, Max-Planck-Institut für
experimentelle Medizin, Göttingen, F.R.G; Département de
Physiologie, Ecole de Médecine, Université de Genève,
Geneva, Switzerland; Centro Studi di Fisiologia del Lavoro
Muscolare, Consiglio Nazionale delle Ricerche, Milan, Italy;
Department of Physiology, State University of New York at
Buffalo, Buffalo NY, USA

INTRODUCTION

Estimations of muscle blood flow distribution by radioactive microsphere trapping have shown a widely uneven distribution in the isolated-
perfused gastrocnemius of the dog (Piiper et al., 1985) and in leg muscles
(vastus lateralis, gastrocnemius and triceps brachii) of intact dogs
(Pendergast et al., 1985). Moreover, the blood flow heterogeneity, already
large at rest, was found to increase both during artificial stimulation
and natural exercise.

The heterogeneity of blood flow observed in dog muscles could not be
attributed to any peculiar pattern of regional muscle fiber type distribution, as found for instance in the rat (Laughlin and Armstrong, 1983;
Laughlin et al., 1984; Mackie and Terjung, 1983).

The aim of the present study (Marconi et al., in press) was to
investigate whether, besides the described spatial inhomogeneity of blood
flow within the muscle, also a time-dependent (temporal) inhomogeneity may
be present. The latter phenomenon, if proved, could explain the rather
paradoxical finding that muscles characterized by considerable fractions
of strongly underperfused mass may work aerobically even at maximal
stimulation rates as shown by the absence of lactic acid accumulation in
blood (Cerretelli et al., 1984; Piiper et al., 1985).

METHODS

The experiments were performed on 5 gastrocnemius muscles (88 ± 16 g,
mean $\pm$ SD) prepared in dogs (20-25 kg body weight) sedated with 2 mg/kg
morphine subcutaneously and anesthetized with 80 mg/kg chloralose and
250 mg/kg urethan intravenously, artificially ventilated with room air.
Anticoagulation was achieved by administration of heparin (7 mg/kg initially and supplemented thoughout the experiment). The surgery, the set-up
and the techniques of the microsphere method were similar to those

described in more detail previously (Piiper et al., 1968; 1985).

In 3 dogs the isolated-perfused gastrocnemius preparation was used according to the 'standard' procedure whereby the muscle received its blood supply only from one artery and the outflowing blood was drained through one main vein whose collateral branches were carefully ligated. In these preparations total blood flow was assessed by venous outflow. In order to avoid manipulations that could lead to functional changes of blood flow distribution, in 2 muscles a simplified (partial) preparation was used. The main artery was cannulated as in the standard procedure but the venous vessels were left intact. This technique had the advantage of leaving the muscle practically intact, but it did not allow the measurement of overall blood flow. An electromagnetic flowmeter probe was inserted around the muscle artery for monitoring the relative muscle blood flow.

The muscle was loaded by 2 to 3 kg-force, using a spring, and stimulated via the centrally severed nerve. Rhythmic tetani (0.2 s duration) were used with a contraction frequency of 20 to 34 per min. The frequency of the impulses (1 ms duration) was 50 Hz and their voltage above that producing maximum response (supramaximal stimulation). The contractions were practically isotonic since the force increased less than 5% from rest to maximal stimulation. During the period of sequential measurements (up to 10 min) the blood flow and the shortening (external power) were close to constant, varying less than ± 5%.

In both preparations, for injection of microspheres a mixing chamber equipped with a magnetic stirrer was inserted into the femoral artery cannula proximal to the muscle. After 9 minutes of steady stimulation, repeated (4 to 5) injections of about 3×10^5 microspheres, 15 μm (or 10 μm) in diameter, labeled with ^{141}Ce, ^{113}Sn, ^{109}Cd, ^{85}Sr, ^{103}Ru or ^{46}Sc, redispersed in 0.5 ml Ringer solution with 10% dextran by an ultrasonic homogenizer, were made via the mixing chamber into the muscle artery at intervals of two minutes.

At the end of each experiment the muscle was fixed in 25% formaldehyde and cut into 150 to 260 pieces of 0.47 ± 0.05 g (mean ± SD). Each piece was weighed and the radioactivity of each differently labeled microsphere species in each piece (containing on the average 1500 microspheres) was determined by a multichannel gamma counter.

Evaluation of the data

The distribution was quantitatively characterized in terms of coefficients of variation (CV), following King et al. (1985).

(1) In 3 muscles 10, 15 (and 25) μm microspheres were injected simultaneously. The average CV of the ratio of 10/15 μm for the individual muscle pieces, 0.12, was taken as index of the inherent error of the method. (There appeared to be no difference in the average spatial distribution of 10 and 15 μm microspheres).

(2) The spatial distribution of microspheres was quantified as the distribution among the pieces, its CV was designated as CV,M(S).

(3) The temporal variability of microsphere distribution was evaluated from the CV resulting from repeated injections of microspheres in each muscle piece, and averaged for the whole muscle: CV,M(T).

(4) The inhomogeneity of blood flow distribution to the pieces, CV,F, both spatial and temporal, was calculated from CV,M, taking into account the CV

of the experimental error, 0.12 (see above):

$$CV,F(S) = \sqrt{CV,M(S)^2 - 0.12^2}$$

$$CV,F(T) = \sqrt{CV,M(T)^2 - 0.12^2}$$

RESULTS

Examples of patterns of microsphere distribution and of their changes with time are presented in fig. 1. In fig. 2 two examples are shown of the variation as a function of time of specific microsphere distribution to single muscle pieces. In fig. 3 the microsphere distribution to individual pieces is shown for 5 consecutive injections in two muscles. The parameters for spatial and temporal variability obtained in the individual muscles, CV,M(S) and CV,M(T), are listed in table 1.

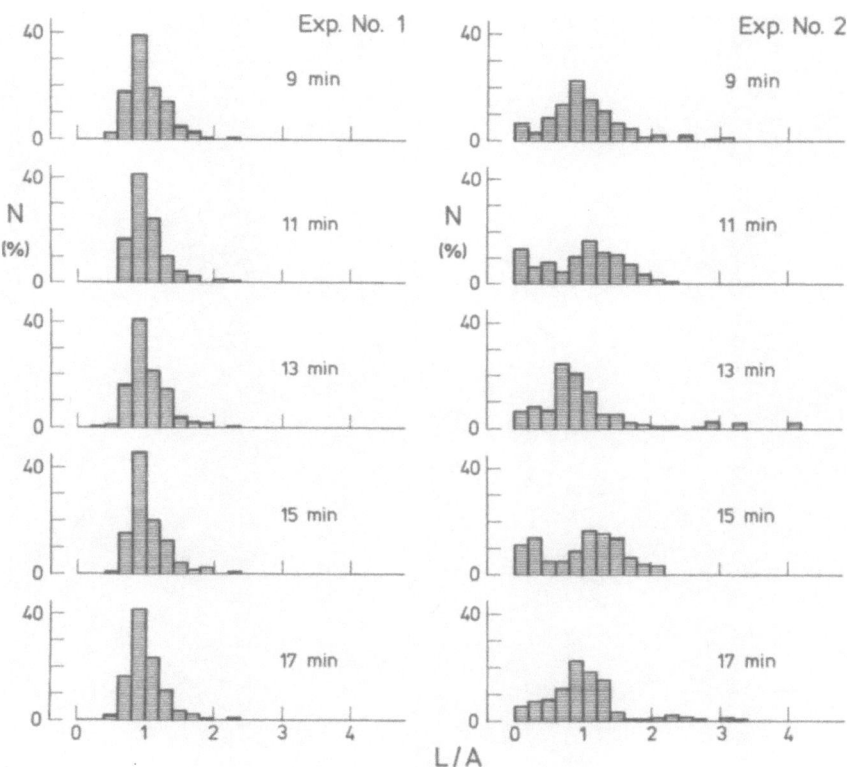

Fig. 1. Spatial distribution of repeatedly injected differently labeled microspheres in two dog gastrocnemius preparations. Abscissa: local/average microsphere number ratio, L/A. Ordinate: frequency distribution of muscle mass, N (cumulative height of columns equal to 100%). Time in minutes, from beginning of stimulation. At 13 min, 10 μm microspheres were injected, at other times, 15 μm microspheres. For additional data, see table 1 (from Marconi et al., in press).

The results may be summarized as follows.

(1) The microspheres were unequally distributed to muscle areas corresponding to the pieces. This is shown for two preparations in fig. 1 and the degree of heterogeneity can be quantified by the CV,M(S) values, averaging 0.45 (table 1). The spatial distribution pattern differs largely between muscles as shown in fig. 1 and by the interindividual scatter in CV,M(S) (CV of CV,M(S) = ± 0.40). The corresponding CV for blood flow distribution (i.e. corrected for methodological scatter) CV,F(S), averages 0.43.

(2) The spatial distribution pattern changes with time, but the interindividual variability is marked (fig. 1).

(3) The variations of microsphere distribution to selected individual pieces are shown for two muscles in fig. 2 and by the CV,M(T) values, averaging 0.23, in table 1. The differences between muscles and between pieces within a muscle are evident from fig. 2 and from the large scatter

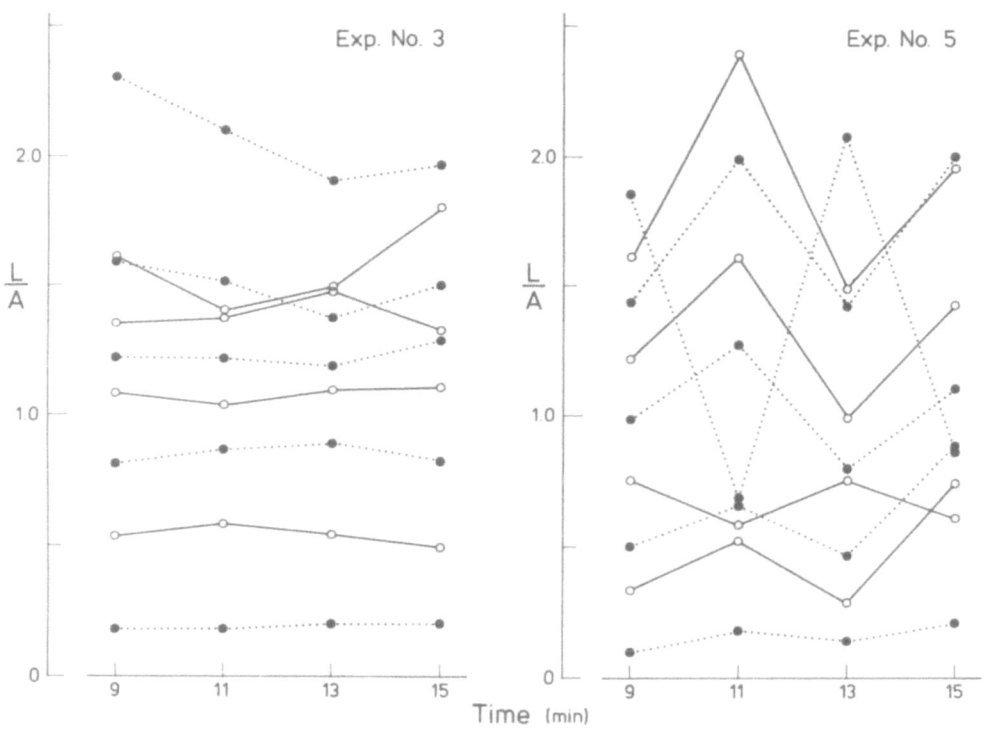

Fig. 2. Variations in time of the distribution of repeatedly injected differently labeled microspheres in two dog gastrocnemius preparations. Abscissa: time from beginning of stimulation. Ordinate: local/average microsphere number ratio (L/A). The 9 pieces were selected to cover the whole L/A range at the first injection. Measurements in the same piece are connected by lines. At 13 min, 10 μm microspheres were injected, at other times, 15 μm microspheres. For additional data, see table 1 (from Marconi et al., in press).

Exp. No. 1

L/A : □ < 0.4 < □ < 0.8 < ▨ < 1.2 < ▧ < 1.6 < ■

Fig. 3a. Pattern of microsphere distribution in Expt.
No. 1. Injections of 5 differently labeled
microspheres during stimulation. Time (in
minutes) from start of stimulation is indicated. Top, proximal insertion of the muscle.
The local/average microsphere concentration
ratio (L/A) is represented by different
shading as explained at the bottom. For additional data, see table 1 (from Marconi et
al., in press).

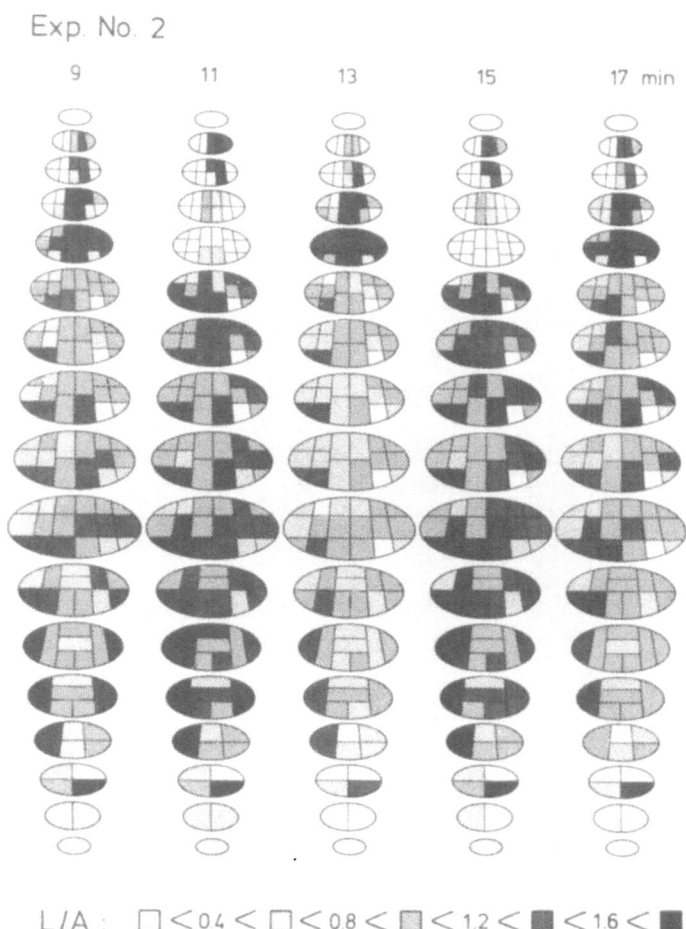

Exp. No. 2

Fig. 3b. Pattern of microsphere distribution in Expt. No. 2. See fig. 3a (from Marconi et al., in press).

Table 1. Spatial and temporal variability of 10-15 μm microsphere distribution in gastrocnemius muscles of 5 dogs. CV, coefficient of variation (= SD/mean). M refers to microspheres; S, to spatial distribution; T, to variation in time; P, to partial preparation; St, to standard preparation.

Exp. No.	1	2	3	4	5	Mean	CV
Preparation	Pa	Pa	St	St	St		
Blood flow [ml/(min·100g)]	20	66	39	39	43	41	
Weight [g]	95	70	112	91	72	88	
No. of pieces	254	164	188	187	152	189	
No. of measurements	5	5	4	4	4		
CV,M(S)	0.26	0.60	0.28	0.48	0.65	0.45	± 0.40
CV,M(T)	0.10	0.28	0.06	0.39	0.34	0.23	± 0.65

of CV,M(T) values, formally yielding ± 0.65 as coefficient of variation (= CV of CV,M(T)). The CV for temporal variations corrected for the methodological scatter, CV,F(T) averages 0.19.

(4) A correlation between the extent of inhomogeneity and the type of preparation (standard or partial) is not evident from the measurements.

DISCUSSION

Validity of procedures

The fundamental assumption of the microsphere embolization method is that the microspheres are distributed in proportion to local blood flow (as prevailing before the injection, i.e. undisturbed). There is little doubt that this is true for whole organs and larger tissue areas, but the limits of the resolving power of the method when small areas are considered have not been adequately established. At asymmetrical branchings of the arterial tree, flow of particles (microspheres, red blood cells) may be more unequally allotted to unequal branches than the total flow (Fung, 1973; Gaehtgens et al., 1979). Such effects may have contributed the more unequal distribution of 25 μm microspheres as compared to 10 μm and 15 μm microspheres. This finding is in accordance with the preferential deposition of larger microspheres (35 μm vs. 15 μm) in areas of high flow, observed in the left ventricle of isolated dog hearts (Yipintsoi et al., 1973). The same tendency has been reported for 15 μm microspheres as compared to a plasma marker with specifically high affinity for endothelium (Little and Bassingthwaighte, 1983; Bassingthwaighte et al., 1987). Thus the blood flow inhomogeneity derived from microsphere deposition may be exaggerated. On the other hand, a further subdivision of the pieces is expected to lead to an increase of inhomogeneity CV.

Use of excessive numbers of microspheres may lead to increased inhomogeneity due to obstruction. It is important to note that the spatial distribution was not broadened after multiple injections. Thus vessel obstruction did not appear to importantly modify the blood flow inhomogeneity derived from microsphere distribution in our experiments.

In preparing the muscle with completely isolated venous outflow it was inevitable to clamp a number of smaller accessory veins. This was expected to produce some perfusion inhomogeneity. However, as shown in table 1, there was no clear-cut difference in the CVM values betwen the 'partial' preparations, with intact vascular supply, and the 'standard' preparations. Thus the contribution of surgical artifacts to the total extent of inhomogeneity must have been small.

The parameters SD and CV have a precisely defined meaning when the quantity varies (or is expected to vary) according to a normal (Gaussian) distribution. The spatial frequency distribution of microspheres (fig. 1) was clearly non-Gaussian, and there was no indication that the temporal variance followed a Gaussian distribution. Thus the use of SD and CV was an expedient to approximately characterize the average width of the frequency distributions.

Local distribution and contractions

The overall spatial frequency distribution of microspheres was similar to that previously described in the same muscle preparation at rest and when stimulated to varied work loads (Piiper et al., 1985). In that study a small increase of the inhomogeneity from lighter to heavier exercise was found. But in those experiments the work load was progressively increased during the course of each experiment. Thus the increasing spatial inequality of microsphere distribution might have been due to the number of injections or to the cumulative number of microspheres injected. In the light of the present experiments these factors appear to be unimportant.

To assess the effects of contractions per se on microsphere distribution, an additional microsphere injection was performed in resting muscles during post-ischemic hyperemia at the time when blood flow had decreased to the mean value prevailing during unoccluded stimulation. The following values were found for the ratio CV,M(S) in hyperemia/CV,M(S) during stimulation: 1.21, 1.56, 0.98; mean ($\pm$ SD) = 1.25 ($\pm$ 0.29). Thus the microsphere distribution inhomogeneity, measured at similar blood flow level, had not decreased after cessation of contractions. This may be interpreted to indicate that the inhomogeneity was not due to locally variable mechanical or metabolic effects of contractions, at least for the pattern chosen for this experiment. Since the O_2 demand during post-ischemic hyperemia certainly was much lower than during stimulation, the arteriovenous O_2 difference must have been much smaller and the venous O_2 content and partial pressure levels much higher. If the local differences in blood flow reflected spatial differences in O_2 demand, they would be expected to disappear or at least to be greatly reduced during post-ischemic hyperemia.

Temporal variations

The variations of local blood flow with time are strikingly different from muscle to muscle and, within a muscle, from piece to piece (fig. 2). In some muscles (3 out of 5) there was indication of periodic flow variations with a period close to 4 minutes (fig. 2, Expt. No. 5). The apparent absence of fluctuations in other experiments (fig. 2, Expt. No. 3) does not mean that they were not present, because one might have missed them due to the slowness of the method. To establish the time course of blood flow fluctuations, evidently methods allowing continuous monitoring are required.

Conclusion

During supramaximal stimulation to rhythmic tetanic contractions, the distribution of embolizing microspheres has shown that the specific local blood flow in the dog gastrocnemius muscle exhibits considerable local and temporal variations whose extent varies within and between muscles.

SUMMARY

The distribution of blood flow in skeletal muscle stimulated to rhythmic isotonic contractions was studied by injections of radioactive microspheres into the arterial supply of gastrocnemius muscles (mean weight 88 g) subsequently cut into 0.5 g pieces for determination of radioactivity. The coefficient of variation (CV = SD/mean) of the ratio of simultaneously injected 10 μm and 15 μm microspheres, 0.12, was taken as the inherent scatter of the method. The average spatial distribution inequality of 10-15 μm microspheres corresponded to a CV of 0.45 and the specific local blood flow inhomogeneity to a CV = 0.43 (= $\sqrt{0.45^2-0.12^2}$), but there were marked differences between muscles. The temporal variability of blood flow in individual muscle pieces was obtained from the comparison of fractional trapping of 4 to 5 differently labeled microspheres injected at intervals of 2 minutes into steadily stimulated muscles. The mean CV for the variations in time was 0.23 and that corrected for methodological scatter, 0.19. There were large differences between muscle pieces within a muscle and between muscles. The presence of considerable spatial and temporal variations of blood flow in exercising muscle during apparent steady state may be important in limiting and/or modulating tissue O_2 supply.

REFERENCES

Bassingthwaighte, J.B., Malone, M.A. Moffett, T.C., King, R.B., Little, S.E., Linke, J.M. and Krohn, K.A., 1987, Validity of microsphere depositions for regional myocardial flows. Am. J. Physiol. 253: H184-H193.

Cerretelli, P., Marconi, C., Pendergast, D., Meyer, M., Heisler, N., and Piiper, J., 1984, Blood flow in exercising muscles by xenon clearance and by microsphere trapping, J. Appl. Physiol. 56: 24-30.

Fung, Y.C., 1973, Stochastic flow in capillary blood vessels, Microvasc. Res. 5: 34-48.

Gaehtgens, P., Pries, A. and Albrecht, K.H., 1979, Model experiments on the effect of bifurcations on capillary blood flow and oxygen transport, Pflügers Arch. 380: 115-120.

King, R.B., Bassingthwaighte, J.B., Hales, J.R.S. and Rowell, L.B., 1985, Stability of heterogeneity of myocardial blood flow in normal awake baboons, Circ. Res. 57: 285-295.

Laughlin, M.H. and Armstrong, R.B., 1983, Rat muscle blood flows as a function of time during prolonged slow treadmill exercise, Am. J. Physiol. 244: H814-H824.

Laughlin, M.H., Mohrman, S.J. and Armstrong, R.B., 1984, Muscular blood flow distribution patterns in the hindlimb of swimming rats, Am. J. Physiol. 246: H398-H403.

Little, S.E. and Bassingthwaighte, J.B., 1983, Plasma-soluble marker for intraorgan regional flows, Am. J. Physiol. 245: H707-H712.

Mackie, B.G. and Terjung, R.L., 1983, Blood flow to different skeletal muscle fiber types during contraction, Am. J. Physiol. 245: H265-H275.

Marconi, C., Heisler, N., Meyer, M., Weitz, H., Pendergast, D.R., Cerretelli, P. and Piiper, J., 1988, Blood flow distribution and its temporal variability in stimulated dog gastrocnemius muscle, <u>Respir. Physiol</u>. (in press).

Pendergast, D.R., Krasney, J.A., Ellis, A., McDonald, B., Marconi, C. and Cerretelli, P., 1985, Cardiac output and muscle blood flow in exercising dogs, <u>Respir. Physiol</u>. 61: 317-326.

Piiper, J., di Prampero, P.E. and Cerretelli, P., 1968, Oxygen debt and high-energy phosphates in gastrocnemius muscle of the dog, <u>Am. J. Physiol</u>. 215: 523-531.

Piiper, J., Pendergast, D.R., Marconi, C., Meyer, M., Heisler, N. and Cerretelli, P., 1985, Blood flow distribution in dog gastrocnemius muscle at rest and during stimulation, <u>J. Appl. Physiol</u>. 48: 2068-2074.

Yipintsoi, T., Dobbs, Jr., W.A., Scanlon, P.D., Knopp, T.J. and Bassingthwaighte, J.B., 1973, Regional distribution of diffusible tracers and carbonized microspheres in the left ventricle of isolated dog hearts, <u>Circ. Res</u>. 33: 573-587.

MUSCLE O$_2$ UPTAKE WHILE PERFUSED AT CONSTANT PRESSURE WITH NORMOXIC BLOOD DURING GLOBAL HYPOXIA

K. Reinhart, D. L. Bredle, C. K. Chapler *, and S. M. Cain

Dept. of Physiology and Biophysics, Univ. of Alabama at
Birmingham, Birmingham, AL, USA, and *Dept. of Physiology,
Queen's Univ., Kingston, Ont., Canada

INTRODUCTION

When the isolated hindlimb muscles of anesthetized dogs were
perfused at constant flow (CF) with normoxic blood while the whole animal
was ventilated with 9% O$_2$ in N$_2$, limb O$_2$ uptake ($\dot{V}O_2$) increased 25%
instead of decreasing with whole body $\dot{V}O_2$ (Bredle et al., 1988a). This
hypoxia-induced increase in limb O$_2$ demand was attributed to the
calorigenic effects of catecholamines liberated at sympathetic nerve
endings and from the adrenal medulla as a result of increased
chemoreceptor activity during hypoxia (Blatteis and Lutherer, 1974; Cain,
1969; Sylvester et al., 1979). The increase in hindlimb $\dot{V}O_2$ was
prevented by β_2-adrenergic receptor blockade (Bredle et al., 1988a) and
was largely dependent on intact innervation to the hindlimb (Bredle et
al., 1988b). Our goal in the present study was two-fold: 1) to ascertain
whether hypoxic vasoconstrictor tone was sufficiently potent to prevent
any increase in limb $\dot{V}O_2$ even if it were perfused with normoxic blood,
and 2) to observe whether any autoregulatory escape would occur if that
were the case. With respect to the second goal, as the ratio of O$_2$
supply to demand decreases, then local vasodilator factors should
accumulate to relieve local vasoconstriction (Cain and Chapler, 1979).
To accomplish these goals, we pump-perfused the limb muscles of
anesthetized dogs with normoxic blood at constant perfusion pressure (CP)
while ventilating them with an hypoxic gas mixture. In this manner,
blood flow varied inversely with the level of sympathetic vasoconstrictor
tone.

METHODS

Six mongrel dogs were anesthetized with pentobarbital sodium (30
mg/kg), paralyzed with succinylcholine chloride, and pump-ventilated to
keep end-tidal PCO$_2$ near 35 Torr. The arterial inflow to and venous
outflow from the left hindlimb muscles were isolated, and the muscles
were pump-perfused with autologous blood via a membrane oxygenator.
Blood flow to the paw was excluded by a tourniquet. Inflow PO$_2$ and PCO$_2$
were maintained near 100 Torr and 35 Torr respectively while the animal
was ventilated with air, 9% O$_2$ in N$_2$, and air again for 40 min periods.
The perfusion pump was controlled by an on-line computer so that flow was
varied to keep perfusion pressure constant at the level observed while
the limb was still autoperfused. The vascular isolation and reactive

hyperemia were verified at the beginning and end of each experiment by briefly stopping and then releasing arterial inflow.

Simultaneous arterial and venous blood samples were drawn for both the whole body and the limb. Blood pH, gas tensions, and O_2 contents were measured with appropriate instruments. Whole body gas exchange was calculated from measurements of expired volume and gas concentrations. Limb $\dot{V}O_2$ was calculated as the product of blood flow and arteriovenous (a-v) O_2 content difference. Limb vascular resistance was calculated by dividing perfusion pressure by the blood flow. The O_2 extraction fraction was calculated as the a-v difference divided by arterial O_2 content.

The results are compared with earlier results (Bredle et al., 1988a) obtained from 8 dogs in which limb blood flow rather than perfusion pressure was kept constant in experiments that were otherwise identical. Data were analyzed between and within groups by ANOVA and Duncans post hoc tests. Significant differences were accepted at p<0.05.

RESULTS

Although perfused with oxygenated blood, limb $\dot{V}O_2$ decreased ~14% in the constant pressure (CP) group by the end of the hypoxic period while

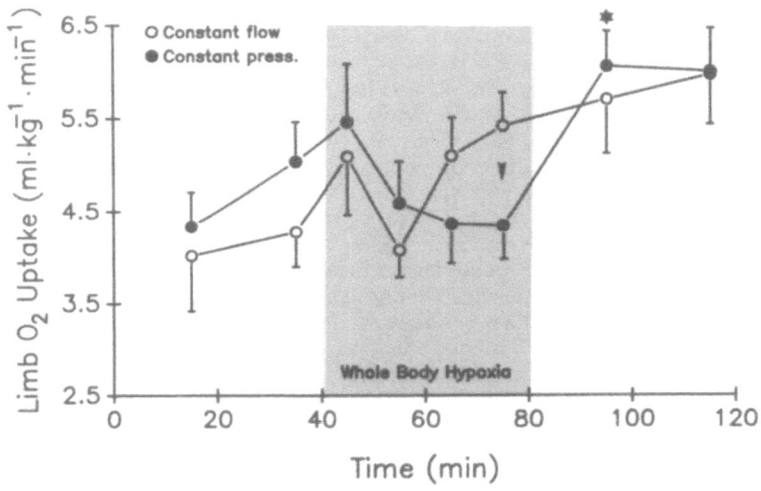

Fig. 1. Hindlimb muscle $\dot{V}O_2$. Values are means ± SE. (✱ sig. diff. from preceding value; sig. diff. between groups)

it had increased ~25% in the CF group (Fig. 1). The difference between the two groups was significant at this point. When the animals were again ventilated with air during the recovery period, there was a sharp and significant rise in limb $\dot{V}O_2$ of the CP group so that it reached the same elevated level as the CF group.

In all cases, limb O_2 delivery (blood flow x arterial O_2 content) was well above any critical level with respect to $\dot{V}O_2$ and never decreased below 14 ml/kg·min on the average. Similarly, average O_2 extraction ratios never exceeded 35% so that there was unquestionably sufficient O_2 reserve left in perfusing blood.

The increase in vasoconstrictor tone reached its peak in both groups after 20 min of hypoxia (Fig. 2). Although vascular resistance was significantly greater at that point in the CF group, the increase from the normoxic level was about the same in both groups, 76% in CF and 70%

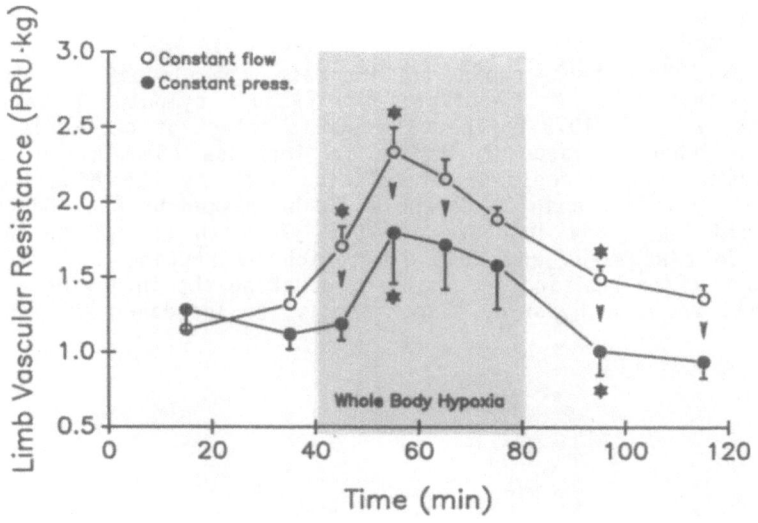

Fig. 2. Hindlimb vascular resistance. (Symbols same as Fig.1).

in CP. No significant change occurred thereafter in hypoxia in either group. Limb resistance fell significantly with the first measurement after the animal was returned to normoxic conditions. The difference between the two groups in the way that they achieved changes in resistanace can be seen in Figures 3 and 4. Whereas perfusion pressure was controlled constant in CP, it increased over 50% during hypoxia in CF. Blood flow, conversely, decreased significantly in CP with whole body hypoxia but not in CF. During recovery, flow increased significantly in CP so that it actually exceeded that in CF.

Limb venous PO_2 decreased significantly in CF but not in CP during whole body hypoxia (Fig. 5).

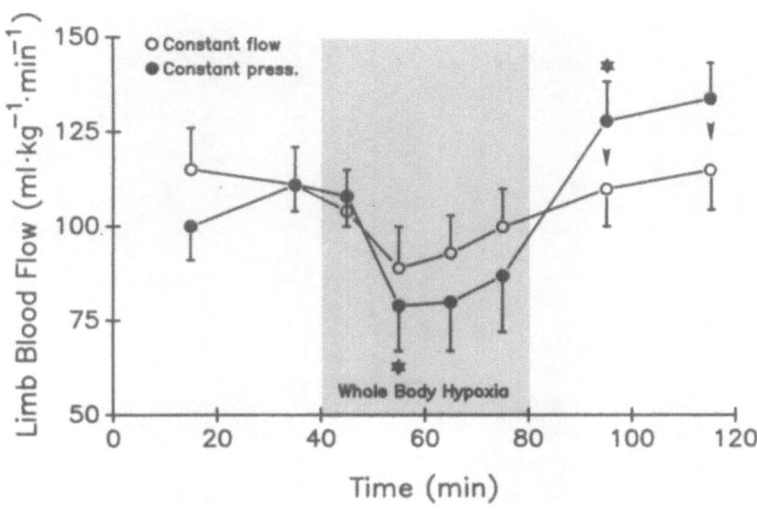

Fig. 3. Hindlimb blood flow. (Symbols same as Fig. 1).

DISCUSSION

Hypoxia has been shown to increase the levels of circulating catecholamines and to activate peripheral sympathetic innervation (Sylvester et al., 1979). The calorigenic effect of catecholamines then causes peripheral tissue O_2 demand to increase (Schmitt et al, 1973; Cain, 1969). In a regional circulation such as the hindlimb muscles however, any increase in O_2 demand would be masked by the reduced supply that would limit any increase in $\dot{V}O_2$. In both CF and CP groups, we thought to circumvent any such limitation by keeping limb O_2 delivery well above usual limiting levels by oxygenating the inflowing blood while the animal was made hypoxic. Surprisingly, an increased $\dot{V}O_2$ was observed

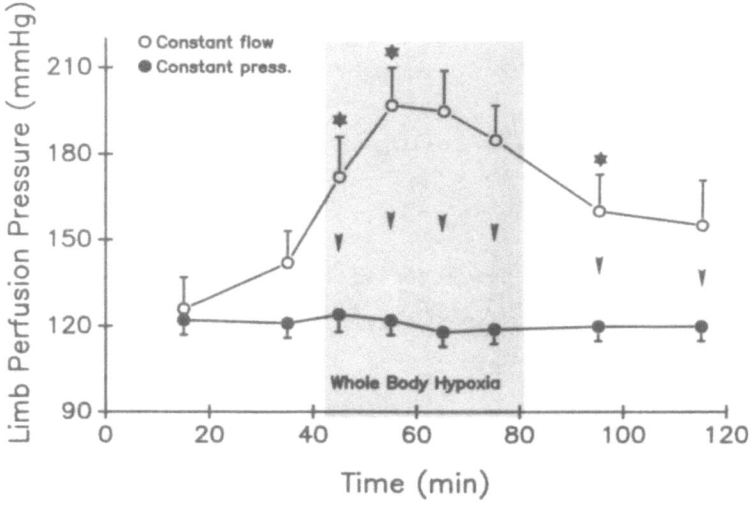

Fig. 4. Hindlimb perfusion pressure. (Symbols same as Fig. 1).

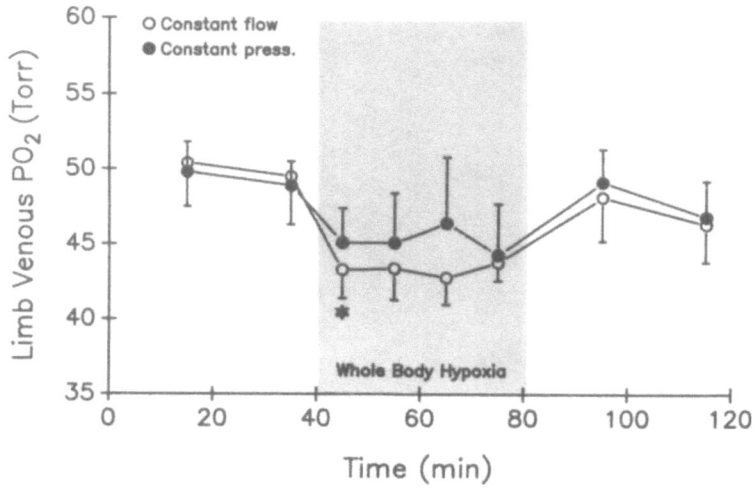

Fig. 5. Hindlimb venous PO_2. (Symbols same as Fig. 1).

only in the CF group whereas the CP group showed a tendency to decrease. We believe that this was a direct result of vasoconstriction acting at the level of exchange vessel control so that O_2 distribution within the muscles was disturbed by increased functional shunting. Whereas this was overcome in CF by forcing flow to stay nearly constant with the perfusion pump, the high venous PO_2 values and low O_2 extraction ratios in CP were evidence that the muscles were unable to satisfy any increase in O_2 demand in spite of adequate oxygen reserve in the perfusing blood. There is no reason to believe that the stimulus to catecholamine calorigenesis was any different between the two groups. The increase in limb $\dot{V}O_2$ of CP immediately upon release of vasoconstrictor tone with return of the animal to normoxic conditions was good evidence that the effect had been present during hypoxia. The exchange vessel controls are thought to be more sensitive than the resistance vessels to metabolic autoregulation (Granger and Shepherd, 1973). As a result, when muscle is confronted by hypoxia, it first increases O_2 extraction and then begins to vasodilate to increase blood flow (Cain and Chapler, 1985). The failure to mobilize more capillaries even to increase O_2 extraction more than minimally, therefore, indicated that preventing catecholamine calorigenesis did not generate metabolic signals to equal those that were generated when the resting O_2 demand of muscle was decreased by hypoxia (Cain and Chapler, 1979; 1985).

This last argument is further supported by the fact that no autoregulatory escape of vascular resistance was seen in CP. Whereas vasoconstriction was overcome by metabolic factors in muscle after 20 min of perfusion with hypoxic blood (Cain and Chapler, 1979), no decrease occurred within 40 min of whole body hypoxia in the present studies. As observed earlier by Costin and Skinner (1970), hypoxic vasoconstriction of skeletal muscle is greatly enhanced if the muscle is perfused with normoxic blood. Although resistance was less in CP than in CF for the first three measurements in hypoxia, the percent increase was not different and the absolute value at the end of the hypoxic period was not different between the two groups. The inability to increase $\dot{V}O_2$ meant that there was a relative tissue hypoxia in CP but this did not contribute to any lessening of sympathetic vasoconstrictor tone in the limb muscles of this group during whole body hypoxia.

We conclude that increased vasoconstrictor tone during whole body hypoxia masked an increase in muscle O_2 demand by interfering with oxygenation of peripheral tissues even though the blood perfusing them was kept normoxic. Because total O_2 delivery was adequate, this interference was at the level of exchange vessel control so that functional shunting of O_2 occurred. Although this might be interpreted as a relative tissue hypoxia, no metabolic autoregulation of tissue oxygenation was evident.

ACKNOWLEDGEMENT

The authors thank W. E. Bradley and J. R. Morgan for invaluable technical support during these experiments. The work was supported by NHLBI Grant Nos. HL 14693, HL 26927, 5T32 HL 0755305 as well as funds from MRC of Canada and Braun Melsungen Stiftung.

REFERENCES

Blatteis, C. M. and Lutherer, L. O., 1974, Reduction by moderate hypoxia of the calorigenic action of catecholamines in dogs. J. Appl. Physiol. 36:337-339.

Bredle, D. L., Chapler, C. K., and Cain, S. M., 1988a, Metabolic and circulatory responses of normoxic skeletal muscle to whole body hypoxia. J. Appl. Physiol. (in press).

Bredle, D. L., Chapler, C. K., and Cain, S. M., 1988b, Central reflex effects of hypoxia on muscle oxygenation. Adv. Exp. Med. Biol. (this volume).

Cain, S. M., 1969, Diminution of lactate rise during hypoxia by PCO_2 and α-adrenergic blockade. Am. J. Physiol. 217:110-116.

Cain, S. M. and Chapler, C. K., 1979, Oxygen extraction by canine hindlimb during hypoxic hypoxia. J. Appl. Physiol. 46:1023-1028.

Cain, S. M. and Chapler, C. K., 1985, Circulatory responses to 2,4-dinitrophenol in dog limb during normoxia and hypoxia. J. Appl. Physiol. 59:698-705.

Costin, J. C. and Skinner, N. S., Jr., 1970, Effects of systemic hypoxia on vascular resistance in dog skeletal muscle. Am. J. Physiol. 218:886-893.

Granger, H. J. and Shepherd, A. P., 1973, Intrinsic microvascular control of tissue oxygen delivery. Microvasc. Res. 5:49-72.

Schmitt, M., Meunier, P., Rochas, A., Chatonnet, J., 1973, Catecholamines and oxygen uptake in dog skeletal muscle in situ. Pflugers Arch. 345:145-158.

Sylvester, J. T., Scharf, S. M., Gilbert, R. D., Fitzgerald, R. S. and Traystman, R. J., 1979, Hypoxic and CO hypoxia in dogs: hemodynamic carotid reflexes, and catecholamines. Am. J. Physiol. 236 (Heart Circ. Physiol. 5):H22-H28.

LOCALIZED HETEROGENEITY OF RED CELL VELOCITY

IN SKELETAL MUSCLE AT REST AND AFTER CONTRACTION

Karel Tyml and Christopher G. Ellis

Department of Medical Biophysics
University of Western Ontario, and
The Robarts Research Institute
London, Ontario, Canada

INTRODUCTION

Heterogeneity of blood flow in the microcirculation is one of the key parameters that determine oxygenation of tissue. The degree of heterogeneity will dictate the degree of mismatch between the convective transport of materials within the microvessels and the diffusive transport from blood to tissue. Although heterogeneity has been recognized as one of the key features of the microcirculation, it has not been measured in sufficient detail (Duling and Damon, 1987).

Recently, we have developed a video-microscopic technique for quantifying spatial heterogeneity of red cell velocity (i.e. variability of velocity among capillaries at one instant of time) in an entire population of capillaries visualized at low magnification (Tyml, 1986). We used the frog sartorius muscle preparation since it offered an important advantage over a mammalian preparation: the much larger size of red cells in frogs allowed using a very low magnification and consequently permitted analysis of large tissue volumes of up to 1 mm³. Using this preparation, we were able to (1) measure the spatial distributions of velocity at rest and after muscle contraction, (2) compute the coefficient variation (CV = SD/mean) from these distributions to provide an index of heterogeneity, and (3) demonstrate for the first time that heterogeneity depends on tissue metabolism (Tyml, 1986).

The analysis of heterogeneity from velocity distributions, however, presents at least two problems. First, it does not show localized changes in heterogeneity such as those present across or along muscle fibres. Second, since distributions at different metabolic states can have different shapes, the computed CV's may not have the same meaning and therefore may not be comparable. The objective of this paper was to address these two problems by providing detailed analysis of velocity in the frog sartorius at resting and after contraction.

METHODS

A sartorius muscle in a mature frog (Rana Pipiens) was prepared for an intravital video-microscopic observation according to a procedure reported previously (Tyml, 1986). The frog was anesthetized with 20% urethane (36 mg/10 g of body weight) and the muscle was exposed by cutting and reflecting the overlying skin. The muscle was covered by glass coverslip, epi-illuminated and viewed under very low magnification (Wild, M420). The total surface area of the muscle visualized on a television monitor was 2.4 × 1.8 mm where a network of capillary segments (i.e. vessels between adjacent branch points) could be identified in a 0.15 mm depth of field. The microcirculation was video-recorded in the resting state and at the peak of hyperemia following 1 min direct electrical stimulation of the muscle (10V amplitude, 6 Hz frequency and 0.1 ms pulse width). Red cell velocity (V_{rbc}) was measured in as many segments as possible (i.e. 67) by the video flying spot technique (Tyml and Ellis, 1982). This technique allows "simultaneous" measurements in segments via repetitive video playbacks.

The red cell velocity in capillary segments found in the surface volume of 2.4 × 1.8 × 0.15 mm was analyzed in 4 ways. First, V_{rbc} distributions at rest and in hyperemia were determined from all capillary segments and the global CV's were computed. Second, velocities were analyzed in 11 optical cross-sections spaced regularly 0.2 mm apart within the above surface volume. Each cross-section was defined as an area of a rectangle whose longer side was a 1.8 mm line drawn across the muscle fibres on the monitor and whose shorter side was the 0.15 mm depth of field. Third, velocities were evaluated in 12 longitudinal muscle strips (strip volume: 2.4 × 0.15 × 0.15 mm) into which the surface volume was divided. Finally, in each cross-section, pairs of capillaries surrounding a muscle fibre were considered. An asymmetry ratio for each pair of velocity measurements of the same direction of flow was determined as the ratio of the smaller to the larger velocity. In this paper, paired t-test was used; $p < 0.05$ was chosen as the level of statistical significance.

RESULTS

Global heterogeneity: Figure 1 shows the resting and hyperemic state distributions. The resting distribution (mean: 0.08 ± 0.06 SD mm/s) was highly skewed; it became more normally shaped during hyperemia (mean: 0.57 ± 0.22 mm/s). In these distributions, CV decreased from 75 to 39% suggesting that, on the scale of the entire visualize surface volume, velocity became more homogeneous after contraction.

Cross-sectional heterogeneity: Figure 2a illustrates the mean velocity in 11 cross sections, based on measurements in 10-32 capillaries per section. The mean resting velocities ranged from 0.04 to 0.12 mm/s. Muscle contraction caused significant increases in mean velocities in all sections (range: 0.45 - 0.76 mm/s). Figure 2b depicts the variability of V_{rbc} in each section. Except for position 1.4, CV decreased in all sections when going from rest to hyperemia, suggesting that V_{rbc} became more homogeneous across muscle fibres. Figures 2c,d summarize the CV data of Fig. 2b. Shown are the CV distributions in both metabolic states. Although the shape of these distributions appears similar, the mean has shifted the left (from 49 to 39%) suggesting a more homogeneous velocity during hyperemia.

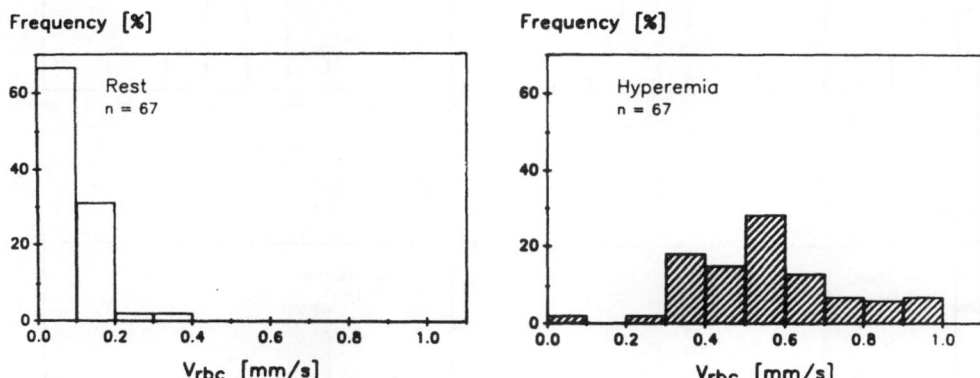

Fig. 1. Spatial distributions of red cell velocity (V_{rbc}) in capillary segments in frog sartorius muscle at rest (left) and at the peak of hyperemia, i.e. 4 min following muscle contraction. V_{rbc} increased significantly from 0.08 ± 0.06 to 0.57 ± 0.22 SD mm/s. The coefficient of variation (CV = SD/mean) decreased from 75 to 39%.

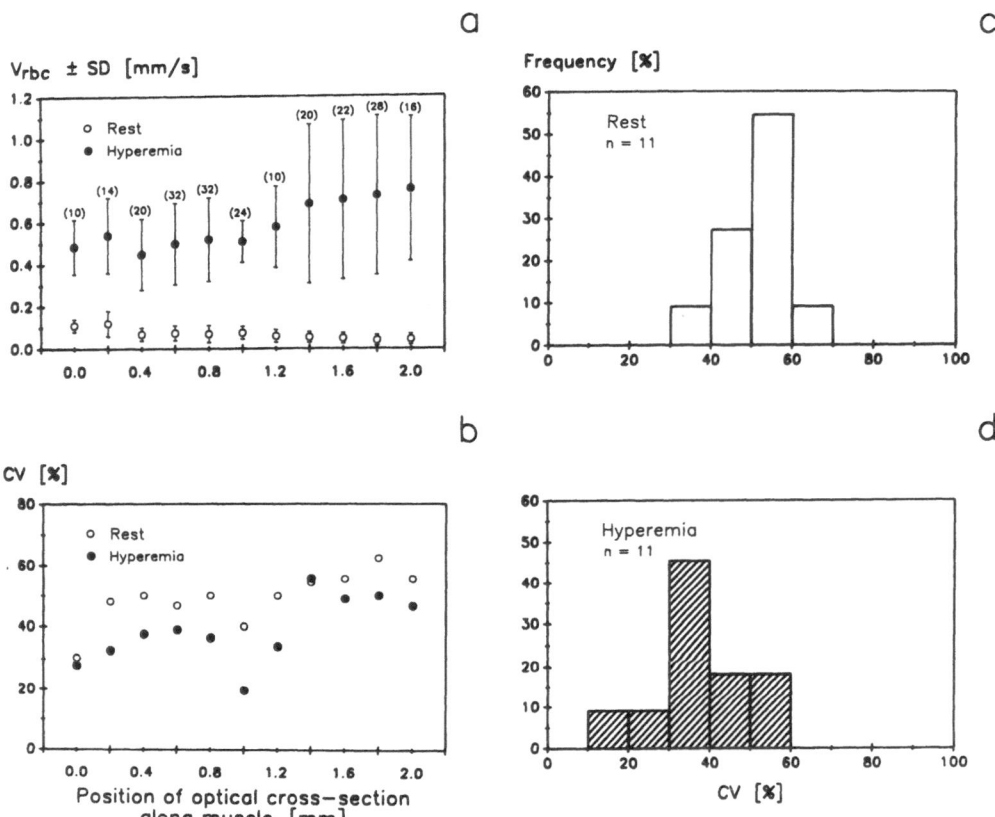

Fig. 2. *Panel a.* Resting and hyperemic velocity in 11 optical cross-sections. In all sections V_{rbc} increased significantly. Numbers in parenthesis show the number of capillaries analyzed per section. *Panel b.* Resting and hyperemic CV's computed from SD and mean values of panel a. *Panels c and d.* Histograms of resting and hyperemic CV's derived from CV's of panel b. The mean value decreased significantly from 49 ± 9 at rest to 39 ± 11 SD % after contraction.

Longitudinal heterogeneity: Figure 3a illustrates the mean velocity in 12 longitudinal strips, based on 4-8 capillary segments per strip. The mean resting velocities ranged from 0.05 to 0.15 mm/s. Contraction increased significantly velocity in all strips (range: 0.44 - 0.70 mm/s). Figure 3b depicts the heterogeneity in velocity for each strip. In the resting state, there was a great scatter among CV values: for positions 0.9 and 1.65, CV was close to 20%, whereas for positions 0 and 0.75 it was nearly 70%. This scatter is also shown in Fig. 3c derived from Fig. 3b. Following muscle contraction, there was an appreciable redistribution of CV values: CV became more comparable among strips and was reduced in all but one strip (position 0.3). The overall CV was reduced from 58 to 34%.

Asymmetry ratio: Figure 4 illustrates mean asymmetry ratios for capillary pairs in 11 cross-sections. The figure shows that, except for position 1.4, velocities around a muscle fibre became more homogeneous. Figure 5 summarizes data from Fig. 4. Based on all 113 pair measurements, the distribution of asymmetry ratio following contraction shifted appreciably to the right (increase in mean: 0.59 ± 0.24 to 0.74 ± 0.24 SD), indicating an increased homogeneity of velocity on capillary-to-capillary basis.

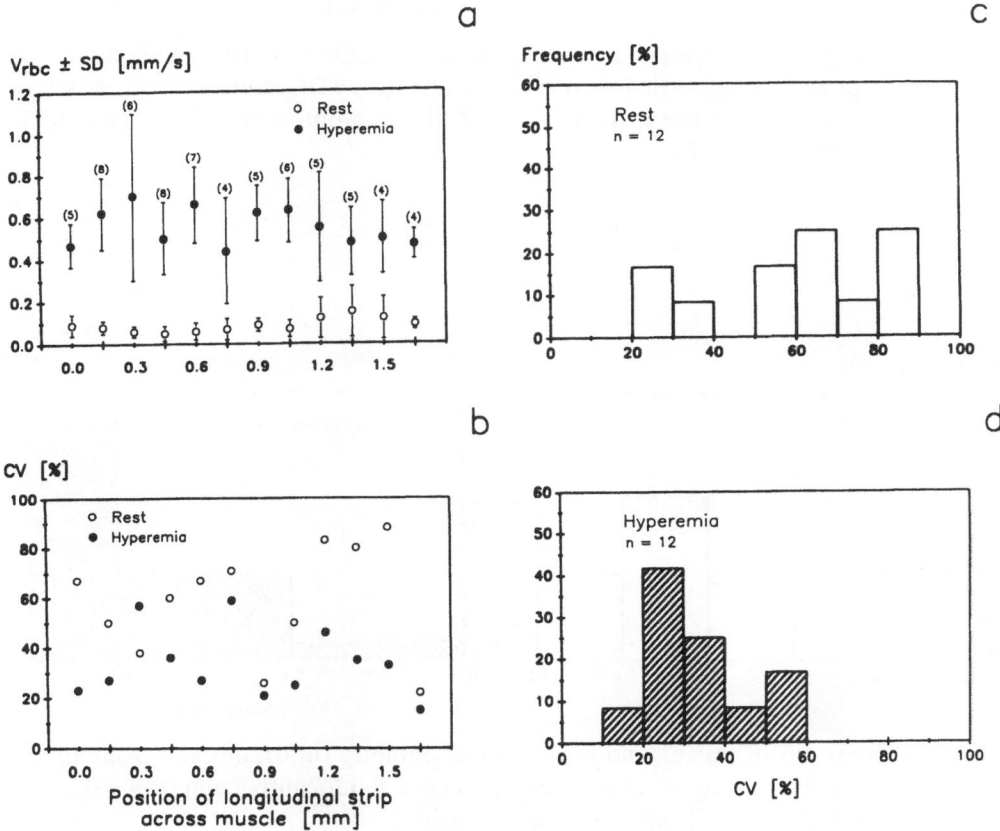

Fig. 3. *Panel a.* Resting and hyperemic velocities in 12 longitudinal strips. In all strips V_{rbc} increased significantly. Numbers in parenthesis show the number of capillaries analyzed per strip. *Panel b.* Resting and hyperemic CV's computed from SD and mean values in panel a. *Panels c and d.* Histograms of resting and hyperemic CV's derived from panel b. The mean decreased significantly from 58 ± 22 at rest to 34 ± 14 SD % during hyperemia.

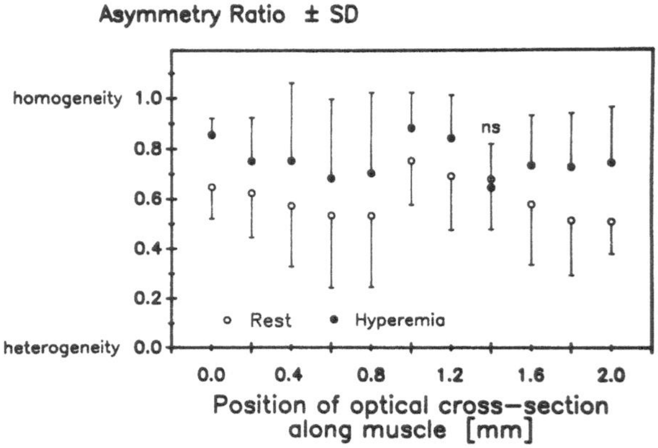

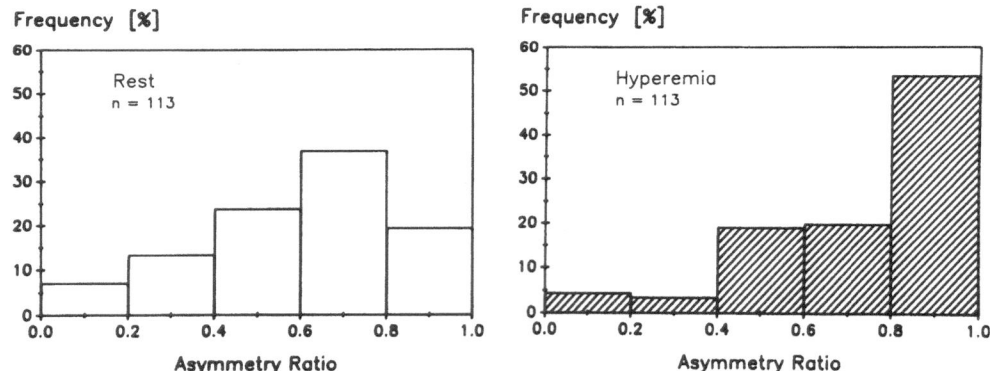

Fig. 4. Resting and hyperemic asymmetry ratio in capillary pairs. In all sections, except at 1.4 mm, ratios increased significantly after contraction. At each position, the number of pairs was half the number of capillaries shown in parenthesis of Fig. 2a.

Fig. 5. Histograms of resting and hyperemic asymmetry ratios, derived from all 11 cross-sections (Fig. 4). The mean increased significantly from 0.59 ± 0.24 at rest to 0.74 ± 0.24 SD after muscle contraction.

DISCUSSION

Referring to Fig. 1, using CV as the conventional index of heterogeneity demonstrated that the global heterogeneity of velocity decreased with muscle contraction. However, this type of index provides only limited information about the flow behaviour in the network. For example, it does not distinguish between the origin of heterogeneity: i.e. within or among microvascular units (Damon and Duling, 1986). The organization of the microvascular network in the frog sartorius is similar to that of the mammalian muscle: microvascular units of 10-20 capillaries are supplied by common terminal arterioles which, in turn, branch off from transverse arterioles that run obliquely or across the muscle fibres (Lund et al., 1987; Koller et al., 1987; Vrielink et al., 1987). The following heterogeneity analysis provides further information on the organizational aspects of the network with respect to the measured V_{rbc} heterogeneity.

Assuming that the supramaximal stimulation resulted in a near-maximal arteriolar dilatation, the hyperemic values of CV (Figs. 1, 2b,d, 3b,d, 4 and 5) reflect the variability in geometry (i.e. length, diameter) of the sartorius microvasculature. This anatomical variability sets the lower limit for the spatial heterogeneity of V_{rbc}. Departures from this limit in the resting state could be mediated by mechanisms operating either within or among microvascular units (i.e. causing intra or inter unit heterogeneity).

The majority of capillary pairs shared common terminal arterioles. Therefore, the asymmetry ratio shown in Fig. 4 is an index of intra-unit heterogeneity. Figures 4 and 5 illustrate that the intra-unit heterogeneity was larger at rest than after contraction. Recently, we have demonstrated that this type of heterogeneity is flow-dependent (Tyml and Mikulash, 1988). Since this dependency most likely reflects hemodynamic events in a single capillary (e.g. passage of red and white cells), we propose that the changes in asymmetry ratio of Figs. 4,5 were linked to differences in blood rheology when comparing rest and hyperemia.

Data of Fig. 2 are based on cross-sectional analysis of 3-4 microvascular units per section. A section-by-section comparison of hyperemic and resting ratios in Fig. 4 reveals a fairly constant departure of resting ratios from hyperemic ratios. This constancy is also apparent in Fig. 2b, although not to the same extent as in Fig. 4. We interpret this to indicate that the change in V_{rbc} heterogeneity in cross-sections was mainly due to the changes in intra-unit heterogeneity, and only to a minor extent to changes in inter-unit heterogeneity (e.g. that caused by diameter variability between terminal arterioles that feed these units).

Figure 3 is based on longitudinal analysis of 3-4 microvascular units. In contrast to the cross-sectional analysis, these units together spanned a long distance (i.e. 2.4 mm) and were therefore affected (in addition to different terminal arterioles) by different transverse arterioles. Referring to Figs. 2d and 3d, the hyperemic values of CV were comparable across and along muscle fibres. However, in contrast to the relative constancy of departures of resting CV's from hyperemic CV's in cross-sections (Fig. 2b), departures along fibres varied greatly (Fig. 3b). This feature is also apparent in Figs. 3c,d, where the spread of CV's was reduced when going from rest to hyperemia. Clearly, the heterogeneity behaviour of microvascular units along fibres was much different from that across fibres. Since inter-unit changes were small when mostly affected by terminal arterioles (see above paragraph), we interpret the large inter-unit changes of Fig. 3 to be due to the variability of diameters of transverse arterioles.

All three types of analysis (asymmetry, cross-sectional, longitudinal) showed the same trend of increased homogeneity after contraction. It appears, therefore, that the increase in homogeneity derived from the global CV (Fig. 1) is meaningful, despite the problem of unequal shapes of velocity distributions. We believe, however, that the three analysis provided additional information regarding mechanism of heterogeneity. For example, analysis of Figs. 2-5 have allowed us to distinguish between the sources of heterogeneity (i.e. rheology vs. vasculature), including the relative contribution of each. The significance of the above analysis may lie in the potential of yielding information of clinical relevance. Clearly, in cases of large perfusion heterogeneity under pathological conditions, therapeutic strategies aiming at an improved perfusion would be greatly helped by the knowledge of either the rheological or vascular component involved.

SUMMARY

Using intravital video microscopy, the present study focussed on a detailed analysis of V_{rbc} heterogeneity in a $2.4 \times 1.8 \times 0.15$ surface volume of a frog sartorius muscle, before and after supramaximal contraction. Heterogeneity of V_{rbc} was evaluated (1) for an entire population of capillaries seen in this volume, (2) for a series of optical cross-sections, (3) along a series of longitudinal muscle strips, and (4) in terms of an asymmetry ratio for pairs of concurrent capillaries surrounding a muscle fibre. All four types of analysis showed an increased V_{rbc} homogeneity after contraction. Velocities became more homogeneous along rather than across muscle fibres. The mean asymmetry ratio became significantly larger during post-contraction hyperemia suggesting that each fibre receives a more uniform blood supply that will contribute to an improved exchange of materials across the capillary wall. The analysis of localized V_{rbc} heterogeneity provides new means of pinpointing the sources of perfusion heterogeneity. It enables, therefore, a specific experimental intervention that is aimed at an improved perfusion under both normal and abnormal conditions.

ACKNOWLEDGEMENTS

We thank Ms. Sarah Wells and Nancy White for excellent technical and secretarial work. This study was supported by the Medical Research Council of Canada and by the Heart and Stroke Foundation of Ontario.

REFERENCES

Damon, D.H., and Duling, B.R. (1986). Distribution of red cells within and among capillary units of striated muscle. Fed. Proc. 45(4): 1157 (Abstract).

Duling, B.R., and Damon, D.H. (1987). An examination of the measurement of flow heterogeneity in striated muscle. Circ. Res. 60(1): 1-13.

Koller, A., Dawant, B., Lin, A., Popel, A.S., and Johnson, P.C. (1987). Quantitative analysis of arteriolar network architecture in cat sartorius muscle. Am. J. Physiol. 253: H154-164.

Lund, N., Damon, D.H., Damon, D.N., and Duling, B.R. (1987). Capillary grouping in hamster tibialis anterior muscles: flow patterns, and physiological significance. Int. J. Microcirc.: Clin. Exp. 5: 359-372.

Tyml, K. (1986). Capillary recruitment and heterogeneity of microvascular flow in skeletal muscle before and after contraction. Microvasc. Res. 32: 84-98.

Tyml, K., and Ellis, C.G. (1982). Evaluation of the flying spot technique as a television method for measuring red cell velocity in microvessels. Int. J. Microcirc.: Clin. Exp. 1: 145-155.

Tyml, K., and Mikulash, K. (1988). Evidence for increased perfusion heterogeneity in skeletal muscle during reduced flow. Microvasc. Res. 35: 316-324.

Vrielink, H.H.E.O., Slaaf, D.W., Tangelder, G.J., and Reneman, R.S. (1987). Does capillary recruitment exist in young rabbit skeletal muscle? Int. J. Microcirc.: Clin. Exp. 6: 321-332.

OTHER ORGANS AND TISSUES

REVERSAL OF CARBON MONOXIDE-CYTOCHROME C OXIDASE BINDING BY

HYPERBARIC OXYGEN IN VIVO

S. D. Brown and C. A. Piantadosi

Department of Medicine and F. G. Hall Hyper- and Hypobaric Center

Duke University Medical Center

Durham, North Carolina 27710

INTRODUCTION

The classic mechanism of CO toxicity is tight binding of CO to hemoglobin which decreases the O_2 carrying capacity of arterial blood and produces tissue hypoxia (1). In vitro effects, such as the shift of the oxygen hemoglobin dissociation curve to the left and CO binding to metallo-enzymes including cytochrome c oxidase are recognized but of uncertain physiological significance (2,3). Not all of the toxic effects of CO however, can be explained on the basis of the classic mechanism (4,5). Observations of CO toxicity at low HbCO may be the result of an effect of CO at the tissue level. Tissue uptake of CO in vivo has been demonstrated for myoglobin (4,5), but binding of CO to cytochrome c oxidase has not been confirmed previously in vivo. The purposes of this study were to document in vivo oxidation-reduction (redox) responses of brain cytochrome c oxidase to CO hypoxia, to identify formation and reversal of the CO-cytochrome a,a_3 ligand and to correlate those findings with the cardiovascular responses, measured HbCO, and arterial pH and blood gas values in a rat model of CO poisoning.

MATERIALS AND METHODS

The experiment was comprised of three parts. First, general physiologic and brain cytochrome a,a_3 redox responses to 0.5 or 1% CO exposures and their responses to hyperbaric oxygen were measured in vivo. Second, HbCO formation in vivo was assessed spectrophotometrically and correlated with HbCO saturations measured by CO-oximetry in a series of graded CO exposures. Third, in vivo spectroscopic scans were made to determine whether CO-mediated cytochrome a,a_3 redox responses were attributable to formation of the cytochrome a,a_3-CO complex, to a hypoxia-mediated increase in the reduction state of cytochrome a or both under conditions specified by the first portion of the experiment.

Animal Preparation and Monitoring

Adult male Sprague Dawley rats (Charles Rivers Laboratories) weighing 150 to 300 gm were anesthetized with intraperitoneal sodium pentobarbital (50mg/kg). Tracheostomies were performed and indwelling catheters were placed in both femoral arteries and one femoral vein. After transfer to the hyperbaric chamber, the rats were paralyzed with intravenous (i.v.) tubocurarine chloride (1.5 mg/kg) to prevent respiratory motion. The rats were ventilated with 90% O_2 balance N_2 via a mechanical rodent ventilator (EDCO Scientific Inc.) to maintain $PaCO_2$ near 35 torr. Additional pentobarbital and curare were given i.v. as necessary to maintain anesthesia. The head was secured in a stereotaxic apparatus and the skull exposed via a longitudinal incision through the scalp. The muscle and fascia were reflected from a point anterior of the nasal suture to well behind the parietoocipital sutures. The rat's arterial blood pressure (Statham, model 23d, strain gauge), bipolar electroencephalogram (EEG) (Grass Instruments, platinum needle electrodes), and rectal temperature were monitored continuously throughout the experiments. A thermostatically controlled heating pad beneath the rat maintained core temperature near 37°C. Arterial blood samples were drawn intermittently to measure PaO_2, $PaCO_2$, pHa, total hemoglobin in grams per 100ml blood (THb), % oxyhemoglobin (HbO_2), % HbCO, % deoxyhemoglobin (Hb) and % methemoglobin (HbMet) and initial O_2 content (Instrument Laboratories, model 813 pH/blood gas analyzer and model 482 CO-oximeter, CO-oximetry). Blood gas analysis was performed in the chamber at 3 ATA by a second blood gas instrument (Radiometer, Denmark) modified to operate at high pressure. Arterial blood samples obtained at 3 ATA for CO-oximetry were sealed at 0° C and analyzed after decompression. O_2 content as ml O_2 per 100ml blood (vol%) to include dissolved O_2 volume was calculated according to the following equation:

Calculated total O_2 vol% =   (O_2 content from CO-oximetry) + (measured PaO_2 (torr) x 0.003 ml dissolved O_2 / torr)

Optical Monitoring

The parietal cortex of the rat was monitored continuously by differential reflectance spectrophotometry within the hyperbaric chamber. The optical measurements were made through the translucent intact skull with a 4 wavelength spectrophotometer. This monitoring approach maintains the normal intracranial circulatory relationships at the expense of some light scattering that decreases the optical signal to noise ratio. The spectrophotometer was of a type described by Jöbsis et al (6) wherein a single incandescent light source (Osram model 58.112) supplies 4 tunable monochromators to produce 2 pairs of monochromatic sample and reference wavelengths (2 to 4 nm bandwidth). The light was pulsed by means of a slotted chopping wheel and conducted to the skull surface via a fiberoptic bundle tipped with an internally reflecting glass rod. This arrangement illuminated a limited area of the rat's right parietal skull. A similar rod with an opaque sheath coupled to the left parietal skull surface collected diffusely reflected light to be measured by a photomultiplier tube (Hamamatsu R928). Differences in intensity of sample and reference (S-R) wavelengths were recorded on a multi-channel chart recorder (Gould, model 560).

The difference spectrum of cytochrome a,a₃ has an absorption maximum at approximately 605 nm that is derived primarily from the reduced iron-porphyrin complex of cytochrome a (3). Absorption at this 605 nm sample wavelength was referenced to 620 nm (6). HbCO was optically monitored by a 569-586 nm S-R pair. Both wavelengths are isosbestic points for Hb and HbO_2 and 569 nm is an absorption peak for HbCO (7). Contributions to the 569-586 nm S-R pair from cytochromes c, c₁ and b are small relative to hemoglobin and were neglected for the purposes of this experiment (8). The optical data were expressed as a percentage of the total labile signal (TLS) measured at each S-R pair. The TLS for 605-620 nm was defined as the difference between maximal oxidation at 1 or 3 ATA and reduction at death. The TLS for 569-586 nm was the difference between control and values obtained while the rats were ventilated with 90% O_2, 1% CO, balance N_2.

Carbon Monoxide and Hyperbaric Exposures

The spectrophotometer was installed in the hyperbaric chamber. Electrical cables from the spectrophotometer to the amplifiers transversed the hull via connectionless, epoxy potted penetrations. Arterial blood samples were obtained for control determinations of blood gases, pH, CO-oximetry while the animal breathed 90% O_2.

The prepared animal was then ventilated with 90% O_2 and either 1% CO for 15 min (or until optical stability) or 0.5% CO for 30 min. CO-oximetry was repeated at the end of the CO exposure periods before the chamber was compressed to 3 ATA at 0.30 ATA per minute while the animal continued to be ventilated with the CO mixture. Arterial blood gas analysis and CO-oximetry were repeated 20 min after compression began. The animals were then ventilated with 90% O_2, balance N_2 for 25 min and blood gas analysis and CO-oximetry was repeated. Rats were then killed with i.v. KCl.

In vivo HbCO formation and cytochrome a,a_3 redox state assessed spectrophotometrically were correlated with measured HbCO levels in graded CO exposures. Six rats were monitored as above while ventilated with 90% O_2 and, in sequence, 0.25% CO, then 0.5% CO, then 1% CO. The rats were killed with i.v. KCl after the 1% CO exposure period to obtain the 605-620 nm TLS. In vivo absorption spectra to detect the cytochrome a_3-CO complex were made by reading transmittance values obtained by scanning from 620 to 568 nm in 4 nm increments with a single monochromator at a 4 nm bandwidth. Spectra were obtained at optical steady states after ventilation with 90% O_2 and either 1% CO for 15 min or 0.5% CO for 30 min at 1 ATA, 12 min after compression began to 3 ATA and 8 min after death at 3 ATA. Transmittance values were converted to log ratios to obtain absorbance difference spectra between oxidized and reduced conditions.

Statistical Methods

Grouped data were expressed as mean ± SEM. Group to group comparisons were made by unpaired t tests. Statistical comparisons for control and experimental data from the same animal were made by paired t test or one way ANOVA. All t tests were done with Bonferroni corrections. P values of <0.05 were accepted as significant.

RESULTS

A representative experimental trace from a rat exposed to 1% CO is shown in Figure 1. Physiologic and optical measurements at control and experimental conditions are summarized in Table 1. Mean arterial pressure (MAP) declined during CO exposure, but stabilized after 8.6±1.3 min of exposure to 90% O_2 + 1% CO and after 14.8±2.8 min of exposure to 90% O_2 + 0.5% CO. Mean arterial pressure (MAP) increased during compression to 3 ATA and stabilized in 7.9±2.1 min in rats breathing 1% CO and in 9.1±0.9 min in rats breathing 0.5% CO. Further treatment of both groups with 90% O_2 at 3 ATA for 25 min recovered MAP to control values. The other physiologic parameters returned to baseline values with HBO except for small persistent elevations of HbCO.

An increase in absorbance at 569 nm relative to 586 nm (the optical HbCO signal) began at 1.2±0.4 and 1.0±0.4 min and stabilized in 19.6±1.1 and 19.5±2.7 min after beginning 1 and 0.5% CO exposure respectively. An increase in absorbance at 605 relative to 620 nm consistent with cytochrome a reduction began after 2.4±0.7 min of CO exposure in both groups and stabilized in 8.7±0.9 and 22.6±2.7 min after 1 and 0.5% CO exposure respectively. Absorbance at 605 nm decreased immediately relative to 620 nm (cytochrome a reoxidation). Upon compression to 3 ATA, the 569-586 nm signal remained stable. The latter observation was confirmed by no change in HbCO measured by CO-oximetry between pre- and 20 min post-compression (p>.90 and .40 for 1% and 0.5% CO). The EEG attenuated in 8 of 10 rats (6.5±0.9 min) in the 1% CO group and 1 of 10 animals at 13 min in the 0.5% CO group. The EEG recovered in all the above rats after compression to 3 ATA. At 3 ATA, absorbance at 605 nm decreased relative to 620 nm consistent with reoxidation of cytochrome a,a_3 to 79±4% TLS in 6.8±1.4 min on 1% CO and 86±3% TLS in 9.3±1.3 min on 0.5% CO. Further treatment of both groups with 90% O_2 at 3 ATA for 25 min recovered the 605-620 nm S-R pair to control values. The mean values at 569-586 nm during recovery exceeded control values (defined as 0) but the difference was not statistically significant.

Table 1. Physiological and Optical Variables

	1% CO, 90% O$_2$ Exposed Rats				0.5% CO, 90% O$_2$ Exposed Rats			
	control	1 ATA+CO	3 ATA+CO	3 ATA	control	1 ATA+CO	3 ATA+CO	3 ATA
MAP mmHg	104±7	62±6*	93±8¶	108±8¶	92±4	75±4*	98±6¶	104±6*
n	10	10	10	10	10	10	10	10
Temp C° §	37.6±0.1	37.6±0.2	38.0±0.3	37.7±0.3	37.0±0.3	37.2±0.2	37.5±0.2	37.3±0.2
n	9	9	9	9	10	10	10	10
pHa §	7.43±0.02	7.34±0.03	7.38±0.02	7.38±0.03	7.44±0.02	–	7.42±0.01	7.46±0.02
n	10	6	8	9	9	0	10	10
PaCO$_2$ mmHg§	33.8±1.6	29.0±4.2	33.6±1.0	37.2±2.5	29.4±2.0	–	27.0±2.3	26.3±2.4
n	10	2	8	9	9	0	10	10
PaO$_2$ mmHg§	346±24	410	1315±55*	1265±65*	363±11	–	1246±77*	1467±23*
n	10	1	9	9	9	0	10	10
CaO$_2$ Vol%	20.2±0.6	7.5±0.2*	11.6±0.2*¶	22.3±0.7*¶	18.2±0.6	9.7±0.4*	14.4±0.4*¶	21.8±0.8*¶
n	8	6	8	8	10	10	10	10
COHb%	3.3±0.1	69.0±0.7*	69.6±0.7*	8.2±1.2*¶	3.0±0.3	53.1±0.7*	52.4±1.0*	5.7±0.5*¶
n	8	6	8	8	10	10	10	10
a,a$_3$ %TLS	87.0±4.2	38.3±3.5*	78.7±4.2¶	99.8±0.2¶	90.8±3.8	50.9±4.7*	84.9±3.0¶	97.3±2.1¶
n	10	10	10	10	8	8	8	8
COHb %TLS	0.0±0.0	91.50±4.2*	91.9±4.7*	-16.2±8.9¶	0.0±0.0	85.0±9.1*	94.3±4.6*	-7.8±9.5¶
n	10	10	10	10	8	8	8	8

* p<.05 compared to control
§ no significant differences within groups by ANOVA
¶ p<.05 compared to previous condition
All values expressed as mean ± standard error of the mean.

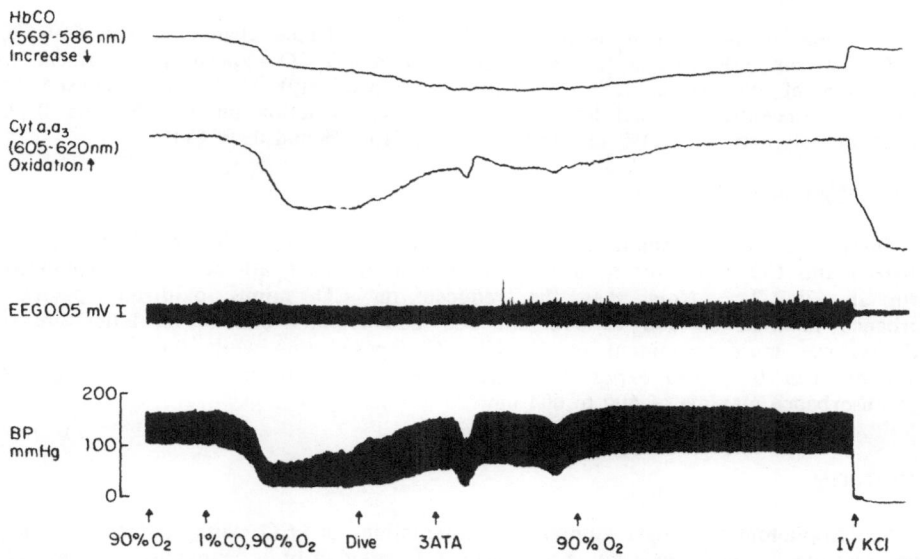

Fig. 1. <u>In vivo</u> optical responses of brain cytochrome a,a3 to 1% CO and hyperbaric exposures.

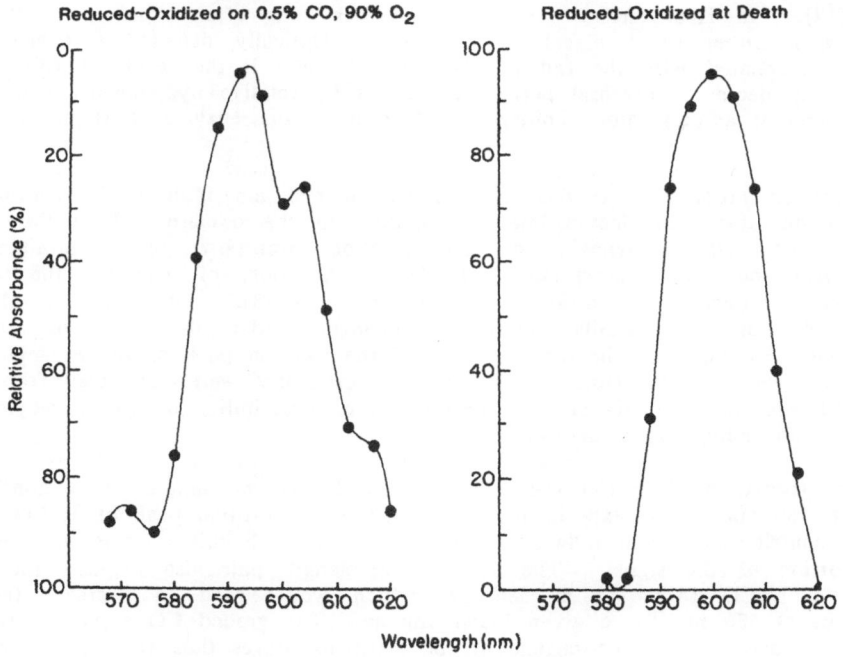

Fig. 2. Reduced minus oxidized difference spectra of rat cortex.

Graded CO Exposure

The relation between measured HbCO and absorbance change at the 569-586 nm and 605-620 nm S-R pair in graded 0.25, 0.5 and 1% CO exposures was linear with a high degree of correlation ($r^2 = .96$ and $.86$ respectively) The rats exposed to the graded CO concentrations had less cytochrome a,a_3 reduction on 1% CO and 0.5% CO than rats acutely exposed to 1% and 0.5% CO ($p<.05$ for 1% and 0.5% CO).

In Vivo Absorbance Spectra

Difference spectra comparing CO exposures at 1 ATA (reduced) minus 3 ATA (oxidized) and CO exposures at 3 ATA (oxidized) minus death at 3 ATA (reduced) are shown in Fig 2. Note that the reduced on CO minus oxidized spectra have absorbance maxima of 588 to 590 nm and 600 to 604 nm. The latter absorbance maximum was more prominent in the 1% CO exposed rats while the former was more prominent after 0.5% CO exposure. The reduced at death minus oxidized spectra had single absorbance maxima of 600 to 604 nm.

DISCUSSION

Spectrophotometric data supporting interaction of CO with cytochrome a,a_3 in blood circulated rats has not been reported previously. Previous studies of the cytochrome a,a_3 CO ligand were conducted using in vitro solutions of cytochrome c oxidase or isolated mitochondrial preparations (3,9). These in vitro observations, conducted under non-physiologic conditions, have raised doubts about the physiologic significance of the CO-cytochrome a,a_3 interaction (4).

Several of the physiological responses to CO hypoxia would increase the likelihood of CO uptake by cytochrome a,a_3 in vivo. Blood pressure fell in a dose dependent fashion during CO exposures. CO produces systemic and cerebral vasodilation (12), though the relative contributions of the hypoxic effects of CO and the tissue effect is unknown. Barbiturate anesthesia potentiates this hypotensive effect (12). In fluorocarbon perfused rats, MAP falls after exposure to 0.5% CO indicating a direct tissue effect of CO (13). Optically detected cytochrome a,a_3 reduction correlated with the fall in MAP and initially is the result of CO hypoxia, and possibly decreased cerebral perfusion induced by relative hypotension, and perhaps altered cerebral autoregulatory ability. Reduction of the enzyme is requisite for CO binding.

Increased absorbance at 605 nm relative to 620 nm indicates cytochrome a,a_3 reduction but does not demonstrate CO binding to the oxidase. Reduction of the oxidase with either hypoxia or an electron transport block produces the characteristic 605 nm absorption peak (3). Addition of CO to the reduced cytochrome produces an additional peak at 590 nm (3). Other known absorbing species do not have similar absorption changes under the conditions of our experiment. We observed the disappearance of the 590 nm peak of the CO-cytochrome a,a_3 ligand when spectra from the 3 ATA CO exposures were subtracted from the 1 ATA CO exposures. This also demonstrates the reversibility of the CO-cytochrome a,a_3 ligand under hyperbaric conditions.

The formation of HbCO was followed reliably by the simple subtraction of 586 nm from 569 nm in our experiments. The 569 nm absorption peak of HbCO in vivo receives contributions from reduced cytochrome b although this appears to comprise a small portion of the signal. The 569-586 wavelength pair also neglects the optical effects of HbCO formation at 586 nm which would tend to decrease the total absorbance at 586 nm for a given blood volume. The graded CO exposures produced the same relative absorbance change as the acute exposures thus further verifying the technique.

The intent of the experiment was not to prove the physiologic significance of CO binding with cytochrome a,a_3 but to demonstrate that binding and its reversal <u>in vivo</u> is a possible explanation for non-hypoxic mechanisms of CO toxicity. The experiments also do not prove the efficacy of hyperbaric oxygen in CO poisoning although the rationale for its use is strengthened inasmuch as reoxidation of cytochrome a,a_3 and reversal of the HbCO ligand occurs despite no change in HbCO level at 3 ATA.

SUMMARY

Cytochrome a,a_3 redox state of the parietal cortex of pentobarbital anesthetized rats was continuously monitored through intact skull with four wavelength differential spectrophotometry during exposure to 90% O_2 plus either 1.0 or 0.5% CO at 1 and 3 (ATA). The formation of HbCO was monitored in the brain by absorbance differences between 569 and 586 nm and correlated positively in graded 0.25 to 1% CO exposures with measured HbCO levels. Exposure to 90% O_2, 1% or 0.5% CO (balance N_2) decreased mean arterial pressure (MAP), calculated arterial O_2 content and cytochrome a,a_3 oxidation measured at 605 nm relative to 620 nm while HbCO rose. After compression to 3 ATA, rats breathing CO mixtures increased MAP and O_2 content with reoxidation of cytochrome a,a_3 while HbCO remained constant. Further treatment of both groups with 90% O_2 at 3 ATA recovered the above parameters to at least control values except small persistent elevations of HbCO. Difference spectra recorded from 568 to 620 nm in parallel experiments showed twin absorbance peaks at 588 to 592 nm and 600 to 605 nm in response to CO. These absorbance maxima were consistent with formation of the cytochrome a_3-CO complex and cytochrome a reduction respectively. These studies indicate that CO binds to reduced cytochrome a_3 in blood circulated rat cortex in CO hypoxia and this effect can be reversed by increasing dissolved arterial O_2 content at 3 ATA.

REFERENCES

1. Bernard C., 1857, Lecons sur less effects des substances toxiques et medicamentenses. Paris: Bailliere et Fils.
2. Roughton FJW, RC Darling, 1944, The effect of carbon monoxide on the oxyhemoglobin dissociation curve. Am J Physiol 141:17-31.
3. Keilin D, and EF Hartree, 1939, Cytochrome and cytochrome oxidase Proc R Soc Lond [Biol.]:127:167.
4. Coburn RF, and HJ Forman, 1987, Carbon monoxide toxicity, in "Handbook of Physiology; Section 3: The Respiratory System," AP Fishman, ed., American Physiologic Society, Bethesda, Maryland.
5. Barlett D, RF Coburn, and EP Radford, 1977, Effects on man and animals, in "Carbon Monoxide, Committee on Medical and Biologic Effects of Environmental Pollutants," National Academy of Sciences, Washington, D.C.
6. Jöbsis FF, JH Keizer, JC La Manna, and M Rosenthal, 1977, Reflectance spectrophotometry of cytochrome a,a_3 in vivo. J Appl Physiol 43:858-872.
7. Van Assendelft, OW, 1970, Light absorbtion spectra of haemoglobin derivatives, in "Spectrophotometry of haemoglobin derivatives," Charles C. Thomas, Publisher, Springfield, Illinois.
8. Piantadosi CA, and FF Jöbsis-VanderVliet, 1984, Spectrophotometry of cerebral cytochrome a,a_3 in bloodless rats, Brain Res, 305:89-94.
9. Chance B, M Erecinska, and M Wagner, 1970, Mitochondrial responses to carbon monoxide toxicity. Ann NY Acad Sci, 174:193-204.
10. Piantadosi CA, 1987, Carbon monoxide, oxygen transport, and oxygen metabolism, J Hyperbaric Med, 2(1):27-44.
11. Sylvia, AL, CA Piantadosi, and FF Jöbsis-VanderVliet, 1985, Energy metabolism and in vivo cytochrome c oxidase redox relationships in hypoxic rat brain, Neurol Res, 7:81-88.

12. Pitt, BR, EP Radford, GH Gurtner and RJ Traystman, 1979, Interaction of carbon monoxide and cyanide on cerebral circulation and metabolism, Arch Environ Health, 34:354-359.

13. Piantadosi CA, PA Lee, and AL Sylvia, 1988, Direct effects of carbon monoxide on cerebral energy metabolism in bloodless rats, J Appl Physiol, 65:878-887.

CYTOPROTECTIVE EFFECT OF ISOTONIC MANNITOL AT LOW OXYGEN TENSION

G. Gronow, P. Prechel, and N. Klause

Department of Physiology
University of Kiel
D-2300 Kiel, FRG

INTRODUCTION

The protective mechanism of mannitol in the postischemic kidney in vivo remains incompletely defined. Mannitol infusion has been reported to prevent postischemic renal failure by vascular effects such as an arterial vasodilation (improving oxgen transport to postischemic renal tissue) as well as by an increased glomerular filtration rate and an enforced osmotic diuresis which, in turn, may have reduced tubular obstructions (Burke et al., 1983; Zager et al., 1985).

On the other hand, at the cellular level, the osmotic effect of a poorly permeable molecule may reduce cellular injury under conditions of reduced oxygen and substrate supply. Extracellular mannitol, for example, prevented disruptive cell swelling, reduced the loss of cell constituents, supported intracellular ion homeostasis, and improved mitochondrial function (Schrier et al., 1984; Gronow et al., 1985).

Recently, it has been pointed out that the reperfusion of ischemic kidneys may cause oxidant stress which contributes to postischemic renal tissue injury (Paller et al., 1988; McCoy et al., 1988). Under conditions of reperfusion and reoxygenation the activity of cellular protective enzymes, or the capacity of oxygen radical scavenging cell constituents may become insufficient, and univalent reduced oxygen (or different species of oxygen free radicals) may then oxidize cellular macromolecules, leading to dysfunction of ion pumping enzymes (Kako et al., 1988) and the liberation of lipid peroxidation products such as malondialdehyde (Paller et al., 1988).

Mannitol, in adition to its known osmodiuretic properties a potent hydroxyl radical scavenger, may reduce in isolated renal microsomes the formation of malondialdehyde (Gutteridge et al., 1984; Cojocel et al., 1985). Furthermore, in the postischemic kidney in vivo, the protective effect of mannitol has been shown to equal that of superoxide dismutase or catalase infusion (Hanssen et al., 1986). However, in the latter experiments no discrimination could be made between a) oxygen radical scavanging effects of mannitol at the cellular level, b) cytoprotection by its osmotic activity, and c) secondary beneficial events due to the above mentioned mannitol-induced vasodilation and increased oxygen delivery to postischemic renal tissue.

To exclude such vascular effects, and to discriminate at the cellular level between osmotic and oxygen radical scavanging effects of mannitol we tested the effect of variable concentrations of hypertonic or isotonic mannitol as well on hypoxic /posthypoxic cellular function as on lipid peroxidation in isolated tubular cells of rat kidney cortex.

MATERIAL AND METHODS

Male Sprague-Dawley rats (body weight 340 - 410g) were anesthetized by the i.p. injection of pentobarbital sodium (80mg /kg body weight). Isolated tubular segments from rat kidney cortex (ITS) were then prepared by mechanical and collagenase treatment as reported earlier (Gronow et al., 1984, 1985). After preincubation and twofold washing ITS were incubated in standard Krebs-Ringer-bicarbonate medium (= KRB, pH 7.35, 0.5 g /100ml bovine albumin, 10 mmol /l lactate, 37°C).

Hypoxia (PO_2 20 min $\leq$ 1 mm Hg) was introduced by gassing the surface of the incubation media with 95% N_2 : 5% CO_2, reoxygenation (30 min) then introduced by gassing the surface with 95% O_2: 5% CO_2. Incubation media were made either hypertonic by the addition of mannitol (up to 200 mmol /l) or kept isotonic by the replacement of NaCl by equiosmolar amounts of mannitol. In studies on hypoxic cellular function (Fig. 2), 10 mmol /l lactate in the incubation medium was replaced by 10 mmol/l glucose, mannose, mannitol, sorbitol, or SO_4^{2-}, respectively. In the reoxygenation period, the addition of 10 mmol /l non-dissociating, osmotically active substances was compensated for by the omission of 5 mmol /l NaCl.

Measurements of intracellular K^+ accumulation, tubular protein, substrate turnover, and the loss of τ-Glutamyltransferase (τGT) have been described earlier (Gronow et al., 1984, 1985). Functions of ITS suspended without mannitol and gassed 50 min with 95% O_2: 5% CO_2 served as oxygenated controls or 100% reference (intracellular K^+ = 394 ±52 nmol /mg protein, τGT-loss = 1.13 ±0.09 mU /mg protein·min , gluconeogenesis, GNG = 4.92 ±0.93 nmol /mg protein·min). Lipid peroxidation was estimated in the supernatant of homogenized and centrifuged (10 min at 600 x g) cells as malondialdehyde equivalents (MDA) according to the thiobarbituric acid assay of Ohkawa et al. (1979). The RA 642-BS-pyrimido-pyrimidin derivative, a scavenger of free hydroxyl radicals, was a gift from Thomae (Biberach, FRG).

All values are expressed per mg tubular protein and are the mean (±SE) of 12 observations. Statistical analysis was employed as paired t-Tests in which a P-value of 0.05 or less was assumed to indicate a significant difference.

RESULTS

Hypertonic Mannitol

Incubation media were made hypertonic by the addition of mannitol. Gluconeogenesis (GNG), intracellular K^+-accumulation (K^+), and τ-Glutamyltransferase loss (τGT) were measured immediately after hypoxia and reoxygenation (Fig.1, left panel). With no mannitol in the incubation medium, GNG (17% of aerobic control values) and K^+ (53% aerobic control) were significantly reduced in reoxygenated ITS, and τGT leakage was markedly elevated (571%). Addition of small amounts of mannitol to the incubation medium significantly improved GNG (39%) and K^+ (95%), the highest values were observed in a range of 10 -30 mmol /l mannitol, which was equivalent to a hyperosmolality of 310 -330 mosmol /kg H_2O).

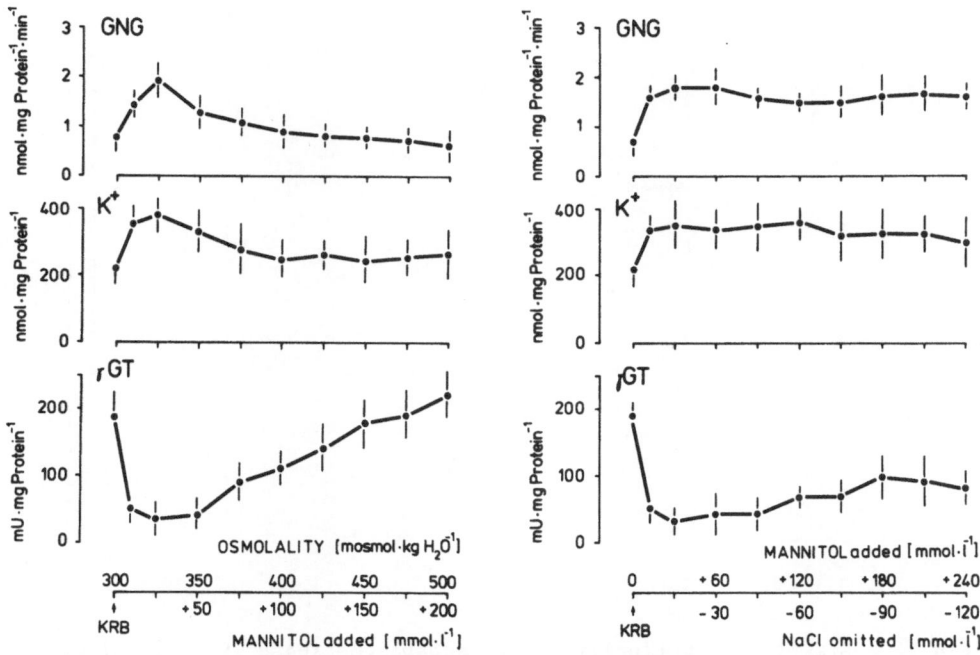

Fig. 1.   Effect of mannitol on gluconeogenesis (GNG), intracellular K^+
accumulation (K^+), and loss of τ-Glutamyltransferase (τGT) in
reoxygenated tubular cells of rat kidney cortex (37°). Left
panel: incubation media (KRB = Krebs-Ringer-bicarbonate) made
hypertonic by the addition of mannitol. Right panel: incubation
medium kept isotonic (300 mosmol /kg H_2O) by the replacement of
equiosmolar amounts of medium NaCl by mannitol (x ±SE, n=12)

According to the improvement in GNG at 10-30 mmol /1 mannitol, τGT-
leakage was also suppressed to near aerobic values. Higher osmolalities
(>375 mosmol /kg H_2O, 75 mmol /1 mannitol added) had no such beneficial
effect, GNG and K^+ declined, and τGT leakage rose with increasing mannitol
concentration. At 100 mmol/1 mannitol, for example, GNG was about 19%, K^+
60%, and τGT-loss about 359% of the aerobic control, respectively.

Isotonic Mannitol

In the next series of experiments, NaCl in the incubation medium was
replaced by equimolar amounts of mannitol (Fig. 1, right panel). In respect
to incubations performed with no mannitol in the medium (KRB-control: GNG
14%, K^+ 55%, τGT 579%), the presence of lower concentrations of mannitol
(10-60 mmol/1) significantly improved GNG (up to 38%), K^+ (up to 89%), and
τGT-loss (about to 33%). Higher concentrations of mannitol did not improve
these values, GNG and K^+ were nearly unchanged, and τGT was slightly elevated. Thus, a cytoprotective effect of low-dose (10 mmol /1) mannitol
replacing 5 mmol/1 NaCl in an isotonic incubation medium (300 mosmol /kg
H_2O) was evident.

In order to further elucidate this "isotonic", low-dose mannitol
effect on reoxygenated renal tubular cells and to test glycolytic energy
production we compared in the next series of experiments the effect of 10
mmol/ mannitol with either 10 mmol /1 glucose, mannose, sorbitol, or $SO_4{}^{2-}$
respectively (Fig. 2): in hypoxia (left panel), lactate formation was high
(about 3 nmol/mg·min) in the presence of metabolic substrates such as glu-

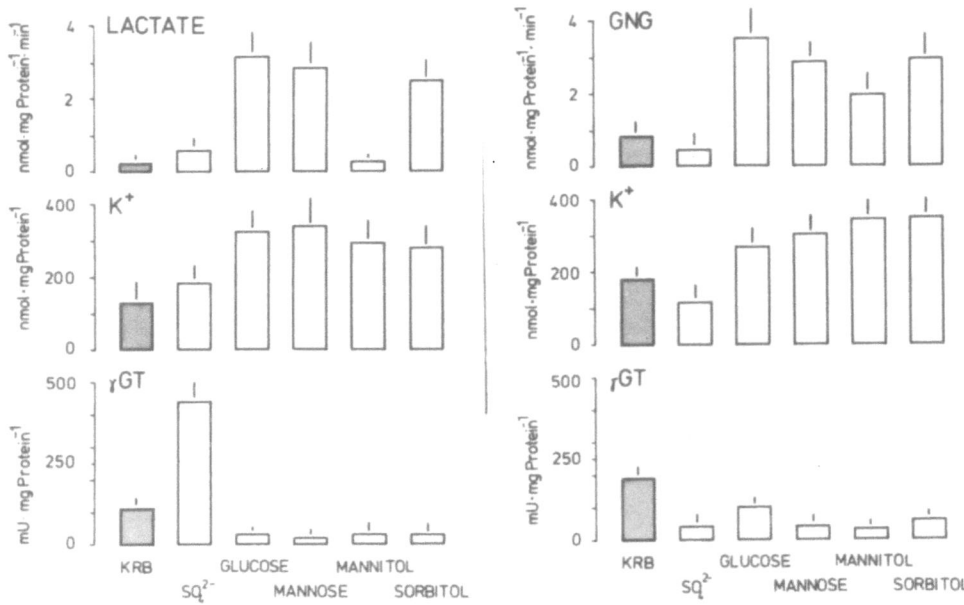

Fig. 2. Effect of osmotically active substances (each 10 mmol /l) on
lactate formation, intracellular K^+ accumulation (K^+), loss of
τ-Glutamyltransferase (τGT), and gluconeogenesis (GNG) in isolated renal tubular cells of rat kidney cortex (37°C). Left
panel: in hypoxia (PO_2 < 1mm Hg, no lactate added). Right panel:
tubules reoxygenated, 10mmol/l lactate added ($\overline{X}$ ±SE, n =12)

cose, mannose, or sorbitol. In contrast, glycolysis in the presence of
sulfate or mannitol was as low (about 0.5 nmol /mg·min) as without mannitol in the incubation medium (= KRB-control).

According to the observed metabolic support via glycolysis, in the
presence of metabolic substrates (glucose, mannose, sorbitol) intracellular K^+ was significantly higher (about 80% of oxygenated control), and
τGT-loss markedly lower (i.e. in the control range) than in KRB-solution
(329%) or with cytotoxic sulfate ions (1243%) in the incubation medium.
Without metabolic support by glycolysis, however, a marked cytoprotective
effect of mannitol (K^+ : 73%, τGT-loss as low as oxygenated control) could
still be observed in respect to cells which were incubated in KRB-solution
and which had also no source for metabolic energy production (control K^+:
33%, τGT: 329%).

After reoxygenation (Fig.2, right panel), functional parameters of
isolated cells paralleled the beneficial mannitol effects observed in hypoxia: K^+ was elevated (81%), GNG significantly supported (35%), and τGT-
loss as low as in aerobic control experiments. With the exception of
substrate-supported gluconeogenesis (60 -70%)in the presence of mannose
and sorbitol mannitol effects in reoxygenated tubules were even comparable
to the effect of metabolic substrates on functional parameters in reoxygenated cells (K^+: 70-80%, τGT-loss as oxygenated control). Reoxygenated
tubular cells incubated without mannitol or metabolic substrates (=KRB)
had a much lower K^+ content (45%), formed less glucose (15%), and lost
more τGT (576%). Thus, the cytoprotective effect of "isotonic" mannitol
observed in hypoxia persisted in the reoxygenation period.

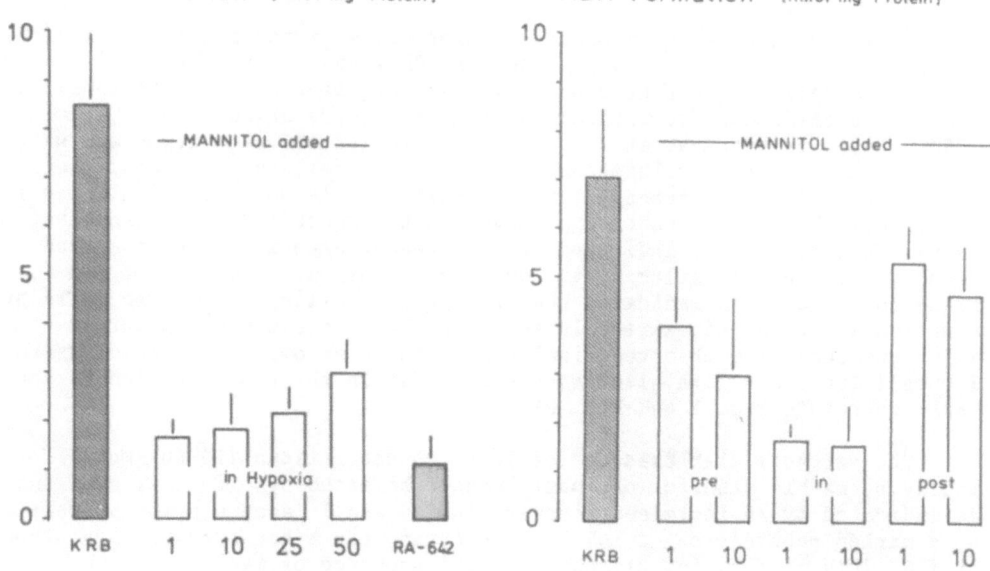

Fig. 3. Effect of hydroxyl radical scavangers mannitol or RA-642 on the formation of malondialdehyde (MDA) equivalents in reoxygenated tubular cells of rat kidney cortex (37°). KRB: no hydroxyl radical scavangers added. Left panel: mannitol (1-50mmol/l) and RA-642 (10 mmol /l) added in hypoxia. Right panel: mannitol present in the incubation medium before (pre), in (in), and after (post) the hypoxic time interval ($\overline{x}$ ±SE, n =12)

Lipid Peroxidation

In the reoxygenation period, lipid peroxidation of tubular membrane lipids was monitored as the formation of malondialdehyde equivalents (MDA, Fig. 3). In respect to cells incubated without mannitol (KRB, 8.5 ±1.4 nmol/mg MDA = 100%), an about fourfold reduction of MDA-formation could be observed at different mannitol concentrations (left panel). The suppression of MDA-formation was maximal at 1 mmol /l mannitol (21% of KRB), and significantly lower at 50 mmol /l mannitol (35% Of KRB). A comparable reduction of MDA-formation could also be observed when the hydroxyl radical scavenger RA-642 was present during hypoxia and subsequent reoxygenation (14% of KRB-control).

In a last series of experiments we tested as to what extent the point in time of mannitol application would affect the observed reduction of MDA-formation in isolated tubular cells (Fig.3, right panel): in respect to cells incubated without mannitol (KRB-control) the presence of 1 or 10 mmol/l mannitol which was added only in hypoxia suppressed MDA-formation in the subsequent reoxygenation period (when ITS were incubated without mannitol) by about 80% (columns marked as "in"). This effect was not as much pronounced (56%) when mannitol was only present in a preincubation period (ITS suspended in hypoxia and during reoxygenation without mannitol, columns marked as "pre"), and the addition of mannitol in the reoxygenation period yielded a significant smaller (30%) mannitol effect in MDA-formation (columns marked as "post").

DISCUSSION

Cellular function of isolated tubular cells of rat renal cortex (ITS), i.e. posthypoxic lactate gluconeogenesis (GNG) and intracellular K^+ accumulation, as well as membrane-bound τGT-loss have been shown to be sensitive parameters of hypoxic stress and subsequent reoxygenation (Gronow et al, 1984, 1985). In hypoxia, at insufficient rates of ATP generation and with no osmotic support of volume regulation by extracellular mannitol, the osmotic activity of intracellular macromolecules is not counterbalanced in renal cells by active transmembraneous ion transport (Macknight and Leaf, 1979). This effect was indicated in the present experiments by a marked reduction of intracellular K^+ at low oxygen tension. A release of membrane-bound τGT then indicated the disruptive swelling of hypoxic cells by osmotically attracted extracellular water. This irreversible membrane injury preceded the observed final reduction in aerobic recovery of anabolic cell functions like gluconeogenesis (GNG) in the reoxygenation period (KRB-control in Fig. 1 and Fig. 2).

The extracellular presence of lower concentrations (10-60 mmol/l) of poorly permeable mannitol obviously counterbalanced hypoxic cell swelling, as indicated by an increased recovery in GNG and K^+ accumulation as well as by a marked suppression of τGT-leakage during and after hypoxic incubation of ITS (Fig. 1 and Fig. 2). However, the presence of mannitol in the incubation medium did not generally prevent hypoxic and posthypoxic dysfunction. At concentrations of more than 90 mmol/l mannitol renal cell functions were impaired by a disproportionally increased transmembraneous solvent drag of osmotically attracted intracellular water. This high extracellular osmolality may have also limited cellular volume regulation in the reoxygenation period, as τGT-loss in "isotonic" mannitol was only about one half of the leakage observed in hypertonic mannitol solution (Fig. 1).

When the effects of four osmotically active substances were compared with the effect of mannitol (Fig. 2), metabolic support by anaerobic substrates was indicated by an increased rate of lactate formation. According to earlier observations (Gronow and Cohen, 1984, Gronow et al., 1985) the presence of anaerobic substrates supported cellular ion homeostasis (K^+) and gluconeogenesis as well as cellular volume regulation (indicated by a low τGT-leakage). However, in the presence of mannitol, glycolysis was as low as in the KRB-control, but a significant improvement in cellular function could be observed: GNG and intracellular K^+ were elevated, and τGT-loss was smaller than in the KRB-control. Thus, the present experiments provide evidence for a non-metabolic effect of isotonic mannitol.

To test the hydroxyl radical scavenging properties of mannitol (Gutteridge et al., 1984; Cojocel et al., 1985) we measured the formation of malondialdehyde (MDA) equivalents in reoxygenated renal tubular cells (Fig. 3). Obviously, a low concentration of mannitol (1 mmol/l) reduced MDA-formation as effective as higher concentrations (up to 50 mmol/l) or the presence of another hydroxyl radical scavanger, RA-642 (left panel). Thus, a predominant protective effect of mannitol by osmotic forces, as has been concluded from in vivo studies (Schrier et al., 1984; Zager et al., 1985) does not appear to be a primary reaction in the present experiments. Instead, in the presence of mannitol, a marked reduction of MDA-formation could be observed in the reoxygenation period (Fig. 3, left panel), indicating a reduced rate of lipid peroxidation even in the presence of small amounts of mannitol. Similar observations have been made in a subcellular preparation. (Cojocel et al., 1985; Paller et al., 1988). Interestingly, in our intact cellular system the observed protective mannitol effect was smallest when one would have expected more pronounced effects, i.e. at higher oxygen tension in the reoxygenation or preincubation period ("pre" and "post" in Fig. 3, right panel).

By contrast, the most pronounced radical scavanging effect of mannitol could be observed when it had only been present in the hypoxic incubation period (Fig. 3, right panel). Thus, under our experimental conditions of extreme hypoxia (PO_2 <1mm Hg) mannitol may have entered hypoxically swelling cells via the intrusion of extracellular water through leaky cell membranes (Macknight and Leaf, 1977). Than, in closer contact to intracellular sites of oxygen radical generation (Xanthine-Oxidase, mitochondria, microsomes, for example) mannitol may have reduced MDA-formation in the subsequent reoxygenation period. According to this hypothesis, mannitol added aerobically (colums marked as "pre" and "post" in Fig. 3) would have penetrated to a lesser extent tighter membranes of sufficiently oxygenated tubular cells.

SUMMARY

The beneficial effect of mannitol infusion in postischemic kidneys remains unresolved. Contradictionary reports may have originated from at least 3 different mechanisms: a) arterial vasodilation, b) osmotic support of hypoxic cellular volume regulation, and c) scavenging of hydroxyl radicals in the reoxygenation period. To exclude vascular effects we tested at 37°C in hypoxic (PO_2 <1 mmHg) and reoxygenated isolated tubular cells of rat kidney cortex cellular function, i.e. intracellular K^+ accumulation (K^+), posthypoxic lactate gluconeogenesis (GNG), loss of membrane-bound τ-Glutamyltransferase (τGT), and formation of a lipid peroxidation product, malondialdehyde (MDA). Mannitol (M) was added to a Ringer incubation medium in variable concentration either without (hypertonic M) or with omission of equiosmolar amounts of NaCl (isotonic M).

K^+ and GNG were significantly supported, and τGT-loss markedly suppressed in a range of 10-50 mmol/l hypertonic and isotonic M. At higher concentrations no improvement (isotonic) or even deleterious effects (hypertonic) of M occurred. Beneficial effects of lower concentrations of M (10 mmol/l) were not correlated to anaerobic glycolysis, and 1 as well as 10 mmol/l M induced a comparable and significant reduction in posthypoxic MDA-formation. This effect was most pronounced when M was only added in hypoxia, indicating leakiness of cellular membranes in hypoxia. It is concluded that a) reported unfavourable effects of M may have been mediated by high (> 60 mmol/l) concentrations of M, and b) that beneficial effects of low-dose (1-10 mmol/l) M were mediated by a support of hypoxic volume regulation (high K^+, low τGT-loss), a better maintenance of cellular membranes (reduced MDA-formation) and an improvement in posthypoxic energy metabolism (GNG).

REFERENCES

Burke, Th. J., Arnold, P. E., Schrier, R. W., 1983, Prevention of ischemic acute renal failure with impermeant solutes, Am.J.Physiol.,244:F646
Cojocel, C., Hannemann, J., Baumann, K., 1985, Cephaloridine-induced lipid peroxidation initiated by reactive oxygen species as a possible mechanism of cephaloridine nephrotoxicity, Biochim.Biophys.Acta, 834:402.
Gronow, G., Cohen, J. J., 1984, Substrate support for renal functions during hypoxia in the perfused kidney. Am.J.Physiol., 247:F618.
Gronow, G., Meya, F., Weiss, Ch., 1984, Studies on tha ability of kidney cells to recover after periods of anoxia. Adv.Exp.Med.Biol.,169:589.
Gronow, G., Benk, P., Franke, H., 1985, Effects of anaerobic substrates on post-anoxic cellular functions in isolated tubular segments of rat kidney cortex. Adv.Exp.Med.Biol., 180:403.

Gutteridge, J. M. C., 1984, Lipid peroxidation initiated by superoxide-dependent hydroxyl radicals using complexed iron and hydrogen peroxide, FEBS letters 172:245.

Hansson, R., Johansson, S., Jonsson, O., Pettersson, S., Scherstén, T., Waldenström, J., 1986, Kidney protection by pretreatment with free radical scavengers and allopurinol: renal function at recirculation after warm ischemia in rabbits. Clin.Sci.,71:245.

Kako, K., Kato, M., Matsuoka, T. Mustapha, A., 1988, Depression of membrane-bound Na^+-K^+- ATPase activity induced by free radicals and by ischemia of kidney.Am.J.Physiol. 254:C330.

MacKnight, A. D. C., Leaf, A., 1977, Regulation of cellular volume, Physiol. Rev., 57:510.

McCoy, R., N., Hill, K. E., Ayon, M. A., Stein, J. H., Burk, R. F., 1988, Oxidant stress following renal ischemia: changes in the glutathione redox ratio. Kidney Int. 33:812.

Ohkawa, H., Ohishi, N., Yagi, K., 1979, Assay for lipid peroxides in animal tissues by thiobarbituric acid reaction, Anal.Biochem. 95:351.

Paller, M. S., 1988, Renal work, glutathione and susceptibility to free radical-mediated postischemic injury. Kidney Int.33:843.

Schrier, R. W., Arnold, P. E., Gordon, J. A., Burke, Th. J., 1984, Protection of mitochondrial function by mannitol in ischemic acute renal failure, Am.J.Physiol. 247:F365.

Zager, R. A., Mahan, J., Merola, A. J., 1985, Effects of mannitol on the postischemic kidney. Lab.Invest. 53:433.

OXYGEN DIFFUSION THROUGH MITOCHONDRIAL MEMBRANES

T. Koyama, M. Kinjo and T. Araiso

Research Institute of Applied Electricity, Hokkaido
University, 060 Saporo, Japan

INTRODUCTION

The membrane viscosity of myocardial mitochondria was measured recently with a time-resolved fluorescence technique. As expected from the low concentration of cholesterol in mitochondrial membranes, the viscosity was relatively small. Treatment with phospholipase A_2 caused an increase in membrane viscosity. To date, the influence of mitochondrial membrane on the oxygen transport to the oxygen-consuming site has not been discussed.

An oxygen gradient can be established in the cell membranes, cell cytosol and also in the phospholipid layer of mitochondrial membranes. Oxygen which diffuses through these gradients is finally consumed by the cytochrome a,a_3 situated on the inner surface of the mitochondrial inner membranes. A gradient can develop in any unstirred medium surrounding an oxygen-consuming structure and can limit the diffusion of oxygen to the oxygen consuming site, cytochrome a,a_3. The requirement of the intact myocardium for a relatively high oxygen tension suggests the existence of an oxygen gradient from the suspending medium to the mitochondrial inner surface (Jones & Kennedy 1986).

Fluorometrically-measured membrane viscosity seems to represent the effective viscosity to the lateral diffusional movement of membrane proteins (Araiso et al. 1986) and we assumed that it might be applicable to the analysis of oxygen diffusion which occurs through the arrayed phospholipid molecules in biomembranes (Koyama and Araiso, 1986). The same assumption is used in the numerical analysis of the diffusion of oxygen through mitochondrial membranes in the present study.

METHODS and CALCULATION

Rat myocardial mitochondria were isolated (Kinjo, Araiso & Koyama, 1988). The suspension of isolated mitochondria was incubated with a hydrofran solution containing a low concentration of diphenyl hexatriene (DPH). The time course of anisotropy decay of fluorescence from DPH molecules entrapped in the mitochondrial membranes was measured using a picosecond time-resolved fluorometer at 37 °C. The membrane viscosity is calculated as the frictional resistance of lateral collision of wobbling phospholipid molecules according to the "wobbling in a cone" model (Kinosita et al. 1977, Kawato et al. 1977, Araiso et al. 1986). Assuming that the membrane viscosity thus measured represents the effective viscosity encountered by oxygen diffusing through the phospholipid bilayer of mitochondrial inner membranes, we analyzed the oxygen gradient in the phospholipid layer of mitochondrial membranes by a radial diffusion model as proposed by Jones and Kennedy (1986) for the myocardial cell.

The mitochondrion is spherical or oval in shape and its minor axis is 2 μm at most. The total inner surface of mitochondria including cristae is 11 $\mu m^2/\mu m^3$ of tissue (11×10^4 cm^2/cm^3) (Page & MacCaster 1973). Let us take as a model of a mitochondrion a long cylinder of which diameter is 2 μm. The cristae infolds are stretched to form a smooth cylinder surface. All the mitochondria are arranged end to end for a long thin cylinder for the ease of calculation. The total length of the cylinder of all the mitochondria contained in 1ml tissue, l, is given by; l = total inner surface/(π x diameter) = 1.75 x 10^8 cm.

The oxygen consumption rate, Kc, by the mitochondrial cylinder is equal to that of the myocardial cell, i.e. a mean value of 10ml O_2/100g cells/min (7.43 x 10^{-7} mol/g/sec) for the whole cardiac cycle (Cat, Millican 1939; Human, Bing et al. 1949,) increasing to 30ml O_2/100 g/min (22.29 x 10^{-7} mol/g/sec) during the systole (Thews 1962). The oxygen consumption rate is assumed to be uniform along the mitochondrial cylinder.

The flux of oxygen, Q, which diffuses through unit length of the mitochondrial cylinder in unit time, is estimated from the oxygen consumption rate of the cell; Q = Kc/l = 7.43 x 10^{-7}/1.75 x 10^8 = 4.25 x 10^{-15} mol/cm/sec on an average and 12.7 x 10^{-15} during the systolic phase.

The mitochondrial cylinder is surrounded by phospholipid bilayers having a thickness of 10 nm; the mitochondrial outer membrane and inner membrane are each 5 nm thick giving a total of 10 nm. The model cylinder is shown in Fig. 1.

The general solution of the differential equation for the radial equation is

$$C = A + B \ln r,$$

where A and B are constants to be determined from the boundary conditions, ln and r represent natural logarhythm and radial distance, respectively.

If the surface of the mitochondrial cylinder (radius = a) is kept at a constant oxygen concentration C_1 and the outer surface of the mitochondrial membrane (radius = b) at C_2, then the flux of oxygen, Q, is given by

$$Q = DS\frac{dC}{dr}\Big|_{r=a} = D2\pi a \frac{(C_2 - C_1)/a}{\ln(b/a)}$$

and

$$Q = \frac{2\pi D (C_2 - C_1)}{1000 \ln(b/a)} \qquad (1),$$

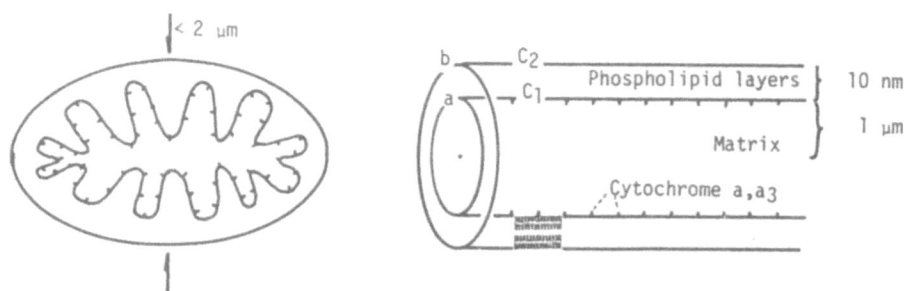

Fig. 1. A scheme of the cylinder model of mitochondria.

where D represents the diffusion coefficient, and 1000 is the denominator giving C in units of mol/l (Jones & Kennedy 1986, cf. Crank 1967).

The value of D is inversely proportional to the viscosity of the surrounding medium. The value of D in water, of which the viscosity is 0.75 cp, is $2 \times 10^{-5} cm^2$/sec at 37 °C. We assumed the diffusion coefficient in mitochondrial membranes, having a viscosity of 43.8 and 51.4 cp (Kinjo, Araiso, Koyama 1988), to be 0.034 and 0.029×10^{-5} cm^2/sec. The absorption coefficient of oxygen in erythrocyte membrane is about 5 times that in water (McCabe 1986). The effective diffusion coefficient should be 0.172 and 0.146×10^{-5} cm^2/sec, respectively.

RESULTS and DISCUSSION

Putting the numerical values given above into equation (1), the oxygen gradient was calculated for various conditions as listed in Table 1. A hypothetical case where the mitochondrial membrane has the diffusion coefficient of 1.5×10^{-5} cm^2/sec, as reported for myocardial tissue (Thews 1962), is included.

Table I. Oxygen gradient across the mitochondrial membrane and related factors.

Condition for Mitochondria	Viscosity (cp)	Effective Diffusion Coefficient. $(10^{-5}$ cm^2/sec)	Oxygen Consumption Rate (ml/100g/min)	$(10^{-7} mol/g/s)$	Oxygen Gradient (nmol)
Intact	43.8	0.172	10	7.43	3.9
			30	22.29	11.6
Phospholipase A_2 treated	51.4	0.146	10	7.43*	4.6
			30	22.29*	13.7
Hypothetical case		1.5	10	7.43	0.44
			30	22.29	1.33

* It was assumed that the phospholipase A_2 treated mitochondria have the same oxygen consumption rate as intact ones.

Since the mitochondrion is a small particle, its surface area to volume ratio is large. The viscosity of the mitochondrial membrane is lower than that of red blood cells. The diffusion of oxygen, therefore, is relatively unhindered in mitochondria. Despite the reduced value of the effective diffusion coefficient, the oxygen gradient of the present cylinder model of mitochondria is only 3.9 nmol for the mean oxygen consumption rate and 11.6 nmol for the systolic consumption rate. These values are small but may be of some importance in hypoxia, since the oxygen concentration for half-activation of cytochrome a,a$_3$ contained in isolated mitochondria is 400 nmol (Kennedy & Jones 1986).

The predicted oxygen gradient may seem small but is fairly large when compared with the value for the hypothetical case shown in Table I. This indicates that the mitochondrial membrane can be a barrier for oxygen diffusion.

The application of phospholipase A_2 caused an increase in membrane viscosity. This results in an increase in the oxygen gradient (see Table I). It is probable that the ischemic activation of phospholipase A_2 destroys biomembranes and reduces the oxygen diffusion into mitochondria.

Jones and Kennedy (1986) proposed a small diffusion coefficient, 2×10^{-6} cm^2/sec, to explain the photometrically determined oxygen gradient across cytosol and cytochrome a,a$_3$. This low value corresponds to a viscosity of 38 cp. If the cytosol has such a high viscosity, the viscosity of the mitochondrial membrane will not constitute any significant barrier. The measurement of the viscosity of cytosol is an important problem but seems to be difficult at present.

SUMMARY

The effect of the mitochondrial membrane on the oxygen supply to the interior of mitochondria was analyzed with a cylinder model of diffusion. This estimation is based on the assumption that cytochrome a,a$_3$ is distributed only on the inner surface of the mitochondrial inner membrane. The diffusion coefficient in the mitochondrial membrane was approximated from the fluorescently-determined viscosity of rat mitochondrial membrane. A pico-second time-resolved fluorometer at 37 °C gave values of 43.8 cp for intact mitochondria and 51.4 cp after phospholipase A$_2$ treatment. Using the mean oxygen consumption rate of 10 ml O$_2$/100 g tissue/sec in beating heart, oxygen gradients of 3.9 and 4.6 nmol was predicted across the intact and phospholipase-A$_2$ treated mitochondrial membranes, respectively. The increased oxygen consumption during systole will yield oxygen gradients of 11.6 and 13.7 nmol. These gradients were much larger than the values estimated in a hypothetical case using the diffusion coefficient for the mitochondrial membrane of 1.5×10^{-5} cm^2/sec. The predicted oxygen gradient suggests a non-uniform distribution of oxygen in the myocardial cell and may be of importance in understanding the relationship between oxygen supply and myocardial function in hypoxia. Phospholipase A$_2$, which is known to be activated in ischemia, destroys the microstructure of myocardial cells, seems deleterious to oxygen transport to cytochrome a,a$_3$.

REFERENCES

Araiso, T., Sindo, Y., Arai, T. and Koyama, T. 1986, Viscosity and order in erythrocyte membranes studied with nanosecond fluorometry. Biorheology 23, 467-483.

Bing, R. J., Hammond, M. M., Handesman, J. C., Powers, S. R., Spencer, F. C., Eckenhoff, J. E., Goodale, W. T., Hafkenshiel, J. H., Kety, S. S. 1949, The measurement of coronary blood flow, oxygen consumption and efficiency of the left ventricle in man. Amer. H. J. 38, 1-24.

Crank, J. 1967, Mathematics of diffusion, Clarendon Press Oxford.

Jones D. P. and Kennedy, F. G. 1986, Analysis of intracellular oxygenation of isolated adult cardiac myocytes. Am. J. Physiol. 250 (Cell Physiol. 19): C384-C390.

Kawato, S., Kinosita, K. Jr. and Ikegami, A. 1977, Dynamic structure of lipid bilayers studied by nanosecond fluorescence technique. Biochemistry 11, 2319-2324.

Kinjo, M. Araiso, T., and Koyama, T. 1988, The effect of phospholipid A$_2$ on dynamic microstructure of phospholipid layer of mitochondria from rat myocardium. J. Appl. Cardiol. in press.

Kinoshita, K. Jr., Kawato, S. and Ikegami, A. 1977, A theroty of fluorescence polarization decay in membranes. Biophys. J. 20, 289-305.

Koyama, T., Araiso, T. 1986, Oxygen diffusion coefficient of cell membranes. In: Longmuir IS ed. Oxygen Transport to Tissue vol. VIII, new York and London, Plkenum Press, 99-106.

McCabe, M. 1986, The solubility of oxygen in erythrocyte ghosts and the flux of oxygen across the red cell membrane. Oxygen Transport to Tissue vol. VIII. Ed. I.S. Longmuir, Plenum Press, New York and London, pp.13-20.

Millican G. A. 1939, Muscle hemoglobin. Physiol. Rev. 19, 505-523.

Page, E. and McCallister, L. P. 1973, Quantitative electron microscopic description of heart muscle cells. Application to normal, hypertrophied and thyroxin-stimulated hearts. Amer. J. Cardiol. 31, 172-181, 1973.

Thews, G. 1962, Die Sauerstoffdrucke im Herzmuskelgewebe. Pflugers Arch. 276, 166-181.

EFFECTS OF DIFFERENT INSPIRATORY/EXPIRATORY (I/E) RATIOS AND PEEP-VENTILATION ON BLOOD GASES AND HEMODYNAMICS IN DOGS WITH SEVERE RESPIRATORY DISTRESS SYNDROME (RDS)

B. Lachmann, W. Schairer, S. Armbruster,
G.J. van Daal, and W. Erdmann

Dept. of Anesthesiology, Erasmus University
Postbus 1738, 3000 DR, Rotterdam
The Netherlands

Despite great advances in the treatment of acute respiratory failure, many patients do not respond to accepted methods of resuscitation. The great majority of these critically ill patients fall into the category of adult respiratory distress syndrome (ARDS).The main immediate therapeutic goal in severe ARDS consists of attempting to overcome hypoxaemia as well as metabolic and respiratory acidosis by means of respiratory therapy, increased inspiratory oxygen concentration and infusion of buffer solution. Some patients, who showed no improvement from this therapeutic regiment, have been treated with extracorporeal membrane oxygenation (ECMO) (1) or with extracorporeal elimination of CO_2 (2) in combination with low frequency ventilation in a few highly specialized intensive care units.

The aim of respiratory therapy in ARDS is to maintain the gas exchange in the lung by opening and stabilizing closed units with a minimal depression of the heart function and the circulation. Numerous clinical and experimental studies were performed to investigate the influence of continuous positive airway pressure (CPAP), positive end-expiratory pressure (PEEP), "super"-PEEP, inspiratory-expiratory ratio (I/E ratio), frequency, inspiratory flow, end-expiratory pause (EIP) and intermittent mandatory ventilation (IMV) on the gas exchange of severely damaged lungs (for review see ref. 3). A significant improvement of the oxygenation could be accomplished by adjusting ventilation in several ways. Up to now it has, however, not been shown that a particular set of data describing the state of the lungs and heart enables making a decision on how respiratory therapy should be performed in an optimal manner.

The failure of respiratory therapy despite a high PEEP and a high inspired oxygen concentration, FIO_2 of 1, in a group of patients with ARDS, the poor results from ECMO and the fact that few intensive care units can perform special, more promising varieties of ECMO give grounds for searching after other methods of respiratory therapy suitable for patients with the most severe respiratory insufficiency.

Method

Altogether 12 dogs with severe respiratory insufficiency were studied. Respiratory insufficiency was produced by bronchial lavage with isotonic saline at body temperature [4]. Variation of the number of lavages enables variation of the severity of functional disturbances. Strict adherence to a certain lavage procedure results in a quite reproducible condition in the lungs. Severe respiratory insufficiency was defined as being present when arterial oxygen tension fell below 60 mm Hg during volume-controlled ventilation with an I/E ratio of 1:2 and an FIO_2 of 1. In order to establish this condition in dogs, it was necessary to repeat the lung lavage ten times on the average, with a volume corresponding to the vital capacity (for details see Lachmann et al. [4]).

Various breathing patterns were produced by a Servo ventilator 900 C (Siemens-Elema AB, Solna, Sweden). Pressure in pressure-controlled ventilation and minute volume in volume-controlled ventilation were kept constant in all of the experimental studies. If not otherwise stated, frequency was 40/min. The effects of different ventilator settings were evaluated by measurements of blood gas and haemodynamic measurements.

Results

Influence of I/E ratio on gas exchange.

The data after lavage are compatible with severe RDS. PaO_2 fell from about
460 mm Hg before lavage to about 55 mm Hg after lavage in the ventilator settings cited above.
After lavage, peak pressure was about 28-30 cm H_2O. Up to this stage, no obvious difference between volume- and pressure-controlled ventilation was observed. When inspiration was prolonged to cover 80% of the cycle, PaO_2 increased. The greatest improvements were found at pressure-controlled ventilation (Fig. 1a). The initial values of arterial CO_2 tension, $PaCO_2$, illustrate that the animals were hyperventilated with 20 ml/kg body weight (BW). The elimination of CO_2 was higher in pressure-controlled ventilation than in volume-controlled ventilation (Fig. 1b).

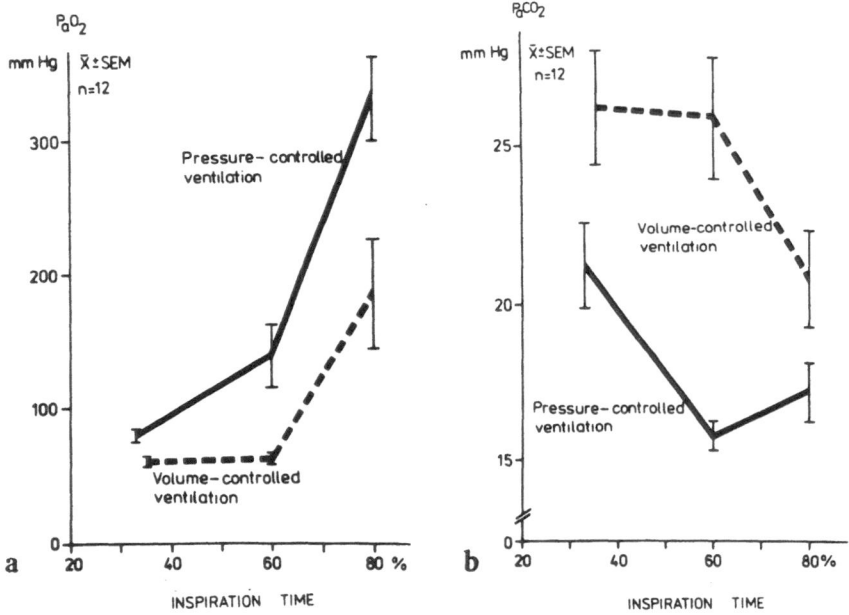

Figure 1. The average PaO_2 (a) and $PaCO_2$ (b) in relation to inspiration time at pressure- and at volume-controlled ventilation in 12 beagles with severe ARDS; SEM, standard error of the mean. Note that despite hyperventilation, oxygenation was poor when inspiration time was short.

The increase in arterial oxygenation corresponds to a drastic reduction in intrapulmonary shunt (Fig. 2). Prolongation of inspiratory time to 80% of the cycle led to a significant increase of mean airway pressure which caused the well-known decrease in cardiac output of about 33% in pressure-controlled ventilation and about 45% in volume-controlled ventilation. A decrease in systemic pressure as well as an increase in pulmonary artery pressure and pulmonary vascular resistance also occurred.

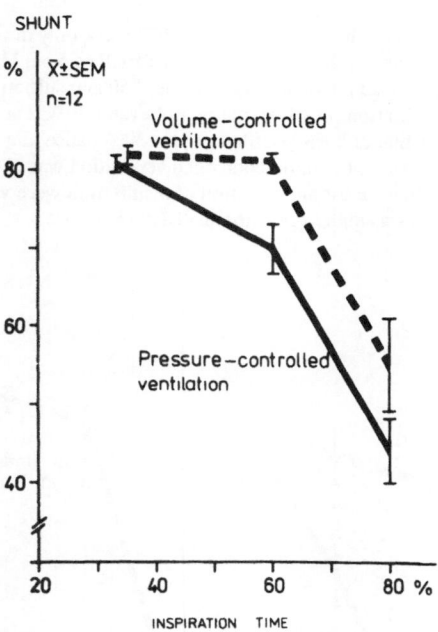

Figure 2. The intrapulmonary shunt was calculated from arterial and mixed venous content of oxygen and related inspiration time at pressure- and volume-controlled ventilation. Same animals as in Fig.1.

The resulting data show clearly that, also without a PEEP, a significant improvement of gas exchange can be reached by prolongation of the inspiratory time to 80% of the cycle. At an I/E ratio of 4:1, a significantly better oxygenation can be reached in the most severely damaged lungs at pressure-controlled ventilation compared to volume-controlled ventilation (Fig. 1). The negative cardiocirculatory consequences are less influenced by a high I/E ratio at pressure-controlled ventilation than at volume-controlled ventilation.

The influence on compliance is difficult to interpret, as at short expirations the conditions are not static at the end of expiration, particularly not when an improvement of compliance leads to a longer time constant of the respiratory system.

Long inspiration time versus PEEP ventilation.

On the basis of the favourable results at pressure-controlled ventilation with an inspiratory time of 80%, it must be asked which advantages this ventilation pattern has over other ventilation patterns that have successfully been clinically applied in connection with PEEP [5,6,7,8].
In these experiments, we compared a high I/E ratio with PEEP not only in regard to gas exchange and airway pressure, but also to haemodynamics. A PEEP of 5 cm H_2O at volume-controlled ventilation efficiently produced adequate oxygenation if 50% insufflation and 30% pause were used (Fig. 3). At 25% insufflation plus 10% pause, not even a PEEP as high as 15 cm H_2O produced as high PaO_2 as that at 50% insufflation plus 30% pause and a PEEP of 5 cm H_2O did. High values of PEEP were at volume-controlled ventilation was 27-30 cm H_2O. It was kept constant in each animal when PEEP and duration of insufflation were varied. When inspiratory time was 80%, PaO_2 was high even without PEEP (Fig. 3).

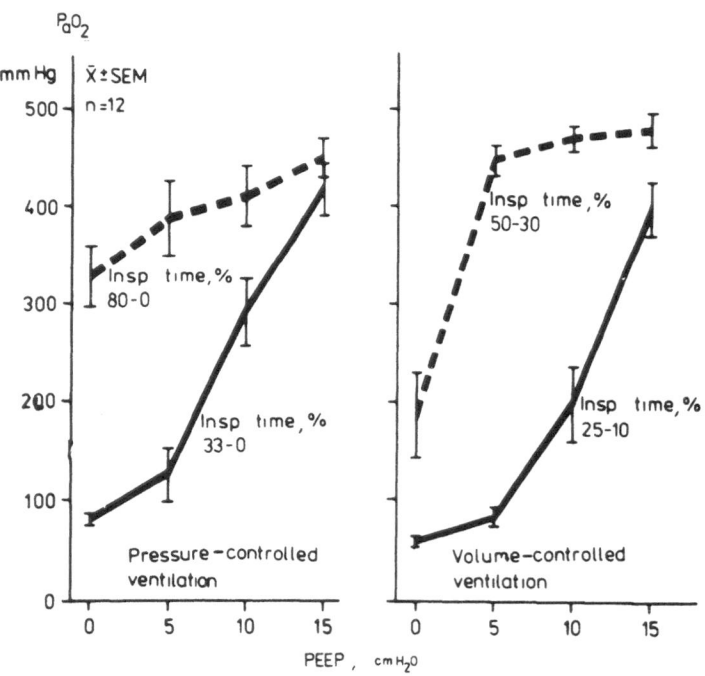

Fig. 3. Effects of PEEP and inspiration time at pressure- and at volume-controlled ventilation on PaO_2. Same animals as in Fig.1.

Further slight increases in PaO_2 were seen when PEEP was present. When PEEP was 10 cm H_2O or higher at pressure-generated ventilation, CO_2 retention was caused by reduced ventilation (Fig. 4). When duration of inspiration was as short as 33%, a higher PEEP was needed to maintain oxygenation. This advantage of PEEP was won at the expense of severe CO_2 retention (Fig. 5) caused by hypoventilation.

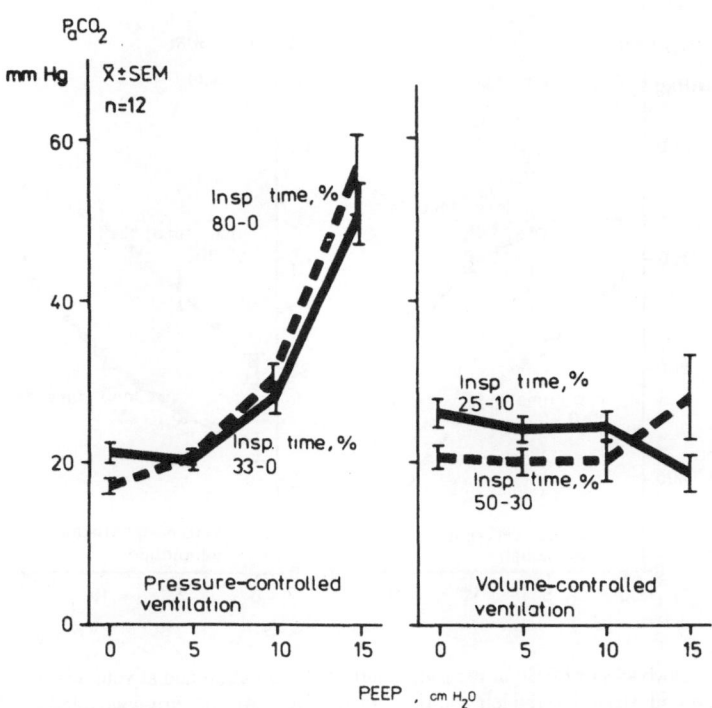

Fig. 4. Effects of PEEP and inspiration time at pressure- and at volume-controlled ventilation on $PaCO_2$. Same animals as in Fig.1.

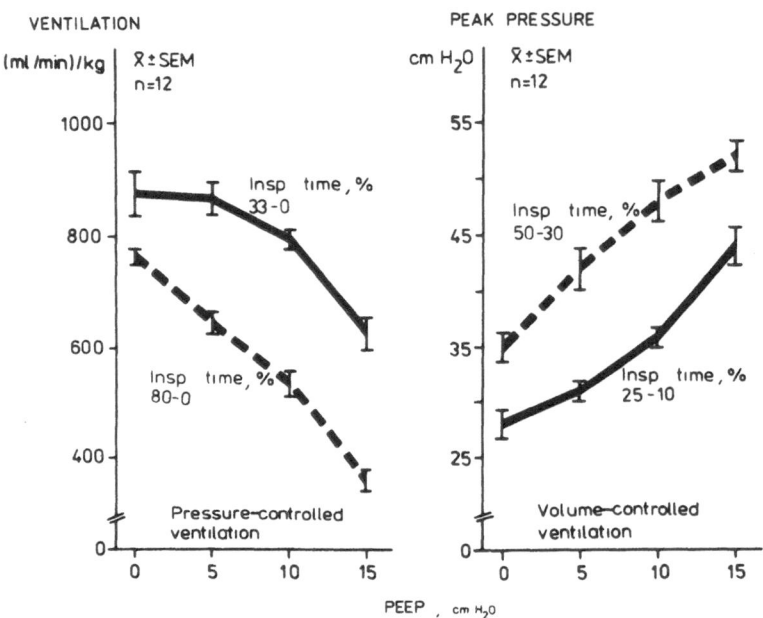

Fig. 5. The drawbacks of PEEP at pressure-controlled ventilation and at volume-controlled ventilation are illustrated to the left and right, respectively. At pressure-controlled ventilation, ventilation decreases with PEEP. At volume-controlled ventilation, peak airway pressures increase with PEEP. Same animals as in Fig.1.

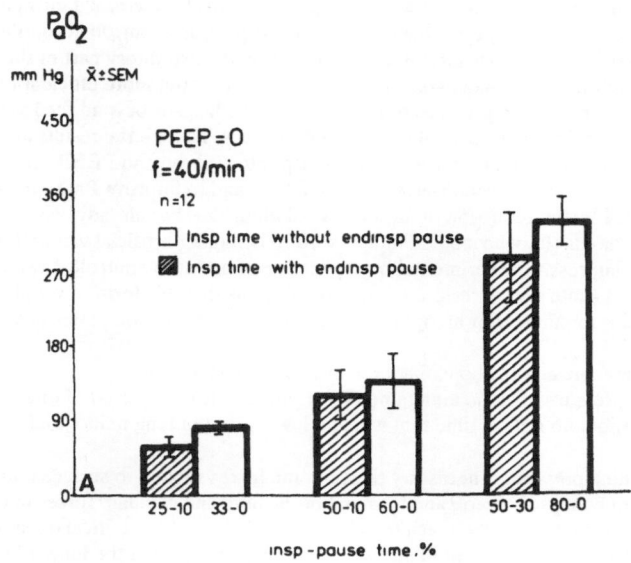

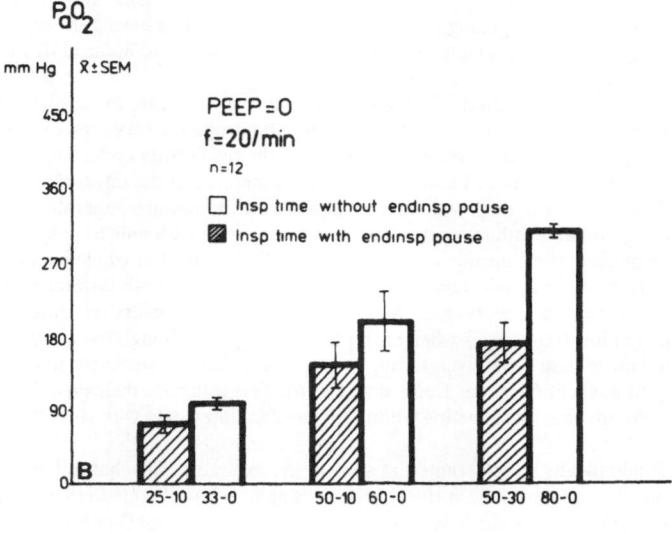

Fig 6. Mean values and standard errors of mean PaO_2 during pressure-controlled ventilation with a frequency of 20 (A) and of 40 (B) per minute dependent on inspiratory time and post-inspiratory pause. Note the high difference of 150 mm Hg in PaO_2 between inspiration time of 50% and pause time of 30% in contrast to 80% inspiration time without pause at a frequency of 20/min (A). During pause time, intrapulmonary pressure fell only by 2-3 cm H_2O.

Discussion

Starting from the PaO_2 tension at pressure-controlled ventilation with an I/E ratio of 4:1 and a PEEP of zero, and searching for other ventilation patterns which produce the same degree of oxygenation (Fig. 3), all such patterns led to a lower cardiac output and a lower oxygen transport. The improvement in oxygenation and the decreased intrapulmonary shunt, from the use of a high I/E ratio and PEEP and at pressure-controlled ventilation, are probably due to prevention of closure of lung units during the expiratory part of the breathing cycle. Of the two manoeuvres, increasing I/E ratio is the more efficient one, probably because the opening pressure of large parts of the lungs is beyond the reach of a practicable PEEP in the severe case of surfactant deficiency. Though the results at pressure-controlled ventilation show that the use of a long inspiratory phase and PEEP acted synergistically to reduce the alveolar-arterial O_2 gradient and to improve PaO_2, the usefulness of PEEP is limited by the reduction of alveolar ventilation. This has already been shown by Herman and Reynolds (9), who studied infants with IRDS under artificial ventilation. Considering the impressive experimental by application of pressure-controlled ventilation with a high I/E ratio in acute severe respiratory failure, it seems that this form of ventilation therapy is superior to others. For an optimal ventilation, this ventilation pattern requires:

1) One must overcome a critical opening pressure during inspiration.
2) This opening pressure must be maintained for a sufficiently long period of time.
3) During expiration, no critical time that would allow closure of lung units should pass.

The critical opening pressure is necessary to overcome forces related to surface tension, e.g. adhesive forces of collapsed alveoli and terminal bronchioli, and capillary forces in the small fluid-filled airways, so that air can reach the alveoli. The height of the critical opening pressure is a function of the content of surface-active phospholipids in the lungs of patients with ARDS. A varying content of surface-active material in different lung areas could also explain the apparent paradox that a large shunt can exist even with high PEEP and with a pressure which exceeds critical opening pressure. Some lung units then have a strong tendency to collapse when expiration is too long and/or when the counterpressure induced by PEEP is not high enough to prevent collapses in these regions.

A sufficiently long application of the critical opening pressure is necessary to ventilate all alveoli despite different retraction forces in the airways. An intrapulmonary pressure that balances closing pressure must be maintained for most of the respiratory cycle. Even a slightly lower pressure will lead to collapse of lung units and to an increase of the intrapulmonary shunt. This was illustrated in 12 dogs with ARDS ventilated with pressure-generated ventilation. Each dog was ventilated with two patterns at two frequencies with only a slight difference. In one pattern, the controlled pressure was applied during the whole inspiratory phase. In the other, insufflation was followed by a pause during which both inspiratory and expiratory lines of the ventilator were shut. Such a pause allows some stress relaxation to occur - particularly at low frequencies when each phase is longer. Although the airway pressure never fell more than 3 cm H_2O during the pause, this fall was sufficient to cause a pronounced drop of PaO_2 at the lower frequency (Fig. 6). This reflects a delicate balance between closing and opening forces acting within the terminal lung units with surfactant deficiency.

The expiration should ideally be interrupted as soon as the expiratory flow has fallen to low values. Monitoring of expiratory flow at the airway opening may be valuable in this regard. It is, however, possible that some units have collapsed even before the total flow has ceased. It may be so that PEEP has a value as an adjuvant to the high I/E ratio at severe RDS, particularly when very short pulmonary time constants warrant higher frequencies than 20/min.

References

1. Zapol WM, Snider M, Hill JD. (1979) Extracorporeal membrane oxygenation in severe acute respiratory failure. JAMA 242: 2193-2196.
2. Gattinoui L, Pesenti A, Rossi GP, Vesconi S, Fox U, Kolobow T, Agostini A, Pelizzola A, Langer M, Uziel L, Longoni F, Damia G. (1980) Treatment of acute respiratory failure with low-frequency positive-pressure ventilation and extracorporeal removal of CO_2. Lancet 2: 292-294.

3. Lachmann B, Danzmann E. (1984) Adult respiratory distress syndrome. In: Van Golde LMG, Batenburgh JJ, Robertson B. (eds). Pulmonary surfactant. Elsevier, Amsterdam: 505-548.
4. Lachmann B, Robertson B, Vogel J. (1980) In vivo lung lavage as an experimental model of the respiratory distress syndrome. Acta Anaesthesiol Scand 24: 231-236.
5. Ashbaugh DG, Petty TL, Bigelow DB, Harris TM. (1969) Continuous positive-pressure breathing (CPPB) in adult respiratory distress syndrome. J Thorac Cardiovasc Surg 57: 31-41.
6. Petty TL, Ashbaugh DG. (1971) The adult respiratory distress syndrome clinical features, factors influencing prognosis and principles of management. Chest 60: 233-239.
7. Petty TL, Newman JH. (1978) Adult respiratory distress syndrome. West J Med 128: 399-407.
8. Suter PM, Fairley HB, Isenberg MD. (1975) Optimum end-expiratory airway pressure in patients with acute pulmonary failure. N Engl J Med 292: 284-289.
9. Herman S, Reynolds EOR. (1971) Methods for improving oxygenation in infants mechanically ventilated for severe hyaline membrane disease. Arch Dis Child 48: 612-617.

IMPROVED ARTERIAL OXYGENATION AND CO_2 ELIMINATION FOLLOWING CHANGES FROM VOLUME-GENERATED PEEP VENTILATION WITH INSPIRATORY/ EXPIRATORY (I/E) RATIO OF 1:2 TO PRESSURE-GENERATED VENTILATION WITH I/E RATIO OF 4:1 IN PATIENTS WITH SEVERE ADULT RESPIRATORY DISTRESS SYNDROME (ARDS)

B. Lachmann, W. Schairer, S. Armbruster,
G.J. van Daal, and W. Erdmann

Dept. of Anesthesiology, Erasmus University
Postbus 1738, 3000 DR, Rotterdam
The Netherlands

In earlier investigations we found that in severe ARDS caused by bilateral lung lavages in adult rabbits and dogs, the arterial oxygenation increases when constant pressure-generated ventilation with decelerating flow and an I/E ratio 4:1 was applied (1,2,3). This specific ventilation pattern was therefore used in 6 patients with severe ARDS who could not be sufficiently oxygenated with PEEP ventilation. Three of them had such severe damage to the lung parenchyma with gas leakage that PEEP pressure higher than 4-8 cm H_2O could not be administered. In two patients, PEEP higher than 12-16 cm H_2O had no influence on oxygenation. In one patient with myocardial re-infarction, a PEEP higher than 8 cm H_2O led to acute cardiac failure. Before the special ventilation pattern was started, all of the patients were ventilated with volume-controlled ventilation with an I/E ratio of 1:2 or 1:1. An FIO_2 of 1 had been used for an average of 72 h.

When the pattern of ventilation was changed to pressure-controlled ventilation with an I/E ratio of 4:1, a significant increase in PaO_2 from an average of 55 mm Hg to 130 mm Hg was observed in 10 min. One hour later, PaO_2 has increased to about 170 mm Hg (Fig. 1a). Despite lowering of the ventilation pressure (Fig. 1c), the arterial oxygenation further increased during the following 23 h. FIO_2 could simultaneously be reduced from 1 to 0.65.

The arterial-end-expiratory CO_2 gradient was about 40 mm Hg before starting the special ventilation. Normalization to 4 mm Hg could be observed within 24 h (Fig. 1b). The thorax-lung compliance improved very little within the first hour, but showed a significant improvement after 24 h (Fig. 1d).

The drastic improvement of the gas exchange and general status of the patient after application of pressure-controlled ventilation with an I/E ratio of 4:1 is further demonstrated in three out of six patients who had very severe respiratory insufficiency.

Case reports

Case I:
A 35-year-old man had a traffic accident and very severe trauma of the thorax and lungs, combined with shock. He was admitted to the intensive care unit, ICU. During volume-controlled ventilation, the chest x-ray showed only a few ventilated lung areas (Fig. 2a). Minute ventilation was 14 l of pure oxygen. PEEP higher than 8 cm H_2O could not be applied because of the large lung parenchyma defects. Arterial oxygen tension was between 30 and 40 mm Hg, and $PaCO_2$ varied between 75 and 91 mm Hg under these conditions. After 10 min of pressure-controlled ventilation with an I/E ratio of 4:1 and a lower peak pressure than before, PaO_2 had increased to a value twice as high as earlier, and $PaCO_2$ was only 46 mm Hg. Arterial oxygenation, CO_2 elimination and compliance improved despite further lowering of

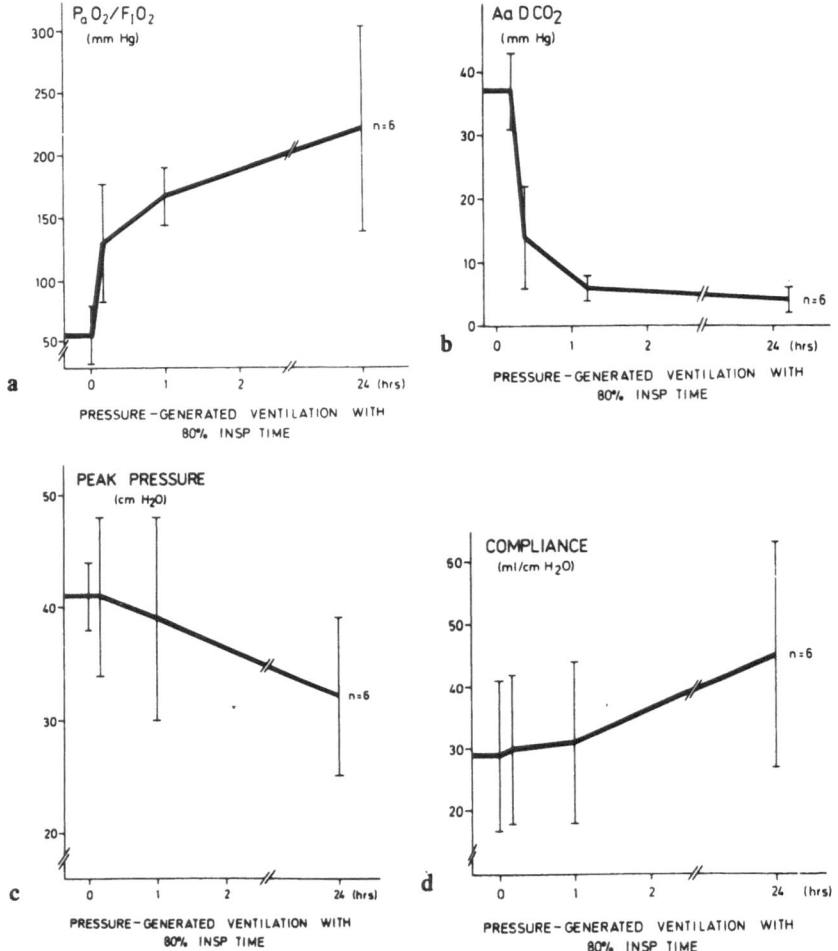

Fig. 1. Course of the quotient PaO_2/FIO_2 (a), arterial-end-expiratory CO_2 gradient ($AaDCO_2$) (b), peak airway pressure (c) and compliance (d) in six patients with most severe ARDS after a change of ventilator therapy from volume-controlled ventilation with I/E ratio of 1:2 or 1:1 to pressure-controlled ventilation with an I/E ratio of 4:1.

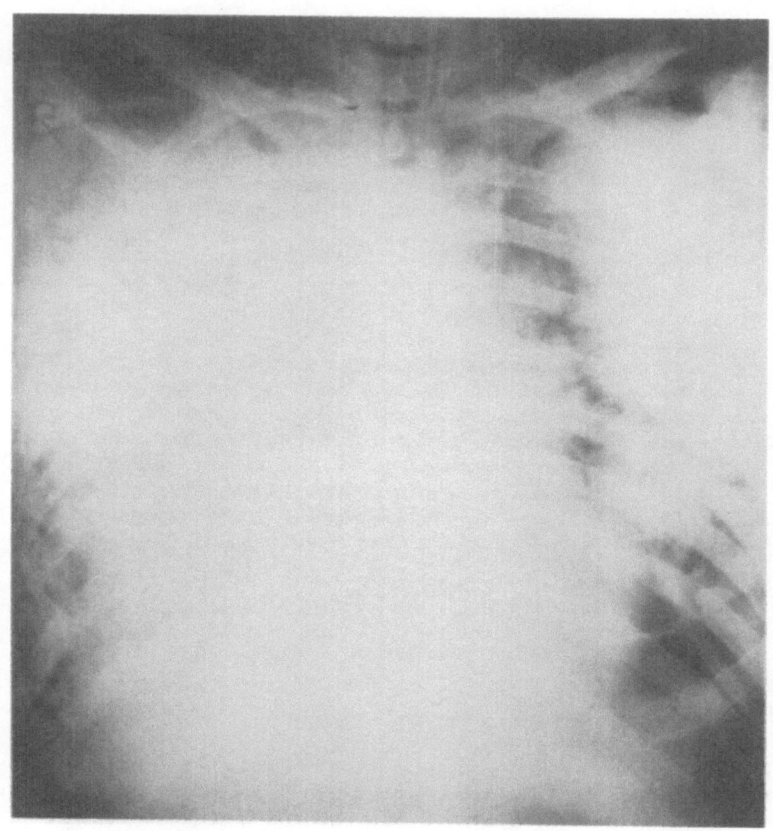

Fig. 2a. Chest x-rays of a 35-year-old man 10 h after blunt trauma of the thorax; artificial ventilation with an I/E ratio of 1:2.

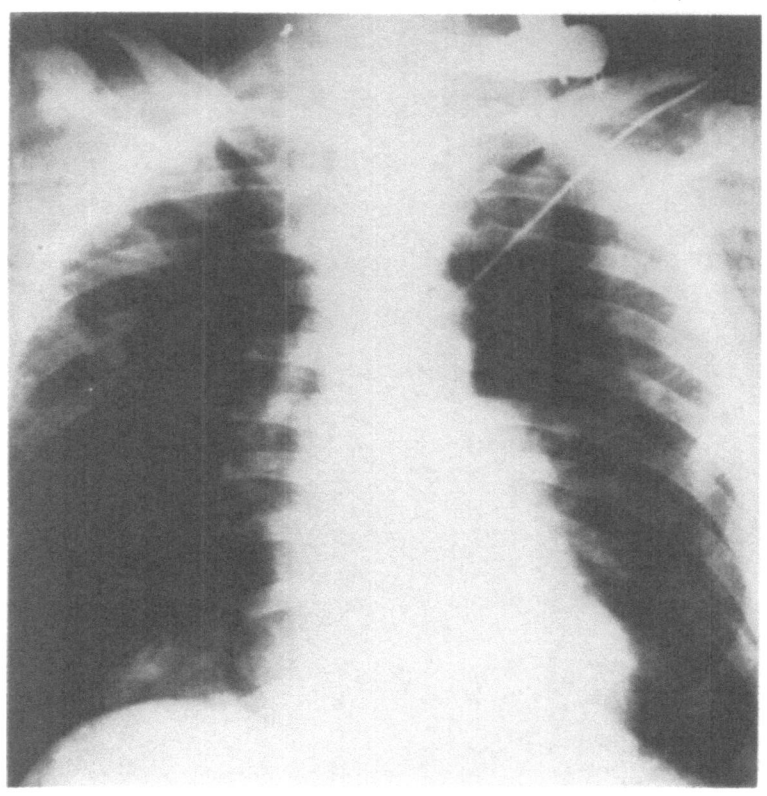

Fig. 2b. Chest x-ray of a the same patient after four days of pressure-controlled ventilation with an I/E ratio of 4:1.

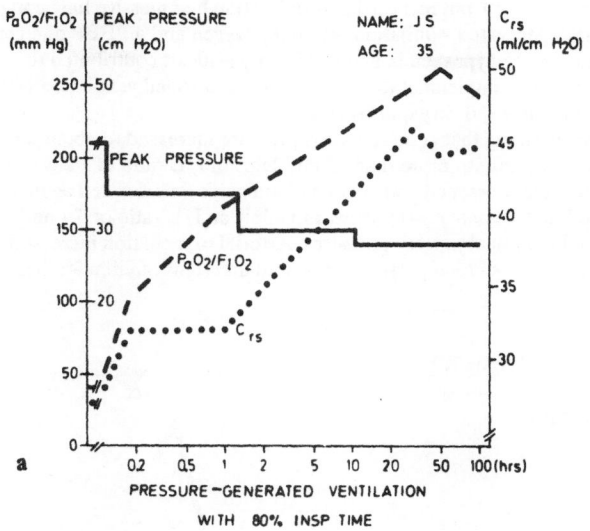

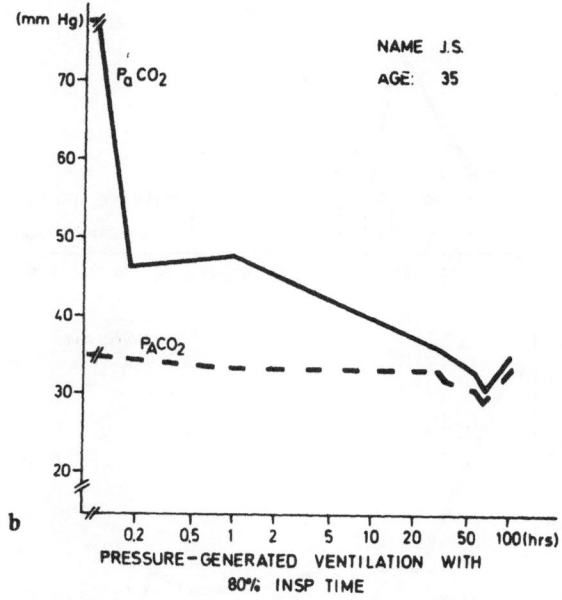

Fig. 3. Evaluation of peak insufflation pressure, compliance and PaO_2/FIO_2 (a) and at end-expiratory $PACO_2$ and $PaCO_2$ (b) at pressure-controlled ventilation with an I/E ratio of 4:1 in the same patient as in Fig. 2. Note the logaritmic time scale. Ventilator setting before starting pressure-controlled ventilation: minute volume = 14 l; peak pressure = 42 cm H_2O; PEEP = 8-10 cm H_2O; FIO_2 = 1.

peak pressure (Fig. 3a,3b). After three days with this breathing pattern, the arterial-end-expiratory CO_2 gradient was nearly zero, indicating that substantial disturbances of ventilation perfusion and of diffusion no longer existed. Chest x-ray after four days was nearly normal (Fig. 2b).

Case II:

A 54-year-old woman suffering from myocardial re-infarction had massive lung congestion. She was treated with volume-generated ventilation with pure oxygen and a PEEP of 4 cm H_2O. Gas exchange deteriorated. Severe hypoxaemia and respiratory acidosis contributed to cardiocirculatory shock. After implementation of pressure-controlled ventilation with an I/E ratio of 4:1, oxygenation improved very quickly (Fig. 4a).

It was then considered alarming that central venous pressure increased to about 20 mm Hg and mean pulmonary artery pressure to more than 35 mm Hg. An I/E ratio of 1:2 was therefore re-initiated. Indices of shock worsened and arterial blood gases deteriorated again dramatically. It was then determined that he only possibility was to apply an I/E ratio of 4:1 and accept high central venous and pulmonary artery pressures. Arterial oxygenation increased with about 100 mm Hg in 30 min, systemic blood pressure improved and signs of circulatory shock ceased.

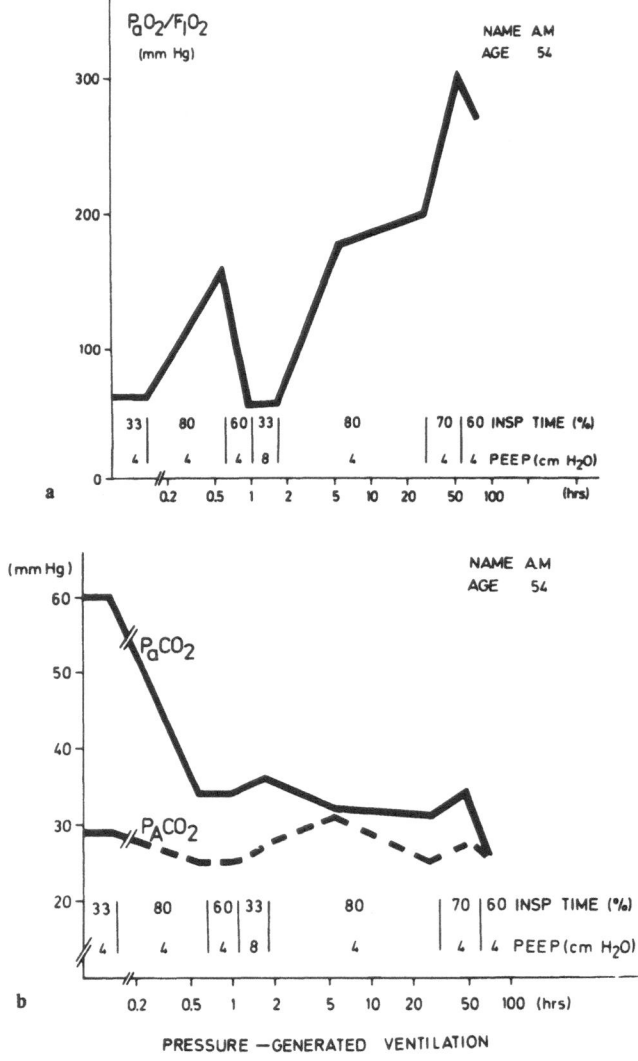

Fig. 4. Course of the PaO_2/FIO_2 (a) and of $PACO_2$ (b) after initiation of pressure-controlled ventilation. Note the logarithmic time scale.

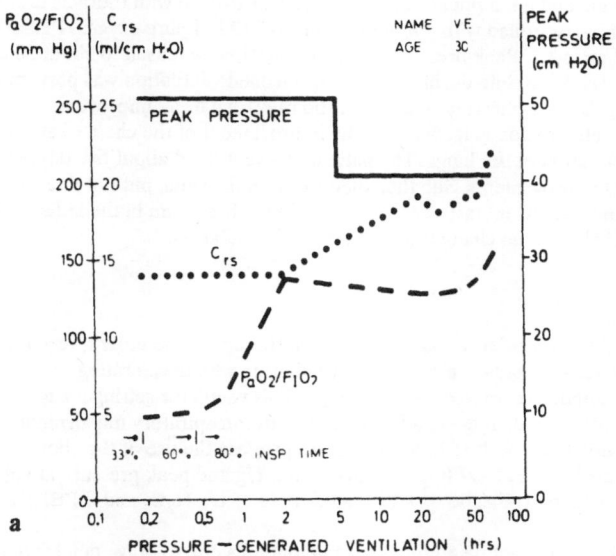

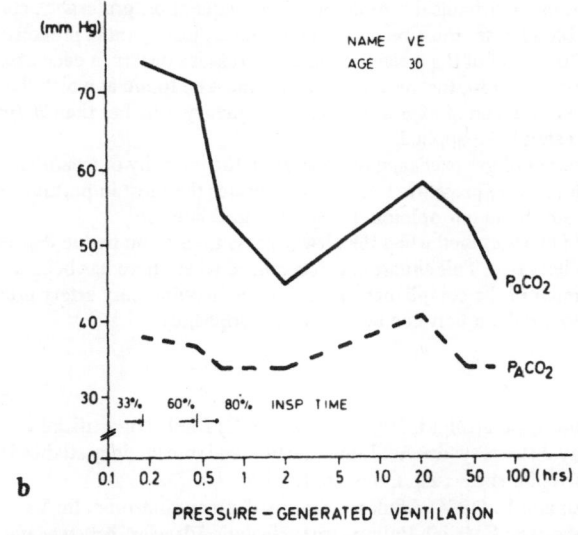

Fig. 5. Course of PaO_2/FIO_2 peak airway pressure, and compliance (a) and of $PaCO_2$ (b) during pressure-controlled ventilation. Note the logarithmic time scale.

Within the next 24 h, further improvement of gas-exchange and circulatory parameters occurred (Fig. 4a, 4b). Inspiratory time could be reduced to 70% and on the second day to 60%. The functional improvement of the lung corresponded to changes of chest x-ray. The patient could be transferred to the medical clinic in good condition a fortnight later.

Case III:

A 30-year-old woman with post-operative complications was admitted to the ICU. She had severe bilateral pneumonia and had already been ventilated for about two weeks. She had developed a septicaemia and got a pneumothorax which was treated with thoracic drainage. Ventilation was volume-controlled with a minute volume of 12 l of pure oxygen. PEEP was 4-8 cm H_2O and the I/E ratio 1:2. Peak pressure during inspiration increased to about 50 cm H_2O. The lengthening of inspiration time during pressure-controlled ventilation was performed step by step in this patient. Only application of an I/E ratio of 4:1 led to an improvement of oxygenation and CO_2 elimination (Fig. 5a, 5b). The improvement of the chest x-ray also indicated successful aeration of the lungs. The patient was ventilated about five days without any further functional improvements. She then died from septicaemia, but not due to respiratory insufficiency. Septic metastases were found in the brain and in the kidneys. The right side of the heart showed no changes due to artificial ventilation.

Discussion

The most important clinical consideration here is that in therapy of the most severe forms of respiratory distress syndrome, pressure-controlled ventilation with decelerating flow and an I/E ratio of 4:1 should be applied to improve gas exchange. This ventilator setting leads, furthermore, to less structural damage and less marked cardiorespiratory impairments then volume-controlled ventilation and PEEP. We therefore propose the use of the above ventilator pattern in RDS patients if the PaO_2/FIO_2 is below 80 mm Hg and peak pressure at volume-controlled ventilation with an I/E ratio of 1:1 is higher than 35 cm H_2O, and if PEEP exceeds 8 cm H_2O.

When the necessary PEEP cannot be attained in some patients due to grave defects in the lung parenchyma leading to excessive gas leakage to the pleural space, pressure-controlled ventilation with a very long inspiration time is the only possibility of maintaining the gas exchange with conventional mechanical ventilation. Very frequent or, preferably, continuous control of the arterial blood gases must be done to find the critical opening pressure that leads to a sudden improvement of the oxygenation. This pressure differs in each case of RDS. By monitoring the expiratory flow, the special frequency must be found at which this flow has not quite fallen to zero at the end of expiration. If this frequency is higher than 20/min, a PEEP of 5-10 cm H_2O should be applied.

When a clear improvement of gas exchange has occurred, the intensity of treatment is lowered by first reducing the inspiratory pressure. Pressure constitutes the most important aetiologic factor in severe lung damage, in our opinion. FIO_2 is reduced later on.

Inspiratory time should be shortened when the airway pressure is to an undue degree transmitted to the capillary bed. This situation is recognized when there has been a considerable improvement of the compliance, increased mean pulmonary artery pressure and a decrease of the pressure gradient between airways and oesophagus.

References

1. Lachmann B, Jonson B, Lindroth M, Robertson B. (1982) Modes of artificial ventilation in severe respiratory distress syndrome. Lung function and morphology studies in rabbits after wash-out of alveolar surfactant. Crit Care Med 10: 724-732.
2. Lachmann B, Danzmann E. (1984) Adult respiratory distress syndrome. In: Van Golde LMG, Batenburgh JJ, Robertson B. (eds). Pulmonary surfactant. Elsevier, Amsterdam: 505-548.
3. Lachmann B, Schairer W, Armbruster S, Van Daal GJ, Erdmann W. (1988) This book.

ELECTROCHEMICAL STUDY OF OXYGEN

BEHAVIOUR WITH LUNG SURFACTANT

Erna Ladanyi

Department of Occupational Health
University of Goettingen
Federal Republic of Germany
D-3400 Goettingen, Windausweg 2

INTRODUCTION

At the very beginning of its way to any tissue inhaled oxygen has to cross the so-called lung surfactant surface layer LSSL. Therefore, the properties of the LSSL especially those influencing its permeability, elasticity and stability are of great interest for oxygen transport studies.

Structure and properties of the LSSL

The physico-chemical behaviour of the LSSL is determined not only by its chemical composition but a great deal also by its molecular structure. LSSL is in fact a film lining the inside of the alveoli being positioned at the interface between the alveolaer air and the aqueous hypophase covering the alveolar cells, as demonstrated with a transmission electron micrograph of a cross-sectioned lung by Weibel and Gil (1968). LSSL and the precursors holding hypophase are the extracellulary, the precursors producing cells the intracellulary part of the lung surfactant system. The LSSL is the active part of the lung surfactant system. As concluded from data derived from lung washings (containing both LSSS and the whole hypophase) this film consists mainly of a great variety of phospholipid molecules (King, 1984) which are supposed to be strongly orientated at the interface the polar heads pointing towards the aqueous phase, the fatty acid chains towards the air. Immuno-gold electronmicroscopy (Ladanyi et al.,1988) of hypophase-isolated LSSL from lung washings (Ladanyi et al., 1987) and of in vitro re-assembled LSSL (Ladanyi et al., 1988) proved that the specific proteins found in the lung washings are also present in the LSSL, at least temporarily.

Role of the LSSL

Although its existence postulated already sixty years ago (von Neergard, 1929) the discussion about the role of LSSL is still in vigour. Control of the surface tension at the air/water interface with maintainement of a normal lung mechanics, defence against bacterial and noxious agents seem to be generally accepted atributes of the LSSL.

We will show in the following that the LSSL plays also a role in the transport of the oxygen to tissues as seen from our electrochemical data.

Some data concerning the electrochemistry of the LSSL

The components of the LSSL behave not only in vivo but also in vitro as typical surface-active agents: they adsorb at interfaces changing the properties of the given interface. The charged mercury electrode/saline interface is a convenient model interface for studying behviour of LSSL. The even spontaneously occuring adsorbtion at the mercury/saline interface results in easily measurable changes of some electrochemical parameters like the double-layer capacitance, the shape of the capacitance-potential curve of the saline (disappearance of the adsorption peak of the chlor ion and appearance of 3 concentration-dependent LSSL-specific maxima on it), the height and the position of the so-called electrocapillary maximum etc.

Electrochemical methods such as d.c. and a.c. polarography using dropping or hanging mercury electrodes proved to be valuable tools in studying such properties of the LSSL in vitro as surface activity (Ladanyi, Zugravu and Tomoaia, 1974; Ladanyi and Stalder, 1979), composition (Ladanyi, 1980), double-layer capacitance (Ladanyi et al.1988), the influence of environmental and occupational noxious agents on different LSSL-properties (Ladanyi et al.,1974; Ladanyi, 1980; Stalder and Ladanyi,1980; Ladanyi and Stalder, 1983; Ladanyi, 1986; Ladanyi,1987; Ladanyi and Stalder, 1987).

Very recent electrochemical data concern the molecularity of the phospholipid layer and the role of the 34 kD specific protein found in lung washings. Double-layer capacitance measurements at the air/water interface of the surface layer in vitro brought a strong evidence for the fact that the phospholipids of that peculiar composition as they are present in lung washings/amnniotic fluid have a tendency to build multilayers of odd numbers (probably trilayers) rather than a monolayer. The role of the proteins is apparently to transform these multilayers in a monolayer and vice-versa by transporting phospholipid molecules from/to the interface. This may occur during the inflation/deflation of the lung since in vitro compression of LSSL results in squeezing out of the protein molecules from the surface layer (Ladanyi et al., 1988).

Oxygen dissolved in saline is reduced at the mercury electrode at a well-defined potential value while the resulting reduction current is proportional to the oxygen concentration. Therefore, any change in the so-called d.c. half-wave- or half-peak-potential value of the oxygen reduction in presence of lung surfactant components will evidence surfactant-determined changes in the ability of the reducing species to reach the electrode. Further more, any change in the current intensity will testify changes in the dissolved oxygen concentration.

788

EXPERIMENTAL

The measurements using a dropping mercury electrode DME as
working electrode and two KCl saturated Ag/AgCl microelectrodes
as reference and auxiliary electrodes, respectively, were
carried out on a Polarecord E 506 (Metrohm, Herisau, Switzerland) by applying linear voltage sweep. The dropping time was
mechanically adjusted to 1.4 sec. The measurements on a hanging
mercury drop electrode HMDE as working electrode, a KCl
saturated Ag/AgCl as reference electrode and a Pt wire as
auxiliary electrode were carried out on a PAR Model 170 (PAR,
Princeton, N.J., USA) polarograph by applying triangular
voltage sweep to the cell. The HMDE had an extremly thin wall
at the tip and was connected to a special device moving it
smoothly up and down through the surface layer into and out of
the vessel. In this way, as the HMDE crossed the surface layer
from the air-side, it became covered with the surface layer.

The biological material consisted of natural lung washings
from humans and rats, of extracted lipids and resuspended
proteins of these lung washings, lipids extracted from human
amniotic fluid and the 34 kD surfactant specific protein.

Saline was prepared of p.a. grade NaCl (Merck, Germany) in
deionized, double-distilled water. The 34 kD surfactant specific protein was a gift of Prof. P. von Wichert
(Universitätspoliklinik Marburg, Germany), the amniotic fluid
extracts were obtained from Prof. M. Shinitzky (The Weizmann
Institute of Science, Rehovot, Israel).

To study the influence of LSSL on oxygen behaviour at the
charged DME and HMDE non-de-aerated saline was used throughout
as supporting electrolyte. In experiments with the DME the
current-potential curves of the saline with and without
aliquots of lung washings were recorded. The HMDE was used to
find out separately the influence of the lipid and of the
protein components of LSSL on the oxygen transport. Therefore,
the HMDE was first let to cross the carefully cleaned surface
layer until it reached the bulk of the saline and subsequently
the current-potential curve of the oxygen reduction at the
uncovered electrode surface was recorded. Then a condensed
layer of lipids was spread over the saline, a new drop of the
HDME was let to cross it into the bulk followed by a recording
on this new drop. Finally, one of the protein solutions was
carefully injected beneath the lipid layer and the current-
potential curve recorded again on an other new drop.

RESULTS AND DISCUSSION

Fig.1 presents the current-potential curves of the oxygen
reduction in saline (curve 1) and in saline with added lung
washing (curve 2). As oxygen is reduced in two steps, the
curves visualize the two reduction waves of the oxygen at a
dropping mercury electrode. The potential value at the half of
the wave height $E_{1/2}$, the so-called half-wave potential is well
defined for oxygen in the given supporting electrolyte (in this
case the saline). As seen, in presence of lung washing $E_{1/2}$ is
shifted with o.1 V towards more negative potential values. This
means that the oxygen, in order to be reduced, has to overcome
a bigger energetic barrier, when lung washing is present.

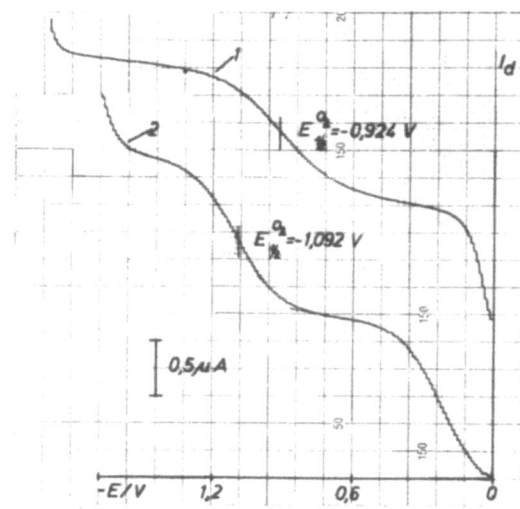

Fig.1.   Current-potential curves of oxygen reduction   in
saline with (curve 2) and without (curve 1) lung washing

Obviously,   the   surface-active components   of   the   lung
washing adsorbed at the mercury electrode surface and formed a
LSSL-like film which hindered the oxygen reduction. The impede-
ment depends on lung washing concentration, as seen in Fig.2
(curve 1).   The less the so-called "original washing" diluted
with saline the bigger the $E_{1/2}$ shift.

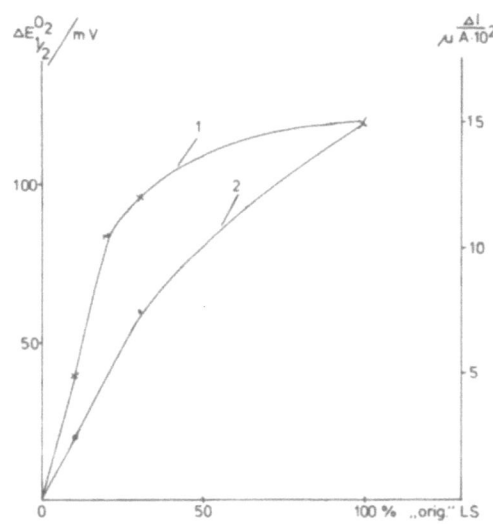

Fig. 2.   Half-wave   potential shift $\Delta E_{1/2}$ (curve 1) of   oxygen
           reduction   and   current   intensity   augmentation ΔI
           (curve 2) depending on lung washing concentration

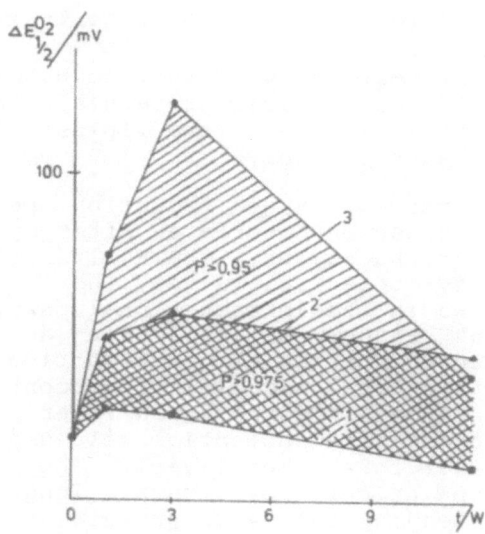

Fig. 3.   Long-time-influence of a single dust-treatment on the
           lung surfactant of laboratory animals compared to
           controls. $\Delta E_{1/2}$ of the oxygen reduction obtained with
           lung washings of animals treated with OX 50 (curve
           1), Bŏhmit (curve 2) and DQ 12 (curve 3)

The curve has the shape of a Langmuir-type isotherme which
shows that the process has a saturation tendency - well known
with adsorption at limited surfaces. In analogy, if the aqueous
hypophase/ alveolar air interface in the lung is tightly
covered with the LSSL, the oxygen transport through the film
can be more difficult. This can be the case at the deflation,
when the alveolar surface is contracted thus the surface
concentration of the LSSL reaches its maximum (when the LSSL is
probably a multilayer). In opposite, if the lung is inflated
and the LSSL has its minimal surface concentration (when the
LSSL is a monolayer) the oxygen exchange with the alveolar
lumen is facilitated. This would make sense, since the inflation requires higher oxygen permeability than the deflation.

        Current intensity, as we know, is proportional to the
oxygen concentration. The more lung washing there is in the
saline the higher is the current intensity (Fig. 2, curve 2).
This means, that the tighter the LSSL is the more oxygen
cumulates in the hypophase. Thus, the hypophase seems to be a
place of storage for the oxygen.

        Fig. 3 shows oxygen reduction shifts $\Delta E_{1/2}$ with lung
washings obtained from animals treated with different dusts
and from controls. As seen, any dust-treatment , even with the
non-silicogene OX 50 (curve 1) causes an augmentation of the
$\Delta E_{1/2}$ compared to the value for the untreated animals. The
highly silicogene DQ 12 (curve 3) causes the biggest and even
after 3 months persisting effect on $E_{1/2}$. The obtained data are
significant. This result could be interpreted in terms of the
stress-theory of Selye: because of the stress caused by the
dust there is an overproduction of surfactant thus there is a

tighter LSSL which defends the alveolar cells from the dust.

It seemed interesting to us to find out which the role of the lipids and of the specific protein(s) in the $E_{1/2}$ shift is. In Fig. 4 the influence of the lipids and of the 34 kD protein are presented separately.

Curve 1 is the 1st oxygen reduction peak obtained in saline, thus on a clean HMDE surface. After spreading amniotic fluid extract over the saline, the $E_{1/2}$ shifts with 0.5 V towards more negative potential values. The lipids alone (which build, as we showed in the introductional part, a multilayer) seem to represent rather a shield than a barrier to the crossing charge needed for the oxygen reduction. In opposite, human 34 kD protein even in a minimal concentration of 0.18 µg/ml, injected beneath of the lipid layer produces a $E_{1/2}$ shift towards more positive potential values (and a change of the lipid multilayer into a monolayer). The same holds when instead of the amniotic fluid extract a lung washing extract is spread and instead of the 34 kD protein the re-dissoluted lung washing rest of the lipid extraction is injected beneath the lipid surface layer.

These results suggest that the LSSL represents a certain energetic barrier to the crossing oxygen. The level of this barrier depends on the composition and structure of the LSSL. Phospholipids and proteins act antagonistic concerning the LSSL permeability to oxygen. Since the protein molecules are squeezed out from the lipid layer at compression (according to our already mentioned immuno-electronmicroscopic results) and compression occurs at deflation of the lung, it can be speculated, that the LSSL actually acts as an one-way gate for the oxygen. Our results which show a surfactant concentration-

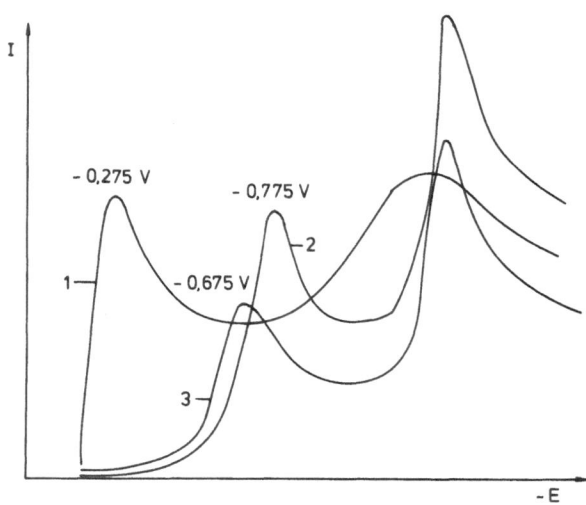

Fig. 4.   Influence of surfactant lipids (curve 2) and specific proteins (curve 3) on the peak-potential of the oxygen reduction in saline (curve 1)

dependent accumulation of oxygen in the lung washings compared to saline are in agreement with this hypothesis.

SUMMARY

At the very beginning of its way to any tissue oxygen has to cross the lung surfactant surface layer LSSL. The influence of LSSL on oxygen was studied in vitro at model interfaces using sensitive d.c. electrochemical techniques. LSSL adsorbed from lung washings at the dropping mercury electrode/saline interface shifted the half-wave potential of oxygen reduction $E_{1/2}$ towards more negative values. The magnitude of $\triangle E_{1/2}$ was dependent on the lung washing concentration and showed a saturation tendency. The intensity of the reduction current was higher in saline samples with surfactant than without and had a rather rising tendency.

Lipids extracted from lung washings or amniotic fluid being spread as a condensed multilayer at the saline/air interface and re-adsorbed on a hanging mercury electrode surface produced a several time bigger shift of oxygen reduction $E_{1/2}$ towards negative potential values than the lung washings did. The surfactant specific 34 kD protein and lung washing proteins injected beneath the lipid layer produced a positive potential shift.

Dust-treatment-caused changes in surfactant quality/ quantity resulted in bigger $\triangle E_{1/2}$ values.

The data suggest that the LSSL represents a certain energetic barrier or even an one-way gate to the crossing oxygen the efficiency of which depends on the composition/structure of the LSSL and that the aqueous hypophase under the LSSL in the alveoli is a place of storage for the oxygen.

REFERENCES

King, R.J., 1984, Isolation and chemical composition of pulmonary surfactant, in " Pulmonary Surfactant", Robertson, B., van Golde, L.M.G., Batenburg,J.J., ed., Elsevier, Amsterdam

Ladanyi, E., Zugravu, E., Tomoaia,M.,1974, Electrochemical Methods in Surface-Activity Studies of Lung Surfactant.I. Polarographic Maximum Suppressing Ability of Lung Surfactant, Int. Arch. Arbeitsmed., 33:245

Ladanyi, E., Stalder,K., 1979, Alternating current-tast-polarographic determination of surface activity of lung surfactant, J. Electroanal.Chem., 99:321

Ladanyi, E., 1980, Polarographische Elektrosorptions-analyse des oberflächenaktiven Systems der lunge (Lung surfactant), Dissertation, Technische Universität Clausthal

Ladanyi, E., Stalder, K., 1980, Changes occuring in the lung surfactant under the action of inhalative occupational substances, Verh. Dtsch. Ges. Arbeitsmed., 20:519

Ladanyi, E., Stalder, K.,1983, Contribution to the medical

importance of dusts resulting by the use of agricultural machins, Verh.Dtsch.Ges.Arbeitsmed., 23:523

Ladanyi, E., 1986, Inhalative Noxen und das Surfactant-System der Lunge, Prax.Klin.Pneumol., 40:465

Ladanyi, E., Möbius, D., Stalder, K., von Wichert, P., 1987, Structure of isolated lung surfactant monolayer, Symposium on Membrane Lipids, 20-21 March,Sintra, Portugal, ACTAS do INSTITUTO de BIOQIMICA (in press)

Ladanyi, E., Stalder, K., 1987, Modellversuche zum Einfluß von Formaldehyd auf das Lungensurfactant, Verh.Dtsch.Ges.Arbeitsmed.,27:545

Ladanyi, E., 1987, Present knowledge in the field of lung surfactant electrochemistry, J.Bioel.Bioenerg., (in Press)

Ladanyi, E., Miller, I., Popovitz-Biro, R., Marikovsky, J., von Wichert, P., Müller, B., Stalder, K., 1988, Molecular structure of the extracellular surface-layer of the human lung surfactant, 3d International Symposium, Basic Research on Lung Surfactant, Marburg, 12-14 September

Neergard, K. von, 1929, Neue Auffassungen über einen Grundbegriff der Atemtechnik. Die Retraktionskraft der Lunge, abhängig von der Oberflächenspannung der Alveolen, Z.Ges.Exp.Med.66:373

Stalder, K., Ladanyi, E., 1980, Changes occuring in the lung surfactant under the action of inhalative occupational substances, Verh.Dtsch.ges.Arbeitsmed., 20:519

Weibel,E., R., Gil,J., 1968, Electron microscopic demonstration of an extracellular duplex lining layer of alveoli. Resp. Phisisiology, 4:42

CHINESE HERBAL MEDICINE INCREASES TISSUE OXYGEN TENSION

Robert A. Linsenmeier, Thomas K. Goldstick, and Shu-Lun Zhang*

Departments of Biomedical Engineering and Chemical Engineering,
Northwestern University, Evanston, IL 60208 USA
*Microcirculation Research Laboratory, Capital Institute of Pediatrics, No. 1
Yue Tan South Street, Nan Li Se Road, Beijing, People's Republic of China

INTRODUCTION

Anisodamine, an alkaloid from the Chinese solancea plant (Anisodus Tanguticus), found only in the foothills of the Himalayas near Tibet, has been employed for centuries as a Chinese herbal medicine. The natural alkaloid was isolated in 1965 and has now been synthesized. It is structurally almost identical to atropine. This synthetic anisodamine, known as compound 654-2, has the identical pharmacological and clinical properties as the naturally occurring one. In the present study, the synthetic compound was used. Anisodamine is said to have been used in China to treat conditions as varied as septic shock, lobar pneumonia, glomerular nephritis, migraine headache, pancreatitis, retinochoroiditis, diabetes, and gangrene. Although its mechanism of action has not yet been fully elucidated, it is known to act as a weak vasodilator in some tissues. It has also been found to inhibit platelet aggregation, reduce coagulation, increase the flexibility of erythrocytes, and block calcium channels. It seemed to us that much of its clinical efficacy might also be explained by an increase in tissue PO_2. Until now this increase has only been measured in skin. As a part of a larger study of retinal tissue oxygen transport, we have attempted to measure the effect of anisodamine on the retinal tissue PO_2 in cats.

METHODS

Experiments were performed on six, dark-adapted, adult cats using methods previously described (Linsenmeier, 1986). Surgery for the placement of cannulas and for eye preparation was done under sodium thiamylal anesthesia. All subsequent experimental measurements were performed under urethane anesthesia (200 mg/kg loading dose followed by 20-40 mg/kg·hr). Animals were paralyzed with pancuronium bromide (0.3 mg/kg·hr) and artificially ventilated. Body temperature was kept close to 38^0C with a regulated heating pad. Arterial PO_2, PCO_2, and pH were measured with a blood gas analyzer (model 158, Corning Medical and Scientific, Medfield, MA) and were maintained within the normal range for cats ($P_aO_2 \approx 100$ mm Hg, $P_aCO_2 \approx 30$ mm Hg and $pH_a \approx 7.40$) by respirator adjustment and the addition of small amounts of 100% oxygen to the inspired gas. The anisodamine studies reported here were performed at the end of a series of other, unrelated, retinal oxygen studies, approximately 18 hours after the initial anesthesia.

A double-barreled microsensor, with one barrel an oxygen microelectrode (Whalen, Riley and Nair, 1967) and the other barrel a voltage microelectrode (Linsenmeier and Yancey, 1987), was introduced into the intact right eye through a 15 gauge (1.83 mm OD) hypodermic needle which penetrated the sclera about 10 mm posterior to the limbus (Figure 1). A silicone rubber boot system allowed the microsensor to move, but prevented the leakage of vitreous. The microsensor was advanced through the vitreous humor with a microdrive (model 607WPC, Kopf Instruments, Tujunga, CA) toward the area centralis until the saline-filled voltage recording barrel indicated contact with the retinal surface (Linsenmeier, 1986). It was then withdrawn into the vitreous so that it was 150 to 300 μm from the retinal surface. The oxygen barrel had a recessed gold cathode, and the overall size of the electrode tip including both electrodes was 3 to 6 μm. The oxygen cathode was about 1 μm in diameter and the recess depth to diameter ratio was always greater than 5. Electrodes were polarized at -0.7 V with respect to a Ag/AgCl wire in the vitreous. They were calibrated in saline at 37°C before and after experiments and had sensitivities in the range of 0.02 to 0.2 pA/mm Hg PO_2 (where one pA equals 10^{-12} Amperes). In the absence of oxygen, the electrical current never exceeded the equivalent of ± 3 mm Hg PO_2. Electrodes were found to be insensitive to anisodamine even at a concentration of 50 mg/100 ml in the calibration bath, over six times the maximum in the cat's blood.

The oxygen measurement system consisted of a polarizing source and a Keithley 614 Electrometer (Keithley Instruments, Cleveland, OH), followed by a buffer amplifier to prevent loading of the electrometer output, a tape recorder (Store 4DS, Racal Recorders, Sarasota, FL), and a chart recorder (440, Gould Inc., Cleveland, OH). Oxygen signals were analyzed off-line by a PDP 11/23 computer system (Digital Equipment Corp., Maynard, MA).

The experimental protocol involved recording preretinal vitreous PO_2 for about half an hour before administering anisodamine. During most of this control period, the cat breathed room air and the PO_2 was very stable. The animal was also subjected to two brief (2 to 3 min) episodes of hyperoxia (100% oxygen inspired) during this control period. A dose of 5 mg/kg synthetic anisodamine HCl (ampules of solution obtained from Beijing Pharmacy, Beijing

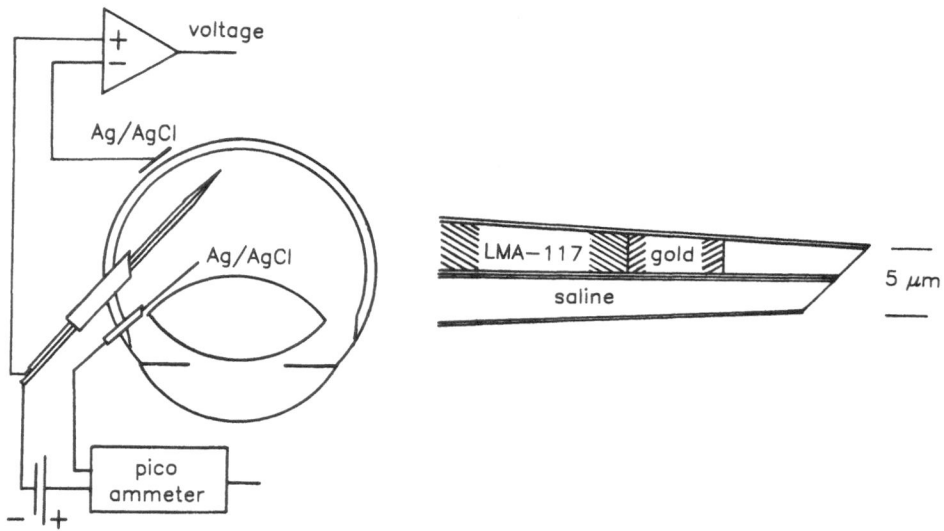

Fig. 1. Schematic diagram of the measurement system and microelectrode employed to obtain the preretinal vitreous PO_2 in cats. Double-barrelled electrodes were inserted into the intact eye through a guide needle and advanced to the retina. The oxygen cathode was polarized at -0.7V with respect to a chlorided silver wire in the vitreous. The second barrel, referenced to an electrode behind the eye, recorded voltage. This was used in these experiments primarily to determine when the electrode touched the retina. The oxygen microelectrode had a gold cathode recessed a distance of at least 5 cathode diameters into the tip. The gold was electroplated onto a low melting-point alloy (LMA) consisting of 44.7% Bi, 22.6% Pb, 19.1% In, 8.3% Sn, and 5.3% Cd. This alloy melts at 117°F and changes size less after casting than the more commonly used Wood's metal. The second barrel was filled with isotonic saline.

Peoples Republic of China, by S.-L.Z.) was then given intravenously over 10 min, as a 3 mg/ml solution. Following the injection, the preretinal PO_2 was recorded for at least an hour, and two more episodes of hyperoxia were induced. In the first two experiments (cats #55 and #56) the injection was given somewhat faster, over 3 and 5 min, but very similar results were obtained so that all results have been considered together.

Inexplicably, in one of the six cats (cat #60) anisodamine appeared to have no effect on either blood pressure or PO_2. A second, repeat administration of anisodamine five hours later to this same cat, with measurements made at that time using a different microsensor, again exhibited no effect. All data from this apparently anisodamine-nonresponder have therefore been omitted.

RESULTS

The mean femoral arterial blood pressure always dropped following the anisodamine infusion (Table I). At the beginning of the infusion the pressure dropped rapidly an average of 26%. It remained at that level for about 20 min and then returned to the previous control value. The rate of return was quite variable. Timed from the start of the 10 min anisodamine infusion, the return to essentially the preanisodamine blood pressure took from 25 to 70 min. From the limited amount of data available, it appears that neither the amount of the drop in pressure nor the recovery time were especially dependent upon the initial blood pressure or the rate of injection, even for an injection as rapid as 3 min compared to the usual 10 min.

Figure 2 shows a typical record of preretinal vitreous PO_2 slightly before, during and after the anisodamine infusion. The cats were always breathing room air when the anisodamine was administered. The preretinal vitreous PO_2 has been shown to be representative of the retinal tissue PO_2 (Alm and Bill, 1972; Linsenmeier et al., 1981) and so the preretinal vitreous PO_2 measurements will be designated simply "retinal PO_2" hereinafter. In general the blood pressure typically remained below the preanisodamine level for at least 25 min, i.e., throughout the time of this record. Nevertheless, the retinal PO_2 rose significantly. An increase was observed in all five anisodamine-responsive cats studied (Table I). The average increase $\pm$SD was 20% $\pm$ 9% which is statistically significant (P < 0.01). The effects on retinal PO_2 lasted at least one hour.

Figure 3 shows the effects of anisodamine on the transient following the change in breathing gas from room air to 100% oxygen. Values for the maximum retinal PO_2 breathing 100% oxygen are also given in Table I. Just as when the cats were breathing room air, anisodamine also increased the retinal PO_2 when they were breathing 100% oxygen, again by about 20% on the average (Table I). But in the 100% oxygen case, the increase was not found to be statistically significant (P > 0.2). This was largely because of the results from one cat (#56) where the value actually decreased, possibly from other causes e.g., movement of the microsensor, adaptation to hyperoxia, etc. One of the

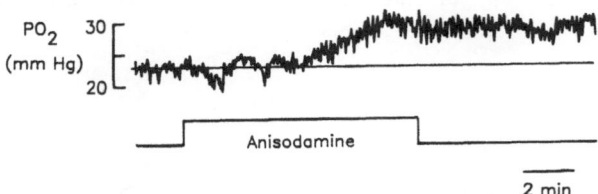

Fig. 2. Typical retinal PO_2 transient following infusion of 5 mg/kg anisodamine over 10 minutes (cat #58). Discontinuities in the lower line mark beginning and end of infusion. The animal was breathing room air throughout this record.

TABLE I

Effect of Anisodamine on Preretinal PO$_2$ Breathing Room Air and 100% Oxygen, and also on Mean Arterial Blood Pressure

Cat No.	PO$_2$, mm Hg Preanisodamine Air	O$_2$*	Postanisodamine Air	O$_2$*	Percent Change with Anisodamine Air	O$_2$	MAP† mm Hg Pre	Min#	Percent Change with Anisodamine
55	25.5	32.6	32.6	42.9	27.8%	31.6%	122	85	-30%
56	26.5	65.7	30.1	53.7	13.6	-18.3	134	100	-25
57	34.4	46.9	43.8	56.1	27.3	19.6	120	100	-17
58	22.9	34.9	28.4	54.0	24.0	54.7	not measured		
59	21.4	38.6	22.9	40.5	7.0	4.9	160	110	-31
Mean	26.1	43.7	31.6	49.4	20.0	18.5	134	99	-26
SD	5.0	13.4	7.7	7.2	9.2	27.5	18	10	7
P					<0.01	NS	<0.01		<0.01

*Maximum PO$_2$ following 2 min of 100% oxygen breathing, the average of two transients.

†MAP = Mean femoral artery blood pressure.

#Min = Minimum MAP following anisodamine administration.

most striking features of the anisodamine effect was the apparent increase in the hyperoxic rate of rise of retinal PO_2, which almost invariably rose much faster following anisodamine.

DISCUSSION

In urethane-anesthetized, dark-adapted cats, intravenous anisodamine administration caused an average transient blood pressure decrease of 26% while simultaneously causing an average retinal PO_2 increase of 20%. Very similar transients in blood pressure have previously been observed in rats (Xiu et al., 1985), rabbits (Fan et al., 1986), and dogs (Guo and Sun, 1982; Shi et al., 1986). Apparently this effect of anisodamine is caused by myocardial depression (Shi et al., 1986) possibly a result of the blocking of calcium channels by anisodamine (Tang et al., 1985) as well as the blocking of α-adrenergic receptors (Varma and Yue, 1986).

The increased retinal PO_2 with anisodamine, in the face of decreased arterial pressure, was rather surprising. However, vasodilatation has previously been observed in some other tissues (Xiu et al., 1985; Fan et al., 1986; Xiu and Intaglietta, 1986). Others have also observed decreased whole blood viscosity (Xiu et al., 1982; Luo et al., 1987), decreased platelet aggregation (Ge et al., 1985; Zhang et al., 1987; Shao et al., 1987), inhibition of thromboxane synthesis (Xiu et al., 1982; Zhang et al., 1987), inhibition of prostaglandin PGI_2 synthesis (Xiao and Chen, 1985), and increased erythrocyte deformability (Luo et al., 1987). Nevertheless it seems to be significant that we for the first time have found an increased PO_2 in neural tissue. This increased retinal PO_2 could have tremendous clinical benefits in treating retinal vascular obstructive disease.

There were two somewhat unexpected results. First, one of the six cats was inexplicably an anisodamine-nonresponder. Perhaps this is a feature of any pharmacological agent. Second, right after the change in breathing gas from room air to 100% oxygen, the rate of increase in retinal PO_2 seemed to be speeded up by anisodamine. Although further study is essential for complete understanding, the rate of gas exchange in the lungs may have influenced this part of the tissue PO_2 transient. Anisodamine is thought to improve lung blood flow (Fan et al., 1986) and perhaps this represents one possible explanation for the faster retinal PO_2 transient.

The most significant finding was that anisodamine increased retinal PO_2 by about 20%. Since all of the other pharmacological actions of anisodamine are thought to be relatively

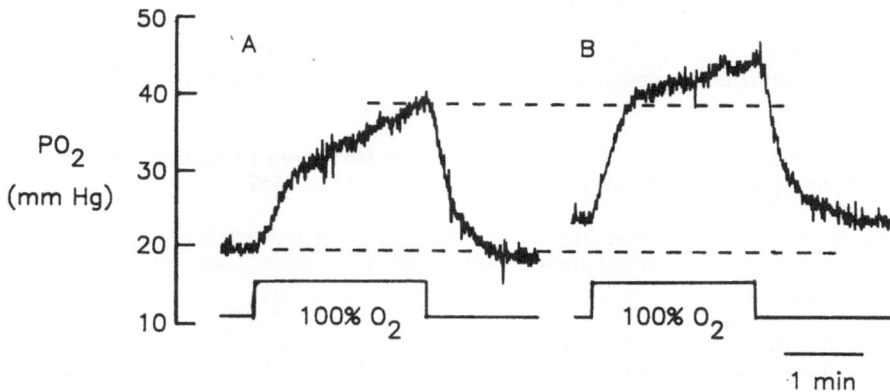

Fig. 3. Typical retinal PO_2 transient following a two-minute episode of breathing 100% oxygen preceded and followed by breathing room air (cat #59). Discontinuities in the lower line mark the beginning and the end of the two-minutes of 100% oxygen breathing.
A. Before anisodamine infusion.
B. After anisodamine infusion.

benign, except possibly for its hypotensive effects, the increase in tissue PO_2 could have great clinical significance in the treatment of tissue hypoxia which undoubtedly plays a role in diabetic vascular disease and sickle cell crises.

SUMMARY

An herbal medicine widely used in China, anisodamine, was investigated to determine its effect on the tissue PO_2 of an important neural tissue, the retina. The anisodamine was injected intravenously over 10 min into anesthetized cats. Although it reduced mean blood pressure an average of 26%, anisodamine simultaneously increased tissue PO_2 an average of 20%. It may therefore be useful in treating clinical conditions thought to be secondary to tissue hypoxia, such as diabetic retinopathy and sickle cell crises.

ACKNOWLEDGEMENTS

The research was supported by U.S. Public Health Service, National Eye Institute grants EY-05034 and EY-07041, and by the Whitaker Foundation. We are grateful for the technical assistance of HD Roh and RD Braun and for the advice and encouragement in the preliminary stages provided by Drs. TA Deutsch, KA Neely and JT Ernest.

REFERENCES

Alm, A and Bill, A, 1972, The oxygen supply to the retina. I. Effects of changes in intraocular and arterial blood pressures, and in arterial PO_2 and PCO_2, on the oxygen tension in the vitreous body of the cat, Acta Physiol. Scand., 84: 261-274.

Fan, Y, Yang, L, Wang, X, Sun, H, Sun, Y, Hu, F, Zhou, W, and Xiao, Y, 1986, Effects of anisodamine and scopolamine on pulmonary circulation in rabbits, Zhongguo Yaoli Xuebao, 7 : 117-121.

Ge, M, Luo, Z, Luo, H, and You, J, 1985, Effect of anisodamine (compound 654-2) on disaggregating platelet microaggregates in the pulmonary circulation of dogs with endotoxemia, Hunan Yixueyuan Xuebao, 10 : 7-11.

Guo, H and Sun, R, 1982, Effects of intravenously injected anisodamine (654) on hemodynamics in normal anesthetized and endotoxin-shock dogs, Kexue Tongbao, 27 : 320.

Linsenmeier, RA, 1986, Effect of light and darkness on oxygen distribution and consumption in the cat retina, J. Gen. Physiol., 88 : 521-542.

Linsenmeier, RA, Goldstick, TK, Blum, RS, and Enroth-Cugell, C, 1981, Estimation of retinal oxygen transients from measurements made in the vitreous humor, Exp. Eye Res., 32: 369-379.

Linsenmeier, RA and Yancey, CM, 1987, Improved fabrication of double-barreled recessed cathode O_2 microelectrodes, J. Appl. Physiol., 63 : 2554-2557.

Luo, G, Shen, C, Yang, X, and Zhang, Y, 1987, Effects of ligustrazine and anisodamine on whole blood viscosity and erythrocyte deformability in rabbits, Zhonghua Yixue Zazhi, 67 : 607-609.

Shao, M, Xu, S, Li, C, and Xue, Q, 1987, Effects of nifedipine and anisodamine on platelet membrane fluidity and platelet aggregation in rats, Zhongguo Yixue Kexueyuan Xuebao, 9 : 381-385.

Shi, J, Miao, Z, Yang, M, Yu, Z, Zhang, S, and Yang, X, 1986, Negative effects of anisodamine on cardiovascular function in anesthetized dogs, Tianjin Yiyao, 14 : 135-138.

Tang, C, Yang, X, Wang, X, Zhao, Q, and Su, J, 1985, Calcium-antagonist effect of anisodamine (654-2), Beijing Yixueyuan Xuebao, 17 : 165-168.

Varma, DR and Yue, TL, 1986, Adrenoceptor blocking properties of atropine-like agents anisodamine and anisodine on brain and cardiovascular tissues of rats, <u>Br. J. Pharmacol.</u>, <u>87</u> : 587-594.

Whalen, WJ, Riley, J, and Nair, P, 1967, A microelectrode for measuring intracellular PO_2, <u>J. Appl. Physiol.</u>, <u>62</u> : 798-801.

Xiao, D and Chen, H, 1985, Changes in the plasma levels of PGE, $PGF_{2\alpha}$ and 6-keto-$PGF_{1\alpha}$ in canine endotoxic shock, <u>Chin. Med. J. (Beijing Engl. Edn.)</u>, <u>98</u> : 501-506.

Xiu, R-J, DeLano, FA, and Zweifach, BW, 1985, Influence of anisodamine on microhemodynamics in skeletal muscle preparations, <u>in</u> "Advances in Chinese Medicinal Materials Research", ed. by HM Chang, HW Yeung, WW Tso, and A Koo, World Scientific Publishing Company, Singapore [international symposium held in 1984], pp. 545-552.

Xiu, R-J, Hammerschmidt, DE, Coppo, PA, and Jacob, HS, 1982, Anisodamine inhibits thromboxane synthesis, granulocyte aggregation, and platelet aggregation, <u>JAMA</u>, <u>247</u> : 1458-1460.

Xiu, R-J and Intaglietta, M, 1986, Computer analysis of the microvascular vasomotion, <u>Chin. Med. J. (Beijing Engl. Edn.)</u>, <u>99</u> : 351-360.

Zhang, S, Chang, AM, Li, CF, Li, ZJ, Yin, ZJ, Zhao, X, and Liang, SL, 1987, Mechanism of the therapeutic effect of anisodamine in disseminated intravascular coagulation: Study of platelet adhesion and aggregation, malondialdehyde, thromboxane B_2, 6-keto-prostaglandin $F_{1\alpha}$, and microcirculation, <u>Exp. Hematol.</u>, <u>15</u> : 65-71.

THE INFLUENCE OF A TEMPORARY CESSATION AND REPERFUSION OF INTESTINAL BLOOD FLOW ON THE LEVEL OF HEPATIC LIPID PEROXIDES.

J. Lutz and A. Augustin

Physiologisches Institut der Universitaet Wuerzburg
Roentgenring 9, D-8700 Wuerzburg, F.R.Germany

INTRODUCTION

The liver is an organ with a very high metabolic rate and thereby also a target for the formation and activity of free oxygen radicals. Most substances to be detoxified reach the liver as they progress from the intestinal circulation through the portal vein. During the continued inflow of such substances and by local processes, a certain amount of lipid peroxides is generated under the influence of simultaneously produced free oxygen radicals. Such lipid peroxides react with thiobarbituric acid in the same way as malondialdehyde. Thus the content of lipid peroxides can be expressed in terms of malondialdehyde.

The content of malondialdehyde or its precursors (MDA) in intestinal tissue is lower than that of the liver. Under the condition of a temporary occlusion of intestinal blood supply, however, the level of MDA in the intestine increases more than sixfold (Augustin and Lutz, 1988). Therefore, we asked the question how the MDA level of the liver would react to these large changes in lipid peroxide generation by the intestine. The influence of treatment with perfluorochemicals and that of free oxygen radical scavengers on this reaction was also examined. In addition we analyzed the effect of antibiotic pretreatment.

MATERIALS AND METHODS

The livers and intestines of 84 male Wistar rats of 254 ± 37 g body weight were tested for their level of lipid peroxides. A reaction with thiobarbituric acid was used in a modification of the methods of Ohkawa et al. (1979) and Yagi (1984). The results were expressed as nmoles of MDA/g wet tissue; values in the figures are given as means $\pm$ SEM.

The portal inflow to the liver was stopped for 90 (or 45) min by a reversible occlusion of the superior mesenteric artery as described by us before (Hamar et al., 1987). Immediately at the end of the occlusion period some of the animals were killed and tissue samples of liver and intestine were taken for the determination of MDA. The remaining animals were killed at different time periods after the beginning of recirculation and again tissue samples from both organs were obtained. Survival from occlusion shock beyond 6 h was enabled by fluid replacement with hydroxyethyl starch in Ringer (HES, 6%, one dose of 50 ml/kg b.wt. at the end of occlusion and a further dose of 25 ml/kg, 2.5 h later) in combination with a coctail of methylprednisolone (30 mg/kg), Totocillin[R] (ampicillin + oxacillin, together 1g/kg), Trasylol[R] (apronitin, 30,000 U/kg) and guanethidine (0.5 mg/kg) at the end of the occlusion period (Lutz et al., 1985).

Treatment regimens with perfluorochemicals (PFC, Fluosol-DA[R], 8 g/b.wt.) together with oxygen breathing (FI $O_2 \sim 0.9$) were also tested, alone as well as with free oxygen radical scavengers (superoxide dismutase, SOD, 10 mg/kg = 30,000 U/kg b.wt.) plus catalase (CAT, 10 mg/kg = 340,000 U/kg) or allopurinol (100 mg/kg b. wt.).

Experiments with antibiotic pretreatment were performed by giving polymyxin B (50 mg/kg) together with neomycin sulfate (500 mg/kg) orally, given in three portions daily during the 5 days preceeding the occlusion.

RESULTS

In control livers the level of MDA was about twice that of the intestine. At the end of the occlusion period, when the intestinal values had increased more than sixfold, the level of lipid peroxides in the liver had decreased to a quarter of the initial level. Thus a large amount of MDA was generated and retained only in the intestine. Two and one half (2.5) h after the end of occlusion and recirculation, the content of MDA in the intestine was further increased to more than 10 fold the initial value, whereas that in the liver remained nearly constant. A longer test period was not possible without special precautions since all animals died. Under the protection of HES in combination with the above mentioned coctail about 50 % of the animals survived; the intestinal content of MDA increased only 1.5 fold 2.5 h after the end of occlusion. The MDA content of the liver remained low. After 15 h there was a diminishing influence of the therapy and after 48 h the MDA level of the intestine had risen again distinctly, whereas that of the liver was in the range of the control value. (Fig. 1).

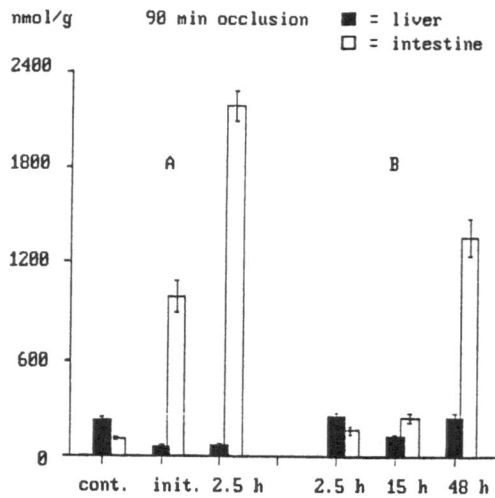

Fig.1. Content of lipid peroxides, expressed in terms of TBA-reactive substances in liver and intestinal tissue, before and at different times after 90 min occlusion of the sup. mes. artery (A). Values at the right (B) were obtained after fluid therapy, since otherwise most animals died after 2.5 h.

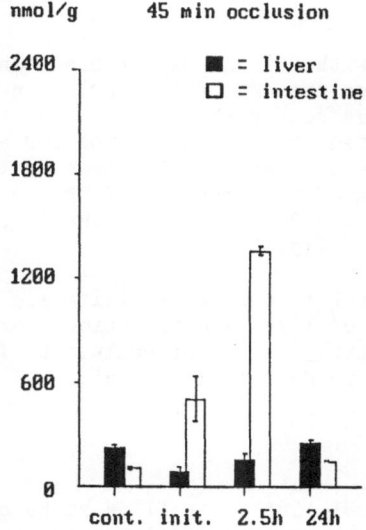

TBA-REACTIVE SUBSTANCES

nmol/g   45 min occlusion

■ = liver
□ = intestine

Fig.2.   Lipid peroxides after 45 min occlusion of the superior mesenteric artery. After   24 h   the levels returned to control
values.

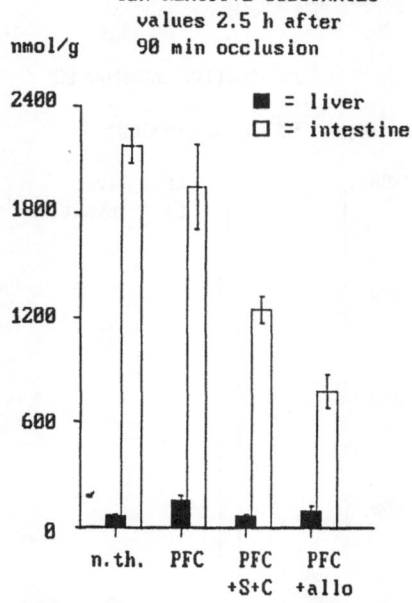

TBA-REACTIVE SUBSTANCES
values 2.5 h after
nmol/g   90 min occlusion

■ = liver
□ = intestine

Fig.3. Lipid peroxides 2.5 h after the end of 90
min occlusion without drugs (n.th.),   under PFC+O$_2$
treatment (PFC),   under PFC+O$_2$ together   with
SOD+CAT (PFC+S+C),   and under PFC+O$_2$ together
with allopurinol (PFC+allo).

When the occlusion period lasted only 45 min, the opposing course of the MDA content in liver and intestine could be observed without a protective treatment; after 24 h the MDA content had returned to the control values (Fig.2).

During treatment with PFC and oxygen breathing, the MDA content of the intestine 2.5 h after the end of the occlusion period was not further elevated above the values seen without this drug; in the liver a certain increase could be detected. During the use of SOD and CAT the MDA content of the intestine increased significantly less (p < 0.01). PFC together with allopurinol attenuated the increase of MDA to less than half the value with PFC alone (p < 0.001), whereas the liver content in all these situations remained low (Fig.3).

Antibiotic pretreatment, besides facilitating very low pre-occlusion values of MDA content of liver and intestine, also prevented the large increase of MDA 2.5 h after the end of occlusion (Fig.4). The survival rate for 48 h increased to 85.7 % (p < 0.05).

DISCUSSION

Granger et al. (1981,1986) as well as Parks et al. (1981,1984) emphasized the role of free oxygen radicals during ischemia and reperfusion of the intestinal circulation. Changes in the permeability of the gut, expressed by a decrease in the osmotic reflection coefficient were demonstrated by these authors, in extension to the results of Haglund and Lundgren (1972), who had shown an increase in the capillary filtration coefficient under these conditions. The large increase of TBA-reactive substances in the intestinal tissue which was found by us after ischemia and reperfusion is a specific sign for the action of oxygen derived free radicals and supports their work.

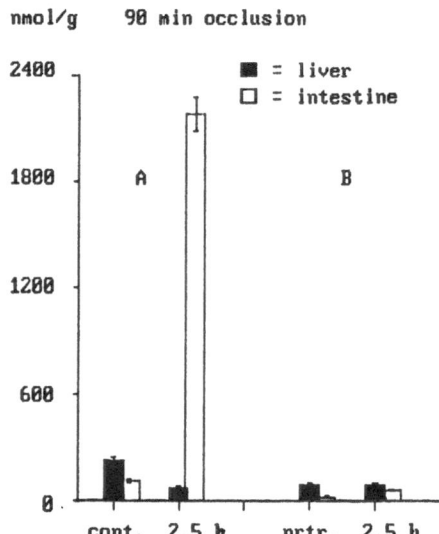

Fig.4. Lipid peroxides before and 2.5 h after the end of occlusion without treatment (A) and after pretreatment (prtr.) with antibiotics (B)

The content of TBA-reactive material in the liver, however, follows a quiet different and less altering course, caused by the different perfusion conditions of this organ. Cessation of portal inflow results in a compensatory increase of hepatic arterial flow, thus keeping the oxygen supply nearly fully compensated (Lutz, 1977, Lutz and Schultze, 1978).

The protective action of free oxygen radical scavengers like SOD and CAT is in accordance with the above mentioned conception. Furthermore, allopurinol acting either as inhibitor of xanthin oxidase or as radical scavenger per se (Moorhouse et al.,1987) showed a convincing effect in reducing the content of MDA. For the liver the results with scavengers fit to those of Adkison et al. (1986), who used the level of serum transaminases and liver oxygen consumption as parameters for the influence of radicals on liver function.

The highly significant reduction of MDA in liver and intestine after antibiotic pretreatment, which was combined with an increase in survival rate, seems to be at variance with results of Haglind et al. (1981), who found only an insignificant protective effect on mortality of such a treatment. The antibiotic used by them, however, consisted of clindamycin in a dose of 40 mg/kg and thus differed qualitatively and quantitatively from that used by us; so the difference might be a methodical one.

CONCLUSION

It was found that during the occlusion of the mesenteric artery the level of lipid peroxides in the liver decreased in spite of an enormous increase in the intestinal tissue. This reveals the importance of an intact vascular transport of lipid peroxides or their precursors for the MDA content in the liver. In the reperfusion period the MDA level of the liver increased only slightly, but even this slight increase was counteracted by the administration of SOD and CAT or allopurinol. The beneficial effect of free oxygen scavengers was most evident in the intestine. The low MDA values in the liver after reflow may be explained by effective degradation of peroxides by this organ. Antibiotic pretreatment was very useful for inhibition of the post-occlusion increase in MDA content of the intestinal tissue as well as that of the liver.

REFERENCES

Adkison, D., Höllwarth, M.E., Benoit, J.N., Parks, D.A., McCord, J.M. and Granger, D.N. (1986). Role of free radicals in ischemia-reperfusion injury to the liver. Acta Physiol. Scand. Suppl. 548, 101-107.

Augustin, A. and Lutz, J. (1988). The effect of a temporary occlusion of the superior mesenteric artery on the level of lipid peroxides in plasma and intestinal tissue. Pflueg. Arch. 411, R108.

Granger, D.N., Rutili G. and McCord, J.M. (1981). Superoxide radicals in feline intestinal ischemia. Gastroent. 81, 22-29.

Granger, D.N., Höllwarth M.E. and Parks, D.A. (1986). Ischemia-reperfusion injury: role of oxygen-derived free radicals. Acta Physiol. Scand. Suppl. 548, 47-63.

Haglind, E., Haglund, U., Lundgren O. and Schersten, T. (1981). Graded intestinal vascular obstruction. II. Effects of antibiotic pretreatment in the rat. Circulat. Shock 8, 41-47.

Haglund, U. and Lundgren, O. (1972). Reactions within consecutive vascular sections of the small intestine of the cat during prolonged hypotension. Act. Physiol. Scand. <u>84</u>, 151-161.

Hamar, J., Dezsi, L., Adam, E., Egri, L., Netzer, K.O., Stark, M. and Lutz, J. (1987). Role of fluid replacement, increased oxygen availability by perfluorochemicals and enhanced RES function in the treatment of mesenteric occlusion shock. Res. Exp. Med. <u>187</u>, 451-459.

Lutz, J. (1977). Hämodynamik und Sauerstoffversorgung der Leber. In: Experimentelle Hepatologie. (Edit. O. Zelder, M. Fischer and H. Hamelmann). Falck, Freiburg p. 1-14.

Lutz, J., Hamar, J., Netzer, K.O. and Stark, M. (1985). Survival from mesenteric occlusion shock influenced by different treatment in rats. Int. J. Microcirc. Clin. Exp. <u>4</u>, 103.

Lutz, J. and Schultze, H.G. (1978). Oxygen consumption and oxygen extraction of the feline liver under different types of induced hypoxia. In: Oxygen transport to tissue III (Edit. I.A. Silver, M. Erecinska and H.I. Bicher), Plenum Pub. Corp. N.Y. p. 537-543.

Moorhouse P.C., Grootveld M., Halliwell, B., Quinlan J.G. and Gutteridge J.M.C. (1987). Allopurinol and oxypurinol are hydroxyl radical scavengers. FEBS letters <u>213</u>, 23-28.

Ohkawa, H., Ohishi, N. and Yagi, K. (1979). Assay for lipid peroxides in animal tissues by thiobarbituric acid reaction. Analytical Biochem. <u>95</u>, 351-358.

Parks, D.A., Bulkley, G.B., Granger, D.N., Hamilton, S.R. and McCord, J.M. (1982). Ischemic injury in the cat small intestine: role of superoxide radicals. Gastroenterol. <u>82</u>, 9-15.

Parks, D.A., Shah, A.K. and Granger, D.N. (1984). Oxygen radicals: effects on intestinal vascular permeability. Amer. J. Physiol. <u>247</u>, G167-G170.

Yagi,K. (1982). Assay for serum lipid peroxide level and its clinical significance. In: Lipid peroxides in biology and medicine. Academic Press Inc. p. 223-242.

OXYGEN GRADIENTS IN TWO REGIONS OF THE EPIPHYSEAL GROWTH PLATE

S. F. Silverton, L. C. Wagerle, M. E. Robiolo,
J. C. Haselgrove and R. E. Forster

Department of Physiology, School of Medicine and
Department of Biochemistry, School of Dental Medicine
University of Pensylvania, Philadelphia, PA, USA

INTRODUCTION

Although oxygen is required for the continued viability
of all cells, controversy exists over oxygen delivery and the
extent of vascularization in the cartilage growth plate. The
anatomy of this region differs among species and also varies in
different bones in the same species. While previous studies of
growth plate metabolism suggest that oxygen gradients exist in
the mineralizing region of cartilage and are important to the
elongation of the bone during normal growth (Brighton and
Heppenstall, 1971), the information necessary to produce a
model of growth plate oxygen demands is not available in the
literature. Measurements of blood flow to the growth plate
have not been reported. Also, the metabolic rates and redox
status of chondrocytes are unknown though some work has been
initiated in our laboratory (Shapiro et al., 1982,1983;,
Silverton et al., 1988).

Our studies have demonstrated that distinct regions of the
growth plate exist which differ appreciably metabolically. Two
of these regions, the resting cartilage zone, directly below
the articular cap, and the region of cellular hypertrophy,
which is adjacent to the bone shaft, exhibit low tissue oxygen
tensions. In the avian growth plate, despite the low oxygen
partial pressures, the resting chondrocyte region shows a
relatively high ATP content and a low NADH, suggesting an
adequate oxygen supply. In comparison, the zone of cellular
hypertrophy which is adjacent to calcified cartilage and bone,
has increased ADP and NADH levels, which could represent a
relative mismatch of oxygen supply and metabolic demands. Our
goals in the following studies were to document blood flow to
the growth plate region in a mammalian species, to measure
oxygen consumption of isolated chondrocytes and to accumulate
data on vascular cross-sectional areas relative to tissue
volume in order to construct an initial model of oxygen
gradients in these two regions of the growth plate.

METHODS

Blood Flow Studies

Animal preparation. Newborn Yorkshire piglets, 2-9 days
of age and weighing 0.77-1.54 kg were lightly anesthetized with
3-4% halothane. Core body temperature was monitored and
maintained at 39°C with a warming blanket. Following local
anesthesia (2% lidocaine) a tracheostomy was performed and
catheters were placed into a femoral artery and vein, a
brachial artery and into the left atrium via a left
thoracotomy. Animals were paralyzed (0.5 mg pancuronium
bromide) and mechanically ventilated.

Blood flow measurements. Blood flow was measured with
the tracer microsphere technique (Heymann et al., 1977).
Microspheres (15.0 $\pm$ 1.0 μM) labeled with ^{141}Ce, ^{85}Sr, or
^{46}Sc (3M, St. Paul, MN) were used. Stock solutions were
sonicated for 5 minutes and shaken vigorously on a vortex mixer
for 2-5 minutes. Approximately 0.7-1.2 x 10^6 microspheres
were placed into specially prepared injection vials (Heymann et
al, 1977) and shaken again on a vortex mixer for 2-5 minutes
before injection. The contents of the injection vial were
flushed into the left atrium over a period of 15-30 seconds.
Reference arterial blood was withdrawn from the brachial artery
for 10 seconds before, during, and for 60 seconds after the
microsphere injection at a rate of 0.97 ml/min (Wagerle et al.,
1988). Each of the isotopes was injected separately, and
multiple experiments were performed. The femoral lines were
utilized for blood gas measurements and drug injections.

At the end of the experiment, the animal was killed by an
injection of euthanasia solution (T-61, American Hoescht) and
the femur and tibia from the non-catheterized hindlimb and
humerus and radius from the non-catheterized forelimb were
removed. Brain samples were also collected. The growth plates
were separated from the metaphyseal bone and the secondary
ossification centers were dissected from the cartilage caps.
Slices of endochondral bone adjacent to the growth plate were
also collected. The tissues were weighed and placed into
counting vials. Reference blood samples were also counted.
Tissues and reference blood were counted for 20 minutes with a
four channel gamma-counter (Beckman Instruments, Fullerton,
CA). Energy windows were set at 125-170, 460-550, and 820-1060
KeV for ^{141}Ce, ^{85}Sr, and ^{46}Sc, respectively. Blood flow
was calculated with the formula

$$F_t = \frac{C_t * F_b * W_t}{C_b * D_t}$$

where F_t and F_b are tissue and reference organ blood flow
in ml/min/100 cm^3, C_t and C_b are radioactivity in tissue
and blood samples, and W_t and D_t are the weight and density
of the tissue samples.

Growth plate and Chondrocyte Preparation

Six to eight week old White Rock chicks were used for
these studies. The animals were sacrificed by cervical
dislocation and the proximal tibial growth cartilage was

exposed. For measurement of vascular and tissue areas, fresh slices of tissue were cut with a razor from the exposed growth plate. The resting and hypertrophic chondrocyte regions were located by direct visualization under the light microscope. The slices in proximity to the articular cartilage contained the resting cells. Slices nearer to the bone contained hypertrophic cells and calcifying cartilage. The fresh slices were visualized under a microscope linked to an image processor and the one dimensional light density profiles were analyzed.

To measure oxygen uptake in chondrocytes, cells were isolated from growth cartilage using a procedure that we have previously published (Kakuta et al., 1986). Briefly, tissue slices from resting, proliferating and hypertrophic cartilage were suspended in Hanks Balanced Salt Solution (HBSS) and digested with 0.1% collagenase at 4°C for 12 hours. Following initial enzyme treatment, the supernatant was decanted and replaced with fresh collagenase; digestion was continued for a further 3 hours at 37°C. To liberate the cells from the matrix, the tissue was gently triturated in cold HBSS 3-4 times. The cells were collected by centrifugation at 10,000 x g for 5 minutes and then resuspended in HBSS. These cells were 95% viable when assessed by trypan blue exclusion and demonstrated high ATP/ADP ratios, suggesting that they are capable of respiring at a normal rate.

Oxygen Uptake by Pd-coproporphyrin Phosphorescence Quenching

Oxygen consumption by cartilage cells was quantitated at physiological oxygen tensions using the technique of Vanderkooi (1987). Briefly, cells were suspended in HBSS containing 20 mM HEPES, 0.2% bovine serum albumin, and 10 μl of a solution containing 1 μM Pd-coproporphyrin in dimethylformamide (final concentration of the porphyrin probe was 4.4 pmoles/ml) in an air-tight glass cuvette. The buffer was gassed with nitrogen to reduce the O_2 tension to the physiological range and the cuvette placed in a fluorometer. The phosphorescence quenching constant, τ, of the heavy metal-heme compound was sampled repetitively and the corresponding O_2 concentration calculated. Protein concentrations were measured with the Lowry (1951) method.

RESULTS

In the neonatal pig, using microsphere technology, blood flow to the growth plate was found to be 40% of the value to the adjacent metaphyseal bone. However, the secondary growth center, which is also next to the growth plate toward the articular cartilage cap, had blood flows which were similar to growth plate values. Table 1 shows the flow values found in bone and compares these measurements to those of brain and inactive muscle (Wagerle et al., 1986).

Vascular and intervascular areas of two distinct regions of the growth plate were quantitated in fresh transverse slices of the chick proximal tibia using an image processor. Figure 1 shows representative one dimensional profiles of tissue sections from a resting cartilage slice and from the zone of cellular hypertrophy. Vascular channels of the resting region of the growth plate were smaller in diameter (see Table 2).

Table 1. Blood flow in several tissues.

blood flow	growth plate	metaphyseal bone	2°growth center	brain	muscle
cc/min 100 cm^3	20±2	50±7	19±2	79±7	21±2

Blood flow values were obtained using radioactive microspheres. The number of animals for bone measurements was 4; for brain and muscle was 41. Values are corrected for the differing tissue densities.

Tissue distances from the periphery of one vascular channel to its neighbor were greater in the resting zone.

Oxygen consumption of isolated chondrocytes as determined by the method of Vanderkooi was constant over a large range of oxygen tensions. However, as the oxygen concentration approached values which have been shown in the literature to be present in the growth plate, oxygen uptake was found to decrease non-linearly. The experiemtnal data was fitted to a Michaelis-Menten function using "Asystant". Manual treatment of the data gave similar results. Shown in Figure 2 is a typical oxygen consumption curve and the computer-fitted curve.

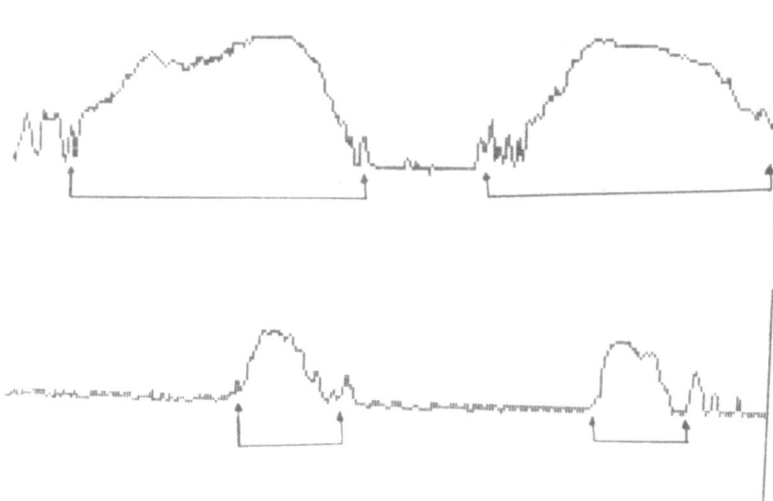

Fig 1. One-dimensional Light Density Profile of Cartilage Slices. The vessels are darker than the intervening tissue. The upper tracing is a profile in the resting chondrocyte region. Lower is from the zone of cellular hypertrophy. Vessel boundaries are marked with paired arrows.

Table 2. Vascular and Tissue Measurements of Two
          Regions of Avian Growth Plate Cartilage.

	Resting Zone	Hypertrophic Zone
vascular channel	82μm±3 (8)	147μm±11 (12)
tissue space	202μm±40 (6)	159μm±16 (6)

Vascular channel radii and tissue spaces were
measured in fresh slices of avian growth plate
using an image processor for quantitation.

The experimental data were used in the Krogh-Erlang
equation to calculate the difference in oxygen tension between
the vascular channel and the periphery of the Krogh cylinder.
The oxygen diffusion coefficient for this model was assumed to
be similar to muscle. The chondrocyte oxygen uptake averaged
8.3 ± 0.8 μM/min-mg protein at 20 μM oxygen concentration. At
a vessel oxygen tension of 20-25mm Hg, the decrease in oxygen
across the tissue cylinder was similar in the resting and
hypertrophic regions, 13.5 mm Hg to 14.8 mm Hg, respectively.

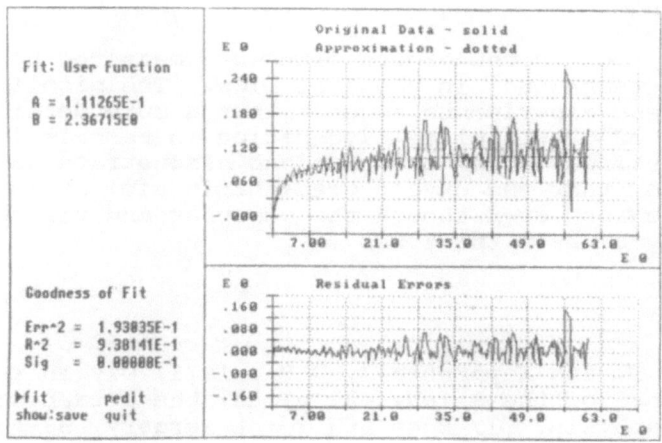

Fig 2. Oxygen Consumption of Isolated Chondrocytes.
        Chondrocytes consumed oxygen at a constant
        rate above 20μM oxygen. At lower oxygen
        tensions computer curve-fitting was employed to
        analyze the data. Using a Michaelis-Menten
        function, A is the maximum velocity and B is the
        oxygen concentration of the half maximal rate.
        The x-axis is μM [O_2], the y-axis is O_2 up-
        in μM/sec.

DISCUSSION

Our primary goal in these studies was to estimate the oxygen gradients which may exist in the growth cartilage. We have measured several parameters in the avian growth plate and have calculated gradients with the Krogh cylinder model. The avian growth plate served as an initial model both because of the relatively deep growth plate (4-5mm) which allowed the geometry to be easily imaged and because of the ease of chondrocyte isolation for oxygen consumption measurements. The diffusion coefficient for oxygen in the resting region and the zone of cellular hypertrophy was not available to us for these first estimates. Nor did we have our own measurements of local oxygen tensions. Both of these parameters would have been helpful. Additionally, the cylinder model makes assumptions which are rather simplistic; the cylinder is infinite, oxygen diffusion occurs only radially. Our calculations also assumed constant oxygen consumption by the cells across the cylinder, whereas our experimental data show a decrease in uptake at lower oxygen tensions. Furthermore, although the literature documents increasing numbers of mitochondria in hypertrophic chondrocytes (Brighton et al., 1973, Hargest et al., 1985), suggesting an increased ability to consume oxygen, our oxygen consumption values were carried out on a mixed population of cells. If resting chondrocytes indeed do consume less oxygen, then the oxygen gradients would be smaller in this region than in the zone of cellular hypertrophy. Such a circumstance would put the hypertrophic cells most removed from the vascular channels at the greatest risk of anoxia. Alternatively, oxygen consumption unrelated to mitochondria could be a feature of hypertrophic chondrocytes, as we have suggested in our metabolic studies (Silverton et al., 1988). Subsequent studies need to be directed at improving the model in the light of these findings.

Blood flow to the growth plate was measured by microsphere techniques in neonatal pigs. The significant flows found in these experiments suggest that a more detailed examination of growth plate oxygenation in mammals is needed. The Krogh cylinder model calculations demonstrate as a first approximation that the oxygen gradient is similar in the two cartilage regions even though the vascular and tissue areas vary greatly between the zones.

REFERENCES

Brighton, C.T. and Heppenstall, R.B., 1971, Oxygen tension in zones of the epiphyseal plate, the metaphysis and diaphysis. J. Bone and Joint Surgery, 53-A:719-728.

Hargest, T.E., Gay, C.V., Schraer, H. and Wasserman, A.J., 1985, Quantitative changes in the vertical distribution of elements in cells and matrix of epiphyseal growth plate cartilage. J. Histochem. Cytochem. 33:275-86.

Heymann, M.A., Payne, B.D., Hoffman, J.I.E., and Rudolph, A.M., 1977, Blood flow measurements with radionuclide labeled particles. Prog. Cardiovasc. Dis. 20:55-79.

Kakuta, S., Golub, E.E., Haselgrove, J.C., Chance, B., Frasca, P. and Shapiro, I.M., 1986, Redox studies of the epiphyseal growth cartilage: pyridine nucleotide

metabolism and the development of mineralization.
J. Bone and Mineral Res., 1:433-440.

Lowry, O.H., Rosebrough, W.J., Farr, A.L. and Randall, R.J.,
1951, Protein measurement with folin phenol reagent.
J. Biol. Chem. 193:265-275.

Shapiro, I.M. Golub, E.E., Haselgrove, J.C., Havery, J. Chance,
B. and Frasca, P., 1982, Initiation of endochondral
calcification is related to changes in the redox state
of hypertrophic chondrocytes. Science 2167:950-952.

Shapiro, I.M., Golub, E.E., May, M. and Rabinowitz, J.S., 1983,
Studies of nucleotides of growth plate cartilage:
Evidence linking changes in cellular metabolism with
cartilage calcification. Bioscience Rep. 3:345-351.

Silverton, S.F., Matsumoto, H., DeBolt, K., Reginato, A., and
Shapiro, I.M., 1988, Pentose phosphate shunt
metabolism by cells of the chick growth cartilage.
Bone (in review).

Vanderkooi, J.M., Grzegroz, M., Green, T.J. and Wilson, D.F.,
1987, An optical method for measurement of dioxygen
concentration based upon quenching of
phosphorescence. J. Biol. Chem. 262:5476-5482.

Wagerle, L.C., Kumar, S.P., and Delivoria-Papadopoulos, M.,
1986, Effect of sympathetic nerve stimulation on
cerebral blood flow in newborn piglets. Pediat.Res.
20:131-135.

Wagerle, L.C., Kumar, S.P., Beik, J. and
Delivoria-Papadopoulos, M., 1988, Blood-brain barrier
to hydrogen ion during acute metabolic acidosis in
piglets. J. Appl. Phsyiol. (in press).

TUMORS

[31]P NUCLEAR MAGNETIC RESONANCE MEASUREMENTS OF ATPase KINETICS IN
MALIGNANT TUMORS

P. Okunieff[1], P. Vaupel[1], and L.J. Neuringer[2]

[1]Dept. Radiation Medicine, Massachusetts General Hospital,
Cancer Center, Harvard Medical School, Boston, MA 02114
[2]Francis Bitter National Magnet Laboratory, Massachusetts
Institute of Technology, Cambridge, MA 02139

INTRODUCTION

The estimation of tumor growth rate is generally made on the basis of
tumor histology, its site of origin, and sequential radiographic or caliper
measurements of its dimensions. There are currently no techniques capable of
measuring the expected growth rate non-invasively, serially and prospective-
ly. An assay, able to measure the growth rate of a partially or fully treated
tumor would be of obvious potential advantage for optimizing therapy. We
propose the use of ^{31}P nuclear magnetic resonance (NMR) magnetization transfer techniques for non-invasive and continuous estimation of tumor growth
kinetics. Specifically, the rate of nucleoside triphosphate (NTP) synthesis
can be quantitatively measured in vivo using ^{31}P-NMR (1-4), and thus it is
possible to accurately measure the metabolic rate of a tumor. Since higher
energy consumption rates in tumors are likely to correlate with more rapid
proliferation and growth (5-9), ^{31}P magnetization transfer NMR should give
an indirect indication of tumor growth rate. The purpose of this study was
to examine the potential for ^{31}P-NMR magnetization transfer experiments, measuring the kinetics of the reactions catalyzed by ATPase and creatine kinase
to indirectly detect the decrease in growth rate which occurs in tumors of
larger size.

MATERIALS AND METHODS

Animal and Tumors. Experimental animals were 8-12 week old C3Hf/Sed mice
derived from our defined flora mouse colony (10). Animals were provided with
sterilized animal pellets and acidified, vitamin K fortified water ad libitum. Early generation isotransplants of a spontaneous fibrosarcoma (FSaII)
were used. Single cell suspensions were prepared by a mechanical procedure
and a defined number of Trypan Blue negative cells were trans- planted
subcutaneously (s.c.) in 5 microliters into the hind foot dorsum.

Tumor Volume Measurements. Tumor volumes were measured three times each
week from the time they were detectable until they reached 1 ml. Volumes were
calculated using an ellipsoid approximation and the three orthogonal diameters measured by calipers [$V = (\pi/6) \times d_1 \times d_2 \times d_3$].

Volume Doubling Times (VDT). Tumor VDT were estimated separately for tumors of microscopic and of macroscopically detectable volumes. Microscopic tumors are defined as those which are not yet detectable by visual examination of the hind foot dorsum. Growth curves were obtained using tumors transplanted s.c. in 5 microliter volumes of tumor cell suspensions, and containing 4.09×10^0 to 7.25×10^5 Trypan Blue negative FSaII cells diluted in Hanks' solution. Twelve sequential dilutions (1 part cell suspension to 2 parts Hanks' solution) were used to obtain the above range of cell inocula (11,12). To assure similar tumor bed and growth conditions each 5 microliter injection volume included 5.5×10^5 lethally irradiated cells [140 Gy delivered to a separate cell suspension immediately preceding tumor injections (11,12)]. Eight to 10 animals were used for each cell suspension dilution (total = 117 tumors). The time for tumors to reach 100 mm^3 was used as the endpoint. Using the interval required for 7.25×10^5 cells to produce a 100 mm^3 tumor as a "standard-time", the interval for 2.42×10^5 cells to grow to 100 mm^3 minus the "standard-time" is the time required for the number of tumor cells to triple. This process was then repeated for each of the smaller tumor cell inocula, and a growth curve was fitted. The VDT data were then extrapolated to estimate the tumor cell density.

Tumor bearing animals from the above experiment were also used to determine the VDT of macroscopically detectable tumors. Tumor volumes become measurable at volumes of 30 to 100 mm^3, and then grow to 1 ml in the subsequent 9 or 10 days independent of the initial inoculum of tumor cells. Various growth models have been used to characterize the shape of the growth curve. The most widely used of these models are the Gompertzian, and logistic growth (13). Both these models have a characteristic slowing of the exponential growth rate as the tumor volume increases. Rather than assuming one of these growth curves would fit the data, the growth curves obtained from the above 117 tumors was fit to a general formula (Equation 1). The VDT was calculated for each inoculum of tumors at each volume measurement time, totalling 49 VDT measurements, each VDT consisting of ≈19 - 20 individual tumor volume measurements. The VDT was calculated using the following equation:

$$VDT = \frac{V \cdot \Delta t}{\Delta V} \qquad \text{[Equation 1]}$$

where V is tumor volume (calculated as the average volume during the time interval Δt), and ΔV is the change in volume during the time interval Δt.

^{31}P Nuclear Magnetic Resonance Spectroscopy.

Spectra were obtained at 8.5 Tesla (corresponding to a resonance frequency of 145.6 MHz for ^{31}P). Details regarding the NMR probe and shimming have been described previously (5,14,15). Tumors in the mouse foot dorsum allow for selective spectroscopy of the tumor without contamination from nearby skeletal muscle (14,16). Spectra were accumulated with a 90° pulse, 10 kHz bandwidth, an 8 sec presaturation pulse, and a 12 sec interpulse interval. The 12 sec interpulse interval allows for near complete relaxation of all resonances. Several spectra were collected of each tumor including presaturation of the γ-NTP, PCr (phosphocreatine), and P$_i$ (inorganic phosphate). Control spectra were obtained with the saturation pulse positioned midway between the γ- and α- NTP resonances. A very high degree of saturation selectivity was possible. Signal-to-noise was sufficient after 64 to 128 averaged free induction decays. Spectral processing included 15-30 Hz matched exponential multiplication or convolution difference. Peak intensities were calculated using three fitting techniques, a Lorentzian linefit, a Gaussian linefit, and maximum peak heights (15-17). The maximum peak heights were the most reproducible parameter for all resonances.

The longitudinal relaxation times (T_1) were measured using the technique of partial saturation. This technique is descibed in detail in a previous publication (5,23). Again, during all or a portion of the interpulse interval a presaturation pulse was applied to the γ–NTP, PCr, or P_i resonance. Standard T_1 measurements were also made using the same technique with a presaturation pulse centered between the γ– and α– NTP resonances. At least six spectra were used to compute each T_1. Interpulse intervals for this set of spectra ranged from 0.6 sec to 12 sec.

RESULTS AND DISCUSSION

Growth Kinetics of Microscopic Tumors. The growth rate of microscopic tumors was exponential (Figure 1). Using an exponential curve fit, the calculated VDT was 19 hr. The correlation coefficient of the exponential fit was 0.98 ($p<0.001$). This VDT is in good agreement with the 16–18 hr cell cycle time and the $\approx$80% growth fraction of small FSaII tumors (18). The similarity between the VDT and cell cycle time suggests that in microscopic tumors essentially all cells are clonogenic, most are cycling, and that the cell loss factor is very small. Extrapolation of the exponential growth curve from 7.25×10^5 cells for "standard–time" yields the number of cells in a 100 mm^3 volume. The density of FSaII cells was estimated to be 2.0×10^5 cells/mm^3 using this algorithm.

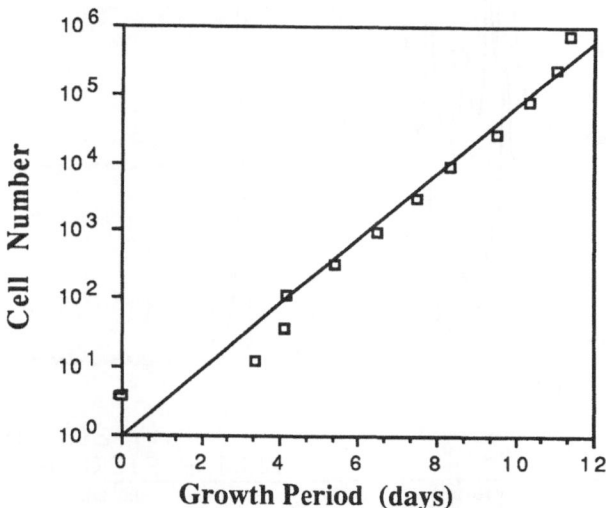

Fig.1. Exponential growth curve for microscopic tumors in the hind foot dorsum of mice (R=0.98, p<0.001).

Growth Kinetics of Measureable Tumors. The VDT for microscopic tumors ($<100 \ mm^3$) was $\approx$0.78 days, small macroscopic FSaII tumors (100–150 mm^3) had a VDT of 2.6 $\pm$ 0.3 (SE) days, and tumors of intermediate size (range: 150 to 300 mm^3; mean: 226 mm^3) had a VDT of 3.2 $\pm$ 0.2 days. Larger tumors (range: 300–600 mm^3; mean: 453 mm^3) had a VDT of 3.8 $\pm$ 0.3 days, which increased to 7.6 $\pm$ 1.3 days in tumors of 1 ml. The increase in VDT from microscopic tumors to the largest tumors is considered a result of a developing hostile tumor micromilieu. Specifically, as a tumor grows, an increasingly limited availability of metabolic and anabolic substrates results in a decreasing number of cells with the required environmental conditions for maximum replication rate.

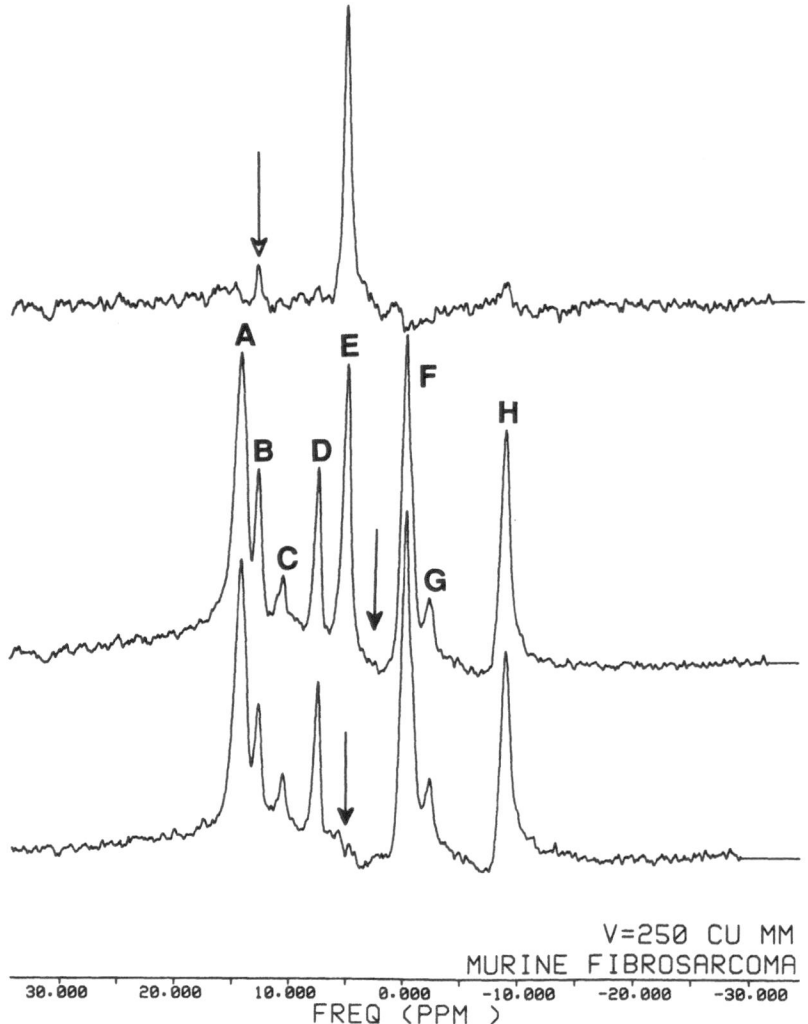

Fig.2. ^{31}P NMR spectra from a 250 mm^3 FSaII tumor in the hind foot dorsum of a C3Hf/Sed mouse. The center spectrum is the control, obtained with a saturation pulse centered between the γ- and α-NTP resonances (solid arrow). The peaks are labelled A: PME; B: P$_i$; C: PDE; D: PCr; E: γ-NTP; F: α-NTP; G: DPDE; H: β-NTP. The lower spectrum was obtained by moving the saturation pulse to coincide with the γ-NTP resonance (solid arrow). The top spectrum is obtained after subtracting the lower spectrum from the center spectrum, and highlights the magnetization transferred from the P$_i$ resonance (open arrow).

Table 1

^{31}P LONGITUDINAL RELAXATION TIMES WITH AND WITHOUT METABOLITE SATURATION

Saturated Resonance	PME	P_i	PCr	γ–NTP	α–NTP	DPDE	β–NTP
Control (none)	3.79 ± 0.41	4.21 ± 0.75	3.26 ± 0.45	1.53 ± 0.23	1.34 ± 0.27	1.79 ± 0.17	1.30 ± 0.15
P_i	2.98 ± 0.49	——	3.21 ± 0.46	1.44 ± 0.16	1.26 ± 0.15	1.69 ± 0.34	1.20 ± 0.13
PCr	3.08 ± 0.07	3.68 ± 0.64	——	1.18 ± 0.16	0.89 ± 0.04	1.23 ± 0.19	0.90 ± 0.07
γ–NTP	3.74 ± 0.39	4.00 ± 0.69	4.10 ± 0.49	——	1.32 ± 0.12	1.62 ± 0.25	1.40 ± 0.10

Values are means (sec) ± SE

NMR Spectroscopy. Figure 2 shows a typical ^{31}P NMR spectrum obtained from a FSaII tumor located in the hind foot dorsum. From left to right the center spectrum shows peaks corresponding to PME [phosphomonoesters including phosphocholine, phosphoethanolamine and sugar phosphates (19–21)], P_i, PDE [glycerophosphocholine, glycerophosphoethanolamine, and perhaps membrane phospholipid degradation products (5,19,22)], PCr, γ–NTP [includes the closely associated β– nucleoside diphosphate resonance(NDP)], α–NTP (includes the α–NDP and nicotinamide adenine dinucleotides), DPDE (di-phosphodiesters), and β–NTP.

The T_1 relaxation times of tumor phosphate resonances are given in Table 1. Each T_1 is the average of 3 to 7 tumor measurements. Not all tumors had sufficiently prominent DPDE resonances to measure T_1. The T_1 relaxation time for PDE is composed of several overlapping resonances which appear to have significantly different T_1s, and was therefore omitted (5). The PME, P_i, and PCr had the longest relaxation times with a range of 2.98 for PME to 4.21 for P_i. No significant effect of saturation on the T_1 relaxation times of these slowly relaxing resonances was detected. There was a trend for the T_1 relaxation times obtained with the PCr resonance saturated to be somewhat lower overall compared to the other measurements. This in some cases reached statistical significance. The biologic meaning of this finding is unclear, since as can be seen in Table 2, the reactions catalyzed by creatine kinase are not associated with significant magnetization transfer for medium or large tumors, and the reaction rate constant of PCr-->NTP is minimal even in small, rapidly growing tumors (Table 2).

The T_1 relaxation times obtained with the appropriate presaturated resonance are used to calculate the rate of P_i utilization in NTP synthesis [obtained by saturating the γ–NTP resonance and measuring the change in the P_i resonance (1–3)]. The reverse reaction rate is also calculated by saturating the P_i resonance and measuring the change in the γ–NTP resonance. The creatine kinase kinetics are similarly observed by saturating the γ–NTP and PCr resonances respectively and calculating the change in the remaining resonance. A typical set of spectra, and the difference spectrum are shown in Figure 2. The center panel shows the control spectrum of a 250 mm^3 FSaII tumor with the presaturation pulse marked by the arrow and the individual

Table 2

HIGH ENERGY PHOSPHATE REACTION RATES

Reaction Measured	Small Tumors (85 – 131 mm^3)	Medium Tumors (168 – 250 mm^3)	Large Tumors (318 – 400 mm^3)
P$_i$ ⟶ NTP	0.071 + 0.003 (n=18)	0.022 + 0.001 (n=12)	0.005 + 0.001 (n=5)
	p<0.01	p<0.01	
PCr ⟶ NTP	0.013 + 0.001 (n=18)	0.003 + 0.001 (n=12)	0.003 + 0.001 (n=5)
	p<0.01	n.s.	
NTP ⟶ P$_i$	0.021 + 0.001 (n=17)	0 + 0.028 (n=6)	0.007 + 0.001 (n=5)
	n.s.	n.s.	
NTP ⟶ PCr	0 + 0.017 (n=12)	0 + 0.064 (n=8)	0 + 0.105 (n=5)
	n.s.	n.s.	

Values are means (sec^{-1}) $\pm$ SE.

resonances labelled. There was no loss of α– or γ–NTP resonance intensity due to the saturation pulse placed between these closely spaced resonances, demonstrating that a high degree of saturation selectivity is possible. The lower spectrum was obtained with presaturation of the γ–NTP resonance. Near complete obliteration of this resonance is observed. The top spectrum was obtained by subtracting the bottom spectrum from the center spectrum and highlights the difference between them. Little or no conversion of PCr to ATP is observed based on the insignificant transfer at the PCr locus. However, NTP is being synthesized from P$_i$ (marked by the open headed arrow). A related reaction is also detected (β–NTP ---> β–NDP) because the β–NDP resonance is a component of the γ–NTP resonance. This chemical reaction and, to a lesser extent, the stereochemical interactions between the γ– and β– NTP resonances are responsible for the small magnetization transfer seen at the β–NTP resonance. Quantitative calculation of the uni–directional reaction kinetics can be made using the equation:

$$\frac{\text{Intensity with Saturation}}{\text{Intensity of Control}} = \frac{(1/T_1)}{(1/T_1) + K} \quad \text{[Equation 2]}$$

where T$_1$ (sec) is the longitudinal relaxation time of the resonance, which represents the metabolite being utilized by the reaction with a rate constant of K. Since the intensities and T$_1$ can be measured, K (sec^{-1}) can be calculated. The constants for the synthesis rate of NTP from PCr and P$_i$ are given along with the rate constants for synthesis of PCr and P$_i$ from NTP in Table 2.

The kinetics associated with the reaction P$_i$ ---> γ–NTP change significantly with tumor size. On the other hand, the lack of significant PCr involvement in energy metabolism despite its high cellular concentrations suggests that PCr is being retained as a sequestered energy source. This pattern of magnetization transfer differs from brain and skeletal muscle, which have substantial magnetization transfer between the PCr and γ-NTP resonances, suggesting the reactions catalyzed by creatine kinase are predominating (3). These normal tissues generally have a meager magnetization transfer between γ-NTP and P_i, which can be enhanced during conditions which increases energy consumption (e.g., muscular work). The PCr in tumor may, therefore, represent a reserve source of energy for mitosis, during which time the production of high energy phosphates are depressed. The decreased P_i ---> γ-NTP reaction rate, which is seen in large compared to small tumors, is compatible with the decreased average metabolic rate predicted to occur in larger tumors. The processes needed for doubling the tumor volume at intervals as short as 19 hr are likely to require the majority of the cellular ATP production. Hence, the 3.5 to 14-fold decline in P_i utilization for NTP synthesis between small and large tumors is also compatible with the observed 5 to 10-fold slower VDT.

CONCLUSION

The decline in the exponential growth rate of tumors of greater size has been documented in virtually all solid experimental tumors. This study for the first time documents a correlation between the rate of NTP synthesis, and the **tumor growth rate**. We showed a significant decrease in tumor growth rate in microscopic vs. measurable tumors, and a continued decrease with each volume doubling of measurable tumors. This decreased rate of tumor growth reflected a similar decrease in the rate of NTP synthesis from P_i. Unlike the situation in normal metabolically active tissues, the creatine kinase reaction is relatively quiescent in tumors, suggesting that the high PCr levels in tumors are a sequestered energy pool. The measurement of the uni-directional reaction kinetics is performed non-invasively in vivo and features a technique which can be repeated serially over time or during treatment. Measuring the rate of ATP synthesis should currently be possible on human tumors in situ using presently available commercial technology.

ACKNOWLEDGEMENTS

This project was supported in part by NIH grants CA48096 and RR00995. We would like to thank Dr. Muneyasu Urano who made the tumor growth kinetics analysis possible and to Mr. Myles Greenberg for expert technical assistance.

REFERENCES

1. J.R. Alger, J.A. Hollander, and R.G. Shulman, In vivo phosphorus-31 nuclear magnetic resonance saturation transfer studies of adenosine-triphosphatase kinetics in Saccharomyces cerevisiae, Biochemistry. 21:2957 (1982).

2. J.R. Alger, and R.G. Shulman, NMR studies of enzymatic rates in vitro and in vivo by magnetization transfer, Q. Rev. Biophys. 17:83 (1984).

3. B.M. Hitzig, J.W. Prichard, H.L. Kantor, W.R. Ellington, J.S. Ingwall, C.T. Burt, S.I. Helman, and J. Koutcher, NMR spectroscopy as an investigative technique in physiology. FASEB J. 1:22 (1987).

4. J.J. Led, and H. Gesmar, The applicability of the magnetization transfer NMR technique to determine chemical exchange rates in extreme cases. The importance of complementary cases, J. Magn. Reson. 49:444 (1982).

5. P. Okunieff, J. Ramsay, T. Tokuhiro, B.M. Hitzig, E. Rummeny, L.J. Neuringer, and H.D. Suit, Estimation of tumor oxygen and metabolic rate using ^{31}P MRS: Correlation of spin-lattice relaxation with tumor growth rate and DNA synthesis, Int. J. Radiat. Oncol. Biol. Phys. 14:1185 (1988).

6. J.P. Freyer, E. Tustanoff, A.J. Franko, and R.M. Sutherland, In situ oxygen consumption rates of cells in V-79 multicellular spheroids during growth, J. Cell. Physiol. 118:53 (1984).

7. G.D. Chiro, J. Hatazawa, D.A. Katz, H.V. Rizzoli, and D.J. De Michele, Glucose utilization by intracranial meningiomas as an index of tumor aggressivity and probability of recurrence: A PET study, Radiology. 164:521 (1987).

8. R. Sutherland, J. Freyer, W. Mueller-Klieser, R. Wilson, C. Heacock, J. Sciandra, and B. Sordat, Cellular growth and metabolic adaptations to nutrient stress environments in tumor microregions, Int. J. Radiat. Oncol. Biol. Phys. 12:611 (1986).

9. E.A. Newsholme, B. Crabtree and M.S.M Ardawi M.S.M., The role of high rates of glycolysis and glutamine utilization in rapidly dividing cells, Bioscience Rep., 5:393 (1985).

10. R.S. Sedlacek, R.P. Orcutt, H.D. Suit, and E.F. Rose, A flexible barrier at cage level for existing colonies: Production and maintenance of a limited stable anaerobic flora in a closed inbred mouse colony, in: "Recent Advances in Germfree Research," S. Sasaki, A. Ozawa, and K. Hashimoto, eds., Tokai University Press, Tokyo (1981).

11. L. Revesz, Effect of tumour cells killed by X-rays upon the growth of admixed viable cells, Nature. 178:1391 (1956).

12. M. Urano, and J. Kahn, Some practical questions in the tumor regrowth assay, in: "Rodent Tumor Models in Experimental Cancer Therapy," R.F. Kallman, ed., Pergamon Press, New York (1987).

13. L. Norton, Cell kinetics in normal tissues and in tumors, in: "Cancer in the Young," A.S. Levine, ed., Masson, New York (1982).

14. P. Okunieff, E. McFarland, E. Rummeny, C. Willett, B.M. Hitzig, L.J. Neuringer, and H. Suit, Effects of oxygen on the metabolism of murine tumors using in-vivo phosphorus-31 NMR, Am. J. Clin. Oncol. 10:475 (1987).

15. P. Okunieff, F. Kallinowski, P. Vaupel, and L.J. Neuringer, Effect of hydralazine-induced vasodilation on the energy metabolism of murine tumors studied by in-vivo ^{31}P-nuclear magnetic resonance spectroscopy, J. Natl. Cancer Inst. 80:745 (1988).

16. P. Okunieff, J.A. Koutcher, L. Gerweck, E. McFarland, M. Urano, M., B.M. Hitzig, L. Neuringer, and H.D. Suit, Tumor size dependent metabolic changes in a murine fibrosarcoma: use of Fourier transform 31P NMR to evaluate tumor energy metabolism, Int. J. Radiat. Oncol. Biol. Phys. 12:793 (1986).

17. P. Okunieff, J.A. Koutcher, and H.D. Suit, Critical Review: Tumor size dependent metabolic changes in a murine fibrosarcoma: use of Fourier transform 31P NMR to evaluate tumor energy metabolism, Invest. Radiol. 22:618 (1987).

18. J. Ramsey, Personal communication (1988).

19. J.S. Cohen, R.C. Lyon, C. Chen, P.J. Faustino, G. Batist, M. Shoemaker, E. Rubalcaba, and K.H. Cowan, Differences in phosphate metabolite levels in drug-sensitive and -resistant human breast cancer cell lines determined by ^{31}P magnetic resonance spectroscopy, Cancer Res. 46: 4087-4090 (1986).

20. J.J.H. Ackerman, T.K. Grove, G.G. Wong, D.G. Gadian, and G.K. Radda, Mapping of metabolites in whole animals by ^{31}P NMR, Nature 283: 167 (1980).

21. W.T. Evanochko, T.T. Sakai, T.C. Ng, R. Krishna, H.D. Kim, R.B. Zeidler, V.K. Ghanta, R.W. Brockman, L.M. Schiffer, P.G. Braunschweiger, and J.D. Glickson, NMR study of in vivo RIF-1 tumors: Analysis of perchloric acid extracts and identification of ^{1}H, ^{31}P, and ^{13}C resonances, Biochim. Biophys. Acta. 805:104 (1984).

22. P.F. Daly, R.C. Lyon, E.J. Straka, and J.S. Cohen, ^{31}P-NMR spectroscopy of human cancer cells proliferating in a basement membrane gel. FASEB J. 2:2596 (1988).

23. R. Freeman, and H.D. Hill, Fourier transform study of NMR spin-lattice relaxation by "progressive saturation", J. Chem. Phys. 54:3367 (1971).

CONTRIBUTION OF DIFFUSION TO THE OXYGEN DEPENDENCE OF ENERGY METABOLISM IN HUMAN NEUROBLASTOMA CELLS

William L. Rumsey, Michael Robiolio, and David F. Wilson

Department of Biochemistry and Biophysics
Medical School
University of Pennsylvania
Philadelphia, PA 19104

INTRODUCTION

The oxygen dependence of cellular energy metabolism is normally expressed as a function of the extracellular oxygen pressure. This means it is a function not only of the oxygen dependence of mitochondrial oxidative phosphorylation, but also of the diffusion induced oxygen pressure differences which develop between the extracellular medium and the mitochondria. Evaluation of the relative contributions of the oxygen dependence of mitochondrial oxidative phosphorylation and diffusion has proven difficult due to technical limitations in the methods available for measuring oxygen in biological samples. In general, workers have chosen to continuously add oxygen to the the suspensions of mitochondria and cells in order to generate "steady states" in which the oxygen pressure was essentially unchanging and the oxygen measurements could be made over longer times. Unfortunately, in this approach the oxygen pressure is higher at the site of addition than in the rest of the suspension and regional differences in oxygen pressure in the medium were present and not adequately evaluated. An optical method for measuring oxygen has been recently developed which has both high sensitivity (10^{-3} M to 10^{-8} M) and rapid response (< 1 msec) (Vanderkooi and Wilson, 1986; Vanderkooi et al, 1987; Wilson et al, 1987; 1988). This method permits accurate measurements under conditions without continuous addition of oxygen and thereby without differences in the oxygen pressure in the extracellular medium.

RESULTS

Measurements have been made of the oxygen pressure required for half-maximal rates of respiration (P_{50}) by suspensions of isolated mitochondria (Wilson et al, 1988). The value of P_{50} is different for different metabolic states of the mitochondria. It is greater than 0.5 torr when the mitochondria are well coupled and in the presence of high levels of ATP. When an uncoupler of mitochondrial oxidative phosphorylation was added, however, the respiratory rate increased 6-10 fold but the P_{50} decreased to less than 0.03 torr.

The oxygen pressure for half maximal P_{50} for suspensions of human neuroblastoma cells (approximately 0.8 torr) was similar to that for coupled mitochondria in the presence of ATP. When uncoupler was added to the cells, the P_{50} decreased only to 0.7 torr (Robiolio et al, 1988), a much smaller decrease than that for suspensions of isolated mitochondria. This decrease in P_{50} occurred despite an increase of 4-5 fold in the cellular respiratory rate. This data indicated that although diffusion did not contribute significantly to the measured value of the P_{50} for oxygen in the respiration of normal neuroblastoma cells, it may have contributed in the case of uncoupler treated cells.

The P_{50} for uncoupled neuroblastoma cells was significantly greater than that for uncoupled mitochondria. This was result is consistent with the possibility that the measured P_{50} for uncoupler treated neuroblastoma cells is diffusion limited. In this case the P_{50} could be a measure of the diffusion induced oxygen pressure difference between the extracellular medium and the mitochondria of the cells.

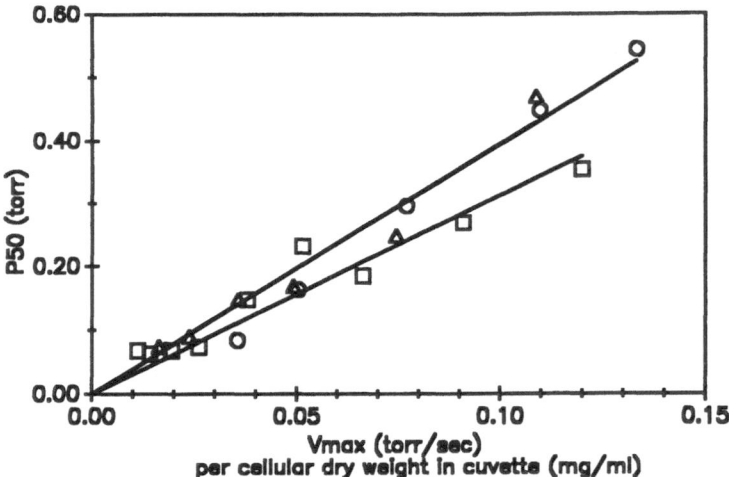

Figure 1. The dependence of V_{max} and P_{50} of uncoupler treated cells on the cellular respiratory rate. The neuroblastoma cells were suspended in Hank's medium containing 0.3% bovine serum albumin and 2 uM Pd-coproporphyrin. The uncoupler, trifluoromethylphenylhydrazone of carbonyl cyanide (FCCP), was added at 1.5 times the concentration required for maximal rates of respiration. The cells were present at approx. 1.8 mg dry weight/ml (o) and twice (▵) or 4 times (◻) that concentration. At each cell concentration, the V_{max} and P_{50} were determined for added amytal concentrations of 0.2 mM, 0.4 mM, and upward by steps of 0.2 mM with each point plotted (maximal plotted concentration of 1.6 mM). A. The measured value of P_{50} is plotted against the measured value of V_{max}. The lines are fitted assuming an intercept of 0. B. The V_{max} values have been normalized to the same concentration of cells by presenting the values as the respiratory rate per mg dry weight of cells.

The diffusion induced oxygen pressure difference in neuroblastoma cells

If the P_{50} for uncoupler treated cells were diffusion limited, decreasing the respiratory rate by additon of inhibitors of the mitochondrial respiratory chain would lower the value of P_{50}. At high respiratory rates the P_{50} should decrease in proportion to the decrease in respiration. At low respiratory rates the P_{50} should level off and approach the value for uncoupled, inhibited mitochondria. The latter region, however, is characterized by a P_{50} value too low to measure (< 0.03 Torr) even by phosphorescence quenching (using Pd-coproporphyrin bound to bovine serum albumin as an oxygen probe).

It can be seen in Figure 1, that inhibition of cellular respiration by amytal, an inhibitor of the NADH dehydrogenase region of the respiratory chain, decreases the P_{50} in proportion to the decrease in respiratory rate. Extrapolation of the data to a respiratory rate of zero gives a P_{50} of less than 0.05 torr. Increasing or decreasing the respiratory rate of the cell suspension by changing the cell concentration had no effect on the measured value of P_{50}.

Comparison of the P_{50} for oxygen in suspensions of normal and uncoupled cells at the same cellular respiratory rates.

Addition of uncoupler to suspensions of neuroblastoma cells caused a 4-5 fold increase in respiratory rate. Subsequent addition of increasing amounts of amytal progressively lowered the cellular respiratory rate, and at an amytal concentration of about 1.2 mM the respiratory rate of the uncoupled cells was equal to that of normal cells. Measured at the same cellular respiratory rate, the P_{50} for oxygen by suspensions of normal cells was 0.8 torr, while that for the uncoupled, inhibited cells was 0.15 torr.

DISCUSSION

The oxygen dependence of mitochondrial oxidative phosphorylation extends well into the range of oxygen pressure under normal physiological conditions (up to at least 30 torr-see Wilson et al., 1979a,b; Wilson and Erecinska, 1985). Most of the oxygen dependence is reflected in changes in the regulatory parameters for oxidative phosphorylation (cytoplasmic [ATP]/[ADP][Pi] and intramitochondrial [NAD$^+$]/[NADH]). A decrease in oxygen pressure to below that required to saturate the cytochrome c oxidase reaction leads to temporary suppression of the rate of ATP synthesis. The excess of ATP use over ATP synthesis lowers the [ATP]/[ADP][Pi], but as the latter decreases the respiratory rate is stimulated (see Wilson et al, 1982). If the oxygen pressure is held steady at the new value, the respiratory rate increases until the rate of ATP synthesis again equals the rate of ATP utilization and a new steady state is attained. Normally these changes are rapid (seconds). Since the rate of ATP utilization is essentially independent of oxygen pressure, the respiratory rate remains nearly constant as the oxygen pressure is lowered until it reaches a level too low to support the required rate of ATP synthesis even at near maximal stimulation. Thus the oxygen dependence of mitochondrial oxidative phosphorylation in cells is observed primarily in the cytoplasmic [ATP]/[ADP][Pi]. The mitochondrial respiratory rate is also, however, dependent on the intramitochondrial [NAD$^+$]/[NADH] (see Erecinska et al, 1978; Wilson et al, 1982) and the latter is determined by the activity of the mitochondrial dehydrogenases.

The activities of the dehydrogenases are highly regulated and, under some conditions, part of the effect of lowering the oxygen pressure is compensated for by decreasing [NAD$^+$]/[NADH] (reduction of the mitochondrial pyridine nucleotide pool). When the oxygen pressure is low enough that the regulatory parameters can no longer stimulate the respiratory rate enough to keep the rate of ATP synthesis equal to its rate of utilization, the respiratory rate falls. This does not occur until after the conditions are reached which will, if sustained for significant lengths of time, result in severe cellular pathology.

Diffusion of oxygen from the extracellular medium to the mitochondria limits the P_{50} for oxygen in uncoupler treated neuroblastoma cells. The oxygen pressure difference between the external medium and the mitochondria is dependent on the mitochondrial respiratory rate. This value can be as high as 0.7 torr for maximal respiratory rates but is only 0.15 torr in normal cells respiring at half maximal rates (P_{50} of 0.8 torr). It appears, at least in human neuroblastoma cells, that diffusion of oxygen from the extracellular medium to the mitochondria does not contribute substantially to the oxygen pressure dependence of cellular energy metabolism. It can, however, limit the respiratory rate at low oxygen pressures when the mitochondria are maximally activated.

In the absence of direct measurements of the oxygen pressure difference between the external medium and the mitochondria, there has been much discussion of the role of oxygen diffusion in determining the oxygen dependence of cellular energy metabolism. Some authors have suggested that oxygen pressure differences of several torr may exist between the cytosol and mitochondria (see for example Jones and Kennedy, 1982; Jones, 1986). This suggestion which has been criticized as inconsistent with reasonable application of diffusion theory to oxygen in biological tissue (Clark and Clark, 1985; Clark et al, 1987; Wilson, 1982; Wilson and Erecinska, 1985), with the measured oxygen dependence of monoamine oxidase (an enzyme in the outer mitochondrial membrane) activity (Katz et al, 1984; Wittenberg and Wittenberg, 1985) and with the absence of local regions of low oxygen in the neighborhood of mitochondria (Gayeski et al, 1987). Our measurements establish that, at least in the case of human neuroblastoma cells, the total oxygen pressure difference from the extracellular medium to the mitochondria is only about 0.15 torr. The difference between the cytoplasm and the mitochondria must be less than the total and is probably only a few tens of millitorr.

In summary: In suspensions of normally respiring human neuroblastoma cells respiration has an oxygen dependence similar to that of suspensions of isolated mitochondria in a comparable metabolic state. When mitochondrial oxidative phosphorylation is uncoupled, respiration at limiting oxygen pressures is indicative of the oxygen pressure difference between the extracellular medium and the mitochondria. Diffusion gives a P_{50} proportional to the cellular respiratory rate, with a value of 0.15 torr for the respiratory rate of normal cells. The oxygen pressure difference between the cytoplasm and the mitochondria is probable only a few tens of millitorr.

Acknowledgements: This research was supported by a grant, GM-21525, from the National Institutes of Health.

REFERENCES

Clark, A., Jr., Clark, P. A. A., Connett, R. J., Gayeski, T. E. J., and Honig, C. R., 1987, How large is the drop in P_{O2} between cytosol and mitochondrion? *Am.J. Physiol.* 252: C583-C587.

Clark, A., Jr., and Clark, P. A. A., 1985, Local oxygen gradients near isolated mitochondria. *Biophys. J.* 48: 931-938.

Erecinska, M., Wilson, D. F., and Nishiki, K., 1978, Homeostatic regulation of cellular energy metabolism: experimental characterization *in vivo* and fit to a model. *Am. J. Physiol.* 234: C82-C89.

Gayeski, T. E. J., Connett, R. J., and Honig, C. R., 1987, The minimum intracellular P_{O2} for maximal cytochrome turnover in red muscle *in situ*. *Am. J. Physiol.* 2523: H906-H915.

Jones, D. P., 1986, Intracellular diffusion gradients of O_2 and ATP. *Am. J. Physiol.* 250: C663-C675.

Jones, D. P. and Kennedy, F. G., 1982, Intracellular oxygen supply during hypoxia. *Am. J. Physiol.* 243: C247-C253.

Katz, I. R., Wittenberg, J. B., and Wittenberg, B. A., 1984, Monoamine oxidase, an intracellular probe of oxygen pressure in isolated cardiac myocytes. *J. Biol. Chem.* 259: 7504-7509.

Robiolio, M., Rumsey, W.L. and Wilson, D.F., 1988, Oxygen diffusion and mitochondrial respiration in neuroblastoma cells. *Amer. J. Physiol.* in press.

Vanderkooi, J. M., Maniara, G., Green, T.J., and Wilson, D. F., 1987, An optical method for measurement of dioxygen based upon quenching of phosphorescence. J. Biol. Chem. 262: 5476-5482.

Vanderkooi, J. M., and Wilson, D. F., 1986, A new method for measuring oxygen in biological systems. In: *Oxygen transport to tissue VIII*, edited by I. A. Longmuir. New York: Plenum, p. 189-193.

Wilson, D. F., 1982, Regulation of *in vivo* mitochondrial oxidative phosphorylation. In: *Membranes and Transport*, edited by A. N. Martonosi. New York: Plenum, vol. 1, p. 349-355.

Wilson, D. F., and Erecinska, M., 1985, Effect of oxygen concentration on cellular metabolism. *Chest* 88: 229-232.

Wilson, D. F., Erecinska, M., Drown, C., and Silver, I. A., 1979a, The oxygen dependence of cellular energy metabolism. *Arch. Biochem. Biophys.* 195: 485-493.

Wilson, D. F., Owen, C. S., and Erecinska, M., 1979b, Quantitative dependence of mitochondrial oxidative phosphorylation on oxygen concentration: a mathematical model. *Arch. Biochem. Biophys.* 195: 494-504.

Wilson, D. F., Rumsey, W. L., Green, T. J., and Vanderkooi, J. M., 1988, The oxygen dependence of mitochondrial oxidative phosphorylation measured by a new optical method for measuring oxygen concentration. *J. Biol. Chem.* 263: 2712-2718.

Wilson, D. F., Vanderkooi, J. M., Green, T. J., Maniara, G., DeFeo, S.P., and Bloomgarden, D. C., 1987, A versatile and sensitive method for measuring oxygen. in *Oxygen Transport to Tissue IX*, edited by I. A. Silver and A. Silver, New York: Plenum, p. 71-77.

Wittenberg, B. A., and Wittenberg, J. B, 1985, Oxygen pressure gradients in isolated cardiac myocytes. *J. Biol. Chem.* 260: 6548-6554.

BLOOD FLOW, TISSUE OXYGENATION, pH DISTRIBUTION, AND ENERGY METABOLISM OF
MURINE MAMMARY ADENOCARCINOMAS DURING GROWTH[*]

P. Vaupel[1], P. Okunieff[1], and L.J. Neuringer[2]

[1]Dept. Radiation Medicine, Massachusetts General Hospital
 Cancer Center, Harvard Medical School, Boston, MA 02114
[2]Francis Bitter National Magnet Laboratory, Massachusetts
 Institute of Technology, Cambridge, MA 02139

INTRODUCTION

Many solid tumors are relatively resistant to non-surgical therapeutic
assaults. A variety of factors are involved in the lack of responsiveness
of these neoplasms including cellular heterogeneity due to genetic differences between cells, and physiological factors created by inadequate and
heterogeneous vascular networks. Thus, properties such as tumor blood flow,
tissue oxygenation, pH distribution and energy metabolism, factors which
generally go hand in hand, markedly influence the therapeutic response.
Since the strategic approach must consider the physiological microenvironment of the cells within a tumor mass and -if possible- should exploit the
metabolic micromilieu in designing therapies, we have measured relevant
parameters in a well established murine tumor cell line with a known size
dependent increase in the radiobiologically hypoxic cell fraction.

MATERIALS AND METHODS

Animals and Tumors. Experimental animals were 10-12 week old C3Hf/Sed
mice derived from our defined flora mouse colony[1]. Animals were provided
with sterilized animal pellets and acidified, vitamin K fortified water ad
libitum. Early generation isotransplants of a spontaneous mammary adenocarcinoma (MCaIV) were used. Single cell suspensions were prepared by a
mechanical procedure and were transplanted subcutaneously (s.c.) into the
hind leg. Tumor volumes on the day of study were calculated by an ellipsoid
approximation using the three diameters ($V = \pi/6 \cdot d_1 \cdot d_2 \cdot d_3$). The parameters
of interest were studied in tumors of different sizes with volumes ranging
from 0.05 to 1.75 ml.

Measurements of Blood Flow in Microregions of Tumors. Flow in microareas of mammary tumors was determined using the H_2 clearance technique as
described by STOSSECK et al.[2] The method is based on the polarographic determination of the amount of hydrogen gas reaching a platinum electrode
(H_2 detector) from a hydrogen gas generating electrode located at a fixed
distance. In this study, two platinum-in-Teflon wires ($\emptyset = 100$ um) placed
approximately 100 um apart were used. The reading device was applied to
the surface of the tumors[3].

---

[*]This study was supported in part by NIH grants CA48096 and RR00995

Measurements of Tissue pO_2 and pH Values with Microelectrodes. The O_2 microelectrodes used during these experiments were of the "gold-in-glass" type as described earlier[4]. The exposed gold tip of these microelectrodes was about 1-5 um in diameter and was coated with a Rhoplex membrane. The pH microelectrodes used in this study were of the Hinke type. In these spear- type glass microelectrodes tip diameters were < 1 um, thus minimizing tissue damage upon penetration. The mice were anesthetized with chlorpromazine (50 ug/g i.m.) and ketamine (40 ug/g i.m.) during measurement of flow, pO_2 and pH distribution. Rectal temperature was maintained close to 37°C by a thermostatically controlled heating pad[4].

^{31}P Nuclear Magnetic Resonance Spectroscopy. Spectra were obtained at 8.5 Tesla (corresponding to a resonance frequency of 145.6 MHz for ^{31}P). Details regarding the NMR probe and shimming have been described previously[5]. Spectra were accumulated with a 60° pulse, a recycle time of 2 sec, a spectral width of 10 kHz, and 128 to 512 averaged free induction decays. Spectral processing included 15-30 Hz matched exponential multiplication, and a convolution difference of 600 Hz. Peak height values were corrected for partial saturation using the T_1 relaxation times[5,6]. The phosphocreatine (PCr) to inorganic phosphate (P_i) chemical shift was used to estimate the nominal intracellular pH of tumor cells[6,7]. All NMR experiments were performed on conscious animals.

RESULTS AND DISCUSSION

Tumor Blood Flow. The average perfusion rate in murine mammary adenocarcinomas significantly decreases with increasing tumor size (see Fig.1). This pattern has also been documented for many other tumor systems[8]. Considering the pathogenetic mechanism(s) directly responsible for the flow decline with increasing tumor mass, a progressive rarefaction of the vascu-

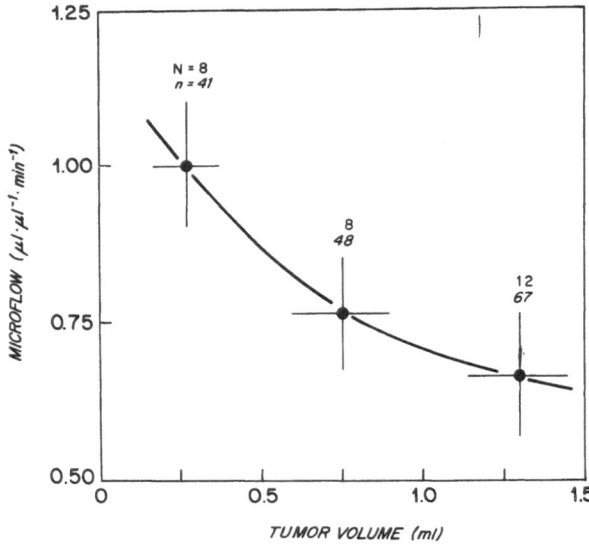

Fig.1. Local tumor blood flow (microflow) in superficial tumor areas as a function of tumor volume. N = number of tumors investigated, n = number of flow measurements performed. Values are means ± SE.

lar network as well as severe structural and functional abnormalities of the tumor microcirculation have to be taken into account. The decrease in tumor blood flow per unit tissue mass is additionally caused by the development of necrosis. The most pronounced flow decline, however, is obvious before any significant necrosis appears indicating that necrosis is not the paramount factor responsible for the size dependent flow drop observed. The decrease in the local flow per unit tissue volume (measured in superficial tumor areas) is in very good agreement with measurements of total tumor blood flow in these murine mammary tumors (see fig. 2; data redrawn from TOZER et al.[9]). Local tumor blood flow varied considerably within a tumor and between tumors of comparable sizes.

Tissue Oxygenation and pH Distribution in Murine Mammary Tumors. As the tumors increased in size, the O_2 partial pressure distribution is shifted to lower values, i.e., the frequency distribution is tilted to the left and more limited in variability than in the subcutis at the site of tumor growth. This pattern becomes evident if the mean or median tissue pO_2 values in the murine tumors are examined as a function of tumor volume (see fig.3). From these data it is concluded that the tissue oxygenation in microareas of the mammary tumors is inadequate and heterogeneous. This is mostly due to heterogeneously distributed restrictions of the tumor microcirculation and, therefore, of the O_2 availability to the cancer cells in vivo. Besides intratumor heterogeneities, there is marked tumor-to-tumor variability in the tissue pO_2 values measured.

Compared to subcutaneous tissue (median pH = 7.3) the pH distributions in the mammary tumors are shifted towards more acidic values (see fig. 4).

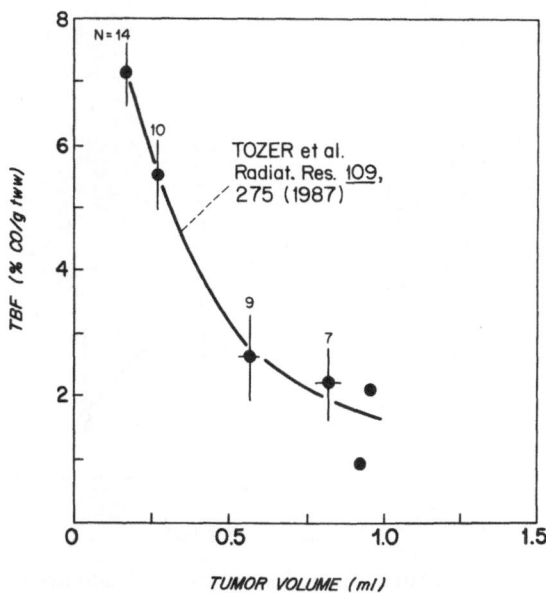

Fig.2. Changes in blood flow (TBF) of MCaIV tumors during growth. Results are expressed as percentages of cardiac output (CO). N = number of tumors investigated, tww = tumor wet weight. Values are means ± SE (redrawn from TOZER et al.[9]).

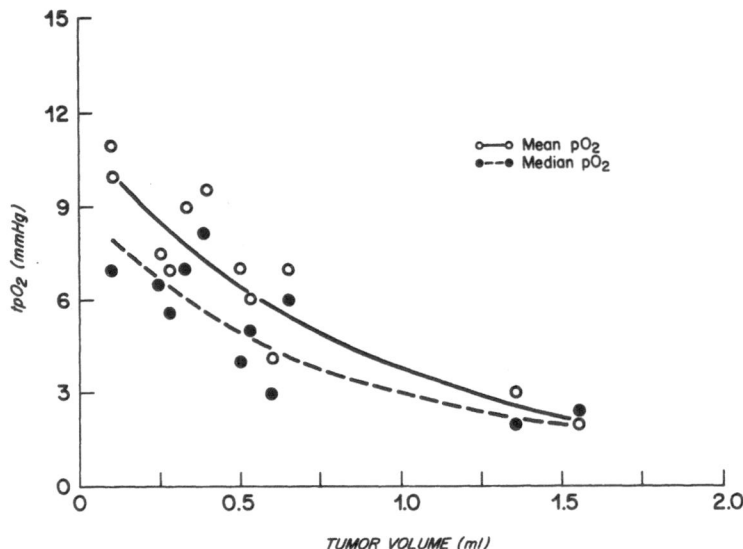

Fig.3. Mean (open circles, solid line) and median (closed circles, broken line) O_2 partial pressures (tpO_2) in mammary tumors as a function of tumor volume.

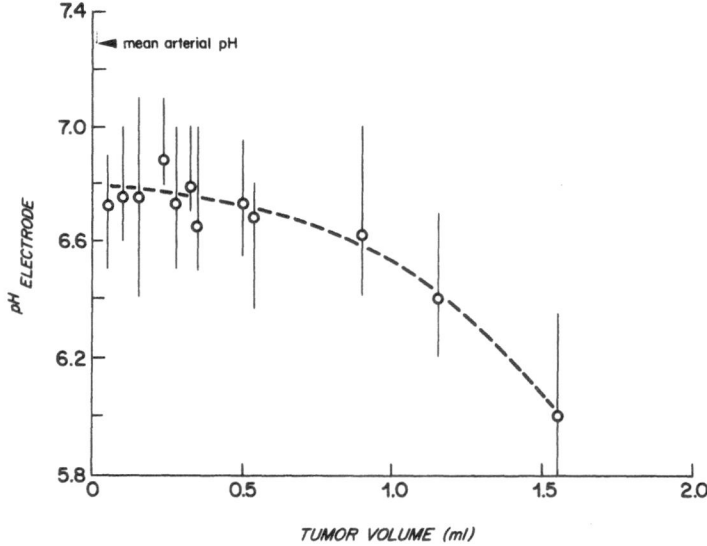

Fig.4. Mean tissue pH values in mammary tumors measured with pH glass microelectrodes. Vertical bars: pH range.

Here again, marked intertumor variations are observed. A trend toward a further acidification of the tumor tissue is found with tumor volumes > 1 ml. pH values measured within individual tumors exhibit very heterogeneous distributions with steep pH gradients in some of them and flat pH profiles in others. The low pH values obtained in murine mammary tumors are comparable to those found earlier in a broad variety of rodent and human tumors[10], and are brought about by a high lactate production rate (due to aerobic and anaerobic glycolysis as well as glutaminolysis) combined with a reduced removal of this metabolite.

Tumor Size Dependent Changes in Tumor Bioenergetics. Typical spectral changes that occur with tumor growth are shown in fig. 5. With increasing tumor volume, PCr and nucleoside triphosphate (NTP) resonances decrease. These changes coincide with an increase of P_i. Tumors with volumes > 0.6 ml often have little NTP and no detectable PCr. Tumor growth also correlates

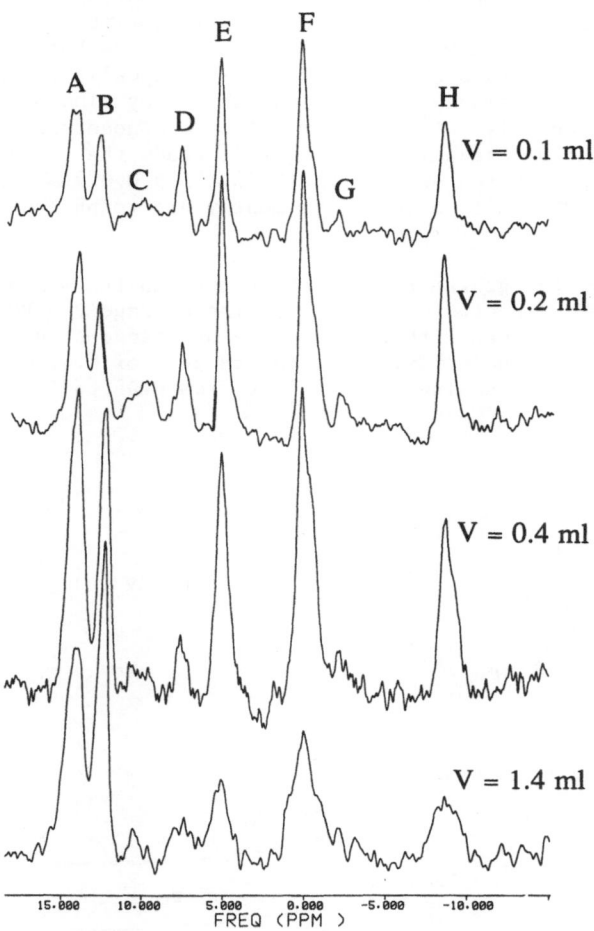

Fig.5. Effect of increasing tumor volume (V) on ^{31}P-NMR spectra of murine mammary tumors. Spectra are individually scaled for better demonstration. A = phosphomonoesters (PME, includes phosphoethanolamine, phosphocholine, and sugar phosphates), B = P_i, C = phosphodiesters (PDE, includes glycerophosphoethanolamine, glycerophosphocholine), D = PCr, E = γ NTP and β NDP, F = α NTP, α NDP, and nicotinamide adenine dinucleotides, G = diphosphodiesters (DPDE), F = β NTP; (NDP = nucleoside diphosphate).

with a significant relative increase in the phosphomonoester (PME) resonance, which has been attributed to an intensified membrane turnover with
tumor growth[11]. Fig. 6 exhibits the growth related changes in the PCr/P_i
and NTP/P_i ratios which start to decline early in tumor growth, i.e., in
tumors without gross necrosis, and reach nadir values in tumors > 0.5 ml.
The PME/NTP ratio increases with tumor growth and then levels off at tumor
volumes > 0.6 ml. At the same time, PME/P_i drops and reaches nadir values
in tumors > 0.5 ml. The changes in the tumor energy metabolism observed
with ^{31}P NMR spectroscopy are in line with data described by TOZER et al.[9]
who found a significant decline of the ATP concentration with tumor growth
(measurements of the tissue ATP levels using an enzymatic assay in MCaIV
tumors of the same size range; see fig. 7).

Nominal Intracellular pH in Murine Mammary Tumors. Using the chemical
shift of the P_i and PCr peaks, the nominal intracellular pH can be estimated
according to the observation of MOON and Richards[7]. Since larger tumors often
have little or no PCr, the PCr peak position can not always be determined
with the needed accuracy. In those tumors pH is estimated using the peak
separation between P_i and αNTP (the αNTP peak, like the PCr peak is a
useful internal standard since it shifts only minimally in the physiologically pH range[7,12]). Because the chemical shift may also be affected by
electrolytes, especially Mg^{2+}, the tumor tissue concentration of free Mg^{2+}
is estimated using NMR techniques[13]. In this study the tissue concentration
of free Mg^{2+} is unaffected during growth (0.135 ± 0.019 mM) indicating that
the chemical P_i- PCr shift observed is mostly due to pH changes in the tissue.

Fig. 8 shows the growth related pH changes in the mammary tumors obtained from ^{31}P-NMR spectroscopy (pH_{NMR}). The average P_i- PCr shift decreases with tumor growth with the majority of change occurring in tumors
> 0.7 ml in volume. An approximate total pH drop of 0.4 to 0.5 pH units
is calculated with increasing tumor volume from 0.05 to 1.75 ml.

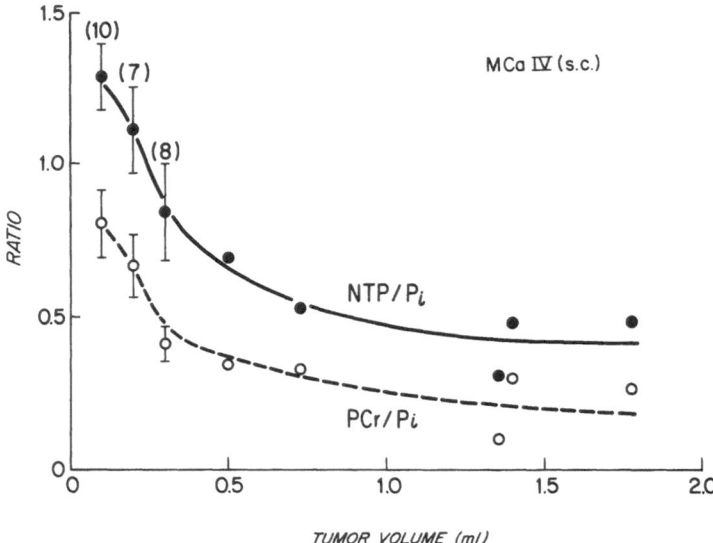

Fig.6. Growth related changes in the PCr/P_i and NTP/P_i ratios. Numbers in
parentheses are numbers of tumors investigated. Values are means
($\pm$ SE).

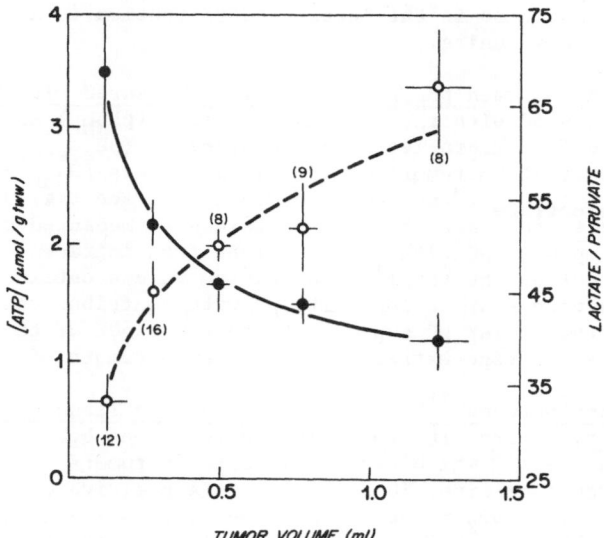

Fig.7. Changes of tissue ATP concentrations (closed circles, solid line), and lactate/pyruvate ratio (open circles, broken line) during tumor growth. Number in parentheses are numbers of tumors investigated. Values are means ± SE, tww = tumor wet weight (redrawn from TOZER et al.[9]).

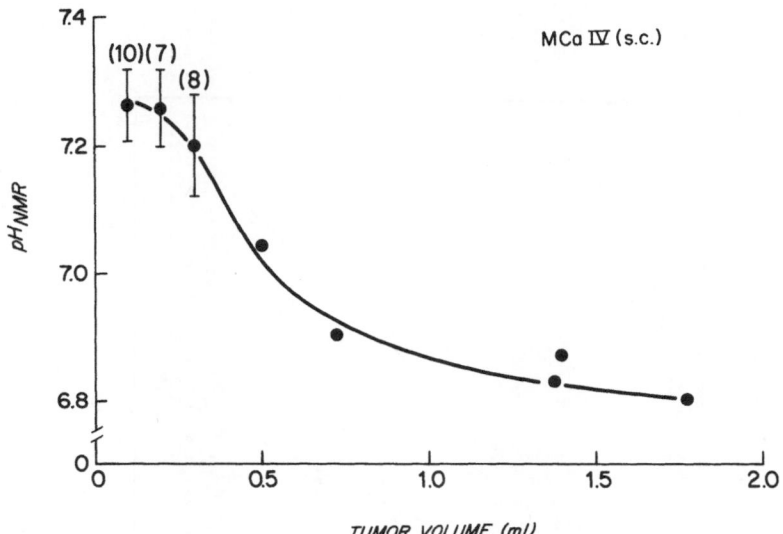

Fig.8. Growth related pH changes in mammary tumors obtained from ^{31}P NMR spectroscopy (pH_{NMR}). Number in parentheses are numbers of tumors investigated. Values are means (± SE).

The drop in pH_{NMR} observed in this study coincides with a 2-fold increase of the lactate levels and of the lactate/pyruvate ratio in MCaIV tumors of the same size range (see fig. 7). Similar pH- lactate- correlations have been described by KALLINOWSKI et al.[14] for ascites tumors. In the latter study, a 2-fold increase in the lactate concentration is accompanied by a pH drop of about 0.45 units.

Correlations Between pH_{NMR} and pH Values Measured with Microelectrodes. If pH values measured with the ^{31}P NMR technique (pH_{NMR}) are compared with pH data obtained from microelectrode measurements ($pH_{electrode}$) there is clear indication that in tumors of identical volumes pH_{NMR} is consistently higher than $pH_{electrode}$ (Δ pH = 0.2 to 0.5 units; see fig. 9). This most probably is due to the fact that the pH values as measured by NMR are best considered a composite pH with contributions from intracellular pH expected to be at least 80% of the total[15], whereas pH values obtained with micro-electrode measurements are a composite pH with contributions from the inter-stitial and intravascular pH expected to be about 50% of the total (estimated value assuming an average extracellular space in tumors of ca. 50%[16]).

Correlations Between ^{31}P NMR Spectroscopy and Tissue pO_2 Measurements in Murine Mammary Tumors. The following correlations between tumor tissue oxygenation and the ^{31}P NMR bioenergetic data in tumors of identical vol-umes are observed (see figs. 10 and 11): (i) a positive correlation between the mean tumor tissue pO_2 values as measured with O_2 sensitive microelec-trodes and the NTP/P_i ratios, and (ii) a positive correlation between the mean pO_2 values and the PCr/P_i ratios. Both ratios are related to tissue oxygenation via the enzyme reactions catalyzed by phosphocreatine kinase and adenylate kinase, respectively. Although the exact relationship between cellular PCr/P_i and NTP/P_i and tissue oxygenation has not been fully elu-cidated, ^{31}P NMR can thus have the potential of providing a means of detec-

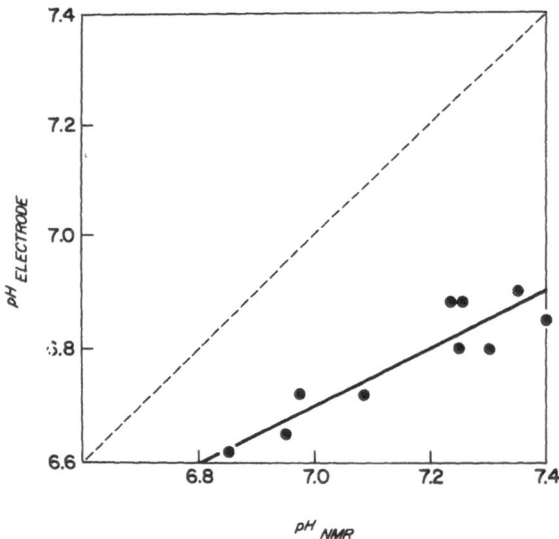

Fig.9. Correlation between pH values measured with the ^{31}P NMR technique (pH_{NMR}) and pH data obtained from microelectrode measurements ($pH_{electrode}$) in tumors of identical volumes. The line of identity (broken line) is shown for comparison.

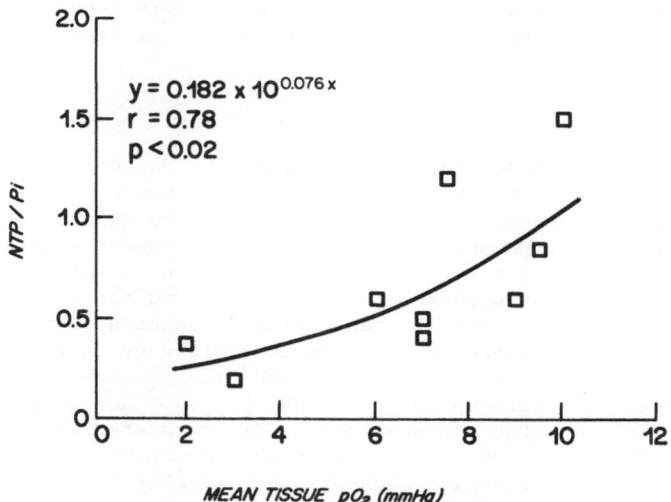

Fig. 10. Correlation between the mean tissue pO_2 values and NTP/P_i ratios in mammary tumors of identical volumes.

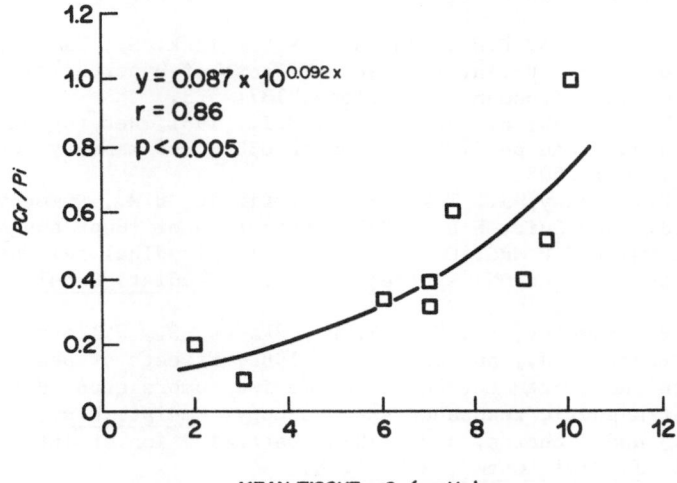

Fig. 11. Correlation between the mean tissue pO_2 values and PCr/P_i ratios in mammary tumors of identical volumes.

ting changes in tumor bioenergetics brought about by changes in tissue oxygenation. Furthermore, the data presented suggest that the microcirculation in isotransplanted MCaIV tumors yields an oxygen limited tumor energy metabolism. PCr/P_i and NTP/P_i ratios are therefore sensitive "detectors" of an inadequate O_2 supply to tumors, which can be evaluated non-invasively and repeatedly. The measurements of the energy parameters thus continues to be worthwhile in studies testing factors able to predict treatment response of human cancer.

CONCLUSIONS

In this study we have characterized the size dependent changes of relevant and interrelated metabolic parameters of a murine mammary carcinoma in vivo. Tumor blood flow, tumor oxygenation and the tissue pH distribution as well as high energy phosphates all declined as tumors enlarged. In each case, the steepest decline occurred at a size range in which gross necrosis is rarely observed, and in which a steep increase in the radiobiologically hypoxic cell fraction is also apparent. From these data we conclude that in this tumor type the energy metabolism is limited by an inadequate O_2 supply which is manifested with tumor growth. The results described include the first correlation between pH_{NMR} and $pH_{electrode}$, and suggest that in the pH range observed, the intracellular pH is consistently higher than the extracellular pH. Using some of the techniques described, the efficacy of some tumor treatment strategies might be specifically "tailored" to achieve a better therapeutic response in this mammary tumor, and might eventually be used to improve the treatment of cancer in patients.

REFERENCES

1. Sedlacek, R.S., and Mason, K.S., 1977, A simple and inexpensive method for maintaining a defined flora mouse colony, Lab. Anim. Sci.,27:667.
2. Stosseck, K., Luebbers, D.W., and Cottin, N., 1974, Determination of local blood flow (microflow) by electrochemically generated hydrogen. Construction and application of the measuring probe, Pfluegers Arch., 348: 225.
3. Bicher, H.I., Hetzel, F.W., Sandhu, T.S., Frinak, S., Vaupel, P., O'Hara, M.D., and O'Brien, T., 1980, Effects of hyperthermia on normal and tumor microenvironment, Radiology, 137: 523.
4. Vaupel, P.W., Frinak, S., and Bicher, H.I., 1981, Heterogeneous oxygen partial pressure and pH distribution in C3H mouse mammary adenocarcinoma, Cancer Res., 41: 2008.
5. Okunieff, P., Ramsay, J., Tokuhiro, T., Hitzig, B.M., Rummeny, E., Neuringer, L.J., and Suit, H.D., 1988, Estimation of tumor oxygen and metabolic rate using ^{31}P MRS: Correlation of longitudinal relaxation with tumor growth rate and DNA synthesis, Int. J. Radiat. Oncol. Biol. Phys., 14: 1185.
6. Okunieff, P., Rummeny, E., Vaupel, P., Skates, S., Willett, C., Hitzig, B.M., Neuringer, L.J., and Suit H.D., 1988, Effects of pentobarbital anesthesia on the energy metabolism of murine tumors studied by in vivo ^{31}P nuclear magnetic resonance spectroscopy, Radiat. Res., in press.
7. Moon, R.B., and Richards, J.H., 1973, Determination of intracellular pH by ^{31}P NMR, J. Biol. Chem., 248: 7276.
8. Vaupel, P., Fortmeyer, H.P., Runkel, S., and Kallinowski, F., 1987, Blood flow, oxygen consumption, and tissue oxygenation of human breast cancer xenografts in nude rats, Cancer Res., 47: 3496.
9. Tozer, G., Suit, H.D., Barlai-Kovach, M., Brunengraber, H., and Biaglow, J., 1987, Energy metabolism and blood perfusion in a mouse mammary adenocarcinoma during growth and following X irradiation, Radiat. Res., 109: 275.
10.Kallinowski, F., and Vaupel, P., 1988, pH distributions in spontaneous and isotransplanted rat tumors, Brit. J. Cancer, in press.

11. Steen, R.G., Tamargo, R.J., McGovern, K.A., Rajan, S.S., Brem, H., Wehrle, J.P., and Glickson, J.D., 1988, In vivo ^{31}P nuclear magnetic resonance spectroscopy of subcutaneous 9L gliosarcoma: Effects of tumor growth and treatment with 1,3-Bis(2-chloroethyl)-1-nitrosourea on tumor bioenergetics and histology, Cancer Res., 48: 676.

12. Okunieff, P.G., Koutcher, J.A., Gerweck, L., McFarland, E., Hitzig, B., Urano, M., Brady, T., Neuringer, L.J., and Suit, H.D., 1986, Tumor size dependent changes in a murine fibrosarcoma: Use of in vivo ^{31}P NMR for non-invasive evaluation of tumor metabolic status, Int. J. Radiat. Oncol. Biol. Phys., 12: 793.

13. Okunieff, P., Kallinowski, F., Vaupel, P., and Neuringer, L.J., 1988, Effects of hydralazine- induced vasodilation on the energy metabolism of murine tumors studied by in vivo ^{31}P nuclear magnetic resonance spectroscopy, J. Natl. Cancer Inst., 80: 745.

14. Kallinowski, F., Tyler, G., Mueller- Klieser, W., and Vaupel, P., 1989, Growth related changes of oxygen consumption rates of tumor cells grown in vitro and in vivo, J. Cell. Physiol., submitted.

15. Evelhoch, J.L., Sapareto, S.A., Jick, D.E.L., and Ackerman, J.J.H.,1984, In vivo metabolic effects of hyperglycemia in murine radiation-induced fibrosarcoma: A ^{31}P NMR investigation, Proc. Natl. Acad. Sci. USA, 81: 6496.

16. Vaupel, P., and Hammersen, F. (eds.), 1983, "Mikrozirkulation in malignen Tumoren", Karger, Basel, Muenchen, Paris, London, New York, Tokyo, Sydney.

INTERRELATIONSHIP AMONG MORPHOLOGY, METABOLISM, AND PROLIFERATION OF TUMOR CELLS IN MONOLAYER AND SPHEROID CULTURE

Stefan Walenta, Anneli Bredel, Ulrich Karbach, Leoni Kunz
Lutz Vollrath[*], and Wolfgang Mueller-Klieser

Department of Applied Physiology and [*]Department of Anatomy
University of Mainz, D-6500 Mainz, FRG

INTRODUCTION

Previous investigations have indicated a positive correlation between the proliferative and metabolic activities of tumor cells in monolayer culture (Freyer et al., 1984; Freyer and Sutherland, 1985; Walenta and Mueller-Klieser, 1987). On the other hand, no difference in the local oxygen consumption has been found between highly proliferating outer cell areas and non-proliferating inner cellular regions in multicellular tumor spheroids (Mueller-Klieser, 1984, 1987). Therefore, the interrelationship among metabolism, proliferation, and cellular morphology was investigated systematically in tumor cells both in monolayer and spheroid culture.

MATERIALS AND METHODS

Measurements were mainly carried out on EMT6/Ro cells derived from a mouse mammary sarcoma. For comparison, additional data on human breast carcinoma cells (MCF-7) were obtained. The respiration rate, the fraction of proliferating cells, the mean cellular volume, and the mean number and volume of mitochondria per cell were measured as a function of time in monolayer and spheroid culture. The respiratory activity of monolayer cells that were trypsinized and suspended in fresh medium was assessed by a photometric method (Mueller-Klieser et al., 1986). Cell numbers and cellular volumes were registered by an automatic cell counter based on the Coulter principle. The local oxygen consumption rates in multicellular spheroids that were grown in suspension cultures (Mueller-Klieser, 1987; Sutherland, 1988) were derived from measurements with O_2-sensitive microelectrodes in these cell aggregates using a special evaluation procedure (Mueller-Klieser, 1984).

The proliferative status of single cells and multicellular spheroids was determined by standard autoradiographic techniques. The labeling index (LI) was obtained as the percentage of labeled cells with regard to the total number of cells considered. The proliferative fraction of cells, i.e., the percentage of proliferating cells related to the total number of viable cells, was derived from the LI in exponentially growing monolayer cells assuming a corresponding proliferating fraction of 100%.

For the quantitative assessment of the cellular morphology, pellets of single cells that were obtained from suspensions by centrifugation and individual multicellular spheroids were investigated by standard electron-microscopic techniques. Sections were randomly made through the packed single cells, whereas only central sections through spheroids were considered for evaluation. Using an automatic image analysing system (Morphomat, Kontron, München, FRG), several characteristic areas were registered in these sections, including the cellular areas of the cells, the nuclei, and the mitochondria. In addition, the mean number of mitochondria per cell area was determined. Estimates could be derived from these data for the cellular volume fraction of nuclei and mitochondria assuming spherical geometry for cells and for mitochondria.

RESULTS AND DISCUSSION

The oxygen consumption rates (Q) per cell and per cellular volume, the labeling index and the proliferative fraction of EMT6 cells as a function of time in monolayer culture are shown in Fig. 1 a+b. It is evident that the transition from the exponential (days 2-3) to the plateau phase (days 5-9) is associated with a decrease in both the proliferative and respiratory activity of the cells. Concomitantly, the mean cellular volume decreases. To evaluate the cell line specificity of these findings, Fig. 2 a+b shows the mean cellular volume and the cellular and volume-related respiration rate of MCF-7 cells. In this case, a decrease of the respective parameters similar to that in EMT6 cells was registered during the transition from the exponential to the plateau growth phase; however, the decrease in mean cellular volume in the second half of the culturing period was preceded by an initial increase of this particular parameter immediately after the initiation of monolayer growth. This is in contrast to a continuous decline of cell volumes found for EMT6 cells. The biphasic behavior of cell volumes during monolayer growth has been registered for several other human cancer cell lines.

The results of the morphometric measurements in EMT6 monolayer cells are shown in Table 1. It is obvious that the decrease in the respiratory activity of tumor cells during monolayer growth is associated with a reduction in the mean cellular volume. This confirms the results of the measurements using the Coulter principle that are also shown in Table 1. The decrease in the oxygen consumption rate per cellular volume, in addition to the registered reduction in Q on a cellular basis, can be explained by a corresponding decrease in the cellular volume fraction of the mitochondria. Obviously, this change is paralleled by a loss in the cytoplasmic volume, whereas the volume of the nucleus remains relatively constant. The spatial distribution of mitochondria within the cells was also altered during monolayer growth. In the exponential growth phase, mitochondria appeared to be rather evenly distributed in one hemisphere of the cell with the nucleus being located in the second hemisphere surrounded by only a narrow seam of cytoplasm. In the plateau phase, the cells exhibit central nuclei with the mitochondria homogeneously distributed in the surrounding cytoplasm. Such a variation in the mitochondrial array within cells may contribute to changes in the cellular oxygen uptake, as has been hypothesized by Jones (1984). In general, the mitochondria did not undergo any obvious structural changes in different monolayer growth phases, although the mean mitochondrial volume decreased during monolayer growth. These findings suggest that the changes in cellular oxygen consumption that are positively correlated with the proliferative activity of cells are mediated through variations in the

848

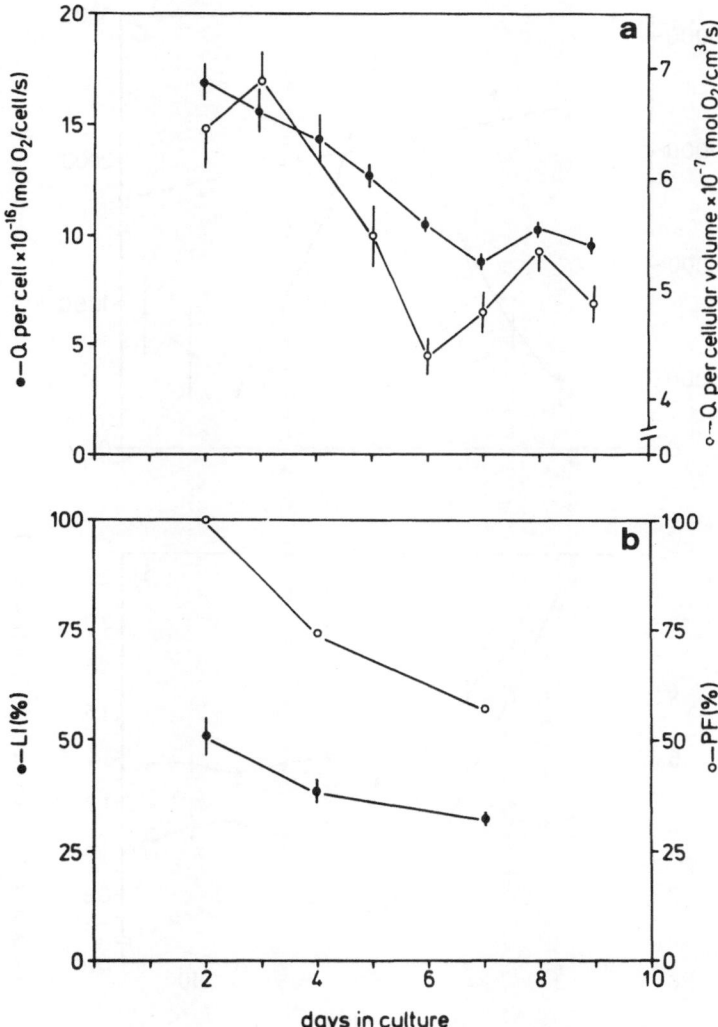

Fig. 1. a: Oxygen consumption rates Q ($\overline{x} \pm$ SEM) per cell (dots) and per
cellular volume (circles) of EMT6 cells as a function of
time in monolayer culture.

b: Labeling index LI ($\overline{x} \pm$ SD; dots) and proliferative fraction
PF (circles) of EMT6 cells as a function of time in mono-
layer culture.

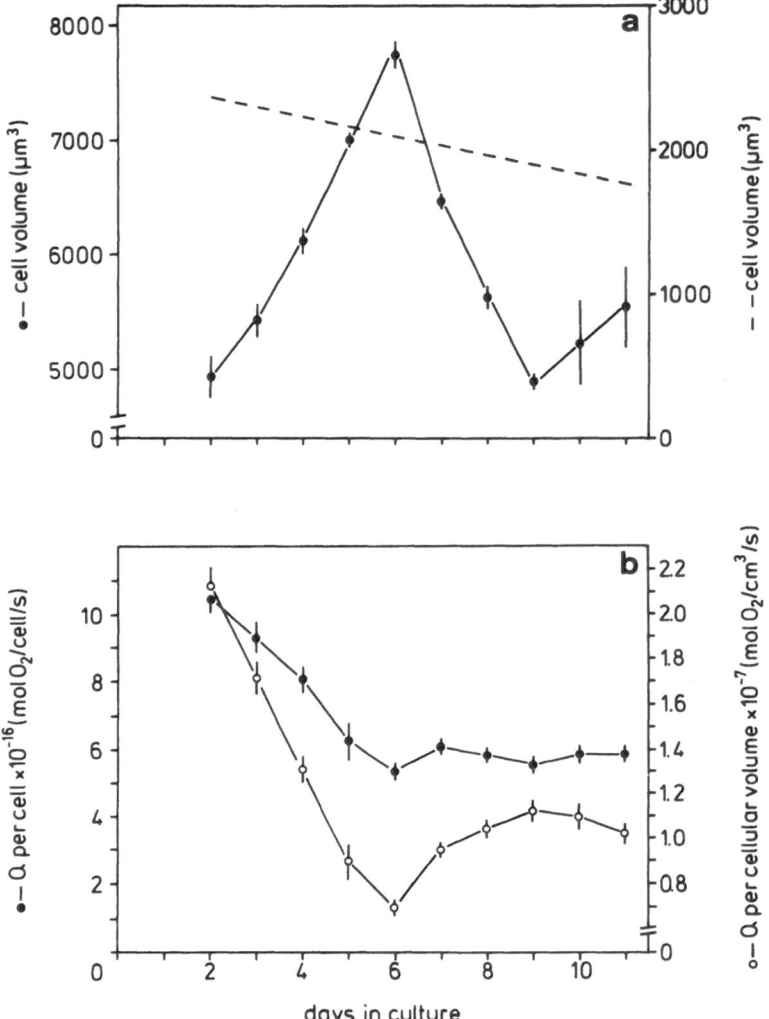

Fig. 2. a: Volume V ($\overline{x} \pm$ SEM) of MCF-7 cells as a function of time in monolayer culture. For comparison, the respective correlation for EMT6 cells is indicated by a dashed line.
b: Oxygen consumption rate Q ($\overline{x} \pm$ SEM) per cell (dots) and per cellular volume V (circles) of MCF-7 cells as a function of time in culture.

Table 1. Results of morphometric measurements in EMT6 monolayer cells in the exponential (2-3 days in culture) and plateau (5-9 days in culture) growth phase. For comparison, the corresponding values for the measured oxygen consumption rate, cellular volume, and labeling index are shown. Values are means $\pm$ SEM except for labeling index ($\bar{x} \pm$ SD).

	exponential	plateau
measured		
cellular areas (μm^2)	101.60 $\pm$ 3.12	70.74 $\pm$ 1.77
number of mitochondria per cell section	34.68 $\pm$ 1.45	20.11 $\pm$ 0.95
mean mitochondrial area (μm^2)	0.13 $\pm$ 0.00	0.10 $\pm$ 0.00
mean area of the nucleus (μm^2)	31.50 $\pm$ 1.42	25.82 $\pm$ 1.09
calculated		
mean cellular volume (μm^3)	3118	1811
cellular volume fraction of mitochondria (%)	0.90	0.50
cellular volume fraction of nucleus (%)	17	22
cellular volume fraction of the cytoplasm (%)	83	78
measured		
mean oxygen consumption rate per cell (mol O_2/cell/s)	$(16.1 \pm 1.0) \times 10^{-17}$	$(10.4 \pm 1.5) \times 10^{-17}$
mean oxygen consumption rate per cell volume (mol O_2/cm^3/s)	$(6.5 \pm 0.35) \times 10^{-8}$	$(4.9 \pm 0.35) \times 10^{-8}$
labeling index (%)	51.9 $\pm$ 3.94	29.8 $\pm$ 1.63
cellular volume (μm^3)	2709 $\pm$ 109	1978 $\pm$ 20

number, volume, and array of the mitochondria rather than through enzyme regulation at a constant pool of mitochondria.

Data similar to those shown in Table 1 were obtained for the morphology of the outer and inner cells in multicellular EMT6 spheroids. These cells correspond to exponential and plateau phase cells from monolayers, respectively. On the other hand, the large and highly proliferative cells in the outermost spheroid regions are very loosely packed compared to the inner spheroid cells. Such a difference in cellular packing density could generate a rather uniform mitochondrial density with regard to a given tissue volume. This may explain the findings of a rather uniform oxygen consumption rate per unit volume in EMT6 spheroids (Mueller-Klieser, 1987).

ABSTRACT

The proliferative activity of tumor cells is positively correlated with the cellular respiration rate. Thus, a decrease in the proliverative fraction during monolayer growth of EMT6 cells is associated with a decrease in the oxygen consumption rate both per cell and per cellular volume. These changes are paralleled by variations in the morphology of the tumor cells. The cellular volume, the cytoplasmic volume, the number and volume of mitochondria is decreased, whereas the volume of the nucleus remains relatively constant during the transition of cells from the exponential to the plateau growth phase. Similar observations have been made in outer, highly proliferative and inner non-proliferating cells in EMT6 spheroids. However, the local oxygen consumption in this threedimensional, tissue-like assemblage of cells is obviously modified by the cellular packing density leading to a rather uniform oxygen uptake in the proliferating and non-proliferating cell areas.

ACKNOWLEDGEMENT

This work was supported by the Deutsche Forschungsgemeinschaft (Mu576/2-3).

REFERENCES

J.P. Freyer, R.M. Sutherland (1985). A reduction in the in situ rates of oxygen and glucose consumption of cells in EMT6/Ro spheroids during growth. J. Cell. Physiol. 124: 516-524.

J.P. Freyer, E. Tustanoff, A.J. Franko, R.M. Sutherland (1984). In situ oxygen consumption rates of cells in V-79 multicellular spheroids during growth. J. Cell. Physiol. 118: 53-61.

D.P. Jones (1984). Effect of mitochondrial clustering on O_2 supply in hepatocytes, Am. J. Physiol. 247: C83-C89.

W. Mueller-Klieser (1984). Method for determination of oxygen consumption rates and diffusion coefficients in multicellular spheroids, Biophys. J. 46: 343-348.

W. Mueller-Klieser (1987). Multicellular spheroids. A review on cellular aggregates in cancer research, J. Cancer Res. Clin. Oncol. 113: 101-122.

W. Mueller-Klieser, R. Zander, P. Vaupel (1986). A new photometric method for oxygen consumption measurements in cell suspensions, J. Applied Physiol. 61: 449–455.

R.M. Sutherland (1988). Cell and environment interactions in tumor microregions: The multicell spheroid model, Science 240: 177–184.

S. Walenta, W. Mueller-Klieser (1987). Oxygen consumption rate of tumor cells as a function of their proliferative status, Adv. Exp. Med. Biol. 215: 389–391.

INFORMATION

TISSUE ENGINEERING

Duane F. Bruley

National Science Foundation
1800 G Street, N.W., Room 1126
Washington, D.C. 20550

Tissue Engineering requires for its implementation cross-disciplinary efforts between diverse groups of bioengineers, biological scientists, and medical researchers. ISOTT has acted as an organized forum to bring these multidisciplinary forces together to develop dialogue regarding the structure, function, maintenance, repair, growth, and destruction of tissue under a variety of physiological and pathological conditions. The growing interest of other societies, academic institutions, the State and Federal governments and private industry has led to heightened awareness of Tissue Engineering.

To explore the prospects for Tissue Engineering a workshop was held on February 26-29, 1988[1]. Approximately 75 people attended the workshop; these included university, governmental agency, and industrial representatives. The presentations included a wide range of tissues, procedures, and background basic science.

For the purpose of this workshop and discussion, the following definition was developed for consideration:

"Tissue Engineering is the application of principles and methods of engineering and life sciences toward fundamental understanding of structure-function relationships in normal and pathological mammalian tissues and the development of biological substitutes to restore, maintain, or improve tissue functions."

The workshop was especially successful because a wide range of senior medical, biological, and engineering scientists expressed interest and shared their experiences and views. It is clear that the workshop participants consider Tissue Engineering as both a forefront of biomedical engineering research and an area that has broad and promising practical application and industrial potential.

---

[1]TISSUE ENGINEERING, A UCLA Symposia, Technology Priority Assessment Workshop. Granlibakken, Lake Tahoe, California, C. Fred Fox and Richard Skalak, February 26-29, 1988.

Three objectives were established for immediate investigations in tissue engineering:

o   Explorations of effects of the environment at both the cellular and tissue level, in-vivo and ex-vivo. This includes the effects of mechanical, electrical, chemical and biochemical stimuli.

o   Investigation of surface phenomena. Many problems in tissue engineering are concerned with cell-to-cell or cell-to-substrate interactions. Studies of surface phenomena including adhesion and detachment are of particular concern.

o   Studies of growth, production, and storage of cells, tissues, and molecular products. An important aspect of the practical application of tissue engineering would be production of cells and tissue on a large scale.

Specific areas that were targeted for investigation include:

o   Microcirculation: endothelial cells, angiogenesis
o   Blood replacements/blood cell activation
o   Vascular prothesis
o   Biomaterial: centered infections and inflammation
o   Skin replacement
o   Artificial organs
o   Cartilage
o   Bone
o   General connective tissue
o   Cell biology and cell mechanics
o   Muscle and soft tissue
o   Brain: cell implantation and prehensile dementia
o   Growth factor and motor nerve regrowth
o   Intractable pain and chromaffin cell transplants

As a result of the workshop, a Tissue Engineering initiative has been implemented at the National Science Foundation. Proposals are welcomed from single investigators or teams involved with cross-disciplinary research.

CONTRIBUTORS

INDEX*

---

*The page numbers given are the first pages of the papers in which these
subjects are covered.